Plant Proteomic Research 3.0

Plant Proteomic Research 3.0

Editors

Setsuko Komatsu
Jesus V. Jorrin-Novo

MDPI • Basel • Beijing • Wuhan • Barcelona • Belgrade • Manchester • Tokyo • Cluj • Tianjin

Editors
Setsuko Komatsu
Fukui University of Technology
Japan

Jesus V. Jorrin-Novo
University of Cordoba
Spain

Editorial Office
MDPI
St. Alban-Anlage 66
4052 Basel, Switzerland

This is a reprint of articles from the Special Issue published online in the open access journal *International Journal of Molecular Sciences* (ISSN 1422-0067) (available at: https://www.mdpi.com/journal/ijms/special_issues/plant-proteomic_3).

For citation purposes, cite each article independently as indicated on the article page online and as indicated below:

LastName, A.A.; LastName, B.B.; LastName, C.C. Article Title. *Journal Name* **Year**, *Volume Number*, Page Range.

ISBN 978-3-0365-0604-3 (Hbk)
ISBN 978-3-0365-0605-0 (PDF)

Contents

About the Editors . . . . **ix**

Setsuko Komatsu and Jesus V. Jorrin-Novo
Plant Proteomic Research 3.0: Challenges and Perspectives
Reprinted from: *Int. J. Mol. Sci.* **2021**, *22*, 766, doi:10.3390/ijms22020766 . . . . **1**

Galina Smolikova, Daria Gorbach, Elena Lukasheva, Gregory Mavropolo-Stolyarenko, Tatiana Bilova, Alena Soboleva, Alexander Tsarev, Ekaterina Romanovskaya, Ekaterina Podolskaya, Vladimir Zhukov, Igor Tikhonovich, Sergei Medvedev, Wolfgang Hoehenwarter and Andrej Frolov
Bringing New Methods to the Seed Proteomics Platform: Challenges and Perspectives
Reprinted from: *Int. J. Mol. Sci.* **2020**, *21*, 9162, doi:10.3390/ijms21239162 . . . . **7**

Maria Tartaglia, Felipe Bastida, Rosaria Sciarrillo and Carmine Guarino
Soil Metaproteomics for the Study of the Relationships Between Microorganisms and Plants: A Review of Extraction Protocols and Ecological Insights
Reprinted from: *Int. J. Mol. Sci.* **2020**, *21*, 8455, doi:10.3390/ijms21228455 . . . . **63**

Dongli He, Rebecca Njeri Damaris, Ming Li, Imran Khan and Pingfang Yang
Advances on Plant Ubiquitylome—From Mechanism to Application
Reprinted from: *Int. J. Mol. Sci.* **2020**, *21*, 7909, doi:10.3390/ijms21217909 . . . . **83**

Zahed Hossain, Farhat Yasmeen and Setsuko Komatsu
Nanoparticles: Synthesis, Morphophysiological Effects, and Proteomic Responses of Crop Plants
Reprinted from: *Int. J. Mol. Sci.* **2020**, *21*, 3056, doi:10.3390/ijms21093056 . . . . **101**

Besma Sghaier-Hammami, Sofiene B.M. Hammami, Narjes Baazaoui, Consuelo Gómez-Díaz and Jesús V. Jorrín-Novo
Dissecting the Seed Maturation and Germination Processes in the Non-Orthodox *Quercus ilex* Species Based on Protein Signatures as Revealed by 2-DE Coupled to MALDI-TOF/TOF Proteomics Strategy
Reprinted from: *Int. J. Mol. Sci.* **2020**, *21*, 4870, doi:10.3390/ijms21144870 . . . . **119**

Olha Lakhneko, Maksym Danchenko, Bogdan Morgun, Andrej Kováč, Petra Majerová and Ľudovit Škultéty
Comprehensive Comparison of Clinically Relevant Grain Proteins in Modern and Traditional Bread Wheat Cultivars
Reprinted from: *Int. J. Mol. Sci.* **2020**, *21*, 3445, doi:10.3390/ijms21103445 . . . . **145**

Li Yu, Boxuan Yuan, Lingling Wang, Yong Sun, Guohua Ding, Ousmane Ahmat Souleymane, Xueyan Zhang, Quanliang Xie and Xuchu Wang
Identification and Characterization of Glycoproteins and Their Responsive Patterns upon Ethylene Stimulation in the Rubber Latex
Reprinted from: *Int. J. Mol. Sci.* **2020**, *21*, 5282, doi:10.3390/ijms21155282 . . . . **167**

Anne Hofmann, Stefanie Wienkoop, Sönke Harder, Fabian Bartlog and Sabine Lüthje
Hypoxia-Responsive Class III Peroxidases in Maize Roots: Soluble and Membrane-Bound Isoenzymes
Reprinted from: *Int. J. Mol. Sci.* **2020**, *21*, 8872, doi:10.3390/ijms21228872 . . . . **189**

Chris J. Hulatt, Irina Smolina, Adam Dowle, Martina Kopp, Ghana K. Vasanth, Galice G. Hoarau, René H. Wijffelsand Viswanath Kiron
Proteomic and Transcriptomic Patterns during Lipid Remodeling in *Nannochloropsis gaditana*
Reprinted from: *Int. J. Mol. Sci.* **2020**, *21*, 6946, doi:10.3390/ijms21186946 . . . . **211**

Yong Lai, Dangquan Zhang, Jinmin Wang, Juncheng Wang, Panrong Ren, Lirong Yao, Erjing Si, Yuhua Kong and Huajun Wang
Integrative Transcriptomic and Proteomic Analyses of Molecular Mechanism Responding to Salt Stress during Seed Germination in Hulless Barley
Reprinted from: *Int. J. Mol. Sci.* **2020**, *21*, 359, doi:10.3390/ijms21010359 . . . . **235**

Ming Li, Ishfaq Hameed, Dingding Cao, Dongli He and Pingfang Yang
Integrated Omics Analyses Identify Key Pathways Involved in Petiole Rigidity Formation in Sacred Lotus
Reprinted from: *Int. J. Mol. Sci.* **2020**, *21*, 5087, doi:10.3390/ijms21145087 . . . . **255**

Quanquan Chen, Ran Huang, Zhenxiang Xu, Yaxin Zhang, Li Li, Junjie Fu, Guoying Wang, Jianhua Wang, Xuemei Du and Riliang Gu
Label-Free Comparative Proteomic Analysis Combined with Laser-Capture Microdissection Suggests Important Roles of Stress Responses in the Black Layer of Maize Kernels
Reprinted from: *Int. J. Mol. Sci.* **2020**, *21*, 1369, doi:10.3390/ijms21041369 . . . . **283**

Sheldon R. Lawrence II, Meghan Gaitens, Qijie Guan, Craig Dufresne and Sixue Chen
S-Nitroso-Proteome Revealed in Stomatal Guard Cell Response to Flg22
Reprinted from: *Int. J. Mol. Sci.* **2020**, *21*, 1688, doi:10.3390/ijms21051688 . . . . **299**

Laura Ceballos-Laita, Elain Gutierrez-Carbonell, Daisuke Takahashi, Andrew Lonsdale, Anunciación Abadía, Monika S. Doblin, Antony Bacic, Matsuo Uemura, Javier Abadía and Ana Flor López-Millán
Effects of Excess Manganese on the Xylem Sap Protein Profile of Tomato (*Solanum lycopersicum*) as Revealed by Shotgun Proteomic Analysis
Reprinted from: *Int. J. Mol. Sci.* **2020**, *21*, 8863, doi:10.3390/ijms21228863 . . . . **319**

Raviraj M. Kalunke, Silvio Tundo, Francesco Sestili, Francesco Camerlengo, Domenico Lafiandra, Roberta Lupi, Colette Larré, Sandra Denery-Papini, Shahidul Islam, Wujun Ma, Stefano D'Amico and Stefania Masci
Reduction of Allergenic Potential in Bread Wheat RNAi Transgenic Lines Silenced for *CM3*, *CM16* and *0.28 ATI* Genes
Reprinted from: *Int. J. Mol. Sci.* **2020**, *21*, 5817, doi:10.3390/ijms21165817 . . . . **347**

Gul Nawaz, Babar Usman, Neng Zhao, Yue Han, Zhihua Li, Xin Wang, Yaoguang Liu and Rongbai Li
CRISPR/Cas9 Directed Mutagenesis of *OsGA20ox2* in High Yielding Basmati Rice (*Oryza sativa* L.) Line and Comparative Proteome Profiling of Unveiled Changes Triggered by Mutations
Reprinted from: *Int. J. Mol. Sci.* **2020**, *21*, 6170, doi:10.3390/ijms21176170 . . . . **365**

Hui Wang, Qingping Zhou and Peisheng Mao
Ultrastructural and Photosynthetic Responses of Pod Walls in Alfalfa to Drought Stress
Reprinted from: *Int. J. Mol. Sci.* **2020**, *21*, 4457, doi:10.3390/ijms21124457 . . . . **389**

Takuya Hashimoto, Ghazala Mustafa, Takumi Nishiuchi and Setsuko Komatsu
Comparative Analysis of the Effect of Inorganic and Organic Chemicals with Silver Nanoparticles on Soybean under Flooding Stress
Reprinted from: *Int. J. Mol. Sci.* **2020**, *21*, 1300, doi:10.3390/ijms21041300 . . . . **409**

Weifeng Luo, Setsuko Komatsu, Tatsuya Abe, Hideyuki Matsuura and Kosaku Takahashi
Comparative Proteomic Analysis of Wild-Type *Physcomitrella Patens* and an OPDA-Deficient
Physcomitrella Patens Mutant with Disrupted *PpAOS1* and *PpAOS2* Genes after Wounding
Reprinted from: *Int. J. Mol. Sci.* **2020**, *21*, 1417, doi:10.3390/ijms21041417 . . . . **425**

**Carol A. Olivares-García, Martín Mata-Rosas, Carolina Peña-Montes, Francisco
Quiroz-Figueroa, Aldo Segura-Cabrera, Laura M. Shannon, Victor M. Loyola-Vargas,
Juan L. Monribot-Villanueva, Jose M. Elizalde-Contreras, Enrique Ibarra-Laclette, Mónica
Ramirez-Vázquez, José A. Guerrero-Analco and Eliel Ruiz-May**
Phenylpropanoids Are Connected to Cell Wall Fortification and Stress Tolerance in Avocado
Somatic Embryogenesis
Reprinted from: *Int. J. Mol. Sci.* **2020**, *21*, 5679, doi:10.3390/ijms21165679 . . . . **443**

Zhi-Lang Qiu, Zhuang Wen, Kun Yang, Tian Tian, Guang Qiao, Yi Hong and Xiao-Peng Wen
Comparative Proteomics Profiling Illuminates the Fruitlet Abscission Mechanism of Sweet
Cherry as Induced by Embryo Abortion
Reprinted from: *Int. J. Mol. Sci.* **2020**, *21*, 1200, doi:10.3390/ijms21041200 . . . . **467**

**Shengjiang Wu, Yushuang Guo, Muhammad Faheem Adil, Shafaque Sehar, Bin Cai,
Zhangmin Xiang, Yonggao Tu, Degang Zhao and Imran Haider Shamsi**
Comparative Proteomic Analysis by iTRAQ Reveals that Plastid Pigment Metabolism
Contributes to Leaf Color Changes in Tobacco (*Nicotiana tabacum*) during Curing
Reprinted from: *Int. J. Mol. Sci.* **2020**, *21*, 2394, doi:10.3390/ijms21072394 . . . . **487**

**Chao Xiang, Xiaoping Yang, Deliang Peng, Houxiang Kang, Maoyan Liu, Wei Li, Wenkun
Huang and Shiming Liu**
Proteome-Wide Analyses Provide New Insights into the Compatible Interaction of Rice with
the Root-Knot Nematode *Meloidogyne graminicola*
Reprinted from: *Int. J. Mol. Sci.* **2020**, *21*, 5640, doi:10.3390/ijms21165640 . . . . **507**

About the Editors

Setsuko Komatsu has been a Professor at Fukui University of Technology, Japan, since 2018. Before this position, she was the Chief of the Field Omics Research Unit at the National Institute of Crop Science and a Professor at the University of Tsukuba, Japan. She was previously employed at Meiji Pharmaceutical University and then also at Keio University School of Medicine. In 1990, she started working on plant proteomics using protein sequencing and mass spectrometry. Her main research interests are within the field of crop proteomics, phytohormone, biochemistry, and molecular biology with a special focus on signal transduction in cells. As the president of the "Asia Oceania Agricultural Proteomics Organization", she is working to promote and further agricultural proteomics.

Jesus V. Jorrin-Novo holds a PhD in Biology from the University of Córdoba (1986) and is a Professor at the Department of Biochemistry and Molecular Biology, University of Córdoba, and Head of the Agroforestry and Plant Biochemistry, Proteomics, and Systems Biology Research Group (AGR-164; http://www.uco.es/investiga/grupos/probiveag/). With a professional life of 36 years, he joined the University as undergraduate student in 1975. His main interest and research are plant biology, mostly focused on crops and, lately, forest tree species. By using a molecular approach, including classic biochemistry and omics techniques, he pursues a deep knowledge of plant development and responses to environmental stresses, in order to understand biodiversity. He has tried to follow what he understands as the main missions of the University: to connect the research and the academy and offer to undergraduate, master's, and PhD students the possibility of active participation in research projects as a means of being trained and educated in science; this without forgetting the translation to the productive sectors and to society. In the mid- and late 2000s, he took an active part in the creation of the SEProt and the EuPA, always trying to promote proteomics research in the plant science field.

 International Journal of
Molecular Sciences

Editorial

Plant Proteomic Research 3.0: Challenges and Perspectives

Setsuko Komatsu [1,*] and Jesus V. Jorrin-Novo [2,*]

[1] Faculty of Environmental and Information Sciences, Fukui University of Technology, Fukui Prefecture, Fukui 910-0028, Japan
[2] Department of Biochemistry and Molecular Biology, ETSIAM, University of Cordoba, UCO-CeiA3, 14014 Cordoba, Spain
* Correspondence: skomatsu@fukui-ut.ac.jp (S.K.); bf1jonoj@uco.es (J.V.J.-N.)

Received: 7 January 2021; Accepted: 11 January 2021; Published: 14 January 2021

Advancements in high-throughput "Omics" techniques have revolutionized plant molecular biology research. Among them, proteomics offers one of the best options for the functional analysis of the genome, and for generating detailed information, which when integrated with that obtained by other classic and "Omics" approaches, will provide a deeper knowledge of the different plant processes. In this regard, responses to stress are key issues in plant biology research. Plants, being sessile in nature, are constantly exposed to environmental challenges resulting in substantial yield loss. To cope with harsh environments, plants have developed a wide range of adaptation strategies involving morpho-anatomical, physiological, and biochemical traits [1], knowledge of which will open new possibilities for crop breeding, improvements, and yield increase. Thus, in recent years, there has been a lot of progress in the understanding of plant responses to environmental cues at the protein level [2].

The generations of proteomics platforms (gel, label, gel free/label free, targeted) which have appeared in the last twenty years are being exploited in describing protein profiles, post-translational modifications and interactions. Nevertheless, the ultimate success of any proteomic strategy lies in various factors including the isolation, separation, visualization, and accurate identification of proteins. Despite recent advancements, more emphasis needs to be given to the protein extraction protocols, especially for very low-abundant, hydrophobic, and large molecular mass proteins. Thus, this amalgamation of diverse mass spectrometry techniques, which when complemented with genome-sequence data and modern bioinformatics analysis with improved sample preparation and fractionation strategies, offers a powerful tool to characterize novel proteins and to follow temporal changes in protein relative abundances under different environmental conditions. Furthermore, post-translational modifications and protein-protein interactions provide deeper insight into protein molecular function.

In this direction, the present "Plant Proteomics 3.0" Special Issue was conceived in an attempt to address recent advancements, as well as the limitations of current proteomic techniques and their diverse applications, to achieve new insights into plant molecular responses to various biotic and abiotic stressors and the molecular bases of other processes. Proteomic focus is also related to translational purposes, including food traceability and allergen detection. In addition, bioinformatic techniques are needed for more confident identification, quantitation, data analysis and networking, especially with non-model, orphan, plants, including medicinal and meditational plants as well as forest tree species. This Issue is the youngest brother of the previous "Plant Proteomic Research" [1] and "Plant Proteomic Research 2.0" [2] Issues. It contains 23 articles, including 4 reviews and 19 original articles. They were contributed by research groups from China, Japan, the USA, Mexico, and various European countries including Spain, Italy, Germany, the Slovak Republic, and Norway.

Four reviews deal with methodological aspects [3,4], mechanisms of ubiquitination [5], and the effect of and responses to nanoparticles treatments [6]. Smolikova et al. [3] summarized the main methodological approaches already employed in seed proteomics, as well as those still waiting for implementation in plant research, such as sample preparation, data acquisition, processing, and post-processing. Tartaglia et al. [4] focused on several soil protein extraction protocols to highlight the methodological challenges for the application of proteomics to soil samples, which enhance the identification of proteins with low abundance or from non-dominant populations. He et al. [5] summarized the latest advances in protein ubiquitination to gain comprehensive and updated knowledge, because ubiquitylation has been widely reported to be involved in many aspects of plant growth and development. Furthermore, Hossain et al. [6] highlighted the current understanding of plant responses to nanoparticles; for this purpose, the synthesis of nanoparticles, their morphophysiological effects on crops, and applications of proteomic techniques are discussed to comprehend the underlying mechanism of nanoparticles stress acclimation.

Shotgun analysis (gel-free proteomics) has supposed classical gel-based analysis (gel-based proteomics) becoming the dominant proteomic platform. One reason is that shotgun analysis affords deeper proteome coverage and allows several thousand proteins to be identified as compared to a few hundred with the gel-based approach. Both gel-based and gel-free are complementary, and the former still a valid approach [7–10]. As evidence of this, out of the 19 original papers, 15 articles report on the use of a gel-free technique. Progress has been fueled by the advancement in mass spectrometry techniques, complemented with genome-sequence data and modern bioinformatic analysis. By using two-dimensional gel electrophoresis coupled with matrix-assisted laser desorption ionization/time of flight, Sghaier-Hammami et al. [7] analyzed changes in the protein profile of non-orthodox *Quercus ilex* seeds during the maturation and germination stages, revealing some differences between orthodox/recalcitrance and viable/non-viable seeds. Lakhneko et al. [8] performed a proteomics study of grains from modern and traditional bread wheat cultivars. They used a detergent-containing buffer, which allowed the extraction of various groups of storage proteins and analyzed them by two-dimensional gel electrophoresis. Yu et al. [9] characterized glycosylated proteins in rubber latex using Pro-Q glycoprotein gel staining. Hofmann et al. [10] determined the alterations in soluble and membrane-bound class III peroxidases of maize under hypoxia by using combined proteomics and transcriptomics approaches.

Proteomics is more integrated with other "Omics" and classic approaches in the systems biology direction. Four original articles [10–13] employed proteomic and transcriptomic analyses. Hulatt et al. [11] analyzedcorrelation between proteomic and transcriptomic data, the regulation of sub-cellular localized proteins in different compartments, gene/protein functional groups, and metabolic pathways in *Nannochloropsis gaditana*. Lai et al. [12] reported that 96 genes were differentially expressed at both transcriptomic and proteomic levels in salt-tolerant and salt-sensitive hulless barley. Li et al. [13] studied at the mRNA and protein levels differences between vertical and floating leaves in sacred lotus, showing that the first is enriched in cell wall and lignin biosynthesis gene products.

In situ and single cell analysis remain some of the main challenges in proteomics studies. In this direction, Chen et al. [14] used shotgun combined with laser-capture microdissection to analyze the black layer of maize kernels, a tissue rich in stress-responsive proteins. Lawrence et al. [15] prepared stomatal guard cells from *Arabidopsis thaliana* using a Scotch-tape method to analyze changes in S-nitrosylation in guard cells during a pathogen challenge. Ceballos-Laita et al. [16] assessed the effects of the exposure of tomato roots to excess Mn on the protein profile of the xylem sap using a shotgun proteomic approach. Furthermore, plant materials are also important for the identification of unique proteins. For example, Kalunke et al. [17] reported characterization of three transgenic lines obtained from the bread wheat cultivar Bobwhite silenced by RNAi in the three ATI genes *CM3*, *CM16* and *0.28*; and these lines show unintended differences in the accumulation of high molecular mass glutenin subunits, which are involved in technological performances, but which do not show differences in terms of yield. Nawaz et al. [18] reported that semi-dwarf rice lines lacking any residual transgene-DNA and off-target effects were generated through CRISPR/Cas9-guided mutagenesis of the *OsGA20ox2* gene in a high yielding Basmati rice line, and the identified proteins were mainly enriched in the carbon metabolism and fixation, glycolysis/gluconeogenesis, photosynthesis, and oxidative phosphorylation pathways.

Proteomic techniques are used in the identification of stress-responsive mechanism in plants under conditions such as salt [14], cold [19], flood [20], and wounding [21]. Wang et al. [19] provided insights into the cold stress responses of early imbibed castor seeds: proteins confer cold tolerance by promoting protein synthesis, protect the cell against damage caused by cold stress, and facilitate resistance or adaptation to cold stress by the increased content of unsaturated fatty acid. Hashimoto et al. [20] indicated that the combined mixture of silver nanoparticles, nicotinic acid, and KNO_3 causes positive effects on soybean seedlings by regulating the protein quality control for the mis-folded proteins in the endoplasmic reticulum, suggesting that it might improve the growth of soybeans under flooding stress. Luo et al. [21] suggested that PpAOS gene, which is involved in 12-Oxo-phytodienoic acid biosynthesis, expression enhances photosynthesis and effective energy utilization in response to wounding in *Physcomitrella patens*.

Olivares-García et al. [22] suggested that in the embryogenic culture of avocados, there is an enhanced phenylpropanoid metabolism for the production of the building blocks of lignin having a role in cell wall reinforcement for tolerating stress response. Qiu et al. [23] indicated that embryo abortion might lead to phytohormone synthesis disorder, which effected signal transduction pathways, and hereby controlled genes involved in cell wall degradation and then caused the abscission of fruitlets in sweet cherry trees. Wu et al. [24] revealed by using iTRAQ that plastid pigment metabolism contributes to leaf color changes in *Nicotiana tabacum* during curing. Xiang et al. [25] hypothesized that RPM1, a disease resistance protein, may confer resistance to *M. graminicola* in rice, and that it may possess broad-spectrum resistance against pathogens including the parasitic nematode *M. graminicola*.

Proteomics is an alive and very active discipline, and it is just one more step on the way to decipher the intricate secrets and complexity of plant systems. The guest editors hope that this special issue will provide readers with a framework for understanding plant proteomics and insights into new research directions within this field. For sure, new methodologies, approaches, equipment, and applications will be continuously appearing, and these will be the subject of "Plant Proteomic Research 4.0". Finally, the guest editors thank all of the authors for their contributions and the reviewers for their critical assessments of these articles. Moreover, they also thank the Assistant Editor, Ms. Chaya Zeng, for giving us the opportunity to serve as guest editors for "Plant Proteomic Research 3.0".

Author Contributions: S.K. and J.V.J.-N. have made substantial, direct, and intellectual contributions to the published articles, and approved them for publication. All authors have read and agreed to the published version of the manuscript.

Conflicts of Interest: The authors declare no conflict of interest.

References

1. Komatsu, S.; Hossain, Z. Preface—Plant Proteomic Research. *Int. J. Mol. Sci.* **2017**, *18*, 88. [CrossRef]
2. Komatsu, S. Plant Proteomic Research 2.0: Trends and Perspectives. *Int. J. Mol. Sci.* **2019**, *20*, 2495. [CrossRef] [PubMed]
3. Smolikova, G.; Gorbach, D.; Lukasheva, E.; Mavropolo-Stolyarenko, G.; Bilova, T.; Soboleva, A.; Tsarev, A.; Romanovskaya, E.; Podolskaya, E.; Zhukov, V.; et al. Bringing New Methods to the Seed Proteomics Platform: Challenges and Perspectives. *Int. J. Mol. Sci.* **2020**, *21*, 9162. [CrossRef] [PubMed]

4. Tartaglia, M.; Bastida, F.; Sciarrillo, R.; Guarino, C. Soil Metaproteomics for the Study of the Relationships Between Microor-ganisms and Plants: A Review of Extraction Protocols and Ecological Insights. *Int. J. Mol. Sci.* **2020**, *21*, 8455. [CrossRef] [PubMed]

5. He, D.; Damaris, R.N.; Li, M.; Khan, I.; Yang, P. Advances on Plant Ubiquitylome—From Mechanism to Application. *Int. J. Mol. Sci.* **2020**, *21*, 7909. [CrossRef]

6. Hossain, Z.; Yasmeen, F.; Komatsu, S. Nanoparticles: Synthesis, Morphophysiological Effects, and Proteomic Responses of Crop Plants. *Int. J. Mol. Sci.* **2020**, *21*, 3056. [CrossRef]

7. Sghaier-Hammami, B.; BM Hammami, S.; Baazaoui, N.; Gómez-Díaz, C.; Jorrín-Novo, J. Dissecting the Seed Maturation and Germination Processes in the Non-Orthodox Quercus ilex Species Based on Protein Signatures as Revealed by 2-DE Coupled to MALDI-TOF/TOF Proteomics Strategy. *Int. J. Mol. Sci.* **2020**, *21*, 4870. [CrossRef]

8. Lakhneko, O.; Danchenko, M.; Morgun, B.; Kováč, A.; Majerová, P.; Škultéty, L'. Comprehensive Comparison of Clinically Relevant Grain Proteins in Modern and Traditional Bread Wheat Cultivars. *Int. J. Mol. Sci.* **2020**, *21*, 3445. [CrossRef]

9. Yu, L.; Yuan, B.; Wang, L.; Sun, Y.; Ding, G.; Souleymane, O.A.; Zhang, X.; Xie, Q.; Wang, X. Identification and Characterization of Glycoproteins and Their Responsive Patterns upon Ethylene Stimulation in the Rubber Latex. *Int. J. Mol. Sci.* **2020**, *21*, 5282. [CrossRef]

10. Hofmann, A.; Wienkoop, S.; Harder, S.; Bartlog, F.; Lüthje, S. Hypoxia-Responsive Class III Peroxidases in Maize Roots: Soluble and Membrane-Bound Isoenzymes. *Int. J. Mol. Sci.* **2020**, *21*, 8872. [CrossRef]

11. Hulatt, C.J.; Smolina, I.; Dowle, A.A.; Kopp, M.; Vasanth, G.K.; Hoarau, G.G.; Wijffels, R.H.; Kiron, V. Proteomic and Transcriptomic Patterns during Lipid Remodeling in *Nannochloropsis gaditana*. *Int. J. Mol. Sci.* **2020**, *21*, 6946. [CrossRef] [PubMed]

12. Lai, Y.; Zhang, D.; Wang, J.; Wang, J.; Ren, P.; Yao, L.; Si, E.; Kong, Y.; Wang, H. Integrative Transcriptomic and Proteomic Analyses of Molecular Mechanism Responding to Salt Stress during Seed Germination in Hulless Barley. *Int. J. Mol. Sci.* **2020**, *21*, 359. [CrossRef] [PubMed]

13. Li, M.; Hameed, I.; Cao, D.; He, D.; Yang, P. Integrated Omics Analyses Identify Key Pathways Involved in Petiole Rigidity Formation in Sacred Lotus. *Int. J. Mol. Sci.* **2020**, *21*, 5087. [CrossRef] [PubMed]

14. Chen, Q.; Huang, R.; Xu, Z.; Zhang, Y.; Li, L.; Fu, J.; Wang, G.; Wang, J.; Du, X.; Gu, R. Label-Free Comparative Proteomic Analysis Combined with Laser-Capture Microdissection Suggests Important Roles of Stress Responses in the Black Layer of Maize Kernels. *Int. J. Mol. Sci.* **2020**, *21*, 1369. [CrossRef]

15. Ii, S.L.; Gaitens, M.; Guan, Q.; Dufresne, C.; Chen, S. S-Nitroso-Proteome Revealed in Stomatal Guard Cell Response to Flg22. *Int. J. Mol. Sci.* **2020**, *21*, 1688. [CrossRef]

16. Ceballos-Laita, L.; Gutierrez-Carbonell, E.; Takahashi, D.; Lonsdale, A.; Abadía, A.; Doblin, M.S.; Bacic, A.; Uemura, M.; Abadía, J.; López-Millán, A.F. Effects of Excess Manganese on the Xylem Sap Protein Profile of Tomato (*Solanum lycopersicum*) as Revealed by Shotgun Proteomic Analysis. *Int. J. Mol. Sci.* **2020**, *21*, 8863. [CrossRef]

17. Kalunke, R.; Tundo, S.; Sestili, F.; Camerlengo, F.; Lafiandra, D.; Lupi, R.; Larré, C.; Denery-Papini, S.; Islam, S.; Ma, W.; et al. Reduction of Allergenic Potential in Bread Wheat RNAi Transgenic Lines Silenced for CM3, CM16 and 0.28 ATI Genes. *Int. J. Mol. Sci.* **2020**, *21*, 5817. [CrossRef]

18. Nawaz, G.; Usman, B.; Zhao, N.; Han, Y.; Li, Z.; Wang, X.; Liu, Y.-G.; Li, R. CRISPR/Cas9 Directed Mutagenesis of OsGA20ox2 in High Yielding Basmati Rice (*Oryza sativa* L.) Line and Comparative Proteome Profiling of Unveiled Changes Triggered by Mutations. *Int. J. Mol. Sci.* **2020**, *21*, 6170. [CrossRef]

19. Wang, H.; Zhou, Q.; Mao, P. Ultrastructural and Photosynthetic Responses of Pod Walls in Alfalfa to Drought Stress. *Int. J. Mol. Sci.* **2020**, *21*, 4457. [CrossRef]

20. Hashimoto, T.; Mustafa, G.; Nishiuchi, T.; Komatsu, S. Comparative Analysis of the Effect of Inorganic and Organic Chemicals with Silver Nanoparticles on Soybean under Flooding Stress. *Int. J. Mol. Sci.* **2020**, *21*, 1300. [CrossRef]

21. Luo, W.; Komatsu, S.; Abe, T.; Matsuura, H.; Takahashi, K. Comparative Proteomic Analysis of Wild-Type Physcomitrella Patens and an OPDA-Deficient Physcomitrella Patens Mutant with Disrupted PpAOS1 and PpAOS2 Genes after Wounding. *Int. J. Mol. Sci.* **2020**, *21*, 1417. [CrossRef] [PubMed]

22. Olivares-García, C.A.; Mata-Rosas, M.; Peña-Montes, C.; Quiroz-Figueroa, F.R.; Segura-Cabrera, A.; Shannon, L.M.; Loyola-Vargas, V.M.; Monribot-Villanueva, J.L.; Elizalde-Contreras, J.M.; Ibarra-Laclette, E.; et al. Phenylpropanoids Are Connected to Cell Wall Fortification and Stress Tolerance in Avocado Somatic Embryogenesis. *Int. J. Mol. Sci.* **2020**, *21*, 5679. [CrossRef] [PubMed]

23. Qiu, Z.; Wen, Z.; Yang, K.; Tian, T.; Qiao, G.; Hong, Y.; Wen, X.P. Comparative Proteomics Profiling Illuminates the Fruitlet Abscission Mechanism of Sweet Cherry as Induced by Embryo Abortion. *Int. J. Mol. Sci.* **2020**, *21*, 1200. [CrossRef] [PubMed]

24. Wu, S.; Guo, Y.; Adil, M.F.; Sehar, S.; Cai, B.; Xiang, Z.; Tu, Y.; Zhao, D.; Shamsi, I.H. Comparative Proteomic Analysis by iTRAQ Reveals that Plastid Pigment Metabolism Contributes to Leaf Color Changes in Tobacco (*Nicotiana tabacum*) during Curing. *Int. J. Mol. Sci.* **2020**, *21*, 2394. [CrossRef] [PubMed]

25. Xiang, C.; Yang, X.; Peng, D.; Kang, H.; Liu, M.; Li, W.; Huang, W.; Liu, S.-M. Proteome-Wide Analyses Provide New Insights into the Compatible Interaction of Rice with the Root-Knot Nematode *Meloidogyne graminicola*. *Int. J. Mol. Sci.* **2020**, *21*, 5640. [CrossRef]

Publisher's Note: MDPI stays neutral with regard to jurisdictional claims in published maps and institutional affiliations.

International Journal of
Molecular Sciences

Review

Bringing New Methods to the Seed Proteomics Platform: Challenges and Perspectives

Galina Smolikova [1], Daria Gorbach [2], Elena Lukasheva [2], Gregory Mavropolo-Stolyarenko [2], Tatiana Bilova [1,3], Alena Soboleva [2,3], Alexander Tsarev [2,3], Ekaterina Romanovskaya [2], Ekaterina Podolskaya [4,5], Vladimir Zhukov [6], Igor Tikhonovich [6,7], Sergei Medvedev [1], Wolfgang Hoehenwarter [8] and Andrej Frolov [2,3,*]

[1] Department of Plant Physiology and Biochemistry, St. Petersburg State University,
199034 St. Petersburg, Russia; g.smolikova@spbu.ru (G.S.); bilova.tatiana@gmail.com (T.B.);
s.medvedev@spbu.ru (S.M.)

[2] Department of Biochemistry, St. Petersburg State University, 199178 St. Petersburg, Russia;
daria.gorba4@yandex.ru (D.G.); elena_lukasheva@mail.ru (E.L.); gm2124@mail.ru (G.M.-S.);
oriselle@ya.ru (A.S.); alexandretsarev@gmail.com (A.T.); rcatherine@mail.ru (E.R.)

[3] Department of Bioorganic Chemistry, Leibniz Institute of Plant Biochemistry, 06120 Halle (Saale), Germany

[4] Institute of Analytical Instrumentation, Russian Academy of Science, 190103 St. Petersburg, Russia;
ek.podolskaya@gmail.com

[5] Institute of Toxicology, Russian Federal Medical Agency, 192019 St. Petersburg, Russia

[6] All-Russia Research Institute for Agricultural Microbiology, 196608 St. Petersburg, Russia;
vladimir.zhukoff@gmail.com (V.Z.); arriam2008@yandex.ru (I.T.)

[7] Department of Genetics and Biotechnology, St. Petersburg State University, 199034 St. Petersburg, Russia

[8] Proteome Analytics Research Group, Leibniz Institute of Plant Biochemistry, 06120 Halle (Saale), Germany;
wolfgang.hoehenwarter@ipb-halle.de

* Correspondence: andrej.frolov@ipb-halle.de; Tel.: +49-(0)-34-555-821350

Received: 10 November 2020; Accepted: 27 November 2020; Published: 1 December 2020

Abstract: For centuries, crop plants have represented the basis of the daily human diet. Among them, cereals and legumes, accumulating oils, proteins, and carbohydrates in their seeds, distinctly dominate modern agriculture, thus play an essential role in food industry and fuel production. Therefore, seeds of crop plants are intensively studied by food chemists, biologists, biochemists, and nutritional physiologists. Accordingly, seed development and germination as well as age- and stress-related alterations in seed vigor, longevity, nutritional value, and safety can be addressed by a broad panel of analytical, biochemical, and physiological methods. Currently, functional genomics is one of the most powerful tools, giving direct access to characteristic metabolic changes accompanying plant development, senescence, and response to biotic or abiotic stress. Among individual post-genomic methodological platforms, proteomics represents one of the most effective ones, giving access to cellular metabolism at the level of proteins. During the recent decades, multiple methodological advances were introduced in different branches of life science, although only some of them were established in seed proteomics so far. Therefore, here we discuss main methodological approaches already employed in seed proteomics, as well as those still waiting for implementation in this field of plant research, with a special emphasis on sample preparation, data acquisition, processing, and post-processing. Thereby, the overall goal of this review is to bring new methodologies emerging in different areas of proteomics research (clinical, food, ecological, microbial, and plant proteomics) to the broad society of seed biologists.

Keywords: data processing; gel-based proteomics; gel-free proteomics; glycation; glycosylation; phosphorylation; post-translational modifications; proteomics

1. Introduction

Seeds represent the basis of the human diet and contribute daily to consumed foods [1]. Hence, the rapidly growing human population requires a secure, continuous supply with foods of appropriate quality and safety [2]. Because of this, modern agriculture aims at sustainable production of high-quality seeds. On the one hand, due to the ongoing climate changes, it can be a challenging task [3]. On the other hand, the improvement of crop plant productivity is desired [4,5]. Hence, the yields of crop biomass and tolerance to environmental stress need to be increased simultaneously. For this, understanding of the fundamental processes accompanying seed development, storage, and germination, is required.

During development, the seed passes through the stages of embryogenesis, filling, and late maturation [6]. Importantly, the last steps of seed maturation are accompanied by development of desiccation tolerance and onset of dormancy—events that affect the tolerance of seeds to storage conditions thus having a great impact on their quality [7]. In turn, environmental stressors, like drought, high salinity, and high light, directly affect maturation of seeds and might compromise their longevity, i.e., ability to maintain viability during storage [8,9]. To a large extent it is underlaid by stress-related metabolic adjustment, i.e., accumulation of osmoprotective amino acids and sugars in plant tissues [10]. In combination with oxidative stress, typically accompanying plant response to abiotic stressors, this might lead to an increased oxidative and glycoxidative modification of seed proteins, that might negatively affect quality of seeds [11]. An understanding of mechanisms underlying these processes is required to ensure sustainable production of seed crops.

Seed germination and seedling development are largely dependent on the metabolic status of the plant, especially availability of reserve substances [12–16]. The chemical nature of these biomolecules differs between plant species. For instance, in cereals, the major seed storage tissue is endosperm, where starch can account for up to 87% of the total dry seed weight [14,17]. In contrast, in legumes, the principle storage organs are cotyledons, which, for example, in soybean, can contain up to 40 and 20% (*w/w*) proteins and oils, respectively [17]. Therefore, due to high constitutive levels of protein biosynthesis, legume seeds represent an excellent platform for the large-scale production of transgenic proteins [16,18,19]. As legume seeds are widely used for human nutrition and animal feed [20], their development is of particular interest to plant biologists. The model plants most commonly used for this are pea (*Pisum sativum*), faba bean (*Vicia faba*), soybean (*Glycine max*), lotus (*Lotus japonicus*), and annual barrel medic (*Medicago truncatula*) [14].

In this context, knowledge of the seed proteome of legume crop plants, as well as its dynamics in response to environmental and biological stressors, may significantly impact the improvement of their quality and nutritional properties [21]. Therefore, such versatile functional genomics tool as proteomics is widely used in seed research [11,22–28]. Currently, it is routinely employed in comprehensive profiling of complex protein extracts and delivers valuable qualitative and quantitative information on protein dynamics in plant organisms. Implementation of cell fractionation techniques allows reasonable simplification of the protein sample matrix and provides better insight into the molecular organization of individual compartments, as well as macromolecular complexes like cytoskeleton, membranes, and cell wall [29–33]. More specifically, the global proteomic approach addresses alterations in multiple biological processes, occurring sequentially and in parallel at the tissue, cell or subcellular level [24,34,35]. Based on the acquired information, development of seeds and seedlings can be systematically characterized [36,37]. Because of this, the molecular basis of seed vigor and its alterations in response to developmental or environmental stimuli have been intensively studied over the past decade [15,28,38–45], although the key proteins associated with seed vigor are still poorly understood. One of the most possible reasons for this is the high complexity of seed proteome compared with highly abundant storage polypeptides and low-abundant regulatory and signaling proteins. For this reason, depletion of storage proteins [46] and enrichment of low-abundant post-translationally modified species [47] is desired.

Thus, here we address the proteomics techniques currently established in plant science and potentially applicable to seed research. Here we discuss methodological aspects, with a special emphasis on sample preparation, data acquisition, processing, and post-processing. We consider both the methods and techniques already successfully introduced in seed proteomics, as well as those prospectively applicable and promising in this field.

2. Seed Proteomics: General Methodology and Applications

The proteome is a complex system, representing a result of interconnected dynamic properties of individual proteins [48]. In agreement with this, proteomics aims at qualitative and/or quantitative characterization of the proteome, with the depth defined by the type and complexity of the specific research aim. Accordingly, so-called discovery proteomics is focused on maximizing sampling depth, annotation of individual proteins, expression patterns of their isoforms, as well as identity and precise localization of post-translational modifications (PTMs) [11,49]. Discovery proteomics is inherently hypothesis-generating, implementing relative or absolute quantification of individual proteins (protein dynamics), in response to internal or external factors [50]. Although to some extent, these tasks can be solved by other analytical techniques (e.g., immunochemical methods, like ELISA or Western blotting) [51], the most efficient way it can be achieved is by mass spectrometry coupled on-line with high-resolution separation techniques [52]. Thereby, protein identity can be assigned by tandem mass spectrometry (MS/MS), i.e., fragmentation of peptide ions, pre-selected in so-called full-MS survey scans (data-dependent acquisition, DDA) or acquired in parallel (data-independent acquisition, DIA) [53,54]. Further post-processing with versatile bioinformatic tools gives access to functions and intracellular localization of individual seed proteins, as well as their involvement in complex protein-protein interaction networks [55].

In general, proteomic analysis can rely either on top-down or bottom-up strategies [56,57]. In the first case, ions of individual proteins with molecular weight below 25 kDa are measured directly, including fragmentation and MS/MS acquisition of protein fragment ions [58,59], whereas the bottom-up strategy employs proteolytic digestion of protein mixtures prior to MS and MS/MS analysis of the resulting proteolytic peptide-specific ions [60]. Thereby, identification of proteins by bottom-up proteomics relies on sequence tags, comprising exact monoisotopic mass, charge, and MS/MS fragmentation patterns acquired for peptide ions [61]. As application of top-down proteomics is restricted by molecular weight and purity requirements for analyte proteins, as well as simply on the protein size and charge abundance [56], its use in seed proteomics is limited [62]. Thus, in the absolute majority of cases, seed proteomics relies on the bottom-up strategy [20,26,63,64].

Obviously, the high complexity of the eukaryotic proteome represents the greatest challenge in proteomics [65]. Therefore, powerful mass spectrometric techniques, used for proteome analysis, need to be complemented by high-resolution and high-throughput separation methods. The major challenge in seed proteomics is the high abundance of storage proteins, which strongly dominate the seed proteome and can be used as protein quality markers [66]. These proteins need to be depleted, before the low-abundance seed polypeptides can be accessed [63]. A variety of depletion techniques, applicable for seed storage proteins, were comprehensively reviewed by Miernyk recently [26]. This can be accomplished by extraction with aqueous (aq.) isopropanol-based solvents [67], or by precipitation in presence of aq. 0.01–0.1% (*w/v*) protamine sulfate [68]. The latter technique proved to be well-compatible with label-free [63] and tag-based [69] quantitative techniques. Alternatively, seed storage globulins (glycinin and ß-conglycenin) can be efficiently removed by 10 mmol/L $CaCl_2$ [70]. Selective enrichment of minor seed proteins can also be achieved by implementation of the combinational peptide ligand libraries technology [71]. Finally, reduction of sample complexity can be achieved by centrifugation-based fractionation of total protein extracts, as it was done to obtain the lipid droplet-enriched protein fraction from tobacco seeds [72]. Recently, Du et al. proposed an alternative approach—absorption on a polyvinylidene fluoride (PVDF) membrane for isolation of lipid body-associated proteins from maize seeds [73].

In general, separation methods employed in proteomics can be attributed to either (i) gel-based or (ii) gel-free strategies [74], clearly different in their methodological setup: gel-based methods assume separation at the protein level, whereas the gel-free approach relies on limited enzymatic hydrolysis prior to separation (Figure 1).

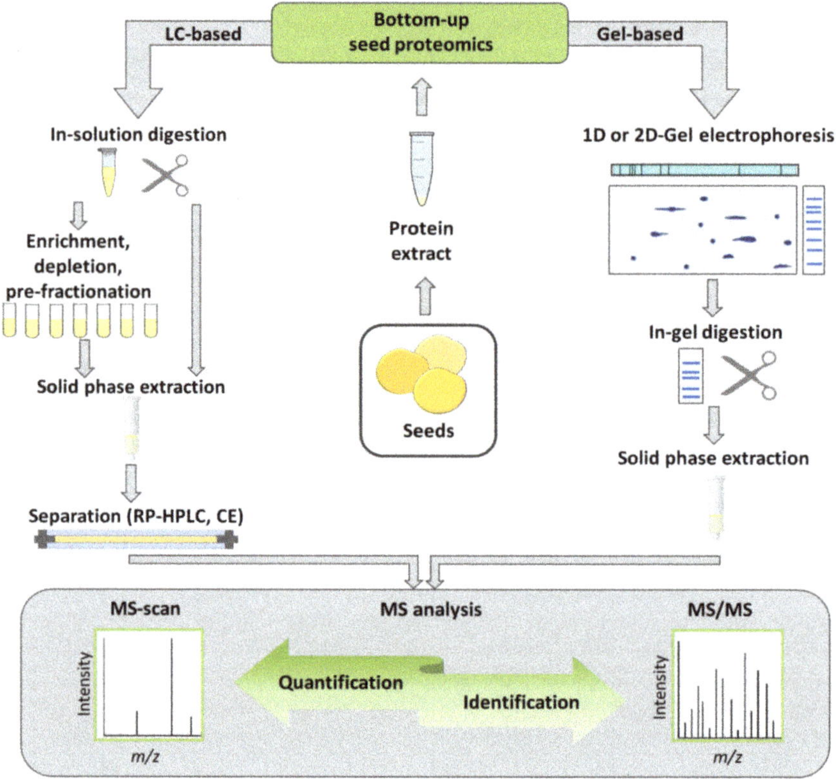

Figure 1. The overview of the experimental workflows for gel-based and gel-free proteomics.

Accordingly, the first group of methods relies on polyacrylamide gel electrophoresis in sodium dodecyl sulfate (SDS-PAGE) or two-dimensional gel electrophoresis (2D-GE), whereas the second one typically employs liquid chromatography (LC). Alternatively, gel-free electrophoretic techniques, such as free-flow or capillary electrophoresis can be used for protein or peptide separation (the methodology is comprehensively reviewed by Dawod et al. [75]), although in plant proteomics these methods are usually applied to analysis of membrane proteins, not to the seed proteome [76].

3. Gel-Based Bottom-Up Proteomics

Due to relatively low analytical resolution (i.e., number of bands reliably detectable in electropherograms) of SDS-PAGE, gel-based proteomics typically relies on 2D-GE [77,78], i.e., a two-step procedure, sequentially employing isoelectrofocusing (IEF) for separation of proteins by isoelectric point, followed by SDS-PAGE for separation by molecular weight (Figure 1). To establish pH gradients, required for IEF, protein samples are solubilized in aqueous buffers supplemented with carrier-ampholytes, chaotropic agents, and detergents [79]. Most often, 7–8 mol/L urea and 2 mol/L thiourea are used as chaotropic agents (i.e., compounds, disrupting the hydrogen bonding network of water) [80] whereas 3-(3-cholamidopropyl)dimethylammonio)-1-propanesulfonate (CHAPS) [81] and Triton X-100 are typically used as detergents (i.e., substances, disrupting intra- and inter-molecule non-polar interactions) [82].

All along with its good compatibility with highly-sensitive visualization and MS-based identification techniques, during the last decades 2D-GE became a powerful and versatile tool for bottom-up proteomics. It allows detection of thousands of proteins in one experiment [83]. Thereby, in both separation steps the resolution of 2D-GE can be adjusted to the specific needs of a proteomics experiment [84]. Thus, selection of a smaller pH range can give better insight into the fractions of basic, neutral or acidic proteins, with longer immobilized pH gradient (IPG) strips giving better spot resolution [85]. At the level of PAGE separation, better resolution can be achieved by increase of gel size. In contrast to gel-free techniques, 2D-GE gives access to the patterns of protein isoforms and relation of isoforms to specific post-translational modifications (PTMs)–phosphorylation, glycosylation, oxidation [86]. Another important feature of 2D-GE is the high reliability of protein quantification, its independence from matrix effects and the availability of orthogonal visualization techniques [87].

On the other hand, 2D-GE has several important limitations. First of all, reproducibility of the method is often compromised, that can be related to inhomogeneity of the gel or cathodic drift, i.e., progressive loss of basic proteins during prolonged application of electric field during electro-focusing [83]. Not less importantly, two-dimensional separation is highly dependent on sample preparation, and inconsistency between replicates at this step might result in high variations in electrophoretic mobility of individual proteins [87]. Finally, it is difficult to resolve the whole proteome in one experiment, as the proteins behind the pI gradient, applied during IEF, remain unseparated [88]. Low abundant and hydrophobic proteins are also often undersampled by 2D-GE.

3.1. Sample Preparation for Gel-Based Proteomics

Protein extraction procedures, conventionally applied in seed proteomics, are usually designed in agreement with this reconstitution scheme (Table A1). To achieve high efficiency of protein extraction, seed tissues are usually shock-frozen in liquid nitrogen and homogenized with a mortar and pistil or/and in a ball mill [26]. Among the variety of available techniques, phenol extraction and trichloroacetic acid (TCA)/acetone precipitation are the most effective methods to achieve comprehensive protein isolation [89]. For quantitative solubilization of seed proteins (especially membrane ones), detergents (e.g., SDS or Triton X-100) need to be added [90,91]. As was explicitly demonstrated, application of detergents is strongly mandatory for reconstitution of seed proteins [92,93]. Technically, TCA/acetone procedure is easier—it relies on direct precipitation of proteins from plant tissues. The efficiency of this approach was demonstrated in the analysis of germinating wheat seeds [94]. Interestingly, as a precipitation agent, acetone can be also used as it is (without TCA) or such extraction can be complemented with a MeOH/CHCl$_3$/H$_2$O procedure—this approach was shown to be advantageous in terms of proteome coverage in the study of quinoa seeds [95]. In contrast, phenol extraction is a two-step procedure, i.e., (i) solubilization of proteins with phenol and (ii) precipitation with methanol afterwards [55]. In comparison to other extraction methods, the TCA/acetone precipitation proved to be highly efficient in the proteome analysis of protein-rich seeds, as was demonstrated for soybean seed protein preparations [96]. This method was also applicable for wheat and rice seeds, which are characterized with a high content (90–95%) of starch and low protein (4–10%) and lipid (about 1%) content [97,98]. On the other hand, although the phenol-based methods might suffer from compromised recovery, this approach provides better purity, in comparison to acetone/TCA extraction. It also ensures reliable removal of anionic polysaccharides and nucleic acids, which might interfere with the 2D-GE procedure [99].

Importantly, while working with whole seeds, special attention needs to be paid to removal of secondary metabolites, especially phenolics, which are typically abundant in seed coats and might react with proteins, affecting electrophoretic separation [100]. This is especially important for the study of berries when it comes to the proteome of the secondary cell wall (seed coat), also rich in anthocyanins, tannins, terpenes [101]. Oxidative damage of proteins by phenolics can be suppressed by the addition of soluble or non-soluble polyvinylpyrollidone (PVP) to the extraction buffer [89] or (even better) by supplementation of PVP directly to plant samples prior to tissue grinding [102]. The effect of this procedure was demonstrated in a proteomics study, dealing with the effect of rhizobia on pea seed productivity [103]. Reducing agents, such as mercaptoethanol, dithiothreitol (DTT), and ascorbic acid are added as well [104], although the latter can promote oxidation and glycation of amino acid side chains in the presence of transition metals, as was shown with different in vitro glycation models [105–107]. It is important to note, that despite its excellent performance, phenol extraction suffers from low sample throughput and requires some experience in handling.

Most other protein extraction methods provide lower protein yields, incomplete protein recovery, and/or compromised purity. Easiest, ground plant tissues can be extracted directly with sample buffer (sodium or potassium phosphate with pH close to neutral) containing urea and usually (but not always) thiourea. This method proved to be applicable for extraction of legume seed proteins [95,96]. To achieve higher recovery of membrane proteins, sample buffer can be supplemented with CHAPS (e.g., for analysis of immature *Medicago truncatula* seeds) [82], Triton X-100 (as was applied in profiling of germinating soybean seeds) [108], Nonidet P40 (NP-40, e.g., used in the analysis of dormant and germinating *Arabidopsis thaliana* seeds) [109], or sodium dodecyl sulfate (SDS) in the presence of 20% (*v/v*) glycerol, successfully applied in the evaluation of persulfide metabolism in developing Arabidopsis seeds [110]. To remove lipids (which can reduce solubility of proteins and affect thereby IEF separation [111]), pre-cleaning with petroleum ether [112] or dichloromethane (DCM) [113] can be implemented in the protocol of isolation proteins from legume seeds. Resulting pre-cleaned extracts can be analyzed directly [114] or further purified, e.g., by precipitation with TCA/acetone, as was applied in the multi-omics study of rice seeds [115]. Alternatively, soluble protein fractions can be extracted by aqueous buffers (typically Tris-HCl or HEPES), further precipitated by acetone, and dried. This method gives access, in particular, to the soluble part of the proteome, which was successfully used to study the reserve mobilization of proteins [108].

It is important to note, that different seed matrices require different isolation protocols to achieve the best possible results. For example, Bose et al. showed that extraction of seed proteins in cereal species requires special considerations in respect of isolation buffer composition: for example for wheat the presence of a chaotropic agent (urea) is advantageous, whereas for rye a Tris-HCl buffer yields better proteome coverage [64]. Additionally, a step-wise extraction procedure to obtain water-soluble (albumins), salt-soluble (globulins), alcohol-soluble prolamins (ethanol extracted), and alcohol-insoluble prolamins (NaOH extracted)—so-called Osborn protocol [116], is efficient in the extraction of proteins from stark-rich seeds [117]. Chaotropic agents (urea, thiourea) and polyvinylpyrrolydone need to be added when cell wall-associated proteins are isolated [71].

3.2. Visualization of Electrophoretic Zones in Gel-Based Proteomics

In an electropherogram, separated proteins can be visualized as patterns of characteristic signals—so-called "spots", visible with the eye or in fluorescence detection mode, respectively [56]. Thus, an appropriate visualization of separated spots becomes critical for successful interpretation of 2D-GE data. Usually, when only major proteins are targeted, corresponding spots are visualized with colloidal Coomassie Blue dye [118], which is completely compatible with MS and, according to our own validation experiments, provides a sensitivity of several dozen nanograms per electrophoretic zone [119]. This technique suits well to the detection of seed proteins and was proven to give reliable quantitative results [120]. When higher sensitivity is desired, various silver staining methods can be employed [121]. This technique is highly sensitive and provides reliable protein detection in the femtomol range [122]. Thereby, formaldehyde-free protocols allow reliable MS-based identification of separated proteins [123], that were successfully applied to detection of proteins in seeds and cold-pressed oils [124]. One must keep in mind, that this method suffers from low linearity [79]. Therefore, fluorescent dyes, like SYPRO™ Ruby [125] and Ruthenium red [126] stains, are advantageous, when quantification is desired. Indeed, such reagents provide sensitivity, close to silver staining, and a linear dynamic range of up to three orders of magnitude [127], being well applicable to gel-based separations of seed proteins [128]. Obviously, due to the high number of detected signals, highly-sensitive methods ultimately require higher resolution. In the best case scenario, 2D-GE separation allows distinction of up to 10,000 spots per experiment [129,130]. Remarkably, as 2D-GE analysis relies on sequential separation by isoelectric point and molecular weight, these two parameters can be easily derived from 2D-electropherograms, and provide, thereby, a further level of protein annotation, in addition to MS- and MS/MS analysis of individual spots [23]. These data can be efficiently applied to cross-validation of results, obtained by gel-free techniques in combination with gel-based/gel-free workflows, increasing, thereby, the overall proteome coverage and identification reliability [120]. It is important to note, that the number of detected spots does not directly correspond to protein identification rates. Indeed, on one hand, polypeptides with similar pI and Mr tend to co-migrate in both separation dimensions. On another, the proteins, represented by several isoforms usually appear as characteristic signal patterns, slightly different in one or both parameters. These signals can also be represented by truncated and post-translationally modified protein variants, which can be in special cases assigned by a combination of Western blotting and tandem mass spectrometry (MS/MS) [131]. Thus, 2D-GE gives valuable information about qualitative and quantitative isoform patterns [34], although the overall numbers of non-redundant proteins, identified in individual 2D-GE, is typically restricted to several hundred [11]. This limitation can be, however, overcome by the implementation of enrichment/depletion and pre-fractionation techniques, which allow detection of overall higher numbers of spots [68].

3.3. Identification of Individual Proteins by Mass Spectrometry

After excision of individual spots, the corresponding proteins can be subjected to limited proteolysis [20]. The digestion protocols typically rely on a well-established procedure, only slightly varying depending on specific experimental setup (Figure 2). This methodology is well-established in plant proteomics [132] and specifically in seed applications [133] as a modification of the procedures, developed earlier for analysis of animal tissues [134]. Thus, when Coomassie dye is used for gel staining, the workflow includes excision of spots, cutting them in small sections, and de-staining by repeated incubation in aqueous methanol- or acetonitrile-containing buffers (Table A2) [135]. After de-staining, gels are dried at room temperature and rehydrated with a protease solution [136]. The dried gel pieces, swollen in protease solution efficiently absorb enzyme, thereby facilitating proteolysis and increasing its efficiency. After completing the digestion, proteolytic peptides can be recovered by repeated dehydration and re-hydration of gel sections. For silver nitrate stained gels, no de-staining procedure prior to mass spectrometry is necessary [137]. However, only formaldehyde-free techniques can be applied, when proteolysis and MS-based characterization of hydrolysates are desired [122]. Importantly, while reduction of disulfide bonds and alkylation of resulted sulfhydryls can be done during sample preparation (Table A2), as can be exemplified by the proteomic characterization of lupin seeds [138]), this step is often implemented after excision of gel slices [31,139,140]. The main limitation of the conventional in-gel digestion workflow is its somewhat limited throughput. However, recently, this was successfully overcome with the introduction of the high-throughput in-gel digestion (HiT-Gel) technique, based on a 96-well plate format (Table A2) [141], which still needs to be established in seed proteomics.

Identification of proteins in tryptic hydrolyzates relies on mass spectrometric (MS) analysis of their components, i.e., proteolytic peptides. Thereby, for reliable identification and quantification of each individual protein, at least one peptide needs to be proteotypic, i.e., uniquely identifying a specific protein and consistently detectable during MS analysis [142]. Thus, mass spectrometric assays, based on such peptides, provide high sensitivity and specificity, being a good alternative to more time-consuming immunochemical techniques [143,144]. As each excised electrophoretic zone contains relatively low numbers of proteins (ideally only one), analysis of the resulting digests often relies on matrix-assisted laser desorption-ionization—time of flight (MALDI-TOF)- or TOF/TOF-MS without introduction of any chromatographic separation step. In the simplest case, MALDI-TOF-MS analysis employs so-called peptide mass fingerprinting, i.e., protein identification by the characteristic MS pattern of proteolytic peptides [145,146]. This approach suffers, however from poor sensitivity and selectivity [56]. Indeed, only highly-abundant proteins generate patterns of proteolytic peptides sufficient for their unambiguous identification. On the other hand, these abundant components suppress the signals of co-migrating peptide ions representing minor proteins, making their annotation challenging. To some extent, this limitation can be overcome by tandem mass spectrometry (MS/MS): MALDI-TOF/TOF instrumentation can be applied to obtain more specific and reliable results [140,147]. However, because of poor precision of precursor isolation (usually relying on time-based selection during separation in the first short field-free region with the precision of not better than ±5 *m/z*), for MALDI-TOF/TOF-MS instruments, identification by several such peptides might be advantageous. Indeed, as it was shown in 2D-GE-MALDI-TOF/TOF-MS experiments with mixtures of six proteins, MS/MS-based identification with three peptides seems to be more reliable [148], that would be in agreement with standards, conventionally accepted in liquid-chromatography-based animal [149] and plant [63] proteomics. In practice, this identification scheme can be approached by the implementation of reversed phase—high performance liquid chromatography (RP-HPLC) separation in off-line mode and/or automated acquisition protocols, relying on MS/MS analysis of multiple selected precursor ions—a strategy, well-applicable to plant proteomics [150]. In general, it needs to be concluded, that the low resolution of the precursor selection algorithm essentially restricts the discovery potential of MALDI-TOF/TOF-MS. Implementation of liquid chromatography (LC) separation prior to MALDI-TOF/TOF-MS analysis allows overcoming this limitation and provides improved identification rates for proteins, co-migrating in one electrophoretic zone (spot) [151]. It is important to note, that in some cases, for example in analysis of seed storage proteins (SSPs), this strategy might not be efficient enough for distinguishing closely related polypeptides. Indeed, such proteins have a high degree of homology in their sequences because they are encoded by paralogous genes. In such cases considering specific signal patterns at 2D-GE protein maps can be helpful for correct protein annotation.

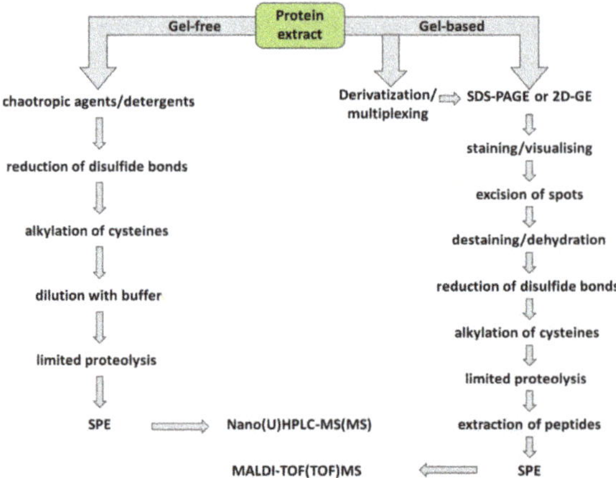

Figure 2. Overview of enzymatic digestion implemented in gel-based and gel-free experimental setup.

One of the most important features of 2D-GE is the possibility for relative quantification of individual electrophoretic zones [152]. This information provides direct access to quantitative profiles of protein abundance (also often referred to as protein expression) and allows clear and straightforward statistical interpretation of data [153]. In this context, protein dynamics, i.e., alterations in protein expression profiles in time or in response to application of certain experimental conditions, can be characterized [23,140]. Thereby, UV or VIS optical density of stained proteins, or intensity of fluorescence signal, can be considered as quantitative parameters. The appropriate integration algorithms are usually implemented in automatized protocols and established software tools, providing convenient access to image acquisition, spot detection, matching, normalization, statistical analysis, and annotation, also often combined with robotized spot cutting and in-gel digestion. The most commonly applied tools rely on numerical approaches (PDQuest (Bio-Rad Laboratories), Delta2D (Decodon), and Melanie (GeneBio) and can be successfully complemented by home-made systems [154].

Unfortunately, although 2D-GE instrumentation is relatively cheap, easy to handle and adequately supported by efficient data analysis software, this method has several intrinsic limitations. The most critical of them is the relatively low inter-gel reproducibility, i.e., difficulties in spot detection and matching in parallel gels [155]. Indeed, each IPG strip and electrophoresis block represents an independent system with separation conditions, different from those of others. Although batch setups for casting and running gels are currently well-established, this inter-system error cannot be eliminated completely because of a local heterogeneity of casting solutions. This fact might also affect dispersion within treatment groups and significance of inter-group comparisons [156]. Fortunately, this essential limitation can be efficiently overcome and adequately corrected by sample multiplexing in terms of the difference gel electrophoresis (DIGE) approach [157]. For this, cyanine dyes (Cy3 and Cy5) can be added to alternating samples, whereas the third dye (Cy2) can be used for standardization [158]. Due to the presence of different fluorophores, each dye can be detected independently without essential cross-talk with limits of detection (LODs) in the higher pg range. Therefore, this method was successfully applied in seed proteomics [20,28].

Although the 2D-GE technique is being continuously improved, some limitations of the method have not yet been overcome. First, as most of the spots are composed of several proteins (typically present in unknown ratios), the reliability of quantitative profiles obtained by 2D-GE is often questionable. Because of this, additional quantitative Western blotting experiments for proteins of interest are desired [31]. Further, even when multiplexing with DIGE is performed, the method might still suffer from inter-replicate variability. This could be overcome by increasing the numbers of technical replicates, that would make experiments more laborious and dependent on personnel performance. Finally, in comparison to gel-free techniques, 2D-GE has lower analytical resolution and maximal achievable linear dynamic range (which for modern LC-MS-based techniques can exceed five orders of magnitude [159]). Whereas analytical resolution of gel-based methods can be essentially increased by sample pre-fractionation (either at the step of sample preparation or isoelectrofocusing), these methods are time-consuming. Therefore, gel-free techniques are being actively introduced into seed proteomics, though one must keep in mind, that due to principally different separation mechanisms, some data—molecular weight and isoelectric point, easily deliverable by 2D-GE, are not accessible by gel-free techniques. It makes 2D-GE-MS identifications quite reliable and useful for discrimination of false-positives generated by the gel-free approach.

4. Gel-Free Bottom-Up Proteomics

The gel-free approach assumes the application of enzymatic proteolysis prior to separation, i.e., the entire protein extract is digested, and the resulting mixture of proteolytic peptides is separated (Figure 1). The analysis typically employs nanoRP-HPLC with electrospray ionization (ESI)-MS coupled on-line [56] or with matrix-assisted laser desorption/ionization (MALDI)-MS coupled off-line [160]. Among these two techniques, LC-ESI-MS is more powerful and may result in identification of up to several hundred thousand peptides [11]. Thus, the LC-MS approach provides impressive analytical resolution and represents a powerful tool, delivering protein identification rates approximately ten-fold higher than conventional 2D-GE methods [161]. In plant science, specifically in seed research, gel-free bottom-up proteomics studies are accomplished as so-called shotgun experiments, assuming comprehensive characterization of complex samples in one LC-MS experiment [162].

Unfortunately, despite of its high analytical power, LC-MS-based proteomics has some limitations that need to be kept in mind when planning experiments and interpreting their results. Thus, in contrast to gel-based ones, this technique is highly sensitive to matrix effects, which are typically manifested as suppression of low-intensity peptide signals by highly abundant co-eluting species [163,164], that might compromise dynamic range of proteome analysis and precision of quantification. Therefore, the separation efficiency of the LC-MS system, prefacing ESI-MS is of critical importance for successful identification of low-abundant peptide quasi-molecular ions and quantification of corresponding proteins.

Another issue is the well-known incompatibility of the ESI mechanism with detergents (e.g., SDS and CHAPS) [165]. As desolvation of low molecular weight analyte ions relies on the solvent evaporation model [166], and detergents suppress evaporation of eluents in droplets, these compounds disrupt the transfer of analytes into the gas phase and dramatically affect the sensitivity of detection. Besides this, conventional detergents (like SDS, Triton X100, and CHAPS) are quantitatively retained in reversed phase chromatography (RPC) and broadly co-elute with proteolytic peptides, further suppressing their signals [167]. Although SDS can be removed by potassium dodecyl sulfate (KDS) precipitation, this procedure requires time-consuming optimization (Table A2) [168] and is therefore still not established for plant material. Thus, the mentioned detergents cannot be employed in sample preparation for LC-MS and isolation of proteins with phenol- and detergent-free aqueous buffers gives access only to the soluble part of the plant proteome [169]. In this case, pre-cleaning of protein extract might be necessary to remove low molecular weight metabolites, which may interfere with enzymatic proteolysis (e.g., by inhibition of proteases). Thereby, preparation of plant aqueous protein extracts requires more purification effort, in comparison, for example, with analogous isolations from mammalian tissues. Indeed, whereas for human blood plasma samples ultrafiltration is sufficient to remove interfering metabolites and salts [170], for plant extracts an additional pre-cleaning by size-exclusion chromatography on PD-10 columns is required [171]. Isolation of seed proteins by this method might be even more challenging: unfortunately, the major storage proteins tend to aggregate in the absence of detergents [93], that might result in the loss of associated low-abundant proteins as well. Therefore, this isolation strategy is hardly applicable in seed proteomics and is generally not recommended. Therefore, the introduction of SDS-PAGE as a sample preparation step can be a good alternative. In terms of this setup, the isolated protein mixture is reconstituted in SDS-sample buffer and separated by SDS-PAGE (Table A2). Afterwards, the whole lanes are excised and cut into several (at least ten) segments for in-gel digestion with protease. After pooling of individual digests, samples can be desalted and analyzed by LC-MS. This protocol proved to be well-applicable to protein-rich legume seeds [108]. However, both of the described strategies have two essential disadvantages: (i) they require time-consuming sample preparation, and (ii) might suffer from the incomplete recovery of the seed proteome. Fortunately, recovery of in-gel digestion can be potentially increased by replacing bis-acrylamide with its disulfide analog—bis-acrylcistamine (BAC) [172]. Treatment of such gels with tris(2-carboxyethyl)phosphin (TCEP) might provide quantitative extraction of proteolytic peptides after digestion. Alternatively, digestion recovery in experiments with seed proteins can be improved when detergent-free buffers containing urea or guanidine chloride are applied [95]. However, despite of this improvement, in the absence of detergents, supplementation of extraction buffers with chaotropic agents does not provide the highest possible proteome coverage.

Importantly, the number of identified proteins can be affected by intrinsic limitations of data-dependent acquisition (DDA) experiments, which are most widely used in LC-MS proteomics [173]. Thus, state of the art quadrupole-orbital trap (Q-Orbitrap), Orbitrap-based tribrids, and quadrupole-time of flight (QqTOF) mass spectrometers are able to perform DDA experiments that comprise up to 20 dependent MS/MS scans, acquired for the most intense signals detected in a survey full-MS scan within a cycle of 0.5–1 s [174,175]. However, despite this high acquisition speed (scan rate up to 20 Hz), the number of peptides that can be sequenced in an MS/MS scan is still limited and usually essentially lower than the number of multi-charged features detected at the MS level. It is especially important in seed proteomics, as seed proteome is strongly dominated by a few major storage proteins, which not only suppress identification of minor polypeptides [63,69] but also involved in the formation of amyloid structures [176]. To some extent, these problems can be addressed by introduction of pre-fractionation steps and method-specific exclusion lists [177]. However, from today's perspective, data-independent acquisition (DIA)-based techniques, ideally in combination with DDA seem to be the best choice [178], especially when post-translational modifications are addressed [179].

Another important issue is the relatively low inter-run reproducibility of spray performance. Because of this, intra- and inter-batch precision of label-free quantification is typically low. Therefore, analytical strategies need to employ inter- and intra-batch normalization [180]. In this context, techniques relying on internal normalization and multiplexing, like metabolic [181], chemical [182,183], and stable isotope/^{18}O [184] labeling are advantageous when precise quantification is desired. Although these methods have already become common in plant proteomics, they are still only applied to green parts (mostly leaf) of the most well-characterized model plants, like *Arabidopsis*. Thus, their use in seed research is still quite limited and uncommon. Finally, probably the most important factors limiting the application of LC-MS-based methods are the high costs of instrumentation and limited availability of well-trained personnel. Due to this and the above-mentioned advantages, 2D-GE remains the main "working horse" in seed proteomics.

Fortunately, current developments in analytical science provide adequate solutions to all (probably, besides the last one mentioned) limitations of the gel-free approach. Typically, these developments were initially introduced in cell, blood plasma, or mammalian tissue proteomics, and so far only a few of these methodological solutions are transferred to plant and seed research. Thus, the most of these improvements were implemented quite recently and mostly in the fields, not related to plant and, especially, seed research. Therefore, with this review, we would like to bring this summarized and critically discussed information to the seed proteomics society with the hope to push their implementation in plant proteomics. In this context, we are convinced that many of these recently proposed method improvements might allow versatile shotgun LC-MS analyses of the total seed proteome in future.

As the first improvement, highly efficient separation can be achieved by nano-scaled ultra-high performance liquid chromatography (UHPLC) [56]. Elution of individual analytes in well-defined chromatographic zones minimizes matrix effects and attenuates suppression of low-intense signals by highly-abundant species [185]. The quantitative reconstitution of membrane proteins without any deleterious effects on the sensitivity of peptide detection is another essential achievement. Thus, at the end of the last decade, Mann and co-workers proposed the so-called filter-aided sample preparation (FASP) method, assuming digestion of proteins in the presence of SDS on membrane filters, and exchanging it by urea afterwards (Table A2) [186]. This technique was successfully transferred to plant and, specifically, to seed biology: it proved to be efficient for digestion of grape leaf [187], cucumber seed proteins [188], and delivered superior protein identification rates in comparison to sample pre-cleaning by SDS-PAGE. Recently, Bose et al. comprehensively compared the performance of FASP and different extraction buffers with seeds of four cereal species [64] and demonstrated 'broad applicability of this technique in seed biology.

An alternative approach relies on immobilization of protein in gel by supplementing with acrylamide/bisacrylamide mixture after its complete solubilization in the appropriate detergent solution (so-called tube-gel digestion protocol) [189]. Immobilization in gel results in unfolding of hydrophobic (e.g., membrane) proteins and promotes their effective digestion. A similar protocol, known as gel-aided sample preparation (GASP), employing immobilization of proteins in the gel, is also applicable to intact cells [190]. These methods are well-suited for the identification of membrane and membrane-bound protein complexes, and GASP protocol was recently applied to the analysis of *Nicotiana benthamiana* leaf proteins [191]. Both FASP and GASP yield up to five-fold higher intensities of peptide signals in comparison to in-solution digestion in absence of detergents. Interestingly, the application of aqueous solutions containing high proportions of acetonitrile results in superior digestion yields when small protein amounts need to be hydrolyzed [192]. Surprisingly, the number of identified peptides was much lower when methanol was used [29]. Recently, the GASP technique was successfully up-scaled to a 96-well format—so-called HiT Gel (High Throughput in Gel Digestion) protocol, successfully applied to *Arabidopsis thaliana* protein extracts [141]. However, this technique has not yet been transferred to seed proteomics.

Further improvement of proteome coverage could be achieved by the application of commercially available detergents, forming insoluble precipitates under low pH values [193]. One such compound, sodium deoxycholate (SDC), is well-known from blood plasma proteomics [170]. Recently, SDC was successfully applied to solubilization and digestion of total barley leaf protein preparation [194] and isolates from germinating maize seeds [195]. Despite of its wide application, precipitation of SDC upon digestion is not always quantitative. The same is the case for sodium laurate [196].

An alternative solubilization/digestion strategy might rely on degradable acid-labile detergents. For example, zwitterionic PPS Silent Surfactant [197] and anionic Protease-MAX [198] surfactants are a good alternative for SDS in digestion buffers. To date, these detergents have only been used rarely in plant proteomics. Thus, Protease-MAX was recently successfully used for the characterization of *Solanum tuberosum* leaf tissue [199], whereas the former detergent, to the best of our knowledge, was not applied to analysis of plant tissues so far. Both these chemicals have not been used yet in seed research. It is important to note that PPS Silent Surfactant and Protease-MAX reagents yield hydrophobic cleavage products, which form films on the surface of aqueous phases and contaminate samples [200]. This ultimately requires additional intensive sample cleaning after proteolysis. The same is known for the anionic acid-labile detergent RapiGest SF Surfactant [201], which is currently most often applied for digestion of human plasma and tissues [202]. Recently, this reagent was successfully employed in the digestion of barley seed protein isolates, resulting in the annotation of 226 polypeptides in shotgun data-independent experiments [203]. Its application to quinoa seeds gave access to new lysine-rich seed storage globulins [162].

Another option available from the company Progenta is anionic, cationic, and zwitterionic acid-labile surfactants I and II (AALS I/II, CALS I/II, and ZALSI/II), among which AALS II, CALS II, and ZALSII have higher protein solubilization potential, i.e., can be applied to hydrophobic (e.g., membrane) proteins [204]. Recently, in a systematic comparison of different degradable surfactants, AALS II (structurally mimicking SDS) showed the best performance in terms of solubilization efficiency and sequence coverage of digested and analyzed standard protein samples [205]. This detergent was successfully optimized for the digestion of protein isolates obtained from seeds of different species [26,55,63]. Importantly, cleavage of this detergent yields relatively hydrophilic products, which can be easily removed from seed protein hydrolysates by solid-phase extraction (SPE) on the reversed-phase [26]. This compound has tunable surfactant properties, i.e., adjustable critical micelle concentration (CMC) [200]. Therefore, recently we selected this surfactant for profiling the total *Arabidopsis* leaf proteome [177,206]. Later, we successfully addressed proteome changes in oilseed rape (*Brassica napus*) seeds during germination [26] and mature pea seeds [55,63].

Remarkably, when we combined application of AALS II with long chromatographic gradients of 1.5–3 h (which allow higher numbers of cycles of DDA experiments per acquisition run), confident identification of thousands of proteins could be achieved not only in *Arabidopsis* leaf tissues [177] but also in pea seed embryos [63]. The effect of the under-sampling phenomenon, known to accompany DDA experiments [173], can be further attenuated by sample enrichment, depletion, or pre-fractionation. For example, removal of legume reserve seed proteins by their precipitation in presence of protamine sulfate or by organic solvents represents a well-established depletion procedure, which provides a deeper insight into the seed proteome [68]. For specific enrichment of seed glyco- and phosphoproteome, adequate protocols were successfully established [207,208]. Other fractionation techniques, like pre-fractionation by hydrophilic interaction liquid chromatography (HILIC) [177], gas-phase fractionation (GPF) at the level of mass analyzer [209], or selective enrichment of glycated proteolytic peptides by boronic acid chromatography (BAC) [171,210] are already established in plant biology and can be easily transferred to the specific field of seed biology. In general, fractionation, enrichment, and depletion procedures dramatically reduce the complexity of peptide mixtures analyzed in individual fractions and thus ensure the complete detection of low-abundant proteins. After combining the results obtained in all experiments, substantial improvement of proteome coverage could be observed.

To achieve the highest possible protein identification rates and precision of label-free quantification, we optimized our proteomics workflow for the highest protein recovery and reproducibility (Figure 3). Thus, in terms of our approach, after shock-freezing in liquid nitrogen, seeds or their parts (e.g., embryos or cotyledons) are ground in a ball mill, and 50–200 mg (depending on species) of frozen powder are taken for phenol extraction [26,89]. After drying, protein isolates are quantitatively reconstituted in the shotgun buffer, containing chaotropic agents (urea and thiourea) and degradable detergent (Progenta Anionic Acid-Labile Surfactant, AALSII) [206]. If protein isolation is performed carefully, and no interphase (typically containing nucleic acids and polysaccharides) is co-precipitated with the target protein fraction, seed protein isolates can be reconstituted completely. If it is not the case, after short centrifugation, protein contents are determined in supernatants. For this, we are using a two-step procedure, relying on protein determination by the 2D-Quant kit (which is based on specific binding of copper to protein [211]) with subsequent cross-validation of results by SDS-PAGE with Coomassie staining according to a well-established protocol [212]. The application of these two orthogonal approaches ensures reliable normalization of protein amounts taken for enzymatic digestion. According to our experience, twice repeated incubation of solubilized proteins with trypsin was sufficient for complete proteolysis of seed samples [56]. The completeness of hydrolysis was confirmed by SDS-PAGE, as was earlier established for human samples [107] and successfully transferred to plant, and specifically seed samples. Having this in hand, AALS can be destroyed, and the digests can be pre-cleaned with RP-SPE either in a cartridge or in a stage-tip format [59,153].

The pre-cleaned samples can be analyzed by nano (U)HPLC-ESI-MS employing either data-independent acquisition (DIA) or data-dependent acquisition (DDA) modes (Figure 3). In QqTOF-MS instruments, DIA relies on the simultaneous fragmentation of multiple peptide ions within a wide range of *m/z*. In its simplest form, this *m/z* range covers the whole measurement mass range (MS^E technology, employed by Waters instruments) [213]. This approach proved to be efficient in discovering seed proteins [112,214]. Alternatively, the full mass range can be split into 10–30 segments or windows of 20–50 *m/z* each wherein all ions are concomitantly fragmented—the so-called sequential window acquisition of all theoretical fragment ion spectra (SWATH) MS technology [215]. The strategy for analysis of SWATH data is based on peptide-centric scoring, which relies on querying chromatographic and mass spectrometric coordinates of the proteins and peptides of interest in the form of so-called peptide query parameters (PQPs) [216]. The SWATH-based DIA approach allows essential coverage of seed proteome and identification of more than 2000 individual proteins in one experiment [217]. In contrast, the DDA scan strategy relies on selective fragmentation of a defined number of the highest intensity signals detected in survey full-MS scans [173]. To increase analytical reproducibility, this approach can be additionally mapped to extracted ion chromatogram (XIC) information [218]. Identification of individual peptide sequences relies on tandem mass spectrometry (MS/MS, Figure 3) whereas quantitative information can be derived from spectrum counting (PSM counting), peak heights, and, in the most precise way, from the integration of extracted ion chromatograms [219]. During the recent decade, this approach remains a working horse of gel-free proteomics [55,220,221].

Unfortunately, both DIA and DDA have advantages and disadvantages, which need to be considered in each specific case when developing an analytical strategy. Thus, DIA delivers rich MS/MS information, although the correct annotation of fragments is critical for this technique. To some extent, targeted extraction of MS/MS spectra might simplify this task [216]. On the other hand, the major challenge of DDA experiments is analytical under-sampling [173]. Indeed, only a limited number of peptide ions can be fragmented in each DDA method cycle, whereas the major part of the proteome remains unidentified. To minimize this limitation, LC-MS analysis can be pre-faced with an additional chromatographic dimension, or implementation of prolonged separation gradients is required [177].

Thereby, the selection of mass spectrometric instrumentation in each case is generally governed by the overall goal of the specific research.

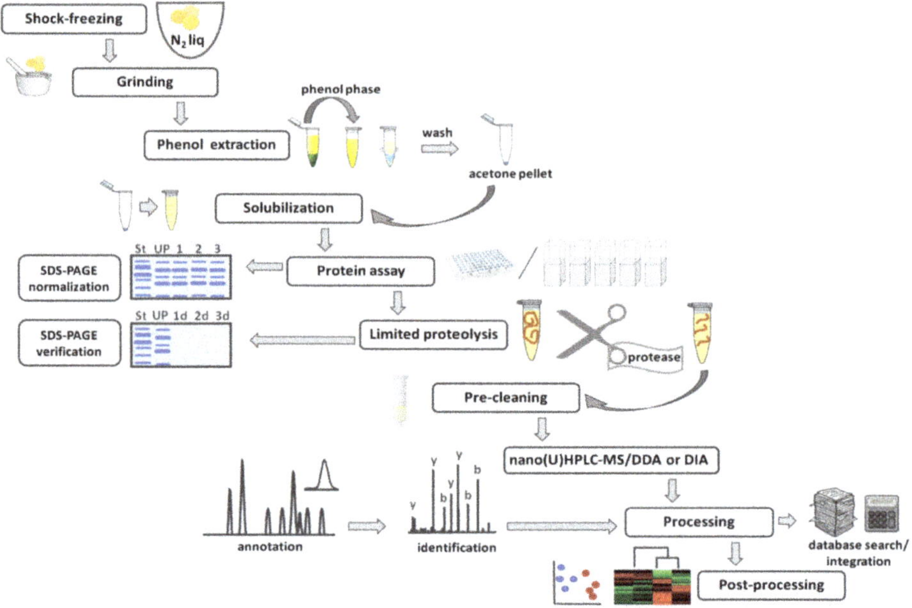

Figure 3. Detailed workflow for LC-based proteomics: protein isolation, sample preparation, and analysis.

In this context, high resolution and high mass accuracy are critical for successful annotation of sequence tags in discovery proteomics. In the simplest case, these requirements are fulfilled only at the MS level. The technical solution for this is the linear ion trap (LIT)-Orbitrap-MS technology, which employs low-frequency orbital traps and long scan times in FT mode. This assumes long MS scans with high (up to 120,000) resolution and mass accuracy about 1–3 ppm, whereas the survey MS/MS scans are acquired with unit resolution [222]. In contrast quadrupole-orbital trap, (Q-Orbitrap) instruments (Q-Exactive and Exploris™ instruments from Thermo Fisher Scientific), orbital trap-based tribrids (Fusion, Lumos und Eclipse mass spectrometers from the same company), and QqTOF mass analyzers rely on high resolution and high mass accuracy both in MS and MS/MS mode [223]. Therefore, for DDA experiments, these instruments represent the best option and give access to higher identification rates in comparison to LIT-Orbitrap-MS [173,224]. Thus, they are widely and successfully applied in seed proteome research, yielding the most comprehensive sequence coverage and giving access to the most representative seed protein datasets.

On the other hand, for targeted proteomics, triple quadrupole (QqQ) mass spectrometers, operating in selected reaction monitoring (SRM) mode, provide the highest sensitivity in the sub-ng/mL range [225]. As QqTOF-MS, from the hardware side, can be considered an extension of QqQ-MS, these instruments are also well suited for analysis in SRM mode [226]. A scan strategy called parallel reaction monitoring (PRM) extends the targeted SRM approach to the Q-Orbitrap mainly because of its two linear quadrupoles configured before the Orbitrap mass analyzer by analogy to the triple quadrupole configuration. It acquires MS/MS spectra of precursor ions isolated in the second quadrupole mass filter as opposed to monitoring individual transitions as it can be accomplished in the selected/multiple reaction monitoring (SRM/MRM) mode, delivering higher specificity with comparable sensitivity [227]. Additionally, it circumvents many of the difficulties associated with SRM development, such as optimization of collision energy and dwell times. That approach has also been demonstrated on the LTQ-Orbitrap with a limit of quantification of 100 amol in a complex plant leaf extract [175].

Although relatively few (in comparison to gel-based) gel-free seed proteomics studies have been published so far, LC-MS based techniques essentially increased the power of plant proteomics in general [57] and the depth of seed proteome characterization in particular [46,63]. Indeed, the first proteomic maps were based on two-dimensional gel electrophoresis (2D-GE) and mass spectrometric (MS) identification of visualized electrophoretic zones (spots) and contained hundreds of proteins [34]. The majority of the identified polypeptides were storage proteins, which strongly dominate the seed proteome and can serve as seed protein quality markers [66]. Removal of these highly abundant storage proteins by isopropanol- [67] or by protamine sulfate-containing solutions [68] could increase coverage of the seed proteome, however, the number of identified proteins never exceeded several hundred. Implementation of liquid chromatography (LC)-MS-based strategies provided a deeper insight into the seed proteome [20,35,64]. Growing numbers of gel-free proteomics studies extend our knowledge providing more detailed information about the seed proteome (abundant and non-abundant proteins, PTMs) in a diversity of plants: barley, rape, wheat, rye, oats, soybeans, pea, etc [26,63,64,203,228]. Thereby depletion of storage proteins and enrichment of low-abundant post-translationally modified species is desired. Indeed, the most representative LC-MS-based study to date of Min et al. performed with soybean seeds, identified 1626 non-redundant proteins [54]. Finally, this year Mergner et al. reported a comprehensive identification of Arabidopsis sees proteins: the authors reported more than ten thousand seed proteins, as a part of the most recent proteomics atlas [229]. This number significantly exceeds the outcomes of gel-based proteomics. Recently we reported the most complete, to the best of our knowledge, legume seed proteome map, comprising about 2000 non-redundant proteins. We compared seed proteomes of yellow- and green-seeded pea cultivars in a comprehensive case study. The analysis revealed a total of 1938 and 1989 non-redundant proteins, respectively [63].

It is important to note that, despite its high efficiency, the gel-free approach has several bottle necks. First of all, it requires expensive instrumentation, highly-competent personnel, and implementation of powerful bioinformatic workflows. Another serious problem is the high complexity of plant matrices, which might result in strong matrix effects, compromising the accuracy of quantification. Therefore, although the linear quantitative dynamic range of gel-based techniques is at least two orders of magnitude lower in comparison to state of the art LC-MS methods, in many cases, 2D-GE might provide more reliable quantification. Therefore, complementation of the gel-free proteomics strategy with the gel-based techniques is always beneficial and gives the best results [140]. Indeed, this complementation may include several important aspects and essentially impact on data reliability due to cross-validation of results. First, simultaneous considering data sets obtained with 2D-GE and nano-LC-MS results in higher protein identification rates and more comprehensive proteome coverage [33,57,230]. Importantly, in contrast to LC-MS, 2D-GE delivers important information on protein molecular weights and isoelectric points [231]. This feature allows direct verification of shotgun proteomics results. Additionally, this method gives information on patterns of isoforms with their post-translational modifications, which can be easily visualized by specific staining [77].

5. Post-Translational Modifications

Despite the importance of data on changes in protein abundance, this information is insufficient for a complete understanding of regulatory events behind seed maturation and germination [11]. Indeed, enzymatic post-translational modifications (PTMs), such as phosphorylation, acetylation, and glycosylation, have a great impact on intracellular signal transduction pathways [232]. Accordingly, knowledge of patterns of regulatory enzymatic PTMs facilitates understanding of biochemical and physiological alterations accompanying seed development, maturation, germination, and ageing [233]. Therefore, a comprehensive analysis of specific modification sites in seed proteins represents an important direction of actual research.

Among post-translational covalent PTMs, reversible protein phosphorylation stands out as the most extensively studied one and essentially contributes to the regulatory network, switching multiple critical cellular functions between the active and inactive state [234]. In this context, knowledge about the changes in the phosphorylation status of individual seed proteins during maturation, germination, and early seedling development might be a good starting point to study their contribution into underlying regulatory events, giving a deep insight into metabolic control of corresponding functions [47]. It is important to note that the phosphorylation state of each individual protein depends on activities of multiple enzymes—kinases and phosphatases, as well as phosphate-binding proteins, which catalyze phosphorylation and dephosphorylation, respectively, at one or several sites [235]. In the last decade, analysis of phosphorylation sites became a powerful tool to address regulatory aspects of seed development and germination [236]. Recently, Meyer et al. reported more than one thousand phosphorylated sites in seed proteins of *Arabidopsis thaliana*, soybean, and rapeseed, and pinpointed RNA biosynthesis and metabolism as the most affected gene ontology categories [47].

In the simplest and most straightforward case, the phosphorylation status of proteins can be derived from 2D-GE experiments, using specific phosphostaining. Indeed, phosphorylation typically affects protein pI values and results, thereby, in the appearance of specific signals for individual phospho isoforms in 2D-GE electropherograms [237]. This approach proved to be the method of choice for the characterization of phosphorylated proteins among seed storage polypeptides—cruciferins, napins, cupins, and vicilins [47,238–241]. Thereby, phosphorylated isoforms can be quantified by chemical dephosphorylation with hydrogen fluoride-pyridine (HF-P) [242–244], i.e., phosphorylation levels can be assessed by the difference in spot intensities, observed in gels, run before and after application of HF-P [245]. Most often, detection or cross-validation of such protein phosphorylation patterns relies on Pro-Q diamond phosphoprotein stain (Pro-Q-DPS), which is known to be a rapid and convenient tool for in-gel detection, mapping, and quantification of polypeptides [246,247]. For example, this method was successfully applied to the characterization of differential phosphorylation patterns accompanying grain development in two different Chinese bread wheat cultivars [248]. Further, this staining protocol was employed in phosphoproteome analysis of germinating seeds and early grown seedlings from *Quercus ilex* L [128]. Implementation of this protocol in parallel to a standard protein quantification procedure (e.g., Coomassie blue, silver, or Sypro Rubi staining) might give direct access to the proteins, whose abundance is regulated via phosphorylation [246]. Moreover, as was shown by Han et al. in their study of rice embryo proteome dynamics during seed germination [249], an extension of this approach on enzyme activity assays allows direct consideration of such regulatory events in the context of accompanying metabolic alterations. Despite its ease in handling and good performance, gel phosphoprotein stains have an important limitation: this method does not allow reliable assessment of phosphorylation status at specific residues when several phosphosites are present in the sequence. Spot excision followed with proteolysis, and MS/MS analysis typically gives access to protein identification, but not localization of phosphorylation sites. In such cases, site-specific antibodies or/and gel-free MS/MS-based techniques [250] need to be applied. Despite of this, the potential of the phosphostaining techniques should not be underestimated. For example, this approach is useful for fast screening of phosphoproteomes isolated from different species or tissues [251].

In contrast to the gel-based phosphostaining strategy, quantification of individual phosphorylation sites gives access to the dynamics of phosphorylation at individual protein residues in plant proteins. In this way, activation or inhibition of specific kinases and/or phosphorylases can be directly addressed either on a relative or absolute basis [252]. Quantitative assessment of these alterations in enzyme activities typically relies on corresponding proteolytic peptides obtained after enzymatic digestion of cellular proteins and enriched by immobilized metal affinity chromatography (IMAC) or metal oxide affinity chromatography (MOAC) [222]. The latter technique is currently recognized as the "working horse" of phosphoproteomics, and in the absolute majority of cases relies on titanium dioxide (TiO_2)

affinity chromatography [253], although the stationary phase can be represented by zirconia (salts of Zr^{4+}) [254], aluminum hydroxide [255] and iron oxide [222] as well. Recently, Ni-functionalized particles [256] and iron (III) stearate Langmuir-Blodgett films [222] were proposed as affinity materials for IMAC, also these techniques are still to be implemented in plant and seed biology. Pre-fractionation of complex protein digests might be another approach, giving access to enhanced phosphopeptide detection. Thus, implementation of strong cation exchange chromatography (SCX), electrostatic repulsion hydrophilic interaction chromatography (ERLIC), and solution isoelectric focusing (sIEF) resulted in enhancement of phosphosite identification [257]. However, despite of their high analytical potential, these methods still need to be optimized for seed research. Regardless of their origin, the obtained eluates or fractions are typically purified by solid-phase extraction and analyzed with nano(U)HPLC-MS and MS/MS [258]. Thereby, these experiments typically rely on label-free techniques, i.e., direct analysis of individual samples without multiplication [259]. However, due to the intrinsic limitation of ESI, all currently available label-free quantification (LFQ) techniques typically suffer from compromised precision and high inter-sample variability [219]. Although appropriate standardization and normalization procedures can improve data quality, this still remains a bottleneck of LFQ analysis [180].

To overcome this limitation, powerful chromatographic techniques can be efficiently complemented with ^{15}N labeling or derivatization-based techniques—tandem mass tags (TMT) or isobaric tags for relative and absolute quantification (iTRAQ) [260]. The first strategy represents an in vivo labeling technique, often also referred to as *Arabidopsis* metabolic labeling [261], functionally analogous to SILAC (stable isotope labeling of amino acids in cell culture). SILAC is commonly used in animal and human cell biology and gives a unique opportunity for fine dissection of regulatory pathways by simultaneous measurement of samples of various biological scenarios and deriving differences in phosphorylation directly from the mass spectra [235]. *Arabidopsis* metabolic labeling, relying on aqueous [262] or solid [263] media allows doing the same at the level of plant organism [264], that is hardly possible even for basic model animals. Although in vivo metabolic labeling was applied to *Arabidopsis* seedlings [265], this method still was not used in seed research. To a large extent, it can be explained by the high costs of ^{15}N-labeled salts and the long time required to obtain seeds, especially for valuable crop plants, like legumes or cereals. However, despite these high costs in vivo labeling of seed phosphoproteome might be a promising direction of new studies.

Techniques based on chemical tags, like iTRAQ and TMT, rely on post-isolation derivatization with multiplexing differentially isotopically labeled agents, subsequent purification with cation exchange chromatography, and nanoRP-(U)HPLC [266]. The quantitative analysis relies on tag-specific signals in the low-*m/z* range of tandem mass spectra, acquired with combined samples [267]. The iTRAQ approach proved to be efficient in analysis of plant proteome [268] and was successfully introduced in seed research. Thus, Zhang et.al. analyzed the regulatory pathways behind dormancy and germination of grassbur twin seeds and proved the regulation of ribosome synthesis and carbohydrate metabolism during seed germination via the phosphoinositide 3-kinase (PI3K) pathway [269]. Moothoo-Padayachie et al. used iTRAQ to explain the mechanisms behind the viability loss upon desiccation of recalcitrant *T. dregeana* seeds [270]. Interestingly, despite its convenience and reliability, TMT still was not used in seed phosphoproteomics.

Another important PTM is glycosylation, which patterns in plants differ significantly from mammals [271]. Besides its known role in the regulation of cell functions and compartmentalization, this modification is believed to impact seed allergenicity, attracting the special attention of food scientists due to a concept of carbohydrate cross-reactive determinants (CCDs) [272]. Seed storage proteins are known to be the most abundant polypeptide fraction, demonstrating a high degree of glycosylation [237]. As glycosylation affects protein molecular weights and isoelectric points, characteristic patterns of differentially glycosylated isoforms can be efficiently characterized by 2D-GE [79]. Individual electrophoretic zones (spots) related to glycosylation can be identified by comparison of electropherograms acquired before and after deglycosylation with protein-N-glycosidase

(PNGase), as was shown by de La Fuente et al. for phaseolin patterns in common bean (*Phaseolus vulgaris* L.) [273]. The relative levels of each glycosylated isoform can be quantified according to the data processing strategy described by Bernal et al. for mapping of patatin isoforms [245] or using similar processing pipelines. Alternatively, glycosylated proteins can be stained directly in-gel. This can be accomplished by the application of a fluorescent hydrazide Pro-Q Emerald 300 dye, available as a Glycoprotein Gel and Blot Stain Kit [274]. The glycoprotein bands or spots can be visualized by fluorescence detection after UV transillumination at 300 nm. Another glycoprotein quantification approach relies on the electrophoretic transfer of the proteins, separated by 2D-GE, to a polyvinylidene difluoride (PVDF)-membrane with subsequent visualization with a concanavalin A (Con A)/horseradish peroxidase (HRP) method [275], as was done by Weiss et al. for fractionated proteome of barley seeds [276]. The interaction of the lectin Con A with glycoproteins can be also employed for affinity chromatography, typically accomplished with agarose-immobilized Con A as a stationary phase, packed in tubes, columns, or grafted on more unconventional solid surfaces like colloidal gold and affinity membranes [277]. This technique can be combined with further analytical methods to probe specific glycoprotein interactomes or enzymatic activities. For example, Ostrowski et al. incubated pea seed glycoproteins with [^{14}C]-indole-3-acetic acid ([^{14}C]-IAA) in the presence of 1-*O*-indole-3-acetyl-β-*D*-glucose (IAGlc) synthase showed that this enzyme impacts on modification of glycosylated proteins with IAA [278]. To address the composition of protein-bound glycans, the oligosaccharide moieties can be cleaved from the polypeptide chains by PNGase, chemically derivatized with a fluorescent tag, and analyzed by hydrophilic interaction liquid chromatography (HILIC) with fluorescent detection, whereas the structure of the glycan moiety can be assessed by MALDI-TOF/TOF-MS [279].

Several other enzymatic PTMs are in the focus of seed biology research due to their involvement in regulation of key physiological processes accompanying seed maturation and germination. Thus, acetylation and succinylation are the PTMs, mainly involved in regulation of metabolism [280]. Recently, the Miernyk's group established a two-level immunoaffinity enrichment procedure for acetylated proteins and peptides for the study of the soybean seed acetylated proteome in a combination of storage protein depletion and linear ion trap (LIT)-Orbitrap-MS [281]. The combination of these techniques allowed identification of 245 acetylated proteins in developing soybean seeds. Later, He et al. applied specific immunoaffinity enrichment of seed protein tryptic digests followed with high resolution (HR)-nanoLC-MS and MS/MS for comprehensive profiling of acetylated and succinylated rice seed proteome that indicated the involvement of these modifications in multiple functions and strong physiological cross-talk between these two PTMs [282]. Another important modification in seed proteins is lipidation (most often—myristoylation, palmitoylation, and prenylation), which is often highly relevant for the interaction of dehydrins with membranes [283,284]. These modifications were detected in the dehydrin/abscisic acid (ABA)-responsive protein of *Fagus silvaticus* seeds by the combination of computational and molecular biology methods [285]. Methylation of K and R residues is widely spread in the plant kingdom [286], although it is clearly under-characterized in seeds. To summarize, among enzymatic modifications, phosphorylation and glycosylation are relatively well studied, whereas the data on acetylation of seed proteins are currently being actively acquired. In contrast, other enzyme-dependent modifications are still to be completely characterized in seed proteome.

6. Data Processing and Post-Processing

Large datasets generated in high-throughput proteomic experiments require powerful computational platforms, which enable reliable annotation of MS and MS/MS signal patterns to sequences of specific proteolytic peptides, give access to precise quantitative assignments, and support further statistical analysis to retrieve biological information. Here we summarize some major recent achievements, facilitating efficient processing and post-processing of proteomics data.

In MS-based proteomics, identification of peptide sequences typically relies on so-called sequence tags (see above) [287]. In the simplest form, a sequence tag might contain only patterns of peptide signals (peptide mass fingerprinting, PMF) [288]. Extension of the approach to MS/MS data (tandem mass spectrometric fragment patterns) provides direct access to sequence information [289]. Moreover, due to the unique character of fragment patterns, tandem mass spectrometry allows analysis of complex protein mixtures [290]. However, interpretation of MS/MS spectra in terms of peptide structure remains a challenging task. Typically, it relies on comparison of MS signals and MS/MS fragment patterns with computed data obtained by in silico digestion of defined proteins and subsequent theoretical fragmentation of matched signals, that is typically done by search machines [291]. Unfortunately, in most cases, MS/MS spectra suffer from poor quality, i.e., from incomplete fragment ion series and presence of signals not related to backbone fragmentation, i.e., other than b-, y- or immonium ions [292]. Moreover, insufficient separation prior to MS might result in overlap of several fragmentation patterns and makes straightforward interpretation difficult and possibly ambiguous.

The first successful approach for high-throughput and reliable identification of peptides relied on a partial unique sequence of at least 2–4 residues derived from characteristic mass increments in MS/MS spectra [293]. The next generation of search engines skipped this intermediate step of identification (tag sequence). Thereby, in one group of engines, having the so-called 'SEQUEST' architecture (PepSearch, SEQUEST, Crux, etc.), matching of theoretical and measured MS/MS spectra relies on cross-correlation analysis [294]. The engines, representing another group (Mascot, OLAV, Andromeda, etc.), estimate the probability of matches between the experimentally measured and calculated fragment m/z values occurring by chance [295,296].

In general, the error tolerance of the measured m/z and the size of the sequence search space are defined by the resolution of computational searches. The values of parent and fragment ion mass tolerance are defined by mass analyzer type and contribute directly to scoring matches [297]. Since search engines perform multiple comparisons on large datasets with defined mass tolerance, this might lead to some amount of false-positive identifications occurring by chance. Therefore, the actual percentage of such false identifications should be corrected by the false discovery rate (FDR) procedures. For this, tandem mass spectra are searched against a decoy database, which represents a false sequence dataset, typically representing a reversed sequence set [298,299] or database, generated by permutation [300–302]. The size of the search sequence space is partially defined by database size. Therefore, the most reliable peptide/protein identification can be achieved when the database size is small (e.g., top-down proteomics). Importantly, the size of the search space denotes not only protein sequences in a primary database but also all considered sequence post-translational modifications (PTMs) [303]. PTMs can be typically recognized by characteristic mass increments, observed in tandem mass spectra [304]. This feature is exploited by the implementation of PTMs in search algorithms [305–307]. Thereby, PTMs can be defined as static (i.e., ultimately present at specifically defined amino acid residues, e.g., carbamidomethyl, formed at all cysteine residues after treatment of proteins with iodoacetamide prior to digestion), or variable (optionally occurring at specific residues, e.g., methionine sulfoxide, often formed during sample preparation). Submission of variable modifications results in significantly increased effective search space and time, as the search engine tests all possible combinations of unmodified and modified residues [297,303].

Another popular method of MS/MS-based peptide identification is *de novo* sequencing. It is designed to extract maximal useful information from experimental spectra translating into proper peptide sequences [308,309]. Since the *de novo* sequencing does not require a sequence database, in broad perspective this technique might be the method of choice to identify proteins which are not present in databases, for example, mutation-modified proteins or proteins of poorly or/and unstudied organisms, whose genomes are not sequenced.

Originally, due to large differences in ionization efficiencies of individual peptides, quantitative perspectives of MS-based proteomics seemed limited. Indeed, high variability in charge state, peptide length, amino acid composition, and PTM patterns result in great differences in ion intensities of peptide signals in spectra, even if they represent the same protein. Thus, for accurate quantitation by intensities of quasi-molecular ions, comparisons between different samples can only be done for the same peptide mass-to-charge ratios (*m/z*) acquired under the same conditions in LC-MS experiments [310]. The strategies of quantitative comparison are usually categorized as label-based (most often—stable-isotope-labeling techniques) and label-free approaches [311]. Label-based techniques rely on the fact that stable ^{13}C, ^{18}O and/or ^{15}N-labeled peptides differ from their non-labeled counterparts only in mass but demonstrate the same chromatographic behavior and ionization efficiency [312]. Label-free quantification, in turn, can rely on feature-based quantification, or spectral counting (SC). Thereby, individual signals of peptide ions are treated as features, which are selected during peak peaking and matched during feature map alignment. The SC quantitation methods count the number of MS/MS spectra that are assigned to the same protein, ignoring intensities of corresponding peptide ions [313]. Label-free methods are characterized by a high analytical resolution and dynamic range. Label free methods are supported by numerous software tools, which make a quantitative comparison of proteins across many samples and biological scenarios facile even for a non-expert (Table A3): they include SuperHirn [314], MaxQuant [315], Progenesis [316], OpenMS [317], Proteome Discoverer [318] and the Trans-Proteomic Pipeline [319]. They implement various protein quantification indices based on both spectral counting and ion signal intensities such as emPAI [320], APEX [321], mSCI [322], TOPn [323], and MS stats [324]. For users working with standard workflows and not developing their own algorithms, Progenesis and MaxQuant, which are suitable for fast data analysis, might represent the best solution. If more flexibility and development of task-specific algorithms are required, open-source packages such as Viper [325], mzMine2v [326], or OpenMS [317] represent the best solution. Large proteomics labs and core facilities might appreciate the modularity and automation provided by pipeline tools (Figure 4) [313].

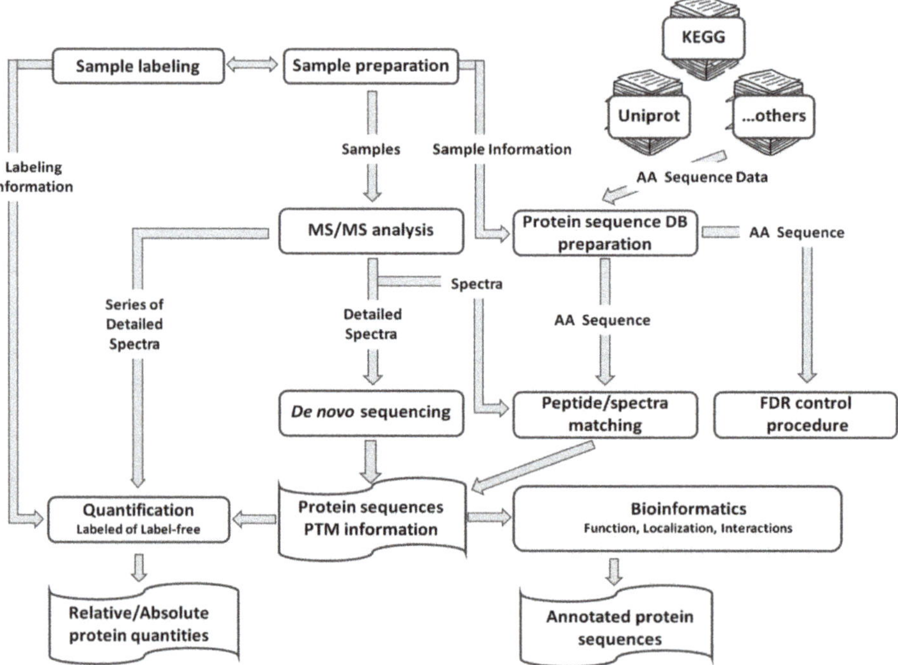

Figure 4. Detailed workflow for LC-based proteomics: protein isolation, sample preparation, and analysis.

Information on protein identities and their relative or absolute abundance is important, but not sufficient for comprehensive understanding of complex interactions within a biological system. To build a generalized image of multiple simultaneous processes occurring in biological systems, bioinformatic tools are required. These tools allow the assignment of an identified protein to a specific subcellular compartment, molecular function, or biological process as well as establishing its protein-protein interactions (PPI) as a network. The information about protein sub-cellular localization might be found in a database of experimentally derived annotations, available at resource SWISS-PROT [327]. However, even the experimental data collections of well-studied organisms (e.g Arabidopsis or yeast) are not nearly complete [328,329]. To overcome this, in silico prediction approaches (e.g., Plant-mPLoc, BUSCA, LOCALIZER, etc.) relying on specific protein sequence features have been developed [330–335]. For example, these algorithms are able to predict the localization of proteins to membranes by the characteristic positions of hydrophobic and hydrophilic residues in their sequence [336]. The information on protein functions can be retrieved from protein sequence databases such as UniProt [337], KEGG [338–340]—especially through KEGG-based tools such as BlastKOALA and GhostKOALA—and plant-specific functional database tools such as MapMan [341]. The two first databases organize functional information in Gene Ontology (GO) tags with the following increasing hierarchy order: enzymatic reaction, metabolic pathway, functional process, biological process [342]. GO is also used by PANTHER web server that combines genomes, gene function classifications, pathways, and statistical analysis tools to analyze large-scale genome-wide experimental data [343]. This functional hierarchy scheme, however, might differ from database to database. For example, in MapMan top hierarchical level comprises 34 functional bins [344]. In general, biological function is indirectly associated with the protein sequence, although similar sequences tend to share functionality [345].

Finally, the analysis of PPIs provides essential information for understanding cellular processes. For example, this information can be obtained by structure-based simulation and statistical/machine learning prediction approaches [346–348]. Thereby, the first group of methods relies on docking and molecular dynamics methods. This type of calculations requires high computational capacity and rich protein structural information and cannot, therefore, be applied to the size of protein sets usually obtained in proteomics studies. Another group of methods relies on protein molecular descriptors (sequence, domain occurrence, co-expression, etc.) usually retrieved from databases. Information on protein interactions (both experimental and predicted) can be obtained from multiple resources, like STRING (PPIs prediction), BioGRID (PPIs and protein-chemical interactions), Reactome/Plant Reactome (PPIs along with metabolic pathways), etc. (Table A3)

7. Future Perspectives

Although seed proteomics represents a well-established field, some of its aspects still need to be elaborated. One of such emerging techniques, only minimally represented in seed research, is peptidomics [349].

Plant peptidomics represents a new rapidly developing field, which became a subject of particular interest after the publication of *A. thaliana* peptidome at the end of the last decade [350]. Plant peptides is a general term for polypeptides of less than 10 kDa which are involved in signaling, regulatory, and defense functions, or produced due to the activity of intracellular proteases [351]. Whereas bioactive peptides, involved in antimicrobial, antifungal, antiviral [352,353], signaling [354–356], and plant-microbial interactions [357] are relatively well-addressed in literature, seed peptidome is much less characterized. Seed storage proteins, representing the main seed polypeptide fraction, are considered as powerhouses of bioactive peptides [358–360]. Thereby, most often, these peptides comprise the sequences of disordered regions in corresponding proteins [361]. Their liberation can be achieved by treatment of the whole seed material or seed protein isolates with different proteases, specificity of which mostly define the biological activity of resulted mixtures [362]. Different variations of this procedure resulted in the characterization of multiple seed-derived antimicrobial peptides (isolated from the seeds of cycad, guava, barnyard grass, legume *Vicia faba*, and cabbage), recently reviewed by Chai et al. [363]. Analogously, antioxidant dipeptides were isolated from perilla (*Perilla frutescens* L) [364], whereas carrot [365] and chia (*Salvia hispanica* L) seed peptides showed inhibited collagenase, hyaluronidase, tyrosinase, and elastase, demonstrating thereby clear anti-ageing effect [366]. In general, discovery and primary analysis of such peptides relied on the bottom-up proteomics setup. Accordingly, several tools conventionally applied to proteomics data can be used here as well: Mascot database [295], KEGG database, BLAST (Basic Local Alignment Search Tool), etc. [367]. On the other hand, some solutions, developed for untargeted metabolomics and routinely used in non-plant peptidomics can be applied as well: XCMS [368], Agilent Mass Profiler Professional [369] and PepEx, which are useful for site visualization and peptide mapping [370]. Peptide structure predictors based on X-ray crystallography and NMR are also applicable in plant peptidomics—Pepstr [371], PEP-FOLD [372], and PepLook [373]. Plant-specific databases, such as The Plant Proteome Database for *A. thaliana*, the Rice Proteome Database, and Promex can be applied for annotation of bioactive peptides in enzymatic hydrolyzates [374]. More specific tools include the ERA database for cyclic peptide sequences [375] and AMPA searching software for antimicrobial proteins [376].

Redox proteomics represents another rapidly developing field of plant proteome research targeting plant response to oxidative stress—the state when production of reactive molecular species, RMS (mostly reactive oxygen and nitrogen species—ROS and RNS, respectively) is overwhelmed with their detoxification with plant antioxidant systems [377–380]. The overall goals of redox proteomics are (i) structural characterization of non-enzymatic PTMs, originating from RMS, and (ii) probing the physiological role of the regulatory pathways behind these modifications. To date, ROS-derived modifications are the best characterized PTMs. Thereby, along with multiple other protein residues, cysteine represents the most redox-sensitive residue: it readily forms methionine sulfoxide and cysteinyl sulfenic acid ($-SOH$), which can be further involved in disulfide formation or oxidation to yield sulfinic ($-SO_2H$) and sulfonic acids ($-SO_3H$) [381]. Sulfenylation, i.e., formation of sulfenic acid is a reversible reaction underlying so-called redox signaling, which appears to involved hundreds, if not thousands site-specifically sulfenylated proteins [382]. Proteomics approaches to characterize the sulfenylation patterns rapidly developed during the current decade. Whereas the earlier approaches relied on a two-step labeling procedure, which was limited to in vitro experiments [383], implementation of direct labeling techniques, specifically targeting sulphenic acid residues with nucleophilic reagents (e.g., DYn-2) gave access to in vivo studies [384]. A step forward was the implementation of compartment-specific genetically encoded probes—e.g., yeast ACTIVATOR PROTEIN 1; YAP1)-based construct YAP-1C, which contains a redox-active residue C598 amenable of formation of mixed disulfides with in vivo sulfenylated cysteinyl residues [385]. Recently, a highly-reactive diazene (1-(pent-4-yn-1-yl)-1H-benzo[c][1,2]thiazin-4(3H)-one 2,2-dioxide)-based probe was introduced [386] and brought a break-through in analysis of sulfenylated proteome [382].

The above mentioned powerful techniques are still waiting for implementation in seed proteomics, although seed development and germination are well-addressed by non-labeling methods of redox proteomics [387]. Thereby, thiol metabolism is addressed with such gel-based proteomics approaches as diagonal electrophoresis [388]. In this context, the main direction of the further research would be the implementation of redox-labeling techniques in the study of seed maturation and seed germination. Another important aspect would be addressing seed quality and tolerance to prolonged storage. On the one hand, accumulation of antinutritive modifications of reserve proteins can be addressed [11]. On the other, as mitochondria are the marker of seed ageing [389], the state of the art redox proteomics can be employed in the study of the mechanisms behind.

As was mentioned above, characterization of plant sulfenylation patterns (which can be defined as sulfenilome) is still needs to be done. Obviously, for this, implementation of labeling techniques combined with LC-MS-based techniques in plant redox proteomics is strongly mandatory. Without any doubt, this will give access to regulatory mechanisms, based on redox signaling.

It is important to note that those structural and functional manifestations of oxidative stress in plant proteome are much more diverse than it usually thought. The corresponding changes in plant protein structure and function need to be considered in the context of the physiological and metabolic response of the plant to oxidative stress. For example, it is well-known that seed maturation is accompanied by accumulation of reserve molecules [390], to a great extent represented by carbohydrates and lipids [391]. Accumulation of fatty acids and carbohydrates at the early steps of seed maturation and their mobilization during germination may affect the patterns on non-enzymatic modifications of seed proteins, which are a highly abundant component of, for example, legume seeds. Indeed in seeds, oxidative stress, accompanying both their maturation and ageing [392–395], develops at the background of high contents of reducing sugars (generated by de-polymerization of reserve polysaccharides) and unsaturated fatty acids (mostly formed during mobilization of reserve lipids) [396–400]. These conditions might result in oxidative degradation of carbohydrates and lipids/fatty acids via metal-catalyzed monosaccharide autoxidation and lipid peroxidation, respectively, yielding in both cases reactive carbonyl compounds (RCCs) [401,402]. These intermediates are readily involved in glyco- and lipoxidation of proteins resulting in the formation of advanced glycoxidation and lipoxidation end products (AGEs and ALEs, respectively), which are potentially pro-inflammatory in mammals and impact on ageing and metabolic diseases [403,404]. Hence, this phenomenon needs to be taken into account when seeds are stored for long periods of time [405], or when the processes, accompanying storage, are simulated in models of accelerated seed ageing [406]. As this human physiology aspect in plant science remains mostly unstudied, these aspects need to be addressed in much more detail in next future. For this, a coordinated effect of plant biochemists, nutritional scientists and human physiologists is required.

Although glyco- and lipoxidation in plants has been addressed by biochemical methods [399,407], a comprehensive analysis of non-enzymatic PTMs in ageing seeds still needs to be done. Most efficiently, such a profiling of glyco- and lipoxidative modifications might rely on LC-MS-based approach with a special focus on in-depth analysis of low-abundant species using enrichment/depletion, pre-fractionation, and result-based exclusion of non-modified peptides from fragmentation [171,177]. Alternatively, modified peptides can be selectively detected by highly sensitive and specific precursor ion scanning and further identified by targeted MS/MS experiments [408,409]. Such experiments might rely on MS/MS analyses with model oxidized, glycated, as well as glyco- and lipoxidized peptides [408–413], which can be obtained by solid-phase peptide synthesis (SPPS) with high yield and purity [414–416]. Thereby, formation and degradation of early and advanced glycation products can be dissected by glycation models based on synthetic peptides. On the one hand, such an approach may give access to the pathways of AGE formation [106,171,417]. On the other hand, it can deliver information about carbonyl compounds involved in glycation [105]. Based on this approach, highly sensitive and reliable LC-MS/MS-based quantification techniques can be developed [418,419], although their application to seed research has still not been accomplished. An alternative approach relies on exhaustive enzymatic hydrolysis followed with MS/MS analysis of resulted amino acid adducts by QqQ-MS using the stable isotope dilution [420] and standard addition [421] approach. This strategy was successfully applied to the assessment of glycation and oxidation in plants [422]. Recently, we optimized this methodology for the analysis of seed protein adducts [93,423]. However, it is just a beginning of a big work, which is still to be done.

Although glycation and lipoxidation are well-known markers of ageing in mammals [424], this was only recently confirmed for plants. Indeed, AGEs were shown to impact ageing not only in plant leaves [425] but also in such specialized structures, as legume root nodules [212], where multiple regulatory proteins and transcription factors were involved in age-dependent modifications. It raises the question about a possible regulatory role of this type of modification in plants. Keeping in mind a pronounced pro-inflammatory effect of glyco- and lipoxidized proteins on organisms of heterotrophic mammalian consumers of seed-derived food, one can conclude that this aspect is highly relevant for food chemistry as well [403]. Taking all this into account, patterns of AGEs, and ALEs in seed proteins of crop plants need to be comprehensively characterized.

8. Conclusions

The seed proteome represents a highly-complex system, only a part of which has been sufficiently described to date. The patterns of PTMs in seed polypeptides are also just partly characterized. Indeed, only phosphorylation and glycosylation were comprehensively addressed so far, whereas the sites of other enzymatic and non-enzymatic modifications remain mostly unknown. Obviously, the depth of its characterization needs to be increased for a better understanding of the processes accompanying seed development and germination. For this, simultaneous consideration of the datasets acquired by LC-MS and 2D-GE is necessary. This would not only increase the proteome coverage but also might allow cross-validation of the obtained data and increase the overall result reliability. On the other hand, due to their potential involvement in cellular regulatory networks, non-enzymatic modifications (first of all, oxidation and glycation) and mechanisms, underlying in vivo physiological effects, need to be addressed. To some extent, the depth of proteome analysis can be increased by the implementation of additional enrichment/depletion or pre-fractionation steps in established workflows. However, for further increase the analytical power of seed proteomics, its methodology needs to be complemented with approaches and techniques currently employed in other fields of proteomics research, like, medical, ecological, food proteomics. We hope that our contribution will help seed biologists to increase the methodological power of seed proteomics and to make a step forward to understanding the fundamental processes of seed development and germination.

Author Contributions: G.S. wrote the Introduction and contributed to other sections, E.L. and A.S. prepared figures and tables and contributed gel- and LC-based proteomics parts, G.M.-S., T.B. and A.T. wrote the data processing part, E.P., V.Z., I.T. and S.M. contributed writing and critical discussion of the manuscript, D.G. and E.R. have done the revision of the manuscript prior to new submission, W.H. and A.F. supervised writing, contributed to all sections, and prepared the final version of the manuscript. All authors have read and agreed to the published version of the manuscript.

Funding: This research was funded by Russian Science Foundation: 50% of the study dealing with general methodology and applications of proteomics in seed science; project № 20-16-00086 to S.M. and 50% of the study dealing with the description of methodological approaches; project № 17-76-30016 to I.T.

Conflicts of Interest: The authors declare no conflict of interest. The funders had no role in the design of the study; in the collection, analyses, or interpretation of data; in the writing of the manuscript, or in the decision to publish the results.

Abbreviations

2D-GE	Two-dimensional gel electrophoresis
AALS I/II	Anionic acid labile surfactant I/II
ABA	Abscisic acid
AGEs	Advanced glycoxidation end products
ALEs	Advanced lipoxidation end products
BAC	Boronic acid chromatography
CALS I/II	Cationic acid labile surfactant I/II
CHAPS	3-(3-cholamidopropyl)dimethylammonio)-1-propanesulfonate
CMC	Critical micelle concentration
DCM	Dichloromethane
DDA	Data-dependent acquisition
DIA	Data-independent acquisition
DIGE	Difference gel electrophoresis
DTT	Dithiothreitol
ESI	Electrospray ionization
FASP	Filter-aided sample preparation

FDR	False discovery rate
GASP	Gel-aided sample preparation
GPF	Gas-phase fractionation
HILIC	Hydrophilic interaction liquid chromatography
HiT-Gel	High-throughput in-gel digestion technique
IEF	Isoelectrofocusing
IPG	Immobilized pH gradient
KDS	Potassium dodecyl sulfate
LC	Liquid chromatography
LIT	Linear ion trap
LODs	Limits of detection
MALDI	Matrix-assisted laser desorption-ionization
MRM	Multiple reaction monitoring
MS	Mass spectrometry
MS/MS	Tandem mass spectrometry
PMF	Peptide mass fingerprinting
PPI	Protein-protein interactions
PQPs	Peptide query parameters
PRM	Parallel reaction monitoring
PTMs	Post-translational modifications
PVP	Polyvinylpyrollidone
RCCs	Reactive carbonyl compounds
RPC	Reversed phase chromatography
SDC	Sodium deoxycholate
SDS	Sodium dodecyl sulfate
SDS-PAGE	Polyacrylamide gel electrophoresis in sodium dodecyl sulfate
SPE	Solid phase extraction
SPPS	Solid phase peptide synthesis
SRM	Selected reaction monitoring
TCA	Trichloroacetic acid
TOF	Time of flight
UHPLC	Ultra-high performance liquid chromatography
XIC	Extracted ion chromatogram
ZALSI/II	Zwitterionic acid labile surfactant I/II

Appendix A

Table A1. Overview of protein extraction techniques.

#	Extraction Technique	Extraction Buffer	Chaotropic Agents	Detergents	Reducing and Chelating Additives	Further Additives	Precipitation (Vprecipitant: Vextract)	Isolate Cleaning	Reconstitution	Ref
1	Phenol extraction	0.5 mol/L Tris-HCl (pH 7.5)	none	none	2% (*v/v*) ME, 50 mmol/L EDTA	1–15% (*w/v*) PVPP, PIC	0.1 mol/L AmAc/MeOH (5:1)	MeOH (3×), acetone (3×)	SDS-PAGE SB, IEF buffer, SB for LC-MS	[89,177]
2	TCA/acetone extraction	10% (*w/v*) TCA in acetone	none	none	2% (*v/v*) ME	1–15% (*w/v*) PVPP, PIC	precipitation at the extraction step	acetone (3×)	SDS-PAGE SB, IEF buffer	[89,94,426]
3	Extraction with urea/thiourea buffer	14 mmol/L Tris-HCl	7 mmol/L urea, 2 mmol/L thiourea	2% (*v/v*) Triton X-100, 58 mmol/L CHAPS	none	PIC, 18 mmol/L ampholytes	none	none	solubilization at the extraction step	[82]
4	Acetone precipitation	20 mmol/L Tris-HCl (pH 7.5)	none	1% (*v/v*) Triton X-100	10 mmol/L EGTA, 1 mmol/L DTT	1 mmol/L PMSF, 250 mmol/L sucrose	precipitation at the extraction step	acetone (3×)	SDS-PAGE SB	[108]
5	Extraction with SDS-Tris buffer	125 mmol/L Tris-HCl	none	4% (*w/v*) SDS	2% (*v/v*) ME	20% (*v/v*) glycerol	none	none	solubilization at the extraction step	[110]
6	Extraction HEPES buffer/delipidation (DCM)	50 mmol/L HEPES buffer	none	none	1 mmol/L EDTA	1 mmol/L PMSF, 0.1 mmol/L nDHGA	acetone (1:5)	none	SDS-PAGE SB, IEF buffer	[113]
7	Extraction with urea/thiourea buffer	6 mmol/L Tris-HCl,4.2 mmol/L Trizma R	7 mmol/L urea, 2 mmol/L thiourea	4% (*w/v*) CHAPS	3% (*w/v*) DTT	PIC, DNAse I, RNAse A	none	none	solubilization at the extraction step	[427]
8	MeOH/CHCl3 precipitation, delipidation (PE)	50 mmol/L Tris-HCl (pH 8.8)	none	1% (*w/v*) SDS	0.07% (*v/v*) ME, 1.5 mmol/L KCl	PIC, delipidation (PE)	MeOH/CHCl$_3$/ddH$_2$O (4:1:3)	SPE	8 mol/L urea in 50 mmol/L ABC	[95]
9	TCA/acetone precipitation, delipidation (PE)	50 mmol/L Tris-HCl (pH 8.8)	none	1% (*w/v*) SDS	0.07% (*v/v*) ME, 1.5 mmol/L KCl	PIC, delipidation (PE)	acetone (1:4)	SPE	8 mol/L urea in 50 mmol/L ABC	[95]
10	Acetone precipitation, delipidation (PE)	50 mmol/L Tris-HCl (pH 8.8)	none	1% (*w/v*) SDS	0.07% (*v/v*) ME, 1.5 mmol/L KCl	PIC, delipidation (PE)	acetone/10% (*w/v*) TCA (1:4)	SPE	8 mol/L urea in 50 mmol/L ABC	[95]

Table A1. *Cont.*

#	Extraction Technique	Extraction Buffer	Chaotropic Agents	Detergents	Reducing and Chelating Additives	Further Additives	Precipitation (Vprecipitant: Vextract)	Isolate Cleaning	Reconstitution	Ref
11	Urea solubilization buffer	8 mol/L urea, 2% (w/v) ampholyte (pH 3–10)	8 mol/L urea	4% (w/v) CHAPS	none	none	none	2D cleanup kit (GE Healthcare)	solubilization at the extraction step	[96]
12	Thiourea/urea solubilization buffer delipidation (hexane)	5 mol/L urea, 2 mol/L thiourea, 0.8% (w/v) ampholytes (pH 3–10)	5 mol/L urea, 2 mol/L thiourea	4% (w/v) CHAPS	65 mmol/L DTT	delipidation (hexane)	none	none	solubilization at the extraction step	[96]
13	Phenol extraction	0.1 mol/L Tris–HCl (pH 8.8)	none	none	10 mmol/L EDTA, 0.4% (v/v) ME	none	AmAc/MeOH (5:1)	0.1mol/L AmAc/MeOH (2×) acetone (2×) MeOH (1×)	8 mol/L urea, 2 mol/L thiourea, 2% (w/v) CHAPS, 2% (v/v) Triton X-100, 50 mmol/L DTT, 0.5% (w/v) ampholytes (pH 3–10)	[96]
14	Modified TCA/acetone precipitation/Urea solubilization extraction	10% (w/v) TCA in acetone	none	none	0.07% (v/v) ME	none	precipitation at the extraction step	acetone (2–3×)	9 mol/L urea, 1% (w/v) CHAPS, 1% (w/v) ampholytes pH (3–10), 1% (w/v) DTT	[96]
15	Phenol extraction	0.5 mol/L Tris-HCl, (pH 7.5)	none	none	2% (v/v) ME, 50 mmol/L EDTA	10% (w/v) PVPP, 1 mmol/L PMSF	0.1 mol/L AmAc/MeOH (2x)	none	IEF buffer, SDS-PAGE SB	[426]
16	Tris/TCA extraction	100 mmol/L Tris, TCA in acetone (pH 8.5)	none	none	5 mmol/L DTT, 1 mmol/L EDTA. 0.07% (v/v) ME	1 mmol/L PMSF	Precipitation at the extraction step	0.07% (v/v) ME in acetone	IEF buffer, SDS-PAGE SB	[426]
17	Tris-base extraction	40 mmol/L Tris	5 mol/L urea, 2 mol/L thiourea	2% (w/v) CHAPS	2% (v/v) ME	5% (w/v) PVP	Precipitation at the extraction step	0.07% (v/v) ME in acetone	IEF buffer, SDS-PAGE SB	[426]
18	TCA/acetone extraction	10% (v/v) TCA in acetone	none	none	20 mmol/L DTT	none	Precipitation at the extraction step	Acetone or 20 mmol/L DTT in acetone, or 10% ddH2O in acetone or 20 mmol/L DTT, 10% ddH2O in acetone	IEF SB	[100]

Table A1. *Cont.*

#	Extraction Technique	Extraction Buffer	Chaotropic Agents	Detergents	Reducing and Chelating Additives	Further Additives	Precipitation (Vprecipitant: Vextract)	Isolate Cleaning	Reconstitution	Ref
19	Phenol extraction	50 mmol/L Tris-HCl (pH 7.5)	none	none	5 mmol/L EDTA, 5 mmol/L DTT	1% (*w/v*) PVPP, 1 mmol/L PMSF	Acetone: supernatant (5:1)	none	Urea buffer (50 mmol/L HEPES (pH 7.8), 8 mol/L urea) SB for LC-MS	[103]
20	TCA/acetone extraction	10% (*v/v*) TCA in acetone	none	none	10 mmol/L DTT	10% (*w/v*) PVPP, 55 mmol/L iodoacetamide, 0.5 mol/L TEAB	Precipitation at the extraction step	Acetone (3×)	TEAB buffer, IEF SB, SB for LC-MS	[102]
21	TCA/acetone and methanol washes and phenol extraction	Phenol (pH 8.0): SDS (1:1)	none	none	none	none	0.1 mol/L AmAc/MeOH	10% (*v/v*) TCA/acetone 0.1 MAmAc 80% MeOH 80% acetone	SDS-PAGE SB, IEF buffer	[111]
22	Tris–HC l/ TCA/acetone extraction	0.1 mol/L Tris–HCl (pH 6.8), 10% (*v/v*) TCA/acetone	none	1% (*w/v*) SDS	0.1 mol/L DTT	none	10% (*v/v*) TCA/acetone	10% (*v/v*) TCA/acetone aqueous 10% TCA (2x) dH2O (1x) acetone (1x)	SDS-PAGE SB	[111]

ABC, ammonium bicarbonate buffer; AmAc, ammonium acetate; CHAPS, 3-[(3-cholamidopropyl)dimethylammonio]-1-propanesulfonate hydrate; DCM, dichloromethane; DTT, dithiothreitol; EDTA, ethylenediaminetetraacetic acid; EGTA, ethylene-bis(oxyethylenenitrilo)tetraacetic acid; HEPES, 4-(2-hydroxyethyl)piperazine-1-ethanesulfonic acid; IEF, isoelectric focusing; LC-MS, liquid chromatography-mass spectrometry; ME, ß-mercaptoethanol; nDHGA, nordihydroguaiaretic acid; PE, petroleum ether; PIC, protein inhibitor cocktail; PMSF, phenylmethylsulfonyl fluoride; PVPP, polyvinylpolypyrrolidone; SB, sample buffer; SDS, sodium dodecyl sulfate; SDS-PAGE, sodium dodecyl sulfate polyacrylamide gel electrophoresis; SPE, solid phase extraction; TCA, trichloroacetic acid; TEAB, triethylammonium bicarbonate; Tris, tris(hydroxymethyl)aminomethane.

Table A2. Digestion strategies used in bottom-up proteomics.

#	Object/Tissue	Methodology						Ref
		Protein Isolation	Detergent/ Chaotropic Agent	Reduction/Alkylation	Protease	Chromatographic System	MS	
		Plant Objects						
1	*Brassica napus* L., seeds	detergent extraction, phenol extraction	none (in-gel digest)	DTT/IA	trypsin	none	MALDI-TOF/TOF-MS	[428]
2	*Lupinus luteus* L., Seeds	delipidation with hexane, acetone precipitation	none (in-gel digest)	DTT/IA	trypsin	RP C18, L-column water-ACN grad., 0.1% (v/v) FA	ESI-IT-MS	[138]
3	*Cicer arietinum* L., plasma membrane aerial parts	chloroform/methanol (5:4) extraction	none (in-gel digest) 4% (w/v) SDS/none	DTT/IA DTT/IA	trypsin trypsin	RP C18, water-ACN grad., 0.1% (v/v) FA	ESI-LTQ-Orbitrap-MS	[31]
4	*Glycine max* L., seeds	delipidation with hexane, extraction with SDS-PAGE SB	none (in-gel digest)	DTT/IA DTT/none	trypsin	none RP, BEH130C18, water-ACN grad. 0.1% (v/v) FA	MALDI-TOF/TOF-MS ESI-QqTOF-MS	[139]
5	*Glycine max* L., seeds	2 steps extraction: protamine sulfate precipitation TCA/acetone precipitation	none (in-gel digest) 4% (w/v) SDS/ 8 mol/L urea	DTT/IA DTT/IA	trypsin trypsin	none RP, C18. water-ACN grad. 0.1% (v/v) FA	MALDI-TOF/TOF-MS ESI-Q-Orbitrap-MS	[140,186]
6	*Arabidopsis thaliana*, aerial parts	1) detergent extraction 2) aq. buffer extraction	none (in-gel digest)	DTT/IA	trypsin	RP, C18 PepMap water-ACN grad. 0.1% (v/v) FA	ESI-Q-Orbitrap-MS	[141]
7	*A. thaliana*, leaves	phenol extraction	0.5% (w/v) AALS/ 7 mol/L urea	TCEP/IA	trypsin	RP, water-ACN grad. 0.1% (v/v) FA	ESI-LIT-Orbitrap ESI-QqTOF-MS	[169]
8	*G. max*, seeds	1) detergent extractionacetone precipitation	none (in-gel digest)	none	trypsin	RP, C18 PepMap column water-ACN grad. 0.1% (v/v) FA	ESI-LIT-MS	[108]
9	*Chenopodium quinoa* W., seeds	1) detergent extraction 2) methanol/chloroform or acetone precipitation	none/8 mol/L urea	DTT/IA	trypsin	RP, Acclaim-C18 column water-ACN grad. 0.1% (v/v) FA	ESI-LIT-Orbitrap-MS	[95]

Table A2. *Cont.*

#	Object/Tissue	Methodology						Ref
		Protein Isolation	Detergent/ Chaotropic Agent	Reduction/Alkylation	Protease	Chromatographic System	MS	
10	*Solanum esculentum* L. roots,	detergent extraction for isolation of cell microsomal fractions by centrifugation	1) methanol ** 2) 0.2% PPS SilentSurfactant/none 3) 0.2% RGS/none 4) none/6 mol/L GdnCl	TCEP/IA	trypsin	RP, BEH C18 water-ACN grad. 0.1% (*v/v*) FA	ESI-QqTOF-MS	[29]
11	*Vitisriparia*, leaves	1) detergent extraction 2) methanol-chloroform extraction	none (in-gel digest) 50% TFE **	DTT/IA DTT/IA	trypsin Lys-C, trypsin	RP, Magic C18AQ resin water-ACN grad. 0.1% (*v/v*) FA	ESI-LIT-Orbitrap-MS	[187]
12	*Cucumis sativus* L., seeds	TCA/acetone (1:9 *w/v*) precipitation	none/8 mol/L urea	DTT/IA	trypsin	RP, C18 water-ACN grad. 0.1% (*v/v*) FA	ESI-Q-Orbitrap-MS	[188]
13	*Hordeum vulgare* L., leaves	detergent extraction	1) none/8 mol/L urea 2) 2% (*w/v*) SDS/8 mol/L urea 3) 1% SDC/none 4) 2% SDC/none	DTT/IA	trypsin	RP, Reprosilpur 120 C18 water-ACN grad. 0.1% (*v/v*) FA	ESI-Q-Orbitrap-MS	[194]
14	*Zea mays* L., seeds	1) detergent extraction 2) TCA/acetone (1:9*w/v*) extraction	none (in-geldigest)	DTT/IA	trypsin	RP, Eksigent C8-CL-120 column water-ACN grad. 0.1% (*v/v*) FA	ESI-QqQ-MS	[429]
15	*Solanum tuberosum* L. leaves,	2 steps: 1) detergent extraction 2) co-immunoprecipitation 10% TCA, 0.07% (*w/v*)	0.1% ProteaseMAX™ surfactant/none	TCEP/MMTS	trypsin	RP, Reprosil C18-AQ, water-ACN grad. 0.1% (*v/v*) FA	ESI-Q-LIT-Orbitrap-MS	[430]
16	*H. vulgare* caryopses,	β-mercaptoethanol in acetone	0.1% RGS/none	DTT/IA	trypsin	RP, C18 water-ACN grad. 0.1% (*v/v*) FA	ESI-QqTOF-MS	[203]
17	*B. napus*, seedling	1) aq. buffer extraction 2) phenol extraction	0.02% (*w/v*) AALS/8 mol/L urea, 2 mol/L thiourea	TCEP/IA	trypsin	RP, Acclaim PepMap 100 C18 column water-ACN grad. 0.1% (*v/v*) FA	ESI-Q-LIT—Orbitrap-MS	[26]

Table A2. *Cont.*

#	Object/Tissue	Methodology						
		Protein Isolation	Detergent/ Chaotropic Agent		Reduction/Alkylation	Protease	Chromatographic System	MS	Ref
		Non-plant objects							
18	Myoglobin, bacteriorhodopsin, BSA *	none	1) 0.1–1.0% RGS/none 2) 1.0% SDC/none 3) 0.1–1.0% SL/none		none	trypsin	none	MALDI-TOF/TOF-MS	[196]
	Rat, liver *	isolation of cell membranes by centrifugation in the gradient of sucrose	1) 1.0% SDS/none 2) 1.0% SL/none 3) 1.0% RGS/none 4) 1.0% SDC/none		DTT/IA	trypsin	RP, C18 PepMap column water-ACN grad. 0.1% (*v/v*) FA	ESI-IT-MS	
19	Rat, liver *	membrane isolation, centrifugation in sucrose gradient	1) 1%(*w/v*) SDS/none 2) 1%(*w/v*)RGS/none 3) none/8mol/L urea 4) 60% (*v/v*) methanol **		DTT/IA	trypsin	RP, C18PepMap column water-ACN grad. 0.1% (*v/v*) FA	ESI-IT-MS	[168]
20	BSA, ubiquitin, myoglobin, PC3 cells *	cell lysis, isolation of cell membranes by centrifugation	none		DTT/IA	trypsin	RP, C18. water-ACN grad. 0.1% (*v/v*) FA	ESI-QqTOF-MS	[189]
21	*Rhodopseudomona spalustris* *	acid extraction and sucrose density fractionation	0, 60 or 80% acetonitrile ** in 50 mmol Tris-HCl, 10 mmol/L CaCl2		DTT/none	trypsin	RP, Vydac C18 water-ACN grad. 0.1% (*v/v*) FA	ESI-IT-MS ESI-FT-ICR-MS	[192]
	Mixture of protein standards *	none	none/6 mol/LGdnHCl						

* not established in seeds or generally in plants, shown as a method to be potentially employed in seed proteomics, ** non detergent or chaotropic reagent used for solubilization; AALS, anionic acid labile surfactant; ACN, acetonitrile; BSA, bovine serum albumin; DTT, dithiothreitol; ESI, electrospray ionization; FA, formic acid; FT-ICR, Fourier transform-ion cyclotron resonance; GdnHCl, guanidinium chloride; IA, iodoacetamide; IT, ion trap; LIT—linear ion trap; MALDI-TOF, matrix assisted laser desorption/ionization—time of flight; MMTS, methyl methanethiosulfonate; MS, mass spectrometry; nano-scaled liquid chromatography; Q, quadrupole mass analyzer; QqTOF, quadrupole-time of flight; RGS, RapiGest SF; RP, reversed phase; SDC, sodium deoxycholate; SDS, sodium dodecyl sulfate; SL, sodium laurate; TCA, trichloroacetic acid; TCEP, *tris*-(2-carboxyethyl)-phosphine; TFE, trifluoroethanol.

Table A3. Software tools for processing and post-processing for proteomics data.

#	Tool	Version	Supported Platform	GUI/CMD	Open Source	Input Formats	Quantification Technique	Ref
1	MaxQuant	v1.6.3.3	Windows, Linux	+/+	+	AB SCIEX (*.wiff), mzXML, Thermo (*.raw), Agilent & Bruker Daltinics (*.d), Uimf (*.uimf)	LFQ/label-based	[431]
2	Peaks	v2.0	Windows, Linux	+/+	−	CID/CAD/HCD/ETD/ECD/EThCD	LFQ/label-based	[432]
3	OpenMS/TOPP	v2.0	Windows, Linux, Mac OS	+/+	+	mzML, mzXML, mzData	LFQ/label-based	[433]
4	Progenesis QI	v2.3	Windows			Agilent & Bruker Daltinics (*.d), AB SCIEX (*.wiff), mzML, mzXML, Thermo& Agilent (*.raw)	LFQ/label-based	[434]
5	Proteome discoverer	v2.2	Windows	+/−	−	mzXML, mzDATA, mzML, MSF	LFQ/label-based	[435]
6	Census	v2.3	Windows, Linux, Mac OS	+/−	−	mzXML	LFQ/label-based possibility to self-define mass tags	[436]
7	SILVER	V3.0	Windows	+/−	+	mzXML, *.raw	SILAC	[437]
8	msInspect	v3.1	Windows, Linux, Mac OS	+/+	+	mzXML	LFQ/label-based	[313,438]
9	mzMine2	v2.6	Windows, Linux, Mac OS	+/−	+	mzML, mzXML, mzData, NetCDF, RAW (Thermo)	LFQ	[326]
10	MassChroQ	v2.2.12	Windows, Linux	−/+	+	mzXML, mzML	LFQ/label-based	[439]
11	Skyline	v4.1	Windows, Linux	+/+	+	.sky, .skyd, mzML, mzXML, major vendor formats	LFQ	[440]

Table A3. *Cont.*

#	Tool	Version	Supported Platform	GUI/CMD	Open Source	Input Formats	Quantification Technique	Ref
12	DIA-Umpire	v2.0	Windows, Linux	−/+	+	mzXML	ICAT, ^{18}O	[441]
13	Viper	v3.49	Windows, Linux	+/−	+	PEK, .CSV (Decon2LS), .mzXML,.mzData	ICAT, ^{18}O	[325]
14	OpenSWATH	v2.2	Windows, Linux, Mac OS	−/+	+	mzML, mzXML, TraML	LFQ/label-based	[215]
15	TPP	v5.1.0	Windows, Linux, Mac OS	+/+	+	mzXML, .RAW (Thermo), wiff, baf (Brucker), pepXML	LFQ	[319,442]
16	moFF	v2.0	Windows, Linux, Mac OS	+/+	+	Thermo (.raw), mzML	LFQ	[443]
17	Mascot Distiller	v2.7	Windows	+/+	−	mzML, mzXML, mzData, major vendors	LFQ/label-based	[444]
18	Corra	v3.1	Linux			mzML, pepXML	LFQ/label-based	[313,445]
19	FlashLFQ	v0.1.61	Windows	−/+	+	MzML, raw	LFQ	[446]
20	Thermo Scientific ProSightPC/ ProSightPD	v4.0/v2.0	Windows	+/−	−	Thermo (*.raw), .PUF, UniProt XML, FASTA, UniProKB	LFQ/abel-based	[447]
21	MassHunter	v10.0	Windows	+/−	−	Agilent (d)	LFQ/label-based	[448]
22	Mercator4.0	v2.0	Web tool, Windows, Linux	+/−	+	FASTA	On-line functional annotation tool sequences	[449]
23	BlastKOALA	-	Web tool	−/−	−	FASTA	Automatic annotation server for genome and metagenome sequences	[341]

Table A3. *Cont.*

#	Tool	Version	Supported Platform	GUI/CMD	Open Source	Input Formats	Quantification Technique	Ref
24	WoLF PSORT	-	Web tool	–/–	–	FASTA	Prediction of sub-cellular localization	[331]
25	BUSCA (Bologna Unified Subcellular Component Annotator)	-	Web tool	–/–	–	FASTA	Prediction of sub-cellular localization	[331,333]
26	eggNOG-mapper	v2	Web tool	–/–	+	FASTA	Functional annotation of large sets of sequences *	[450]
27	PANTHER	v.14.0	Web server	–/–	–	FASTA, gene ID(.txt)	Large-scale genome-wide experimental data **	[343]
28	STRING	v11.0	Web server	–/–	–	protein name (.txt), gene ID (.txt)	Protein-protein association networks	[451]

* based on fast orthology assignments using pre-computed eggNOG v5.0 clusters and phylogenies, ** system that combines genomes, gene function classifications, pathways and statistical analysis tools; ^{18}O, ^{18}O labeling by ^{18}O-enriched water; CMD, command line; GUI, graphical user interface; ICAT, isotope-coded affinity tag; LFQ, label free quantification; SILAC, stable isotope labeling by/with amino acids in cell culture.

References

1. FAO. *Seeds Toolkit—Module 5: Seed Marketing*; Food and Agriculture Organization of the United Nations: Rome, Italy, 2018; ISBN 9789251309537.
2. FAO. *Statistical Book. Part 3. Feeding the World*; Food and Agriculture Organization of the United Nations: Rome, Italy, 2013; ISBN 978-92-5-107396-4.
3. FAO; IFAD; UNICEF; WFP; WHO. *The State of Food Security and Nutrition in the World 2019. Building Climate Resilience for Food Security and Nutrition*; Food and Agriculture Organization of the United Nations: Rome, Italy, 2018; ISBN 9789251305713.
4. Bradford, K.J.; Dahal, P.; Van Asbrouck, J.; Kunusoth, K.; Bello, P.; Thompson, J.; Wu, F. The dry chain: Reducing postharvest losses and improving food safety in humid climates. *Trends Food Sci. Technol.* **2018**, *71*, 84–93. [CrossRef]
5. Miller, J.K.; Herman, E.M.; Jahn, M.; Bradford, K.J. Strategic research, education and policy goals for seed science and crop improvement. *Plant Sci.* **2010**, *179*, 645–652. [CrossRef]
6. Bewley, J.D.; Bradford, K.J.; Hilhorst, H.W.M.; Nonogaki, H. *Seeds: Physiology of Development, Germination and Dormancy*, 3rd ed.; Springer: New York, NY, USA, 2013; ISBN 978-1-4614-4692-7.
7. Leprince, O.; Pellizzaro, A.; Berriri, S.; Buitink, J. Late seed maturation: Drying without dying. *J. Exp. Bot.* **2017**, *68*, 827–841. [CrossRef] [PubMed]
8. Finch-Savage, W.E.; Bassel, G.W. Seed vigour and crop establishment: Extending performance beyond adaptation. *J. Exp. Bot.* **2016**, *67*, 567–591. [CrossRef]
9. Marques, A.; Buijs, G.; Ligterink, W.; Hilhorst, H. Evolutionary ecophysiology of seed desiccation sensitivity. *Funct. Plant Biol.* **2018**, *45*, 1083. [CrossRef]
10. Szarka, A.; Tomasskovics, B.; Bánhegyi, G. The Ascorbate-glutathione-α-tocopherol. Triad in Abiotic Stress Response. *Int. J. Mol. Sci.* **2012**, *13*, 4458–4483. [CrossRef]
11. Frolov, A.; Mamontova, T.; Ihling, C.; Lukasheva, E.; Bankin, M.; Chantseva, V.; Vikhnina, M.; Soboleva, A.; Shumilina, J.; Mavropolo-Stolyarenko, G.; et al. Mining seed proteome: From protein dynamics to modification profiles. *Biol. Commun.* **2018**, *63*, 43–58. [CrossRef]
12. Aguirre, M.; Kiegle, E.; Leo, G.; Ezquer, I. Carbohydrate reserves and seed development: An overview. *Plant Reprod.* **2018**, *31*, 263–290. [CrossRef]
13. Baud, S. Seeds as oil factories. *Plant Reprod.* **2018**, *31*, 213–235. [CrossRef]
14. Gallardo, K.; Thompson, R.; Burstin, J. Reserve accumulation in legume seeds. *Comptes Rendus Biol.* **2008**, *331*, 755–762. [CrossRef]
15. Wang, W.-Q.Q.; Liu, S.-J.J.; Song, S.-Q.Q.; Møller, I.M. Proteomics of seed development, desiccation tolerance, germination and vigor. *Plant Physiol. Biochem.* **2015**, *86*, 1–15. [CrossRef] [PubMed]
16. Miernyk, J.A.; Hajduch, M. Seed proteomics. *J. Proteomics* **2011**, *74*, 389–400. [CrossRef] [PubMed]
17. Copeland, L.O.; McDonald, M.B. The chemistry of seeds. *Princ. Seed Sci. Technol.* **1999**, 40–58. [CrossRef]
18. Robić, G.; Farinas, C.S.; Rech, E.L.; Miranda, E.A. Transgenic soybean seed as protein expression system: Aqueous extraction of recombinant β-glucuronidase. *Appl. Biochem. Biotechnol.* **2010**, *160*, 1157–1167. [CrossRef]
19. Schmidt, M.A.; Herman, E.M. Proteome rebalancing in soybean seeds can be exploited to enhance foreign protein accumulation. *Plant Biotechnol. J.* **2008**, *6*, 832–842. [CrossRef]
20. Rathi, D.; Gayen, D.; Gayali, S.; Chakraborty, S.; Chakraborty, N. Legume proteomics: Progress, prospects, and challenges. *Proteomics* **2016**, *16*, 310–327. [CrossRef]
21. Shewry, P.R.; Casey, R. Seed Proteins. In *Seed Proteins*; Shewry, P.R., Casey, R., Eds.; Springer: Dordrecht, The Netherlands, 1999; pp. 1–10. ISBN 978-94-010-5904-6.
22. Gallardo, K.; Job, C.; Groot, S.P.C.; Puype, M.; Demol, H.; Vandekerckhove, J.; Job, D. Proteomic analysis of arabidopsis seed germination and priming. *Plant Physiol.* **2001**, *126*, 835–848. [CrossRef]
23. Gallardo, K.; Le Signor, C.; Vandekerckhove, J.; Thompson, R.D.; Burstin, J. Proteomics of medicago truncatula seed development establishes the time frame of diverse metabolic processes related to reserve accumulation. *Plant Physiol.* **2003**, *133*, 664–682. [CrossRef]
24. Catusse, J.; Job, C.; Job, D. Transcriptome- and proteome-wide analyses of seed germination. *Comptes Rendus Biol.* **2008**, *331*, 815–822. [CrossRef]

25. Rajjou, L.; Lovigny, Y.; Groot, S.P.C.; Belghazi, M.; Job, C.; Job, D. Proteome-wide characterization of seed aging in Arabidopsis: A comparison between artificial and natural aging protocols. *Plant Physiol.* **2008**, *148*, 620–641. [CrossRef]

26. Frolov, A.; Didio, A.; Ihling, C.; Chantzeva, V.; Grishina, T.; Hoehenwarter, W.; Sinz, A.; Smolikova, G.; Bilova, T.; Medvedev, S. The effect of simulated microgravity on the *Brassica napus* seedling proteome. *Funct. Plant Biol.* **2018**, *45*, 440. [CrossRef]

27. Vanderschuren, H.; Lentz, E.; Zainuddin, I.; Gruissem, W. Proteomics of model and crop plant species: Status, current limitations and strategic advances for crop improvement. *J. Proteomics* **2013**, *93*, 5–19. [CrossRef] [PubMed]

28. Miernyk, J.A. Seed proteomics. In *Plant Proteomics. Methods in Molecular Biology (Methods and Protocols)*; Humana Press: Totowa, NJ, USA, 2014; Volume 1072, pp. 361–377.

29. Mbeunkui, F.; Goshe, M.B. Investigation of solubilization and digestion methods for microsomal membrane proteome analysis using data-independent LC-MSE. *Proteomics* **2011**, *11*, 898–911. [CrossRef] [PubMed]

30. Balmer, Y.; Vensel, W.H.; DuPont, F.M.; Buchanan, B.B.; Hurkman, W.J. Proteome of amyloplasts isolated from developing wheat endosperm presents evidence of broad metabolic capability. *J. Exp. Bot.* **2006**, *57*, 1591–1602. [CrossRef] [PubMed]

31. Barua, P.; Subba, P.; Lande, N.V.; Mangalaparthi, K.K.; Prasad, T.S.K.; Chakraborty, S.; Chakraborty, N. Gel-based and gel-free search for plasma membrane proteins in chickpea (*Cicer arietinum* L.) augments the comprehensive data sets of membrane protein repertoire. *J. Proteomics* **2016**, *143*, 199–208. [CrossRef] [PubMed]

32. Yadeta, K.A.; Elmore, J.M.; Coaker, G. Advancements in the analysis of the *Arabidopsis* plasma membrane proteome. *Front Plant Sci* **2013**, *4*, 86. [CrossRef] [PubMed]

33. Komatsu, S.; Wang, X.; Yin, X.; Nanjo, Y.; Ohyanagi, H.; Sakata, K. Integration of gel-based and gel-free proteomic data for functional analysis of proteins through Soybean Proteome Database. *J. Proteomics* **2017**, *163*, 52–66. [CrossRef]

34. Bourgeois, M.; Jacquin, F.; Savois, V.; Sommerer, N.; Labas, V.; Henry, C.; Burstin, J. Dissecting the proteome of pea mature seeds reveals the phenotypic plasticity of seed protein composition. *Proteomics* **2009**, *9*, 254–271. [CrossRef]

35. Komatsu, S.; Hashiguchi, A. Subcellular proteomics: Application to elucidation of flooding-response mechanisms in soybean. *Proteomes* **2018**, *6*, 13. [CrossRef]

36. Wang, W.Q.; Møller, I.M.; Song, S.Q. Proteomic analysis of embryonic axis of *Pisum sativum* seeds during germination and identification of proteins associated with loss of desiccation tolerance. *J. Proteomics* **2012**, *77*, 68–86. [CrossRef]

37. Hajduch, M.; Casteel, J.E.; Hurrelmeyer, K.E.; Song, Z.; Agrawal, G.K.; Thelen, J.J. Proteomic analysis of seed filling in *Brassica napus* developmental characterization of metabolic isozymes using high-resolution. *Plant Physiol.* **2006**, *141*, 32–46. [CrossRef] [PubMed]

38. Zhang, H.; Wang, W.Q.; Liu, S.J.; Møller, I.M.; Song, S.Q. Proteome analysis of poplar seed vigor. *PLoS ONE* **2015**, *10*, e0132509. [CrossRef] [PubMed]

39. Ventura, L.; Donà, M.; Macovei, A.; Carbonera, D.; Buttafava, A.; Mondoni, A.; Rossi, G.; Balestrazzi, A. Understanding the molecular pathways associated with seed vigor. *Plant Physiol. Biochem.* **2012**, *60*, 196–206. [CrossRef] [PubMed]

40. Catusse, J.; Meinhard, J.; Job, C.; Strub, J.M.; Fischer, U.; Pestsova, E.; Westhoff, P.; Van Dorsselaer, A.; Job, D. Proteomics reveals potential biomarkers of seed vigor in sugarbeet. *Proteomics* **2011**, *11*, 1569–1580. [CrossRef]

41. Rajjou, L.; Duval, M.; Gallardo, K.; Catusse, J.; Bally, J.; Job, C.; Job, D. Seed germination and vigor. *Annu. Rev. Plant Biol.* **2012**, *63*, 507–533. [CrossRef]

42. Hatzig, S.V.; Frisch, M.; Breuer, F.; Nesi, N.; Ducournau, S.; Wagner, M.-H.; Leckband, G.; Abbadi, A.; Snowdon, R.J. Genome-wide association mapping unravels the genetic control of seed germination and vigor in Brassica napus. *Front. Plant Sci.* **2015**, *6*, 221. [CrossRef]

43. Xin, X.; Lin, X.-H.; Zhou, Y.-C.; Chen, X.-L.; Liu, X.; Lu, X.-X. Proteome analysis of maize seeds: The effect of artificial ageing. *Physiol. Plant.* **2011**, *143*, 126–138. [CrossRef]

44. Smolikova, G.N.; Shavarda, A.L.; Alekseichuk, I.V.; Chantseva, V.V.; Medvedev, S.S. The metabolomic approach to the assessment of cultivar specificity of *Brassica napus* L. seeds. *Russ. J. Genet. Appl. Res.* **2016**, *6*, 78–83. [CrossRef]

45. Narula, K.; Sinha, A.; Haider, T.; Chakraborty, N.; Chakraborty, S. Seed Proteomics: An overview. In *Agricultural Proteomics Volume 1*; Salekdeh, G.H., Ed.; Springer International Publishing: Cham, Switzerland, 2016; Volume 1, pp. 31–52. ISBN 978-3-319-43273-1.

46. Min, C.W.; Gupta, R.; Kim, S.W.; Lee, S.E.; Kim, Y.C.; Bae, D.W.; Han, W.Y.; Lee, B.W.; Ko, J.M.; Agrawal, G.K.; et al. Comparative biochemical and proteomic analyses of soybean seed cultivars differing in protein and oil content. *J. Agric. Food Chem.* **2015**, *63*, 7134–7142. [CrossRef]

47. Meyer, L.J.; Gao, J.; Xu, D.; Thelen, J.J. Phosphoproteomic analysis of seed maturation in arabidopsis, rapeseed, and soybean. *Plant Physiol.* **2012**, *159*, 517–528. [CrossRef]

48. Larance, M.; Lamond, A.I. Multidimensional proteomics for cell biology. *Nat. Rev. Mol. Cell Biol.* **2015**, *16*, 269–280. [CrossRef]

49. Plomion, C.; Meddour, H.; Kohler, A.; Jacob, D.; Bastien, C.; Dreyer, E.; De Daruvar, A.; Guehl, J.; Schmitter, J.; Martin, F.; et al. Mapping the proteome of poplar and application to the discovery of drought-stress responsive proteins. *Proteomics* **2006**, *6*, 6509–6527. [CrossRef] [PubMed]

50. Bantscheff, M.; Schirle, M. Quantitative mass spectrometry in proteomics: A critical review. *Anal. Bioanal. Chem.* **2007**, *389*, 1017–1031. [CrossRef] [PubMed]

51. Jimenez-lopez, J.C.; Foley, R.C.; Brear, E.; Clarke, V.C.; Lima-cabello, E.; Florido, J.F.; Singh, K.B.; Alché, J.D.; Smith, P.M.C. Characterization of narrow-leaf lupin (*Lupinus angustifolius* L.) recombinant major allergen IgE-binding proteins and the natural β -conglutin counterparts in sweet lupin seed species. *Food Chem.* **2018**, *244*, 60–70. [CrossRef] [PubMed]

52. Luthria, D.L.; Maria John, K.M.; Marupaka, R.; Natarajan, S. Recent update on methodologies for extraction and analysis of soybean seed proteins. *J. Sci. Food Agric.* **2018**, *98*, 5572–5580. [CrossRef]

53. Alvarez, S.; Roy Choudhury, S.; Sivagnanam, K.; Hicks, L.M.; Pandey, S. Quantitative proteomics analysis of *Camelina sativa* seeds overexpressing the AGG3 gene to identify the proteomic basis of increased yield and stress tolerance. *J. Proteome Res.* **2015**, *14*, 2606–2616. [CrossRef]

54. Hart-Smith, G.; Reis, R.S.; Waterhouse, P.M.; Wilkins, M.R. Improved quantitative plant proteomics via the combination of targeted and untargeted data acquisition. *Front. Plant Sci.* **2017**, *8*, 1669. [CrossRef]

55. Mamontova, T.; Afonin, A.M.; Ihling, C.; Soboleva, A.; Lukasheva, E.; Sulima, A.S.; Shtark, O.Y.; Akhtemova, G.A.; Povydysh, M.N.; Sinz, A.; et al. Profiling of seed proteome in pea (*Pisum sativum* L.) lines characterized with high and low responsivity to combined inoculation with nodule bacteria and arbuscular mycorrhizal fungi. *Molecules* **2019**, *24*, 1603. [CrossRef]

56. Soboleva, A.; Schmidt, R.; Vikhnina, M.; Grishina, T.; Frolov, A. Maillard Proteomics: Opening new pages. *Int. J. Mol. Sci.* **2017**, *18*, 2677. [CrossRef]

57. Jorrín-Novo, J.V.; Pascual, J.; Sánchez-Lucas, R.; Romero-Rodríguez, M.C.; Rodríguez-Ortega, M.J.; Lenz, C.; Valledor, L. Fourteen years of plant proteomics reflected in proteomics: Moving from model species and 2DE-based approaches to orphan species and gel-free platforms. *Proteomics* **2015**, *15*, 1089–1112. [CrossRef]

58. Catherman, A.D.; Skinner, O.S.; Kelleher, N.L. Top down proteomics: Facts and perspectives. *Biochem. Biophys. Res. Commun.* **2014**, *445*, 683–693. [CrossRef] [PubMed]

59. Chmelik, J.; Zidkova, J.; Rehulka, P.; Petry-Podgorska, I.; Bobalova, J. Influence of different proteomic protocols on degree of high-coverage identification of nonspecific lipid transfer protein 1 modified during malting. *Electrophoresis* **2009**, *30*, 560–567. [CrossRef] [PubMed]

60. Gillet, L.C.; Leitner, A.; Aebersold, R. Mass spectrometry applied to bottom-up proteomics: Entering the high-throughput era for hypothesis testing. *Annu. Rev. Anal. Chem.* **2016**, *9*, 449–472. [CrossRef] [PubMed]

61. Mørtz, E.; O'Connor, P.B.; Roepstorff, P.; Kelleher, N.L.; Wood, T.D.; McLafferty, F.W.; Mann, M. Sequence tag identification of intact proteins by matching tanden mass spectral data against sequence data bases. *Proc. Natl. Acad. Sci. USA* **1996**, *93*, 8264–8267. [CrossRef]

62. Hummel, M.; Wigger, T.; Brockmeyer, J. Characterization of mustard 2S albumin allergens by bottom-up, middle-down and top-down proteomics: A consensus set of isoforms of Sin a 1. *J. Proteome Res.* **2015**, *14*, 1547–1556. [CrossRef]

63. Mamontova, T.; Lukasheva, E.; Mavropolo-Stolyarenko, G.; Proksch, C.; Bilova, T.; Kim, A.; Babakov, V.; Grishina, T.; Hoehenwarter, W.; Medvedev, S.; et al. Proteome map of pea (*Pisum sativum* L.) embryos containing different amounts of residual chlorophylls. *Int. J. Mol. Sci.* **2018**, *19*, 4066. [CrossRef]

64. Bose, U.; Broadbent, J.A.; Byrne, K.; Hasan, S.; Howitt, C.A.; Colgrave, M.L. Optimisation of protein extraction for in-depth profiling of the cereal grain proteome. *J. Proteomics* **2019**, *197*, 23–33. [CrossRef]

65. Nilsen, T.W.; Graveley, B.R. Expansion of the eukaryotic proteome by alternative splicing. *Nature* **2010**, *463*, 457–463. [CrossRef]

66. Bourgeois, M.; Jacquin, F.; Cassecuelle, F.; Savois, V.; Belghazi, M.; Aubert, G.; Quillien, L.; Huart, M.; Marget, P.; Burstin, J. A PQL (protein quantity loci) analysis of mature pea seed proteins identifies loci determining seed protein composition. *Proteomics* **2011**, *11*, 1581–1594. [CrossRef]

67. Natarajan, S.S.; Krishnan, H.B.; Lakshman, S.; Garrett, W.M. An efficient extraction method to enhance analysis of low abundant proteins from soybean seed. *Anal. Biochem.* **2009**, *394*, 259–268. [CrossRef]

68. Kim, Y.J.Y.C.; Wang, Y.; Gupta, R.; Kim, S.T.S.W.; Min, C.W.; Kim, Y.J.Y.C.; Park, K.H.; Agrawal, G.K.; Rakwal, R.; Choung, M.-G.G.; et al. Protamine sulfate precipitation method depletes abundant plant seed-storage proteins: A case study on legume plants. *Proteomics* **2015**, *15*, 1760–1764. [CrossRef]

69. Min, C.W.; Park, J.; Bae, J.W.; Agrawal, G.K.; Rakwal, R.; Kim, Y.; Yang, P.; Kim, S.T.; Gupta, R. In-Depth Investigation of low-abundance proteins in matured and filling stages seeds of glycine max employing a combination of protamine sulfate precipitation and TMT-Based quantitative proteomic analysis. *Cells* **2020**, *9*, 1517. [CrossRef] [PubMed]

70. Krishnan, H.B.; Oehrle, N.W.; Natarajan, S.S. A rapid and simple procedure for the depletion of abundant storage proteins from legume seeds to advance proteome analysis: A case study using *Glycine max*. *Proteomics* **2009**, *9*, 3174–3188. [CrossRef] [PubMed]

71. Righetti, P.G.; Boschetti, E. Low-abundance plant protein enrichment with peptide libraries to enlarge proteome coverage and related applications. *Plant Sci.* **2020**, *290*, 110302. [CrossRef] [PubMed]

72. Kretzschmar, F.K.; Doner, N.M.; Krawczyk, H.E.; Scholz, P.; Schmitt, K.; Valerius, O.; Braus, G.H.; Mullen, R.T.; Ischebeck, T. Identification of low-abundance lipid droplet proteins in seeds and seedlings. *Plant Physiol.* **2020**, *182*, 1326–1345. [CrossRef]

73. Du, C.; Liu, A.; Niu, L.; Cao, D.; Liu, H.; Wu, X.; Wang, W. Proteomic identification of lipid-bodies-associated proteins in maize seeds. *Acta Physiol. Plant.* **2019**, *41*, 70. [CrossRef]

74. Tan, B.C.; Lim, Y.S.; Lau, S. Proteomics in commercial crops: An overview. *J. Proteomics* **2017**, *169*, 176–188. [CrossRef]

75. Dawod, M.; Arvin, N.E.; Kennedy, R.T. Recent advances in protein analysis by capillary and microchip electrophoresis. *Analyst* **2017**, *142*, 1847–1866. [CrossRef]

76. Kota, U.; Goshe, M.B. Advances in qualitative and quantitative plant membrane proteomics. *Phytochemistry* **2011**, *72*, 1040–1060. [CrossRef]

77. Rogowska-Wrzesinska, A.; Le Bihan, M.-C.; Thaysen-Andersen, M.; Roepstorff, P. 2D gels still have a niche in proteomics. *J. Proteomics* **2013**, *88*, 4–13. [CrossRef]

78. Ostergaard, O.; Finnie, C.; Laugesen, S.; Roepstorff, P.; Svennson, B. Proteome analysis of barley seeds: Identification of major proteins from two-dimensional gels (pI 4-7). *Proteomics* **2004**, *4*, 2437–2447. [CrossRef] [PubMed]

79. Rabilloud, T.; Lelong, C. Two-dimensional gel electrophoresis in proteomics: A tutorial. *J. Proteomics* **2011**, *74*, 1829–1841. [CrossRef] [PubMed]

80. Xu, X.; Fan, R.; Zheng, R.; Li, C.; Yu, D. Proteomic analysis of seed germination under salt stress in soybeans. *J. Zhejiang Univ. Sci. B* **2011**, *12*, 507–517. [CrossRef] [PubMed]

81. Rabilloud, T.; Chevallet, M.; Luche, S.; Lelong, C. Two-dimensional gel electrophoresis in proteomics: Past, present and future. *J. Proteomics* **2010**, *73*, 2064–2077. [CrossRef] [PubMed]

82. Gallardo, K.; Kurt, C.; Thompson, R.; Ochatt, S. *In vitro* culture of immature *M. truncatula* grains under conditions permitting embryo development comparable to that observed *in vivo*. *Plant Sci.* **2006**, *170*, 1052–1058. [CrossRef]

83. Magdeldin, S.; Enany, S.; Yoshida, Y.; Xu, B.; Zhang, Y.; Zureena, Z.; Lokamani, I.; Yaoita, E.; Yamamoto, T. Basics and recent advances of two dimensional- polyacrylamide gel electrophoresis. *Clin. Proteomics* **2014**, *11*, 16. [CrossRef]

84. Rune, M. *Mass Spectrometry Data Analysis in Proteomics*; Humana Press: Totowa, NJ, USA, 2006; Volume 367, ISBN 1-59745-275-0.

85. Friedman, D.B.; Hoving, S.; Westermeier, R. Chapter 30 isoelectric focusing and two-dimensional gel electrophoresis. In *Methods in Enzymology*; Elsevier Inc.: Amsterdam, The Netherlands, 2009; pp. 515–540.

86. Lee, P.Y.; Saraygord-Afshari, N.; Low, T.Y. The evolution of two-dimensional gel electrophoresis—From proteomics to emerging alternative applications. *J. Chromatogr. A* **2020**, *1615*, 460763. [CrossRef]

87. Jorrin-Novo, J.V.; Komatsu, S.; Sanchez-Lucas, R.; Rodríguez de Francisco, L.E. Gel electrophoresis-based plant proteomics: Past, present, and future. Happy 10th anniversary Journal of Proteomics! *J. Proteomics* **2019**, *198*, 1–10. [CrossRef]

88. Meleady, P. Two-Dimensional gel electrophoresis and 2D-DIGE. In *Difference Gel Electrophoresis. Methods in Molecular Biology*; Humana Press: New York, NY, USA, 2018; pp. 3–14.

89. Isaacson, T.; Damasceno, C.M.B.; Saravanan, R.S.; He, Y.; Catalá, C.; Saladié, M.; Rose, J.K.C. Sample extraction techniques for enhanced proteomic analysis of plant tissues. *Nat. Protoc.* **2006**, *1*, 769–774. [CrossRef]

90. Wang, W.; Vignani, R.; Scali, M.; Cresti, M. A universal and rapid protocol for protein extraction from recalcitrant plant tissues for proteomic analysis. *Electrophoresis* **2006**, *27*, 2782–2786. [CrossRef]

91. Wang, X.; Li, X.; Deng, X.; Han, H.; Shi, W.; Li, Y. A protein extraction method compatible with proteomic analysis for the euhalophyte Salicornia europaea. *Electrophoresis* **2007**, *28*, 3976–3987. [CrossRef] [PubMed]

92. Soboleva, A.; Vikhnina, M.; Grishina, T.; Frolov, A. Probing protein glycation by chromatography and mass spectrometry: Analysis of glycation adducts. *Int. J. Mol. Sci.* **2017**, *18*, 2557. [CrossRef] [PubMed]

93. Antonova, K.; Vikhnina, M.; Soboleva, A.; Mehmood, T.; Heymich, M.; Leonova, T.; Bankin, M.; Lukasheva, E.; Gensberger-Reigl, S.; Medvedev, S.; et al. Analysis of chemically labile glycation adducts in seed proteins: Case study of methylglyoxal-derived hydroimidazolone 1 (MG-H1). *Int. J. Mol. Sci.* **2019**, *20*, 3659. [CrossRef] [PubMed]

94. Damerval, C.; de Dominique, V.; Zivy, M.; Thiellement, H. Technical improvements in two-dimensional electrophoresis increase the level of genetic variation detected in wheat-seedling proteins. *Electrophoresis* **1986**, *7*, 52–54. [CrossRef]

95. Capriotti, A.L.; Cavaliere, C.; Piovesana, S.; Stampachiacchiere, S.; Ventura, S.; Zenezini Chiozzi, R.; Laganà, A. Characterization of quinoa seed proteome combining different protein precipitation techniques: Improvement of knowledge of nonmodel plant proteomics. *J. Sep. Sci.* **2015**, *38*, 1017–1025. [CrossRef]

96. Natarajan, S.; Xu, C.; Caperna, T.J.; Garrett, W.M. Comparison of protein solubilization methods suitable for proteomic analysis of soybean seed proteins. *Anal. Biochem.* **2005**, *342*, 214–220. [CrossRef]

97. Fu, Y.; Zhang, H.; Mandal, S.N.; Wang, C.; Chen, C.; Ji, W. Quantitative proteomics reveals the central changes of wheat in response to powdery mildew. *J. Proteomics* **2016**, *130*, 108–119. [CrossRef]

98. Baslam, M.; Kaneko, K.; Mitsui, T. iTRAQ-based proteomic analysis of rice grains. In *Plant Proteomics: Methods and Protocols, Methods in Molecular Biology*; Jorrin-Novo, J.V., Valledor, L., Castillejo, M.A., Rey, M.-D., Eds.; Springer: Berlin/Heidelberg, Germany, 2020; Volume 213, pp. 405–414.

99. Görg, A.; Weiss, W. Two-dimensional electrophoresis with immobilized pH gradients. In *Proteome Research: Two-Dimensional Gel Electrophoresis and Identification Methods*; Springer-Verlag: Berlin/Heidelberg, Germany, 2000; pp. 57–106.

100. Vâlcu, C.-M.; Schlink, K. Reduction of proteins during sample preparation and two-dimensional gel electrophoresis of woody plant samples. *Proteomics* **2006**, *6*, 1599–1605. [CrossRef]

101. Negri, A.S.; Prinsi, B.; Scienza, A.; Morgutti, S.; Cocucci, M.; Espen, L. Analysis of grape berry cell wall proteome: A comparative evaluation of extraction methods. *J. Plant Physiol.* **2008**, *165*, 1379–1389. [CrossRef]

102. Dong, Y.; Wang, Q.; Zhang, L.; Du, C.; Xiong, W. Dynamic Proteomic characteristics and network integration revealing key proteins for two kernel tissue developments in popcorn. *PLoS ONE* **2015**, *10*, e0143181. [CrossRef]

103. Ranjbar Sistani, N.; Kaul, H.; Desalegn, G.; Wienkoop, S. Rhizobium impacts on seed productivity, quality, and protection of *Pisum sativum* upon disease stress caused by didymella pinodes: Phenotypic, proteomic, and metabolomic traits. *Front. Plant Sci.* **2017**, *8*, 1961. [CrossRef] [PubMed]

104. Cremer, F.; van de Walle, C. Method for extraction of proteins from green plant tissues for two-dimensional polyacrylamide gel electrophoresis. *Anal. Biochem.* **1985**, *147*, 22–26. [CrossRef]

105. Milkovska-Stamenova, S.; Schmidt, R.; Frolov, A.; Birkemeyer, C. GC-MS method for the quantitation of carbohydrate intermediates in glycation systems. *J. Agric. Food Chem.* **2015**, *63*, 5911–5919. [CrossRef]

106. Frolov, A.; Schmidt, R.; Spiller, S.; Greifenhagen, U.; Hoffmann, R. Arginine-derived advanced glycation end products generated in peptide–glucose mixtures during boiling. *J. Agric. Food Chem.* **2014**, *62*, 3626–3635. [CrossRef] [PubMed]

107. Greifenhagen, U.; Frolov, A.; Blüher, M.; Hoffmann, R. Plasma Proteins modified by advanced glycation end products (AGEs) reveal site-specific susceptibilities to glycemic control in patients with type 2 diabetes. *J. Biol. Chem.* **2016**, *291*, 9610–9616. [CrossRef] [PubMed]

108. Han, C.; Yin, X.; He, D.; Yang, P. Analysis of Proteome profile in germinating soybean seed, and its comparison with rice showing the styles of reserves mobilization in different crops. *PLoS ONE* **2013**, *8*, e56947. [CrossRef]

109. Nishimura, N.; Tsuchiya, W.; Moresco, J.J.; Hayashi, Y.; Satoh, K.; Kaiwa, N.; Irisa, T.; Kinoshita, T.; Schroeder, J.I.; Yates, J.R.; et al. Control of seed dormancy and germination by DOG1-AHG1 PP2C phosphatase complex *via* binding to heme. *Nat. Commun.* **2018**, *9*, 1–14. [CrossRef]

110. Lorenz, C.; Brandt, S.; Borisjuk, L.; Rolletschek, H.; Heinzel, N.; Tohge, T.; Fernie, A.R.; Braun, H.-P.; Hildebrandt, T.M. The role of persulfide metabolism during *Arabidopsis* seed development under light and dark conditions. *Front. Plant Sci.* **2018**, *9*, 1381. [CrossRef]

111. Wang, W.; Vignani, R.; Scali, M.; Sensi, E.; Tiberi, P.; Cresti, M. Removal of lipid contaminants by organic solvents from oilseed protein extract prior to electrophoresis. *Anal. Biochem.* **2004**, *329*, 139–141. [CrossRef]

112. Murad, A.M.; Rech, E.L. NanoUPLC-MSE proteomic data assessment of soybean seeds using the Uniprot database. *BMC Biotechnol.* **2012**, *12*, 82. [CrossRef]

113. Satour, P.; Youssef, C.; Châtelain, E.; Vu, B.L.; Teulat, B.; Job, C.; Job, D.; Montrichard, F. Patterns of protein carbonylation during *Medicago truncatula* seed maturation. *Plant. Cell Environ.* **2018**, *41*, 2183–2194. [CrossRef] [PubMed]

114. Tang, H.; Ming, Z.; Liu, R.; Xiong, T.; Grevelding, C.G.; Dong, H.; Jiang, M. Development of Adult worms and granulomatous pathology are collectively regulated by *t*- and *b*-cells in mice infected with *Schistosoma japonicum*. *PLoS ONE* **2013**, *8*, e54432.

115. Rakwal, R.; Hayashi, G.; Shibato, J.; Deepak, S.A.; Gundimeda, S.; Simha, U.; Padmanaban, A.; Gupta, R.; Han, S.; Kim, S.T.; et al. Progress toward rice seed OMICS in Low-level gamma radiation environment in iitate Village, Fukushima. *J. Hered.* **2018**, *109*, 206–211. [CrossRef] [PubMed]

116. Pinheiro, C.; Sergeant, K.; Machado, C.M.; Renaut, J.; Ricardo, C.P. Two Traditional maize inbred lines of contrasting technological abilities are discriminated by the seed flour proteome. *J. Proteome Res.* **2013**, *12*, 3152–3165. [CrossRef]

117. Niu, L.; Ding, H.; Zhang, J.; Wang, W. Proteomic analysis of starch biosynthesis in maize seeds. *Starch - Stärke* **2019**, *71*, 1800294. [CrossRef]

118. Neuhoff, V.; Arold, N.; Taube, D.; Ehrhardt, W. Dependence of Particle and fiber properties. *Electrophoresis* **1988**, *9*, 255–262. [CrossRef]

119. Frolov, A.; Blüher, M.; Hoffmann, R. Glycation sites of human plasma proteins are affected to different extents by hyperglycemic conditions in type 2 diabetes mellitus. *Anal. Bioanal. Chem.* **2014**, *406*, 5755–5763. [CrossRef]

120. Romero-Rodríguez, M.C.; Jorrín-Novo, J.V.; Castillejo, M.A. Toward characterizing germination and early growth in the non-orthodox forest tree species Quercus ilex through complementary gel and gel-free proteomic analysis of embryo and seedlings. *J. Proteomics* **2019**, *197*, 60–70. [CrossRef]

121. Chevalier, F. Standard dyes for total protein staining in gel-based proteomic analysis. *Materials* **2010**, *3*, 4784–4792. [CrossRef]

122. Mortz, E.; Krogh, T.N.; Vorum, H.; Görg, A. Improved silver staining protocols for high sensitivity protein identification using matrix-assisted laser desorption/ionization-time of flight analysis. *Proteomics* **2001**, *1*, 1359–1363. [CrossRef]

123. Chevallet, M.; Luche, S.; Strub, J.; Van Dorsselaer, A.; Rabilloud, T. Sweet silver: A formaldehyde-free silver staining using aldoses as developing agents, with enhanced compatibility with mass spectrometry. *Proteomics* **2008**, *8*, 4853–4861. [CrossRef] [PubMed]

124. Puumalainen, T.J.; Puustinen, A.; Poikonen, S.; Turjanmaa, K.; Palosuo, T.; Vaali, K. Proteomic identification of allergenic seed proteins, napin and cruciferin, from cold-pressed rapeseed oils. *Food Chem.* **2015**, *175*, 381–385. [CrossRef] [PubMed]

125. Berggren, K.; Chernokalskaya, E.; Steinberg, T.H.; Kemper, C.; Lopez, M.F.; Diwu, Z.; Haugland, R.P.; Patton, W.F. Background-free, high sensitivity staining of proteins in one- and two-dimensional sodium dodecyl sulfate-polyacrylamide gels using a luminescent ruthenium complex. *Electrophoresis* **2000**, *21*, 2509–2521. [CrossRef]

126. Rabilloud, T.; Strub, J.; Luche, S.; Van Dorsselaer, A.; Lunardi, J. A comparison between Sypro Ruby and ruthenium II tris (bathophenanthroline disulfonate) as. *Proteomics* **2001**, *1*, 699–704. [CrossRef]

127. Zhou, G.; Li, H.; DeCamp, D.; Chen, S.; Shu, H.; Gong, Y.; Flaig, M.; Gillespie, J.W.; Hu, N.; Taylor, P.R.; et al. 2D Differential In-gel electrophoresis for the identification of esophageal scans cell cancer-specific protein markers. *Mol. Cell. Proteomics* **2002**, *1*, 117–123. [CrossRef]

128. Romero-Rodriguez, M.; Abril, N.; Sánchez-Lucas, R.; Jorrin Novo, J. V Multiplex staining of 2-DE gels for an initial phosphoproteome analysis of germinating seeds and early grown seedlings from a non-orthodox specie: *Quercus ilex* L. subsp. ballota [Desf.] Samp. *Front. Plant Sci.* **2015**, *6*, 620. [CrossRef]

129. Klose, J.; Kobalz, U. Two-dimensional electrophoresis of proteins: An updated protocol and implications for a functional analysis of the genome. *Electrophoresis* **1995**, *16*, 1034–1059. [CrossRef]

130. Gygi, S.P.; Corthals, G.L.; Zhang, Y.; Rochon, Y.; Aebersold, R. Evaluation of two-dimensional gel electrophoresis-based proteome analysis technology. *Proc. Natl. Acad. Sci. USA* **2000**, *97*, 9390–9395. [CrossRef]

131. Chassaigne, H.; Trégoat, V.; Nørgaard, J.V.; Maleki, S.J.; van Hengel, A.J. Resolution and identification of major peanut allergens using a combination of fluorescence two-dimensional differential gel electrophoresis, Western blotting and Q-TOF mass spectrometry. *J. Proteomics* **2009**, *72*, 511–526. [CrossRef]

132. Jorrin-Novo, J.V. Plant proteomics methods and protocols. In *Plant Proteomics*; Jorrin-Novo, J.V., Komatsu, S., Weckwerth, W., Wienkoop, S., Eds.; Humana Press: Totowa, NJ, USA, 2014; Volume 1072, ISBN 978-1-62703-630-6.

133. Scippa, G.S.; Rocco, M.; Ialicicco, M.; Trupiano, D.; Viscosi, V.; Di Michele, M.; Arena, S.; Chiatante, D.; Scaloni, A. The proteome of lentil (*Lens culinaris* Medik.) seeds: Discriminating between landraces. *Electrophoresis* **2010**, *31*, 497–506. [CrossRef]

134. Talamo, F.; D'Ambrosio, C.; Arena, S.; Del Vecchio, P.; Ledda, L.; Zehender, G.; Ferrara, L.; Scaloni, A. Proteins from bovine tissues and biological fluids: Defining a reference electrophoresis map for liver, kidney, muscle, plasma and red blood cells. *Proteomics* **2003**, *3*, 440–460. [CrossRef] [PubMed]

135. Schönberg, A.; Rödiger, A.; Mehwald, W.; Galonska, J.; Christ, G.; Helm, S.; Thieme, D.; Majovsky, P.; Hoehenwarter, W.; Baginsky, S. Identification of STN7/STN8 kinase targets reveals connections between electron transport, metabolism and gene expression. *Plant J.* **2017**, *90*, 1176–1186. [CrossRef] [PubMed]

136. Rosenfeld, J.; Capdevielle, J.; Guillemot, J.C.; Ferrara, P. In-gel digestion of proteins for internal analysis after one- or two-dimensional gel electrophoresis. *Anal. Biochem.* **1992**, *203*, 173–179. [CrossRef]

137. Blum, H.; Beier, H.; Gross, H.J. Improved silver staining of plant proteins, RNA and DNA in polyacrylamide gels. *Electrophoresis* **1987**, *8*, 93–99. [CrossRef]

138. Ogura, T.; Ogihara, J.; Sunairi, M.; Takeishi, H.; Aizawa, T.; Olivos-Trujillo, M.R.; Maureira-Butler, I.J.; Salvo-Garrido, H.E. Proteomic characterization of seeds from yellow lupin (*Lupinus luteus* L.). *Proteomics* **2014**, *14*, 1543–1546. [CrossRef] [PubMed]

139. Gomes, L.S.; Senna, R.; Sandim, V.; Silva-Neto, M.A.C.; Perales, J.E.A.; Zingali, R.B.; Soares, M.R.; Fialho, E. Four conventional soybean [*Glycine max* (L.) Merrill] seeds exhibit different protein profiles as revealed by proteomic analysis. *J. Agric. Food Chem.* **2014**, *62*, 1283–1293. [CrossRef] [PubMed]

140. Min, C.W.; Lee, S.H.; Cheon, Y.E.; Han, W.Y.; Ko, J.M.; Kang, H.W.; Kim, Y.C.; Agrawal, G.K.; Rakwal, R.; Gupta, R.; et al. In-depth proteomic analysis of *Glycine max* seeds during controlled deterioration treatment reveals a shift in seed metabolism. *J. Proteomics* **2017**, *169*, 125–135. [CrossRef]

141. Swart, C.; Martínez-Jaime, S.; Gorka, M.; Zander, K.; Graf, A. Hit-gel: Streamlining *in-gel* protein digestion for high-throughput proteomics experiments. *Sci. Rep.* **2018**, *8*, 8582. [CrossRef]

142. Mallick, P.; Schirle, M.; Chen, S.S.; Flory, M.R.; Lee, H.; Martin, D.; Ranish, J.; Raught, B.; Schmitt, R.; Werner, T.; et al. Computational prediction of proteotypic peptides for quantitative proteomics. *Nat. Biotechnol.* **2007**, *25*, 125–131. [CrossRef]

143. Vandemoortele, G.; Staes, A.; Gonnelli, G.; Samyn, N.; De Sutter, D.; Vandermarliere, E.; Timmerman, E.; Gevaert, K.; Martens, L.; Eyckerman, S. An extra dimension in protein tagging by quantifying universal proteotypic peptides using targeted proteomics. *Sci. Rep.* **2016**, *6*, 27220. [CrossRef]

144. Keerthikumar, S.; Mathivanan, S. Proteotypic Peptides and Their Applications. In *Methods in Molecular Biology*; Humana Press: New York, NY, USA, 2017; Volume 1549, pp. 101–107.

145. Bhattacharya, A. Proteotypic Peptides. In *Encyclopedia of Systems Biology*; Dubitzky, W., Wolkenhauer, O., Cho, K.-H., Yokota, H., Eds.; Springer: New York, NY, USA, 2013; p. 1800. ISBN 978-1-4419-9863-7.

146. Clauser, K.R.; Baker, P.; Burlingame, A.L. Role of accurate mass measurement (10 ppm) in protein identification strategies employing MS or MS/MS and database searching. *Anal. Chem.* **1999**, *71*, 2871–2882. [CrossRef] [PubMed]

147. Çakir, B.; Gülseren, İ. Identification of novel proteins from black cumin seed meals based on 2D gel electrophoresis and MALDI-TOF/TOF-MS analysis. *Plant Foods Hum. Nutr.* **2019**, *74*, 414–420. [CrossRef] [PubMed]

148. Wu, W.W.; Wang, G.; Baek, S.J.; Shen, R.-F. Comparative study of three proteomic quantitative methods, DIGE, cICAT, and iTRAQ, using 2D gel- or LC−MALDI TOF/TOF. *J. Proteome Res.* **2006**, *5*, 651–658. [CrossRef] [PubMed]

149. Wu, W.; Dai, R.-T.; Bendixen, E. Comparing SRM and SWATH methods for quantitation of bovine muscle proteomes. *J. Agric. Food Chem.* **2019**, *67*, 1608–1618. [CrossRef]

150. Que, Y.; Xu, L.; Lin, J.; Ruan, M.; Zhang, M.; Chen, R. Differential protein expression in sugarcane during sugarcane- sporisorium scitamineum interaction revealed by 2-DE and MALDI-TOF-TOF/MS. *Comp. Funct. Genomics* **2011**, *2011*, 1–10. [CrossRef]

151. Vadivel, A.K.A. Gel-based proteomics in plants: Time to move on from the tradition. *Front. Plant Sci.* **2015**, *6*, 369.

152. Maaß, S.; Becher, D. Methods and applications of absolute protein quantification in microbial systems. *J. Proteomics* **2016**, *136*, 222–233. [CrossRef]

153. Nedenskov Jensen, K.; Jessen, F.; Jørgensen, B.M. Multivariate Data analysis of two-dimensional gel electrophoresis protein patterns from few samples. *J. Proteome Res.* **2008**, *7*, 1288–1296. [CrossRef]

154. Marengo, E.; Robotti, E.; Antonucci, F.; Cecconi, D.; Campostrini, N.; Righetti, P.G. Numerical approaches for quantitative analysis of two-dimensional maps: A review of commercial software and home-made systems. *Proteome Sci.* **2005**, *5*, 654–666. [CrossRef]

155. Voss, T.; Haberl, P. Observations on the reproducibility and matching efficiency of two-dimensional electrophoresis gels: Consequences for comprehensive data analysis. *Electrophoresis* **2000**, *21*, 3345–3350. [CrossRef]

156. Arentz, G.; Weiland, F.; Oehler, M.K.; Hoffmann, P. State of the art of 2D DIGE. *Proteomics Clin. Appl.* **2015**, *9*, 277–288. [CrossRef] [PubMed]

157. Beckett, P. The Basics of 2D DIGE. In *Difference Gel Electrophoresis (DIGE). Methods in Molecular Biology (Methods and Protocols)*; Humana Press: New York, NY, USA, 2012; Volume 854, pp. 9–19.

158. Morgan, M.E.; Minden, J.S. Difference gel electrophoresis: A single gel method for detecting changes in protein extracts. *Electrophoresis* **1997**, *18*, 2071–2077.

159. TripleTOF®6600 System. Available online: https://sciex.com/products/mass-spectrometers/qtof-systems/tripletof-systems/tripletof-6600-system (accessed on 29 November 2020).

160. Krokhin, O.V.; Ens, W.; Standing, K.G. MALDI QqTOF MS Combined with off-line HPLC for Characterization of protein primary structure and post-translational modifications. *J. Biomol. Tech.* **2005**, *16*, 427–438.

161. Köcher, T.; Pichler, P.; Swart, R.; Mechtler, K. Analysis of protein mixtures from whole-cell extracts by single-run nanoLC-MS/MS using ultralong gradients. *Nat. Protoc.* **2012**, *7*, 882–890. [CrossRef] [PubMed]

162. Burrieza, H.P.; Rizzo, A.J.; Moura Vale, E.; Silveira, V.; Maldonado, S. Shotgun proteomic analysis of quinoa seeds reveals novel lysine-rich seed storage globulins. *Food Chem.* **2019**, *293*, 299–306. [CrossRef]

163. Frolov, A.; Henning, A.; Böttcher, C.; Tissier, A.; Strack, D.; Bo, C.; Tissier, A.; Strack, D.; Böttcher, C.; Tissier, A.; et al. An UPLC-MS/MS Method for the simultaneous identification and quantitation of cell wall phenolics in *Brassica napus* seeds. *J. Agric. Food Chem.* **2013**, *61*, 1219–1227. [CrossRef]

164. Matthiesen, R.; Bunkenborg, J. Introduction to mass spectrometry-based proteomics. In *Methods in Molecular Biology*; Humana: New York, NY, USA, 2020; pp. 1–58.

165. Vissers, J.P.C.; Chervet, J.P.; Salzmann, J.P. Sodium dodecyl sulphate removal from tryptic digest samples for on-line capillary liquid. *J. Mass Spectrom.* **1996**, *31*, 1021–1027. [CrossRef]

166. Cole, R.B. *Electrospray and MALDI Mass Spectrometry: Fundamentals, Instrumentation, Practicalities, and Biological Applications*, 2nd ed; John Wiley & Sons, Inc.: Hoboken, NJ, USA, 2012; ISBN 9780471741077.

167. Chen, E.I.; Cociorva, D.; Norris, J.L.; Yates, J.R. Optimization of mass spectrometry-compatible surfactants for shotgun proteomics. *J. Proteome Res.* **2007**, *6*, 2529–2538. [CrossRef]

168. Lin, Y.; Jiang, H.; Yan, Y.; Peng, B.; Chen, J.; Lin, H.; Liu, Z. Shotgun analysis of membrane proteomes by an improved SDS-assisted sample preparation method coupled with liquid chromatography–tandem mass spectrometry. *J. Chromatogr. B* **2012**, *911*, 6–14. [CrossRef]

169. Bilova, T.; Greifenhagen, U.; Paudel, G.; Lukasheva, E.; Brauch, D.; Osmolovskaya, N.; Tarakhovskaya, E.; Balcke, G.U.; Tissier, A.; Vogt, T.; et al. Glycation of plant proteins under environmental stress—Methodological approaches, potential mechanisms and biological role. In *Abiotic and Biotic Stress in Plants—Recent Advances and Future Perspectives*; InTech: London, UK, 2016. [CrossRef]

170. Greifenhagen, U.; Frolov, A.; Blüher, M.; Hoffmann, R. Site-specific analysis of advanced glycation end products in plasma proteins of type 2 diabetes mellitus patients. *Anal. Bioanal. Chem.* **2016**, *408*, 5557–5566. [CrossRef]

171. Bilova, T.; Lukasheva, E.; Brauch, D.; Greifenhagen, U.; Paudel, G.; Tarakhovskaya, E.; Frolova, N.; Mittasch, J.; Balcke, G.U.G.U.; Tissier, A.; et al. A Snapshot of the plant glycated proteome: Structural, functional and mechanistic aspect. *J. Biol. Chem.* **2016**, *291*, 7621–7636. [CrossRef]

172. Takemori, N.; Takemori, A.; Ishizaki, J.; Hasegawa, H. Enzymatic protein digestion using a dissolvable polyacrylamide gel and its application to mass spectrometry-based proteomics. *J. Chromatogr. B* **2014**, *967*, 36–40. [CrossRef]

173. Kalli, A.; Smith, T.; Sweredoski, M.J.; Hess, S. Evaluation and optimization of mass spectrometric settings during data-dependent acquisition mode: Focus on LTQ-orbitrap mass analyzers. *J. Proteome Res.* **2013**, *12*, 3071–3086. [CrossRef] [PubMed]

174. Robinson, A.E.; Binek, A.; Venkatraman, V.; Searle, B.C.; Holewinski, R.J.; Rosenberger, G.; Parker, S.J.; Basisty, N.; Xie, X.; Lund, P.J.; et al. Lysine and arginine protein post-translational modifications by enhanced dia libraries: Quantification in murine liver disease. *J. Proteome Res.* **2020**, *19*, 4163–4178. [CrossRef] [PubMed]

175. Majovsky, P.; Naumann, C.; Lee, C.; Lassowskat, I.; Trujillo, M.; Dissmeyer, N.; Hoehenwarter, W. Targeted Proteomics analysis of protein degradation in plant signaling on an LTQ-orbitrap mass spectrometer. *J. Proteome Res.* **2014**, *13*, 4246–4258. [CrossRef] [PubMed]

176. Antonets, K.S.; Belousov, M.V.; Sulatskaya, A.I.; Belousova, M.E.; Kosolapova, A.O.; Sulatsky, M.I.; Andreeva, E.A.; Zykin, P.A.; Malovichko, Y.V.; Shtark, O.Y.; et al. Accumulation of storage proteins in plant seeds is mediated by amyloid formation. *PLoS Biol.* **2020**, *18*, e3000564. [CrossRef] [PubMed]

177. Paudel, G.; Bilova, T.; Schmidt, R.; Greifenhagen, U.; Berger, R.; Tarakhovskaya, E.; Stöckhardt, S.; Balcke, G.U.; Humbeck, K.; Brandt, W.; et al. Osmotic stress is accompanied by protein glycation in *Arabidopsis thaliana*. *J. Exp. Bot.* **2016**, *67*, 6283–6295. [CrossRef]

178. Hart-Smith, G. Combining Targeted and untargeted data acquisition to enhance quantitative plant proteomics experiments. In *Plant Proteomics. Methods in Molecular Biology*; Humana: New York, NY, USA, 2020; pp. 169–178.

179. Juarez-Escobar, J.; Elizalde-Contreras, J.M.; Loyola-Vargas, V.M.; Ruiz-May, E. A Phosphoproteomic analysis pipeline for peels of tropical fruits. In *Plant Proteomics. Methods in Molecular Biology*; Humana: New York, NY, USA, 2020; pp. 179–196.

180. Wilm, M. Quantitative proteomics in biological research. *Proteomics* **2009**, *9*, 4590–4605. [CrossRef]

181. Ćeranić, A.; Doppler, M.; Büschl, C.; Parich, A.; Xu, K.; Koutnik, A.; Bürstmayr, H.; Lemmens, M.; Schuhmacher, R. Preparation of uniformly labelled 13C- and 15N-plants using customised growth chambers. *Plant Methods* **2020**, *16*, 46. [CrossRef]

182. Liu, W.; Li, L.; Zhang, Z.; Dong, M.; Jin, W. iTRAQ-based quantitative proteomic analysis of transgenic and non-transgenic maize seeds. *J. Food Compos. Anal.* **2020**, *92*, 103564. [CrossRef]

183. Chen, L.; Wang, Z.; Li, M.; Ma, X.; Tian, E.; Sun, A.; Yin, Y. Analysis of the natural dehydration mechanism during middle and late stages of wheat seeds development by some physiological traits and iTRAQ-based proteomic. *J. Cereal Sci.* **2018**, *80*, 102–110. [CrossRef]

184. Nelson, C.J.; Hegeman, A.D.; Harms, A.C.; Sussman, M.R. A Quantitative analysis of arabidopsis plasma membrane using trypsin-catalyzed 18 O labeling. *Mol. Cell. Proteomics* **2006**, *5*, 1382–1395. [CrossRef] [PubMed]

185. Picotti, P.; Bodenmiller, B.; Aebersold, R. Proteomics meets the scientific method. *Nat. Methods* **2013**, *10*, 24–27. [CrossRef] [PubMed]

186. Wiśniewski, J.R.; Zougman, A.; Nagaraj, N.; Mann, M. Universal sample preparation method for proteome analysis. *Nat. Methods* **2009**, *6*, 359–362. [CrossRef] [PubMed]

187. George, I.S.; Fennell, A.Y.; Haynes, P.A. Protein identification and quantification from riverbank grape, *Vitis riparia*: Comparing SDS-PAGE and FASP-GPF techniques for shotgun proteomic analysis. *Proteomics* **2015**, *15*, 3061–3065. [CrossRef]

188. Zhang, N.; Zhang, H.-J.; Sun, Q.-Q.; Cao, Y.-Y.; Li, X.; Zhao, B.; Wu, P.; Guo, Y.-D. Proteomic analysis reveals a role of melatonin in promoting cucumber seed germination under high salinity by regulating energy production. *Sci. Rep.* **2017**, *7*, 503. [CrossRef]

189. Lu, X.; Zhu, H. Tube-gel digestion. *Mol. Cell. Proteomics* **2005**, *4*, 1948–1958. [CrossRef]

190. Fischer, R.; Kessler, B.M. Gel-aided sample preparation (GASP)—A simplified method for gel-assisted proteomic sample generation from protein extracts and intact cells. *Proteomics* **2015**, *15*, 1224–1229. [CrossRef]

191. Muñoz-Talavera, A.; Gómez-Lim, M.Á.; Salazar-Olivo, L.A.; Reinders, J.; Lim, K.; Escobedo-Moratilla, A.; López-Calleja, A.C.; Islas-Carbajal, M.C.; Rincón-Sánchez, A.R. Expression of the biologically active insulin analog SCI-57 in *Nicotiana benthamiana*. *Front. Pharmacol.* **2019**, *10*. [CrossRef]

192. Strader, M.B.; Tabb, D.L.; Hervey, W.J.; Pan, C.; Hurst, G.B. Efficient and specific trypsin digestion of microgram to nanogram quantities of proteins in organic–aqueous solvent systems. *Anal. Chem.* **2006**, *78*, 125–134. [CrossRef]

193. Zhang, Y.; Fonslow, B.R.; Shan, B.; Baek, M.-C.; Yates, J.R. Protein Analysis by shotgun/bottom-up proteomics. *Chem. Rev.* **2013**, *113*, 2343–2394. [CrossRef] [PubMed]

194. Wang, W.-Q.; Jensen, O.N.; Møller, I.M.; Hebelstrup, K.H.; Rogowska-Wrzesinska, A. Evaluation of sample preparation methods for mass spectrometry-based proteomic analysis of barley leaves. *Plant Methods* **2018**, *14*, 72. [CrossRef]

195. Kołodziejczyk, I.; Dzitko, K.; Szewczyk, R.; Posmyk, M.M. Exogenous melatonin expediently modifies proteome of maize (*Zea mays* L.) embryo during seed germination. *Acta Physiol. Plant.* **2016**, *38*, 146. [CrossRef]

196. Lin, Y.; Huo, L.; Liu, Z.; Li, J.; Liu, Y.; He, Q.; Wang, X.; Liang, S. Sodium laurate, a novel protease- and mass spectrometry-compatible detergent for mass spectrometry-based membrane proteomics. *PLoS ONE* **2013**, *8*, e59779. [CrossRef] [PubMed]

197. Norris, J.L.; Porter, N.A.; Caprioli, R.M. Mass Spectrometry of intracellular and membrane proteins using cleavable detergents. *Anal. Chem.* **2003**, *75*, 6642–6647. [CrossRef] [PubMed]

198. Zakowick, H.; Schagat, T.; Yoder, D.; Niles, A.L. Measuring cell health and viability sequentially by same-well multiplexing using the GloMax®-Multi Detection System. *Promega Notes* **2008**, *99*, 25–28.

199. *Technical Bulletin: ProteaseMAX(TM) Surfactant, Trypsin Enhancer*; Promega Corporation: Madison, WI, USA, 2015; ISBN 6082744330.

200. Li, M.; Powell, M.J.; Razunguzwa, T.T.; O'doherty, G.A. A general approach to anionic acid-labile surfactants with tunable properties. *J. Org. Chem.* **2010**, *75*, 6149–6153. [CrossRef]

201. Ross, A.R.S.; Lee, P.J.; Smith, D.L.; Langridge, J.I.; Whetton, A.D.; Gaskell, S.J. Identification of proteins from two-dimensional polyacrylamide gels using a novel acid-labile surfactant. *Proteomics* **2002**, *2*, 928–936. [CrossRef]

202. Jagadeeshaprasad, M.G.; Batkulwar, K.B.; Meshram, N.N.; Tiwari, S.; Korwar, A.M.; Unnikrishnan, A.G.; Kulkarni, M.J. Targeted quantification of N-1-(carboxymethyl) valine and N-1-(carboxyethyl) valine peptides of β-hemoglobin for better diagnostics in diabetes. *Clin. Proteomics* **2016**, *13*, 7. [CrossRef]

203. Kaspar-Schoenefeld, S.; Merx, K.; Jozefowicz, A.M.; Hartmann, A.; Seiffert, U.; Weschke, W.; Matros, A.; Mock, H. Label-free proteome profiling reveals developmental-dependent patterns in young barley grains. *J. Proteomics* **2016**, *143*, 106–121. [CrossRef]

204. Increase Analytical Accuracy LC/MS: Solvents, Blends, Standards, Surfactants; ThermoFisher Scientific, USA. Available online: https://beta-static.fishersci.com/content/dam/fishersci/en_US/documents/programs/scientific/brochures-and-catalogs/guides/lcms-solvents-guide.pdf (accessed on 29 November 2020).

205. Waas, M.; Bhattacharya, S.; Chuppa, S.; Wu, X.; Jensen, D.R.R.; Omasits, U.; Wollscheid, B.; Volkman, B.F.; Noon, K.R.; Gundry, R.L. Combine and conquer: Surfactants, solvents, and chaotropes for robust mass spectrometry based analyses of membrane proteins. *Anal. Chem.* **2014**, *86*, 1551–1559. [CrossRef] [PubMed]

206. Frolov, A.; Bilova, T.; Paudel, G.; Berger, R.; Balcke, G.U.G.U.; Birkemeyer, C.; Wessjohann, L.A.L.A. Early responses of mature *Arabidopsis thaliana* plants to reduced water potential in the agar-based polyethylene glycol infusion drought model. *J. Plant Physiol.* **2017**, *208*, 70–83. [CrossRef] [PubMed]

207. Hogrebe, A.; von Stechow, L.; Bekker-Jensen, D.B.; Weinert, B.T.; Kelstrup, C.D.; Olsen, J. V Benchmarking common quantification strategies for large-scale phosphoproteomics. *Nat. Commun.* **2018**, *9*, 1045. [CrossRef] [PubMed]

208. Dam, S.; Thaysen-Andersen, M.; Stenkjær, E.; Lorentzen, A.; Roepstorff, P.; Packer, N.H.; Stougaard, J. Combined N-glycome and N-glycoproteome analysis of the *Lotus japonicus* seed globulin fraction shows conservation of protein structure and glycosylation in legumes. *J. Proteome Res.* **2013**, *12*, 3383–3392. [CrossRef]

209. Canterbury, J.D.; Merrihew, G.E.; Maccoss, M.J.; Goodlett, D.R.; Shaffer, S.A. Comparison of Data acquisition strategies on quadrupole ion trap instrumentation for shotgun proteomics. *J. Am. Soc. Mass Spectrom.* **2014**, *25*, 2048–2059. [CrossRef]

210. Frolov, A.; Hoffmann, R. Analysis of Amadori peptides enriched by boronic acid affinity chromatography. *Ann. N. Y. Acad. Sci.* **2008**, *1126*, 253–256. [CrossRef]

211. Manual 2D Quant Kit. Available online: https://www.gelifesciences.com/gehcls_images/GELS/RelatedContent/Files/1314729545976/litdoc28954714AE_20110830215136.pdf (accessed on 29 November 2020).

212. Matamoros, M.A.; Kim, A.; Peñuelas, M.; Ihling, C.; Griesser, E.; Hoffmann, R.; Fedorova, M.; Frolov, A.; Becana, M. Protein carbonylation and glycation in legume nodules. *Plant Physiol.* **2018**, *177*, 1510–1528. [CrossRef]

213. Plumb, R.S.; Johnson, K.A.; Rainville, P.; Smith, B.W.; Wilson, I.D.; Castro-Perez, J.M.; Nicholson, J.K. UPLC/MSE; a new approach for generating molecular fragment information for biomarker structure elucidation. *Rapid Commun. Mass Spectrom.* **2006**, *20*, 1989–1994. [CrossRef]

214. Uvackova, L.; Skultety, L.; Bekesova, S.; McClain, S.; Hajduch, M. The MSE-proteomic analysis of gliadins and glutenins in wheat grain identifies and quantifies proteins associated with celiac disease and baker's asthma. *J. Proteomics* **2013**, *93*, 65–73. [CrossRef]

215. Ludwig, C.; Gillet, L.; Rosenberger, G.; Amon, S.; Collins, B.C.; Aebersold, R. Data-independent acquisition-based SWATH-MS for quantitative proteomics: A tutorial. *Mol. Syst. Biol.* **2018**, *14*, e8126. [CrossRef]

216. Gillet, L.C.; Navarro, P.; Tate, S.; Ro, H.; Selevsek, N.; Reiter, L.; Bonner, R.; Aebersold, R.; Röst, H.; Selevsek, N.; et al. Targeted data extraction of the MS / MS spectra generated by data-independent acquisition: A new concept for consistent and accurate proteome analysis. *Mol. Cell. Proteomics* **2012**, *11*, 1–17. [CrossRef] [PubMed]

217. Zhu, F.-Y.; Chen, M.-X.; Su, Y.-W.; Xu, X.; Ye, N.-H.; Cao, Y.-Y.; Lin, S.; Liu, T.-Y.; Li, H.-X.; Wang, G.-Q.; et al. SWATH-MS Quantitative analysis of proteins in the rice inferior and superior spikelets during grain filling. *Front. Plant Sci.* **2016**, *7*, 1926. [CrossRef] [PubMed]

218. Bateman, N.W.; Goulding, S.P.; Shulman, N.J.; Gadok, A.K.; Szumlinski, K.K.; MacCoss, M.J.; Wu, C.C. Maximizing Peptide identification events in proteomic workflows using data-dependent acquisition (DDA). *Mol. Cell. Proteomics* **2014**, *13*, 329–338. [CrossRef] [PubMed]

219. Zhu, W.; Smith, J.W.; Huang, C.-M. Mass spectrometry-based label-free quantitative proteomics. *J. Biomed. Biotechnol.* **2010**, *2010*, 1–6. [CrossRef]

220. Nogueira, F.C.S.; Palmisano, G.; Soares, E.L.; Shah, M.; Soares, A.A.; Roepstorff, P.; Campos, F.A.P.; Domont, G.B. Proteomic profile of the nucellus of castor bean (*Ricinus communis* L.) seeds during development. *J. Proteomics* **2012**, *75*, 1933–1939. [CrossRef] [PubMed]

221. Fercha, A.; Capriotti, A.L.; Caruso, G.; Cavaliere, C.; Samperi, R.; Stampachiacchiere, S.; Laganà, A. Comparative analysis of metabolic proteome variation in ascorbate-primed and unprimed wheat seeds during germination under salt stress. *J. Proteomics* **2014**, *108*, 238–257. [CrossRef]

222. Gladilovich, V.; Greifenhagen, U.; Sukhodolov, N.; Selyutin, A.; Singer, D.; Thieme, D.; Majovsky, P.; Shirkin, A.; Hoehenwarter, W.; Bonitenko, E.; et al. Immobilized metal affinity chromatography on collapsed Langmuir-Blodgett iron(III) stearate films and iron(III) oxide nanoparticles for bottom-up phosphoproteomics. *J. Chromatogr. A* **2016**, *1443*, 181–190. [CrossRef]

223. Kelstrup, C.D.; Bekker-jensen, D.B.; Arrey, T.N.; Hogrebe, A.; Harder, A.; Olsen, J. V Performance evaluation of the Q exactive HF - X for shotgun proteomics. *J. Proteome Res.* **2018**, *17*, 727–738. [CrossRef]
224. Scheltema, R.A.; Hauschild, J.; Lange, O.; Hornburg, D.; Denisov, E.; Damoc, E.; Kuehn, A.; Makarov, A.; Mann, M.; Exactive, Q. The Q exactive HF, a benchtop mass spectrometer with a pre-filter, high-performance quadrupole and an ultra-high-field orbitrap. *Mol Cell Proteomics* **2014**, *13*, 3698–3708. [CrossRef]
225. Shi, T.; Su, D.; Liu, T.; Tang, K.; Camp, D.G.; Qian, W.-J.; Smith, R.D. Advancing the sensitivity of selected reaction monitoring-based targeted quantitative proteomics. *Proteomics* **2012**, *12*, 1074–1092. [CrossRef]
226. Nakamura, K.; Hirayama-Kurogi, M.; Ito, S.; Kuno, T.; Yoneyama, T.; Obuchi, W.; Terasaki, T.; Ohtsuki, S. Large-scale multiplex absolute protein quantification of drug-metabolizing enzymes and transporters in human intestine, liver, and kidney microsomes by SWATH-MS: Comparison with MRM/SRM and HR-MRM/PRM. *Proteomics* **2016**, *16*, 2106–2117. [CrossRef] [PubMed]
227. Peterson, A.C.; Russell, J.D.; Bailey, D.J.; Westphall, M.S.; Coon, J.J. Parallel reaction monitoring for high resolution and high mass accuracy quantitative, targeted proteomics. *Mol. Cell. Proteomics* **2012**, *11*, 1475–1488. [CrossRef] [PubMed]
228. Zargar, S.M.; Mahajan, R.; Nazir, M.; Nagar, P.; Kim, S.T.; Rai, V.; Masi, A.; Ahmad, S.M.; Shah, R.A.; Ganai, N.A.; et al. Common bean proteomics: Present status and future strategies. *J. Proteomics* **2017**, *169*, 239–248. [CrossRef] [PubMed]
229. Mergner, J.; Frejno, M.; List, M.; Papacek, M.; Chen, X.; Chaudhary, A.; Samaras, P.; Richter, S.; Shikata, H.; Messerer, M.; et al. Mass-spectrometry-based draft of the *Arabidopsis* proteome. *Nature* **2020**, *579*, 409–414. [CrossRef] [PubMed]
230. Rodríguez de Francisco, L.; Romero-Rodríguez, M.C.; Navarro-Cerrillo, R.M.; Miniño, V.; Perdomo, O.; Jorrín-Novo, J.V. Characterization of the orthodox *Pinus occidentalis* seed and pollen proteomes by using complementary gel-based and gel-free approaches. *J. Proteomics* **2016**, *143*, 382–389. [CrossRef] [PubMed]
231. Nguyen, T.-P.; Cueff, G.; Hegedus, D.D.; Rajjou, L.; Bentsink, L. A role for seed storage proteins in *Arabidopsis* seed longevity. *J. Exp. Bot.* **2015**, *66*, 6399–6413. [CrossRef] [PubMed]
232. Li, X.; Wang, W.; Chen, J. From pathways to networks: Connecting dots by establishing protein-protein interaction networks in signaling pathways using affinity purification and mass spectrometry. *Proteomics* **2015**, *15*, 188–202. [CrossRef]
233. He, D.; Yang, P. Proteomics of rice seed germination. *Front. Plant Sci.* **2013**, *4*, 246. [CrossRef]
234. Derouiche, A.; Cousin, C.; Mijakovic, I. Protein phosphorylation from the perspective of systems biology. *Curr. Opin. Biotechnol.* **2012**, *23*, 585–590. [CrossRef]
235. Von Stechow, L.; Francavilla, C.; Olsen, J.V. Recent findings and technological advances in phosphoproteomics for cells and tissues. *Expert Rev. Proteomics* **2015**, *12*, 469–487. [CrossRef]
236. Yin, X.; Wang, X.; Komatsu, S. Phosphoproteomics: Protein phosphorylation in regulation of seed germination and plant growth. *Curr. Protein Pept. Sci.* **2018**, *19*, 401–412. [CrossRef]
237. Mouzo, D.; Bernal, J.; López-Pedrouso, M.; Franco, D.; Zapata, C. Advances in the biology of seed and vegetative storage proteins based on two-dimensional electrophoresis coupled to mass spectrometry. *Molecules* **2018**, *23*, 2462. [CrossRef]
238. Agrawal, G.K.; Thelen, J.J. Large scale identification and quantitative profiling of phosphoproteins expressed during seed filling in oilseed rape. *Mol. Cell. Proteomics* **2006**, *5*, 2044–2059. [CrossRef]
239. López-Pedrouso, M.; Alonso, J.; Zapata, C. Evidence for phosphorylation of the major seed storage protein of the common bean and its phosphorylation-dependent degradation during germination. *Plant Mol. Biol.* **2014**, *84*, 415–428. [CrossRef]
240. Irar, S.; Oliveira, E.; Pagès, M.; Goday, A. Towards the identification of late-embryogenic-abundant phosphoproteome in *Arabidopsis* by 2-DE and MS. *Proteomics* **2006**, *6*, S175–S185. [CrossRef]
241. Wan, L.; Ross, A.R.S.; Yang, J.; Hegedus, D.D.; Kermode, A.R. Phosphorylation of the 12 S globulin cruciferin in wild-type and abi1-1 mutant *Arabidopsis thaliana* (thale cress) seeds. *Biochem. J.* **2007**, *404*, 247–256. [CrossRef]
242. Kuyama, H.; Toda, C.; Watanabe, M.; Tanaka, K.; Nishimura, O. An efficient chemical method for dephosphorylation of phosphopeptides. *Rapid Commun. Mass Spectrom.* **2003**, *17*, 1493–1496. [CrossRef]
243. Kita, K.; Okumura, N.; Takao, T.; Watanabe, M.; Matsubara, T.; Nishimura, O.; Nagai, K. Evidence for phosphorylation of rat liver glucose-regulated protein 58, GRP58/ERp57/ER-60, induced by fasting and leptin. *FEBS Lett.* **2006**, *580*, 199–205. [CrossRef]

244. Woo, E.M.; Fenyo, D.; Kwok, B.H.; Funabiki, H.; Chait, B.T. Efficient identification of phosphorylation by mass spectrometric phosphopeptide fingerprinting. *Anal. Chem.* **2008**, *80*, 2419–2425. [CrossRef]

245. Bernal, J.; López-Pedrouso, M.; Franco, D.; Bravo, S.; García, L.; Zapata, C. Identification and mapping of phosphorylated isoforms of the major storage protein of potato based on two- dimensional electrophoresis. In *Advances in Seed Biology*; InTech: London, UK, 2017.

246. Sinha, A.; Haider, T.; Narula, K.; Ghosh, S.; Chakraborty, N.; Chakraborty, S. Integrated seed proteome and phosphoproteome analyses reveal interplay of nutrient dynamics, carbon–nitrogen partitioning, and oxidative signaling in chickpea. *Proteomics* **2020**, *20*, 1900267. [CrossRef]

247. Han, C.; Yang, P. Two dimensional gel electrophoresis-based plant phosphoproteomics. In *Phospho-Proteomics. Methods in Molecular Biology*; Springer: New York, NY, USA, 2016.

248. Guo, G.; Lv, D.; Yan, X.; Subburaj, S.; Ge, P.; Li, X.; Hu, Y.; Yan, Y. Proteome characterization of developing grains in bread wheat cultivars (*Triticum aestivum* L.). *BMC Plant Biol.* **2012**, *12*, 147. [CrossRef]

249. Han, C.; Wang, K.; Yang, P. Gel-Based Comparative phosphoproteomic analysis on rice embryo during germination. *Plant Cell Physiol.* **2014**, *55*, 1376–1394. [CrossRef]

250. Li, M.; Yin, X.; Sakata, K.; Yang, P.; Komatsu, S. Proteomic analysis of phosphoproteins in the rice nucleus during the early stage of seed germination. *J. Proteome Res.* **2015**, *14*, 2884–2896. [CrossRef]

251. Wang, Y.; Tong, X.; Qiu, J.; Li, Z.; Zhao, J.; Hou, Y.; Tang, L.; Zhang, J. A phosphoproteomic landscape of rice (*Oryza sativa*) tissues. *Physiol. Plant.* **2017**, *160*, 458–475. [CrossRef]

252. Forzani, C.; Carreri, A.; De La Fuente Van Bentem, S.; Lecourieux, D.; Lecourieux, F.; Hirt, H. The *Arabidopsis* protein kinase Pto-interacting 1-4 is a common target of the oxidative signal-inducible 1 and mitogen-activated protein kinases. *FEBS J.* **2011**, *278*, 1126–1136. [CrossRef]

253. Chen, Y.; Hoehenwarter, W. Rapid and reproducible phosphopeptide enrichment by tandem metal oxide affinity chromatography: Application to boron deficiency induced phosphoproteomics. *Plant J.* **2019**, *98*, 370–384. [CrossRef]

254. Chang, I.F.; Hsu, J.L.; Hsu, P.H.; Sheng, W.A.; Lai, S.J.; Lee, C.; Chen, C.W.; Hsu, J.C.; Wang, S.Y.; Wang, L.Y.; et al. Comparative phosphoproteomic analysis of microsomal fractions of *Arabidopsis thaliana* and *Oryza sativa* subjected to high salinity. *Plant Sci.* **2012**, 131–142. [CrossRef]

255. Fíla, J.; Matros, A.; Radau, S.; Zahedi, R.P.; Čapková, V.; Mock, H.-P.; Honys, D. Revealing phosphoproteins playing role in tobacco pollen activated in vitro. *Proteomics* **2012**, *12*, 3229–3250. [CrossRef]

256. Kurdyukov, D.A.; Chernova, E.N.; Russkikh, Y.V.; Eurov, D.A.; Sokolov, V.V.; Bykova, A.A.; Shilovskikh, V.V.; Keltsieva, O.A.; Ubyivovk, E.V.; Anufrikov, Y.A.; et al. Ni-functionalized submicron mesoporous silica particles as a sorbent for metal affinity chromatography. *J. Chromatogr. A* **2017**, *1513*, 140–148. [CrossRef]

257. Yeh, T.-T.; Ho, M.-Y.; Chen, W.-Y.; Hsu, Y.-C.; Ku, W.-C.; Tseng, H.-W.; Chen, S.-T.; Chen, S.-F. Comparison of different fractionation strategies for in-depth phosphoproteomics by liquid chromatography tandem mass spectrometry. *Anal. Bioanal. Chem.* **2019**, *411*, 3417–3424. [CrossRef]

258. Vu, L.D.; Zhu, T.; Verstraeten, I.; van de Cotte, B.; Gevaert, K.; De Smet, I. Temperature-induced changes in the wheat phosphoproteome reveal temperature-regulated interconversion of phosphoforms. *J. Exp. Bot.* **2018**, *69*, 4609–4624. [CrossRef] [PubMed]

259. Qiu, J.; Hou, Y.; Tong, X.; Wang, Y.; Lin, H.; Liu, Q.; Zhang, W.; Li, Z.; Nallamilli, B.R.; Zhang, J. Quantitative phosphoproteomic analysis of early seed development in rice (*Oryza sativa* L.). *Plant Mol. Biol.* **2016**, *90*, 249–265. [CrossRef] [PubMed]

260. Li, J.; Silva-Sanchez, C.; Zhang, T.; Chen, S.; Li, H. Phosphoproteomics technologies and applications in plant biology research. *Front. Plant Sci.* **2015**, *6*, 430. [CrossRef] [PubMed]

261. Marcus, K. (Ed.) *Quantitative Methods in Proteomics*; Methods in Molecular Biology; Humana Press: Totowa, NJ, USA, 2012; Volume 893, ISBN 978-1-61779-884-9.

262. Bindschedler, L.V.; Palmblad, M.; Cramer, R. Hydroponic isotope labelling of entire plants (HILEP) for quantitative plant proteomics; an oxidative stress case study. *Phytochemistry* **2008**, *69*, 1962–1972. [CrossRef] [PubMed]

263. Hebeler, R.; Oeljeklaus, S.; Reidegeld, K.A.; Eisenacher, M.; Stephan, C.; Sitek, B.; Stühler, K.; Meyer, H.E.; Sturre, M.J.G.; Dijkwel, P.P.; et al. Study of early leaf senescence in arabidopsis thaliana by quantitative proteomics using reciprocal 14 N/ 15 N labeling and difference gel electrophoresis. *Mol. Cell. Proteomics* **2008**, *7*, 108–120. [CrossRef]

264. Minkoff, B.B.; Stecker, K.E.; Sussman, M.R. Rapid Phosphoproteomic effects of abscisic acid (ABA) on wild-type and aba receptor-deficient *A. thaliana* mutants. *Mol. Cell. Proteomics* **2015**, *14*, 1169–1182. [CrossRef]

265. Lewandowska, D.; ten Have, S.; Hodge, K.; Tillemans, V.; Lamond, A.I.; Brown, J.W.S. Plant SILAC: Stable-isotope labelling with amino acids of *Arabidopsis* seedlings for quantitative proteomics. *PLoS ONE* **2013**, *8*, e72207. [CrossRef]

266. Wong, M.M.; Bhaskara, G.B.; Wen, T.-N.; Lin, W.-D.; Nguyen, T.T.; Chong, G.L.; Verslues, P.E. Phosphoproteomics of *Arabidopsis* highly ABA-induced1 identifies AT-Hook–Like10 phosphorylation required for stress growth regulation. *Proc. Natl. Acad. Sci. USA* **2019**, *116*, 2354–2363. [CrossRef]

267. Li, H.; Han, J.; Pan, J.; Liu, T.; Parker, C.E.; Borchers, C.H. Current trends in quantitative proteomics—An update. *J. Mass Spectrom.* **2017**, *52*, 319–341. [CrossRef]

268. Fan, S.; Meng, Y.; Song, M.; Pang, C.; Wei, H.; Liu, J.; Zhan, X.; Lan, J.; Feng, C.; Zhang, S.; et al. Quantitative Phosphoproteomics analysis of nitric oxide–responsive phosphoproteins in cotton leaf. *PLoS ONE* **2014**, *9*, e94261. [CrossRef]

269. Zhang, G.-L.; Zhu, Y.; Fu, W.-D.; Wang, P.; Zhang, R.-H.; Zhang, Y.-L.; Song, Z.; Xia, G.-X.; Wu, J.-H. iTRAQ Protein profile differential analysis of dormant and germinated grassbur twin seeds reveals that ribosomal synthesis and carbohydrate metabolism promote germination possibly through the PI3K pathway. *Plant Cell Physiol.* **2016**, *57*, 1244–1256. [CrossRef] [PubMed]

270. Moothoo-Padayachie, A.; Macdonald, A.; Varghese, B.; Pammenter, N.W.; Govender, P. Sershen uncovering the basis of viability loss in desiccation sensitive *Trichilia dregeana* seeds using differential quantitative protein expression profiling by iTRAQ. *J. Plant Physiol.* **2018**, *221*, 119–131. [CrossRef]

271. Schoberer, J.; Strasser, R. Plant glyco-biotechnology. *Semin. Cell Dev. Biol.* **2018**, *80*, 133–141. [CrossRef]

272. Aalberse, R.; Koshte, V.; Clemens, J. Immunoglobulin E antibodies that crossreact with vegetable foods, pollen, and *Hymenoptera venom*. *J. Allergy Clin. Immunol.* **1981**, *68*, 356–364. [CrossRef]

273. La Fuente, M.D.; López-pedrouso, M.; Alonso, J.; Santalla, M.; De Ron, A.M.; Álvarez, G.; Zapata, C. In-depth characterization of the phaseolin protein diversity of common bean (*Phaseolus vulgaris* L.) Based on two-dimensional electrophoresis and mass spectrometry. *Food Technol. Biotechnol.* **2012**, *50*, 315–325.

274. Mehta-D'souza, P. Detection of glycoproteins in polyacrylamide gels using Pro-Q Emerald 300 Dye, a fluorescent periodate schiff-base stain. In *Protein Gel Detection and Imaging. Methods in Molecular Biology*; Humana Press: New York, NY, USA, 2018; pp. 115–119.

275. Weiss, W.; Postel, W.; Görg, A. Qualitative and quantitative changes in barley seed protein patterns during the malting process analyzed by sodium dodecyl sulfate-polyacrylamide gel electrophoresis with respect to malting quality. *Electrophoresis* **1992**, *13*, 787–797. [CrossRef]

276. Weiss, W.; Postel, W.; Görg, A. Application of sequential extraction procedures and glycoprotein blotting for the characterization of the 2-D polypeptide patterns of barley seed proteins. *Electrophoresis* **1992**, *13*, 770–773. [CrossRef]

277. Madera, M.; Mann, B.; Mechref, Y.; Novotny, M.V. Efficacy of glycoprotein enrichment by microscale lectin affinity chromatography. *J. Sep. Sci.* **2008**, *31*, 2722–2732. [CrossRef]

278. Ostrowski, M.; Hetmann, A.; Jakubowska, A. Indole-3-acetic acid UDP-glucosyltransferase from immature seeds of pea is involved in modification of glycoproteins. *Phytochemistry* **2015**, *117*, 25–33. [CrossRef]

279. Wang, T.; Hu, X.-C.; Cai, Z.-P.; Voglmeir, J.; Liu, L. Qualitative and quantitative analysis of carbohydrate modification on glycoproteins from seeds of *Ginkgo biloba*. *J. Agric. Food Chem.* **2017**, *65*, 7669–7679. [CrossRef]

280. Rao, R.S.P.; Thelen, J.J.; Miernyk, J.A. Is Lys-Nε{open}-acetylation the next big thing in post-translational modifications? *Trends Plant Sci.* **2014**, *19*, 550–553. [CrossRef] [PubMed]

281. Smith-Hammond, C.L.; Swatek, K.N.; Johnston, M.L.; Thelen, J.J.; Miernyk, J.A. Initial description of the developing soybean seed protein Lys-Nε-acetylome. *J. Proteomics* **2014**, *96*, 56–66. [CrossRef] [PubMed]

282. He, D.; Wang, Q.; Li, M.; Damaris, R.N.; Yi, X.; Cheng, Z.; Yang, P. Global Proteome analyses of lysine acetylation and succinylation reveal the widespread involvement of both modification in metabolism in the embryo of germinating rice seed. *J. Proteome Res.* **2016**, *15*, 879–890. [CrossRef] [PubMed]

283. Traverso, J.A.; Meinnel, T.; Giglione, C. Expanded impact of protein N-myristoylation in plants. *Plant Signal. Behav.* **2008**, *3*, 501–502. [CrossRef] [PubMed]

284. Running, M.P. The role of lipid post–translational modification in plant developmental processes. *Front. Plant Sci.* **2014**, *5*, 50. [CrossRef]

285. Kalemba, E.M.; Litkowiec, M. Functional characterization of a dehydrin protein from *Fagus sylvatica* seeds using experimental and in silico approaches. *Plant Physiol. Biochem.* **2015**, *97*, 246–254. [CrossRef]

286. Friso, G.; van Wijk, K.J. Update: Post-translational protein modifications in plant metabolism. *Plant Physiol.* **2015**, *169*, 1469–1487. [CrossRef]

287. Mann, M.; Højrup, P.; Roepstorff, P. Use of mass spectrometric molecular weight information to identify proteins in sequence databases. *Biol. Mass Spectrom.* **1993**, *22*, 338–345. [CrossRef]

288. Henzel, W.J.; Watanabe, C.; Stults, J.T. Protein identification: The origins of peptide mass fingerprinting. *J. Am. Soc. Mass Spectrom.* **2003**, *14*, 931–942. [CrossRef]

289. Eng, J.K.; McCormack, A.L.; Yates, J.R. An approach to correlate tandem mass spectral data of peptides with amino acid sequences in a protein database. *J. Am. Soc. Mass Spectrom.* **1994**, *5*, 976–989. [CrossRef]

290. Cottrell, J.S. Protein identification using MS/MS data. *J. Proteomics* **2011**, *74*, 1842–1851. [CrossRef] [PubMed]

291. Stein, S.E.; Scott, D.R. Optimization and testing of mass spectral library search algorithms for compound identification. *J. Am. Soc. Mass Spectrom.* **1994**, *5*, 859–866. [CrossRef]

292. Collisionally induced dissociation of protonated peptide ions and the interpretation of product ion spectra. In *Protein Sequencing and Identification Using Tandem Mass Spectrometry*; Wiley Online Books; John Wiley & Sons, Inc.: Hoboken, NJ, USA, 2005; pp. 64–116. ISBN 9780471721987.

293. Mann, M.; Wilm, M. Error-tolerant identification of peptides in sequence databases by peptide sequence tags. *Anal. Chem.* **1994**, *66*, 4390–4399. [CrossRef] [PubMed]

294. Tabb, D.L. The SEQUEST family tree. *J. Am. Soc. Mass Spectrom.* **2015**, *26*, 1814–1819. [CrossRef] [PubMed]

295. Perkins, D.N.; Pappin, D.J.C.; Creasy, D.M.; Cottrell, J.S. Probability-based protein identification by searching sequence databases using mass spectrometry data. *Electrophoresis* **1999**, *20*, 3551–3567. [CrossRef]

296. Cox, J.; Neuhauser, N.; Michalski, A.; Scheltema, R.A.; Olsen, J.V.; Mann, M. Andromeda: A peptide search engine integrated into the MaxQuant environment. *J. Proteome Res.* **2011**, *10*, 1794–1805. [CrossRef] [PubMed]

297. Eng, J.K.; Searle, B.C.; Clauser, K.R.; Tabb, D.L. A Face in the Crowd: Recognizing peptides through database search. *Mol. Cell. Proteomics* **2011**, *10*, R111.009522. [CrossRef]

298. Moore, R.E.; Young, M.K.; Lee, T.D. Qscore: An algorithm for evaluating SEQUEST database search results. *J. Am. Soc. Mass Spectrom.* **2002**, *13*, 378–386. [CrossRef]

299. Peng, J.; Elias, J.E.; Thoreen, C.C.; Licklider, L.J.; Gygi, S.P. Evaluation of multidimensional chromatography coupled with tandem mass spectrometry (LC/LC–MS/MS) for large-scale protein analysis: The yeast proteome. *J. Proteome Res.* **2003**, *2*, 43–50. [CrossRef]

300. Reiter, L.; Claassen, M.; Schrimpf, S.P.; Jovanovic, M.; Schmidt, A.; Buhmann, J.M.; Hengartner, M.O.; Aebersold, R. Protein identification false discovery rates for very large proteomics data sets generated by tandem mass spectrometry. *Mol. Cell. Proteomics* **2009**, *8*, 2405–2417. [CrossRef]

301. Savitski, M.M.; Wilhelm, M.; Hahne, H.; Kuster, B.; Bantscheff, M. A scalable approach for protein false discovery rate estimation in large proteomic data sets. *Mol. Cell. Proteomics* **2015**, *14*, 2394–2404. [CrossRef] [PubMed]

302. Elias, J.E.; Gygi, S.P. Target-decoy search strategy for increased confidence in large-scale protein identifications by mass spectrometry. *Nat. Methods* **2007**, *4*, 207–214. [CrossRef] [PubMed]

303. Sinitcyn, P.; Rudolph, J.D.; Cox, J. Computational Methods for understanding mass spectrometry–based shotgun proteomics data. *Annu. Rev. Biomed. Data Sci.* **2018**, *1*, 207–234. [CrossRef]

304. Angel, T.E.; Aryal, U.K.; Hengel, S.M.; Baker, E.S.; Kelly, R.T.; Robinson, E.W.; Smith, R.D. Mass spectrometry-based proteomics: Existing capabilities and future directions. *Chem. Soc. Rev.* **2012**, *41*, 3912. [CrossRef]

305. Hansen, B.T.; Davey, S.W.; Ham, A.-J.L.; Liebler, D.C. P-Mod: An algorithm and software to map modifications to peptide sequences using tandem MS data. *J. Proteome Res.* **2005**, *4*, 358–368. [CrossRef]

306. Kertesz, V.; Connelly, H.M.; Erickson, B.K.; Hettich, R.L. PTMSearchPlus: Software tool for automated protein identification and post-translational modification characterization by integrating accurate intact protein mass and bottom-up mass spectrometric data searches. *Anal. Chem.* **2009**, *81*, 8387–8395. [CrossRef]

307. Na, S.; Paek, E. Software eyes for protein post-translational modifications. *Mass Spectrom. Rev.* **2015**, *34*, 133–147. [CrossRef]

308. Medzihradszky, K.F.; Chalkley, R.J. Lessons in *de novo* peptide sequencing by tandem mass spectrometry. *Mass Spectrom. Rev.* **2015**, *34*, 43–63. [CrossRef]

309. Jeong, K.; Kim, S.; Pevzner, P.A. UniNovo: A universal tool for de novo peptide sequencing. *Bioinformatics* **2013**, *29*, 1953–1962. [CrossRef]

310. Hu, J. The importance of experimental design in proteomic mass spectrometry experiments: Some cautionary tales. *Briefings Funct. Genomics Proteomics* **2005**, *3*, 322–331. [CrossRef]

311. Schulze, W.X.; Usadel, B. Quantitation in mass-spectrometry-based proteomics. *Annu. Rev. Plant Biol.* **2010**, *61*, 491–516. [CrossRef]

312. Oda, Y.; Huang, K.; Cross, F.R.; Cowburn, D.; Chait, B.T. Accurate quantitation of protein expression and site-specific phosphorylation. *Proc. Natl. Acad. Sci. USA* **1999**, *96*, 6591–6596. [CrossRef] [PubMed]

313. Nahnsen, S.; Bielow, C.; Reinert, K.; Kohlbacher, O. Tools for label-free peptide quantification. *Mol. Cell. Proteomics* **2013**, *12*, 549–556. [CrossRef] [PubMed]

314. Mueller, L.N.; Rinner, O.; Schmidt, A.; Letarte, S.; Bodenmiller, B.; Brusniak, M.-Y.; Vitek, O.; Aebersold, R.; Müller, M. SuperHirn – a novel tool for high resolution LC-MS-based peptide/protein profiling. *Proteomics* **2007**, *7*, 3470–3480. [CrossRef] [PubMed]

315. Cox, J.; Mann, M. MaxQuant enables high peptide identification rates, individualized p.p.b.-range mass accuracies and proteome-wide protein quantification. *Nat. Biotechnol.* **2008**, *26*, 1367–1372. [CrossRef]

316. Progenesis QI User Guide: Analysis Workflow Guidelines for DDA Data. Available online: http://www.nonlinear.com/progenesis/qi-for-proteomics/v4.1/user-guide/ (accessed on 29 November 2020).

317. Bertsch, A.; Gröpl, C.; Reinert, K.; Kohlbacher, O. OpenMS and TOPP: Open source software for LC-MS data analysis. In *Data Mining in Proteomics. Methods in Molecular Biology (Methods and Protocols)*; Hamacher, M., Eisenacher, M., Stephan, C., Eds.; Humana Press: Totowa, NJ, USA, 2011; pp. 353–367. ISBN 978-1-60761-987-1.

318. Available online: https://www.thermofisher.com (accessed on 29 November 2020).

319. Available online: http://tools.proteomecenter.org/software.php (accessed on 29 November 2020).

320. Ishihama, Y.; Oda, Y.; Tabata, T.; Sato, T.; Nagasu, T.; Rappsilber, J.; Mann, M. Exponentially modified protein abundance index (emPAI) for estimation of absolute protein amount in proteomics by the number of sequenced peptides per protein. *Mol. Cell. Proteomics* **2005**, *4*, 1265–1272. [CrossRef]

321. Braisted, J.C.; Kuntumalla, S.; Vogel, C.; Marcotte, E.M.; Rodrigues, A.R.; Wang, R.; Huang, S.-T.; Ferlanti, E.S.; Saeed, A.I.; Fleischmann, R.D.; et al. The APEX quantitative proteomics tool: Generating protein quantitation estimates from LC-MS/MS proteomics results. *BMC Bioinform.* **2008**, *9*, 529. [CrossRef]

322. Sun, A.; Zhang, J.; Wang, C.; Yang, D.; Wei, H.; Zhu, Y.; Jiang, Y.; He, F. Modified spectral count index (mSCI) for Estimation of protein abundance by protein relative identification possibility (RIPpro): A new proteomic technological parameter. *J. Proteome Res.* **2009**, *8*, 4934–4942. [CrossRef]

323. Silva, J.C.; Gorenstein, M.V.; Li, G.-Z.; Vissers, J.P.C.; Geromanos, S.J. Absolute quantification of proteins by LCMS E. *Mol. Cell. Proteomics* **2006**, *5*, 144–156. [CrossRef]

324. Clough, T.; Key, M.; Ott, I.; Ragg, S.; Schadow, G.; Vitek, O. Protein quantification in label-free LC-MS experiments. *J. Proteome Res.* **2009**, *8*, 5275–5284. [CrossRef]

325. Monroe, M.E.; Tolić, N.; Jaitly, N.; Shaw, J.L.; Adkins, J.N.; Smith, R.D. VIPER: An advanced software package to support high-throughput LC-MS peptide identification. *Bioinformatics* **2007**, *23*, 2021–2023. [CrossRef] [PubMed]

326. Pluskal, T.; Castillo, S.; Villar-Briones, A.; Orešič, M. MZmine 2: Modular framework for processing, visualizing, and analyzing mass spectrometry-based molecular profile data. *BMC Bioinform.* **2010**, *11*, 395. [CrossRef] [PubMed]

327. Bairoch, A.; Apweiler, R. The SWISS-PROT protein sequence data bank and its supplement TrEMBL in 1999. *Nucleic Acids Res.* **1999**, *27*, 49–54. [CrossRef] [PubMed]

328. Simpson, J.C.; Pepperkok, R. Localizing the proteome. *Genome Biol.* **2003**, *4*, 240. [CrossRef] [PubMed]

329. Huh, W.-K.; Falvo, J.V.; Gerke, L.C.; Carroll, A.S.; Howson, R.W.; Weissman, J.S.; O'Shea, E.K. Global analysis of protein localization in budding yeast. *Nature* **2003**, *425*, 686–691. [CrossRef]

330. Yu, C.-S.; Chen, Y.-C.; Lu, C.-H.; Hwang, J.-K. Prediction of protein subcellular localization. *Proteins Struct. Funct. Bioinforma.* **2006**, *64*, 643–651. [CrossRef] [PubMed]

331. Horton, P.; Park, K.-J.; Obayashi, T.; Fujita, N.; Harada, H.; Adams-Collier, C.J.; Nakai, K. WoLF PSORT: Protein localization predictor. *Nucleic Acids Res.* **2007**, *35*, W585–W587. [CrossRef]

332. Briesemeister, S.; Rahnenführer, J.; Kohlbacher, O. YLoc—An interpretable web server for predicting subcellular localization. *Nucleic Acids Res.* **2010**, *38*, W497–W502. [CrossRef]

333. Savojardo, C.; Martelli, P.L.; Fariselli, P.; Profiti, G.; Casadio, R. BUSCA: An integrative web server to predict subcellular localization of proteins. *Nucleic Acids Res.* **2018**, *46*, W459–W466. [CrossRef]

334. Sperschneider, J.; Catanzariti, A.-M.; DeBoer, K.; Petre, B.; Gardiner, D.M.; Singh, K.B.; Dodds, P.N.; Taylor, J.M. LOCALIZER: Subcellular localization prediction of both plant and effector proteins in the plant cell. *Sci. Rep.* **2017**, *7*, 44598. [CrossRef]

335. Chou, K.-C.; Shen, H.-B. Plant-mPLoc: A Top-Down Strategy to Augment the Power for Predicting Plant Protein Subcellular Localization. *PLoS ONE* **2010**, *5*, e11335. [CrossRef] [PubMed]

336. Käll, L.; Krogh, A.; Sonnhammer, E.L. A combined transmembrane topology and signal peptide prediction method. *J. Mol. Biol.* **2004**, *338*, 1027–1036. [CrossRef] [PubMed]

337. Bateman, A.; Martin, M.J.; O'Donovan, C.; Magrane, M.; Alpi, E.; Antunes, R.; Bely, B.; Bingley, M.; Bonilla, C.; Britto, R.; et al. UniProt: The universal protein knowledgebase. *Nucleic Acids Res.* **2017**, *45*, D158–D169.

338. Kanehisa, M.; Goto, S.; Sato, Y.; Kawashima, M.; Furumichi, M.; Tanabe, M. Data, information, knowledge and principle: Back to metabolism in KEGG. *Nucleic Acids Res.* **2014**, *42*, D199–D205. [CrossRef] [PubMed]

339. Ogata, H.; Goto, S.; Sato, K.; Fujibuchi, W.; Bono, H.; Kanehisa, M. KEGG: Kyoto encyclopedia of genes and genomes. *Nucleic Acids Res.* **1999**, *27*, 29–34. [CrossRef]

340. Klie, S.; Nikoloski, Z. The Choice between mapman and gene ontology for automated gene function prediction in plant science. *Front. Genet.* **2012**, *3*. [CrossRef]

341. Kanehisa, M.; Sato, Y.; Morishima, K. BlastKOALA and GhostKOALA: KEGG tools for functional characterization of genome and metagenome sequences. *J. Mol. Biol.* **2016**, *428*, 726–731. [CrossRef]

342. Kanehisa, M.; Sato, Y.; Kawashima, M.; Furumichi, M.; Tanabe, M. KEGG as a reference resource for gene and protein annotation. *Nucleic Acids Res.* **2016**, *44*, D457–D462. [CrossRef]

343. Mi, H.; Muruganujan, A.; Huang, X.; Ebert, D.; Mills, C.; Guo, X.; Thomas, P.D. Protocol update for large-scale genome and gene function analysis with the PANTHER classification system (v.14.0). *Nat. Protoc.* **2019**, *14*, 703–721. [CrossRef]

344. Usadel, B.; Poree, F.; Nagel, A.; Lohse, M.; Czedik-Eysenberg, A.; Stitt, M. A guide to using MapMan to visualize and compare omics data in plants: A case study in the crop species, maize. *Plant. Cell Environ.* **2009**, *32*, 1211–1229. [CrossRef]

345. Lohse, M.; Nagel, A.; Herter, T.; May, P.; Schroda, M.; Zrenner, R.; Tohge, T.; Fernie, A.R.; Stitt, M.; Usadel, B. Mercator: A fast and simple web server for genome scale functional annotation of plant sequence data. *Plant. Cell Environ.* **2014**, *37*, 1250–1258. [CrossRef] [PubMed]

346. Saha, I.; Klingström, T.; Forsberg, S.; Wikander, J.; Zubek, J.; Kierczak, M.; Plewczynski, D. Evaluation of machine learning algorithms on protein-protein interactions. In *Advances in Intelligent Systems and Computing*; Springer: Cham, Switzerland, 2014; pp. 211–218. ISBN 9783319023083.

347. Hayashi, T.; Matsuzaki, Y.; Yanagisawa, K.; Ohue, M.; Akiyama, Y. MEGADOCK-Web: An integrated database of high-throughput structure-based protein-protein interaction predictions. *BMC Bioinform.* **2018**, *19*, 62. [CrossRef] [PubMed]

348. Liu, S.; Liu, C.; Deng, L. Machine Learning approaches for protein–protein interaction hot spot prediction: Progress and comparative assessment. *Molecules* **2018**, *23*, 2535. [CrossRef] [PubMed]

349. Schrader, M. Origins, technological development, and applications of peptidomics. In *Methods in Molecular Biology*; Springer: Berlin/Heidelberg, Germany, 2018.

350. Ohyama, K.; Ogawa, M.; Matsubayashi, Y. Identification of a biologically active, small, secreted peptide in *Arabidopsis* by in silico gene screening, followed by LC-MS-based structure analysis. *Plant J.* **2008**, *55*, 152–160. [CrossRef] [PubMed]

351. Farrokhi, N.; Whitelegge, J.P.; Brusslan, J.A. Plant peptides and peptidomics. *Plant Biotechnol. J.* **2008**, *6*, 105–134. [CrossRef]

352. Segonzac, C.; Monaghan, J. Modulation of plant innate immune signaling by small peptides. *Curr. Opin. Plant Biol.* **2019**, *51*, 22–28. [CrossRef]

353. Pan, S.; Agyei, D.; Jeevanandam, J.; Danquah, M.K. Bio-active peptides: Role in plant growth and defense. In *Natural Bio-Active Compounds*; Springer: Singapore, 2019; pp. 1–29.

354. Gancheva, M.S.; Malovichko, Y.V.; Poliushkevich, L.O.; Dodueva, I.E.; Lutova, L.A. Plant peptide hormones. *Russ. J. Plant Physiol.* **2019**, *66*, 171–189. [CrossRef]

355. Stührwohldt, N.; Schaller, A. Regulation of plant peptide hormones and growth factors by post-translational modification. *Plant Biol.* **2019**, *21*, 49–63. [CrossRef]

356. Oh, E.; Seo, P.J.; Kim, J. Signaling peptides and receptors coordinating plant root development. *Trends Plant Sci.* **2018**, *23*, 337–351. [CrossRef]

357. Pan, H.; Wang, D. Nodule cysteine-rich peptides maintain a working balance during nitrogen-fixing symbiosis. *Nat. Plants* **2017**, *3*, 17048. [CrossRef]

358. Ayala-Niño, A.; Rodríguez-Serrano, G.M.; González-Olivares, L.G.; Contreras-López, E.; Regal-López, P.; Cepeda-Saez, A. Sequence Identification of bioactive peptides from amaranth seed proteins (*Amaranthus hypochondriacus* spp.). *Molecules* **2019**, *24*, 3033. [CrossRef] [PubMed]

359. Sharma, S.; Verma, H.N.; Sharma, N.K. Cationic bioactive peptide from the seeds of *Benincasa hispida*. *Int. J. Pept.* **2014**, *2014*, 1–12. [CrossRef] [PubMed]

360. Maruyama, N.; Mikami, B.; Utsumi, S. The development of transgenic crops to improve human health by advanced utilization of seed storage proteins. *Biosci. Biotechnol. Biochem.* **2011**, *75*, 823–828. [CrossRef] [PubMed]

361. Prak, K.; Maruyama, Y.; Maruyama, N.; Utsumi, S. Design of genetically modified soybean proglycinin A1aB1b with multiple copies of bioactive peptide sequences. *Peptides* **2006**, *27*, 1179–1186. [CrossRef]

362. Pu, C.; Tang, W. The antibacterial and antibiofilm efficacies of a liposomal peptide originating from rice bran protein against: Listeria monocytogenes. *Food Funct.* **2017**, *8*, 4159–4169. [CrossRef]

363. Chai, T.-T.; Tan, Y.-N.; Ee, K.-Y.; Xiao, J.; Wong, F.-C. Seeds, fermented foods, and agricultural by-products as sources of plant-derived antibacterial peptides. *Crit. Rev. Food Sci. Nutr.* **2019**, *59*, S162–S177. [CrossRef]

364. Yang, J.; Hu, L.; Cai, T.; Chen, Q.; Ma, Q.; Yang, J.; Meng, C.; Hong, J. Purification and identification of two novel antioxidant peptides from perilla (*Perilla frutescens* L. Britton) seed protein hydrolysates. *PLoS ONE* **2018**, *13*, e0200021. [CrossRef]

365. Ye, N.; Hu, P.; Xu, S.; Chen, M.; Wang, S.; Hong, J.; Chen, T.; Cai, T. Preparation and characterization of antioxidant peptides from carrot seed protein. *J. Food Qual.* **2018**, *2018*, 1–9. [CrossRef]

366. Aguilar-Toalá, J.E.; Liceaga, A.M. Identification of chia seed (*Salvia hispanica* L.) peptides with enzyme inhibition activity towards skin-aging enzymes. *Amino Acids* **2020**, *52*, 1149–1159. [CrossRef]

367. Dallas, D.C.; Guerrero, A.; Parker, E.A.; Robinson, R.C.; Gan, J.; German, J.B.; Barile, D.; Lebrilla, C.B. Current peptidomics: Applications, purification, identification, quantification, and functional analysis. *Proteomics* **2015**, *15*, 1026–1038. [CrossRef]

368. Smith, C.A.; Want, E.J.; O'Maille, G.; Abagyan, R.; Siuzdak, G. XCMS: Processing mass spectrometry data for metabolite profiling using nonlinear peak alignment, matching, and identification. *Anal. Chem.* **2006**, *78*, 779–787. [CrossRef] [PubMed]

369. Meyrand, M.; Dallas, D.C.; Caillat, H.; Bouvier, F.; Martin, P.; Barile, D. Comparison of milk oligosaccharides between goats with and without the genetic ability to synthesize αs1-casein. *Small Rumin. Res.* **2013**, *113*, 411–420. [CrossRef] [PubMed]

370. Guerrero, A.; Dallas, D.C.; Contreras, S.; Chee, S.; Parker, E.A.; Sun, X.; Dimapasoc, L.; Barile, D.; German, J.B.; Lebrilla, C.B. Mechanistic Peptidomics: Factors that dictate specificity in the formation of endogenous peptides in human milk. *Mol. Cell. Proteomics* **2014**, *13*, 3343–3351. [CrossRef] [PubMed]

371. Kaur, H.; Garg, A.; Raghava, G.P.S. Raghava PEPstr: A *de novo* method for tertiary structure prediction of small bioactive peptides. *Protein Pept. Lett.* **2007**, *14*, 626–631. [CrossRef]

372. Thévenet, P.; Shen, Y.; Maupetit, J.; Guyon, F.; Derreumaux, P.; Tufféry, P. PEP-FOLD: An updated de novo structure prediction server for both linear and disulfide bonded cyclic peptides. *Nucleic Acids Res.* **2012**, *40*, W288–W293. [CrossRef]

373. Beaufays, J.; Lins, L.; Thomas, A.; Brasseur, R. *In silico* predictions of 3D structures of linear and cyclic peptides with natural and non-proteinogenic residues. *J. Pept. Sci.* **2012**, *18*, 17–24. [CrossRef]

374. Wienkoop, S.; Staudinger, C.; Hoehenwarter, W.; Weckwerth, W.; Egelhofer, V. ProMEX—A mass spectral reference database for plant proteomics. *Front. Plant Sci.* **2012**, *3*, 3. [CrossRef]

375. Colgrave, M.L.; Poth, A.G.; Kaas, Q.; Craik, D.J. A new "era" for cyclotide sequencing. *Biopolymers* **2010**, *94*, 592–601. [CrossRef]

376. Fesenko, I.A.; Arapidi, G.P.; Skripnikov, A.Y.; Alexeev, D.G.; Kostryukova, E.S.; Manolov, A.I.; Altukhov, I.A.; Khazigaleeva, R.A.; Seredina, A.V.; Kovalchuk, S.I.; et al. Specific pools of endogenous peptides are present in gametophore, protonema, and protoplast cells of the moss Physcomitrella patens. *BMC Plant Biol.* **2015**, *15*, 1–16. [CrossRef]

377. Del Río, L.A. ROS and RNS in plant physiology: An overview. *J. Exp. Bot.* **2015**, *66*, 2827–2837. [CrossRef]

378. Turkan, I. ROS and RNS: Key signalling molecules in plants. *J. Exp. Bot.* **2018**, *69*, 3313–3315. [CrossRef] [PubMed]

379. Birben, E.; Sahiner, U.M.; Sackesen, C.; Erzurum, S.; Kalayci, O. Oxidative stress and antioxidant defense. *World Allergy Organ. J.* **2012**, *5*, 9–19. [CrossRef] [PubMed]

380. Mock, H.P.; Dietz, K.J. Redox proteomics for the assessment of redox-related posttranslational regulation in plants. *Biochim. Biophys. Acta Proteins Proteomics* **2016**, *1864*, 967–973. [CrossRef] [PubMed]

381. Dietz, K.J. Peroxiredoxins in plants and cyanobacteria. *Antioxid. Redox Signal.* **2011**, *15*, 1129–1159. [CrossRef] [PubMed]

382. Huang, J.; Willems, P.; Wei, B.; Tian, C.; Ferreira, R.B.; Bodra, N.; Martínez Gache, S.A.; Wahni, K.; Liu, K.; Vertommen, D.; et al. Mining for protein S-sulfenylation in *Arabidopsis* uncovers redox-sensitive sites. *Proc. Natl. Acad. Sci. USA* **2019**, *116*, 21256–21261. [CrossRef]

383. Wang, H.; Wang, S.; Lu, Y.; Alvarez, S.; Hicks, L.M.; Ge, X.; Xia, Y. Proteomic analysis of early-responsive redox-sensitive proteins in *Arabidopsis*. *J. Proteome Res.* **2012**, *11*, 412–424. [CrossRef] [PubMed]

384. Akter, S.; Carpentier, S.; Van Breusegem, F.; Messens, J. Identification of dimedone-trapped sulfenylated proteins in plants under stress. *Biochem. Biophys. Reports* **2017**, *9*, 106–113. [CrossRef]

385. De Smet, B.; Willems, P.; Fernandez-Fernandez, A.D.; Alseekh, S.; Fernie, A.R.; Messens, J.; Van Breusegem, F. *In vivo* detection of protein cysteine sulfenylation in plastids. *Plant J.* **2019**, *97*, 765–778. [CrossRef]

386. Akter, S.; Fu, L.; Jung, Y.; Lo Conte, M.; Lawson, J.R.; Lowther, W.T.; Sun, R.; Liu, K.; Yang, J.; Carroll, K.S. Chemical proteomics reveals new targets of cysteine sulfinic acid reductase. *Nat. Chem. Biol.* **2018**, *14*, 995–1004. [CrossRef]

387. Nietzel, T.; Mostertz, J.; Ruberti, C.; Née, G.; Fuchs, P.; Wagner, S.; Moseler, A.; Müller-Schüssele, S.J.; Benamar, A.; Poschet, G.; et al. Redox-mediated kick-start of mitochondrial energy metabolism drives resource-efficient seed germination. *Proc. Natl. Acad. Sci. USA* **2020**, *117*, 741–751. [CrossRef]

388. Ratajczak, E.; Staszak, A.; Wojciechowska, N.; Bagniewska-Zadworna, A.; Dietz, K. Regulation of thiol metabolism as a factor that influences the development and storage capacity of beech seeds. *J. Plant Physiol.* **2019**, *239*, 61–70. [CrossRef] [PubMed]

389. Ratajczak, E.; Małecka, A.; Ciereszko, I.; Staszak, A. Mitochondria are important determinants of the aging of seeds. *Int. J. Mol. Sci.* **2019**, *20*, 1568. [CrossRef] [PubMed]

390. Smolikova, G.; Kreslavski, V.; Shiroglazova, O.; Bilova, T.; Sharova, E.; Frolov, A.; Medvedev, S. Photochemical activity changes accompanying the embryogenesis of pea (*Pisum sativum*) with yellow and green cotyledons. *Funct. Plant Biol.* **2018**, *45*, 228. [CrossRef] [PubMed]

391. Landry, E.J.; Fuchs, S.J.; Hu, J. Carbohydrate composition of mature and immature faba bean seeds. *J. Food Compos. Anal.* **2016**, *50*, 55–60. [CrossRef]

392. Bailly, C.; Kraner, I. Analyses of ROS and antioxidants in relation to seed longevity and germination. In *Seed Dormancy:Methods and Protocols*; Kermode, A.R., Ed.; Springer: Berlin/Heidelberg, Germany, 2011; pp. 343–367.

393. Oracz, K.; Bouteau, H.E.-M.; Farrant, J.M.; Cooper, K.; Belghazi, M.; Job, C.; Job, D.; Corbineau, F.; Bailly, C. ROS production and protein oxidation as a novel mechanism for seed dormancy alleviation. *Plant J.* **2007**, *50*, 452–465. [CrossRef] [PubMed]

394. Leymarie, J.; Vitkauskaité, G.; Hoang, H.H.; Gendreau, E.; Chazoule, V.; Meimoun, P.; Corbineau, F.; El-Maarouf-Bouteau, H.; Bailly, C. Role of reactive oxygen species in the regulation of *Arabidopsis* seed dormancy. *Plant Cell Physiol.* **2012**, *53*, 96–106. [CrossRef] [PubMed]

395. Bailly, C. Active oxygen species and antioxidants in seed biology. *Seed Sci. Res.* **2004**, *14*, 93–107. [CrossRef]

396. Wolff, I.A. Seed lipids. *Science* **1966**, *154*, 1140–1149. [CrossRef]

397. Ponquett, R.T.; Ross, G. Lipid autoxidation and seed ageing: Putative relationships between seed longevity and lipid stability. *Seed Sci. Res.* **1992**, *2*, 51–54. [CrossRef]

398. Ouzouline, M.; Tahani, N.; Demandre, C.; El Amrani, E.; Benhassaine-Kesri, G.; Serghini Caid, H. Effects of accelerated aging upon the lipid composition of seeds from two soft wheat varieties from Morocco. *Grasas y Aceites* **2009**, *60*, 367–374.

399. Murthy, U.M.N.; Kumar, P.P.; Sun, W.Q.; Murthy, N.U.M.; Kumar, P.P.; Sun, W.Q. Mechanisms of seed ageing under different storage conditions for *Vigna radiata* (L.) Wilczek: Lipid peroxidation, sugar hydrolysis, Maillard reactions and their relationship to glass state transition. *J. Exp. Bot.* **2003**, *54*, 1057–1067. [CrossRef] [PubMed]

400. Hoekstra, F.A.; Golovina, E.A.; Tetteroo, F.A.; Wolkers, W.F. Induction of desiccation tolerance in plant somatic embryos: How exclusive is the protective role of sugars? *Cryobiology* **2001**, *43*, 140–150. [CrossRef] [PubMed]

401. Wolff, S.P.; Dean, R.T. Glucose autoxidation and protein modification. The potential role of 'autoxidative glycosylation' in diabetes. *Biochem. J.* **1987**, *245*, 243–250. [CrossRef]

402. Ayala, A.; Muñoz, M.F.; Argüelles, S. Lipid Peroxidation: Production, metabolism, and signaling mechanisms of malondialdehyde and 4-hydroxy-2-nonenal. *Oxid. Med. Cell. Longev.* **2014**, *2014*, 1–31. [CrossRef] [PubMed]

403. Vistoli, G.; De Maddis, D.; Cipak, A.; Zarkovic, N.; Carini, M.; Aldini, G. Advanced glycoxidation and lipoxidation end products (AGEs and ALEs): An overview of their mechanisms of formation. *Free Radic. Res.* **2013**, *47*, 3–27. [CrossRef] [PubMed]

404. Møller, I.M.; Rogowska-Wrzesinska, A.; Rao, R.S.P. Protein carbonylation and metal-catalyzed protein oxidation in a cellular perspective. *J. Proteomics* **2011**, *74*, 2228–2242. [CrossRef]

405. Strelec, I.; Hlevnjak, M. Accumulation of Amadori and Maillard products in wheat seeds aged under different storage conditions. *Croat. Chem. Acta* **2008**, *81*, 131–137.

406. Wettlaufer, S.H.; Leopold, A.C. Relevance of Amadori and Maillard products to seed deterioration. *Plant Physiol.* **1991**, *97*, 165–169. [CrossRef]

407. Colville, L.; Bradley, E.L.; Lloyd, A.S.; Pritchard, H.W.; Castle, L.; Kranner, I. Volatile fingerprints of seeds of four species indicate the involvement of alcoholic fermentation, lipid peroxidation, and Maillard reactions in seed deterioration during ageing and desiccation stress. *J. Exp. Bot.* **2012**, *63*, 6519–6530. [CrossRef]

408. Greifenhagen, U.; Nguyen, V.D.V.D.; Moschner, J.; Giannis, A.; Frolov, A.; Hoffmann, R. Sensitive and site-specific identification of carboxymethylated and carboxyethylated peptides in tryptic digests of proteins and human plasma. *J. Proteome Res.* **2015**, *14*, 768–777. [CrossRef]

409. Schmidt, R.; Böhme, D.; Singer, D.; Frolov, A. Specific tandem mass spectrometric detection of AGE-modified arginine residues in peptides. *J. Mass Spectrom.* **2015**, *50*, 613–624. [CrossRef] [PubMed]

410. Frolov, A.; Hoffmann, P.; Hoffmann, R. Fragmentation behavior of glycated peptides derived fromD-glucose,D-fructose andD-ribose in tandem mass spectrometry. *J. Mass Spectrom.* **2006**, *41*, 1459–1469. [CrossRef] [PubMed]

411. Zhang, Q.; Frolov, A.; Tang, N.; Hoffmann, R.; van de Goor, T.; Metz, T.O.; Smith, R.D.R.D. Application of electron transfer dissociation mass spectrometry in analyses of non-enzymatically glycated peptides. *Rapid Commun. Mass Spectrom.* **2007**, *21*, 661–666. [CrossRef] [PubMed]

412. Fedorova, M.; Frolov, A.; Hoffmann, R. Fragmentation behavior of Amadori-peptides obtained by non-enzymatic glycosylation of lysine residues with ADP-ribose in tandem mass spectrometry. *J. Mass Spectrom.* **2010**, *45*. [CrossRef]

413. Ehrlich, H.; Hanke, T.; Frolov, A.; Langrock, T.; Hoffmann, R.; Fischer, C.; Schwarzenbolz, U.; Henle, T.; Born, R.; Worch, H. Modification of collagen in vitro with respect to formation of Nε-carboxymethyllysine. *Int. J. Biol. Macromol.* **2009**, *44*, 51–56. [CrossRef]

414. Frolov, A.; Singer, D.; Hoffmann, R. Sites-specific synthesis of Amadori-modified peptides on solid phase. *J. Pept. Sci.* **2006**, *12*, 389–395. [CrossRef]

415. Frolov, A.; Hoffmann, R. Separation of Amadori peptides from their unmodified analogs by ion-pairing RP-HPLC with heptafluorobutyric acid as ion-pair reagent. *Anal. Bioanal. Chem.* **2008**, *392*, 1209–1214. [CrossRef]

416. Frolov, A.; Singer, D.; Hoffmann, R. Solid-phase synthesis of glucose-derived Amadori peptides. *J. Pept. Sci.* **2007**, *13*, 862–867. [CrossRef]

417. Greifenhagen, U.; Frolov, A.; Hoffmann, R. Oxidative degradation of N ε-fructosylamine-substituted peptides in heated aqueous systems. *Amino Acids* **2015**, *47*, 1065–1076. [CrossRef]

418. Spiller, S.; Frolov, A.; Hoffmann, R. Quantification of specific glycation sites in human serum albumin as prospective type 2 diabetes mellitus biomarkers. *Protein Pept. Lett.* **2018**, *24*, 887–896. [CrossRef]

419. Soboleva, A.; Modzel, M.; Didio, A.; Płóciennik, H.; Kijewska, M.; Grischina, T.; Karonova, T.; Bilova, T.; Stefanov, V.; Stefanowicz, P.; et al. Quantification of prospective type 2 diabetes mellitus biomarkers by stable isotope dilution with bi-labeled standard glycated peptides. *Anal. Methods* **2017**, *9*, 409–418. [CrossRef]

420. Thornalley, P.J.; Battah, S.; Ahmed, N.; Karachalias, N.; Agalou, S.; Babaei-Jadidi, R.; Dawnay, A. Quantitative screening of advanced glycation endproducts in cellular and extracellular proteins by tandem mass spectrometry. *Biochem. J.* **2003**, *375*, 581–592. [CrossRef] [PubMed]

421. Smuda, M.; Henning, C.; Raghavan, C.T.; Johar, K.; Vasavada, A.R.; Nagaraj, R.H.; Glomb, M.A. Comprehensive analysis of maillard protein modifications in human lenses: Effect of age and cataract. *Biochemistry* **2015**, *54*, 2500–2507. [CrossRef] [PubMed]

422. Rabbani, N.; Al-Motawa, M.; Thornalley, P.J. Protein glycation in plants—An under-researched field with much still to discover. *Int. J. Mol. Sci.* **2020**, *21*, 3942. [CrossRef]

423. Leonova, T.; Popova, V.; Tsarev, A.; Henning, C.; Antonova, K.; Rogovskaya, N.; Vikhnina, M.; Baldensperger, T.; Soboleva, A.; Dinastia, E.; et al. Does protein glycation impact on the drought-related changes in metabolism and nutritional properties of mature pea (*Pisum sativum* L.) seeds? *Int. J. Mol. Sci.* **2020**, *21*, 567. [CrossRef]

424. Dammann, P.; Sell, D.R.; Begall, S.; Strauch, C.; Monnier, V.M. Advanced Glycation End-Products as Markers of Aging and longevity in the long-lived Ansell's mole-rat (*Fukomys anselli*). *J. Gerontol. Ser. A* **2012**, *67A*, 573–583. [CrossRef]

425. Bilova, T.; Paudel, G.; Shilyaev, N.; Schmidt, R.; Brauch, D.; Tarakhovskaya, E.; Milrud, S.; Smolikova, G.; Tissier, A.; Vogt, T.; et al. Global proteomic analysis of advanced glycation end products in the *Arabidopsis* proteome provides evidence for age-related glycation hot spots. *J. Biol. Chem.* **2017**, *292*, 15758–15776. [CrossRef]

426. Vilhena, M.B.; Franco, M.R.; Schmidt, D.; Carvalho, G.; Azevedo, R.A. Evaluation of protein extraction methods for enhanced proteomic analysis of tomato leaves and roots. *An. Acad. Bras. Cienc.* **2015**, *87*, 1853–1863. [CrossRef]

427. Galland, M.; He, D.; Lounifi, I.; Arc, E.; Clément, G.; Balzergue, S.; Huguet, S.; Cueff, G.; Godin, B.; Collet, B.; et al. An integrated "multi-omics" comparison of embryo and endosperm tissue-specific features and their impact on rice seed quality. *Front. Plant Sci.* **2017**, *8*, 1984. [CrossRef]

428. Yin, X.; He, D.; Gupta, R.; Yang, P. Physiological and proteomic analyses on artificially aged *Brassica napus* seed. *Front. Plant Sci.* **2015**, *6*, 112. [CrossRef]

429. Kołodziejczyk, I.; Dzitko, K.; Szewczyk, R.; Posmyk, M.M. Exogenous melatonin improves corn (*Zea mays* L.) embryo proteome in seeds subjected to chilling stress. *J. Plant Physiol.* **2016**, *193*, 47–56. [CrossRef] [PubMed]

430. DeBlasio, S.L.; Johnson, R.S.; MacCoss, M.J.; Gray, S.M.; Cilia, M. Model system-guided protein interaction mapping for virus isolated from phloem tissue. *J. Proteome Res.* **2016**, *15*, 4601–4611. [CrossRef] [PubMed]

431. Tyanova, S.; Temu, T.; Carlson, A.; Sinitcyn, P.; Mann, M.; Cox, J. Visualization of LC-MS/MS proteomics data in MaxQuant. *Proteomics* **2015**, *15*, 1453–1456. [CrossRef]

432. PEAKS AB Software, Version 2.0. Available online: https://www.bioinfor.com/peaks-ab-software/ (accessed on 29 November 2020).

433. Cox, B.; Kislinger, T.; Wigle, D.A.; Kannan, A.; Brown, K.; Okubo, T.; Hogan, B.; Jurisica, I.; Frey, B.; Rossant, J.; et al. Integrated proteomic and transcriptomic profiling of mouse lung development and Nmyc target genes. *Mol. Syst. Biol.* **2007**, *3*, 109. [CrossRef]

434. Röst, H.L.; Sachsenberg, T.; Aiche, S.; Bielow, C.; Weisser, H.; Aicheler, F.; Andreotti, S.; Ehrlich, H.-C.; Gutenbrunner, P.; Kenar, E.; et al. OpenMS: A flexible open-source software platform for mass spectrometry data analysis. *Nat. Methods* **2016**, *13*, 741–748. [CrossRef] [PubMed]

435. Proteome Discoverer. User Guide. Software Version 2.2; Thermo Fisher Scientific Inc. 2017. Available online: https://assets.thermofisher.com/TFS-Assets/CMD/manuals/Man-XCALI-97808-Proteome-Discoverer-User-ManXCALI97808-EN.pdf (accessed on 29 November 2020).

436. Park, S.K.; Yates, J.R. Census for proteome quantification. In *Current Protocols in Bioinformatics*; John Wiley & Sons, Inc.: Hoboken, NJ, USA, 2010; Volume 29, pp. 13.12.1–13.12.11.

437. Chang, C.; Zhang, J.; Han, M.; Ma, J.; Zhang, W.; Wu, S.; Liu, K.; Xie, H.; He, F.; Zhu, Y. SILVER: An efficient tool for stable isotope labeling LC-MS data quantitative analysis with quality control methods. *Bioinformatics* **2014**, *30*, 586–587. [CrossRef]

438. Bellew, M.; Coram, M.; Fitzgibbon, M.; Igra, M.; Randolph, T.; Wang, P.; May, D.; Eng, J.; Fang, R.; Lin, C.; et al. A suite of algorithms for the comprehensive analysis of complex protein mixtures using high-resolution LC-MS. *Bioinformatics* **2006**, *22*, 1902–1909. [CrossRef]

439. Valot, B.; Langella, O.; Nano, E.; Zivy, M. MassChroQ: A versatile tool for mass spectrometry quantification. *Proteomics* **2011**, *11*, 3572–3577. [CrossRef]

440. MacLean, B.; Tomazela, D.M.; Shulman, N.; Chambers, M.; Finney, G.L.; Frewen, B.; Kern, R.; Tabb, D.L.; Liebler, D.C.; MacCoss, M.J. Skyline: An open source document editor for creating and analyzing targeted proteomics experiments. *Bioinformatics* **2010**, *26*, 966–968. [CrossRef]

441. Tsou, C.-C.; Avtonomov, D.; Larsen, B.; Tucholska, M.; Choi, H.; Gingras, A.-C.; Nesvizhskii, A.I. DIA-Umpire: Comprehensive computational framework for data-independent acquisition proteomics. *Nat. Methods* **2015**, *12*, 258–264. [CrossRef]

442. Deutsch, E.W.; Mendoza, L.; Shteynberg, D.; Farrah, T.; Lam, H.; Tasman, N.; Sun, Z.; Nilsson, E.; Pratt, B.; Prazen, B.; et al. A guided tour of the Trans-Proteomic Pipeline. *Proteomics* **2010**, *10*, 1150–1159. [CrossRef] [PubMed]

443. Argentini, A.; Goeminne, L.J.E.; Verheggen, K.; Hulstaert, N.; Staes, A.; Clement, L.; Martens, L. moFF: A robust and automated approach to extract peptide ion intensities. *Nat. Methods* **2016**, *13*, 964–966. [CrossRef] [PubMed]

444. An introduction to Mascot Distiller. Available online: http://www.matrixscience.com/distiller.html (accessed on 29 November 2020).

445. Brusniak, M.-Y.; Bodenmiller, B.; Campbell, D.; Cooke, K.; Eddes, J.; Garbutt, A.; Lau, H.; Letarte, S.; Mueller, L.N.; Sharma, V.; et al. Corra: Computational framework and tools for LC-MS discovery and targeted mass spectrometry-based proteomics. *BMC Bioinform.* **2008**, *9*, 542. [CrossRef] [PubMed]

446. Millikin, R.J.; Solntsev, S.K.; Shortreed, M.R.; Smith, L.M. Ultrafast peptide label-free quantification with FlashLFQ. *J. Proteome Res.* **2018**, *17*, 386–391. [CrossRef] [PubMed]

447. ProSightPC 4.0 Quick Start Guide. Revision A XCALI-97800; Thermo Fisher Scientific Inc. 2016. Available online: https://assets.thermofisher.com/TFS-Assets/CMD/Product-Guides/QS-XCALI-97800-ProSightPC-QSXCALI97800-EN.pdf (accessed on 29 November 2020).

448. Sanford, H.; Harnos, S. Agilent MassHunter Qualitative Data Analysis; Software B.07.00. 2017. Available online: https://www.agilent.com/cs/library/eseminars/public/Session_3_Qualitative_Analysis_Basics.pdf (accessed on 29 November 2020).

449. Schwacke, R.; Ponce-Soto, G.Y.; Krause, K.; Bolger, A.M.; Arsova, B.; Hallab, A.; Gruden, K.; Stitt, M.; Bolger, M.E.; Usadel, B. MapMan4: A Refined protein classification and annotation framework applicable to multi-omics data analysis. *Mol. Plant* **2019**, *12*, 879–892. [CrossRef] [PubMed]

450. Huerta-Cepas, J.; Forslund, K.; Coelho, L.P.; Szklarczyk, D.; Jensen, L.J.; von Mering, C.; Bork, P. Fast Genome-wide functional annotation through orthology assignment by eggNOG-mapper. *Mol. Biol. Evol.* **2017**, *34*, 2115–2122. [CrossRef]

451. Szklarczyk, D.; Gable, A.L.; Lyon, D.; Junge, A.; Wyder, S.; Huerta-Cepas, J.; Simonovic, M.; Doncheva, N.T.; Morris, J.H.; Bork, P.; et al. STRING v11: Protein-protein association networks with increased coverage, supporting functional discovery in genome-wide experimental datasets. *Nucleic Acids Res.* **2019**, *47*, D607–D613. [CrossRef]

Publisher's Note: MDPI stays neutral with regard to jurisdictional claims in published maps and institutional affiliations.

International Journal of
Molecular Sciences

MDPI

Review

Soil Metaproteomics for the Study of the Relationships Between Microorganisms and Plants: A Review of Extraction Protocols and Ecological Insights

Maria Tartaglia [1], Felipe Bastida [2], Rosaria Sciarrillo [1] and Carmine Guarino [1,*]

[1] Department of Science and Technology, University of Sannio, via de Sanctis snc, 82100 Benevento, Italy; mtartaglia@unisannio.it (M.T.); sciarrillo@unisannio.it (R.S.)
[2] CEBAS-CSIC, Department of Soil and Water Conservation, Campus Universitario de Espinardo, 30100 Murcia, Spain; fbastida@cebas.csic.es
* Correspondence: guarino@unisannio.it; Tel.: +39-824-305145

Received: 21 September 2020; Accepted: 9 November 2020; Published: 11 November 2020

Abstract: Soil is a complex matrix where biotic and abiotic components establish a still unclear network involving bacteria, fungi, archaea, protists, protozoa, and roots that are in constant communication with each other. Understanding these interactions has recently focused on metagenomics, metatranscriptomics and less on metaproteomics studies. Metaproteomic allows total extraction of intracellular and extracellular proteins from soil samples, providing a complete picture of the physiological and functional state of the "soil community". The advancement of high-performance mass spectrometry technologies was more rapid than the development of ad hoc extraction techniques for soil proteins. The protein extraction from environmental samples is biased due to interfering substances and the lower amount of proteins in comparison to cell cultures. Soil sample preparation and extraction methodology are crucial steps to obtain high-quality resolution and yields of proteins. This review focuses on the several soil protein extraction protocols to date to highlight the methodological challenges and critical issues for the application of proteomics to soil samples. This review concludes that improvements in soil protein extraction, together with the employment of ad hoc metagenome database, may enhance the identification of proteins with low abundance or from non-dominant populations and increase our capacity to predict functional changes in soil.

Keywords: soil metaproteomics; protein extraction; rhizosphere crosstalk; polluted soil

---

1. Introduction

The term "rhizosphere" was coined in 1904 by the agronomist and plant physiologist Hiltner to identify the portion of soil near the roots that, as he first postulated, welcomed an abundant microbial community influenced by the root-released substances. Over the years, the definition of the rhizosphere has expanded and, to date, it is described as composed of the endorizosphere, the rhizoplane, and the ectorhizosphere [1,2]. The substances released from the roots greatly influence the rhizosphere microbial pool and guide the complex plant–microorganisms interactions. The amount and the composition of the root exudates depend on the plant species and the physiological state and can be altered by the environmental conditions and the biotic or abiotic stresses to which the plant is subjected [3]. The roots exude low- and high-molecular-weight organic compounds, mainly amino acids, peptides, proteins, monosaccharaides, disaccharides and polysaccharides, mucilage, phenolic compounds, secondary metabolites, and some important factors for microbial attraction such as malic acid and citric acid. Root exudates model the set of microorganisms present in the rhizosphere through species–species specific attraction/repulsion chemotaxis [4,5]. In fact, exudates can favor positive interactions such as symbiosis, mycorrhizae association, and recruitment of plant growth-promoting bacteria (PGPR) and can inhibit negative interactions such as herbivory, parasitism, competition, and allelopathy [1,6].

The rhizosphere of a polluted soil is generally deeply altered, as it has undergone particular and specific adaptations that are a consequence of the degree and type of pollution. The ecto–endophytic properties are altered and conditioned by polluting substances and by the soil stationary conditions (matrix composition and structure alteration, pH variation, etc.). Further complexity is added to the system by the engineering of the rhizosphere (PGPR, mycorrhizae, etc.) in bioremediation processes.

To understand the complex communication between plant and soil microorganisms, metagenomics and metatranscriptomics are not enough, and a metaproteomic approach must be used in order to assess the real functionality of soil microbes in the rhizosphere. In fact, metagenomics is subject to various limitations, such us the molecular extraction techniques, bioinformatics of annotation, and data processing, but above all for the lack of correlation between the presence of a functional gene and the actual activity of that gene [7]. Moreover, recent studies suggest that a great part of DNA in soil belongs to dead or dormant cells [8]. Even metatranscriptomics cannot provide us accurate information of gene functionality, considering the various post-transcriptional regulations to which mRNAs are subject. Hence, from a quantitative point of view, metagenomics and metatranscriptomics can give us a relative abundance of a functional gene and its mRNA, but not its effective functionality, intended as a protein expression [7–9]. The metaproteomics approach allows us to analyze the taxonomic data, can offer a functional characterization of the biomass present in the rhizosphere, and has great potential as it can help identify the metabolic dynamics between and within species [10–14]. Although the analysis of the total rhizosphere proteome can provide fundamental information to understand the environment–plant–microorganism interactions, soil protein extraction is very complex due to the nature of the soil matrix. The methodological challenges in protein extractions from soil are many; the matrix is complex, chemically and physically heterogeneous, and very variable. Moreover, to extract intracellular proteins, a passage of cell lysis is necessary. Furthermore, the low quantity of proteins present binds humic substances, making their extraction even more complex [15,16]. Metaproteomics analysis of polluted soils has great potential to shed light on the impact of diversity and microbial functionality established in a disturbed soil, assigning proteins to specific phylogenetic and functional groups. This aspect is amplified at the rhizosphere level where particular co-metabolic systems and relationships between different organisms persist, particularly in the dynamic prospects of pollutant degradation. In addition, the metaproteomic application becomes an important tool when the rhizosphere is engineered by adding bacterial and/or fungal cocktails, biostimulants, and nanomaterials to enhance the effect of bioremediation. In fact, the metaproteomic analysis would allow us to obtain a clear vision of the co-metabolic perspectives generated by the engineering of the rhizosphere and it would offer novel information on the genetic and metabolic processes occurring during soil bioremediation. However, soil proteins are associated with humic substances and clay particles [17–19], making it hard to obtain a sufficiently clean protein extract for further processing and downstream analyses. To overcome these challenges, sophisticated protocols for sample preparation must be tested. This review provides the reader with detailed information on the different protocols for protein extraction from soil samples, focusing on those interested in rhizosphere soil, offering a panorama that is at least exhaustive of the current extraction methodological trends and of the current quali–quantitative limits. It also intends to provide an element of discussion on the future prospects and the need and usefulness of metaproteomic studies on polluted rhizospheres and bio-mediated soils.

2. Methods: Selected Studies for This Review

The study of protein expression in environmental samples is a hot topic. On the Web of Science (http://www.webofknowledge.com/), using "soil" and "metaproteomics" as topics it was possible to find 126 records: 95 articles, 28 reviews, two proceedings papers, one book chapter (plus one data paper, one reprint, and one editorial material) published from 1990 to date (June 2020). According to their pertinence, 71 were selected. The same research on Scopus (https://www.scopus.com/) yielded 109 documents: 73 articles, 20 reviews, eight book chapters, three short surveys, two conference papers, one data paper, one letter, and one undefined (June 2020). Of the 73 articles, only those that were truly relevant and in English were taken into consideration. A cross-comparison between these 54 elected papers and the 71 obtained from the search on Web Of Science showed that 49 papers were shared between the searches, four were present in Scopus but not in Web Of Science and vice versa; 22 articles were present in Web Of Science but not in Scopus, with a final count of 75 articles taken into consideration. The same search on PubMed (https://pubmed.ncbi.nlm.nih.gov/) provided 83 results, 31 of which were rejected because they were reviews, non-related or non-English articles. Of the remaining articles, only five were not already present in the search results from the Web of Science and Scopus and were added to the total papers taken into consideration.

3. Soil Matrix Components that Interfere with Protein Extraction

Metaproteomics has great potential to provide new information on rhizospheric interactions, but the success of this approach depends on the quantity and quality of the extracted proteins. The soil, in fact, is one of the most complex matrices for the extraction of proteins due to its physical, structural, and biological complexity, its heterogeneity, and due to the presence of a mixture of organic macromolecules that interfere with the metaproteomic workflow [14,20]. Humic substances, co-extracted with proteins, interfere with the quantification of proteins with spectrophotometric techniques and appear as brown streaks on SDS-PAGE and 2D-PAGE gels, preventing the detection of protein bands/spots. They also cause peptide signal suppression in LC-MS/MS measurements [19,21–23]. The more or less strong bonds between proteins and the soil matrix are influenced by the specific characteristics of the soil under analysis, such as pH, composition, and mineral content [23–26]. Most extraction protocols involve the use of liquid phenol to remove humic substances [21,23,27–29] and polyvinylpolypyrrolidone (PVPP) [11,23]. However, it is necessary to take into account the limited safety in the use of phenol, and the poor results obtained with PVPP. In order to remove, or in any case limit, the interference of humic acids, various strategies have been tested such as the precipitation of humic substance with trivalent aluminum ions, the use of salt solution of inorganic divalent or trivalent cations to desorb the immobilized proteins from humic substance, and the addition of $CaCl_2$ [19,30,31]. However, the fact that many proteins are linked to humic substances is functional to their stability. For instance, many extracellular enzymes are known to be stabilized within humic substances [32]. Therefore, any method that attempts to remove humics will likely discard proteins as well and interfere with their biological significance and function. In addition, it is difficult to cleave proteins from humics. Another interfering substance is keratin that is not a problem until its concentration remains low compared to the proteins of interest. If its concentration, on the other hand, is too high, it masks the signal of the peptides of interest and shows artifacts mainly at 54–57 kDa and 65–68 kDa [33,34]. To minimize keratin contamination, all materials should be handled with nitrile gloves, and latex must be avoided because it contains significant quantities.

4. Soil Protein Extraction

Different approaches and protocols have been used over the years and are present in literature in order to obtain feasible protein extraction in quantitative and qualitative terms (Supplementary Materials Table S1). The general workflow is almost common to all protocols and involves sample collection and storage, cell lysis, protein extraction, and a series of washes (Figure 1). The protein extract thus obtained is ready for quantification, SDS page or 2DE electrophoresis, and mass spectrometry (MS) [15,35]. Protein extraction, obviously, begins with the collection of the soil sample that is sifted to remove stones and debris and make the sample homogeneous. To inactivate the proteases present in the soil, the sample must be stored at −80 °C until use. To extract both extracellular and intracellular proteins, a first cell lysis step is necessary. In a first work that recognized the importance of protein extraction from environmental samples, Ogunseitan [36] tested two extraction techniques.

Both methods provided as a first step a sample wash with ice-cold 50 mM Tris-HCl, pH 7.4, followed by two types of physical cell lysis. With the "boiling method", cell lysis was obtained by placing the samples in a boiling water bath for 10 min in 1:1 water and extraction buffer. In the "freeze–thaw method", instead, hot–cold cycles were used in order to free intracellular proteins before the extraction. A series of centrifugations and the use of a filter to remove debris and DNA allowed protein separation by electrophoresis. The proteins extracted with both methods were between 14 kDa and 97 kDa. In quantitative terms, on the other hand, with the first method, the estimated protein concentration was 1–5 μg/g, and with the freeze–thaw method, 20–50 μg protein/g [36]. Murase et al. [37] restricted the pool of interest to extracellular proteins only and therefore developed a protein extraction from soil that did not include cell lysis. Extracellular proteins, derived from the death and disruption of cells of organisms, or as extracellular enzymes that are continuously excreted by microorganisms and plants, are present in very small quantities. For this reason, the extraction was carried out on a starting sample of 100 g. The work showed how the buffer pH affects the extraction quality; in fact, a higher pH of the phosphate buffer resulted in a greater quantity of soil organic matter (SOM) included in the prepared sample, resulting in a high background with unclear protein bands on SDS—PAGE [37]. The protocol proposed by Ogunseitan [36] was taken up and modified in a subsequent work in which the protein extraction was carried out on soil contaminated with cadmium. The protocol and the extraction (with a protease inhibitor cocktail to prevent proteolysis) have been optimized to obtain a better quantitative yield of the extraction. Cell lysis was obtained with two different techniques: the first involved four repeated cycles of freezing in liquid nitrogen and thawing at 25 °C. The "bead beating method" instead, used 0.2 g of glass beads (Sigma, 150–212 m in diameter) per gram of soil stirred in a Fast Prep machine. The results obtained from the subsequent protein quantization showed that 32% more protein was extracted when using the snap freeze method when compared with the amount of protein obtained from the bead beating one [38]. In these early papers, SDS-PAGE showed proteins between 90 and 110 kDa, and few between 36 and 51 kDa. Schulze et al. [39] utilized MS-based proteomics to analyze proteins isolated from dissolved organic matter (DOM) using suction plates installed at different depths in a deciduous forest. In this procedure, only extracellular proteins were separated from the inorganic material and were separated from the organic molecules by gel filtration and subject to SDS-PAGE. After digestion with trypsin, the peptides obtained were separated with Nanoflow Liquid Chromatography (LC) before analysis with tandem mass spectrometry (MS/MS). This made it possible to identify the type, functions, and origin of the extracellular proteins extracted; furthermore, sampling at different depths made it possible to establish that there was a progressive reduction in the number of proteins identified as the soil depth increases [39]. Subsequently, an extraction method was developed in which an incubation with NaOH can break the microbial cells and separate the proteins linked to minerals and humic acids, while the consequent extraction, a classic protein extraction from plant material [40], involves the use of phenol to separate proteins from the humic organic matter. At the same time, phenol allows the removal of DNA and RNA that negatively influence the subsequent separation of proteins on gel [41]. However, with this extraction technique, it was not possible to completely eliminate the humic substance that appeared as a brown streak on the polyacrylamide gel of the SDS-PAGE [21]. To focus on the problem of the close interaction between proteins and humic acids, a study was done to test the efficiency of three different extractants, 0.1 M sodium pyrophosphate (pH 7), 67 mM phosphate buffer (pH 6), and 0.5 M potassium sulphate (pH 6.6), in recovering total protein quantity and the β-glucosidase enzyme from two natural forest soils [42]. The extracellular proteins were extracted with sodium pyrophosphate at a 1:5 *w/v* ratio, at 37 °C for 24 h under shaking, according to a previous protocol by the same authors [43] and with phosphate buffer and K_2SO_4 at a 1:3 *w/v* ratio at room temperature for 1 h under shaking, modifying a previous protocol [37]. The total proteins instead, were extracted by the chloroform fumigation–extraction method [44]. Despite that polyvinylpyrrolidone (PVPP) and a filtration column were used, the colorimetric assay to establish protein concentration was difficult due to the presence of co-extracted interfering substances as already found in previous studies [30]. The SDS-PAGE showed a high intracellular protein concentration using sodium pyrophosphate, while a higher concentration of extracellular protein was shown using potassium sulphate as an extractant. However, the need to develop protocols with better purification phases was still evident [42]. Chen et al. [45] devised a sequential extraction method, through sequentially extracting soil in 0.05 M citrate, pH 8, or citrate plus SDS buffer (1% *w/v* SDS, 0.1 M Tris-HCl, pH 6.8, 20 mM dithiothreitol (DTT)) at room temperature, followed by a classic phenol extraction and a proteic precipitation with five volumes of cold methanol plus 0.1 M ammonium acetate overnight at −20 °C. DTT, not used in previous protocols, was useful for reducing intra-molecular and inter-molecular disulphide bonds between cysteine residues by helping protein solubilization and aggregate disruption. The protein pellet obtained after centrifugation was washed with cold methanol and twice with cold acetone, air-dried, and solubilized for SDS-PAGE and 2DE. Although the results showed that the SDS buffer was more powerful than other extractants tested and phenol extraction was effective to remove interfering substances, this approach was not considered applicable for deep proteome studies, due to the reduced number of proteins extracted and to the low resolution of 2-DE separation [45]. A new protocol for deep proteome characterization of microorganisms in soil was developed by Chourey et al. [22]. This protocol involves the use of an alkaline-SDS buffer with a high concentration of DTT, followed by a 10 min boiling bath in SDS buffer. The combination of a detergent-rich buffer and thermal action induced cell lysis and the inhibition of protease activity. Cell lysis was followed by the precipitation of proteins with the addition of trichloroacetic acid (TCA) overnight at −10 °C. The protein pellet obtained was then subjected to a series of cold acetone washes. The pellet was then solubilized in a guanidine buffer prior to liquid chromatography-mass spectrometric characterization [22]. This approach tested on soils inoculated with bacteria identified 500–1000 proteins by 2D-LC–MS/MS. This protocol was well described in detail in later work [46]. The same procedure [22] has been utilized in different subsequent studies by Bastida et al. [47–49]. The combined protocol of Chen et al. [45] was refined and modified in a later work by Wang et al. [50], increasing the extracted protein content and shortening the extraction time, replacing the sequential citrate and SDS buffer extractions with a single treatment of the two buffers. Through this modified protocol, an average of 1000 spots on a 2-D gel of rice soil samples was detected, of which 122 proteins were identified by MALDI-TOF/TOF–MS [50]. Keiblinger et al. [23] compared four different protein extraction procedures: SDS extraction without phenol, NaOH in combination with phenol extraction, SDS buffer followed by phenol extraction, and the combination of SDS/phenol with prior washing steps. Proteins were analyzed by two-dimensional liquid chromatography/tandem mass spectrometry (2D-LCeMS/MS) and, in both soil types, extraction with the SDS buffer followed by phenol resulted in an increase in the numbers of extracted proteins [23]. Lin et al. [51] adopted the protocol of Wang et al. [50], testing it to obtain a metaproteomic profile of sugarcane rhizosphere soil. The combination of the extractions with the citrate buffer and the SDS buffer, followed by phenolic extraction, allowed the resolution of 143 protein spots with high resolution and repeatability, 109 of which were successfully analyzed by MALDI TOF-TOF MS [51].

In all the protocols seen so far, the main obstacle to performing good soil protein extraction seems to be the co-extraction of humic acids, which are arduous to separate from proteins and make it difficult, if not impossible, to quantify proteins with spectrophotometric approaches and to obtain a good resolution on polyacrylamide gels, for both the SDS-PAGE and 2DE. In fact, soils with high microbial biomass are usually associated with a higher organic matter content (including humics). In contrast, soils with low organic matter usually also have lower microbial biomass, which also limits protein extraction. Electrostatic attraction drives encapsulation of positively charged proteins, at pH 5 to 8, by soil humic and fulvic acids derived from terrestrial and mixed terrestrial–aquatic sources [52]. To obtain a complete soil metaproteomic analysis, it is necessary to directly extract both extracellular and intracellular proteins, and therefore, as already mentioned, the passage of cell lysis prior to protein extraction is fundamental. However, the disadvantages of the direct extraction method includes incomplete lysis and co-extraction of humic substances [14,45]. In order to overcome these issues, Nicora et al. [53] tried a pre-treatment of the soil sample with an amino acid mixture to evaluate the ability to produce a non-adsorptive surface on soil samples. Furthermore, the compatibility of this procedure with the subsequent protein extraction, tryptic digestion, and mass spectrometric analysis was assessed. The results showed the possibility of increasing protein identifications through blockage of binding sites for a variety of soil and sediment textures using proteins from lysed *Escherichia coli* cells, although the need for a better extraction buffer was still underlined [53]. Given the heterogeneity of the proposed methods, three different protocols [21,22,38], were tested to compare their effectiveness in cell lysis, humic acid cleaning, and protein purification [47]. Although the quantity of extracted proteins is limited with all the protocols tested, the comparison showed that the extraction technique greatly influences the type of extracted proteins and consequently the taxonomic and functional information obtained from them. Importantly, this study quantified the yield of protein extraction through the analyses of amino acids and not through colorimetric approaches, thus avoiding interferences related to humic substances. In fact, Chourey's method allowed a greater extraction and identification of different proteins, especially bacterial, while Singleton's method allowed a greater extraction of fungal proteins [47]. However, the need for a robust analysis of the different extraction protocols for soils with different characteristics was still evident. The study of all proteins recovered directly from environmental samples must take into account the high soil heterogeneity that makes the possibility to develop a universal extraction protocol suited for all different matrix types difficult. Mattarozzi et al. [23] tested three different protocols in a metaproteomics study based on liquid chromatography-high resolution mass spectrometry (LCHRMS) in serpentine soil contaminated by heavy metals, mainly Ni, Co, and Cr. Three different protocols were applied to the homogenized soil samples: the first involved the use of the NoviPure Soil Protein Kit (Qiagen), the second was in accordance with Chourey et al. [22], and the third involved phenol-SDS extraction [23]. The comparison between these methods established that the protocol proposed by Chourey was particularly efficient for protein extraction, allowing a greater extraction of membrane proteins. The NoviPure Soil Protein Kit instead, probably thanks to the strong mechanical action in lysing the cells, allowed a greater extraction of intracellular proteins. The phenol-SDS method allowed extracting proteins equally distributed in the intracellular and membrane compartment. In quantitative terms, on the other hand, the extraction with the NoviPure Soil Protein Kit yielded the lowest values for the colorimetric assay. Given the higher number of unique identified proteins obtained by mixing the extracts of the three methods, [54], it is recommended that different extraction approaches be combined to improve metaproteomic analysis. Liu et al. [55] applied the SDS–phenol method with a previous treatment of forest soil samples ground in liquid N_2. The subsequent analysis of the protein pellet obtained allowed the detection of 640 proteins by nano-LC-MS/MS [55]. In improving the extraction protocol, Qian and Hettich [56] took into consideration the insoluble nature of humic acids in acid solution, and the greater molecular weight of humic acids compared to peptides. The co-extract of humic substances, which interferes with the correct protein extraction from soil samples, is treated with the combination of a detergent and precipitation in trichloroacetic acid (TCA) and a filtration phase at pH 2~3 with a 10-filter kDa. According to the authors, this approach should improve extraction in qualitative terms, without adversely affecting it in quantitative terms and without interfering with subsequent analyses [18]. An experiment was conducted to identify the best protein extraction approach for the extracellular metaproteome that is diluted in soil samples. Various methods for protein concentration, extraction, precipitation, and solubilization have been tested; however, most methods failed to provide pure samples and therefore negatively influenced protein migration on gels and gel background clarity. The combination of phenol-based and TCA-acetone methods has been proposed as a strategy to capture the entire extracellular proteome [56]. The main limiting factor of this approach is the cultured growth of the microbial material prior to protein extraction. This strategy, which certainly increases the quantity and quality of the extracted protein sample, is focused only on the microbial communities and ignores all the extracellular proteins present in the rhizosphere due to the interaction with the roots and their exudates. Despite the progress made, protein extraction needs to be improved in order to efficiently characterize microbial functions in soil. Since soil characteristics, such as organic C, clay, and humic acid contents, represent the main factors that hinder good protein extraction, especially for extracellular proteins, in a recent work, the efficiency of two different extracellular protein extraction methods ofr different soils, without detergents or ultra-sonication, was compared [57]. The first one with a pH 6.5 buffer [58] and the second with an extraction protocol, in accordance with previous works [26,59] based on 0.1 M sodium pyrophosphate buffer, at pH 7.1. The work showed that the type of buffer not only influenced the quantity, but also the type of extracted protein. The absence of a standardized method for the extraction of soluble proteins from the soil has been discussed in a recent work, in which the capacities of the different extractors was tested on three soil types (Cambisol, Ferralsol, and Histosol) [60]. To evaluate the effective yield of these protocols, soluble 14C-labeled proteins were added to the soil samples. For protein extraction, deionized water, Na-pyrophosphate (0.01, 0.05, 0.1 M; pH 7.0), Na-citrate (0.01, 0.05, 0.1, 0.5 M; pH 8.0), Tris-SDS (0.01, 0.05, 0.1 M SDS with 0.05 M Tris; pH 7.0), phosphate buffer K (0.01, 0.05, 0.1, 0.5 M; pH 6.0 and 8.0), $CaCl_2$, NaOH and K_2SO_4 (0.01, 0.05, 0.1, 0.5 M), methanol, and ethanol (25%, 50%, 75%, 100% *v/v*) were used. The results obtained by Greenfield and colleagues have shown that protein extraction is strongly influenced by solvent and soil characteristics. Although no type of extraction permitted complete incubated protein extraction, the maximum recovery was observed with NaOH (0.1 M; 61–80%) and Na-pyrophosphate (0.05 M, pH 7.0; recovery 45–75%) [60]. However, this study focused only on hydrophilic proteins, and is therefore not conclusive for total protein extraction from this type of matrix. In a 2018 study by Callister et al. [61] an extraction protocol from human tissue [61] was modified and adapted for soil samples. To extract a good quantity of protein, 60 g of silt-loam soil was used. Cell lysis was entrusted to the mechanical action of a mix of beads (0.9–2.0 mm stainless steel beads, 0.1 mm zirconia beads, and 0.1 mm garnet beads). Protein extraction occurred using 2:1 chloroform: methanol (*v/v*) in a 5:1 ratio over sample volume in MilliQ water, with vigorous mixing. After sonication, incubation at −80 °C and centrifugation, the protein-containing interphase was taken and precipitated in ice-cold 100% methanol. The pellet obtained after centrifuging was frozen and lyophilized before being resolubilized in an SDS-Tris buffer containing DTT, then incubated for 30 min at 300 rpm, 50 °C. The samples were centrifuged and, in order to remove humic substances, the protein extracts were again precipitated in 20% trichloroacetic acid (TCA); the obtained pellet after the centrifugation underwent three washes in ice-cold acetone [62]. Although the protein pellet obtained was suitable for 1D LC-MS/MS and 2D LC-MS/MS with a good depth of coverage, the reproducibility was poor; moreover, the protocol is particularly laborious and requires more time than the others proposed. Protein sample preparation for mass spectrometry is a key step for a good metaproteomics approach; in gel-free metaproteomic workflows, protein purification is critical and starts with a precipitation stage. Precipitation in TCA is the most used and induces the formation of stable molten protein globules, which are difficult to resolve in aqueous buffer for proteolytic digestion before the MS analysis. To overcome this problem Eddhif and colleagues proposed a washing step of TCA-precipitated proteins in ethanol/HCl. This approach, tested on several matrices including soil, allowed an increase in protein recovery in aqueous buffer from 50 to 96% [63,64].

While a limited number of protocols are currently available for the extraction and purification of soil proteins, several protocols for the isolation of soil DNA are available. In these works, the addition of Al^{3+} was proposed and the pH adjustment of the extraction solution to 5.5-6 to effectively remove humic substances [34,65–68]. This soil DNA extraction protocol has been adapted for protein extraction from soil [19], evaluating the most relevant parameters of the protocol, such as buffer pH and detergent concentration, to obtain the maximum recovery of proteins and minimum levels of humic substance co-extraction. To evaluate the performance of this approach, several soil samples with variable physico–chemical characteristics were subjected to the classic extraction with phenol, the NoviPure Soil Protein Kit, and the adapted protocol by Mandalakis et al. [19], that provided the flocculation of humic substances with trivalent aluminum ions. The latter has proven to be a non-toxic, inexpensive, and viable approach for the purification of proteins from soils [19]. In one of the most recent works, an optimization of already cited protocols [21,23], with modifications at several steps, was done to improve the metaproteomic analysis of rhizospheric samples [69]. The use of polyvinylpolypyrrolidone (PVPP) to clean samples from humic acids and liquid nitrogen to minimize proteolysis was proposed again in this protocol. The proteins were extracted with a double extraction, first with 10% TCA/acetone and then with phenol (pH 8.0) and SDS buffer. Although the mass spectrometry proteomics data showed a total of 696 proteins, the images on the SDS-PAGE showed a poor quantity and quality of the bands obtained [69].

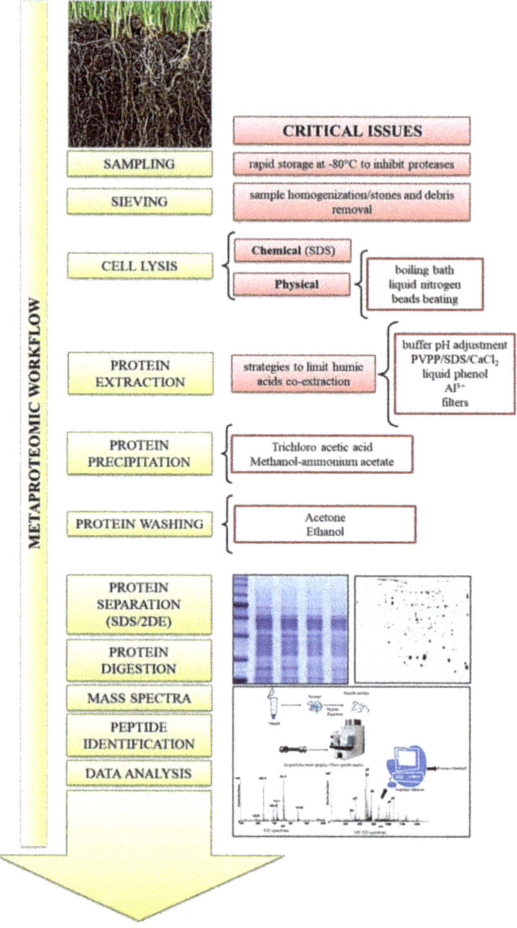

Figure 1. Metaproteomic workflow and critical issues.

4.1. Kit for Protein Extraction from Soil

Given the potential of the proteomics approach in the analysis of environmental sample functional status and the various critical issues during the protein extraction from complex matrices, numerous protein extraction kits have been developed and are now available. Since 2013, a kit for soil protein extraction has been available (NoviPure® by MO BIO Laboratories, Carlsbad, CA, USA); although this kit does not involve toxic reagents and its use is practical and fast, it is rather expensive and does not allow adapting the extractive protocol to the specific characteristics of the soil under consideration. Despite the limitations, protein extraction is entrusted to the NoviPure Soil Protein Kit only, while in some papers the kit performances are compared to classic extraction techniques (Table 1). From the comparison, the use of the kit seems to provide comparable results, if not inferior in quantitative and qualitative terms, compared to traditional extraction techniques [19,54,70–72].

Table 1. Use of the NoviPure Soil Protein kit, comparison with other extraction techniques, and the starting matrix.

Reference	Matrix	NoviPure Soil Protein Kit/Comparative Extraction Techniques
Hansen et al. [71]	Activated Sludge	Kuhn et al. [23] Barr et al. [45] optimized Chourey et al. [22] optimized NoviPure Soil Protein Kit
Butterfield et al. [73]	Grassland sub-root soil	NoviPure Soil Protein Kit + Amicon®Ultra-4 Centrifugal Filter Units (30 KDa)
Mattarozzi et al. [54]	Serpentine soil	Chourey et al. [46] Keiblinger et al. [16] NoviPure Soil Protein Kit
Hori et al. [74]	*Pinus contorta* litter	NoviPure Soil Protein Kit
Mandalakis et al. [19]	Agricultural surface soil	Phenol-based extraction Optimized Al3+ -based method NoviPure Soil Protein Kit TRIzol
Cheng et al. [70]	Stony Corals	Phenol-based extraction Trichloroacetic acid (TCA)-acetone; Glass bead-assisted extraction NoviPure Soil Protein Kit
Yao et al. [75]	Tropical Soil	NoviPure Soil Protein Kit modified by Butterfield et al. [73]
Bona et al. [76]	*Vitis vinifera* rhizosphere	NoviPure Soil Protein Kit
Zhou et al. [77]	Entisol	NoviPure Soil Protein Kit
Ouyang et al. [78]	Vegetable garden surface soil	NoviPure Soil Protein Kit
Mattarozzi et al. [72]	Rhizospheric maize soil	Benndorf et al. [21] NoviPure Soil Protein Kit

4.2. Co-Extraction for Multiomics Approach

In order to obtain significant data from the "omics" approaches, DNA, RNA, proteins, and/or metabolites must be extracted simultaneously from the same sample with a co-extraction protocol that allows the extraction of all the biomolecules of interest without altering them [79]. The method is based on the co-extraction of DNA, RNA, and proteins [80], which provides for the use of phenol-chloroform, which has been taken up and adapted in various works to allow the co-extraction from soil samples with sufficient yields to allow downstream analysis [81–83]. Given the potential of the multiomic approach for soil samples, to reduce time and costs, Nicora et al. [84] proposed a 3-in-1 method for the simultaneous extraction of metabolites, proteins, and lipids (MPLEx) protocol [84]. The protocol, previously used only for lipid extraction, involves the use of organic solvents, chloroform, methanol, and water. Chloroform is not miscible with water and allows the three-phase chemical separation of the components: the aqueous phase contains the hydrophilic metabolites, the interface the proteins, and in the lower chloroform phase the lipid layer. The MPLEx protocol can be used on most soils; however, in highly organic soils such as peat, soil debris remains in the intermediate layer, negatively affecting protein extraction [84].

5. Metaproteomics Applications

Among many other factors, root exudates modulate rhizosphere microbial activity, actively take part in the geochemical and nutrient cycles in the soil, and in the decomposition of organic matter and mineralization processes, affecting the plant growth and health status. Furthermore, given the ability of some microorganisms to biodegrade or detoxify polluting substances accumulated in soils, they play a fundamental role in soil bioremediation, and metaproteomics can reveal new pathways and populations involved in bioremediation. Most of these bacteria, very active in the soil, are not culturable, but vary greatly in quantity and functionality in the presence of host plants. Given the urgent need to clean up soils polluted by human activities, the search for effective and easily applicable strategies is a topic of great interest. Consequently, the metaproteomics approach, with protein extraction directly from rhizospheric soil samples, has wide potential applications (Figure 2; Table 2), being able to provide not only taxonomic but also functional information, providing a complete picture of functional–taxonomic relationships in soil in a given time [85].

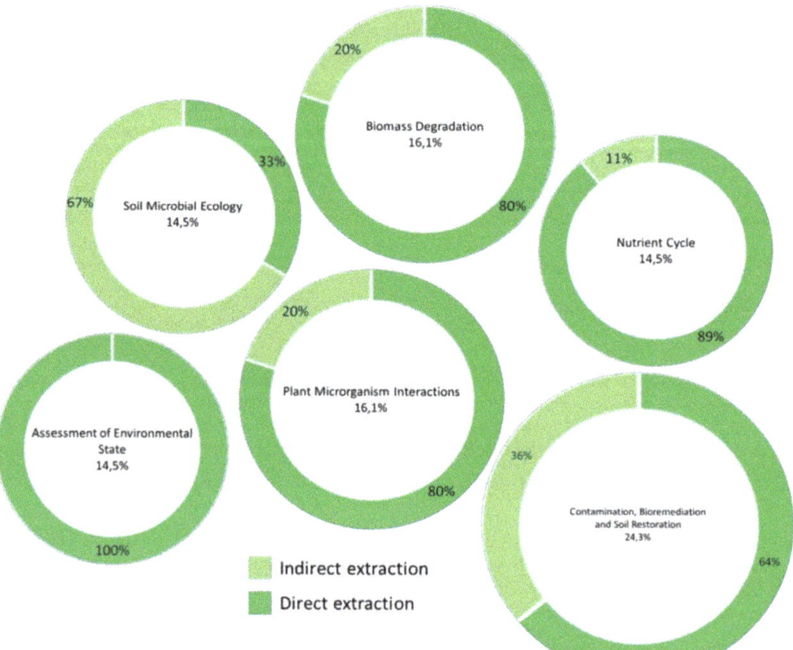

Figure 2. Percentage of metaproteomics applications out of the total number of works taken into consideration with frequency of direct and indirect approaches in extraction.

Table 2. Examples of possible application areas of metaproteomics approaches for soil.

Nutrient Cycle	Biomass Degradation	Soil Microbial Ecology	Plant-Microorganism Interactions	Assessment of Environmental State	Contamination, Bioremediation, and Soil Restoration
Bastida et al., (2016) [48]	Aylward et al., (2012) [86]	Bastida et al., (2016 a) [87]	Bao et al., (2014) [88]	Bastida et al., (2014) [47]	Bastida et al., (2015a,b) [89,90]
Bastida et al., (2019) [49]	Butterfield et al., (2016) [73]	Fernandez-Martinez et al., (2019) [91]	Bona et al., (2019) [76]	Bastida et al., (2018) [57]	Benndorf et al., (2009) [92]
Canizares et al., (2011) [93]	Hori et al., (2018) [74]	Festa et al., (2017) [94]	Knief et al., (2012) [95]	Chen et al., (2019) [96]	Chen et al., (2020) [97]
Chen et al., (2019) [96]	Keiblinger et al., (2012 b) [98]	Maron et al., (2008) [99]	Lin et al., (2013) [51]	Liu et al., (2017) [55]	Chourey et al., (2013) [100]
Orellana et al., (2019) [101]	Liu et al., (2015) [102]	Martinez-Alonso et al., (2019) [103]	Manikandan et al., (2017) [104]	Liu et al., (2019) [105]	Christensen et al., (2019) [106]
Starke et al., (2016) [107]	Schneider et al., (2012) [108]	Mattarozzi et al., (2020) [72]	Renu et al., (2019) [69]	Starke (2017) [109]	Guazzaroni et al., (2013) [110]
Tan et al., (2019) [111]	Zhang et al., (2018) [112]	Sekhon et al., (2016) [113]		Wang et al., (2017) [114]	Halter et al., (2011) [115]
Yao et al., (2018) [75]		Sidibè et al., (2016) [116]	Wu et al., (2011) [117]		Lechner et al., (2018) [118]
Zecchin et al., (2018) [119]		Williams et al., (2010) [120]	Zampieri et al., (2016) [11]		Lünsmann et al., (2016) [121]
					Mattarozzi et al., (2017) [54]
					Ouyang et al., (2019) [78]
					Singleton et al., (2003) [38]
					Sukul et al., (2017) [122]

In this review, we focused on metaproteomics applications related to the interactions between plants and microorganisms in the rhizosphere, and the presence of contaminants and bioremediation processes.

5.1. Plant–Microorganism Interactions

Soil rhizospheric metaproteomics is a promising tool to understand the complex plant–microorganism interactions and the rhizosphere in situ biological/functional properties and how they can adapt to environmental factors. In a metaproteomics study on rice crop phyllosphere and rhizosphere, Knief et al. [95] used the All Prep DNA/RNA/Protein MiniKit (Qiagen) and identified 4600 total proteins. The comparison among differentially expressed proteins between the rhizosphere and phyllosphere showed that those involved in nitrogen fixation and methanol use processes were particularly abundant in the rhizosphere sample [117]. In the rice rhizosphere, Wu et al. [117] identified 122 proteins, mainly involved in energy and secondary metabolism, protein and nucleotide biosynthesis, signal transduction, and resistance. In order to evaluate how the consecutive *Rehmannia glutinosa* monoculture alters the rhizosphere, the results of the metaproteomics analysis of rhizosphere samples taken in subsequent years were compared. Of the 152 spots obtained by 2-DE, only 49 proteins were identified by LIFT-MALDI TOF-TOF-MS, and of these, 33 were differentially expressed. According to the KEGG (Kyoto Encyclopedia of Genes and Genomes) database, these were mainly involved in primary metabolism, in secondary metabolism, and in the response to stress [117]. The metaproteomics approach was used in a later work by Lin et al. [51] to evaluate the effect of yield decline in ratoon sugarcane. Of the 109 protein spots identified by LC-MALDI TOF/TOF, only 33 were differentially expressed. Of these proteins, those derived from plants were mainly involved in primary metabolism and stress responses, while microbial ones were mainly involved in membrane transport and signal transduction [51]. Bao et al. [88] combined a metaproteomic study with catalyzed reporter deposition–fluorescence in situ hybridization (CARD-FISH) to analyze methanotrophs localized in rice root tissues. Further, the metaproteomics approach has been used on a truffle-ground soil already analyzed from a metagenomics point of view [88]. These authors identified proteins from fungi, bacteria, and Viridiplantae and analyzed their function according to the KEGG database [88]. Microbial activities in the rhizosphere affect plant health; consequently, the composition and functions of the soil biome are fundamental for the development and engineering of sustainable agriculture. Manikandan et al. [104] compared protein expression in wilt-infected and healthy tomato rhizospheres. The protein extracts were identified and were derived from plants, bacteria, and fungi. Among these, 11 proteins were differentially expressed and were involved in plant defense responses [104]. Starke et al., [107], with a combination of genomic and proteomics approaches, identified the bacteria involved in the short-term assimilation and tracked the flow of plant-derived N. Interestingly, this study provides one of the first insights into the combined use of protein and stable isotope probing (SIP) in soil rhizosphere to reveal the involvement of microbial populations and functional pathways in relation to the nitrogen cycle [107]. Moreover, in order to identify the repertoire of secreted microbial proteins and use to cooperate or compete and gain information about the interaction pathways between roots and rhizosphere communities, Bona et al. [89] focused attention on the metaproteomics analysis of the rhizosphere of *Vitis vinifera* cv. Pinot Noir. The extracted protein analysis allowed the identification of 150 bacterial genera and the understanding of the functional state of the rhizosphere [89]. In 2019, the first metaproteomics study on the maize rhizosphere was reported and identified 696 proteins with different functions from 393 different species. Most of these proteins belonged to three functional groups: carbon metabolism, glycine biosynthesis, and IMP (Inosine Mono Phosphate) biosynthesis [69].

5.2. Rhizosphere Under Environmental Factors

Several works in literature aimed to elucidate how the natural environmental conditions or their alterations influence and alter the rhizosphere. The long-term effects of deforestation on soil microbial communities have been explored through protein extraction from soil, carried out according to [22], allowing the identification of 715 proteins in a deforested soil sample and 472 from a forest soil sample [89]. The taxonomic and functional analyses allowed the conclusion that if on the one hand deforestation causes a loss of total microbial biomass; on the other, it induces an increase in diversity [89]. In a subsequent work by Liu et al. [55], a temperature rise of +4 °C in a sample of forest soil was applied in order to quantify the effects of global warming on the rhizosphere. The analysis of the extracted proteins allowed the establishment that the bacterial proteins far exceeded those of a fungal nature, and that the differences after the exposure to the increase in temperature were few in taxonomic terms, confirming the ability to adapt and microbial resilience, while from a functional point of view, there was a marked increase in the proteins involved in soil respiration. Confirming the crucial importance of climatic conditions in determining the abundance of soil microbial proteins and the metabolic activities of the main phyla present there, Bastida and colleagues assessed the metaproteome of soils from different climatic zones associated with distinct vegetation patterns and under an increasing temperature [123,124]. The results showed that the increase in temperature corresponded to an increase in the proteins derived from Actinobacteria and Planctomycetes, and decrease of those of Proteobacteria. From a functional point of view, on the other hand, all proteins related to cellular energetic processes and soil respiration were over-expressed [123]. The protocol proposed by Chourey et al. [22], suitable for semiarid soils, was used to evaluate the effects of water/salinity stress on the soil microbial community of a grapefruit agro-ecosystem model. The metaproteomics approach made it possible to identify functional changes and adaptations to stress in the pool of microorganisms present in the rhizosphere whose phylogenetic diversity remained constant [109].

5.3. Contaminants, Bioremediation, and Soil Restoration

The possibility of using soil proteomic analysis as an indicator of cadmium contamination was proposed by Singleton and colleagues in a 2003 paper that highlighted the reduction in the number of total proteins extracted from soil samples containing cadmium. However, no subsequent analysis was performed on the protein extracts obtained [38]. Benndorf and colleagues, modifying their own previews of protein extraction protocols, analyzed the metaproteome of soil's anoxic microbial communities and their ability to biodegrade petroleum hydrocarbons, especially benzene that is particularly recalcitrant in the absence of oxygen. Through the separation of cells from soil granules, cell lysis by ultrasonication, and purification of proteins by phenol extraction, 240 protein spots on 2D-gels were found, and several proteins were successfully identified. These included some enzymes, such as enoyl-CoA hydratase, involved in the anoxic degradation of xenobiotics [92]. Combining different extraction techniques, as previously seen, the metabolic and functional effects in metal-tolerant and metal hyperaccumulator plants, growing in soil contaminated by heavy metals (Ni, Co, Cr), were analyzed [54]. Of the 800 proteins identified, most were involved in stress response or metal uptake, highlighting the ability of the rhizosphere to plastically adapt to microenvironment conditions [54]. Christensen et al., combined metaproteomics and metagenomics to evaluate the presence, distribution, and diversity of Hg-methylating bacteria, responsible for the microbial conversion in the soil of methyl mercury (MeHg), a toxic contaminant [106]. Chourey et al. [100], using an improvement of their previously published protocol validated the usefulness of the metaproteomics approach, identifying the main microbial communities and the pathways activated following contamination by uranium and nitrate. Metagenomics and metaproteomics approaches were applied by Guazzaroni and colleagues to reconstruct the degradation pathway of naphthalene. The results of the multiomics approach indicated the biodegradation capacity of microbial communities, which are extremely suitable for naphthalene. Furthermore, the bio-stimulation of polluted mesocosms induced the release of recalcitrant compounds from the soils that seemed to remodel the composition in terms of the variety and abundance of the soil microbial community [110]. In a recent work, Chen et al. [97] analyzed the effect of dibutyl phthalate on the microbiome of agricultural soils, demonstrating through metabolomics, metaproteomics, and enzyme activity assays, the presence of this compound in the soil induces an up-regulation of a series of microbial proteins involved in the metabolism and transport of sugars. Consequently, there was a clear decrease in glucose in the soil that could lead to negative effects on the rhizosphere [97]. In a 2016 paper, Bastida et al. [48] evaluated the bioremediation effect of compost amendment in hydrocarbon-contaminated soils. Metaproteomics showed that the addition of compost, inducing an increase in microbial catabolic activity, promoted a net reduction of polycyclic aromatic hydrocarbons [48]. The same authors pointed out the correlation between the organic amendment and phyla-lifestyles, as well as the main biochemical degradation pathways in this polluted soil. By combining the metagenomics and metaproteomics approaches, the biocatalytic potential of the microbial community in a soil contaminated with used cooking oil was evaluated, confirming its lipolytic activity [122].

6. Conclusions

Soil metaproteomics is a research field of great interest given the information obtained from it, not only to understand the physiological and functional states of the interaction between microorganisms and plants present in the rhizosphere, but also to be able to "engineer" the rhizosphere, modifying the microbial pool and/or resident plants, in response to environmental stress, climate change, and in order to promote the bioremediation of contaminated soils. Indeed, more than any other approach, the analysis of the rhizosphere metaproteome allows us to evaluate the responses of the rhizosphere microbiome to environmental changes and also to determine how the rhizosphere can respond to targeted engineering interventions to enhance its capabilities. To evaluate the phylogenetic and functional correlations of a contaminated soil, the proteomics approach represents a valid means to test the effective potential of different bioremediation and soil restoration approaches. In spite of the limitations of metaproteomics, which are certainly shared by many other soil approaches based on the extraction of biomarkers in soil (i.e., DNA or RNA extraction, fatty acids, etc.), the evolution of this field has been enormous in recent years and is providing important insights into the relationships between the phylogeny and functionality of microbial communities. One of the major limitations to the progress of the metaproteomics approach for soil is the lack of a universal extraction protocol. More effort should be addressed to improve protein extraction methods while it is acknowledged that extensive soil heterogeneity in chemical, physical, and biological properties makes this task difficult. The first major distinction between the extraction methods proposed over the years is that between direct and indirect extractions. In the direct one, cell lysis occurs directly in the native soil sample. This guarantees a greater understanding of the interactions present in the rhizosphere but is more influenced by the matrix characteristics and is subject to the interfering substances co-extraction. The indirect method involves the physical separation of microbial cells from the soil and only after cell lysis and protein extraction. Although in a way the problem of interfering substances is solved, there is the risk of isolating only some microbial species, losing those strictly adhering to soil particles or that are non-culturable, and the whole pool of proteins released by the roots and microorganisms in the rhizosphere. Therefore, in order to fully exploit the enormous potential of the metaproteomics approach for soil, it is necessary to develop and use a type of direct and in-situ extraction. As seen in the literature, the various protocols used to date for cell lysis and direct and indirect extractions from the soil significantly affect the quality and quantity of extracted proteins. In the absence of an optimal protocol, suitable for all soil types, a strategy to solve the problem could be to combine several extractions, including the pool of extracted proteins, before subsequent analysis. Another important aspect to improve is the quantification of the extracted proteins. Several studies concluded that colorimetric methods are not valid, and thus alternative approaches (i.e., quantifying amino acids after hydrolysis) must be taken into consideration to quantify and normalize the amount of proteins to be used in downstream electrophoresis and mass-spectrometry analyses. Finally, improvement of the overall metaproteomic workflow, not only the protein extraction methods, but also protein/peptide separation, mass spectrometry approaches, and bioinformatics coupled to metagenomics, will provide a more complete picture of the microbially mediated processes taking place in soil, also including the identification of low abundant proteins which are of paramount importance for soil sustainability and bioremediation.

Supplementary Materials: The following are available online at http://www.mdpi.com/1422-0067/21/22/8455/s1.

Funding: F. Bastida is grateful to the Spanish Ministry of Science, Innovation and Universities and FEDER funds for the project AGL2017–85755-R.

Conflicts of Interest: The authors declare no conflict of interest.

References

1. Bakker, P.A.H.M.; Berendsen, R.L.; Doornbos, R.F.; Wintermans, P.C.A.; Pieterse, C.M.J. The rhizosphere revisited: Root microbiomics. *Front. Plant Sci.* **2013**, *4*, 165. [CrossRef] [PubMed]
2. McNear, D.H., Jr. The Rhizosphere—Roots, Soil and Everything In Between. *Nat. Educ. Knowl.* **2013**, *4*, 1.
3. Vives-Peris, V.; De Ollas, C.; Gómez-Cadenas, A.; Pérez-Clemente, R.M. Root exudates: From plant to rhizosphere and beyond. *Plant Cell Rep.* **2020**, *39*, 3–17. [CrossRef] [PubMed]
4. Bais, H.P.; Weir, T.L.; Perry, L.G.; Gilroy, S.; Vivanco, J.M. The role of root exudates in rhizosphere interactions with plants and other organisms. *Annu. Rev. Plant Biol.* **2006**, *57*, 233–266. [CrossRef] [PubMed]
5. Saleh, D.; Sharma, M.; Seguin, P.; Jabaji, S. Organic acids and root exudates of *Brachypodium distachyon*: Effects on chemotaxis and biofilm formation of endophytic bacteria. *Can. J. Microbiol.* **2020**, *66*, 562–575. [CrossRef] [PubMed]
6. Olanrewaju, O.S.; Ayangbenro, A.S.; Glick, B.R.; Babalola, O.O. Plant health: Feedback effect of root exudates-rhizobiome interactions. *Appl. Microbiol. Biotechnol.* **2019**, *103*, 1155–1166. [CrossRef]
7. White, R.A.; Rivas-Ubach, A.; Borkum, M.I.; Köberl, M.; Bilbao, A.; Colby, S.M.; Hoyt, D.W.; Bingol, K.; Kim, Y.-M.; Wendler, J.P.; et al. The state of rhizospheric science in the era of multi-omics: A practical guide to omics technologies. *Rhizosphere* **2017**, *3*, 212–221. [CrossRef]
8. Carini, P.; Marsden, P.J.; Leff, J.W.; Morgan, E.; Strickland, M.S.; Fierer, N. Relic DNA is abundant in soil and obscures estimates of soil microbial diversity. *Nat. Microbiol.* **2017**, *2*, 16242. [CrossRef]
9. Prosser, J.I. Dispersing misconceptions and identifying opportunities for the use of 'omics' in soil microbial ecology. *Nat. Rev. Genet.* **2015**, *13*, 439–446. [CrossRef]
10. Tang, H.; Li, S.; Ye, Y. A Graph-Centric Approach for Metagenome-Guided Peptide and Protein Identification in Metaproteomics. *PLoS Comput. Biol.* **2016**, *12*, e1005224. [CrossRef]
11. Zampieri, E.; Chiapello, M.; Daghino, S.; Bonfante, P.; Mello, A. Soil metaproteomics reveals an inter-kingdom stress response to the presence of black truffles. *Sci. Rep.* **2016**, *6*, 25773. [CrossRef] [PubMed]
12. Gutleben, J.; De Mares, M.C.; Van Elsas, J.D.; Smidt, H.; Overmann, J.; Sipkema, D. The multi-omics promise in context: From sequence to microbial isolate. *Crit. Rev. Microbiol.* **2018**, *44*, 212–229. [CrossRef] [PubMed]
13. Li, S.; Tang, H.; Ye, Y. A Meta-proteogenomic Approach to Peptide Identification Incorporating Assembly Uncertainty and Genomic Variation. *Mol. Cell. Proteom.* **2019**, *18*, S183–S192. [CrossRef] [PubMed]
14. Bastida, F.; Moreno, J.L.; Nicolás, C.; Hernandez, T.; Garcia, C. Soil metaproteomics: A review of an emerging environmental science. Significance, methodology and perspectives. *Eur. J. Soil Sci.* **2009**, *60*, 845–859. [CrossRef]
15. Abiraami, T.V.; Singh, S.; Nain, L. Soil metaproteomics as a tool for monitoring functional microbial communities: Promises and challenges. *Rev. Environ. Sci. Bio Technol.* **2019**, *19*, 73–102. [CrossRef]
16. Keiblinger, K.M.; Wilhartitz, I.C.; Schneider, T.; Roschitzki, B.; Schmid, E.; Eberl, L.; Riedel, K.; Zechmeister-Boltenstern, S. Soil metaproteomics—Comparative evaluation of protein extraction protocols. *Soil Biol. Biochem.* **2012**, *54*, 14–24. [CrossRef]
17. Arenella, M.; Giagnoni, L.; Masciandaro, G.; Ceccanti, B.; Nannipieri, P.; Renella, G. Interactions between proteins and humic substances affect protein identification by mass spectrometry. *Biol. Fertil. Soils* **2013**, *50*, 447–454. [CrossRef]
18. Qian, C.; Hettich, R.L. Optimized Extraction Method To Remove Humic Acid Interferences from Soil Samples Prior to Microbial Proteome Measurements. *J. Proteome Res.* **2017**, *16*, 2537–2546. [CrossRef]
19. Mandalakis, M.; Panikov, N.S.; Polymenakou, P.N.; Sizova, M.V.; Stamatakis, A. A simple cleanup method for the removal of humic substances from soil protein extracts using aluminum coagulation. *Environ. Sci. Pollut. Res.* **2018**, *25*, 23845–23856. [CrossRef]
20. Starke, R.; Jehmlich, N.; Bastida, F. Using proteins to study how microbes contribute to soil ecosystem services: The current state and future perspectives of soil metaproteomics. *J. Proteom.* **2019**, *198*, 50–58. [CrossRef]

21. Benndorf, D.; Balcke, G.U.; Harms, H.; Von Bergen, M. Functional metaproteome analysis of protein extracts from contaminated soil and groundwater. *ISME J.* **2007**, *1*, 224–234. [CrossRef] [PubMed]

22. Chourey, K.; Jansson, J.; Verberkmoes, N.; Shah, M.; Chavarria, K.L.; Tom, L.M.; Brodie, E.L.; Hettich, R.L. Direct Cellular Lysis/Protein Extraction Protocol for Soil Metaproteomics. *J. Proteome Res.* **2010**, *9*, 6615–6622. [CrossRef] [PubMed]

23. Kuhn, R.; Benndorf, D.; Rapp, E.; Reichl, U. Metaproteome analysis of sewage sludge from membrane bioreactors. *Proteomics* **2011**, *11*, 2738–2744. [CrossRef] [PubMed]

24. Giagnoni, L.; Magherini, F.; Landi, L.; Taghavi, S.; Modesti, A.; Bini, L.M.; Nannipieri, P.; Van Der Lelie, D.; Renella, G. Extraction of microbial proteome from soil: Potential and limitations assessed through a model study. *Eur. J. Soil Sci.* **2011**, *62*, 74–81. [CrossRef]

25. Keiblinger, K.M.; Fuchs, S.; Zechmeister-Boltenstern, S.; Riedel, K. Soil and leaf litter metaproteomics—A brief guideline from sampling to understanding. *FEMS Microbiol. Ecol.* **2016**, *92*. [CrossRef]

26. Nannipieri, P. Role of Stabilised Enzymes in Microbial Ecology and Enzyme Extraction from Soil with Potential Applications in Soil Proteomics. *Nucl. Acids Proteins Soil* **2006**, *8*, 75–94. [CrossRef]

27. Chen, S.; Rillig, M.C.; Wang, W. Improving soil protein extraction for metaproteome analysis and glomalin-related soil protein detection. *Proteomics* **2009**, *9*, 4970–4973. [CrossRef]

28. Huang, Q.; Wang, Y.; Li, B.; Chang, J.; Chen, M.; Li, K.; Yang, G.; He, G. TaNAC29, a NAC transcription factor from wheat, enhances salt and drought tolerance in transgenic Arabidopsis. *BMC Plant Biol.* **2015**, *15*, 1–15. [CrossRef]

29. Taylor, E.B.; Williams, M.A. Microbial Protein in Soil: Influence of Extraction Method and C Amendment on Extraction and Recovery. *Microb. Ecol.* **2009**, *59*, 390–399. [CrossRef]

30. Criquet, S.; Farnet, A.; Ferre, E. Protein measurement in forest litter. *Biol. Fertil. Soils* **2002**, *35*, 307–313. [CrossRef]

31. Jin, P.; Song, J.; Yang, L.; Jin, X.; Wang, X.C. Selective binding behavior of humic acid removal by aluminum coagulation. *Environ. Pollut.* **2018**, *233*, 290–298. [CrossRef] [PubMed]

32. Burns, R.G.; Deforest, J.L.; Marxsen, J.; Sinsabaugh, R.L.; Stromberger, M.E.; Wallenstein, M.D.; Weintraub, M.N.; Zoppini, A. Soil enzymes in a changing environment: Current knowledge and future directions. *Soil Biol. Biochem.* **2013**, *58*, 216–234. [CrossRef]

33. Ochs, D. Protein contaminants of sodium dodecyl sulfate-polyacrylamide gels. *Anal. Biochem.* **1983**, *135*, 470–474. [CrossRef]

34. Bell, A.W.; HUPO Test Sample Working Group; Deutsch, E.W.; Au, C.E.; Kearney, R.E.; Beavis, R.; Sechi, S.; Nilsson, T.; Bergeron, J.J.M.; Beardslee, T.A.; et al. A HUPO test sample study reveals common problems in mass spectrometry–based proteomics. *Nat. Methods* **2009**, *6*, 423–430. [CrossRef]

35. Chiapello, M.; Zampieri, E.; Mello, A. A Small Effort for Researchers, a Big Gain for Soil Metaproteomics. *Front. Microbiol.* **2020**, *11*, 88. [CrossRef]

36. Ogunseitan, O.A. Direct extraction of proteins from environmental samples. *J. Microbiol. Methods* **1993**, *17*, 273–281. [CrossRef]

37. Murase, A.; Yoneda, M.; Ueno, R.; Yonebayashi, K. Isolation of extracellular protein from greenhouse soil. *Soil Biol. Biochem.* **2003**, *35*, 733–736. [CrossRef]

38. Singleton, I.; Merrington, G.; Colvan, S.; Delahunty, J. The potential of soil protein-based methods to indicate metal contamination. *Appl. Soil Ecol.* **2003**, *23*, 25–32. [CrossRef]

39. Schulze, W.X.; Gleixner, G.; Kaiser, K.; Guggenberger, G.; Mann, M.; Schulze, E.-D. A proteomic fingerprint of dissolved organic carbon and of soil particles. *Oecologia* **2004**, *142*, 335–343. [CrossRef]

40. Wang, W.; Scali, M.; Vignani, R.; Spadafora, A.; Sensi, E.; Mazzuca, S.; Cresti, M. Protein extraction for two-dimensional electrophoresis from olive leaf, a plant tissue containing high levels of interfering compounds. *Electrophoresis* **2003**, *24*, 2369–2375. [CrossRef]

41. Kirby, K.S. A new method for the isolation of ribonucleic acids from mammalian tissues. *Biochem. J.* **1956**, *64*, 405–408. [CrossRef] [PubMed]

42. Masciandaro, G.; Macci, C.; Doni, S.; Maserti, B.; Leo, A.C.-B.; Ceccanti, B.; Wellington, E. Comparison of extraction methods for recovery of extracellular β-glucosidase in two different forest soils. *Soil Biol. Biochem.* **2008**, *40*, 2156–2161. [CrossRef]

43. Masciandaro, G.; Ceccanti, B.; Gallardo, J. Organic matter properties in cultivated versus set-aside arable soils. *Agric. Ecosyst. Environ.* **1998**, *67*, 267–274. [CrossRef]

44. Hofman, J.; Dušek, L. Biochemical analysis of soil organic matter and microbial biomass composition—A pilot study. *Eur. J. Soil Biol.* **2003**, *39*, 217–224. [CrossRef]

45. Barr, J.J.; Hastie, M.L.; Fukushima, T.; Plan, M.R.; Tyson, G.; Gorman, J.J.; Bond, P.L. Metaproteomic analysis of laboratory scale phosphorus removal reactors reveals functional insights of aerobic granular sludge. IWA Biofilm Conference: Processes in Biofilms, 27–30 October 2011, Shanghai, China.

46. Chourey, K.; Hettich, R.L. Utilization of a Detergent-Based Method for Direct Microbial Cellular Lysis/Proteome Extraction from Soil Samples for Metaproteomics Studies. *Adv. Struct. Safety Stud.* **2018**, 293–302. [CrossRef]

47. Bastida, F.; Hernández, T.; García, C. Metaproteomics of soils from semiarid environment: Functional and phylogenetic information obtained with different protein extraction methods. *J. Proteom.* **2014**, *101*, 31–42. [CrossRef]

48. Bastida, F.; Jehmlich, N.; Lima, K.; Morris, B.; Richnow, H.; Hernández, T.; Von Bergen, M.; García, C. The ecological and physiological responses of the microbial community from a semiarid soil to hydrocarbon contamination and its bioremediation using compost amendment. *J. Proteom.* **2016**, *135*, 162–169. [CrossRef]

49. Bastida, F.; Jehmlich, N.; Martínez-Navarro, J.; Bayona, V.; García, C.; Moreno, J. The effects of struvite and sewage sludge on plant yield and the microbial community of a semiarid Mediterranean soil. *Geoderma* **2019**, *337*, 1051–1057. [CrossRef]

50. Wang, H.-B.; Zhang, Z.-X.; Li, H.; He, H.-B.; Fang, C.-X.; Zhang, A.-J.; Li, Q.-S.; Chen, R.-S.; Guo, X.-K.; Lin, H.-F.; et al. Characterization of Metaproteomics in Crop Rhizospheric Soil. *J. Proteome Res.* **2011**, *10*, 932–940. [CrossRef]

51. Lin, W.; Wu, L.; Lin, S.; Zhang, A.; Zhou, M.; Lin, R.; Wang, H.; Chen, J.; Zhang, Z.; Lin, R. Metaproteomic analysis of ratoon sugarcane rhizospheric soil. *BMC Microbiol.* **2013**, *13*, 1–13. [CrossRef]

52. Tomaszewski, J.E.; Schwarzenbach, R.P.; Sander, M. Protein Encapsulation by Humic Substances. *Environ. Sci. Technol.* **2011**, *45*, 6003–6010. [CrossRef] [PubMed]

53. Nicora, C.D.; Anderson, B.J.; Callister, S.J.; Norbeck, A.D.; Purvine, S.O.; Jansson, J.K.; Mason, O.U.; David, M.M.; Jurelevicius, D.; Smith, R.D.; et al. Amino acid treatment enhances protein recovery from sediment and soils for metaproteomic studies. *Proteomics* **2013**, *13*, 2776–2785. [CrossRef] [PubMed]

54. Mattarozzi, M.; Manfredi, M.; Montanini, B.; Gosetti, F.; Sanangelantoni, A.M.; Marengo, E.; Careri, M.; Visioli, G. A metaproteomic approach dissecting major bacterial functions in the rhizosphere of plants living in serpentine soil. *Anal. Bioanal. Chem.* **2017**, *409*, 2327–2339. [CrossRef] [PubMed]

55. Liu, N.; Keiblinger, K.M.; Schindlbacher, A.; Wegner, U.; Sun, H.; Fuchs, S.; Lassek, C.; Riedel, K.; Zechmeister-Boltenstern, S. Microbial functionality as affected by experimental warming of a temperate mountain forest soil—A metaproteomics survey. *Appl. Soil Ecol.* **2017**, *117*, 196–202. [CrossRef]

56. Speda, J.; Johansson, M.A.; Carlsson, U.; Karlsson, M. Assessment of sample preparation methods for metaproteomics of extracellular proteins. *Anal. Biochem.* **2017**, *516*, 23–36. [CrossRef]

57. Bastida, F.; Jehmlich, N.; Torres, I.F.; García, C. The extracellular metaproteome of soils under semiarid climate: A methodological comparison of extraction buffers. *Sci. Total. Environ.* **2018**, *619*, 707–711. [CrossRef]

58. Tabatabai, M. Soil Enzymes. In *Agronomy Monographs*; Wiley: Hoboken, NJ, USA, 2015; pp. 903–947.

59. Bastida, F.; Algora, C.; Hernández, T.; García, C. Feasibility of a cell separation-proteomic based method for soils with different edaphic properties and microbial biomass. *Soil Biol. Biochem.* **2012**, *45*, 136–138. [CrossRef]

60. Greenfield, L.M.; Hill, P.W.; Paterson, E.; Baggs, E.M.; Jones, D.L. Methodological bias associated with soluble protein recovery from soil. *Sci. Rep.* **2018**, *8*, 1–6. [CrossRef]

61. Folch, J.; Ascoli, I.; Lees, M.; Meath, J.A.; LeBaron, N. Preparation of lipide extracts from brain tissue. *J. Biol. Chem.* **1951**, *191*, 833–841.

62. Callister, S.J.; Fillmore, T.L.; Nicora, C.D.; Shaw, J.B.; Purvine, S.O.; Orton, D.J.; White, R.A.; Moore, R.J.; Burnet, M.C.; Nakayasu, E.S.; et al. Addressing the challenge of soil metaproteome complexity by improving metaproteome depth of coverage through two-dimensional liquid chromatography. *Soil Biol. Biochem.* **2018**, *125*, 290–299. [CrossRef]

63. Eddhif, B.; Lange, J.; Guignard, N.; Batonneau, Y.; Clarhaut, J.; Papot, S.; Geffroy-Rodier, C.; Poinot, P. Study of a novel agent for TCA precipitated proteins washing - comprehensive insights into the role of ethanol/HCl on molten globule state by multi-spectroscopic analyses. *J. Proteom.* **2018**, *173*, 77–88. [CrossRef] [PubMed]

64. Eddhif, B.; Guignard, N.; Batonneau, Y.; Clarhaut, J.; Papot, S.; Geffroy-Rodier, C.; Poinot, P. TCA precipitation and ethanol/HCl single-step purification evaluation: One-dimensional gel electrophoresis, bradford assays, spectrofluorometry and Raman spectroscopy data on HSA, Rnase, lysozyme - Mascots and Skyline data. *Data Brief* **2018**, *17*, 938–953. [CrossRef] [PubMed]
65. Braid, M.D.; Daniels, L.M.; Kitts, C.L. Removal of PCR inhibitors from soil DNA by chemical flocculation. *J. Microbiol. Methods* **2003**, *52*, 389–393. [CrossRef]
66. Dong, D.; Yan, A.; Liu, H.; Zhang, X.; Xu, Y. Removal of humic substances from soil DNA using aluminium sulfate. *J. Microbiol. Methods* **2006**, *66*, 217–222. [CrossRef] [PubMed]
67. Zhang, J.; Dai, J.; Wang, R.; Li, F.; Wang, W. Adsorption and desorption of divalent mercury (Hg2+) on humic acids and fulvic acids extracted from typical soils in China. *Colloids Surf. A Physicochem. Eng. Asp.* **2009**, *335*, 194–201. [CrossRef]
68. Zhang, Y.-R.; Xue, T.; Wang, R.-Q.; Dai, J.-L. FTIR Spectroscopic Structural Characterization of Forest Topsoil Humic Substances and Their Adsorption and Desorption for Mercury. *Soil Sci.* **2013**, *178*, 436–441. [CrossRef]
69. Renu; Gupta, S.K.; Rai, A.K.; Sarim, K.M.; Sharma, A.; Budhlakoti, N.; Arora, D.; Verma, D.K.; Singh, D.P. Metaproteomic data of maize rhizosphere for deciphering functional diversity. *Data Brief* **2019**, *27*, 104574. [CrossRef]
70. Cheng, H.-M.; Zhao, H.; Yang, T.; Ruan, S.; Wang, H.; Xiang, N.; Zhou, H.; Li, Q.X.; Diao, X. Comparative evaluation of five protocols for protein extraction from stony corals (Scleractinia) for proteomics. *Electrophoresis* **2018**, *39*, 1062–1070. [CrossRef]
71. Hansen, S.H.; Stensballe, A.; Nielsen, P.H.; Herbst, F.-A. Metaproteomics: Evaluation of protein extraction from activated sludge. *Proteomics* **2014**, *14*, 2535–2539. [CrossRef]
72. Mattarozzi, M.; Di Zinno, J.; Montanini, B.; Manfredi, M.; Marengo, E.; Fornasier, F.; Ferrarini, A.; Careri, M.; Visioli, G. Biostimulants applied to maize seeds modulate the enzymatic activity and metaproteome of the rhizosphere. *Appl. Soil Ecol.* **2020**, *148*, 103480. [CrossRef]
73. Butterfield, C.N.; Li, Z.; Andeer, P.F.; Spaulding, S.; Thomas, B.C.; Singh, A.; Hettich, R.L.; Suttle, K.B.; Probst, A.J.; Tringe, S.G.; et al. Proteogenomic analyses indicate bacterial methylotrophy and archaeal heterotrophy are prevalent below the grass root zone. *PeerJ.* **2016**, *4*, e2687. [CrossRef] [PubMed]
74. Hori, C.; Gaskell, J.; Cullen, D.; Sabat, G.; Stewart, P.E.; Lail, K.; Peng, Y.; Barry, K.; Grigoriev, I.V.; Kohler, A.; et al. Multi-omic Analyses of Extensively DecayedPinus contortaReveal Expression of a Diverse Array of Lignocellulose-Degrading Enzymes. *Appl. Environ. Microbiol.* **2018**, *84*, e01133-18. [CrossRef] [PubMed]
75. Yao, Q.; Li, Z.; Song, Y.; Wright, S.J.; Guo, X.; Tringe, S.G.; Tfaily, M.M.; Paša-Tolić, L.; Hazen, T.C.; Turner, B.L.; et al. Community proteogenomics reveals the systemic impact of phosphorus availability on microbial functions in tropical soil. *Nat. Ecol. Evol.* **2018**, *2*, 499–509. [CrossRef] [PubMed]
76. Bona, E.; Massa, N.; Novello, G.; Boatti, L.; Cesaro, P.; Todeschini, V.; Magnelli, V.; Manfredi, M.; Marengo, E.; Mignone, F.; et al. Metaproteomic characterization of Vitis vinifera rhizosphere. *FEMS Microbiol. Ecol.* **2019**, *95*, 1–16. [CrossRef] [PubMed]
77. Zhou, Z.-F.; Wei, W.-L.; Shi, X.-J.; Liu, Y.-M.; He, X.-H.; Wang, M.-X. Twenty-six years of chemical fertilization decreased soil RubisCO activity and changed the ecological characteristics of soil cbbL-carrying bacteria in an entisol. *Appl. Soil Ecol.* **2019**, *141*, 1–9. [CrossRef]
78. Ouyang, W.-Y.; Su, J.-Q.; Richnow, H.H.; Adrian, L. Identification of dominant sulfamethoxazole-degraders in pig farm-impacted soil by DNA and protein stable isotope probing. *Environ. Int.* **2019**, *126*, 118–126. [CrossRef]
79. Roume, H.; Heintz-Buschart, A.; Muller, E.E.; Wilmes, P. Sequential Isolation of Metabolites, RNA, DNA, and Proteins from the Same Unique Sample. In *Methods in Enzymology*; Elsevier BV: Amsterdam, The Netherlands, 2013; Volume 531, pp. 219–236.
80. Griffiths, R.I.; Whiteley, A.S.; O'Donnell, A.G.; Bailey, M.J. Rapid Method for Coextraction of DNA and RNA from Natural Environments for Analysis of Ribosomal DNA- and rRNA-Based Microbial Community Composition. *Appl. Environ. Microbiol.* **2000**, *66*, 5488–5491. [CrossRef]
81. Gunnigle, E.; Ramond, J.-B.; Frossard, A.; Seeley, M.; Cowan, D.A. A sequential co-extraction method for DNA, RNA and protein recovery from soil for future system-based approaches. *J. Microbiol. Methods* **2014**, *103*, 118–123. [CrossRef]
82. Feinstein, L.M.; Sul, W.J.; Blackwood, C.B. Assessment of Bias Associated with Incomplete Extraction of Microbial DNA from Soil. *Appl. Environ. Microbiol.* **2009**, *75*, 5428–5433. [CrossRef]

83. Thorn, C.E.; Bergesch, C.; Joyce, A.; Sambrano, G.; McDonnell, K.; Brennan, F.; Heyer, R.; Benndorf, D.; Abram, F. A robust, cost-effective method for DNA, RNA and protein co-extraction from soil, other complex microbiomes and pure cultures. *Mol. Ecol. Resour.* **2019**, *19*, 439–455. [CrossRef]

84. Nicora, C.D.; Burnum-Johnson, K.E.; Nakayasu, E.S.; Casey, C.P.; White, R.A.; Chowdhury, T.R.; Kyle, J.E.; Kim, Y.-M.; Smith, R.D.; Metz, T.O.; et al. The MPLEx Protocol for Multi-omic Analyses of Soil Samples. *J. Vis. Exp.* **2018**, e57343. [CrossRef] [PubMed]

85. Von Bergen, M.; Jehmlich, N.; Taubert, M.; Vogt, C.; Bastida, F.; Herbst, F.-A.; Schmidt, F.; Richnow, H.-H.; Seifert, J. Insights from quantitative metaproteomics and protein-stable isotope probing into microbial ecology. *ISME J.* **2013**, *7*, 1877–1885. [CrossRef] [PubMed]

86. Aylward, F.; Burnum, K.E.; Scott, J.J.; Suen, G.; Tringe, S.G.; Adams, S.M.; Barry, K.W.; Nicora, C.D.; Piehowski, P.D.; Purvine, S.O.; et al. Metagenomic and metaproteomic insights into bacterial communities in leaf-cutter ant fungus gardens. *ISME J.* **2012**, *6*, 1688–1701. [CrossRef] [PubMed]

87. Bastida, F.; Torres, I.F.; Moreno, J.L.; Baldrian, P.; Ondoño, S.; Ruiz-Navarro, A.; Hernández, T.; Richnow, H.H.; Starke, R.; García, C.; et al. The active microbial diversity drives ecosystem multifunctionality and is physiologically related to carbon availability in Mediterranean semi-arid soils. *Mol. Ecol.* **2016**, *25*, 4660–4673. [CrossRef] [PubMed]

88. Bao, Z.; Okubo, T.; Kubota, K.; Kasahara, Y.; Tsurumaru, H.; Anda, M.; Ikeda, S.; Minamisawa, K. Metaproteomic Identification of Diazotrophic Methanotrophs and Their Localization in Root Tissues of Field-Grown Rice Plants. *Appl. Environ. Microbiol.* **2014**, *80*, 5043–5052. [CrossRef] [PubMed]

89. Bastida, F.; García, C.; Von Bergen, M.; Moreno, J.L.; Richnow, H.H.; Jehmlich, N. Deforestation fosters bacterial diversity and the cyanobacterial community responsible for carbon fixation processes under semiarid climate: A metaproteomics study. *Appl. Soil Ecol.* **2015**, *93*, 65–67. [CrossRef]

90. Bastida, F.; Selevsek, N.; Torres, I.F.; Hernández, T.; García, C. Soil restoration with organic amendments: Linking cellular functionality and ecosystem processes. *Sci. Rep.* **2015**, *5*, 15550. [CrossRef] [PubMed]

91. Fernández-Martínez, M.A.; Severino, R.D.S.; Moreno-Paz, M.; Gallardo-Carreño, I.; Blanco, Y.; Warren-Rhodes, K.; García-Villadangos, M.; Ruiz-Bermejo, M.; Barberán, A.; Wettergreen, D.; et al. Prokaryotic Community Structure and Metabolisms in Shallow Subsurface of Atacama Desert Playas and Alluvial Fans After Heavy Rains: Repairing and Preparing for Next Dry Period. *Front. Microbiol.* **2019**, *10*, 1641. [CrossRef]

92. Benndorf, D.; Vogt, C.; Jehmlich, N.; Schmidt, Y.; Thomas, H.; Woffendin, G.; Shevchenko, A.; Richnow, H.-H.; Von Bergen, M. Improving protein extraction and separation methods for investigating the metaproteome of anaerobic benzene communities within sediments. *Biodegradation* **2009**, *20*, 737–750. [CrossRef] [PubMed]

93. Cañizares, R.; Benitez, E.; Ogunseitan, O.A. Molecular analyses of β-glucosidase diversity and function in soil. *Eur. J. Soil Biol.* **2011**, *47*, 1–8. [CrossRef]

94. Festa, S.; Coppotelli, B.; Madueño, L.; Loviso, C.L.; Macchi, M.; Tauil, R.M.N.; Valacco, M.P.; Morelli, I.S. Assigning ecological roles to the populations belonging to a phenanthrene-degrading bacterial consortium using omic approaches. *PLoS ONE* **2017**, *12*, e0184505. [CrossRef] [PubMed]

95. Knief, C.; Delmotte, N.; Chaffron, S.; Stark, M.; Innerebner, G.; Wassmann, R.; Von Mering, C.; Vorholt, J.A. Metaproteogenomic analysis of microbial communities in the phyllosphere and rhizosphere of rice. *ISME J.* **2012**, *6*, 1378–1390. [CrossRef] [PubMed]

96. Chen, J.; Arafat, Y.; Din, I.U.; Yang, B.; Zhou, L.; Wang, J.; Letuma, P.; Wu, H.; Qin, X.; Wu, L.; et al. Nitrogen Fertilizer Amendment Alter the Bacterial Community Structure in the Rhizosphere of Rice (*Oryza sativa* L.) and Improve Crop Yield. *Front. Microbiol.* **2019**, *10*, 2623. [CrossRef] [PubMed]

97. Chen, W.; Wang, Z.; Xu, W.; Tian, R.; Zeng, J. Dibutyl phthalate contamination accelerates the uptake and metabolism of sugars by microbes in black soil. *Environ. Pollut.* **2020**, *262*, 114332. [CrossRef] [PubMed]

98. Keiblinger, K.M.; Schneider, T.; Roschitzki, B.; Schmid, E.; Eberl, L.; Hämmerle, I.; Leitner, S.; Richter, A.; Wanek, W.; Riedel, K.; et al. Effects of stoichiometry and temperature perturbations on beech leaf litter decomposition, enzyme activities and protein expression. *Biogeosciences* **2012**, *9*, 4537–4551. [CrossRef]

99. Maron, P.-A.; Maitre, M.; Mercier, A.; Lejon, D.P.H.; Nowak, V.; Ranjard, L. Protein and DNA fingerprinting of a soil bacterial community inoculated into three different sterile soils. *Res. Microbiol.* **2008**, *159*, 231–236. [CrossRef] [PubMed]

100. Chourey, K.; Nissen, S.; Vishnivetskaya, T.; Shah, M.; Pfiffner, S.; Hettich, R.L.; Löffler, F.E. Environmental proteomics reveals early microbial community responses to biostimulation at a uranium- and nitrate-contaminated site. *Proteomics* **2013**, *13*, 2921–2930. [CrossRef] [PubMed]

101. Orellana, L.H.; Hatt, J.K.; Iyer, R.; Chourey, K.; Hettich, R.L.; Spain, J.C.; Yang, W.H.; Chee-Sanford, J.C.; Sanford, R.A.; Löffler, F.E.; et al. Comparing DNA, RNA and protein levels for measuring microbial dynamics in soil microcosms amended with nitrogen fertilizer. *Sci. Rep.* **2019**, *9*, 1–11. [CrossRef] [PubMed]

102. Liu, D.; Li, M.; Mingxiao, L.; Zhao, Y.; Chaowei, Z.; Song, C.; Zhu, C. Metaproteomics reveals major microbial players and their biodegradation functions in a large-scale aerobic composting plant. *Microb. Biotechnol.* **2015**, *8*, 950–960. [CrossRef] [PubMed]

103. Martinez-Alonso, E.; Pena-Perez, S.; Serrano, S.; Garcia-Lopez, E.; Alcazar, A.; Cid, C. Taxonomic and functional characterization of a microbial community from a volcanic englacial ecosystem in Deception Island, Antarctica. *Sci. Rep.* **2019**, *9*, 1–14. [CrossRef] [PubMed]

104. Manikandan, R.; Karthikeyan, G.; Raguchander, T. Soil proteomics for exploitation of microbial diversity in Fusarium wilt infected and healthy rhizosphere soils of tomato. *Physiol. Mol. Plant Pathol.* **2017**, *100*, 185–193. [CrossRef]

105. Liu, D.; Keiblinger, K.M.; Leitner, S.; Wegner, U.; Zimmermann, M.; Fuchs, S.; Lassek, C.; Riedel, K.; Zechmeister-Boltenstern, S. Response of Microbial Communities and Their Metabolic Functions to Drying–Rewetting Stress in a Temperate Forest Soil. *Microorganism* **2019**, *7*, 129. [CrossRef] [PubMed]

106. Christensen, G.A.; Gionfriddo, C.M.; King, A.J.; Moberly, J.G.; Miller, C.L.; Somenahally, A.C.; Callister, S.J.; Brewer, H.; Podar, M.; Brown, S.D.; et al. Determining the Reliability of Measuring Mercury Cycling Gene Abundance with Correlations with Mercury and Methylmercury Concentrations. *Environ. Sci. Technol.* **2019**, *53*, 8649–8663. [CrossRef] [PubMed]

107. Starke, R.; Kermer, R.; Ullmann-Zeunert, L.; Baldwin, I.T.; Seifert, J.; Bastida, F.; Von Bergen, M.; Jehmlich, N. Bacteria dominate the short-term assimilation of plant-derived N in soil. *Soil Biol. Biochem.* **2016**, *96*, 30–38. [CrossRef]

108. Schneider, T.D.; Keiblinger, K.M.; Schmid, E.; Sterflinger-Gleixner, K.; Ellersdorfer, G.; Roschitzki, B.; Richter, A.; Eberl, L.; Zechmeister-Boltenstern, S.; Riedel, K. Who is who in litter decomposition? Metaproteomics reveals major microbial players and their biogeochemical functions. *ISME J.* **2012**, *6*, 1749–1762. [CrossRef]

109. Starke, R.; Bastida, F.; Abadía, J.; García, C.; Nicolás, E.; Jehmlich, N. Ecological and functional adaptations to water management in a semiarid agroecosystem: A soil metaproteomics approach. *Sci. Rep.* **2017**, *7*, 10221. [CrossRef]

110. Guazzaroni, M.E.; Herbst, F.-A.; Lores, I.; Tamames, J.; Peláez, A.I.; López-Cortés, N.; Alcaide, M.; Del Pozo, M.V.; Vieites, J.M.; Von Bergen, M.; et al. Metaproteogenomic insights beyond bacterial response to naphthalene exposure and bio-stimulation. *ISME J.* **2013**, *7*, 122–136. [CrossRef]

111. Tan, H.; Kohler, A.; Miao, R.; Liu, T.; Zhang, Q.; Zhang, B.; Jiang, L.; Wang, Y.; Xie, L.; Tang, J.; et al. Multi-omic analyses of exogenous nutrient bag decomposition by the black morel Morchella importuna reveal sustained carbon acquisition and transferring. *Environ. Microbiol.* **2019**, *21*, 3909–3926. [CrossRef]

112. Zhang, B.; Thomas, B.W.; Beck, R.; Liu, K.; Zhao, M.; Hao, X. Labile soil organic matter in response to long-term cattle grazing on sloped rough fescue grassland in the foothills of the Rocky Mountains, Alberta. *Geoderma* **2018**, *318*, 9–15. [CrossRef]

113. Sekhon, S.S.; Kim, M.; Um, H.-J.; Kobayashi, F.; Iwasaka, Y.; Shi, G.; Chen, B.; Cho, S.-J.; Min, J.; Kim, Y.-H. Proteomic Analysis of Microbial Community Inhabiting Asian Dust Source Region. *Clean Soil Air Water* **2015**, *44*, 25–28. [CrossRef]

114. Wang, X.; Yang, T.; Lin, B.; Tang, Y. Effects of salinity on the performance, microbial community, and functional proteins in an aerobic granular sludge system. *Chemosphere* **2017**, *184*, 1241–1249. [CrossRef] [PubMed]

115. Halter, D.; Cordi, A.; Gribaldo, S.; Gallien, S.; Goulhen-Chollet, F.; Heinrich-Salmeron, A.; Carapito, C.; Pagnout, C.; Montaut, D.; Séby, F.; et al. Taxonomic and functional prokaryote diversity in mildly arsenic-contaminated sediments. *Res. Microbiol.* **2011**, *162*, 877–887. [CrossRef] [PubMed]

116. Sidibé, A.; Simao-Beaunoir, A.-M.; Lerat, S.; Giroux, L.; Toussaint, V.; Beaulieu, C. Proteome Analyses of Soil Bacteria Grown in the Presence of Potato Suberin, a Recalcitrant Biopolymer. *Microbes Environ.* **2016**, *31*, 418–426. [CrossRef] [PubMed]

117. Wu, L.; Wang, H.; Zhang, Z.; Lin, R.; Zhang, Z.; Lin, W. Comparative Metaproteomic Analysis on Consecutively Rehmannia glutinosa-Monocultured Rhizosphere Soil. *PLoS ONE* **2011**, *6*, e20611. [CrossRef] [PubMed]

118. Lechner, U.; Türkowsky, D.; Dinh, T.T.H.; Al-Fathi, H.; Schwoch, S.; Franke, S.; Gerlach, M.-S.; Koch, M.; Von Bergen, M.; Jehmlich, N.; et al. Desulfitobacterium contributes to the microbial transformation of 2,4,5-T by methanogenic enrichment cultures from a Vietnamese active landfill. *Microb. Biotechnol.* **2018**, *11*, 1137–1156. [CrossRef] [PubMed]

119. Zecchin, S.; Mueller, R.C.; Seifert, J.; Stingl, U.; Anantharaman, K.; Von Bergen, M.; Cavalca, L.; Pester, M. Rice Paddy Nitrospirae Carry and Express Genes Related to Sulfate Respiration: Proposal of the New Genus "Candidatus Sulfobium" . *Appl. Environ. Microbiol.* **2017**, *84*, e02224-17. [CrossRef]

120. Williams, M.A.; Taylor, E.B.; Mula, H.P. Metaproteomic characterization of a soil microbial community following carbon amendment. *Soil Biol. Biochem.* **2010**, *42*, 1148–1156. [CrossRef]

121. Lünsmann, V.; Kappelmeyer, U.; Benndorf, R.; Martinez-Lavanchy, P.M.; Taubert, A.; Adrian, L.; Duarte, M.; Pieper, D.H.; Von Bergen, M.; Müller, J.A.; et al. In situ protein-SIP highlightsBurkholderiaceaeas key players degrading toluene by para ring hydroxylation in a constructed wetland model. *Environ. Microbiol.* **2016**, *18*, 1176–1186. [CrossRef]

122. Sukul, P.; Schäkermann, S.; Bandow, J.E.; Kusnezowa, A.; Nowrousian, M.; Leichert, L.I.O. Simple discovery of bacterial biocatalysts from environmental samples through functional metaproteomics. *Microbiome* **2017**, *5*, 28. [CrossRef]

123. Bastida, F.; Crowther, T.W.; Prieto, I.; Routh, D.; García, C.; Jehmlich, N. Climate shapes the protein abundance of dominant soil bacteria. *Sci. Total. Environ.* **2018**, 18–21. [CrossRef]

124. Bastida, F.; Torres, I.F.; Andrés-Abellán, M.; Baldrian, P.; López-Mondéjar, R.; Větrovský, T.; Richnow, H.H.; Starke, R.; Ondoño, S.; García, C.; et al. Differential sensitivity of total and active soil microbial communities to drought and forest management. *Glob. Chang. Biol.* **2017**, *23*, 4185–4203. [CrossRef] [PubMed]

Publisher's Note: MDPI stays neutral with regard to jurisdictional claims in published maps and institutional affiliations.

International Journal of
Molecular Sciences

Review

Advances on Plant Ubiquitylome—From Mechanism to Application

Dongli He [1], Rebecca Njeri Damaris [1], Ming Li [1] and Pingfang Yang [1,*]

[1] State Key Laboratory of Biocatalysis and Enzyme Engineering, School of Life Sciences, Hubei University, Wuhan 430062, China; hedongli@hubu.edu.cn (D.H.); njerirebecca09@gmail.com (R.N.D.); limit@wbgcas.cn (M.L.)

[2] Department of Basic and Translational Sciences, School of Dental Medicine, University of Pennsylvania, Philadelphia, PA 19014, USA; drimran@upenn.edu

* Correspondence: yangpf@hubu.edu.cn

Received: 18 July 2020; Accepted: 17 October 2020; Published: 24 October 2020

Abstract: Post-translational modifications (PTMs) of proteins enable modulation of their structure, function, localization and turnover. To date, over 660 PTMs have been reported, among which, reversible PTMs are regarded as the key players in cellular signaling. Signaling mediated by PTMs is faster than re-initiation of gene expression, which may result in a faster response that is particularly crucial for plants due to their sessile nature. Ubiquitylation has been widely reported to be involved in many aspects of plant growth and development and it is largely determined by its target protein. It is therefore of high interest to explore new ubiquitylated proteins/sites to obtain new insights into its mechanism and functions. In the last decades, extensive protein profiling of ubiquitylation has been achieved in different plants due to the advancement in ubiquitylated proteins (or peptides) affinity and mass spectrometry techniques. This obtained information on a large number of ubiquitylated proteins/sites helps crack the mechanism of ubiquitylation in plants. In this review, we have summarized the latest advances in protein ubiquitylation to gain comprehensive and updated knowledge in this field. Besides, the current and future challenges and barriers are also reviewed and discussed.

Keywords: ubiquitylation; plant; method; machinery; function; crosstalk; database

1. Introduction

Plants are constantly exposed to dynamic environmental conditions due to their sessile nature, which compels their cells to evolve and acquire the ability to change and survive from their endogenous status rapidly. The internal signal transduction ultimately induces modulation of cellular proteins in response to external stimuli (e.g., light or temperature stress). These post translational modifications (PTMs) impact protein's location, stability and activity, eventually triggering a faster response [1]. PTMs greatly modify the classic central dogma and thus attract considerable attention. Therefore, PTMs have become a major concern in protein research, especially the reversible PTMs, such as phosphorylation, ubiquitylation, acetylation, methylation and O-GlcNacylation. To date, there are more than 660 PTMs reported in the Uniprot database http://www.uniprot.org/docs/ptmlist) [2]. Particularly, the frequent PTMs crosstalk within or between proteins forwards the signaling cascades accurately in a mandatory manner [3,4].

Ubiquitylation is one of the most prevalent PTMs, which was originally identified as a modulator of cellular protein turnover and homeostasis [5]. However, with the advancement of technology, its functions have been extended far beyond what was initially known. Henceforth, ubiquitylation is found widely involved in pivotal processes such as protein turnover, genomic integrity, signaling processing and cell transport to direct cell proliferation, differentiation or survival through communication systems [6,7]. Increasing evidence indicates that ubiquitylation participates in almost all events in the entire life cycle of the plant from seed germination to flowering, senescence, as well as the pathogen responses among others. For instance, during rice seed germination, He et al. (2020) detected 2576 lysine ubiquitylated (Kub) sites in 1171 proteins and the ubiquitylation was supposed to modulate the protein function more than just providing 26S degradation signals in the early stage of rice seed germination [6]. Guo et al. (2017) identified 3263 Kub sites in 1611 proteins in the ethylene treated Petunia Corollas and the global proteome and ubiquitylome were negatively correlated, indicating the involvement of ubiquitylation in protein degradation of petunias petal senescence [8]. Wang et al. (2020) quantified 1926 unique ubiquitylation sites corresponding to 1053 proteins in the de-etiolating seedling leaves of *Zea mays*, reflecting the role of ubiquitylation in photosynthesis and light signaling [9].

Ubiquitylation is second to glycosylation, for having the most complex modifications due to its diverse target-binding ways. This has created great challenges in the identification of the primary modification and functional elucidation. However, the small ubiquitin (Ub) protein with only 76 amino acid is highly conserved throughout all eukaryotic cells, which has formed a universal language in organisms from yeast to humans [7]. This review will discuss the updates on protein ubiquitylation, from its coding mechanism to research methods, functions, crosstalk and related databases. We will also highlight the recent applications of ubiquitylation in the plant kingdom.

2. The Ubiquitylation Machinery and Code

Ubiquitylation describes the process of the conjugation of Ub to substrates, which is sequentially catalyzed by a Ub—activating enzyme (E1), a Ub -conjugating enzyme (E2) and a Ub ligase (E3) [10]. Typically, it forms an iso-peptide bond between the C- terminus of Ub and an ε-amino group of a lysine residue of a substrate but it also can be targeted to other amino acids like Cys, Ser and Thr residues [11,12] or a protein's N-terminus methionine [13]. Glutamic acid (E), aspartic acid (D) and Alanine (A; neutral) were highly enriched around the Kub sites, however, the basophilic residues histidine (H), arginine (R) and lysine (K) were found to be excluded from the adjacent positions [6,8,9]. The flexibility of conjugation dictates the diversity of ubiquitylation. Ub can be attached to a protein at one residue (mono-ub) or multiple residues (multi-ub), more Ub molecules might be added by E2s/E3s and form polymeric chains (poly-Ub, Figure 1) through selective conjugation to its seven lysine residues (K6, K11, K27, K29, K33, K48 and K63) as well as its N-terminal methionine (M1) [14]. This results in distinct structures and functions.

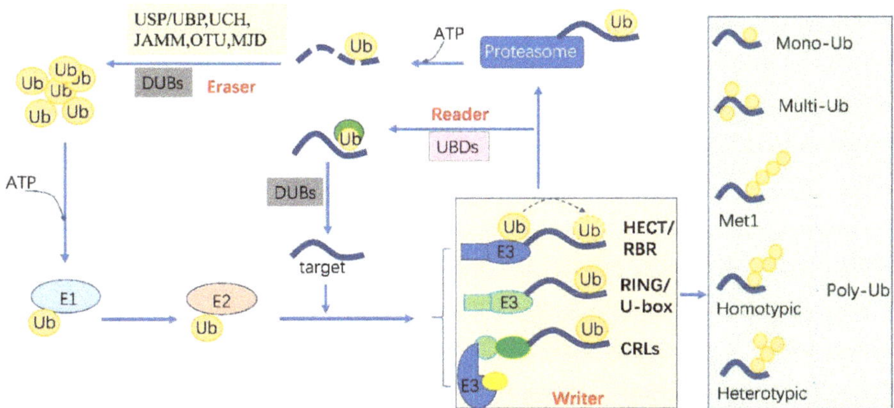

Figure 1. The protein ubiquitylation cascade and its components. Free Ub (Ub) molecules are activated through 3 sequential reactions catalyzed by a Ub-activating enzyme (E1), Ub-conjugating enzyme (E2) and Ub ligase (E3) in an ATP-dependent manner. Based on the transferring mode of Ub Figure 2, the E3s (writer of ubiquitylation) are classed into HECT, RBR, RING, U-box and Cullin-RING E3 ligases. Ub linkage can form into mono-, multimono- (multi-) or poly- ubiquitylation. Ubiquitylation sites are recognized by the proteins carrying Ub binding domains (UBDs, reader of ubiquitylation, including the cap of proteasome) and then direct the targets to be recycled by the deubiquitylases (DUBs, eraser, USP/UBP, UCH, JAMM, OTU and MJD in plants) or 26S proteasome-mediated degradation. HECT, homologous to E6-associated protein C-terminus; RBR, RING-in-between-RING; U-box, a modified RING motif without the full complement of Zn^{2+}-binding ligands; USP/UBP, ubiquitin specific proteases/ubiquitin-specific processing enzymes; UCH, ubiquitin carboxyl-terminal hydrolases; OUT, ovarian tumor proteases; JAMM, JAB1/MPN/MOV34 domain associated metalloisopeptidase; MJD, Machado-Joseph family proteins.

The essential Ub are usually encoded redundantly in the eukaryotic genome with mono-Ub unit fusing to ribosomal or organized as poly-Ub units in a tandem linear [15]. These Ub proteins contain 1–7 Ub units. In Arabidopsis, 12 functional Ub have been identified [16]. These fused- or poly-Ub are initially processed into a single Ub molecule by deubiquitylases (DUBs) before being conjugated to its substrates. The DUBs are also responsible for recycling Ub from the substrates (Figure 1). In plants, there are five different DUB families [17]. Approximately 50 DUBs have been identified in Arabidopsis [17] and 100 of these in human beings [18].

Ubiquitylation and deubiquitylation are tightly regulated in vivo. First, the ATP dependent E1 enzyme captures the Ub through its active-site Cys residue and forms a thioester bond between the C terminus of Ub; then the Ub is transferred onto a Cys residue of E2 [19]. E1 and E2 are relatively conserved in eukaryotes. However, the E3 ligases are diverse among organisms and in different biological processes, where they can selectively recruit specific targets. In humans, there are about 9 E1s and 40 E2s [10] and 600 E3s [18]. These enzymes are more complex in plants. Arabidopsis has 2 E1s and 47 E2s, while approximately 1500 potential E3s; Rice has 6 E1s, 49 E2s and more than 1300 E3s [20,21]. E3 ligases are characterized as ubiquitylation writers with different domains, such as HECT(homologous to E6-associated protein C-terminus), RING, UBOX(a modified RING motif without the full complement of Zn^{2+}-binding ligands), RBR(RING-in-between-RING) and so forth [22]. The ubiquitylated protein can be read by proteins with Ub- binding domains (UBDs, such as Ub-interacting motif (UIMs) and proteasomal receptor) and then be directed to the downstream biological process. Finally, the DUBs or the regulatory cap of the proteasome will erase the Ub from the substrates, thereafter, the free Ub can be recycled [23,24]. The whole cycle constitutes a powerful ubiquitylation language and performs essential signaling functions in all eukaryotes (Figure 1).

3. Methods of Ubiquitylation Detection and Application in Plant Ubiquitylome

The low site-specific stoichiometry, short lifespan, reversible modification, condition-specific expression and complex Ub conjugation architectures bring considerable obstacles in developing deep and accurate catalogs of ubiquitylation. Despite these challenges, greater improvement has been achieved in the identification and verification methods in recent years, especially in plants.

Based on the binding properties between UBDs protein and ubiquitin, three classic methods have been developed to purify the ubiquitylated proteins, including the single-step enrichment, Tandem Affinity Purification (TAP) protocol and two-step affinity tandem Ub binding entities (TUBEs, Figure 2A–C). The single-step enrichment was established to purify the ubiquitylated proteins using affinity matrices through UBDs and the monoclonal anti-ubiquitin antibodies, directly [25]. This method has successfully identified hundreds of ubiquitylated proteins but many false positives were also identified due to non-specific binding under the nondenaturing conditions. TAP protocol greatly avoid this shortcoming, with an initial production of a stable Arabidopsis transgenic line expressing poly-UBQ gene encoding Ub monomers N-terminally tagged with hexahistidine and then purified with sequential Ub-affinity and strong denaturing nickel chelate-affinity chromatography [26]. Saracco et al. (2009) reported that although only 54 non-redundant targets expressed by 90 possible isoforms were identified by mass spectrometry due to the high stringency of TAP, the accuracy was highly improved. Two-step affinity took advantage of the same Arabidopsis transgenic line of TAP, adopting TUBEs developed by Lopitz-Otsoa et al. [27], which drastically improved the purification stringency and yielded about 950 ubiquitylation substrates in the whole Arabidopsis seedlings [28].

The above methods brought great leap for ubiquitylation identification in plants but have not revealed the exact modified site, through site-directed mutagenesis of the lysine residues, therefore, few ubiquitylated sites have been verified [28]. For the lysine ubiquitylation, when the modified protein is digested by trypsin, the remains becomes a specific C-terminal remnant of Lys-ε-Gly-Gly (K-ε-GG, DiGly). Searching spectra for the typical DiGly footprint, Maor et al. (2007) successfully identified 85 precise DiGly footprints on 56 proteins in Arabidopsis [25]. The development of antibodies that recognize DiGly remnant was the first breakthrough that made the proteome-wide investigation of the exact ubiquitylation sites by LC-MS/MS possible (Figure 2D) [29]. Through affinity chromatography with K-ε-GG specific antibody, the tryptic ubiquitylated peptides were efficiently enriched, thereafter, analyzed by high quality MS/MS, which paved a way for real ubquitylome and as a result, thousands of proteins were identified in different species [6,30,31]. K-ε-GG antibody can also unbiasedly recognize the epitope on the Ub itself and provide the information for poly-linkage sites. However, the K-ε-GG antibody cannot capture modifications occurring at the N-terminal or other residues, moreover, neither can it differentiate the tryptic cleavage of other small related protein modifiers [32], such as small Ub-related modifier (SUMO).

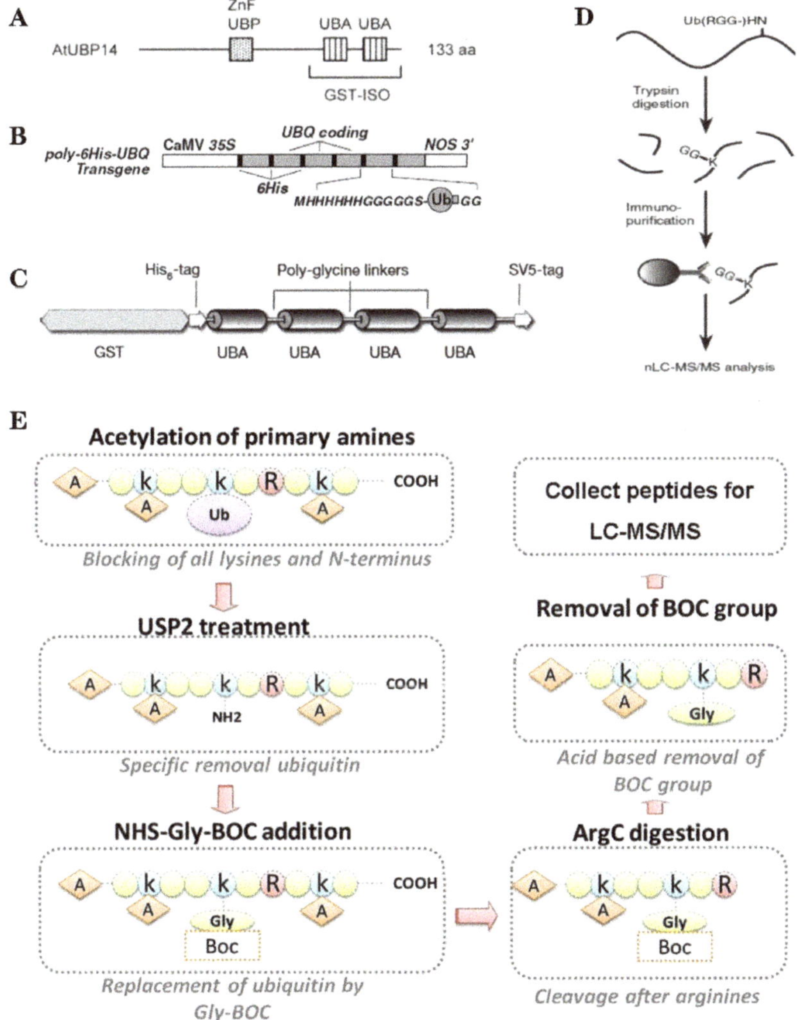

Figure 2. The classic methods of ubiquitylome in plant. (**A**), a single enrichment step approach using UBA (Ub-associated) motif [25]. (**B**), tandem affinity purification (TAP) approach using poly His-tag-UBQ (Ub) motif [26]. (**C**), two-step affinity tandem Ub-binding entities (TUBE) [33]. (**D**), Affinity chromatography using Lys-ε-Gly-Gly (K-ε-GG) specific antibody [29]. (**E**), the Ub COFRADIC (combined fractional diagonal chromatography) pipeline [34].

Ub combined fractional diagonal chromatography (COFRADIC) method was established as a complementary alternative to the K-ε-GG antibody [35]. This protocol first blocks all primary amines (lysines and N termini) via chemical acetylation and removes Ubs with a plant specific DUB USP2cc, then attaches a chemical handle to these free primary amines, subsequently isolating peptides via two consecutive reverse-phase HPLC (RP-HPLC) runs. Afterwards, the handle is removed by ArgC that cleaves the sequence after arginine. As USP2cc specifically recognizes the last five amino acids [36], COFRADIC can successfully avoid the false positive with other Ub-like modifications. USP2cc only recognizes Ub independent of other affected residues, allowing the identification of ubiquitylation on other residues in addition to lysine. Walton et al. (2016) used COFRADIC to identify 16 proteins with

N-terminal ubiquitylation in Arabidopsis (Figure 2E) [34]. However, due to the biochemistry-associated bias, COFRADIC generates peptides that are probably too long or too short to be identified by MS/MS and that are difficult for branched Ub chains detection. Therefore, combining the K-ε-GG antibody affinity and the COFRADIC method may provide a deeper insight into ubiquitylome.

When a Ub is added to the N-terminal methionine (M) of another Ub, it forms a linear poly-Ub chains that is not detectable by all the above methods. This is because the KGG- antibody does not recognize the characteristic GGMQIFVK peptides while other N-terminally tagged Ub constructs prevent linear poly-Ub chain assembly (e.g., TAP). Kliza et al. (2017) established a new method of identifying linear poly-Ub-modified proteins, in which a lysine-less internally streptavidin tagged Ub (INT-Ub.7KR) was first constructed, followed by stable isotope labeling of amino acids in cell culture (SILAC)-based mass spectrometry [37]. Using this method, several known linear poly-Ub targets were successfully validated in T-REx HEK293T cells, which provided an effective strategy for liner poly-Ub detection. This method could be modified and be applied in plant proteomics research in the future.

Thanks to the remarkable development of the LC-MS/MS for proteome analysis with high sensitivity and resolution, it is now possible to quickly identify and quantify ubiquitylated proteins in high throughput. Precise relative quantification of ubiquitylated peptides and sites is a big challenge. SILAC has been successfully used for comparison of ubiquitylation dynamics in animal cells. However, its application is rather difficult in plants due to the low efficiency of the in vivo protein labelling. The optimized label-free methods have been proven to be effective in some plants such as petunias, rice and maize [6,8,9], although the accuracy and repeatability require further improvements. In plants, comparative proteome, the advanced tag label methods, for example, isobaric tags for relative and absolute quantitation (iTRAQ) or tandem mass tag (TMT), have been widely used but the K-ε-GG antibody cannot recognize the di-Gly remnant when its N-terminus is derivatized with iTRAQ or TMT. Rose et al. (2016) labelled the ubiquitylated proteins with TMT10 after elution with K-ε-GG antibody and 9000 ubiquitylated peptides were quantified using up to 7 mg labelled sample [38]. To improve the sensitivity and throughput, Namrata et al. (2016) recently developed a rapid and multiplexed protocol termed UbiFast, in which the K-ε-GG antibody is first labelled with TMT and then used to isolate the ubiquitylated peptides. UbiFast facilitated quantification of 10,000 ubiquitylation sites from only 500 μg peptides, which makes large scale comparative ubquitylome more accurate and sensitive [39].

4. Multiple Functions Played by Protein Ubiquitylation in Plants

Using the protein/peptide affinity plus high-quality MS/MS technology, thousands of ubiquitylated substrates can be identified with one experiment. These substrates carry different types of Ub-chain and may possibly exert diverse effects on the targets (Table 1, Figure 3). Mono-Ub/multi-Ub may precisely change the protein activity and interaction [40], notably, mono-ub-dependent protein degradation has recently been reported. Poly-Ub performs diverse functions, with homotypic K48-linked poly-Ub being the predominant (>50%) linkage and mainly guides proteins to the ubiquitin proteasome system (UPS) for degradation [41]. Also, some K11-, K27-, K29- and K33-linked poly-Ub direct proteins for degradation [42]. In Arabidopsis cell-free system, the K29-chain were confirmed to regulate the degradation of DELLA proteins, repressors in GA signaling pathways [43]. K63-linkage is another abundant poly-Ub despite it being proteasome-independent. K63-linked poly-Ub chain formation is critical for vacuolar targeting of the auxin efflux carrier component 2 (PIN2) [44]. Along with the development of the specific antibody, some heterogenous poly-Ub chains were also discovered, such as K63- poly-Ub that may serve as a "seed" for K48-poly-Ub and form the K48/K63 branched chains, which then directed the targets for degradation [45]. However, to date, the evidence of branched poly-Ub chains in plants is not discovered. In a study conducted by Swatek et al. (2019), they demonstrated that the leader protease (Lbpro) of foot-and-mouth disease virus can incompletely remove ubiquitin from substrates and generate DiGly-modified proteins. This Ub-clipping methodology may provide new insight into the combinatorial complexity and architecture of the ubiquitin (including the mono-, multi- or poly- Ub) code in plants [46].

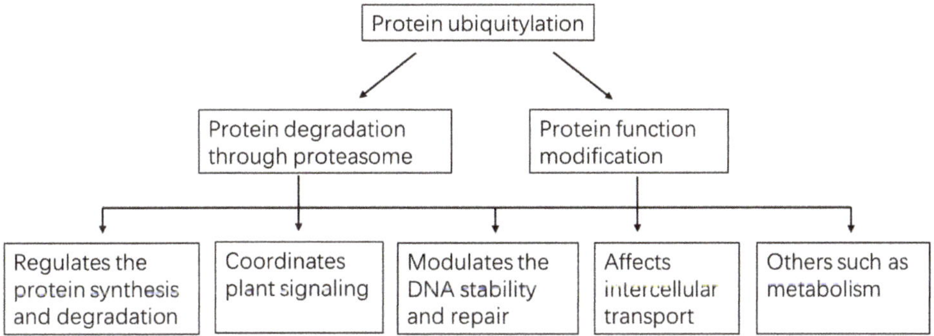

Figure 3. A global view of functions played by protein ubiquitylation in plants.

4.1. Ubiquitylation Regulates Protein Synthesis and Degradation

Ribosomes are the main molecular machinery that generates nascent polypeptide within the cells. The cytoplasmic ribosomes consist of two sub units (40S and 60S) with folded ribosomal RNA (5.8S, 18S and 28S rRNA), approximately 80 ribosomal proteins (RPs) are present in eukaryotes. During the ribosomal synthesis, RPs are initially incorporated into the ribosome in accordance with the processing of pre-rRNA transcripts into mature rRNAs before being exported to the cytoplasm. In this highly efficient process, many RPs are "unemployed" to become potential ubiquitylation substrates for nuclear proteasome-mediated degradation [47]. Stress conditions may cause massive amounts of RPs to become "unemployed" and ubiquitylated and this has been confirmed in de-etiolating maize seedlings and senescing petunias petal [8,9].

Ubiquitylation distinguishes itself from other PTMs as it is widely involved in protein stability. The bulk of protein degradation in living cells depends on UPS. The proteasome complex constitutes of a 14-subunit 20S core protease and an 18-or-more subunit 19S regulatory particle, which can recognize ubiquitylated proteins and degrade them into small peptides, while removing the Ub for recycling [48]. Protein ubiquitylation performs important function of protein's quality control in cells, clearing those misfolded or damaged proteins, which occurs through tagging substrates with K11-, K48- or K11/K48-branched poly-Ub chains [49]. After treating Arabidopsis seedlings with proteasome inhibitor MG132, more than half of the ubiquitylated proteins increased at the ubiquitylation level, which might be later degraded by 26S proteasome [28]. At the same time, 26S proteasome can be modified by ubiquitylation and the abundance of most ubiquitylated 26S subunits increased significantly (average of 3.9-fold). This increase can be attributed to turnover of inhibited proteasome complexes [28]. Besides, other UPS components, including E1, E2 and E3 enzymes and de-ubiquitylating enzymes were also found ubiquitylated, possibly repressing UPS activity [6,50].

Autophagy is another foremost mechanism of protein degradation and remobilization, which maintains cellular homeostasis by recycling cytoplasmic components through forming a double-membrane autophagosome that subsequently degrades them via the lysosome. It is a highly conserved cellular process prevailing in all eukaryotes [51]. Series of autophagy-related proteins (ATGs) have been discovered in plants, so far. In Arabidopsis, the ATG1–ATG13 kinase complex are key positive regulators to induce the autophagic vesiculation [52]. The ubiquitylation system has been reported to control the protein stability of ATG1–ATG13 complex [53]. Qi et al. (year) discovered the RING-type E3 ligases seven in absentia of Arabidopsis thaliana (SINAT) proteins regulate autophagy by interacting with ATG proteins and modulating the stability of the ATG1–ATG13 kinase complex under either normal nutrient or some starvation conditions [54]. During the senescence of petunias petal, ubiquitylation of ATG8b (Lys-11) was first found to be up-regulated by ethylene [8]. Ubiquitylation is also known for the removal of misfolded or aggregated proteins in autophagy. In another study, the NBR1 (neighbor of BRCA1) protein was reported as an adaptor protein recruited to Ub-positive protein aggregates and degraded by autophagy in the animal cells [55].

Table 1. Potential function of various ubiquitin linkage (The red ones have not been recorded in plants).

Ub code	Function	Substrate	Reference
Mono-/multi-ubiquitylation	changes the protein activity and interaction	e.g., lysine 119 of H2B	[56]
K29-chains	proteasomal degradation	e.g., DELLA proteins	[43]
K48-chains	proteasomal degradation	e.g., Aux/IAA	[57]
K63-chains	endocytic sorting, DNA repair, degradation but proteasome-independent	e.g., PIN2	[44]
M1	inflammation signaling	e.g., NF-κB	[58]
K6-chains	Mitophagy	e.g., mitofusin-2 (Mfn2)	[59]
K11-chains	proteasomal degradation/cell cycle regulation	e.g., anaphase-promoting complex APC/C	[60]
K27-chains	proteasomal degradation	e.g., NS4B	[61]
K33-chains	proteasomal degradation	e.g., ERCC1 (nucleotide excision repair protein)	[62]
M1/K63-linked	transcription factor activation (K63-poly-ub as a prerequisite for the formation of M1-poly-Ub)	e.g., canonical IκB kinase IKKα and IKKβ	[63]
K11/K48-linked	proteasomal degradation, cell-cycle and quality control	e.g., anaphase-promoting complex APC/C	[49]
K48/K63-chains	proteasomal degradation (K63- poly-ub serving as a "seed" for K48-poly-ub)	e.g., proapoptotic regulator TXNIP	[45]

4.2. Ubiquitylation Coordinates Plant Signaling

Because of its reversible nature, rapid kinetics and the versatility of outcomes, ubiquitylation can facilitate the coordination of external and internal environmental signals in space and time. In an earliest study, Shanklin et al. (1987) reported that ubiquitylation mediated the conversion of phytochrome forms between red light-absorbing form (Pr) and far-red light-absorbing form (Pfr.) [64]. The light was then revealed to induce degradation of phytochrome interacting factor PIF1 by the $CUL4^{COP1–SPA}$ E3 ligase initiated photomorphogenesis [65]. Thereafter, the involvement of ubiquitylation in various plant signaling pathways has been broadly reported such as calcium signaling, 14-3-3 proteins and G-proteins mediated signaling [66,67].

The proteasome-mediated degradation acts as central regulator in most phytohormone signaling pathways in plants, such as gibberellic acid (GA), abscisic acid (ABA), auxin, brassinosteroids (BRs), ethylene, salicylic acid (SA), jasmonic acid (JA) among others. DELLA (SLR1) protein, one known repressor of the GA signaling, can be degraded by the 26S proteasome via SCF^{GID2} in a GA-dependent manner [43]. In Arabidopsis, the negative regulator *ABI3*-interacting protein2 (AIP2) of ABA signaling can promote abscisic acid insensitive 3 (ABI3) polyubiquitylation for degradation [68], while RING-type E3 ligase keep on going (KEG) can ubiquitinate ABI5, the upstream transcriptional factor of ABI3 and regulate its abundance [69]. Auxin or indole-3-acetic acid (IAA) can trigger nuclear cascades by modulating the recruitment of most short-lived AUXIN/INDOLE-3-acetic acid (AUX/IAA) transcriptional repressors through multimeric SCF-type E3 Ub ligases, which then leads to ubiquitylation and proteasome-dependent degradation of AUX/IAA [57,70]. The membrane-bound receptor complex BRI1/BAK1 of Brassinosteroids (BRs) is regulated by K63 poly-ubiquitylation [71]. The ethylene insensitive2 (EIN2) plays a central role in the signaling of gaseous hormone ethylene. This process is tightly regulated through proteasomal degradation [72]. In the absence of ethylene, EIN3 is targeted for degradation by Cullin1-RING Finger E3 Ligase CRL1s [73].

4.3. Ubiquitylation Modulates DNA Stability and Repair

Sessile plants may encounter serious genotoxic stress capable of evoking complex DNA damage response (DDR) to protect genomic integrity. Ub-dependent signaling could regulate the DDR like double-strand break response (DSB), nucleotide excision repair (NER), together with other PTMs. During template switching and the DSB response, K63 Poly-Ub assembles as a scaffold for the signaling complex. The BRCA1/BARD1 Ub ligase was observed to assemble K6 linkages in vitro and in overexpression systems. However, a detailed mechanism between K6 linkages and DDR is still elusive [74]. K6- and K33-linked polyubiquitylation were detected having undergone a bulk increase in response to DNA damaged by UV [75]. Proliferating cell nuclear antigen (PCNA) can recruit DNA trans lesion polymerases or the translocase for DDR [76]. Upon rice seed imbibition, Kub164 of PCNA was induced significantly, meanwhile, four Rad23 (Radiation) family proteins and a DDI1(DNA-damage inducible) were rapidly ubiquitylated, which is evidence for widespread DDR during the initiation of seed germination [6].

Monoubiquitylation plays a central role in modulating nucleosome/chromatin structure (histones H2A, H2B, H3 and H4) and DNA accessibility for gene-specific transcription [77–79]. Histone H2A was the first protein to be identified as a substrate for ubiquitylation [80]. In human cells, about 10% of H2A molecules are monoubiquitylated on K119 by a family of multi-subunit E3 ligases known as poly-comb repressive complex 1 (PRC1) [81]. H2A monoubiquitylation can induce repressive histone H3K27 trimethylation and in turn recruits more PRC1 complexes, which allows H2A ubiquitylation to spread along whole chromosomes and further restricts gene expression [82]. Beside inhibiting transcription, monoubiquitylation of other histones result in divergent consequences in controlling gene expression, DNA methylation or DNA repair, for example ubiquitylation of K120 on H2B stimulates gene expression [83]; monoubiquitylation of proximal Lys residues on H3 enables epigenetic inheritance of DNA methylation status [84]; monoubiquitylation of K91 of H4 is important for DNA damage signaling [85]. De-ubiquitylation on these histones occurring at the right time and place promotes rapid changes in transcriptional programs [86].

4.4. Ubiquitylation Affects Intercellular Transport

Ubiquitylation has an extended role in controlling cellular transmembrane transport, protein trafficking, precise location and protein stability. Epidermal growth factor receptor (EGFR) is the first protein identified as evidence of ubiquitylation role in protein transport. This modification occurs on multiple Lys residues during endocytosis internalization [87]. The endosomal sorting complex required for transport (ESCRT) machinery can traffic the flow of ubiquitylated proteins from endosomes to lysosomes. The five distinct complexes (ESCRT-0, -I, -II, -III and the Vps4 complex) have a clear division of tasks to guarantee a smooth transport of ubiquitylated cargos along the endosomal trafficking pathway [88]. The internalized receptors were first recognized by the ESCRT-0 complex, which was associated with multiple Ub subunits and then handed over sequentially to other complexes [89]. Endoplasmic-reticulum-associated protein degradation (ERAD) eliminates defective membrane or luminal proteins of the ER. Substrates of ERAD are ubiquitylated by ER-localized E3s and are delivered to the proteasome for degradation. Membrane-localized E3 ligases are also found in the Golgi apparatus, mitochondria, chloroplast or peroxisomes [6,90]. Ubiquitylation is capable of controlling the trafficking machinery. Contrarily, monoubiquitylation of COPII coat protein controls secretory protein transfer from the ER to the Golgi apparatus [91]. Vesicle transport can move proteins among different locations within the cell organelles. Soluble N-ethylmaleimide-sensitive factor attachment protein receptors (SNARE) are required for the fusion of transport vesicles with the correct target membrane. VAMP8 (v-SNARE) ubiquitylation is required for vesicle fusion in cytokinesis [92]. The authors have identified 2 v-SNARE and 7 t-SNARE proteins that were rapidly ubiquitylated during rice seed germination [6].

5. Crosstalk between Ubiquitylation and Other PTMs

Plant proteins are subjected to a wide array of PTMs. These modifications can occur on the same amino acid residue (s) of a substrate at various temporal points or different amino acid residues of the same substrate. Different PTMs influence each other in coordinating multiple signal transduction events in an orchestrated manner. PTM crosstalk greatly improves the ability of sessile plants for rapid response to different external and internal cues. Among these PTMs, several are reversible including ubiquitylation, phosphorylation, SUMOylation, S-nitrosylation and glycosylation, which may endow proteins with opposing biochemical activities.

As two of the most prevalent PTMs in eukaryotic proteomes, crosstalk between phosphorylation and ubiquitylation can take various forms. For example, phosphorylation can change the stability, function site or substrate recognition of the E3 enzyme and ubiquitylation can directly modify phosphorylation receptor kinase activity [93]. The crosstalk will ultimately cause different outcomes, such as substrate degradation (e.g., FLS2 in Arabidopsis, the leucine-rich repeat receptor-like kinase Flagellin-Sensing 2) [94], changing the enzymatic activity (e.g., plant U-box (PUB) type E3 Ub ligase PUB12/13) [95] or subcellular localization (e.g., BRI1) [71]. Different phosphorylation sites of the same substrate can lead to opposite effects, for PUB2 in Medicago, phosphorylation at Ser316 suppresses the E3 ligase activity of PUB2; however, phosphorylation at Ser421 promotes its E3 ligase activity [96]. The most cited crosstalk between phosphorylation and ubiquitylation is phosphodegrons, in which one or more phosphorylation site(s) function in a short linear motif to promote the subsequent ubiquitylation and degradation of a substrate. The irreversible and robust phosphodegrons are critical for cell cycle progression. The Cdc4 protein was the first F-box protein to be described, which is capable of forming a WD40 fold that recognizes phosphorylated peptides [97]. Using the sequential enrichment method, Swaney et al. (2016) globally analyzed the phosphorylation and ubiquitylation cross-talk in protein degradation in yeast *Saccharomyces cerevisiae*, 466 proteins with 2100 phosphorylation sites co-occurring with 2189 ubiquitylation sites were identified. Further analysis showed that distinct phosphorylation sites are often in conjunction with ubiquitylation in a highly conserved manner [67].

In addition to the above-mentioned crosstalk, ubiquitylation widely interacts with many other PTMs. In Arabidopsis, SNC1 is acetylated at the N terminal, functioning as N-degron for its ubiquitylation leading to its degradation, which results in a decrease in plant's immunity [98]. Redox modifications, for example, S-nitrosylation, disulfides and S-glutathionylation can directly regulate enzymes constituting the ubiquitylation system and affect enzymatic activity [99].

Ub itself can be modified by various PTMs. For instance, phosphorylation has been observed in most of Ub serine, threonine and tyrosine residues [100] and the most frequently modified residues are Ser57 and Ser65 in yeast and mammalian cells, respectively. For example, the PINK1 kinase can phosphorylate S65 of Ub at the mitochondrial surface and activate ubiquitin ligase Parkin in mammals, which will promote the ubiquitylation of numerous mitochondrial outer membrane proteins, finally resulting in the mitochondrial autophagy (mitophagy) [101]. Phosphorylation also inhibits certain E2s, E3s and DUBs and modulates the ubiquitylation cycle [102]. Only 0.03 and 0.01% acetylation of Ub occurred on K6 and K48 in human cells. Overexpression of acetyl-mimetic Ub (K6Q) can repress K11-, K48- and K63-linked Ub chain elongation on substrates but stabilize the monoubiquitylation of histone H2B. We also have detected phosphorylated and acetylated Ub in the germinating rice seed [103,104]. These modifications may change the structure of the Ub-chain and its related signaling in vivo.

6. Related Databases Developed for Plant Ubiquitylation

Owing to the advancement of high throughput of high-quality mass spectrometry (MS)-based proteomics, numerous ubiquitylation sites have been identified, resulting in an enormous collection of ubiquitylation data. Accordingly, some professional databases have been developed to collect the massive ubiquitylated data and analyze the protein ubiquitylation networks (Table 2).

Table 2. Databases Developed for Plant Ubiquitylation.

Name	Website	Aims	Updated Time
UbiProt	http://ubiprot.org.ru	experimentally obtained	2007
PEIMAN	http://bs.ipm.ir/softwares/PEIMAN	predict and compute the ubiquitylation and other PTMs	2015
plantsUPS	http://bioinformatics.cau.edu.cn/plantsUPS	comparative analysis of UPS in higher plants	2009
COFRADIC method	http://bioinformatics.psb.ugent.be/webtools/Ub_viewer	Arabidopsis	2016
PTMCode	https://ptmcode.embl.de	integrative information of known and predicted PTMs	2015
dbPTM	http://dbptm.mbc.nctu.edu.tw	comprehensively functional and structural analyses for PTMs	2019
Plant PTM Viewer	https://www.psb.ugent.be/webtools/ptm-viewer	tools to analyze the potential role of PTMs	2019
CKSAAP_UbSite	http://systbio.cau.edu.cn/cksaap_ubsite/i	software to predict ubiquitylation sites	2013
UbiPred	http://flipper.diff.org/app/tools/info/2503	predict ubiquitylation sites	2010
UbPred	http://www.ubpred.org	random forest-based predictor of potential ubiquitination sites	2010

Keeping in view of the importance, in 2007, Chernorudskiy et al. created UbiProt (http://ubiprot.org.ru) which provided retrievable information about overall characteristics of a particular ubiquitylated protein, related ubiquitylation and deubiquitylation machinery and related literature references. However, this database could only collect 1104 ubiquitylated substrates from 12 species [105]. Considering the expensive and time-consuming nature of experimental methods, a computational method PEIMAN (Post-translational modification Enrichment Integration and Matching Analysis) was developed to study, analyze, predict, count and compute ubiquitylation and other PTMs [106]. Around 2.2% of genomic genes in most species belong to the UPS and thus Du et al. (2009) constructed the plants UPS database (http://bioinformatics.cau.edu.cn/plantsUPS/) that enabled the comparative analysis of UPS in higher plants [20]. This database collected 24 E1,417 E2 and 7624 E3 from seven plant species distributed in 11 UPS-involved gene families. However, due to some reasons, these databases are not updated time to time. Recently, Van-Nui et al. established UbiNet (http://csb.cse.yzu.edu.tw/UbiNet/), which has accumulated 43,948 experimentally verified ubiquitylation sites from 14,692 ubiquitylated proteins of humans and also provides a comprehensive map of intracellular ubiquitylation networks [107]. In Arabidopsis, an ubquitylome using modified COFRADIC technology was shared by the author with information concerning the respective splice variant, the modified sequence, the sequence window, as well as the position of the site within the protein (http://bioinformatics.psb.ugent.be/webtools/Ub_viewer/) [34].

Meanwhile, some databases provide integrative information of multiple PTMs and facilitate analysis of the related crosstalk. Among them, PTMCode (https://ptmcode.embl.de/) provides a resource of known and predicted PTMs functional associations between protein and PTMs within and between interacting proteins, it was updated in 2015 and currently contains 316,546 modified sites from 69 different PTM types [108]. Another database, dbPTM (http://dbptm.mbc.nctu.edu.tw/), an informative resource for PTMs, was established in 2012, now it has been updated to version dbPTM3.0, collecting 908,917 experimental PTM sites of over 130 PTM types. In addition, dbPTM also provides comprehensive functional and structural analyses for PTMs [109]. Patrick et al. (2019) integrated 19 types of protein modifications in plant proteins from five different species and established the Plant PTM Viewer, which comprises approximately 370, 000 PTM sites and remains open for submission, this repository will help to assume the role and potential interplay of PTMs in specific proteins [110].

In parallel, several useful softwares were also developed during the establishment of these databases. These include CKSAAP_UbSite (http://protein.cau.edu.cn/cksaap_ubsite/), UbiPred (http://iclab.life.nctu.edu.tw/ubipred) and UbPred (http://www.ubpred.org). These web servers may facilitate the prediction of ubiquitylation sites in proteins according to extract sequence features of amino acids and the class-balanced accuracy reaches over 70%.

7. Future Challenges of Plant Protein Ubiquitylation

Along with the great achievement, more concerns have also been raised in the study of plant protein ubiquitylation and addressing these challenges will bring new insights into the ubiquitylation structure and function. First, the discovery of particular connections of ubiquitylation in plants. Although the linkage-specific antibodies, affinity purification and mass spectrometry provide powerful approaches in investigating the designated Ub modified sites (N-terminal or lysine residues), they cannot be utilized for exploration of specific modification sites (like Cys, Ser and Thr residues). Therefore, as a greater challenge, exponential expansion of knowledge is needed to understand the complex poly-Ub chain structure (homo-/hetero- and branched-/linear- chain) and its corresponding functions. Limited by the relatively long lifespan and complex cell structure, the development and application of the technologies in plants are especially difficult. Moreover, understanding the detailed mechanism of the ubiquitylation system and uncovering its potential functions requires further exploration. Interactions between an E3 ligase and its cognate substrate are weak and dynamic and the ubiquitylation is unstable and reversible, which fail to identify the full spectrum of related E3 ligase and its substrates. Therefore, methods of verification and reconstruction of these modifications in vivo or in vitro are urgently required. The establishment of a comprehensive database is something to achieve in the near future. Recently, databases have been established to collect and process the massive ubiquitylation data, however, many of them are not promptly updated. Extensive integration of existing information (including ubiquitylation and the other PTMs) will provide a reliable public platform for scientific research, which will bring new insights into deciphering the PTM language (not only ubiquitylation) underlying the proteome and facilitate in developing innovative technologies in agriculture.

Author Contributions: Conceptualization and Original Draft Preparation, D.H. and P.Y., Review & Editing, P.Y., R.N.D., I.K. and M.L. All authors have read and agreed to the published version of the manuscript.

Funding: This work was supported by the National Natural Science Foundation of China (NSFC, No. 31671775). and China Scholarship Council (CSC No. 201908420055).

Conflicts of Interest: The authors declare no conflict of interest.

References

1. Millar, A.H.; Heazlewood, J.L.; Giglione, C.; Holdsworth, M.J.; Bachmair, A.; Schulze, W.X. The scope, functions and dynamics of posttranslational protein modifications. *Annu. Rev. Plant Biol.* **2019**, *70*, 119–151. [CrossRef]

2. Consortium, U. UniProt: A worldwide hub of protein knowledge. *Nucleic Acids Res.* **2019**, *47*, D506–D515. [CrossRef] [PubMed]

3. Hunter, T. The age of crosstalk: Phosphorylation, ubiquitination and beyond. *Mol. Cell* **2007**, *28*, 730–738. [CrossRef] [PubMed]

4. Zhang, Y.; Zeng, L. Crosstalk between Ubiquitination and Other Posttranslational Protein Modifications in Plant Immunity. *Plant Commun.* **2020**, *1*, 100041. [CrossRef]

5. Goldstein, G.; Scheid, M.; Hammerling, U.; Schlesinger, D.; Niall, H.; Boyse, E. Isolation of a polypeptide that has lymphocyte-differentiating properties and is probably represented universally in living cells. *Proc. Natl. Acad. Sci. USA* **1975**, *72*, 11–15. [CrossRef] [PubMed]

6. He, D.; Li, M.; Damaris, R.N.; Bu, C.; Xue, J.; Yang, P. Quantitative ubiquitylomics approach for characterizing the dynamic change and extensive modulation of ubiquitylation in rice seed germination. *Plant J.* **2020**, *101*, 1440–1447. [CrossRef] [PubMed]

7. Oh, E.; Akopian, D.; Rape, M. Principles of ubiquitin-dependent signaling. *Annu. Rev. Cell Dev. Biol.* **2018**, *34*, 137–162. [CrossRef] [PubMed]

8. Guo, J.; Liu, J.; Wei, Q.; Wang, R.; Yang, W.; Ma, Y.; Chen, G.; Yu, Y. Proteomes and Ubiquitylomes Analysis Reveals the Involvement of Ubiquitination in Protein Degradation in Petunias. *Plant Physiol.* **2017**, *173*, 668–687. [CrossRef]

9. Wang, Y.-F.; Chao, Q.; Li, Z.; Lu, T.-C.; Zheng, H.-Y.; Zhao, C.-F.; Shen, Z.; Li, X.-H.; Wang, B.-C. Large-scale Identification and Time-course Quantification of Ubiquitylation Events During Maize Seedling De-etiolation. *Genom. Proteom. Bioinform.* **2020**. [CrossRef]

10. Ye, Y.; Rape, M. Building ubiquitin chains: E2 enzymes at work. *Nat. Rev. Mol. Cell Biol.* **2009**, *10*, 755–764. [CrossRef]

11. Pao, K.-C.; Wood, N.T.; Knebel, A.; Rafie, K.; Stanley, M.; Mabbitt, P.D.; Sundaramoorthy, R.; Hofmann, K.; van Aalten, D.M.; Virdee, S. Activity-based E3 ligase profiling uncovers an E3 ligase with esterification activity. *Nature* **2018**, *556*, 381–385. [CrossRef] [PubMed]

12. Wang, X.; Herr, R.A.; Hansen, T.H. Ubiquitination of substrates by esterification. *Traffic* **2012**, *13*, 19–24. [CrossRef] [PubMed]

13. Breitschopf, K.; Bengal, E.; Ziv, T.; Admon, A.; Ciechanover, A. A novel site for ubiquitination: The N-terminal residue and not internal lysines of MyoD, is essential for conjugation and degradation of the protein. *EMBO J.* **1998**, *17*, 5964–5973. [CrossRef] [PubMed]

14. Peng, J.M.; Schwartz, D.; Elias, J.E.; Thoreen, C.C.; Cheng, D.M.; Marsischky, G.; Roelofs, J.; Finley, D.; Gygi, S.P. A proteomics approach to understanding protein ubiquitination. *Nat. Biotechnol.* **2003**, *21*, 921–926. [CrossRef]

15. Callis, J.; Carpenter, T.; Sun, C.-W.; Vierstra, R.D. Structure and evolution of genes encoding polyubiquitin and ubiquitin-like proteins in Arabidopsis thaliana ecotype Columbia. *Genetics* **1995**, *139*, 921–939.

16. Callis, J. The ubiquitination machinery of the ubiquitin system. *Arab. Book Am. Soc. Plant Biol.* **2014**, *12*, e0174. [CrossRef]

17. Isono, E.; Nagel, M.-K. Deubiquitylating enzymes and their emerging role in plant biology. *Front. Plant Sci.* **2014**, *5*, 56. [CrossRef]

18. Rape, M. Ubiquitylation at the crossroads of development and disease. *Nat. Rev. Mol. Cell Biol.* **2018**, *19*, 59–70. [CrossRef]

19. Schulman, B.A.; Harper, J.W. Ubiquitin-like protein activation by E1 enzymes: The apex for downstream signalling pathways. *Nat. Rev. Mol. Cell Biol.* **2009**, *10*, 319–331. [CrossRef]

20. Du, Z.; Zhou, X.; Li, L.; Su, Z. plantsUPS: A database of plants' Ubiquitin Proteasome System. *BMC Genom.* **2009**, *10*, 227. [CrossRef]

21. Stone, S.L.; Hauksdottir, H.; Troy, A.; Herschleb, J.; Kraft, E.; Callis, J. Functional analysis of the RING-type ubiquitin ligase family of Arabidopsis. *Plant Physiol.* **2005**, *137*, 13–30. [CrossRef] [PubMed]

22. Wenzel, D.M.; Lissounov, A.; Brzovic, P.S.; Klevit, R.E. UBCH7 reactivity profile reveals parkin and HHARI to be RING/HECT hybrids. *Nature* **2011**, *474*, 105–108. [CrossRef] [PubMed]

23. Clague, M.J.; Barsukov, I.; Coulson, J.M.; Liu, H.; Rigden, D.J.; Urbe, S. Deubiquitylases from Genes to Organism. *Physiol. Rev.* **2013**, *93*, 1289–1315. [CrossRef] [PubMed]

24. Komander, D.; Clague, M.J.; Urbe, S. Breaking the chains: Structure and function of the deubiquitinases. *Nat. Rev. Mol. Cell Biol.* **2009**, *10*, 550–563. [CrossRef] [PubMed]

25. Maor, R.; Jones, A.; Nühse, T.S.; Studholme, D.J.; Peck, S.C.; Shirasu, K. Multidimensional protein identification technology (MudPIT) analysis of ubiquitinated proteins in plants. *Mol. Cell Proteom.* **2007**, *6*, 601–610. [CrossRef]

26. Saracco, S.A.; Hansson, M.; Scalf, M.; Walker, J.M.; Smith, L.M.; Vierstra, R.D. Tandem affinity purification and mass spectrometric analysis of ubiquitylated proteins in Arabidopsis. *Plant J.* **2009**, *59*, 344–358. [CrossRef]

27. Lopitz-Otsoa, F.; Rodriguez, M.S.; Aillet, F. *Properties of Natural and Artificial Proteins Displaying Multiple Ubiquitin-Binding Domains*; Portland Press Ltd.: London, UK, 2010.

28. Kim, D.Y.; Scalf, M.; Smith, L.M.; Vierstra, R.D. Advanced proteomic analyses yield a deep catalog of ubiquitylation targets in Arabidopsis. *Plant Cell* **2013**, *25*, 1523–1540. [CrossRef]

29. Xu, G.; Paige, J.S.; Jaffrey, S.R. Global analysis of lysine ubiquitination by ubiquitin remnant immunoaffinity profiling. *Nat. Biotechnol.* **2010**, *28*, 868. [CrossRef]

30. Kim, W.; Bennett, E.J.; Huttlin, E.L.; Guo, A.; Li, J.; Possemato, A.; Sowa, M.E.; Rad, R.; Rush, J.; Comb, M.J. Systematic and quantitative assessment of the ubiquitin-modified proteome. *Mol. Cell* **2011**, *44*, 325–340. [CrossRef]

31. Wagner, S.A.; Beli, P.; Weinert, B.T.; Nielsen, M.L.; Cox, J.; Mann, M.; Choudhary, C. A Proteome-wide, Quantitative Survey of In Vivo Ubiquitylation Sites Reveals Widespread Regulatory Roles. *Mol. Cell Proteom.* **2011**, *10*. [CrossRef]

32. Vierstra, R.D. The expanding universe of ubiquitin and ubiquitin-like modifiers. *Plant Physiol.* **2012**, *160*, 2–14. [CrossRef] [PubMed]
33. Hjerpe, R.; Aillet, F.; Lopitz-Otsoa, F.; Lang, V.; England, P.; Rodriguez, M.S. Efficient protection and isolation of ubiquitylated proteins using tandem ubiquitin-binding entities. *EMBO Rep.* **2009**, *10*, 1250–1258. [CrossRef] [PubMed]
34. Walton, A.; Stes, E.; Cybulski, N.; Van Bel, M.; Inigo, S.; Durand, A.N.; Timmerman, E.; Heyman, J.; Pauwels, L.; De Veylder, L.; et al. It's Time for Some "Site"-Seeing: Novel Tools to Monitor the Ubiquitin Landscape in Arabidopsis thaliana. *Plant Cell* **2016**, *28*, 6–16. [CrossRef]
35. Stes, E.; Laga, M.; Walton, A.; Samyn, N.; Timmerman, E.; De Smet, I.; Goormachtig, S.; Gevaert, K. A COFRADIC protocol to study protein ubiquitination. *J. Proteome Res.* **2014**, *13*, 3107–3113. [CrossRef] [PubMed]
36. Renatus, M.; Parrado, S.G.; D'Arcy, A.; Eidhoff, U.; Gerhartz, B.; Hassiepen, U.; Pierrat, B.; Riedl, R.; Vinzenz, D.; Worpenberg, S. Structural basis of ubiquitin recognition by the deubiquitinating protease USP2. *Structure* **2006**, *14*, 1293–1302. [CrossRef] [PubMed]
37. Kliza, K.; Taumer, C.; Pinzuti, I.; Franz-Wachtel, M.; Kunzelmann, S.; Stieglitz, B.; Macek, B.; Husnjak, K. Internally tagged ubiquitin: A tool to identify linear polyubiquitin-modified proteins by mass spectrometry. *Nat. Methods* **2017**, *14*, 504–512. [CrossRef] [PubMed]
38. Rose, C.M.; Isasa, M.; Ordureau, A.; Prado, M.A.; Beausoleil, S.A.; Jedrychowski, M.P.; Finley, D.J.; Harper, J.W.; Gygi, S.P. Highly multiplexed quantitative mass spectrometry analysis of ubiquitylomes. *Cell Syst.* **2016**, *3*, 395–403.e394. [CrossRef] [PubMed]
39. Udeshi, N.D.; Mani, D.C.; Satpathy, S.; Fereshetian, S.; Gasser, J.A.; Svinkina, T.; Olive, M.E.; Ebert, B.L.; Mertins, P.; Carr, S.A. Rapid and deep-scale ubiquitylation profiling for biology and translational research. *Nat. Commun.* **2020**, *11*, 1–11. [CrossRef]
40. Schnell, J.D.; Hicke, L. Non-traditional functions of ubiquitin and ubiquitin-binding proteins. *J. Biol. Chem.* **2003**, *278*, 35857–35860. [CrossRef] [PubMed]
41. Hershko, A.; Ciechanover, A.; Varshavsky, A. The ubiquitin system. *Nat. Med.* **2000**, *6*, 1073–1081. [CrossRef]
42. Jin, L.; Williamson, A.; Banerjee, S.; Philipp, I.; Rape, M. Mechanism of ubiquitin-chain formation by the human anaphase-promoting complex. *Cell* **2008**, *133*, 653–665. [CrossRef]
43. Wang, F.; Zhu, D.; Huang, X.; Li, S.; Gong, Y.; Yao, Q.; Fu, X.; Fan, L.-M.; Deng, X.W. Biochemical insights on degradation of Arabidopsis DELLA proteins gained from a cell-free assay system. *Plant Cell* **2009**, *21*, 2378–2390. [CrossRef] [PubMed]
44. Leitner, J.; Petrášek, J.; Tomanov, K.; Retzer, K.; Pařezová, M.; Korbei, B.; Bachmair, A.; Zažímalová, E.; Luschnig, C. Lysine63-linked ubiquitylation of PIN2 auxin carrier protein governs hormonally controlled adaptation of Arabidopsis root growth. *Proc. Natl. Acad. Sci. USA* **2012**, *109*, 8322–8327. [CrossRef]
45. Ohtake, F.; Tsuchiya, H.; Saeki, Y.; Tanaka, K. K63 ubiquitylation triggers proteasomal degradation by seeding branched ubiquitin chains. *Proc. Natl. Acad. Sci. USA* **2018**, *115*, E1401–E1408. [CrossRef]
46. Swatek, K.N.; Usher, J.L.; Kueck, A.F.; Gladkova, C.; Mevissen, T.E.; Pruneda, J.N.; Skern, T.; Komander, D. Insights into ubiquitin chain architecture using Ub-clipping. *Nature* **2019**, *572*, 533–537. [CrossRef]
47. Shcherbik, N.; Pestov, D.G. Ubiquitin and ubiquitin-like proteins in the nucleolus: Multitasking tools for a ribosome factory. *Genes Cancer* **2010**, *1*, 681–689. [CrossRef]
48. Book, A.J.; Smalle, J.; Lee, K.H.; Yang, P.Z.; Walker, J.M.; Casper, S.; Holmes, J.H.; Russo, L.A.; Buzzinotti, Z.W.; Jenik, P.D.; et al. The RPN5 Subunit of the 26s Proteasome Is Essential for Gametogenesis, Sporophyte Development and Complex Assembly in Arabidopsis. *Plant Cell* **2009**, *21*, 460–478. [CrossRef]
49. Meyer, H.-J.; Rape, M. Enhanced protein degradation by branched ubiquitin chains. *Cell* **2014**, *157*, 910–921. [CrossRef]
50. Deshaies, R.J.; Joazeiro, C.A.P. RING Domain E3 Ubiquitin Ligases. *Annu. Rev. Biochem.* **2009**, *78*, 399–434. [CrossRef]
51. McEwan, D.G.; Dikic, I. The three musketeers of autophagy: Phosphorylation, ubiquitylation and acetylation. *Trends Cell Biol.* **2011**, *21*, 195–201. [CrossRef]
52. Li, F.; Chung, T.; Vierstra, R.D. AUTOPHAGY-RELATED11 plays a critical role in general autophagy-and senescence-induced mitophagy in Arabidopsis. *Plant Cell* **2014**, *26*, 788–807. [CrossRef] [PubMed]
53. Suttangkakul, A.; Li, F.; Chung, T.; Vierstra, R.D. The ATG1/ATG13 protein kinase complex is both a regulator and a target of autophagic recycling in Arabidopsis. *Plant Cell* **2011**, *23*, 3761–3779. [CrossRef] [PubMed]

54. Qi, H.; Xia, F.-N.; Xie, L.-J.; Yu, L.-J.; Chen, Q.-F.; Zhuang, X.-H.; Wang, Q.; Li, F.; Jiang, L.; Xie, Q. TRAF family proteins regulate autophagy dynamics by modulating AUTOPHAGY PROTEIN6 stability in Arabidopsis. *Plant Cell* **2017**, *29*, 890–911. [CrossRef] [PubMed]

55. Kirkin, V.; Lamark, T.; Sou, Y.-S.; Bjørkøy, G.; Nunn, J.L.; Bruun, J.-A.; Shvets, E.; McEwan, D.G.; Clausen, T.H.; Wild, P. A role for NBR1 in autophagosomal degradation of ubiquitinated substrates. *Mol. Cell* **2009**, *33*, 505–516. [CrossRef] [PubMed]

56. Fierz, B.; Chatterjee, C.; McGinty, R.K.; Bar-Dagan, M.; Raleigh, D.P.; Muir, T.W. Histone H2B ubiquitylation disrupts local and higher-order chromatin compaction. *Nat. Chem. Biol.* **2011**, *7*, 113. [CrossRef]

57. Winkler, M.; Niemeyer, M.; Hellmuth, A.; Janitza, P.; Christ, G.; Samodelov, S.L.; Wilde, V.; Majovsky, P.; Trujillo, M.; Zurbriggen, M.D. Variation in auxin sensing guides AUX/IAA transcriptional repressor ubiquitylation and destruction. *Nat. Commun.* **2017**, *8*, 1–13. [CrossRef]

58. Tokunaga, F.; Sakata, S.-I.; Saeki, Y.; Satomi, Y.; Kirisako, T.; Kamei, K.; Nakagawa, T.; Kato, M.; Murata, S.; Yamaoka, S. Involvement of linear polyubiquitylation of NEMO in NF-κB activation. *Nat. Cell Biol.* **2009**, *11*, 123–132. [CrossRef]

59. Michel, M.A.; Swatek, K.N.; Hospenthal, M.K.; Komander, D. Ubiquitin linkage-specific affimers reveal insights into K6-linked ubiquitin signaling. *Mol. Cell* **2017**, *68*, 233–246.e235. [CrossRef]

60. Min, M.; Mevissen, T.E.; De Luca, M.; Komander, D.; Lindon, C. Efficient APC/C substrate degradation in cells undergoing mitotic exit depends on K11 ubiquitin linkages. *Mol. Biol. Cell* **2015**, *26*, 4325–4332. [CrossRef]

61. Zhang, Y.; Zhang, H.; Zheng, G.-L.; Yang, Q.; Yu, S.; Wang, J.; Li, S.; Li, L.-F.; Qiu, H.-J. Porcine RING finger protein 114 inhibits classical swine fever virus replication via K27-linked polyubiquitination of viral NS4B. *J. Virol.* **2019**, *93*, e01248-19. [CrossRef]

62. Yang, L.; Ritchie, A.-M.; Melton, D.W. Disruption of DNA repair in cancer cells by ubiquitination of a destabilising dimerization domain of nucleotide excision repair protein ERCC1. *Oncotarget* **2017**, *8*, 55246. [CrossRef]

63. Emmerich, C.H.; Ordureau, A.; Strickson, S.; Arthur, J.S.C.; Pedrioli, P.G.; Komander, D.; Cohen, P. Activation of the canonical IKK complex by K63/M1-linked hybrid ubiquitin chains. *Proc. Natl. Acad. Sci. USA* **2013**, *110*, 15247–15252. [CrossRef] [PubMed]

64. Shanklin, J.; Jabben, M.; Vierstra, R.D. Red light-induced formation of ubiquitin-phytochrome conjugates: Identification of possible intermediates of phytochrome degradation. *Proc. Natl. Acad. Sci. USA* **1987**, *84*, 359–363. [CrossRef] [PubMed]

65. Xu, X.; Paik, I.; Zhu, L.; Huq, E. Illuminating progress in phytochrome-mediated light signaling pathways. *Trends Plant Sci.* **2015**, *20*, 641–650. [CrossRef]

66. Sorel, M.; Mooney, B.; de Marchi, R.; Graciet, E. Ubiquitin/Proteasome System in Plant Pathogen Responses. *Annu. Plant Rev. Online* **2018**, 65–116. [CrossRef]

67. Swaney, D.L.; Beltrao, P.; Starita, L.; Guo, A.L.; Rush, J.; Fields, S.; Krogan, N.J.; Villen, J. Global analysis of phosphorylation and ubiquitylation cross-talk in protein degradation. *Nat. Methods* **2013**, *10*, 676–682. [CrossRef]

68. Gao, D.Y.; Xu, Z.S.; He, Y.; Sun, Y.W.; Ma, Y.Z.; Xia, L.Q. Functional analyses of an E3 ligase gene AIP2 from wheat in Arabidopsis revealed its roles in seed germination and pre-harvest sprouting. *J. Integr. Plant Biol.* **2014**, *56*, 480–491. [CrossRef]

69. Bueso, E.; Rodriguez, L.; Lorenzo-Orts, L.; Gonzalez-Guzman, M.; Sayas, E.; Munoz-Bertomeu, J.; Ibanez, C.; Serrano, R.; Rodriguez, P.L. The single-subunit RING-type E3 ubiquitin ligase RSL1 targets PYL4 and PYR1 ABA receptors in plasma membrane to modulate abscisic acid signaling. *Plant J.* **2014**, *80*, 1057–1071. [CrossRef]

70. Chapman, E.J.; Estelle, M. Mechanism of auxin-regulated gene expression in plants. *Annu. Rev. Genet.* **2009**, *43*, 265–285. [CrossRef]

71. Martins, S.; Dohmann, E.M.N.; Cayrel, A.; Johnson, A.; Fischer, W.; Pojer, F.; Satiat-Jeunemaitre, B.; Jaillais, Y.; Chory, J.; Geldner, N.; et al. Internalization and vacuolar targeting of the brassinosteroid hormone receptor BRI1 are regulated by ubiquitinationpavr. *Nat. Commun.* **2015**, *6*, 1–11. [CrossRef]

72. Ju, C.; Chang, C. Mechanistic insights in ethylene perception and signal transduction. *Plant Physiol.* **2015**, *169*, 85–95. [CrossRef]

73. Qiao, H.; Chang, K.N.; Yazaki, J.; Ecker, J.R. Interplay between ethylene, ETP1/ETP2 F-box proteins and degradation of EIN2 triggers ethylene responses in Arabidopsis. *Genes Dev.* **2009**, *23*, 512–521. [CrossRef]

74. Kulathu, Y.; Komander, D. Atypical ubiquitylation—The unexplored world of polyubiquitin beyond Lys48 and Lys63 linkages. *Nat. Rev. Mol. Cell Biol.* **2012**, *13*, 508–523. [CrossRef] [PubMed]

75. Elia, A.E.; Boardman, A.P.; Wang, D.C.; Huttlin, E.L.; Everley, R.A.; Dephoure, N.; Zhou, C.; Koren, I.; Gygi, S.P.; Elledge, S.J. Quantitative Proteomic Atlas of Ubiquitination and Acetylation in the DNA Damage Response. *Mol. Cell* **2015**, *59*, 867–881. [CrossRef] [PubMed]

76. Ciccia, A.; Elledge, S.J. The DNA Damage Response: Making It Safe to Play with Knives. *Mol. Cell* **2010**, *40*, 179–204. [CrossRef]

77. Zhang, Y. Transcriptional regulation by histone ubiquitination and deubiquitination. *Genes Dev.* **2003**, *17*, 2733–2740. [CrossRef] [PubMed]

78. Shilatifard, A. Chromatin modifications by methylation and ubiquitination: Implications in the regulation of gene expression. *Annu. Rev. Biochem.* **2006**, *75*, 243–269. [CrossRef]

79. Pinder, J.B.; Attwood, K.M.; Dellaire, G. Reading, writing and repair: The role of ubiquitin and the ubiquitin-like proteins in DNA damage signaling and repair. *Front. Genet.* **2013**, *4*, 45. [CrossRef]

80. Goldknopf, I.L.; French, M.F.; Musso, R.; Busch, H. Presence of protein A24 in rat liver nucleosomes. *Proc. Natl. Acad. Sci. USA* **1977**, *74*, 5492–5495. [CrossRef]

81. McGinty, R.K.; Henrici, R.C.; Tan, S. Crystal structure of the PRC1 ubiquitylation module bound to the nucleosome. *Nature* **2014**, *514*, 591–596. [CrossRef]

82. Blackledge, N.P.; Farcas, A.M.; Kondo, T.; King, H.W.; McGouran, J.F.; Hanssen, L.L.; Ito, S.; Cooper, S.; Kondo, K.; Koseki, Y. Variant PRC1 complex-dependent H2A ubiquitylation drives PRC2 recruitment and polycomb domain formation. *Cell* **2014**, *157*, 1445–1459. [CrossRef]

83. Pavri, R.; Zhu, B.; Li, G.; Trojer, P.; Mandal, S.; Shilatifard, A.; Reinberg, D. Histone H2B monoubiquitination functions cooperatively with FACT to regulate elongation by RNA polymerase II. *Cell* **2006**, *125*, 703–717. [CrossRef] [PubMed]

84. Ishiyama, S.; Nishiyama, A.; Saeki, Y.; Moritsugu, K.; Morimoto, D.; Yamaguchi, L.; Arai, N.; Matsumura, R.; Kawakami, T.; Mishima, Y. Structure of the Dnmt1 reader module complexed with a unique two-mono-ubiquitin mark on histone H3 reveals the basis for DNA methylation maintenance. *Mol. Cell* **2017**, *68*, 350–360.e357. [CrossRef] [PubMed]

85. Tessadori, F.; Giltay, J.C.; Hurst, J.A.; Massink, M.P.; Duran, K.; Vos, H.R.; van Es, R.M.; Study, D.D.D.; Scott, R.H.; van Gassen, K.L. Germline mutations affecting the histone H4 core cause a developmental syndrome by altering DNA damage response and cell cycle control. *Nat. Genet.* **2017**, *49*, 1642. [CrossRef] [PubMed]

86. Mevissen, T.E.; Komander, D. Mechanisms of deubiquitinase specificity and regulation. *Annu. Rev. Biochem.* **2017**, *86*, 159–192. [CrossRef] [PubMed]

87. Levkowitz, G.; Waterman, H.; Zamir, E.; Kam, Z.; Oved, S.; Langdon, W.Y.; Beguinot, L.; Geiger, B.; Yarden, Y. C-Cbl/Sli-1 regulates endocytic sorting and ubiquitination of the epidermal growth factor receptor. *Genes Dev.* **1998**, *12*, 3663–3674. [CrossRef] [PubMed]

88. Isono, E.; Kalinowska, K. ESCRT-dependent degradation of ubiquitylated plasma membrane proteins in plants. *Curr. Opin. Plant Biol.* **2017**, *40*, 49–55. [CrossRef] [PubMed]

89. Schmidt, O.; Teis, D. The ESCRT machinery. *Curr. Biol.* **2012**, *22*, R116–R120. [CrossRef]

90. Dobzinski, N.; Chuartzman, S.G.; Kama, R.; Schuldiner, M.; Gerst, J.E. Starvation-dependent regulation of golgi quality control links the TOR signaling and vacuolar protein sorting pathways. *Cell Rep.* **2015**, *12*, 1876–1886. [CrossRef]

91. Gorur, A.; Yuan, L.; Kenny, S.J.; Baba, S.; Xu, K.; Schekman, R. COPII-coated membranes function as transport carriers of intracellular procollagen I. *J. Cell Biol.* **2017**, *216*, 1745–1759. [CrossRef]

92. Mukai, A.; Mizuno, E.; Kobayashi, K.; Matsumoto, M.; Nakayama, K.I.; Kitamura, N.; Komada, M. Dynamic regulation of ubiquitylation and deubiquitylation at the central spindle during cytokinesis. *J. Cell Sci.* **2008**, *121 Pt 8*, 1325–1333. [CrossRef]

93. Vu, L.D.; Gevaert, K.; De Smet, I. Protein Language: Post-Translational Modifications Talking to Each Other. *Trends Plant Sci.* **2018**, *23*, 1068–1080. [CrossRef] [PubMed]

94. Gómez-Gómez, L.; Boller, T. FLS2: An LRR receptor–like kinase involved in the perception of the bacterial elicitor flagellin in Arabidopsis. *Mol. Cell* **2000**, *5*, 1003–1011. [CrossRef]

95. Lu, D.; Lin, W.; Gao, X.; Wu, S.; Cheng, C.; Avila, J.; Heese, A.; Devarenne, T.P.; He, P.; Shan, L. Direct ubiquitination of pattern recognition receptor FLS2 attenuates plant innate immunity. *Science* **2011**, *332*, 1439–1442. [CrossRef]

96. Liu, J.; Deng, J.; Zhu, F.; Li, Y.; Lu, Z.; Qin, P.; Wang, T.; Dong, J. The MtDMI2-MtPUB2 negative feedback loop plays a role in nodulation homeostasis. *Plant Physiol.* **2018**, *176*, 3003–3026. [CrossRef] [PubMed]

97. Orlicky, S.; Tang, X.; Willems, A.; Tyers, M.; Sicheri, F. Structural basis for phosphodependent substrate selection and orientation by the SCFCdc4 ubiquitin ligase. *Cell* **2003**, *112*, 243–256. [CrossRef]

98. Xu, F.; Huang, Y.; Li, L.; Gannon, P.; Linster, E.; Huber, M.; Kapos, P.; Bienvenut, W.; Polevoda, B.; Meinnel, T. Two N-terminal acetyltransferases antagonistically regulate the stability of a nod-like receptor in Arabidopsis. *Plant Cell* **2015**, *27*, 1547–1562. [CrossRef] [PubMed]

99. Pajares, M.; Jiménez-Moreno, N.; Dias, I.H.; Debelec, B.; Vucetic, M.; Fladmark, K.E.; Basaga, H.; Ribaric, S.; Milisav, I.; Cuadrado, A. Redox control of protein degradation. *Redox Biol.* **2015**, *6*, 409–420. [CrossRef] [PubMed]

100. Yau, R.; Rape, M. The increasing complexity of the ubiquitin code. *Nat. Cell Biol.* **2016**, *18*, 579–586. [CrossRef]

101. Okatsu, K.; Oka, T.; Iguchi, M.; Imamura, K.; Kosako, H.; Tani, N.; Kimura, M.; Go, E.; Koyano, F.; Funayama, M. PINK1 autophosphorylation upon membrane potential dissipation is essential for Parkin recruitment to damaged mitochondria. *Nat. Commun.* **2012**, *3*, 1–10. [CrossRef]

102. Herhaus, L.; Dikic, I. Expanding the ubiquitin code through post-translational modification. *EMBO Rep.* **2015**, *16*, 1071–1083. [CrossRef] [PubMed]

103. He, D.L.; Wang, Q.; Li, M.; Damaris, R.N.; Yi, X.L.; Cheng, Z.Y.; Yang, P.F. Global Proteome Analyses of Lysine Acetylation and Succinylation Reveal the Widespread Involvement of both Modification in Metabolism in the Embryo of Germinating Rice Seed. *J. Proteome Res.* **2016**, *15*, 879–890. [CrossRef]

104. Li, M.; Yin, X.J.; Sakata, K.; Yang, P.F.; Komatsu, S. Proteomic Analysis of Phosphoproteins in the Rice Nucleus During the Early Stage of Seed Germination. *J. Proteome Res.* **2015**, *14*, 2884–2896. [CrossRef] [PubMed]

105. Chernorudskiy, A.L.; Garcia, A.; Eremin, E.V.; Shorina, A.S.; Kondratieva, E.V.; Gainullin, M.R. UbiProt: A database of ubiquitylated proteins. *BMC Bioinform.* **2007**, *8*, 126. [CrossRef] [PubMed]

106. Nickchi, P.; Jafari, M.; Kalantari, S. PEIMAN 1.0: Post-translational modification Enrichment, Integration and Matching ANalysis. *Database* **2015**, *2015*. [CrossRef] [PubMed]

107. Nguyen, V.-N.; Huang, K.-Y.; Weng, J.T.-Y.; Lai, K.R.; Lee, T.-Y. UbiNet: An online resource for exploring the functional associations and regulatory networks of protein ubiquitylation. *Database* **2016**, *2016*. [CrossRef]

108. Minguez, P.; Letunic, I.; Parca, L.; Garcia-Alonso, L.; Dopazo, J.; Huerta-Cepas, J.; Bork, P. PTMcode v2: A resource for functional associations of post-translational modifications within and between proteins. *Nucleic Acids Res.* **2015**, *43*, D494–D502. [CrossRef]

109. Huang, K.-Y.; Lee, T.-Y.; Kao, H.-J.; Ma, C.-T.; Lee, C.-C.; Lin, T.-H.; Chang, W.-C.; Huang, H.-D. dbPTM in 2019: Exploring disease association and cross-talk of post-translational modifications. *Nucleic Acids Res.* **2019**, *47*, D298–D308. [CrossRef]

110. Willems, P.; Horne, A.; Van Parys, T.; Goormachtig, S.; De Smet, I.; Botzki, A.; Van Breusegem, F.; Gevaert, K. The Plant PTM Viewer, a central resource for exploring plant protein modifications. *Plant J.* **2019**, *99*, 752–762. [CrossRef] [PubMed]

Publisher's Note: MDPI stays neutral with regard to jurisdictional claims in published maps and institutional affiliations.

International Journal of
Molecular Sciences

MDPI

Review

Nanoparticles: Synthesis, Morphophysiological Effects, and Proteomic Responses of Crop Plants

Zahed Hossain [1], Farhat Yasmeen [2] and Setsuko Komatsu [3,*]

[1]   Department of Botany, University of Kalyani, West Bengal 741235, India; zahed_kly@yahoo.com
[2]   Department of Botany, Women University, Swabi 23340, Pakistan; fyasmeen@wus.edu.pk
[3]   Department of Environmental and Food Science, Fukui University of Technology, Fukui 910-8505, Japan
*   Correspondence: skomatsu@fukui-ut.ac.jp; Tel.: +81-776-29-2466

Received: 21 March 2020; Accepted: 23 April 2020; Published: 26 April 2020

Abstract: Plant cells are frequently challenged with a wide range of adverse environmental conditions that restrict plant growth and limit the productivity of agricultural crops. Rapid development of nanotechnology and unsystematic discharge of metal containing nanoparticles (NPs) into the environment pose a serious threat to the ecological receptors including plants. Engineered nanoparticles are synthesized by physical, chemical, biological, or hybrid methods. In addition, volcanic eruption, mechanical grinding of earthquake-generating faults in Earth's crust, ocean spray, and ultrafine cosmic dust are the natural source of NPs in the atmosphere. Untying the nature of plant interactions with NPs is fundamental for assessing their uptake and distribution, as well as evaluating phytotoxicity. Modern mass spectrometry-based proteomic techniques allow precise identification of low abundant proteins, protein–protein interactions, and in-depth analyses of cellular signaling networks. The present review highlights current understanding of plant responses to NPs exploiting high-throughput proteomics techniques. Synthesis of NPs, their morphophysiological effects on crops, and applications of proteomic techniques, are discussed in details to comprehend the underlying mechanism of NPs stress acclimation.

Keywords: nanoparticles; crop; proteomics; plant-nanoparticles interaction; nanoparticles synthesis

1. Introduction

Rapid advancement in nanotechnology has taken the food industry to a new height [1]. Nanoparticles (NPs) are ultrafine particles with a size of less than 100 nm in at least one dimension [2]. Owing to having unique physical and chemical properties, such as high surface area and nanoscale size, these microscopic particles have the potential to improve the quality of food processing, packaging, storage, transportation, functionality, and other safety aspects of food [2]. Moreover, in recent years, nanotechnology has gained tremendous attention in agriculture sector as promising agents for plant growth, fertilizers, and pesticides, ensuring sustainable crop production [3]. The engineered nanomaterials have a wide range of applications in the healthcare industry, including drug delivery [4], cellular imaging and diagnosis [5], cancer therapy [6], antimicrobials [7], biosensors [8], anti-diabetic agents [9], and cosmetics [10]. Nevertheless, unsystematic release of nano-containing biosolids and agrochemicals is a serious threat to the environment, including plants [11].

Among metal based NPs, iron NPs are widely used in environmental remediation, biomedical, diagnostic field, and drug delivery because of their unique properties, such as excellent biodegradability, low cytotoxicity, and ability to attach with multiple targeted ligands or antibodies [12,13]. Few studies have been conducted to assess the impact of iron NPs on plants [14,15]. Kim et al. [14] reported that exposure of iron NPs triggered root elongation in *Arabidopsis thaliana* by nZVI-mediated OH radical-induced cell wall loosening. Conversely, iron–ion/NPs did not affect physiological parameters in lettuce plant [15]. Similar to iron, copper NPs have diverse applications, such as electro metallic agent, wood preservative, bioactive, and lubricant [16]. However, unmanaged discharge of copper NPs into the environment poses an increasing threat to plants [17]. Hence, there is urgent need of in-depth research for understanding the various pathways involved in NPs stress response mechanisms in plants. Most of the phytotoxicity research so far conducted is focused on effects of NPs on seed germination and, at very early growth stages, of the plants [18]. Techniques, including cytotoxicity study [19], transcriptomics [20], and proteomics [21] have been widely used for analyzing uptake, bioaccumulation, biotransformation, and risks of NPs for food crops. Moreover, NP-mediated phytotoxicity as well as their ecotoxicity was conducted on mammalian cells [22]. These high-throughput genome-based omics techniques have been used extensively to dissect plant responses to NPs [23]. Although transcriptional analysis was performed in a variety of organisms including microbes, humans, mammalian cell lines, and other model organisms [24], information about plant–NPs interactions and NP-mediated phytotoxicity is still limited.

The high-throughput techniques used in proteomics focus on revealing structure and conformation of proteins, protein–protein, and protein–ligand interactions. Proteomics offer several advantages over the genome or transcriptome-based technologies as it directly deals with the functional molecules rather than DNA or mRNA [25]. Gel-based or gel-free proteomic techniques, protein chips/microarrays, and protein biomarkers have been widely used for reliable identification and accurate quantitation of stress responsive proteins for dissecting plant stress signaling pathways [26]. Improved protein extraction protocol and advancement in mass spectrometry have made proteomics a rapid, sensitive, and reliable technique for identification and characterization of differentially modulated proteins to assess the possible impact of NPs on crops. Alternative to single omics approach, multi-omics techniques, such as combination of transcriptomics, proteomics, and metabolomics offer more advantages in identifying the underlying response mechanisms of plants towards the environmental contaminants, including NPs [27]. This review highlights the various methods used for synthesis of NPs, their morphophysiological impact on crop plants, and applications of proteomic techniques to comprehend the underlying mechanism of NPs stress acclimation.

2. Methods for NPs Synthesis

The size, concentration, and stability of NPs primarily determine their effects on plants [23]. The characteristics of NPs largely depend on their mode of synthesis. There are various physical, chemical, and biological methods for the synthesis of economically important NPs [28]. Although the methods of NPs synthesis are diverse, there is a bare necessity to develop some ecofriendly processes so that they may be less hazardous to the environment (Table 1).

2.1. Physical Methods for NPs Synthesis

These methods are being used for the synthesis of various economically important NPs, such as silver, copper, iron, titanium, and others. The method of tube furnace was used for the synthesis of spherical silver NPs [29]; while laser ablation resulted in the formation of triangular bipyramidal nanocrystals of silver [30]. NPs synthesized by Ytterbium fiber laser ablation were spherical in shape and polycrystalline in nature [31]. Iron NPs with the globular shape were produced using the thermal dehydration method [32]; whereas irregular shape was attained with thermal decomposition approach [33]. Furthermore, copper NPs with spherical shaped and uniform diameters were synthesized using the thermal decomposition approach [34]. The topographic map indicated that NPs synthesized through sodium borohydrate as the reducing agent produced the NPs with irregular surfaces [35], while the polyol method synthesized pure crystalline copper NPs with cubic surface [36]. When tween 80 was added as modification in the polyol method, it resulted in the formation of crystalline copper NPs [37]. The physical approaches mainly synthesized the NPs with uniform morphological characteristics, which ultimately affected their response towards the environment as well as to the living ecosystem.

2.2. Chemical Methods for NPs Synthesis

The chemical reduction using a variety of organic/inorganic reducing agents, electrochemical techniques, physicochemical reduction, and radiolysis is a well-accepted approach for the synthesis of NPs [38]. The process of reduction through various chemicals led to the synthesis of the diverse shape of properties of NPs, such as silver nitrate reduction with sodium borohydrate resulted in the mixture of spherical and rod shaped silver NPs [39]; however, iron NPs were spherical when iron salt was reduced with sodium borohydrate [40]. The reduction of copper salts with sodium borohydrate produced spherical [41] and irregular NPs [35]. Sonochemical and thermal reduction of copper hydrazine carboxylate produced a network of irregular shaped copper NPs [42]. Wet chemical synthesis involving stoichiometric reaction also produced spherical copper NPs [43]. Moreover, wet chemical method produced nanowires of silver [44]; while spherical silver NPs were produced on ascorbic acid as a reducing agent [45]. Mesoporous silica resulted in the formation of iron NPs having uniform pore size and large surface area [46]. The zinc NPs with crystalline shaped morphology were obtained using ammonium carbamate as a precipitating agent [47]; while refluxing zinc acetate precursor in diethylene and triethylene glycol synthesized oval to rod shaped NPs [48]. Due to the usage of various chemicals for NPs synthesis, there is growing concern about the possible release and effect of NPs in the surrounding environment.

2.3. Biological and Green Methods for NPs Synthesis

In biological and green methods, living organisms, such as bacteria, viruses, and plants, are used as capping and reducing agents. The crystal lattice structure of synthesized copper NPs was achieved through *Morganella* [49]. Silver NPs with spherical and cubic shaped having crystalline nature were synthesized using extracts of *Litchi chinensis* [50], *Eucalyptus macrocarpa* [51], and *Rhazya stricta* [52]. Iron NPs were synthesized using leaf extract of barberry, *Elaeagnus angustifolia*, sattron, *Ziziphus jujube* [53], grape tree [54], and green tea [55]. The involvement of *Albizia lebbeck* bioactive compounds in the stabilization of zincoxide NPs were confirmed through various techniques and revealed irregular spherical morphology [56]; while crystalline hexagonal stage was obtained through the seed extract of *Ricinus communis* [57]. Leaf extract of *Aloe vera* also synthesized highly stable and spherical zinc oxide NPs [58]. Copper NPs were produced using extracts of *Ocimum sanctum* leaf [59], *Cassia alata* flower [60], *Capparis zelynica* leaf [61], and *Syzygium aromaticum* solution [62]. Studies have shown that green synthesis methods exploiting plants or microorganisms are relatively safe, inexpensive, and environment-friendly.

Table 1. Mode of synthesis and characteristics of commercially important nanoparticles (NPs).

NPs	Mode of Synthesis	Size (nm)	Characters	Ref *
	Litchi chinensis leaf extract	41–55	Crystalline nature	[50]
	Tube furnace	6.2–21.5	Spherical shape	[29]
	Laser ablation	20–50	Pentagonal one dimensional (1-D) nanorods, nanowires, cubic/triangular-bipyramidal nanocrystals	[30]
Silver NPs	Carboxymethylated chitosan with ultraviolet light irradiation	2–8	Cubic crystal structure	[40]
	Eucalyptus macrocarpa leaf extract	10–100	Spherical and cubic shaped	[51]
	Sodium borohydride	2–4	Nanorods	[63]
	Silver nitrate with sodium borate	20–50	Mixture of spherical and rod NPs	[39]
	Wet chemical method	20	Nanowires	[44]
	Ascorbic acid as a reducing agent	31	Spherical shaped	[45]
	Silver nitrate and methanolic *Rhazya stricta* root extract	20	Spherical shaped	[52]

Table 1. *Cont.*

NPs	Mode of Synthesis	Size (nm)	Characters	Ref *
Iron NPs	Leaf extract of barberry, *Elaeagnus angustifolia*, *Ziziphus jujube*	40	Spherical shaped	[53]
	Sodium borohydride	44.87	Spherical shaped	[40]
	Ferric chloride precursor with sodium borohydride	6	Spherical in shape	[40]
	Grape tree leaf extract	10–30	Spherical and non-agglomerated	[54]
	Green tea extract	40–60	Amorphous in nature, chain morphology	[55]
	Mesoporous silica	10–300	Uniform pore size, large surface area, high accessible pore volume	[46]
	Thermal dehydration	6–10	globular-shape crystallites	[32]
	Thermal decomposition	50	Irregular and not spherical	[33]
Zinc oxide NPs	*Albizia lebbeck*	66.25	Irregular spherical morphology	[56]
	Chamomile flower extract	48.2	Pure crystalline	[64]
	Ricinus communis seed extract	20	Crystalline hexagonal	[57]
	Ammonium carbamate	10–15	Crystallite rod-shape	[47]
	Aloe vera leaf extract	25–40	Highly stable and spherical	[58]
	Refluxing zinc acetate precursor in diethylene/triethylene glycol	15–100	Oval to rod shape	[48]
Copper NPs	Alcothermal method	6	High dispersion, narrow size distribution	[9]
	Sodium borohydride	17.25	Spherical shaped	[41]
	Thermal decomposition	15–30	Nearly spherical with relatively uniform diameters	[34]
	Biosynthesis by *Morganella*	15–20	Crystal lattice structure	[49]
	Sodium borohydride	15	Pure crystalline metallic phase with face centered cubic, rich in dents, irregular surface	[35]
	Polyol method	45	Pure crystalline with face centered cubic structure	[36]
	Ocimum sanctum leaf extract	77	Different organic molecules, high crystallinity	[59]
	Wet chemical synthesis involving stoichiometric reaction	9	Spherical	[43]
	Polyol method by copper acetate hydrate in tween 80	580	Crystalline nature	[37]
	Reduction of copper (II) acetate in water and 2-ethoxyethanol using hydrazine under reflux	6–23	Spherical	[40]
	Thermal reduction	200–250	Irregular particles	[42]
	Sonochemical reduction	50–70	Irregular network of small NPs	[42]
	Cassia alata flower extract	110–280	Aggregates with rough, particles, spherical	[60]
	Capparis zeylanica leaf extract	50–100	Cubical structure	[61]
	Syzygium aromaticum extract	5–40	Spherical and granular nature	[62]
Titanium oxide NPs	Ytterbium fiber laser ablation	25	Spherical and polycrystalline	[31]
	Taguchi method	18.11	Spherical	[65]
	Sol-gel method	15	Crystalline and nearly spherical	[66]

* Ref means references.

3. Morphological and Physiological Effects of NPs on Crops

The most advanced interdisciplinary tool with the larger potential in agriculture for increased crop productivity is the nanotechnology in which NPs with varying size, concentration, and surface charge influenced the growth and development of diverse plant species [67]. A variety of NPs have been tested against germination of seeds, growth of shoot/root, and crop production [68]. NPs exert species-specific toxicity, plant organ specificity, as well as stress dependency (Table 2).

3.1. Plant Species Specificity of NPs

The impact of NPs depends on the type of plant species used. The aqueous suspension of aluminum oxide NPs improved the root growth of radish [69] but reduced in cucumber [70]. The aqueous suspension of titanium oxide NPs increased root length of wheat [71] but inhibited in cucumber [72]. The iron NPs aqueous suspension increased root length of *Arabidopsis thaliana* [14] and restricted in lettuce [15]. The aqueous suspension of titanium oxide NPs inhibited root elongation in cucumber [69] and carrot [70], but enhanced the growth of maize [1], wheat [73], and spinach [74–77]. The carbon-nanotubes suspension increased germination rate, fresh biomass, and seedling length in *Solanum lycopersicum* [78], *Allium cepa* [79], and wheat [80], while reduced in *Cucurbita pepo* [81], rice [72], and lettuce [79]. These studies have increased our understanding of phytotoxicity and plant responses towards NPs.

3.2. Plant Organ Specific Effects of NPs

The carbon nanotubes, copper-oxide NPs, and titanium-dioxide NPs increased resistance to fungal infection by altering the level of endogenous hormones [82]. The direct application of silver NPs reduced seedling biomass of wheat [83], zucchini [81], mung bean [83], and cabbage [84]; while it regulated the seedling growth in maize [84] and *Vigna radiata* [83]. The hydroponic applications of silver NPs enhanced root elongation in rice [85]; while it reduced in zucchini [81]. Changes in the morphological characteristics of treated plants depend on the types of NPs used. Silver NPs and aluminum-oxide NPs reduced [86] and improved [87], respectively, growth of wheat. The iron NPs enhanced germination ratio and plant growth [88]; while copper NPs inhibited the growth of wheat [89]. The flowering and yield of rice reduced on carbon nanotubes exposure [73]; while enhanced under cerium-oxide NPs treatment [90]. Silver NPs [84] and cerium-oxide NPs [91] improved the growth of maize; while aluminum-oxide NPs [70], titanium-oxide NPs [1], and copper NPs [80] treatments led to growth reduction. Keeping in view these studies, NPs might be involved in the alteration of growth in plants.

3.3. Stress Dependency of NPs

Various modes of applications determine the effects of NPs on growth and productivity of plants. Direct application of aluminum oxide NPs improved root length of wheat [87]; while reduced in maize in hydroponic condition [70]. Exposure of aluminum oxide NPs improved survival percentage and weight/length of root including hypocotyl of soybean under flooding stress [92,93]. There are some NPs with the capability to keep the same effects on the plant, though, applied through various ways, e.g., titanium-oxide NPs improved the growth of spinach when applied through seed treatment [94] and foliar spray [95]. Similarly, soil or direct application of iron NPs increased the growth [96] and yield [97] of wheat. The alteration in the morphology of plants is dependent on the mode of application and the type of NPs exposure is dependent on the mode of application.

Table 2. Mode of applications and morphophysiological responses of crops upon NPs treatments.

NPs	Species	Mode of application	Morphophysiological responses	Ref *
Silver NPs	Rice	Hydroponic application	Enhanced root length	[85]
	Wheat	Direct application	Reduced seedling growth	[86]
	Zucchini	Direct application	Reduced seedling biomass	[81]
	Wheat	Direct application	Reduced seedling biomass	[83]
	Mung bean	Direct application	Reduced seedling biomass	[83]
	Cabbage	Direct application	Decreased root length	[84]
	Maize	Direct application	Increased root length	[84]
	Eruca sativa	Direct application	Increased root length	[98]
	Ajwain	Direct application	Improved water use efficiency, nutrient uptake, reduced fertilizer requirement	[99]
	Zucchini	Hoagland solution	Reduced rate of transpiration	[81]
	Mung bean	Direct application	Regulated seedling growth	[83]
Aluminum oxide NPs	Wheat	Direct application	Enhanced root growth	[87]
	Maize	Hydroponic application	Reduced root elongation	[70]
	Soybean	Direct application	Improved survival and root growth	[92]
	Maize	Direct application	Increased root length	[69]
	Soybean	Flooding	Increased root length	[93]
	Radish	Aqueous suspension	Improved root growth	[69]
	Cucumber	Aqueous suspension	Reduced root growth	[70]

Table 2. *Cont.*

NPs	Species	Mode of application	Morphophysiological responses	Ref *
Titanium oxide NPs	Wheat	Aqueous suspension	Increased root length	[71]
	Rose	Water-agar plates with suspension	Enhanced plant resistance to fungal infection by altering endogenous hormones content	[82]
	Cucumber	Aqueous suspension	Restricted root growth	[69]
	Carrot	Aqueous suspension	Restricted root growth	[70]
	Wheat	Aqueous suspension	Reduced biomass	[100]
	Spinach	Seed treatment	Enhanced growth	[74]
	Spinach	Seed treatment	Significantly affected the plant growth	[94]
	Spinach	Foliar spray	Increased seedling growth	[95]
	Chickpea	Foliar spray	Improved redox status	[101]
	Spinach	Seed treatment	Increased dry weight and chlorophyll content	[94]
	Narbon bean	Seed treatment	Reduced seed germination and root length	[1]
	Maize	Seed treatment	Reduced seed germination and root length	[1]
	Wheat	Aqueous suspension	Increased shoot length	[73]
	Spinach	Aqueous suspension	Increased fresh and dry biomass	[74]
	Spinach	Aqueous suspension	Improved growth related to nitrogen fixation	[75]
	Spinach	Aqueous suspension	Improved light absorbance and carbon dioxide assimilation	[76]
Iron NPs	Lettuce	Aqueous suspension	High concentration inhibited germination	[15]
	Wheat	Direct application	Enhanced seed germination and plant growth	[88]
	Pumpkin	Direct application	No toxic effect	[102]
	Wheat	Direct application	Increased shoot and root biomass	[96]
	Wheat	Soil applied	Increased spike length, number of grains per spike, 1000 grain weight	[103]
	Various plants	Direct application	Development of thicker roots	[104]
Copper/ Copper oxide NPs	Wheat	Direct application	Reduced root and seedling growth	[89]
	Rose	Water-agar plates with suspension	Increased plant resistance to fungal infection by altering endogenous hormones content	[82]
	Pumpkin	Aqueous suspension	Reduced biomass	[81]
	Wheat	Direct application	Reduced seed germination	[103]
	Wheat	Direct application	Increased plant growth and biomass	[97]
	Maize	Aqueous suspension	Reduced seedling growth	[80]
	Mung bean	Agar culture media	Reduced seedling growth	[89]
	Wheat	Agar culture media	Reduced seedling growth	[89]
	Zucchini	Aqueous suspension	Reduced biomass and root growth	[81]
	Rice	Aqueous suspension	Decreased seed germination and seedlings growth	[105]
	Barley	Aqueous suspension	Restricted shoot and root growth	[106]
	Maize	Aqueous suspension	Suppressed root elongation	[80]
	Barley	Aqueous suspension	Decreased plasto globule and starch granule	[107]
	Maize	Aqueous suspension	Reduced shoot and root biomass	[108]
Zinc oxide NPs	*Pleuroziumschreberi*	NPs suspension	Reduced L-ascorbic acid content	[109]
	Wheat	NPs suspension	Reduced biomass	[100]
	Soybean	Direct application	Increased root growth	[91]
	Soybean	Direct application	Decreased root growth	[91]
	Ryegrass	Direct application	Reduced biomass, shrunken root tips, broken epidermis/root caps	[69]
	Soybean	Direct application	Increased root growth	[110]
	Maize	Aqueous suspension	Highly reduced root growth	[69]
	Ryegrass	Hoagland solution	Reduced biomass, shrank root tips, broken epidermis/root cap, highly vacuolated and collapsed cortical cells	[69]

Table 2. *Cont.*

NPs	Species	Mode of application	Morphophysiological responses	Ref *
Carbon nanotubes	Rose	Water-agar plates with suspensions	Increased plant resistance to fungal infection by altering endogenous hormones content	[82]
	Tomato	Aqueous suspension	Enhanced seed germination, fresh biomass, stem length	[78]
	Onion	Direct application	Increased root length	[79]
	Rice	Direct application	Delayed flowering and decreased yield	[72]
	Pumpkin	Aqueous suspension	Reduced biomass	[81]
	Wheat	Direct application	Increased root length	[80]
	Tomato	Aqueous suspension	Increased germination rate, fresh biomass, stem length	[78]
	Rice	MS medium	Delayed flowering and decreased yield	[72]
	Tomato	Aqueous suspension	Reduced root length	[79]
	Lettuce	Aqueous suspension	Reduced root length at longer exposure	[79]
Cerium oxide NPs	Wheat	Direct application	Enhanced shoot growth, biomass, grain yield	[18]
	Lettuce	Direct application	Inhibited root growth	[69]
	Maize	Direct application	Increased stem and root growth	[91]
	Maize	Aqueous suspension	Increased root and stem growth	[91]
	Tomato	Aqueous suspension	Reduced shoot growth	[91]
	Maize	Aqueous suspension	Reduced biomass	[91]
	Sorghum	Foliar spray	Increased leaf carbon assimilation rates, pollen germination, seed yield	[111]
	Rice	Direct application	Enhanced growth	[112]
	Onion	Foliar spray	Improved yield, plant growth, nutrient content	[113]
Gold NPs	Lettuce	Aqueous suspension	Enhanced root elongation	[104]
	Cucumber	Aqueous suspension	Improved germination	[104]
Nd2O3NPs	Pumpkin	Aqueous suspension	Increased antioxidant capacity	[114]

* Ref means references.

4. Applications of Proteomic Techniques to Assess the Impact of NPs on Crops

With the advancements in mass spectrometry, proteomics has become a powerful technology for the identification and characterization of stress-induced proteins. Detailed proteome analysis of plant organelles generates comprehensive information about the intrinsic mechanisms of plant stress responses towards NPs. Proteomic analyses of various crops exposed to different NPs are summarized in Table 3.

4.1. Proteomic Analysis of Silver NPs Challenged Crops

Silver NPs are considered as a promising antibacterial agent due to their strong biocidal effect against microorganisms [115]. These NPs are synthesized through different physical, chemical, and biological methods and well-defined parameters of size and shape [28]. The effects of silver NPs were initially analyzed using proteomic techniques in *Chlamydomonas* [116], *Escherichia coli* [117], and *Bacillus thuringiensis* [118]. Currently, various crop plants were exposed to silver NPs and their effects were analyzed using gel-based or gel-free proteomic techniques. Our gel-free proteomic study revealed restricted growth of soybean seedlings under silver NPs treatment [119]. Proteins related to secondary metabolism, cell organization, and hormone metabolism were mostly influenced by silver NPs exposure. In contrast, silver NPs of 15 nm in size significantly improved the soybean growth under flooding stress by enhancing proteins linked to amino acid synthesis [120]. In wheat, the accumulation of different cellular compartmental proteins on silver NPs exposure in shoot and root was mainly involved in metabolism and cell defense [86]. Silver NPs with chemical exposure increased the proteins related to photosynthesis and protein synthesis, while decreased the glycolysis, signaling, and cell wall related proteins in wheat [121]. Large numbers of proteins involved in the primary metabolism were increased in soybean [119]. Silver NPs treatment increased the proteins related to protein degradation, while decreased protein synthesis related proteins in soybean; indicating that it might improve the growth of soybean under flooding stress through protein quality control [122]. Proteins related to the oxidative stress, signaling, transcription, protein degradation, cell wall synthesis, cell division, and apoptosis were found to be increased in silver NPs exposed rice [118]. In *Eruca sativa*, proteins associated with the endoplasmic reticulum and vacuole were differentially modulated under silver NPs exposure [86]. These findings indicate that silver NPs primarily influence various metabolic processes in wheat and protein quality control in soybeans; thus, improving plant growth.

4.2. Proteomic Analysis of Aluminum Oxide NPs Stressed Crops

Aluminum oxide NPs are mostly used in military and commercial products [123]. Extensive usage of aluminum oxide NPs leads towards their possible leakage into environment, which ultimately interacts with living organisms including plants [124]. Proteomic analysis of soybean root treated with aluminum oxide NPs revealed an increase in the number of proteins related to protein synthesis, transport, and development during the recovery from flooding [92]. A study by Mustafa et al. [120] revealed that proteins associated with the ascorbate-glutathione cycle, as well as ribosomal proteins, were differentially influenced by aluminum oxide NPs. Moreover, high abundance of proteins involved in oxidation-reduction, stress signaling, hormonal pathways related to growth and development, were evident in aluminum oxide NPs challenged soybean [119]. A separate study has shown growth promoting effects of aluminum oxide NPs in the soybean under flooding stress by regulating energy metabolism and cell death [125].

4.3. Proteomic Analysis of Crops Exposed to Copper NPs and Iron NPs

Among the various metal-based NPs, copper NPs are by far the most well studied NPs whose toxicity has been tested in wide range of crops. They have wide applications in electronics, air/liquid filtration, ceramics, wood preservation, bioactive coatings, and films/textiles [16]. At the cellular level, copper acts as structural and catalytic component of many proteins involved in various metabolic processes. In wheat seedlings, abundance of proteins associated with glycolysis and tricarboxylic acid cycle was found to be increased; while, photosynthesis and tetrapyrrole synthesis related proteins were decreased on exposure to copper nanoparticles [97]. Wheat grains obtained after NPs exposure were analyzed through gel-free proteomic technique, which indicated an increase in proteins involved in starch degradation and glycolysis [96].

Similar to copper NPs, iron NPs have extensive industrial, commercial, and biomedical applications [12]. Because of their high reactivity and magnetic property, iron NPs have been used as remediation agents for environmental applications [13]. Iron NPs have known stimulatory effects on the seed germination and plant growth of wheat [96]. Authors exploited gel-free/label-free proteomic technique to elucidate the impact of iron NPs on shoot growth of drought tolerant and salt tolerant wheat varieties. A study revealed that differentially expressed proteins in both varieties were mainly associated with photosynthesis. Notably, proteins related to light reaction were enhanced in the salt tolerant variety compared to drought tolerant wheat on iron NPs exposure. A separate study on grain analysis of wheat indicated an increase in the number of proteins related to starch degradation, glycolysis, and the tricarboxylic acid cycle [103].

4.4. Proteomic Analysis of Other NPs Challenged Crops

One of the most commonly used nanomaterials in agriculture and the energy sector is titanium dioxide NPs [126]. They have diverse applications in personal skincare products, water-treatment agents, and bactericidal agents owing to their high stability and anticorrosive/photocatalytic properties [127,128]. The toxicological effects of nanometer titanium dioxide on a unicellular green alga *Chlamydomonas reinhardtii* were accessed by monitoring the changes in the physiology and cyto-ultrastructure [129]. Authors reported nano titanium dioxide mediated inhibition in photosynthetic efficiency and cell growth, with increased contents of carotenoids and chlorophyll b.

In addition, various NPs are being extensively utilized to improve the growth and productivity of crop plants. However, application of zinc oxide NPs had marked effects on soybean seedling growth, rigidity of roots, and root cell viability [119]. Gel-free proteomic analysis revealed down regulation oxidation-reduction cascade associated proteins, including GDSL motif lipase 5, SKU5 similar 4, galactose oxidase, and quinone reductase in zinc oxide NPs exposed roots. A separate study on cerium oxide NPs treatment in maize indicated enhanced accumulation of heat shock proteins (HSP70) and increased activity of ascorbate peroxidase and catalase [130]. This up regulated antioxidant defense system might help maize plants to overcome NPs-induced oxidative stress damages.

All of these studies indicate that NPs have the potential to modulate plant metabolic processes, and impact of NPs could be either positive or negative, depending on the plant species and type of nanoparticles used, their size, composition, concentration, and physical/chemical properties.

Table 3. Summary of proteomic analyses of various crops exposed to different NPs.

NPs	Plant	Organ	Proteomic Technique	Protein Response	Ref *
Silver NPs	Soybean	Root	Gel-free (nanoLC–MS/MS)	Decreased proteins associated with secondary metabolism, cell organization, and hormone metabolism.	[119]
	Eruca sativa	Root	Gel-based (2-DE, nanoLC–nESI-MS/MS)	Altered endoplasmic reticulum and vacuolar proteins involved in sulfur metabolism.	[98]
	Wheat	Root	Gel-based (2-DE, LC–MS/MS)	Altered proteins involved in metabolism and cell defense.	[86]
	Soybean	Root	Gel-free (nanoLC–MS/MS)	Altered proteins associated with stress, cell metabolism, signaling.	[125]
	Soybean	Root, Hyp **	Gel-free (nanoLC–MS/MS)	Decreased protein synthesis with increased amino acid synthesis.	[93]
	Soybean	Root, Hyp **	Gel-free (nanoLC–MS/MS)	Increased protein degradation related proteins. Decreased protein synthesis associated proteins.	[122]
	Wheat	Shoot	Gel-free (nanoLC–MS/MS)	Increased proteins related to photosynthesis and protein synthesis. Decreased proteins linked to glycolysis, signaling, cell wall.	[121]
	Tobacco	Root, Leaf	Gel-based (2-DE, MALDI-TOF/TOF MS)	Altered abundance of root proteins involved in abiotic/biotic and oxidative stress responses. In leaf, proteins associated with photosynthesis markedly changed.	[131]
Aluminum oxide NPs	Soybean	Root, Hyp **	Gel-free (nanoLC–MS/MS)	Increased proteins related to protein synthesis, transport, and development during post- flooding recovery period.	[92]
	Soybean	Root, Hyp **	Gel-free (nanoLC–MS/MS)	Regulated the ascorbate/glutathione pathway and increased ribosomal proteins.	[120]
	Soybean	Root, Leaf	Gel-free (nanoLC–MS/MS)	Increased proteins involved in oxidation, stress signaling, and hormonal pathways.	[119]
	Soybean	Root, Hyp **	Gel-free (nanoLC–MS/MS)	Decreased energy metabolism and changed proteins related to glycolysis compared to flooding stress.	[125]
Copper NPs	Wheat	Shoot	Gel-free (nanoLC–MS/MS)	Increased proteins related to glycolysis and tricarboxylic acid cycle.	[97]
	Wheat	Seed	Gel-free (nanoLC–MS/MS)	Increased proteins involved in starch degradation and glycolysis.	[103]
Iron NPs	Wheat	Shoot	Gel-free (nanoLC–MS/MS)	Decreased proteins linked to photosynthesis and protein metabolism.	[96]
	Wheat	Seed	Gel-free (nanoLC–MS/MS)	Increased proteins related to starch degradation, glycolysis, tricarboxylic acid cycle.	[103]
Zinc oxide NPs	Soybean	Root, Leaf	Gel-free (nanoLC–MS/MS)	Decreased proteins involved in oxidation-reduction, stress signaling, and hormonal pathways.	[119]
Cerium oxide NPs	Maize	Shoot	Gel-free (nanoLC–ESI-MS/MS)	Increased accumulation of heat shock protein. Increased ascorbate/ peroxidase/ catalase activity.	[130]

* Ref means reference; ** Hyp stands for Hypocotyl. Abbreviations: 2-DE, two-dimensional gel electrophoresis; nESI, nanoelectro spray ionization; MALDI-TOF, matrix-assisted laser desorption ionization time-of-flight.

5. NPs Uptake and Mode of Action

The phytotoxicity of NPs largely depends on the particle size, concentration and chemistry of NPs, in addition to the chemical milieu of the subcellular sites at which the NPs are deposited [23].

Plants, being an indispensable component of terrestrial ecosystems, serve as a potential route for the factory discharged-NPs to enter the plant root system and their transportation to other parts of the plants, resulting bioaccumulation in the food chain [132]. The physico-chemical properties of soil matrix (viz. mineral composition, pH, ionic strength, dissolved organic matter, etc.) as well as the of metal based NPs (viz. size, surface charge, surface coating, etc.) are the determining factors for NPs mobility [133]. Primary-lateral root junctions are the prime sites through which NPs could enter xylem via cortex and finally reach the central cylinder [23]. Study on the uptake pathways of zinc oxide NPs by maize roots reveals that majority of the total zinc oxide NPs undergo dissolution in the exposure medium, and the released Zn^{2+} ions are only taken up by the roots [134]. Only a small fraction of zinc oxide NPs adsorbed on the root surface can cross the root cortex as a result of speedy cell division and root tip elongation, apart from their entry to vascular system through the gap of the Casparian strip at the sites of the primary–lateral root junction.

Once NPs enter the root cells, these ultrafine particles upon dissolution discharge metal ions that interact with the functional groups of proteins (carboxyl and sulfhydryl groups) causing altered protein activity. The released redox-active metal ions could trigger reactive oxygen species (ROS) generation through the Fenton and Haber–Weiss reactions [135]. In these reactions, the hydrogen peroxide (H_2O_2) is decayed by the metal ions leading to the formation of more toxic ROS, namely hydroxyl radical (•OH) and hydroxyl anion (OH^-). Elevated ROS generation was documented in leaves of soybean exposed to zinc oxide NPs and silver NPs [119] as well as in copper oxide NPs challenged rice [105]. These NPs mediated excess ROS formation disturbs the cellular redox system in favor of oxidized forms, causing oxidative damage to vital cellular components including nucleic acids, lipids, and proteins [135].

Cellular compartments with extremely high oxidizing metabolic activity or with an intense rate of electron flow, such as mitochondria, chloroplasts, and peroxisomes, constitute a major source of ROS production in plants [136]. Investigations have revealed that zinc oxide NPs mediated deregulation of photosynthetic efficiency in plants is due to the down regulation of chlorophyll synthesis genes and structural genes of photosystem I [137,138]. To protect cells against such oxidative damages, plants have developed robust multi-component antioxidant defense system comprising of both enzymatic and non-enzymatic machineries [119,139]. The enzymatic antioxidant defense system chiefly includes ROS scavenging enzymes of the ascorbate–glutathione cycle, which operates in nearly all plant cell organelles [140]. The orchestrated action of key antioxidant enzymes viz. superoxide dismutase (SOD), ascorbate peroxidase (APX), catalase (CAT), monodehydroascorbate reductase (MDHAR), dehydroascorbate reductase (DHAR), and glutathione reductase (GR) is an adaptive strategy of plant to cope with the NPs induced oxidative stress damages.

Moreover, NPs exposure often leads to disruption of cellular redox homeostasis and cause cell membrane damage through lipid peroxidation [105,106,108]. Among the ROS, hydroxyl radical (•OH) is known to be the most reactive, capable of stealing hydrogen atom from a methylene ($-CH_2-$) group present in polyunsaturated fatty acid side chain of membrane lipids and, thus, initiates lipid peroxidation [141]. Since, •OH is derived from H_2O_2 as a consequence of one electron reduction, H_2O_2 scavenging peroxides play essential roles in protecting lipid membranes from NPs mediated oxidative stress. Among ROS, a recent study revealed down regulation of *ascorbate peroxidase* (*APX1*) in zinc oxide NPs challenged maize leaves with concomitant increased malondialdehyde (MDA) level, an indicative of oxidative stress induced damage to the lipid membrane [108]. The NPs-induced higher membrane damage is in accordance with the previous reports in rice [105] and Syrian barley [106].

Apart from enzymatic component of ascorbate-glutathione cycle, plants have evolved a second line of defense to cope with the NPs induced oxidative stress. The thioredoxin (Trx) family protein is one of them, engage in mitigating oxidative damages by providing reducing power to reductases, detoxifying lipid hydroperoxides or repairing oxidized proteins. They also act as regulators of scavenging mechanisms and key components of signaling pathways in the plant antioxidant network [142]. In addition, these proteins are necessary for their potential roles as facilitators and regulators of protein folding and chaperone activity [143]. Furthermore, plant quinone reductases (QRs) are involved redox reactions and act as detoxification enzymes of free radicals. Soybean seedlings exposed to zinc oxide NPs and silver NPs treatments exhibited significantly declined abundance of Trx and QR proteins [119]. Severe oxidative burst evident in zinc oxide NPs and silver NPs challenged soybean might be the result of such declined protein abundance affecting optimum growth of seedlings. Enzymes of shikimate pathway involved in the synthesis of amino acids (phenylalanine, tryptophan, and tyrosine) were also found to be affected under NPs exposure. These aromatic amino acids not only act as substrates for the protein synthesis, but are also linked with formation of secondary products, including lignin, suberin, and phytoalexins. The abundance of 3-deoxy-D-arabino-heptulosonate-7-phosphate (DAHP) synthase, the first enzyme of the shikimate pathway, was reported to be decreased in soybean under silver NPs treatment [119]. The reduced shoot length of silver NPs exposed soybean seedlings might be the result of such marked decline in DAHP synthase level. In a nutshell, low abundance of proteins involved in oxidation-reduction, shikimate pathway might limit the growth of the silver NPs challenged soybean seedlings up to a certain level. Summarizing all these findings, a comprehensive model of cellular responses to NPs is presented in Figure 1.

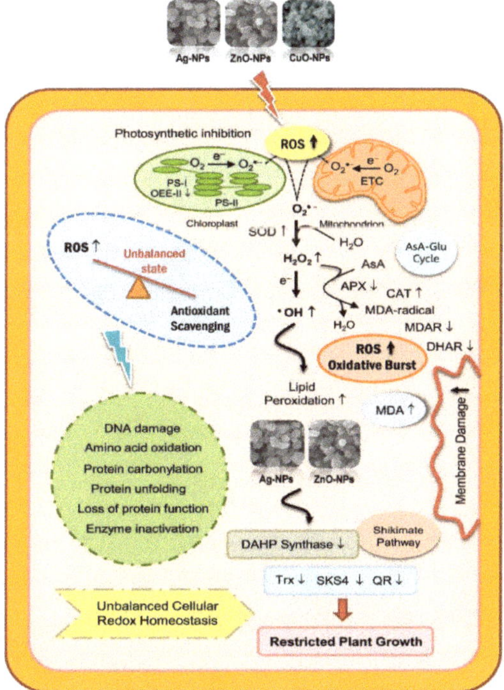

Figure 1. Schematic illustration of diverse cellular responses to nanoparticles (NPs). Exposure to metal based-NPs triggers oxidative stress through enhanced reactive oxygen species (ROS) generation, disruption of redox homeostasis, impaired photosynthetic activity, mitochondrial dysfunction, lipid peroxidation, and membrane damage. Upward arrows indicate increased and downward arrows indicate decreased protein abundance in response to NPs stress, respectively. Abbreviations: APX, ascorbate peroxidase; AsA, reduced ascorbate; CAT, catalase; DAHP, 3-deoxy-D-arabino-heptulosonate-7-phosphate; DHAR, dehydroascorbate reductase; ETC, electron transport chain; H_2O_2, hydrogen peroxide; MDA, malondialdehyde; MDA-radical, monodehydroascorbate radical; MDAR, monodehydroascorbate reductase; ·OH, hydroxyl radical; OEE, oxygen-evolving enhancer; PS, photosystem; QR, quinone reductase; ROS, reactive oxygen species; SOD, superoxide dismutase; Trx, thioredoxin.

6. Conclusions

Nanotechnology has gained tremendous momentum in recent times because of the wide applications of NPs in agriculture, cosmetic industry, cellular imaging, medical diagnosis, biosensing, drug delivery, and cancer therapy. Nevertheless, unintended release of such commercially manufactured nanomaterials into the environment has raised global concern. Hence, considerable attention is now being paid to the methods and strategies of NPs synthesis, plant-nanomaterials interactions, and their environmental fate. As compared to traditional physical and chemical processes, green synthesis of NPs using microorganisms and plants is an environment-friendly, cost effective, safe, biocompatible, green alternative approach for large scale production of NPs. Morphophysiological as well as proteomic studies on NPs-induced phytotoxicity reveal that particle size, concentration, and chemistry of NPs, as well as the type of plant species used, are the key factors determining the type and magnitude of the cellular responses. However, more initiatives must be taken to find out whether the metal-based NPs exert phytotoxicity exclusively due to their high surface area and nanoscale size or due to the released metal ions. Moreover, there is a need for more comprehensive omics approach integrating genomics, transcriptomics, proteomics, and metabolomics, so that the impact of the applied NPs on plants can be assessed well in time.

Author Contributions: Conceptualization, S.K.; writing-original draft preparation, Z.H., F.Y., and S.K.; writing-review and editing, Z.H. and S.K.; visualization, Z.H. and S.K.; supervision, S.K.; All authors have read and agreed to the published version of the manuscript.

Funding: This research was supported by grant (No. 2) from Fukui University of Technology, Japan.

Conflicts of Interest: The authors declare no conflict of interest.

References

1. Singh, T.; Shukla, S.; Kumar, P.; Wahla, V.; Bajpai, V.K.; Rather, I.A. Application of nanotechnology in food Science: Perception and overview. *Front. Microbiol.* **2017**, *8*, e1501. [CrossRef] [PubMed]
2. Bajpai, V.K.; Kamle, M.; Shukla, S.; Mahato, D.K.; Chandra, P.; Hwang, S.K.; Kumar, P.; Huh, Y.S.; Han, Y.-K. Prospects of using nanotechnology for food preservation, safety, and security. *J. Food Drug Anal.* **2018**, *26*, 1201–1214. [CrossRef] [PubMed]
3. Khot, L.R.; Sankaran, S.; Maja, J.M.; Ehsani, R.; Schuster, E.W. Applications of nanomaterials in agricultural production and crop protection: A review. *Crop. Prot.* **2012**, *35*, 64–70. [CrossRef]
4. Albrecht, M.A.; Evans, C.W.; Raston, C.L. Green chemistry and the health implications of nanoparticles. *Green Chem.* **2006**, *8*, 417–432. [CrossRef]
5. Xu, D.; Liu, M.; Zou, H. A new strategy for fabrication of water dispersible and biodegradable fluorescent organic nanoparticles with AIE and ESIPT characteristics and their utilization for bioimaging. *Talanta* **2017**, *174*, 803–808. [CrossRef] [PubMed]
6. Hassan, H.F.; Mansour, A.M.; Abo-Youssef, A.M.; Elsadek, B.E.; Messiha, B.A. Zinc oxide nanoparticles as a novel anticancer approach; in vitro and in vivo evidence. *Clin. Exp. Pharmacol. Physiol.* **2017**, *44*, 235–243. [CrossRef] [PubMed]
7. Nirmala, M.; Anukaliani, A. Synthesis and characterization of undoped and TM (Co, Mn) doped ZnO nanoparticles. *Mater. Lett.* **2011**, *65*, 2645–2648. [CrossRef]
8. Thapa, A.; Soares, A.C.; Soares, J.C. Carbon nanotube matrix for highly sensitive biosensors to detect pancreatic cancer biomarker CA19-9. *ACS Appl. Mater. Interfaces* **2017**, *9*, 25878–25886. [CrossRef] [PubMed]
9. El-Gharbawy, R.M.; Emara, A.M.; Abu-Risha, S.E. Zinc oxide nanoparticles and a standard antidiabetic drug restore the function and structure of beta cells in Type-2 diabetes. *Biomed. Pharmacother.* **2016**, *84*, 810–820. [CrossRef]
10. Rosi, N.L.; Mirkin, C.A. Nanostructures in biodiagnostics. *Chem. Rev.* **2005**, *105*, 1547–1562. [CrossRef]
11. Auffan, M.; Rose, J.; Bottero, J.Y.; Lowry, G.V.; Jolivet, J.P.; Wiesner, M.R. Towards a definition of inorganic nanoparticles from an environmental, health and safety perspective. *Nat. Nanotechnol.* **2009**, *4*, 4–41. [CrossRef] [PubMed]
12. Teja, A.S.; Koh, P.Y. Synthesis, properties, and applications of magnetic iron oxide nanoparticles. *Prog. Cryst. Growth Charact.* **2009**, *55*, 22–45. [CrossRef]

13. Yan, W.; Lien, H.I.; Koel, B.E.; Zhang, W. Iron nanoparticles for environmental clean-up: Recent development and future outlook. *Environ. Sci. Processes Impacts* **2013**, *15*, 63–77. [CrossRef]

14. Kim, J.; Lee, Y.; Kim, E.; Gu, S.; Sohn, E.J.; Soe, Y.S.; An, H.J.; Chang, Y.S. Exposure of iron nanoparticles to *Arabidopsis thaliana* enhances root elongation by triggering cell wall loosening. *Environ. Sci. Technol.* **2014**, *48*, 3477–3485. [CrossRef] [PubMed]

15. Trujillo-Reyes, J.; Majumdar, S.; Botez, C.E.; Peralta-Videa, J.R.; Gardea-Torresdey, J.L. Exposure studies of core–shell Fe/Fe$_3$O$_4$ and Cu/CuO NPs to lettuce (*Lactuca sativa*) plants: Are they a potential physiological and nutritional hazard? *J. Hazard. Mater.* **2014**, *267*, 255–263. [CrossRef] [PubMed]

16. Yang, C.W.; Zen, J.M.; Kao, Y.L.; Hsu, C.T.; Chung, T.C.; Chang, C.C.; Chou, C.C. Multiple screening of urolithic organic acids with copper nanoparticle-plated electrode: Potential assessment of urolithic risks. *Anal. Biochem.* **2009**, *395*, 224–230. [CrossRef]

17. Jia, H.; Chen, S.; Wang, X.; Shi, C.; Liu, K.; Zhang, S.; Li, J. Copper oxide nanoparticles alter cellular morphology via disturbing the actin cytoskeleton dynamics in Arabidopsis roots. *Nanotoxicology* **2020**, *14*, 127–144. [CrossRef]

18. Rico, C.M.; Majumdar, S.; Duarte-Gardea, M.; Peralta-Videa, J.R.; Gardea-Torresdey, J.L. Interaction of nanoparticles with edible plants and their possible implications in the food chain. *J. Agric. Food Chem.* **2011**, *59*, 3485–3498. [CrossRef]

19. Patlolla, A.K.; Berry, A.; May, L.; Tchounwou, P.B. Genotoxicity of silver nanoparticles in *Vicia faba*: A pilot study on the environmental monitoring of nanoparticles. *Int. J. Environ. Res. Public Health* **2012**, *9*, 1649–1662. [CrossRef]

20. Kaveh, R.; Li, S.; Ranjbar, S.; Tehrani, R.; Brueck, C.L.; Aken, B. Changes in *Arabidopsis thaliana* gene expression in response to silver nanoparticles and silver ions. *Environ. Sci. Technol.* **2013**, *47*, 10637–10644. [CrossRef]

21. Hossain, Z.; Mustafa, G.; Komatsu, S. Plant responses to nanoparticle stress. *Int. J. Mol. Sci.* **2015**, *16*, 26644–52663. [CrossRef] [PubMed]

22. Lee, S.; Kim, S.; Kim, S.; Lee, I. Assessment of phytotoxicity of ZnO NPs on a medicinal plant, *Fagopyrum esculentum*. *Environ. Sci. Pollut. Res.* **2013**, *20*, 848–854. [CrossRef] [PubMed]

23. Dietz, K.J.; Herth, S. Plant nanotoxicology. *Trends Plant Sci.* **2011**, *16*, 582–589. [CrossRef] [PubMed]

24. Asharani, P.V.; Mun, G.L.K.; Hande, M.P.; Valiyaveettil, S. Cytotoxicity and genotoxicity of silver nanoparticles in human cells. *ACS Nano* **2009**, *3*, 279–290. [CrossRef]

25. Barkla, B.J.; Vera-Estrella, R.; Pantoja, O. Progress and challenges for abiotic stress proteomics of crop plants. *Proteomics* **2013**, *13*, 1801–1815. [CrossRef]

26. Abdelhamid, H.N.; Wu, H.F. Proteomics analysis of the mode of antibacterial action of nanoparticles and their interactions with proteins. *Trends Anal. Chem.* **2015**, *65*, 30–46. [CrossRef]

27. Wang, W.; Luo, M.; Fu, Y.; Wang, S.; Efferth, T.; Zu, Y. Glycyrrhizic acid nanoparticles inhibit LPS-induced inflammatory mediators in 264.7 mouse macrophages compared with unprocessed glycyrrhizic acid. *Int. J. Nanomed.* **2013**, *8*, 1377.

28. Iravani, S.; Golghar, B. Green synthesis of silver nanoparticles using *Pinus ldarica* bark extract. *Biol. Med. Res. Int.* **2014**, *2013*, 1–5.

29. Jung, J.; Oh, H.; Noh, H.; Ji, J.; Kim, S. Metal nanoparticle generation using a small ceramic heater with a local heating area. *J. Aerosol. Sci.* **2006**, *37*, 1662–1670. [CrossRef]

30. Tsuji, T.; Kakita, T.; Tsuji, M. Preparation of nano-size particle of silver with femtosecond laser ablation in water. *Appl. Surface Sci.* **2003**, *206*, 314–320. [CrossRef]

31. Boutinguiza, M.; Comesaña, R.; Lusquiños, F.; Riveiro, A.; Pou, J. Production of nanoparticles from natural hydroxylapatite by laser ablation. *Nanoscale Res. Lett.* **2011**, *6*, 255. [CrossRef] [PubMed]

32. Zboril, R.; Mashlan, M.; Barcova, K.; Vujtek, M. Thermally induced solid-state syntheses of γ-Fe$_2$O$_3$ nanoparticles and their transformation to α-Fe$_2$O$_3$ via ε-Fe$_2$O$_3$. *Hyperfine Interact.* **2002**, *139*, 597. [CrossRef]

33. Chin, S.F.; Pang, S.C.; Tan, C.H. Green synthesis of magnetite nanoparticles (via thermal decomposition method) with controllable size and shape. *J. Mater. Environ. Sci.* **2011**, *3*, 299–302.

34. Das, D.; Nathb, B.C.; Phukonc, P.; Dolui, S.K. Synthesis and evaluation of antioxidant and antibacterial behavior of CuO nanoparticles. *Colloids Surf. B Biointerfaces* **2013**, *101*, 430–433. [CrossRef] [PubMed]

35. Soomro, R.A.; Sherazi, S.T.; Sirajuddin, H.; Memon, N.; Shah, M.R.; Kalwar, N.H.; Hallam, K.R.; Shah, A. Synthesis of air stable copper nanoparticles and their use in catalysis. *Adv. Mat. Lett.* **2014**, *5*, 191–198. [CrossRef]

36. Park, B.K.; Jeong, S.; Kim, D.; Moon, J.; Lim, S.; Kim, J.S. Synthesis and size control of monodisperse copper nanoparticles by polyol method. *J. Colloid Interface Sci.* **2007**, *311*, 417–424. [CrossRef]

37. Ramyadevi, J.; Jeyasubramanian, K.; Marikani, A.; Rajakumar, G.; Rahuman, A. Synthesis and antimicrobial activity of copper nanoparticles. *Mater. Lett.* **2012**, *71*, 114–116. [CrossRef]

38. Peyser, L.A.; Vinson, A.E.; Bartko, A.P.; Dickson, R.M. Photoactivated fluorescence from individual silver nanoclusters. *Science* **2001**, *291*, 103–106. [CrossRef]

39. Orendoff, C.J.; Gearheart, L.; Jana, N.R.; Murphy, C.J. Aspect ratio dependence on surface enhanced Raman scattering using silver and gold nanorod substrates. *Phys. Chem. Chem. Phys.* **2006**, *8*, 165–170. [CrossRef]

40. Huang, K.; Ehrman, S.H. Synthesis of iron nanoparticles via chemical reduction with palladium ion seeds. *Langmuir* **2007**, *23*, 1419–1426. [CrossRef]

41. Selvarani, M.; Prema, P. Evaluation of antibacterial efficacy of chemically synthesized copper and zerovalent iron nanoparticles. *Asian J. Pharm. Clin. Res.* **2013**, *6*, 223–227.

42. Dhas, N.A.C.; Raj, P.; Gedanken, A. Synthesis, characterization, and properties of metallic copper nanoparticles. *Chem. Mater.* **1998**, *10*, 1446–1452. [CrossRef]

43. Ruparelia, J.P.; Chatterjee, A.K.; Duttagupta, S.P.; Mukherji, S. Strain specificity in antimicrobial activity of silver and copper nanoparticles. *Acta Biomater.* **2008**, *4*, 707–716. [CrossRef]

44. Fu, H.; Yang, X.; Jiang, X.; Yu, A. Bimetallic Ag–Au nanowires: Synthesis, growth mechanism, and catalytic properties. *Langmuir* **2013**, *29*, 7134–7142. [CrossRef] [PubMed]

45. Qin, Y.; Ji, X.; Jing, J.; Liu, H.; Wu, H.; Yang, W. Size control over spherical silver nanoparticles by ascorbic acid reduction. *Colloids Surf. A* **2010**, *372*, 172–176. [CrossRef]

46. Kim, J.; Kim, H.S.; Lee, N.; Kim, T.; Kim, H.; Yu, T.; Song, I.C.; Moon, W.K.; Hyeon, T. Multifunctional uniform nanoparticles composed of a magnetite nanocrystal core and a mesoporous silica shell for magnetic resonance and fluorescence imaging and for drug delivery, *Angewandte Chemie. Int. Ed.* **2008**, *47*, 8438–8441. [CrossRef] [PubMed]

47. Wang, L.; Muhammed, M. Synthesis of zinc oxide nanoparticles with controlled morphology. *J. Mater. Chem.* **1999**, *9*, 2871–2878. [CrossRef]

48. Pranjali, P.M.; Patil, P.M.; Dhanavade, M.J.; Badiger, M.V.; Shadija, P.G.; Lokhande, A.C.; Bohara, R.A. Synthesis and characterization of zinc oxide nanoparticles by using polyol chemistry for their antimicrobial and antibiofilm activity. *Biochem. Biophys. Rep.* **2019**, *17*, 71–80.

49. Ramanathan, R.; Field, M.R.; O'Mullane, A.P.; Smooker, P.M.; Bhargava, S.K.; Bansal, V. Aqueous phase synthesis of copper nanoparticles: A link between heavy metal resistance and nanoparticle synthesis ability in bacterial systems. *Nanoscale* **2013**, *5*, 2300–2306. [CrossRef]

50. Iqbal, M.J.; Ali, S.; Rashid, S.; Kamran, M.; Malik, M.F.; Sughra, K.; Zeeshan, N.; Afroz, A.; Saleem, J.; Saghir, M. Biosynthesis of silver nanoparticles from leaf extract of *Litchi chinensis* and its dynamic biological impact on microbial cells and human cancer cell lines. *Cell Mol. Biol.* **2018**, *64*, 42–47. [CrossRef]

51. Poinern, G.E.J.; Chapman, P.; Shah, M.; Fawcett, D. Green biosynthesis of silver nanocubes using the leaf extracts from Eucalyptus macrocarpa. *Nano Bull.* **2013**, *2*, 1–7.

52. Shehzad, A.; Qureshi, M.; Jabeen, S.; Ahmad, R.; Alabdalall, A.H.; Aljafary, M.A.; Al-Suhaimi, E. Synthesis, characterization and antibacterial activity of silver nanoparticles using *Rhazya stricta*. *Peer J.* **2018**, *6*, 60–86. [CrossRef]

53. Gholami, A.; Khosravi, R.; Khosravi, A.; Samadi, Z. Data on the optimization of the synthesis of green iron nanoparticles using plants indigenous to South Khorasan. *Data Brief* **2018**, *21*, 1779–1783. [CrossRef]

54. Machado, S.; Pinto, S.L.; Grosso, J.P.; Nouws, H.P.A.; Albergaria, J.T.; Delerue-Matos, C. Green production of zero-valent iron nanoparticles using tree leaf extracts. *Sci. Total Environ.* **2013**, *445–446*, 1–8. [CrossRef]

55. Shahwana, T.; Abu Sirriah, S.; Nairata, M.; Boyacıb, E.; Eroglub, A.E.; Scottc, T.B.; Hallam, K.R. Green synthesis of iron nanoparticles and their application as a fentonlike catalyst for the degradation of aqueous cationic and anionic dyes. *Chem. Eng. J.* **2011**, *172*, 258–266.

56. Umar, H.; Kavaz, D.; Rizaner, N. Biosynthesis of zinc oxide nanoparticles using *Albizia lebbeck* stem bark, and evaluation of its antimicrobial, antioxidant, and cytotoxic activities on human breast cancer cell lines. *Int. J. Nanomedicine* **2018**, *14*, 87–100. [CrossRef]

57. Shobha, N.; Nanda, N.; Giresha, A.S.; Manjappa, P.; Dharmappa, K.K.; Nagabhushana, B.M. Synthesis and characterization of Zinc oxide nanoparticles utilizing seed source of *Ricinus communis* and study of its antioxidant, antifungal and anticancer activity. *Mater. Sci. Eng. C. Mater. Biol. Appl.* **2019**, *97*, 842–850. [CrossRef]

58. Sangeetha, G.; Rajeshwari, S.; Venckatesh, R. Green synthesis of zinc oxide nanoparticles by *aloe barbadensis miller* leaf extract: Structure and optical properties. *Mater. Res. Bull.* **2011**, *46*, 2560–2566. [CrossRef]

59. Kulkarni, V.D.; Kulkarni, P.S. Green synthesis of copper nanoparticles using *Ocimum sanctum* leaf extract. *Int. J. Chem. Stud.* **2013**, *1*, 1–4.

60. Jayalakshmi, Y.A. Green synthesis of copper oxide nanoparticles using aqueous extract of flowers of *cassia alata* and particles characterisation. *Int. J. Nanomater. Biostruct.* **2014**, *4*, 66–71.

61. Saranyaadevi, K.; Subha, V.; Ravindran, R.S.; Renganathan, S. Synthesis and characterization of copper nanoparticle using *Capparis zeylanica* leaf extract. *Int. J.Chem.Tech. Res.* **2014**, *6*, 4533–4541.

62. Subhankari, I.; Nayak, P.L. Synthesis of copper nanoparticles using *Syzygium aromaticum* (Cloves) aqueous extract by using green chemistry. *World J. Nano Sci. Technol.* **2013**, *2*, 14–17.

63. Aslani, F.; Bagheri, S.; Julkapli, N.M.; Juraimi, A.S.; Hashemi, F.S.G.; Baghdadi, A. Effects of engineered nanomaterials on plants growth: An overview. *Sci. World J.* **2014**, *2014*, 641759. [CrossRef] [PubMed]

64. Ogunyemi, S.O.; Abdallah, Y.; Zhang, M.; Fouad, H.; Hong, X.; Ibrahim, E.; Masum, M.M.I.; Hossain, A.; Mo, J.; Li, B. Green synthesis of zinc oxide nanoparticles using different plant extracts and their antibacterial activity against *Xanthomonas oryzae* pv. oryzae. *Artif. Cells Nanomed. Biotechnol.* **2019**, *47*, 341–352. [CrossRef]

65. Taran, M.; Rad, M.; Alayi, M. Biosynthesis of TiO_2 and ZnO nanoparticles by *Halomonas elongata* IBRC-M 10214 in different conditions of medium. *Bioimpacts* **2017**, *8*, 81–89. [CrossRef]

66. Dodoo-Arhin, D.; Buabeng, F.P.; Mwabora, J.M.; Amaniampong, P.N.; Agbe, H.; Nyankson, E.; Obada, D.O.; Asiedu, N.Y. The effect of titanium dioxide synthesis technique and its photocatalytic degradation of organic dye pollutants. *Heliyon* **2018**, *4*, e00681. [CrossRef]

67. Ma, X.; Geiser-Lee, J.; Deng, Y.; Kolmakov, A. Interactions between engineered nanoparticles (ENPs) and plants: Phytotoxicity, uptake and accumulation. *Sci. Total Environ.* **2010**, *408*, 3053–3061. [CrossRef] [PubMed]

68. Budhani, S.; Egboluche, N.P.; Arslan, Z.; Yu, H.; Deng, H. Phytotoxic effect of silver nanoparticles on seed germination and growth of terrestrial plants. *J. Environ. Sci. Health C Environ. Carcinog. Ecotoxicol. Rev.* **2019**, *37*, 330–355. [CrossRef]

69. Lin, D.; Xing, B. Phytotoxicity of nanoparticles: Inhibition of seed germination and root growth. *Environ. Pollut.* **2007**, *150*, 243–250. [CrossRef]

70. Yang, L.; Watts, D.J. Particles surface characteristics may play an important role in phytotoxicity of alumina nanoparticles. *Toxicol. Lett.* **2005**, *158*, 122–132. [CrossRef]

71. Larue, C.; Laurette, J.; Herlin-Boime, N.; Khodja, H.; Fayard, B.; Flank, A.; Brisset, F.; Brisset, M. Accumulation, translocation, and impact of TiO_2 nanoparticles in wheat (*Triticum aestivum* spp.): Influence of diameter and crystal phase. *Sci. Total Environ* **2012**, *43*, 197–208. [CrossRef]

72. Lin, S.; Reppert, J.; Hu, Q.; Hudson, J.S.; Reid, M.L.; Ratnikova, T.A.; Rao, A.M.; Luo, H.; Key, P.C. Uptake, translocation, and transmission of carbon nanomaterials in rice plants. *Small* **2009**, *5*, 1128–1132. [CrossRef]

73. Rafique, C.; Arshad, N.; Khakhar, M.F.; Qazi, I.A.; Hamza, A.; Vivic, N. Growth response of wheat to titania nanoparticles application. *NJES* **2014**, *7*, 42–46.

74. Hong, F.H.; Zhou, J.; Liu, C.; Yang, F.; Wu, C.; Zheng, L.; Yang, P. Effect of nano-TiO_2 on photochemical reaction of chloroplasts of spinach. Biol. *Trace Elem. Res.* **2005**, *105*, 269–279. [CrossRef]

75. Yang, F.; Liu, C.; Gao, F.; Su, M.; Wu, X.; Zheng, L.; Hong, F.; Yang, P. The improvement of spinach growth by nano-anatase TiO_2 treatment is related to nitrogen photoreduction. *Biol. Trace Elem. Res.* **2007**, *119*, 77–88. [CrossRef]

76. Linglan, M.; Chao, L.; Chunxiang, Q.; Sitao, Y.; Jie, L.; Fengqing, G.; Fashui, H. Rubisco activase mRNA expression in spinach: Modulation by nanoanatase treatment. *Biol. Trace Elem. Res.* **2008**, *122*, 168–178. [CrossRef]

77. Lu, C.M.; Zhang, C.Y.; Wen, J.Q.; Wu, G.R.; Tao, M.X. Research of the effect of nanometer materials on germination and growth enhancement of Glycine max and its mechanism. *Soybean Sci.* **2010**, *21*, 168–172.

78. Khodakovskaya, M.; Dervishi, E.; Mahmood, M.; Xu, Y.; Li, Z.; Watanabe, F.; Biris, A.S. Carbon nanotubes are able to penetrate plant seed coat and dramatically affect seed germination and plant growth. *ACS Nano* **2009**, *3*, 3221–3227. [CrossRef]

79. Canas, J.E.; Long, M.; Nations, S.; Vadan, R.; Dai, L.; Luo, M. Effects of functionalized and nonfunctionalized single-walled carbon-nanotubes on root elongation of select crop species. *Nanomater. Environ.* **2008**, *27*, 1922–1931.

80. Wang, X.; Han, H.; Liu, X.; Gu, X.; Chen, K.; Lu, D. Multi-walled carbon nanotubes can enhance root elongation of wheat (*Triticum aestivum*) plants. *J. Nanopart. Res.* **2012**, *14*, 841. [CrossRef]

81. Stampoulis, D.; Sinha, S.K.; White, J.C. Assay-dependent phytotoxicity of nanoparticles to plants. *Environ. Sci. Technol.* **2009**, *43*, 9473–9479. [CrossRef] [PubMed]

82. Hao, Y.; Fang, P.; Ma, C.; White, J.C.; Xiang, Z.; Wang, H.; Zhang, Z.; Rui, Y.; Xing, B. Engineered nanomaterials inhibit *Podosphaera pannosa* infection on rose leaves by regulating phytohormones. *Environ. Res.* **2018**, *170*, 1–6. [CrossRef]

83. Singh, D.; Kumar, A. Effects of nano silver oxide and silver ions on growth of *Vigna radiata*. *Bull. Environ. Contam. Toxicol.* **2015**, *95*, 379–384. [CrossRef] [PubMed]

84. Pokhrel, L.R.; Dubey, B. Evaluation of developmental responses of two crop plants exposed to silver and zinc oxide nanoparticles. *Sci. Total Environ.* **2013**, *45*, 321–332. [CrossRef] [PubMed]

85. Yang, R.; Diao, Y.; Abayneh, B. Removal of Hg(0) from simulated flue gas over silver-loaded rice husk gasification char. *R. Soc. Open Sci.* **2018**, *5*, 180248. [CrossRef]

86. Vannini, C.; Domingo, G.; Onelli, E.; Mattia, F.D.; Bruni, I.; Marsoni, M.; Bracale, M. Phytotoxicand genotoxic effects of silver nanoparticles exposure on germinating wheat seedlings. *J. Plant Physiol.* **2014**, *171*, 1142–1148. [CrossRef] [PubMed]

87. Riahi-Madvar, A.; Rezaee, F.; Jalali, V. Effects of alumina nanoparticles on morphological properties and antioxidant system of *Triticum aestivum*. *Iran. J. Plant Physiol.* **2012**, *3*, 595–603.

88. Feizi, H.; Moghaddam, P.R.; Shahtahmassebi, N.; Fotovot, A. Assessment of concentration of nano and bulk iron oxide particles on early growth of wheat (*Triticum aestivum* L.). *Annu. Rev. Res. Biol.* **2013**, *3*, 752–761.

89. Lee, W.M.; An, Y.J.; Yoon, H.; Kweon, H.S. Toxicity and bioavailability of copper nanoparticles to the terrestrial plants mung bean (*Phaseolus radiatus*) and wheat (*Triticum aestivum*): Plant agar test for water-insoluble nanoparticles. *Nanomater. Environ.* **2008**, *27*, 1915–1921. [CrossRef]

90. Divva, K.; Vijayan, S.; Nair, S.J.; Jisha, M.S. Optimization of chitosan nanoparticle synthesis and its potential application as germination elicitor of *Oryza sativa* L. *Int. J. Biol. Macromol.* **2018**, *124*, 1053–1059. [CrossRef]

91. López-Moreno, M.; De la Rosa, G.; Hernandez-Viezcas, J.; Castillo-Michel, H.; Botez, C.; Peralta-Videa, J.; Gardea-Torresdey, J. Evidence of the differential biotransformation and genotoxicity of ZnO and CeO nanoparticles on soybean (*Glycine max*) plants. *Environ. Sci. Technol.* **2010**, *44*, 7315–7320. [CrossRef] [PubMed]

92. Yasmeen, F.; Raja, N.I.; Mustafa, G.; Sakata, K.; Komatsu, S. Quantitative proteomic analysis of post flooding recovery in soybean roots exposed to aluminum oxide NPs. *J. Proteom.* **2016**, *143*, 136–150. [CrossRef] [PubMed]

93. Mustafa, G.; Komatsu, S. Insights into the response of soybean mitochondrial proteins to various sizes of aluminum oxide nanoparticles under flooding stress. *J. Proteome Res.* **2016**, *15*, 4464–4475. [CrossRef] [PubMed]

94. Zheng, L.; Hong, F.; Lu, S.; Liu, C. Effect of nano-TiO$_2$ on strength of naturally aged seeds and growth of spinach. *Biol. Trace Elem. Res.* **2005**, *104*, 83–91. [CrossRef]

95. Gao, F.Q.; Hong, F.H.; Liu, C.; Zheng, L.; Su, M.Y.; Wu, X.; Yang, F.; Wu, C.; Yang, P. Mechanism of nano-anatase TiO$_2$ on promoting photosynthetic carbon reaction of spinach e inducing complex of Rubisco activase. *Biol. Trace Elem. Res.* **2006**, *111*, 239–253. [CrossRef]

96. Yasmeen, F.; Raja, N.I.; Razzaq, A.; Komatsu, S. Gel-free/label-free proteomic analysis of wheat shoot in stress tolerant varieties under iron nanoparticles exposure. *Biochim. Biophys. Acta* **2016**, *1864*, 1586–1598. [CrossRef]

97. Yasmeen, F.; Raja, N.I.; Ilyas, N.; Komatsu, S. Quantitative proteomic analysis of shoot in stress tolerant wheat varieties on copper nanoparticle exposure. *Plant Mol. Biol. Rep.* **2018**, *36*, 326–340. [CrossRef]

98. Vannini, C.; Domingo, G.; Onelli, E.; Prinsi, B.; Marsoni, M.; Espen, L.; Bracale, M. Morphological and proteomic responses of *Eruca sativa* exposed to silver nanoparticles or silver nitrate. *PLoS ONE* **2013**, *8*, e68752. [CrossRef]

99. Seghatoleslami, M.; Faiza, H.; Mousav, G.; Berahmand, A. Effect of magnetic field and silver nanoparticles on yield and water use efficiency of *Carum copticum* under water stress conditions. *Pol. J. Chem. Technol.* **2015**, *17*, 110–114. [CrossRef]

100. Jacob, D.L.; Borchardt, J.D.; Navaratnam, L.; Otte, M.L.; Bezbaruah, A.N. Uptake and translocation of Ti from nanoparticles in crops and wetland plants. *Int. J. Phytoremediation* **2013**, *15*, 142–153. [CrossRef]

101. Mohammadi, R.; Maali-Amiri, R.; Abbasi, A. Effect of TiO2 nanoparticles on chickpea response to cold stress. *Biol. Trace Elem. Res.* **2013**, *152*, 403–410. [CrossRef] [PubMed]

102. Zhu, H.; Han, J.; Xiao, J.Q.; Jin, Y. Uptake, translocation, and accumulation of manufactured iron oxide by pumpkin plants. *J. Environ. Monit.* **2008**, *10*, 713–717. [CrossRef]

103. Yasmeen, F.; Raja, N.I.; Razzaq, A.; Komatsu, S. Proteomic and physiological analyses of wheat seeds exposed to copper and iron nanoparticles. *Biochim. Biophys. Acta Proteins Proteom.* **2017**, *1865*, 28–42. [CrossRef] [PubMed]

104. Barrena, R.; Casals, E.; Colón, J.; Font, X.; Sánchez, A.; Puntes, V. Evaluation of the ecotoxicity of model nanoparticles. *Chemosphere* **2009**, *75*, 850–857. [CrossRef] [PubMed]

105. Shaw, A.K.; Hossain, Z. Impact of nano-CuO stress on rice (*Oryza sativa* L.) seedlings. *Chemosphere* **2013**, *93*, 906–915. [CrossRef] [PubMed]

106. Shaw, A.K.; Ghosh, S.; Kalaji, H.M.; Bosa, K.; Brestic, M.; Zivcak, M.; Hossain, Z. Nano-CuO stress induced modulation of antioxidative defense and photosynthetic performance of Syrian barley (*Hordeum vulgare* L.). *Environ. Exp. Bot.* **2014**, *102*, 37–47. [CrossRef]

107. Rajput, V.D.; Minkina, T.; Fedorenko, A.; Mandzhieva, S.; Sushkova, S.; Lysenko, V.; Duplii, N.; Azarov, A.; Aleksandrovich, V. Destructive effect of copper oxide nanoparticles on ultrastructure of chloroplast, plastoglobules and starch grains in spring barley (*Hordeum sativum*). *Int. J. Agric. Biol.* **2019**, *21*, 171–174.

108. Adhikari, S.; Adhikari, A.; Ghosh, S.; Roy, D.; Azahar, I.; Basuli, D.; Hossain, Z. Assessment of ZnO-NPs toxicity in maize: An integrative microRNAomic approach. *Chemosphere* **2020**, *249*, 126197. [CrossRef]

109. Motyka, O.; Štrbová, K.; Olšovská, E.; Seidlerová, J. Influence of nano ZnO exposure to plants on l-ascorbic acid levels: Indication of nanoparticle-induced oxidative stress. *J. Nanosci. Nanotechnol.* **2019**, *19*, 3019–3023. [CrossRef]

110. López-Moreno, M.L.; De la Rosa, G.; Hernández-Viezcas, J.A.; Peralta-Videa, J.R.; Gardea-Torresdey, J.L. X-ray absorption spectroscopy (XAS) corroboration of the uptake and storage of CeO(2) nanoparticles and assessment of their differential toxicity in four edible plant species. *J. Agric. Food Chem.* **2010**, *58*, 3689–3693. [CrossRef]

111. Djanaguiraman, M.; Belliraj, N.; Bossmann, S.H.; Prasad, P.V.V. High-temperature stress alleviation by selenium nanoparticle treatment in grain sorghum. *ACS Omega* **2018**, *3*, 2479–2491. [CrossRef] [PubMed]

112. Divya, K.; Smitha, V.; Jisha, M.S. Antifungal, antioxidant and cytotoxic activities of chitosan nanoparticles and its use as an edible coating on vegetables. *Int. J.Biol. Macromol.* **2018**, *114*, 572–577. [CrossRef] [PubMed]

113. Abd El-Aziz, M.E.; Morsi, S.M.M.; Salama, D.M.; Abdel-Aziz, M.S.; AbdElwahed, M.S.; Shaaban, E.A.; Youssef, A.M. Preparation and characterization of chitosan/polyacrylicacid/copper nanocomposites and their impact on onion production. *Int. J. Biol. Macromol.* **2019**, *123*, 856–865. [CrossRef] [PubMed]

114. Chen, G.; Ma, C.; Mukherjee, A.; Musante, C.; Zhang, J.; White, J.C.; Dhankher, O.P.; Xing, B. Tannic acid alleviates bulk and nanoparticle Nd2O3 toxicity in pumpkin: A physiological and molecular response. *Nanotoxicology* **2016**, *10*, 1243–1253. [CrossRef] [PubMed]

115. Weir, E.; Lawlor, A.; Whelan, A.; Regan, F. The use of nanoparticles in antimicrobial materials and their characterization. *Analyst* **2008**, *133*, 835–845. [CrossRef]

116. Simon, D.F.; Domingos, R.F.; Hauser, C.; Hutchins, C.M.; Zerges, W.; Wilkinson, K.J. Transcriptome sequencing (RNA-seq) analysis of the effects of metal nanoparticle exposure on the transcriptome of *Chlamydomonas reinhardtii*. *Appl. Environ. Microbiol.* **2013**, *79*, 4774–4785. [CrossRef]

117. Lok, C.N.; Ho, C.M.; Chen, R.; He, Q.Y.; Yu, W.Y.; Sun, H.; Tam, P.K.; Chiu, J.F.; Che, C.M. Proteomic analysis of the mode of antibacterial action of silver nanoparticles. *J. Proteome Res.* **2006**, *5*, 916–924. [CrossRef] [PubMed]

118. Mirzajani, F.; Askari, H.; Hamzelou, S.; Schober, Y.; Rompp, A.; Ghassempour, A.; Spengler, B. Proteomics study of silver nanoparticles toxicity on *Bacillus thuringiensis*. *Ecotoxicol. Environ. Saf.* **2014**, *100*, 122–130. [CrossRef] [PubMed]

119. Hossain, Z.; Mustafa, G.; Sakata, K.; Komatsu, S. Insights into the proteomic response of soybean towards Al_2O_3, ZnO, and Ag nanoparticles stress. *J. Hazard. Mater.* **2016**, *304*, 291–305. [CrossRef] [PubMed]

120. Mustafa, G.; Sakata, K.; Komatsu, S. Proteomic analysis of soybean root exposed to varying sizes of silver nanoparticles under flooding stress. *J. Proteom.* **2016**, *148*, 113–125. [CrossRef]

121. Jhanzab, H.M.; Razzaq, A.; Bibi, Y.; Yasmeen, F.; Yamaguchi, H.; Hitachi, K.; Tsuchida, K.; Komatsu, K. Proteomic analysis of the effect of inorganic and organic chemicals on silver nanoparticles in wheat. *Int. J. Mol. Sci.* **2019**, *20*, 825. [CrossRef] [PubMed]

122. Hashimoto, T.; Mustafa, G.; Nishiuchi, T.; Komatsu, S. Comparative analysis of the effect of inorganic and organic chemicals with silver nanoparticles on soybean under flooding stress. *Int. J. Mol. Sci.* **2020**, *21*, 1300. [CrossRef] [PubMed]

123. Handy, R.D.; Kamme, F.; Lead, J.R.; Hasselov, M.; Owen, R.; Crane, M. The ecotoxicology and chemistry of manufactured nanoparticles. *Ecotoxicology* **2008**, *17*, 287–314. [CrossRef] [PubMed]

124. Shabnam, N.; Kim, H. Non-toxicity of nano alumina: A case on mung bean seedlings. *Ecotoxicol. Environ. Saf.* **2018**, *165*, 423–433. [CrossRef]

125. Mustafa, G.; Sakata, K.; Komatsu, S. Proteomic analysis of flooded soybean root exposed to aluminum oxide nanoparticles. *J. Proteom.* **2015**, *128*, 280–297. [CrossRef]

126. Hou, J.; Wang, L.; Wang, C.; Zhang, S.; Liu, H.; Li, S.; Wang, X. Toxicity and mechanisms of action of titanium dioxide nanoparticles in living organisms. *J. Environ. Sci.* **2019**, *75*, 40–53. [CrossRef]

127. Riu, J.; Maroto, A.; Rius, F.X. Nanosensors in environmental analysis. *Talanta* **2006**, *69*, 288–301. [CrossRef]
128. Tan, X.; Wang, X.; Chen, C.; Sun, A. Effect of soil humic and fulvic acids, pH and ionic strength on Th(IV) sorption to TiO_2 nanoparticles. *Appl. Radiat. Isot.* **2007**, *65*, 375–381. [CrossRef]
129. Chen, L.Z.; Zhou, L.N.; Liu, Y.D.; Deng, S.Q.; Wu, H.; Wang, G.H. Toxicological effects of nanometer titanium dioxide (nano-TiO_2) on *Chlamydomonas reinhardtii*. *Ecotoxicol. Environ. Saf.* **2012**, *84*, 155–162. [CrossRef]
130. Zhao, L.; Peng, B.; Hernandez-Viezcas, J.A.; Rico, C.; Sun, Y.; Peralta-Videa, J.R.; Tang, X.; Niu, G.; Jin, L.; Varela-Ramirez, A. Stress response and tolerance of *Zea mays* to CeO2 nanoparticles: Cross talk among H_2O_2, heat shock protein, and lipid peroxidation. *ACS Nano* **2012**, *6*, 9615–9622. [CrossRef]
131. PeharecŠtefanić, P.; Jarnević, M.; Cvjetko, P.; Biba, R.; Šikić, S.; Tkalec, M.; Cindrić, M.; Letofsky-Papst, I.; Balen, B. Comparative proteomic study of phytotoxic effects of silver nanoparticles and silver ions on tobacco plants. *Environ. Sci. Pollut. Res. Int.* **2019**, *26*, 22529–22550. [CrossRef] [PubMed]
132. Du, W.; Sun, Y.; Ji, R.; Zhu, J.; Wu, J.; Guo, H. TiO2 and ZnO nanoparticles negatively affect wheat growth and soil enzyme activities in agricultural soil. *J. Environ. Monit.* **2011**, *13*, 822e828. [CrossRef] [PubMed]
133. Chen, H. Metal based nanoparticles in agricultural system: Behavior, transport, and interaction with plants. *Chem. Spec. Bioavailab.* **2018**, *30*, 123–134. [CrossRef]
134. Lv, J.; Zhang, S.; Luo, L.; Zhang, J.; Yang, K.; Christie, P. Accumulation, speciation and uptake pathway of ZnO nanoparticles in maize. *Environ. Sci. Nano* **2015**, *2*, 68–77. [CrossRef]
135. Halliwell, B.; Gutteridge, J.M.C. *Free Radicals in Biology and Medicine*, 2nd ed.; Clarendon Press: Oxford, UK, 1989.
136. Tripathy, B.C.; Oelmüller, R. Reactive oxygen species generation and signaling in plants. *Plant Signal Behav.* **2012**, *7*, 1621–1633. [CrossRef]
137. Wang, X.; Yang, X.; Chen, S.; Li, Q.; Wang, W.; Hou, C.; Gao, X.; Wang, L.; Wang, S. Zinc oxide nanoparticles affect biomass accumulation and photosynthesis in Arabidopsis. *Front. Plant Sci.* **2016**, *6*, 1243. [CrossRef]
138. Wang, X.P.; Li, Q.Q.; Pei, Z.M.; Wang, S.C. Effects of zinc oxide nanoparticles on the growth, photosynthetic traits, and antioxidative enzymes in tomato plants. *Biol. Plant.* **2018**, *62*, 801e808. [CrossRef]
139. Hossain, Z.; Nouri, M.Z.; Komatsu, S. Plant cell organelle proteomics in response to abiotic stress. *J. Prot. Res.* **2012**, *11*, 37–48. [CrossRef]
140. Mittova, V.; Volokita, M.; Guy, M.; Tal, M. Activities of SOD and the ascorbate–glutathione cycle enzymes in subcellular compartments in leaves and roots of the cultivated tomato and its wild salt-tolerant relative *Lycopersicon pennelli*. *Physiol. Plant.* **2000**, *110*, 42–51. [CrossRef]
141. Barber, D.J.W.; Thomas, J.K. Reactions of radicals with lecithin bilayers. *Radiat. Res.* **1978**, *74*, 51–58. [CrossRef]
142. Dos Santos, C.V.; Rey, P. Plant thioredoxins are key actors in oxidative stress response. *Trends Plant Sci.* **2006**, *11*, 329–334. [CrossRef] [PubMed]
143. Berndt, C.; Lillig, C.H.; Holmgren, A. Thioredoxins and glutaredoxins as facilitators of protein folding. *Biochim. Biophys. Acta* **2008**, *1783*, 641–650. [CrossRef] [PubMed]

 International Journal of
Molecular Sciences

Article

Dissecting the Seed Maturation and Germination Processes in the Non-Orthodox *Quercus ilex* Species Based on Protein Signatures as Revealed by 2-DE Coupled to MALDI-TOF/TOF Proteomics Strategy

Besma Sghaier-Hammami [1,2,*], Sofiene B.M. Hammami [3], Narjes Baazaoui [4], Consuelo Gómez-Díaz [5] and Jesús V. Jorrín-Novo [2,*]

[1] Centre de Biotechnologie de Borj-Cédria, Laboratoire des Plantes Extrêmophiles, BP 901, Hammam-Lif 2050, Tunisia

[2] Agroforestry and Plant Biochemistry, Proteomics and Systems Biology, Department of Biochemistry and Molecular Biology, University of Cordoba, UCO-CeiA3, 14014 Cordoba, Spain

[3] Institut National Agronomique de Tunisie (INAT), Laboratoire LR13AGR01, Université de Carthage, 43 Avenue Charles Nicolle, Tunis 1082, Tunisia; sofiene.hammami@inat.u-carthage.tn

[4] King Khalid University, Abha 61421, Saudia Arabia; baazaouinarjes@gmail.com

[5] Unidad de Proteómica, Servicio Central de Apoyo a la Investigación (SCAI), Universidad de Córdoba, 14014 Córdoba, Spain; bc2godic@uco.es

* Correspondence: besma.sghaierhammami@cbbc.rnrt.tn (B.S.-H.); bf1jonoj@uco.es (J.V.J.-N.); Tel.: +216-79325855 (B.S.-H.); +34-957-218-609 (J.V.J.-N.)

Received: 19 June 2020; Accepted: 7 July 2020; Published: 9 July 2020

Abstract: Unlike orthodox species, seed recalcitrance is poorly understood, especially at the molecular level. In this regard, seed maturation and germination were studied in the non-orthodox *Quercus ilex* by using a proteomics strategy based on two-dimensional gel electrophoresis coupled to matrix-assisted laser desorption ionization/time of flight (2-DE-MALDI-TOF). Cotyledons and embryo/radicle were sampled at different developmental stages, including early (M1–M3), middle (M4–M7), and late (M8–M9) seed maturation, and early (G1–G3) and late (G4–G5) germination. Samples corresponding to non-germinating, inviable, seeds were also included. Protein extracts were subjected to 2-dimensional gel electrophoresis (2-DE) and changes in the protein profiles were analyzed. Identified variable proteins were grouped according to their function, being the energy, carbohydrate, lipid, and amino acid metabolisms, together with protein fate, redox homeostasis, and response to stress are the most represented groups. Beyond the visual aspect, morphometry, weight, and water content, each stage had a specific protein signature. Clear tendencies for the different protein groups throughout the maturation and germination stages were observed for, respectively, cotyledon and the embryo axis. Proteins related to metabolism, translation, legumins, proteases, proteasome, and those stress related were less abundant in non-germinating seeds, it related to the loss of viability. Cotyledons were enriched with reserve proteins and protein-degrading enzymes, while the embryo axis was enriched with proteins of cell defense and rescue, including heat-shock proteins (HSPs) and antioxidants. The peaks of enzyme proteins occurred at the middle stages (M6–M7) in cotyledons and at late ones (M8–M9) in the embryo axis. Unlike orthodox seeds, proteins associated with glycolysis, tricarboxylic acid cycle, carbohydrate, amino acid and lipid metabolism are present at high levels in the mature seed and were maintained throughout the germination stages. The lack of desiccation tolerance in *Q. ilex* seeds may be associated with the repression of some genes, late embryogenesis abundant proteins being one of the candidates.

Keywords: *Quercus ilex*; acorns; non-orthodox seeds; seed maturation; seed germination; proteomics; embryo axis; cotyledon; inviable seeds

---

1. Introduction

Protein signatures, experimentally afforded by proteomics, determine the phenotype, characteristics and properties of biological systems and processes. We have applied this principle to study seed development and germination in the non-orthodox *Quercus ilex*, the most representative and emblematic forest tree species of the Mediterranean forest and of the agrosilvopastoral ecosystem "dehesa" [1]. Despite being a priority, reforestation and breeding programs with this species are hindered by its long lifespan, lack of natural regeneration, and its biological properties. These latter, impede a viable and feasible germplasm conservation and propagation, which is also linked to the lack of knowledge of its biology, especially at the molecular level [1].

Seed formation and germination are unique characteristics of spermatophytes which support propagation in mostplants, and they are crucial not only for plant development but for human food supply, the conservation of genetic resources, and breeding programs [2].

Depending on the seed's characteristics, properties, maturation, and germination, plant species are classified as recalcitrant (non-orthodox), intermediate or orthodox [3]. The main difference is their sensitivity to water loss and desiccation. Recalcitrant seeds are those that lose viability when dried to a water content of less than 12%, and that lack a dehydration stage during seed ripening [4]. Intermediate and recalcitrant seeds are unable to survive drying and chilling as they rapidly lose their germination capacity and viability during storage [5]. Therefore, for these species, long-term germplasm conservation using conventional seed storage methods is not possible and the development of an alternative strategy remains a real challenge.

The genetic, genomic, and molecular bases of seed development and germination have been extensively studied and acquired in the case of orthodox seeds but are poorly understood in the recalcitrant. Seed maturation and germination, and the transition from one stage to the other, implies complex physiological and biochemical processes. They are induced by environmental factors and under the control of phytohormones and other signaling molecules. They were further mediated by genetic and epigenetic gene expression programs [6–9]. This results in stage-specific transcriptomic, proteomic and metabolomic profiles. They are characterized by structural and functional signatures including operating metabolic pathways, reactive oxygen species (ROS) production, synthesis, accumulation, and mobilization of reserve proteins, and the production of antioxidant and other stress-related proteins, among others [10,11]. Recalcitrant seeds are able to germinate immediately after shedding, thus implying an active metabolism in the mature stage, a property that is absent in quiescent orthodox seeds [12].

As a continuation of our previous work published on *Q. ilex* acorns and seed germination [13–15], we have undertaken a proteomic analysis of seed development and germination. Proteomics has been previously employed in the study of seeds [16], and a number of papers that have focused on germination and viability in Quercus spp. have been previously published [17,18]. The novelty of the present work was the study of the maturation and germination processes as a continuum, and the independent analysis of the cotyledon and the embryo axis throughout 14 stages, nine for maturation and five for germination. From the stage-specific protein profiles visualized, the molecular basis of seed development in this non-orthodox *Q. ilex* species can be hypothesized. This and similar studies would help in establishing clonal propagation methods by means of somatic embryogenesis through the monitoring and molecular comparison of somatic, mostly inviable, and zygotic, mostly viable, embryos [19], as well as the establishment of ex situ germplasm cryoconservation protocols [18].

2. Results

2.1. Acorn Maturation and Germination Stages

The present study is a continuation of our previous work [13], in which the protein profiles of the Holm oak seed cotyledon and the embryo axis at the mature, harvesting, stage were compared. In this paper, the changes in protein profiles were analyzed throughout different acorn developmental (middle,

M4–M6, and late, M8–M9 stages, with M9 corresponding to the mature acorn) and germination (early, G1–G3, and late, G4–G5 stages) phases (Figures S1 and S2).

Apart from the visual aspect, morphometry, and weight (Table 1), the different stages were characterized by tissue water content (Figure 1).

Table 1. Morphometric parameters (length, maximum diameter) and weight of the acorns at the different maturation stages (M1 to M9). M9 corresponds to the mature fruit. The data are presented as the mean ± Standard error of six replicates. Different letters indicate a significant difference with $p < 0.05$ (Tukey test).

Stages	Length (mm)	Maximum Diameter (mm)	Weight (g)
M1	5.00 ± 0.28f	3.07 ± 0.16g	0.03 ± 0.0003e
M2	5.03 ± 0.24f	3.55 ± 0.32g	0.05 ± 0.0003de
M3	8.61 ± 0.36ef	7.16 ± 0.34f	0.19 ± 0.04de
M4	10.88 ± 0.67de	8.95 ± 0.31e	0.40 ± 0.06de
M5	14.28 ± 0.78cd	10.65 ± 0.39d	0.83 ± 0.068cd
M6	18.04 ± 1.04cd	11.86 ± 0.42cd	1.33 ± 0.11cd
M7	25.73 ± 1.56b	13.42 ± 0.53bc	2.72 ± 0.35b
M8	36.16 ± 1.48a	15.77 ± 0.41a	5.24 ± 0.17a
M9	33.12 ± 0.88a	14.95 ± 1.20ab	3.50 ± 0.30b

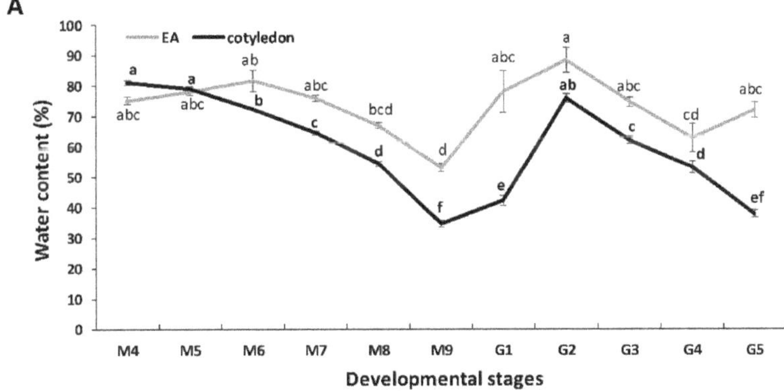

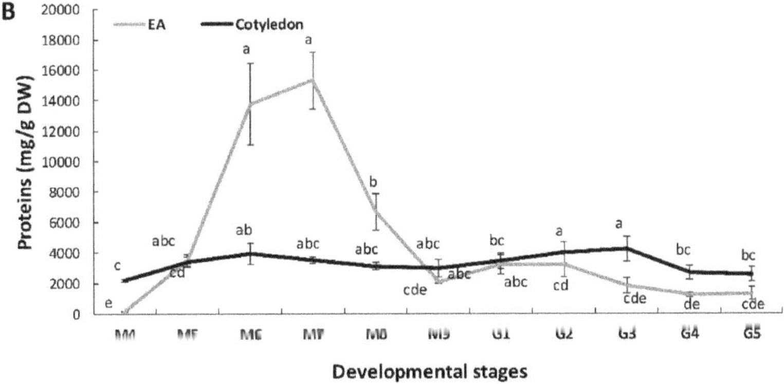

Figure 1. Water (**A**) and protein content (**B**) in seed parts, cotyledon and embryo axis (EA)/radicle, at the different stages of maturation (M1–M9) and germination (G1–G5) processes. Values correspond to the mean of six replicates, with bars corresponding to the Standard error. Different letters indicate significant difference with $p < 0.05$ (Tukey test).

In cotyledons, the water content decreased from 80% at M4 to 35% at the mature, M9, stage (Figure 1A). Upon imbibition, water content increased at G1 up to 76%, then decreasing again to 35% at G5. The water content in the embryo axis remained nearly constant, 75% throughout the maturation stages and a decrease to 53% was observed only at M9. Upon imbibition, water content increased to 89% and remained above 62% at the germination stages analysis (Figure 1A).

2.2. Protein Extraction, 1- and 2-D Gel Electrophoresis

Proteins were extracted by a trichloroacetic acid-acetone/Phenol/chloroform method and quantified by Bradford (Figure 1B). During maturation, the protein content was much higher, in relative terms (per g of tissue dry weight), in the embryo axis than in the cotyledon. In the embryo axis, an initial increase, with maxima at M6 and M7 (around 15 g/g DW) was followed by an ulterior decline to values of below 2 g/g DW at M9. During all germination stages, the amount of proteins remained constant, with similar values to M9, mature acorns. For the cotyledon, the protein content remained nearly constant at all stages from M4 to G5, with values of around 3 g/g DW. For the non-germinating (NG) acorns, cotyledon and the embryo axis presented a similar protein content corresponding to around 4 g/g DW.

The protein extracts from seeds at different developmental stages (from M4 to G5) and tissues (embryo axis, cotyledon) were analyzed by 1-D electrophoresis to discriminate the most important stages on which changes occurred during seeds development (Figure 2A, Figure S3, and Table S1).

The number of bands resolved was 38–50 (cotyledon), and 23–58 (the embryo axis). The highest number of bands corresponded to the M9 stage in cotyledon, whereas in the embryo axis they corresponded to M6–M9 (Table S2). During maturation stages, thirty-six out of the 50 different bands were variable in cotyledon, while 32 out of 58 were variable in the embryo axis. During germination, the number of bands decreased in the embryo axis /radical; however, it remained constant in cotyledon. Twenty out of 53 protein bands were qualitatively variable in cotyledon, while 15 out of 52 were variable in the embryo axis. The NG acorns presented more protein bands in embryo axis than in cotyledon (Table S2).

Based on the results obtained by 1-DE, a 2-DE analysis (Figure 2B, Figure S4) was then performed on the last four development stages (M6 to M9, representing middle and late stages) and on G3 and G5 germination ones (representing early and late stages). It is there where most of the changes in protein profiles were observed. Some variability in the 2-DE pattern among replicates was observed and, consequently, the consistent spots, i.e., those present in all three replicates (representing about 70% of the total), were considered in the statistical analysis (Table S3). During maturation, the number of consistent spots ranged between 446–506 (cotyledon), and 326–470 (the embryo axis), with the maximum value corresponding to M6–M8 for cotyledon, and M9 for the embryo axis. The number of spots decreased during germination compared to the M9 stage both in cotyledon and the embryo axis (Table 2).

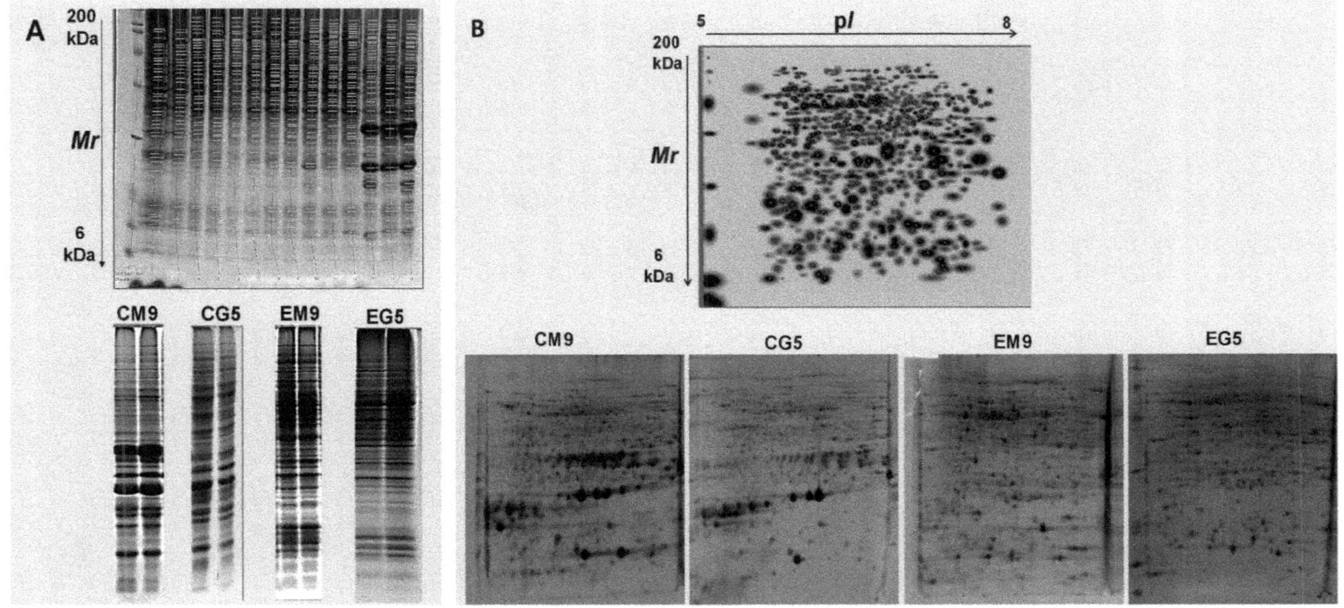

Figure 2. Protein profiles of seed parts, cotyledons and embryo, as visualized by (**A**): 1-D (60 μg of proteins were separated, molecular marker (Mr) is given on the left) and by (**B**): 2-D gel electrophoresis (250 μg of proteins were separated in the first dimension on an immobilized, linear, 5–8 pH gradient and in the second dimension on a 12% acrylamide-SDS gel). Sypro Ruby (SYPRO Ruby Protein Stains Bio-Rad) staining was used to visualize gels following the indications of the instruction manual. Representative images, corresponding to specific stages (M9 and G5) were included. For 2-DE, the master gel is presented. C: cotyledon; E: embryo axis.

Table 2. Total number of consistent spots matched between the cotyledon and embryo axis during maturation (M6 to M9) and germination (G3 and to G5) stages. DP: differential protein spots during maturation and germination. NG: non-germinated acorns. The data are presented as the mean ± standard error of three replicates. Different letters indicate significantly difference with $p < 0.05$ (Tukey test).

Matched Spots	M6	M7	M8	M9	G3	G5	NG	DP
Cotyledon	504 ± 4.3a	506.6 ± 1.5a	489 ± 1ab	456.6 ± 23b	248.6 ± 46c	230.6 ± 17c	234.6 ± 23c	558
Embryo axis	325.3 ± 0.53c	382 ± 2bc	422.6 ± 2.8b	470 ± 1a	385.3 ± 25b	243 ± 53d	388.3 ± 62b	591

By comparing the different maturation and germination stages in each part of the seed tissue, a total of 558 and 591 spots showed significant changes (±1.5-fold, $p < 0.05$, Table 2) in abundance in cotyledon and the embryo axis, respectively. The NG acorns presented more spots in embryo axis (388) than in cotyledon (234) (Table 2).

Some of the spots (10% of identified proteins) showed qualitative differences (presence/absence) between samples and can be proposed as markers of viability, maturation and germination stages, and tissues (cotyledons and embryo). The list of those markers included:

(i) Markers of viability, which were absent in non-germinating seeds, spot 7503 (with the identified protein being an Alcohol dehydrogenase), spot 6425 (lactate/malate dehydrogenase), spots 308, 1312, 2313 and 2315 (with the identified protein being an Enoyl-ACP reductase precursor) and spot 8630 (with the identified protein being a dehydrin protein).

(ii) Markers of developmental stages, which were abundant only in germination, spots 5123 and 6102 (with the identified protein being an Osmotin), so they were chosen as biomarkers for germination. However, spots 2406 and 2603 (with the identified protein as ATP synthase alpha/beta family protein) were only abundant in late maturation, so they were chosen as biomarkers for this phase. The spots 4834, 5819, 6801 and 2406 (with the identified protein being a Transketolase) were highly abundant at middle maturation stages, and spot 8630 (with the identified protein being dehydrin) was highly abundant only at late maturation stages.

(iii) Markers of storage tissues, which were abundant only in cotyledon, spots 8705, 6125, 7232 and 7307 (with the identified protein being a Legumin precursor), spots 17, 1013, 2005, 3009, 6002, 8006, 8008 and 8110 (with the identified protein being a RmlC-like cupins superfamily protein) and spots 5212, 4220 and 4218 (with the identified protein being a 11S globulin isoform 3).

2.3. Statistical Analysis of the Data

2-DE spot intensity data were subjected to multivariant, Venn (Figure 3) and principal component analysis (PCA) (Figure 4) during maturation and germination stages, and for seed parts, the cotyledons and the embryo axis.

During cotyledon's development, the different maturation stages (M6–M9) had 407 common protein spots, while the germination stages (G3–G5) shared only few common proteins (145) (Figure 3). Contrarily, during the embryo axis development, the maturation (M6–M9) and germination (G1–G3) stages shared a higher number of proteins, 316 and 405, respectively (Figure 3). The embryo axis of the non-germinating acorn (ENG) and that of EG3 have 74 common spots, but the cotyledons of NG and CG3 only have 20.

In the first PCA performed, comparing stages, the first two components accounted for 57% variability (35% for PC1, and 22% for PC2), and 62% variability (46% for PC1, and 17% for PC2) for the cotyledon and the embryo axis, respectively, with the first seven components accounting for 99.99% of the biological variability of the dataset in all the developmental stages (Table S4).

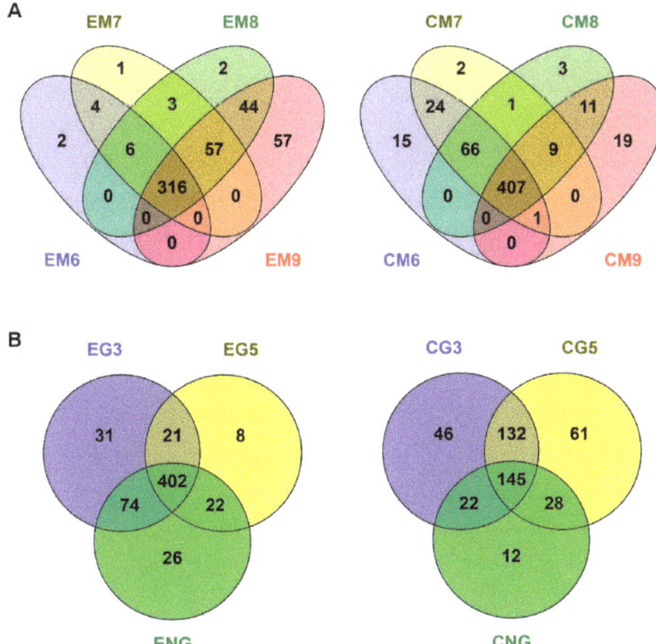

Figure 3. Venn diagrams represent the number of protein spots generated from the 2-DE analyses in cotyledon and embryonic axis during maturation (**A**): from M6 to M9, and during germination (**B**): G3 and G5. The diagrams were plotted using Venny software. The overlapping region between any two groups represents the number of common protein spots between development stages and/or tissues. C: cotyledon; E: embryo; M: maturation; G: germination.

All cotyledon and embryo axis maturation stages were clearly separated from the germination stages by PC1 (Figure 4A,B). PC2 discriminated, on one hand, cotyledons from non-germinating and germinating ones (CG3 and CG5), and on the other hand, early maturation (CM6–CM8) from late (CM9) stages (Figure 4A). For the embryo axis, PC2 separated EM6–EM7 from EM8–EM9. However, at the germination stage, PC2 grouped the embryo axis of non-germinating acorns (ENG) with EG3 and separated it from EG5 (Figure 4B). Based on PC1, 228 and 41 spots decreased in intensity from maturation to germination in the cotyledon and the embryo axis, respectively. Conversely, six spots in the cotyledon and 119 in the embryo axis increased in intensity from maturation to germination (Figure S5).

With regard to the second two PCA analyses, when comparing seed parts (Figure 4C,D), the first two components accounted for 68% variability (50% for PC1, and 18% for PC2), and 57% variability (35% for PC1, and 22% for PC2) for the maturation and germination stages, respectively. The seven principal components accounted for 99.99% of the biological variability of the dataset in all developmental stages (Table S4). It is clear that PC1 discriminated the maturation of the embryo axis from that of cotyledon, and similarly for the germination stages (Figure 4C,D). The PC2 discriminated germination and maturation stages similarly to what was observed in the previous analyses (i) and (ii). Based on PC1, 185 spots were more abundant during maturation in cotyledon than in the embryo axis, and, inversely, 162 spots were more abundant during germination in embryo axis than in cotyledon (Figure S5).

Similar results in differences and distances between samples were obtained when performing a hierarchical clustering analysis (Figure S6).

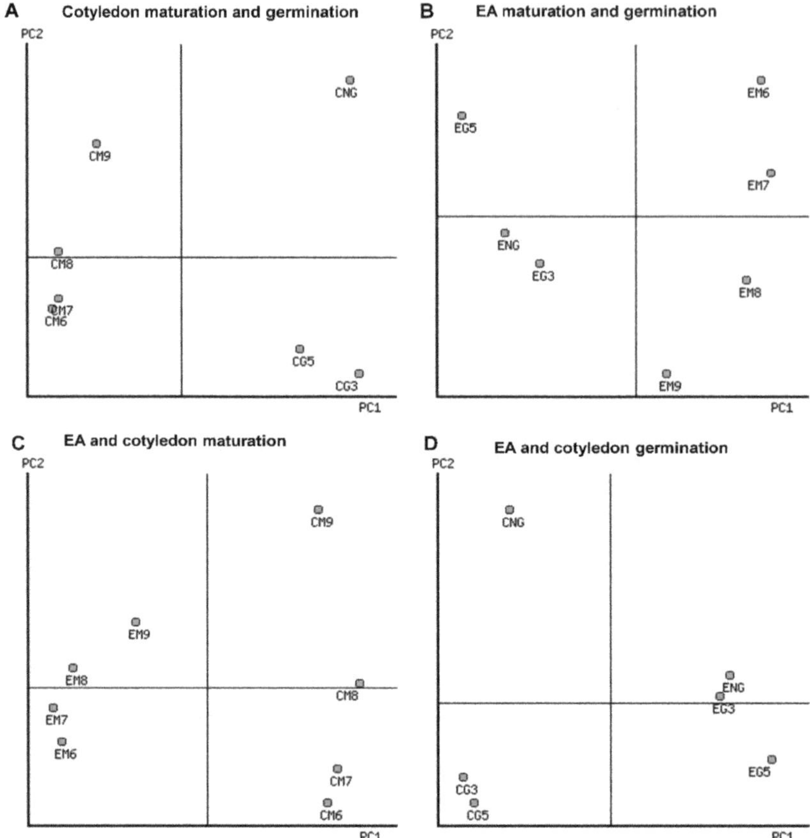

Figure 4. Principal component analysis of 2-DE consistent spots. The plots correspond to cotyledons
(**A**), the embryo axis (**B**), maturation (**C**), and germination (**D**). More details are included in Table S4.

*2.4. Protein Identification, Functional Classification of the Differentially Abundant Proteins, and Changes in
Abundance Throughout the Maturation and Germination Stages*

The selection of spots that could be used as biomarkers for each developmental stage would be of
great interest. The loading of each particular spot to PC 1, 2, and 3 was determined from the factor
matrix generated during the PCA. A total of 160 (for cotyledon) and 57 (for embryo axis) spots showed
the highest correlation (above |0.9|) with each PC determined (Table S5). These spots can be used
to differentiate the stages studied and, once identified, to improve the explanation of the variability
contained in those groups.

Variable spots were excised from the gel and subjected to matrix-assisted laser desorption
ionization/time of flight (MALDI-TOF) analysis. For identification, the *Quercus ilex* homemade and
the National Center for Biotechnology Information (NCBI) databases were used. Only proteins
with a high confidence in the (peptide mass fingerprinting) PMF matches, based on a score of over
71, were considered significant ($p < 0.05$). Out of the total variable spots, 178 (cotyledon) and 143
(embryo) were successfully identified (Table S6). In general, for the proteins identified, *Mr* values were
overestimated under the electrophoresis conditions employed, or, in some cases, as reserve proteins,
that may be the result of protein aggregation. On the other hand, the experimental and theoretical *pI*
were very close. Some of the proteins identified were present in spots having different *Mr* and *pI* and
they corresponded to isoforms or proteoforms as a result of post-translational modifications (PTMs).

That is the case of phosphoglycerate mutase, enolase, and NAD(P)-linked oxidoreductase superfamily protein in embryos, and reserve proteins, pyruvate kinase, pyruvate dehydrogenase, in cotyledons (Table S6).

The 178 proteins identified in the cotyledon could be classified into five functional and 10 sub-functional groups, based on their putative function according to the Kyoto Encyclopedia of Genes and Genomes (KEGG). The most abundant categories were associated essentially with energy metabolism (47%), storage proteins (13%), cell defense and rescue (11%) and protein biosynthesis and destination (10%). The 143 proteins identified in the embryo axis could be classified into seven functional and eight sub-functional groups. The most abundant categories were associated with energy metabolism (21%), cell defense and rescue (31%) and protein biosynthesis and destination (12%) (Table S6).

To understand the changes in the proteome of the different tissue parts of non-orthodox seeds during development and germination, the protein abundance of cotyledon and the embryo axis were compared using the Genesis software package. In addition, the different expression dynamics of each protein identified was represented by a heat map (Figure 5A,B). The trends of those proteins in the two different tissue parts during all the developmental stages are summarized in Figure 6.

As a general rule, in cotyledon, a similar tendency was observed for most of the groups. Protein abundance was maximum at M6, and then decreased at M9 (mature stage) to low values, which were maintained at the germination stages. The exception was in the reserve proteins, whose abundance increased from M6, presenting maximum values at M8 and M9. Some of them were mobilized and hence decreased in abundance at M9 and germination stages. More tendencies were observed in embryo proteins. Some of the groups, such as those of the energetic and carbohydrate metabolism, increased from M6 or M8, having maximum values at the germination stages. Protein abundance in the reserve proteins, and stress responses, including the detoxification of reactive oxygen species, increased from M6 to M8 or M9, then decreased to low values throughout the germination stages for those with maximum values at M8, or remaining at the M9 high levels. For other groups (amino acid metabolism, lipid metabolism, protein biosynthesis and proteolysis) no specific trends were obtained.

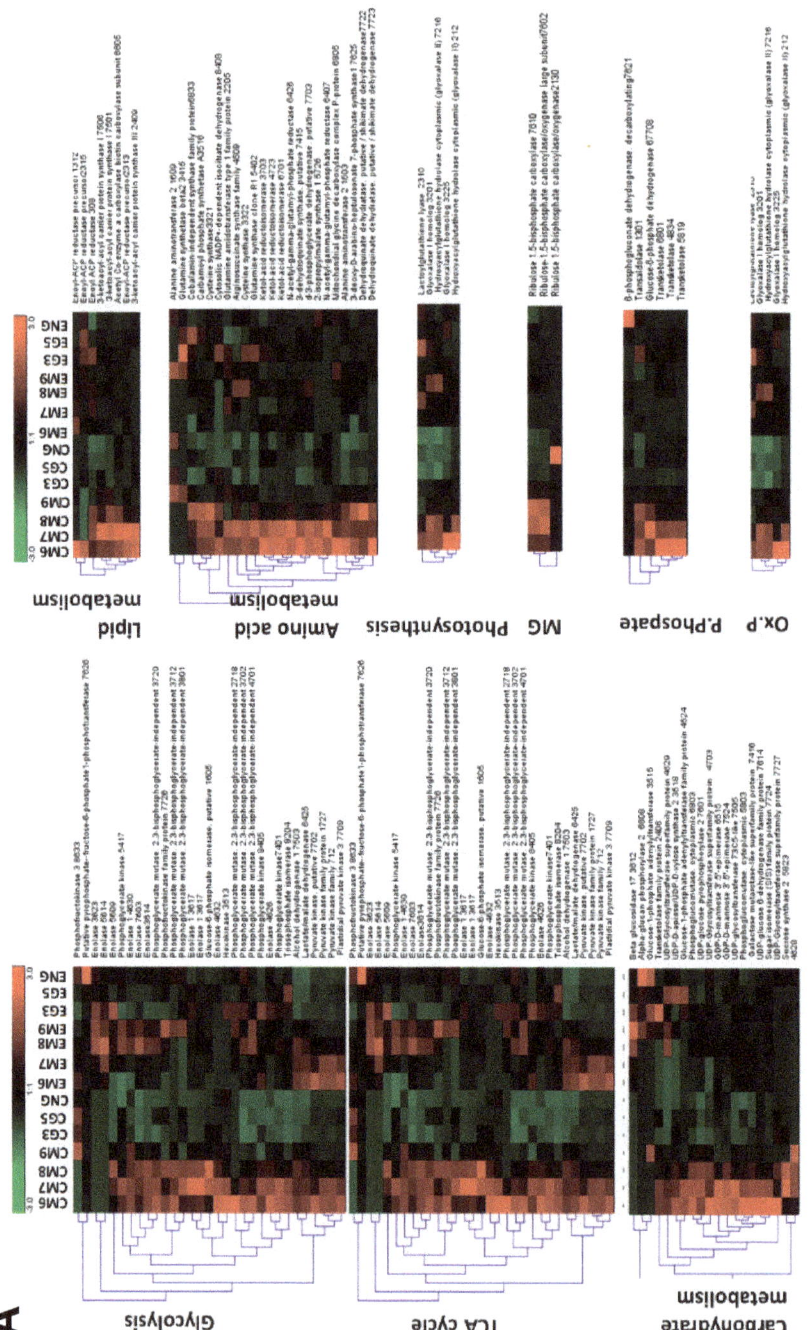

Figure 5. *Cont.*

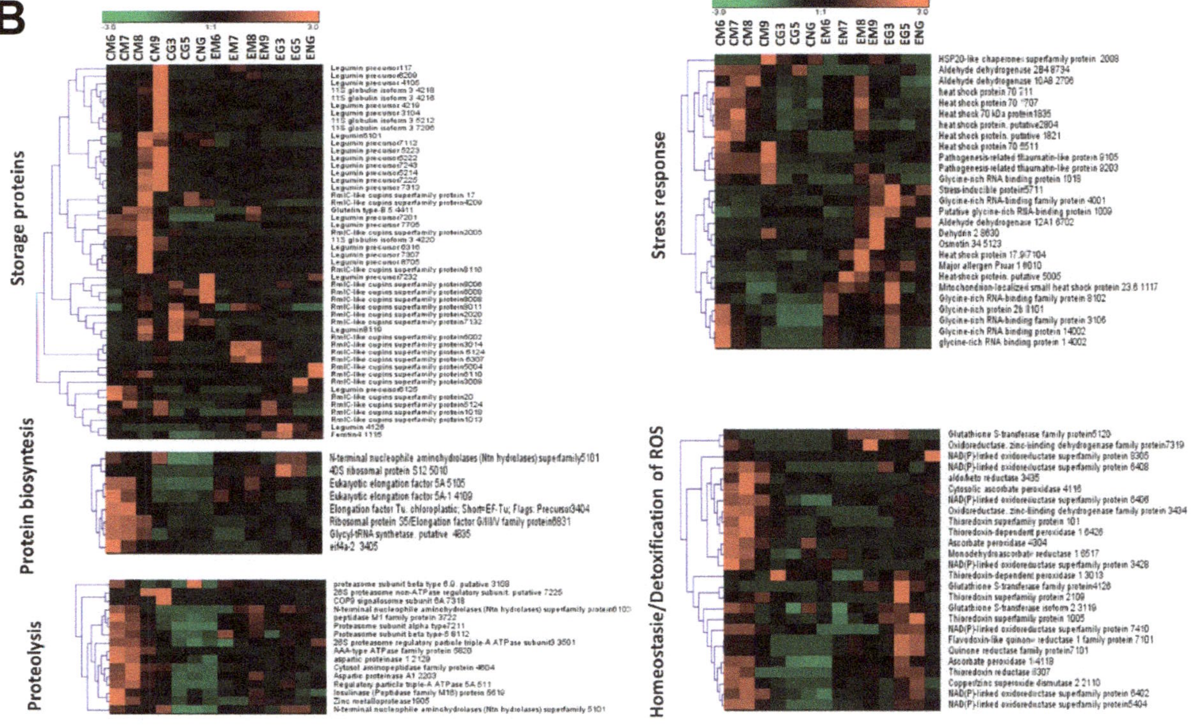

Figure 5. (**A**,**B**) Two-way hierarchical cluster of differentially variable proteins showing changes in abundance in cotyledon and embryo axis along the maturation and germination processes according to the Genesis software package (Eisen et al. 1998). A heatmap representation of the clustered spots shows the protein values according to the level of normalized experiments, which are indicated from −3 (minimum positive expression: clear color) to 3 (maximum positive expression: dark color); black indicates zero expression. Proteins were classified according to the Kyoto Encyclopedia of Genes and Genomes (KEGG).

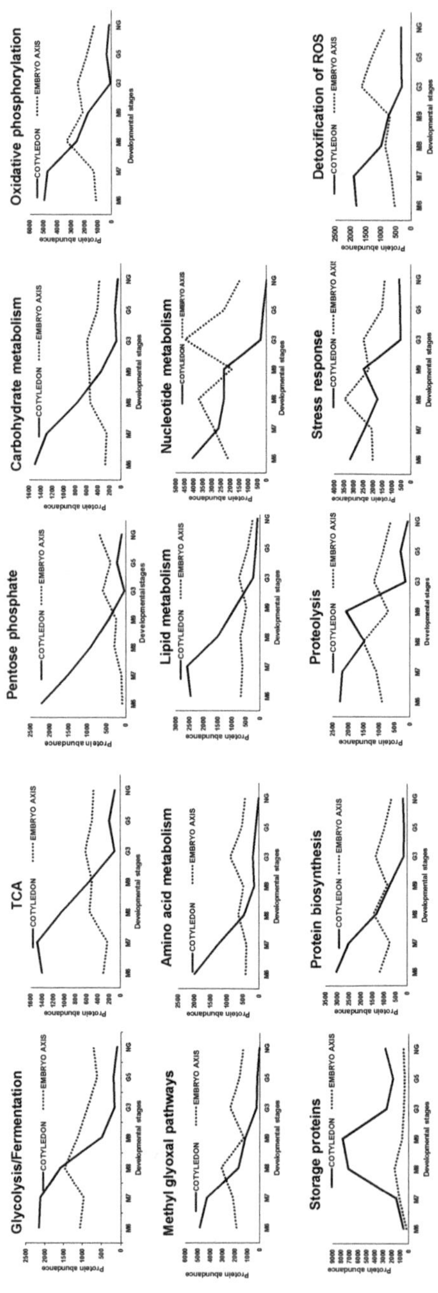

Figure 6. Time course trends of the different protein groups along the maturation and germination stages. Values correspond to the mean abundance of all the proteins within each group (see Figure 5 and Table S6). Data from non-germinating, unviable seeds have been also included. Black line: cotyledon; point rond line: embryo axis.

3. Discussion

As a continuation of the previously published work on proteomics of mature acorns from the non-orthodox *Quercus ilex* tree [13], the protein signatures of cotyledons and the embryo axis throughout seed development and germination phases have been established from data obtained using a 2-DE coupled to MALDI-TOF/TOF MS proteomics strategy.

The protein profiles at six maturation (M4–M9) and five germination (G1–G5) stages, with M9 corresponding to the mature acorns at shedding, were analyzed. Non-germinating seeds after a two-week period upon imbibition were also included in the study. Acorns started to be harvested once they were clearly visible on the tree (M1). Then, the different stages (M1–M9) were chosen based on periodical surveys and visual differences (Figure S1), resulting in differences in their morphometry (width, length), weight, and water content (Table 1, Figure 1A). Almost identical stages were employed in the transcriptomic profiling of developing cork oak acorns [20].

Water content is crucial for viability in recalcitrant seeds, especially in the embryo axis [4], which is protected from dehydration by the waterproof pericarp, seed coat, and the surrounding cotyledon [21]. The lowest water content values corresponded to the mature shedding stage (around 35 and 55% for the cotyledon and the embryo axis, respectively) (Figure 1A). Upon imbibition, germination starts, and seed tissues become hydrated (around 75 and 85% for cotyledon and the embryo axis, respectively, at G2). Even though the relative water content of seeds was above the percentage of a viable one (around 35%), around 30% of the acorns did not germinate in the two weeks after imbibition. On the other hand, and as synchronized germination does not occur in recalcitrant *Quercus* spp. [22] the stages and parameters corresponded to the most representative fruits on each date.

Although 1-D gel electrophoresis is not very resolutive, it is simple and useful when there are a large number of samples, as in the present work (around 30 samples including stages and replicates). This technique allowed us to discriminate between samples, and to group them [15]. In the present research and based on differences in the 1-D protein band profiles (Figure 2A, Figure S3 and Table S2), the last four developmental stages M6 to M9, corresponding to filling and maturation. This was deduced from the increase in weight and decrease in water content. The two germination stags, G3 (early) and G5 (late), were selected for 2-DE-MS analysis.

2-D gel electrophoresis resulted in 300–500 consistent spots resolved in the 5–8 pH range, with maximum values at M6–M8 in cotyledons, and M9 in the embryo axis (Figure 2B, Figure S4 and Table 2). This is in good agreement with the total protein amount in the extracts as determined by the Bradford assay (Figure 1B). Similar comparative values in protein abundance between stages have been shown in different seed proteomics studies [23]. Storage proteins (legumin, 11S globulin, RmlC-like cupins superfamily protein) are those most abundant in the seed [15,24], and, according to the data presented, they are accumulated earlier in cotyledons than in the embryo axis, although the relative abundance is much higher in the former. In a related paper, it has been shown that storage proteins, mostly 7S globulins, are those most commonly accumulated in embryo and endosperm, although with higher values in endosperm at the mature drying stage [25]. The synthesis and accumulation pattern of reserve proteins seems to be quite similar in different plant species, independently of their botanical group (monocot, dicot) and seed characteristic (orthodox or recalcitrant), although there must be differences in the accumulation kinetics for individual seeds [26]. As it is well known for different plant systems [27], the degradation and mobilization of reserve proteins occurs at early germination stages, as revealed by a significant decrease in the number of resolved spots in both cotyledon and the embryo axis (Table 2).

In non-germinating seeds, and according to the protein content values (3.2 and 3.3 g/g DW in the embryo axis and cotyledon, respectively), the 1-D (40 and 46 resolved bands for, respectively, cotyledon and embryo axis) and 2-D (234 and 388 resolved spots for, respectively, cotyledon and embryo axis) profiles, we can speculate that the loss of viability may be associated with the inability to synthesize and accumulate some reserve proteins among others [28].

The 2-DE profile was stage specific, with 558 (cotyledon) and 591 (the embryo axis) variable spots in all the samples (Table 2). In the PCA analysis, PC1 clearly separated the maturation and germination phases, including non-germinating acorns and tissues, and PC2 stages (Figure 4), together explained 56–67% of the variability. From PC2 data, we can clearly differentiate CM6-CM8 from the CM9 stages and EM6/EM7 from EM8/EM9 stages. According to the variable spots, the cotyledon and the embryo axis behaved differently. Thus, in the former, in the transition from maturation to germination, 228 spots decreased in abundance and only six increased. In contrast, in the embryo axis, 119 increased and 41 decreased (Figure S5). So, most metabolic activity should occur in cotyledon at early stages and in the embryo axis at the late ones. Thus, 185 spots were more abundant in cotyledon than in embryo axis at maturation, while 162 were more abundant in the embryo than in the cotyledon at germination. The embryo axis had a large number of common spot proteins in the late maturation and early germination stages (Figure 3), pointing to, unlike orthodox seeds, a continuum between maturation and germination.

Out of the total variable spots, 211 (cotyledon) and 181 (the embryo axis) were subjected to MALDI-TOF/TOF MS with 178 and 143, respectively, were identified (Table S6). Identified proteins were classified into five (embryo) and seven (cotyledon) functional groups, with 10 and 8 pathways represented within the metabolic group. Next, the trends observed for each group and pathway will be discussed.

An important metabolic activity at specific stages of maturation and germination phases does occur according to the protein signatures, and takes place, as a general tendency, earlier in the cotyledon (M6–M7) than in the embryo axis (M8–M9). Thus, maximum protein abundance was observed in cotyledons during the middle seed filling stages (M6–M7) for glycolytic enzymes (e.g., enolase, and phosphoglycerate kinase), which moved during the late stages, M8–M9, in the embryo axis. The coexistence of anaerobic as well as aerobic energy metabolism can be deduced from the presence in the set of variable proteins of those corresponding to enzymes of TCA cycle, oxidative phosphorylation (ATP synthetase and fermentative alcohol dehydrogenase and lactic dehydrogenase). They were found in both tissues with a similar evolution pattern to that of glycolytic enzymes in cotyledon and maximum at late (M9) and middle germination stages in the embryo (EG3).

The NAD(P)H/NAD(P) balance and homeostasis is one of the key features in metabolism. It is mostly determined by a balance between catabolic oxidative reactions and fermentation reactions and to a lesser extent (based on protein abundance data) of the pentose-phosphate pathway. The abundance of pentose-phosphate pathway enzymes (Glucose-6-phosphate dehydrogenase 6, Transketolase) was much greater in cotyledon (maximum at middle M6) than in embryo (maximum at early G3).

As previously discussed, the presence of RubisCO was deduced from data on the large subunit in both seed parts with higher values in cotyledon than in embryo. Maximum values that were found at M7 in cotyledons, and M8–M9, in the embryo, are controversial. This may be because the seeds are photosynthetically active at early development or later on at the seedling stage, which is manifested at the G5 peak in cotyledon. It can be speculated that the presence of RubisCO may contribute to CO_2 reassimilation after decarboxylation reactions [13,29].

Methyl glyoxal and other reactive and toxic aldehydes are by-products of the glycolytic route and are thus assumed to be accumulated at the maturation and germination stages, which must be detoxified, this taking place through the methyl glyoxal pathway [25]. Enzymes of this pathway were identified in both cotyledon and embryo (Glyoxalase I and II, and lactoyl glutathione lyase) with peaks at M6 (cotyledon) and M9-early germination stages, in the embryo, coincident, in both cases, with a maximum of glycolytic enzymes.

Seed germination and early seedling growth is mostly supported by reserves stored in cotyledons [30]. In the case of Holm oak cotyledon, reserve nutrients are stored as starch and proteins with lipids being the minor components [15]. Seed filling and storage accumulation occur at middle stages just prior to drying and maturation and paradoxically *"Final grain weight showed few or no significant correlations with enzyme activities, sugar levels, or starch content during grain filling, or with starch content at maturity"* [31].

Seed reserve proteins play a vital role as they are the source of amino acids and precursors of other N compounds, they are present mostly in cotyledons and to a lesser extent in the embryo axis. Those of *Q. ilex* mostly belong to the legumin family, also with some representatives of the cupin (two different RmlC-like cupins superfamily protein in cotyledon and embryo), globulin (11S globulin isoform 3 in cotyledons), and glutelin (Glutelin type-B 5 in the embryo) ones [13]. Two and five different types of legumins were identified in the embryo and cotyledon, respectively, being present as precursors mature forms or degradation products. Legumins, cupins, and globulins started to accumulate at middle M6 in both seed parts, with the maximum at late (M8–M9) maturation stages, they decreased at germination. This pattern is characteristic of most seeds, independently of their orthodox or recalcitrant character [32]. The exception was observed for a couple of members of the cupin family in cotyledons that showed a clear peak at germination stages. This observation is in the direction of proposing alternative roles for this family of proteins [33].

The protein biosynthesis and degradation processes are behind protein accumulation. In the present work, and within the set of variable proteins, three in the embryo and four in the cotyledons related to protein biosynthesis have been identified. They are assumed to have a specific role in seed maturation and germination (Table S6). This is the case of a Glycyl-tRNA synthetase, present in both tissues, whose gene inactivation in *Arabidopsis thaliana* led to plant embryo development arrest [34]. Ribosomal proteins and elongation factors have also been implicated in seed development and dormancy in *N. tabacum* and *A. thaliana* [35]. The evolution pattern of these proteins is similar to that reported for other groups in cotyledons, with the maximum at M6–M8. In the embryo axis, they were present throughout all the stages analyzed with accumulation kinetics and maximum values depending on the protein.

Extensive protein degradation occurred at late seed development and germination, as revealed by the identification of reserve protein degradation products, up to 10 different ones for legumins and cupins. These degradation products are observed at late maturation (M8–M9) and germination stages, in cotyledons, and, to a lesser extent, in the embryo axis. Degradation products of the RubisCO large subunit also appeared in the cotyledon at germination stages.

According to the set of variable proteins identified, protein degradation is carried out either by unspecific proteases or by the proteasome complex [36]. Among the former, a cytosol aminopeptidase family protein, Zinc metalloprotease, and Aspartic proteinase A1, were identified in the embryo axis, and Aspartic proteinase 1, Peptidase M1 family protein, AAA-type ATPase family protein, in cotyledons. Even though most of them are not well characterized, their presence in seeds has been previously reported. Among these proteins, an Aspartic proteinase 1, the COP9 signalosome, and the AAA-type ATPase family protein regulator have been implicated in the dormancy, viability, and germination processes in Arabidopsis seed [37]. While having maximum values at M7 in cotyledons, protease abundance decreased at late maturation 9, having very low levels at germination or almost none in non-germinated seeds. Maximum values in the embryo axis took place at M8, keeping it more or less constant at germination stages, except in non-germinating seeds.

Other pathways are less well represented in the set of variable proteins, thus preventing any discussion on their role and relevance in seed development, but just mentioning them and verifying their presence in other seeds. Those proteins were linked to carbohydrate, amino acids, and lipid metabolism.

Within the carbohydrate metabolism in the embryo axis, the list included enzymes implicated in the synthesis of starch, the most abundant compound in acorns [15] including phosphorylases and Glucose-1-phosphate adenylyl transferase [38]. Other enzymes identified that were previously reported in seeds were UDP-Glycosyl transferase (an activity involved in the glycosylation of different metabolites; [39], beta glucosidases, UDP-glucose pyrophosphorylase 2 (whose gene inactivation impeded seed set in rice); [40], UDP-D-apiose/UDP-D-xylose synthase 2 (related to cell wall formation [41]. The set in cotyledons comprises the sugar isomerase (SIS) family protein [42], the UDP-glucose 6-dehydrogenase family protein [43], glucose-1-phosphate adenylyl transferase, sucrose synthase 2 [44], GDP-D-mannose 3′,5′-epimerase [45]. The evolution in cotyledons was similar to the previous ones, with the maximum at M7–M9 with later a decay/degradation at germination. In the embryo axis, the peak is displaced at M8–9 and the values are kept at germination stages.

Amino acids play a key role in seed central metabolism, that is used for the synthesis of storage proteins and others, catabolic fuel, precursor of vitamins, hormones, and secondary metabolites [46]. Up to four proteins in the embryo and 18 in the cotyledon, corresponding to enzymes of the amino acid metabolism, have been identified. They showed a similar trend to the other groups in cotyledons, with maximum values at M6–M8 and later a decay at low levels at mature and germination stages. The protein abundance values in the embryo were lower than in cotyledon and a general trend was not observed. Some of them, like alanine aminotransferase, have been reported to play a role in seed dormancy [47].

The list of proteins identified was related to: Arg (embryo, arginosuccinate synthase, [48]; cotyledon carbamoyl phosphate synthetase A: spot, [49]; Glx (embryo, glutamine synthase [50]; cotyledon glutamine amidotransferase type 1 family protein: N-acetyl-gamma-glutamyl-phosphate reductase, and glutamate-1-semialdehyde 2,1-aminomutase 1); Cys (embryo cysteine synthase, [51]; ser (cotyledon d-3-phosphoglycerate dehydrogenase [52]; eu, Ile (cotyledon putative, 2-isopropylmalate synthase 1 [53]; ketol-acid reductoisomerase, Phe/Tyr (cotyledon 3-deoxy-D-arabino-heptulosonate 7-phosphate synthase 1: dehydroquinate dehydratase, putative/shikimate dehydrogenase;3-dehydroquinate synthase, putative [54].

A relevant datum of the fatty acid composition in *Q. ilex* seed is that its profile is quite similar to that of olive oil, with oleic acid being the most abundant (around 60%), followed by linoleic and palmitic (around 18%) and finally stearic (around 4%). Up to three different proteins of enzymes implicated in fatty acid metabolism have been identified, including Enoyl-ACP reductase and 3-ketoacyl-acyl carrier protein synthase, in cotyledon and embryo, plus acetyl co-enzyme, a carboxylase biotin carboxylase subunit, only in cotyledon. The evolution of the proteins throughout maturation and germination, in both the embryo and the cotyledon, is similar to the above discussed for other groups.

Desiccation tolerance in orthodox seeds, and seed viability and longevity depend, to a great extent, on activation, on defense and stress-related genes. Thus, it is not surprising that cell defense and rescue was the functional group most represented in the preformed protein analysis, with two main sub-groups within it, corresponding to redox homeostasis and ROS detoxification, and a more heterogeneous group, including HSPs, and glycine-rich RNA-binding, among others.

Heat-shock proteins (HSPs) are ubiquitous and act as molecular chaperones, thus favoring a competent protein folding and preventing protein aggregation, among other roles, with different families classified according to their size. Differently from vegetative tissues, they are constitutively present in seeds, with differences related to the accumulated type, that have been implicated in desiccation tolerance and with some of them having relevant roles [55]. In the present study, HSPs were identified in the cotyledon (70 type, previously reported in vegetative tissues of Quercus spp., [56], and the embryo axis (70, 20, and 17.9)). The relative abundance was similar in both parts, with a maximum at M6–M8 in cotyledons and a later decay/degradation to M9 and germination, and M7–M9 in the embryo axis, with some of them having high values at germination. This pattern is similar to the one reported for most of the seeds studied [7].

Another family of stress-related proteins well represented in *Q. ilex* seeds was the Glycine-rich RNA-binding one, also involved in signaling and development [57]. Based on relative abundance, some major proteoforms showed some characteristics not observed for other stress-related proteins and groups of them. First, they accumulated at earlier stages than M6, and, second, they were present in embryos in a larger amount throughout the M7–M9 and germination stages.

Within the drought stress-related proteins, two more were identified, including Dehydrin 2 (embryo), and Annexin 1 (embryo and cotyledon). Dehydrin genes have been studied in detail in *Quercus robur*, and their presence confirmed in seeds, being implicated in maturation [58]. Dehydrin transcripts were also detected in germinating *Q. ilex* seeds, with transcript abundance being maximum at mature acorn, then decreasing at later germination stages [14]. The expression of annexins at high levels has been reported for a number of seeds [59].

Proteins related to responses to biotic stresses and plant immunity are linked to specific developmental stages, as is the case for seed maturation and germination, the key stages in the plant biological cycle [7]. This is without discarding the fact that the presence of an endogenous microflora also keeps defense genes activated. In this regard, unpublished metabolomic data by the authors' laboratory, have revealed the existence, in cotyledons, of fungal and bacteria-derived metabolites. The pathogenesis-related proteins identified in embryo included osmotin 34 [60] and Major allergen Pruar 1 [61].

Seed viability and longevity relies on antioxidants and ROS scavengers [7]. ROS species are produced as a consequence of metabolic reactions in which participate oxygen, and, because of its reactivity, its concentration should be kept below toxicity levels. However, they also act as signaling molecules implicated in development and stress response gene expression regulation. It has been well documented how these species are produced during the seed maturation and germination phases and some hypotheses on their role in signaling and protein redox states have been proposed [62].

The set of antioxidant and redox-related enzymes identified include a Cu/Zn superoxide dismutase, ascorbate and dehydroascorbate peroxidase, Thioredoxins, NAD(P) linked oxidoreductase (the embryo axis and cotyledons), and glutathione-S-transferase (only in embryo). All of them have been previously reported in seeds, being related to seed development and longevity [62]. They have a different evolution pattern; thus, GST appeared mostly in embryo at M7–M9 and at germination stages. Thioredoxins proteoforms had maxima in cotyledons at M7 (101), M9 (3103), later decaying at germination stages, while in embryo they corresponded to G3. NAD(P)-linked oxidoreductase had the standard pattern observed in cotyledons and embryo, maximum at M7–M8 and ulterior decay for the former, and M8–M9 for the latter. Finally, the SOD pattern was similar for cotyledon, and had more or less high constant levels throughout all the stages.

Taking all data together enables us to discuss the differences between viable, germinating, and non-viable, non-germinating seeds, cotyledon and the embryo axis, and orthodox and recalcitrant seeds.

The simple visualization of the gels showed that the number of spots was much lower in non-viable, non-germinating seeds than in viable, germinating ones at M9, so the loss of viability may be associated with the inability to synthesize and accumulate some proteins [63].

This result can be explained by different mechanisms that mediate gene silencing or impedes mRNA translation. Thus, several epigenetic marks keep the chromatin condensed, thus impeding gene transcription. Epigenetic processes mediating chromatin structure have been implicated in seed development, and repressed chromatin states through histone posttranslational modifications are responsible for abortion, small size, decreased set, and other abnormal seed phenotypes [64]. Small non-coding RNAs, including short interfering (siRNA) and micro (miRNA) regulate seed development and germination through mRNA cleavage and inactivation [65].

The data reveal that the main differences occurred in the embryo axis. In this direction, proteins related to translation, legumins, proteases, proteasome, and stress-related ones were less abundant in non-germinating seeds. Although there are exceptions to this rule (e.g., fermentative alcohol and lactic dehydrogenase, some ATP synthase, methylglyoxal metabolizing enzymes), the content of enzyme proteins in the different pathways (glycolysis, carbohydrate, amino acid, and lipid metabolism) was lower in non-germinating cotyledons and the embryo axis.

The functional groups and metabolic pathways identified in cotyledons and in the embryo were quite similar. There were some differences in the central, energy, and methyl glyoxal pathways, revealing the particularities of each tissue. Thus, the presence of some proteins was only detected in the embryo axis (alcohol and lactic DH, some enzymes of the TCA) or in the cotyledons (enzymes of the Pentose-phosphate pathway). The set of variable proteins identified in cotyledons and in the embryo axis was different for the carbohydrate, and amino acid pathways. Cotyledons were enriched in reserve proteins and protein-degrading enzymes, clearly pointing to the function of this seed part as a reservoir of nutrients. On the other hand, the embryo axis was enriched in cell defense and rescue proteins, including HSP and antioxidants. All these differences reflect the particularities of each group. The singularity of each seed part, with similar conclusions, has been reported for other plant systems, using transcriptomics and other approaches, including cotton [66] and vetch [67].

To summarize, based on the qualitative differences between some of the variable spots we can propose the proteins within them as biomarkers of seed viability, stage, and tissue. As biomarkers of seed germination, we found that Osmotin proteins were abundant in germinating acorns and absent along the maturation stage. Dehydrin, alcohol dehydrogenase, lactate/malate dehydrogenase and enoyl-ACP reductase precursor were only detected in viable seeds, so they can be considered as biomarkers of seed viability. As markers of storage tissue, we found the legumin precursor and 11S globulin, which were highly abundant in cotyledons. Beta glucosidase and ATP synthase, however, were highly abundant in the embryonic axis and could be considered as biomarkers of the embryonic axis. As markers for stage development, we found that transketolase was highly abundant at middle maturation stages, while dehydrin was highly abundant only at late maturation stages.

From a biological and metabolically point of view, both cotyledons and the embryo axis are characterized by an active metabolism throughout the filling and maturation phases that occur even at the mature acorn stage, with seed development and germination being a continuous process. According to the protein abundance values, we can even conclude that there is a greater metabolic activity in the middle stages of maturation in cotyledon (M6–M7), then moving total ones (M8–M9) in the embryo, considering it as being partly independent. Most of the proteins decreased in cotyledon at M9 and in the germination phase, while the opposite happened in the embryo axis. In cotyledon, most of the energy is assumed to be driven towards the synthesis of reserve nutrients, and, once accumulated, it ceases, moving to the degradation and mobilization of reserves. In the embryo axis, the energy is driven towards synthesis, growth and development [68].

Orthodox and recalcitrant seeds have common characteristics in terms of their changes in gene expression, protein profiles, and metabolism, so seed maturation and germination processes are similar. Just as an example, fermentation is the major pathway supporting ATP and energy demand for reserve synthesis, this being determined by the hypoxic conditions due to low oxygen uptake. Even RubisCO large subunit has been detected in both cotyledons and embryo, although it does not implicate active photosynthesis. Both orthodox and recalcitrant seeds accumulate reserve, antioxidant, and stress-related proteins, in either cotyledon or in the embryo axis.

The main difference between orthodox and recalcitrant seeds is the existence in the former but not in the latter of a desiccation stage associated with the acquisition of drought tolerance, a bridge, as defined by [12], between the maturation and germination phases. This transition and the drought tolerance implicated imply changes in gene expression and metabolic switches that have not been observed in the present proteomics study with *Q. ilex* [69]. Recalcitrant seeds are thus being described as those shedding with an active metabolism, which is supported by the data presented in this work, while orthodox ones maintain almost no metabolic activity [70]. Thus, proteins associated with glycolysis, TCA cycle, cell wall metabolism, are present at high levels at the maturation stage, while associated genes are down-regulated during seed desiccation, and then up-regulated during germination, in orthodox seeds [12,71]. As pointed out by [5],high metabolic rates at the maturing stage is a signature of recalcitrant seeds that lack the quiescent stage at the end of seed development, which characterizes orthodox ones.

However, the difference is never qualitative or clear as black and white. Thus, genes encoding proteins of the translation machinery, DNA repair, proteolysis, and energy metabolism, are highly abundant in the stored mRNA population, and they serve for germination [68]. Despite the absence of proteins, as is the case for HSPs, transcripts are accumulated in the late maturation stages [7].

Desiccation tolerance acquisition has been associated with the accumulation of osmolytes, disaccharides and oligosaccharides, storage proteins, late embryogenesis abundant proteins and HSP, and antioxidative defenses. Out of them, LEAs, and the synthesis of osmolytes, and di- and oligosaccharides have not been visualized in the present proteomics analysis, but they are a general characteristic of orthodox seeds [7], so it is possible to speculate on these absent proteins in *Q. ilex* as being one of the causes of recalcitrance. The presence of the galactinol synthase transcript, implicated in the synthesis raffinose, related to desiccation tolerance, has been noted in *Q. ilex* seeds, with its abundance being maximum at M9 and then decaying at later germination stages [14]. In this direction, it has been suggested that the lack of desiccation tolerance of *A. marina* seeds might be related to the absence of desiccation-related LEAs [72], a fact also reported for other recalcitrant species [73]. LEA transcripts have been identified in *Q. ilex* leaves when they have been induced in response to drought. Other processes could be responsible for their loss of viability, such as the overproduction of ROS, and the decrease in antioxidant defenses [74].

However, the process is complex and there is no simple answer. Aspects such as hormonal balance, and signal transduction pathways, leading to structural and metabolic gene up-regulation have to play a key role; however, this information has remained elusive to the proteomics platform employed in this work so we had to shift to other proteomics and other -omics strategies. Thus, by using GC–MS metabolomics, low levels of abscisic acid and high levels of gibberellins were detected in mature *Q. ilex* acorns [14], which favor seed germination [75].

Finally, epigenetic processes and epigenetic marks mediating gene expression can be, to some extent, responsible for the differences between stages, tissues, viable and non-viable, recalcitrant and orthodox seeds. Epigenetics in orthodox seeds has been studied to some extent [6,76] but has just started to be investigated in recalcitrant ones [77,78].

4. Material and Methods

4.1. Plant Material

Acorns were collected from Holm oak trees at the Cerro Muriano (Córdoba, Spain; 38°0′ 0″ north, 4°46′0″ west) location. They were sampled at nine developmental stages throughout a six-month period (June to November 2014), named M1 to M9, according to days after full bloom (DAB) (Figure S1), with M9 corresponding to mature fruits. Only healthy acorns were picked, taken to the laboratory, abundantly washed with tap water, and blot dried with filter paper. Maximum diameter, length and weight were determined, selecting for ulterior analysis those that had similar morphometric parameters (Table 1, Figure S1).

Based on morphometric parameters and water content, the different stages were grouped as follows: early (M1–M3), middle (M4–M7), late (M8–M9) maturation, and early (G1–G3) and late (G4–G5) germination. They corresponded to the Phases I/II (cell division and expansion), III (dry matter accumulation), and IV (water content reduction, desiccation for orthodox seeds). A similar classification has been established previously [79].

Seed parts (cotyledons and embryo axis) were macerated in a mortar under liquid nitrogen until a fine powder was obtained. Powder was stored at −70 °C until protein extraction. Acorns from three different trees, in a number of five (M9) to fifty (M1), depending on the developmental stages, were processed, each tree being considered as a replicate.

Seeds were germinated on humid perlite (50 seeds/replicate, three replicates) in a plant growth chamber under 12/12 h photoperiod (25/20 °C) for two weeks (Figure S7). The perlite was humidified every three days. A 70% percentage of germination was obtained and, as corresponds to non-orthodox species, no total synchronized germination was observed. Only synchronized seeds were sampled at different times, depending on the length of the radicle, corresponding to G1 (2 days, radicle protrusion), G2 (3 days, 1 cm), G3 (4 days, 1.5 cm), G4 (7 days, 3 cm) and G5 (10 days, 5 cm) (Figure S2B).

As an additional sample, embedded acorns that did not germinate in the two weeks were included (NG seeds).

Relative water content was determined for all the tissues at the different developmental stages, from maturation to germination [14].

4.2. Protein Extraction and Gel Electrophoresis

Three independent extractions (biological replicates) for each seed tissue (cotyledon and embryo axes, 0.5 g each) at maturation M4 to M9, and germination G1 to G5 stages were performed. Tissue powder was extracted using the TCA-acetone-phenol protocol [80,81]. The final pellet was dissolved in 40–200 µL of 7 M urea, 2 M Thiourea, 4% CHAPS, Triton X100 2% and 100 mM Dithiothreitol (DTT) (Bio-Lyte 3/10, Bio-Rad, #1631113). Once the pellet was solubilized and the insoluble material eliminated by centrifugation, the protein content was quantified by the method of [82] using bovine serum albumin as a standard.

Proteins (60 µg per replicate) were separated by one-dimensional SDS–PAGE according to [83] and stained with Coomassie Brilliant Blue–R250 [84]. The molecular weight was determined relative to protein markers (SDS–PAGE Standards, Bio–Rad, 161–0304). Band intensity was analyzed by the Quantity-one software (Bio-Rad, Hercules, CA, USA). Normalized band intensity values based on total intensity bands were calculated for each sample line and used for the statistical evaluation of differential abundance.

Two-dimensional electrophoresis was performed in the 5–8 pH range, as this was where most of the proteins were focused [15]. Immobilized pH gradient (IPG) strips (17 cm, 5–8 pH linear gradient; Bio–Rad, #1632004) were passively rehydrated for 16 h with 250 µg protein in 300 µLof IEF solubilization buffer (7 M urea; 2 M Thiourea; 4% [*w/v*] CHAPS; 0.2% [*v/v*] IPG buffer 5–8, 100 mM DTT; and 0.01% [*w/v*] bromophenol blue). The strips were loaded onto a Bio–Rad Protean IEF Cell system (Bio-Rad, Hercules, CA, USA) and proteins electro-focused [80]. Strips were equilibrated, subjected to second dimension SDS–PAGE and gels Sypro Ruby (SYPRO®Ruby Protein Stains, Bio-Rad, Hercules, CA, USA) stained [85]. Images were analyzed with the PDQuest™ software (Bio-Rad, Hercules, CA, USA), using a guided protein spot detection method [86]. Normalized spot volumes based on total quantity in valid spots were calculated for each 2–DE gel and used for the statistical evaluation of differential protein abundance. Experimental M_r values were calculated by mobility comparisons with low-range molecular weight standards (Bio-Rad, Hercules, CA, USA) run in a separate marker lane on the SDS–gel, while pI was determined by using a 5–8 linear scale over the total dimension of the IPG strips.

4.3. Mass Spectrometry Analysis and Protein Identification

Spots were automatically excised (Investigator ProPic, Genomic Solutions), transferred to Multi well 96 plates and digested with modified porcine trypsin (sequencing grade; Promega, Madison, WI 53711-5399, USA) by using a ProGest (Genomics Solution) digestion station. The digestion protocol used was that of [87] with minor variations. Gel plugs were distained by incubation (twice for 30 min) with 200 mM ammonium bicarbonate in 40% acetonitrile (ACN) at 37 °C, then subjected to three consecutive dehydration/rehydration cycles with pure ACN and 25 mM ammonium bicarbonate in 40% ACN, respectively, and finally dried at room temperature for 10 min. Trypsin (20 µL) at a final concentration of 12.5 ng/µL in 25 mM ammonium bicarbonate was added to the dry gel pieces and the digestion proceeded at 37 °C overnight.

Peptides were extracted from gel plugs by adding 10 µL of 1% TFA (15 min incubation), and then purified by ZipTip. Peptides were deposited onto a MALDI plate using the dry droplet method (ProMS, Genomic Solutions) and the α–cyano hydroxycinnamic acid as a matrix at 5 µg/µL concentration in 70% ACN, 0.1% TFA. Samples were analyzed in a 4700 Proteomics Analyzer MALDI–TOF/TOF Mass Spectrometer (Applied Biosystems, Canada), in the m/z range of 800–4000, with an accelerating voltage of 20 kV, in reflectron mode and with a delayed extraction set to 120 ns. Spectra were calibrated using the trypsin autolysis peaks at $m/z = 842.509$ and $m/z = 2211.104$ as internal standards. The three most abundant ions were then subjected to MS/MS analysis, providing information that could be used to determine the peptide sequence.

A combined search (MS plus MS/MS) was performed using GPS Explorer[TM]software v 3.5 (Applied Biosystems, Canada) over NCBInr and homemade *Quercus* protein database [88] using the MASCOT v 1.9 search engine (Matrix Science Ltd., London, UK). The following parameters were allowed: taxonomy restrictions to Viridiplantae, one missed cleavage, 10 ppm mass tolerance in MS and 0.5 Da for MS/MS data, cysteine carbamidomethylation as a fixed modification and methionine oxidation as a variable modification. The confidence in the peptide mass fingerprinting matches (p <0.05) was based on the MOWSE score and confirmed by the accurate overlapping of the matched peptides with the major peaks of the mass spectrum.

4.4. Statistical Analysis

For the statistical and cluster analyses of protein abundance values, the web-based software National Institute of Aging (NIA) array analysis tool was utilized [89]. This software tool selects statistically valid protein spots based on the analysis of variance (ANOVA). After uploading the data table and indication of biological replications, the data were analyzed statistically using the following settings: error model max (average, actual), 0.01 proportions of highest variance values to be removed before variance averaging, 10 degrees of freedom for the Bayesian error model, 0.05 FDR threshold, and zero permutations.

The entire data set was analyzed by principal component analysis (PCA) using the following settings: covariance matrix type, four principal components, 2–fold change threshold for clusters, and 0.5 correlation threshold for clusters. PCA results were represented as a biplot, with consistent proteins in those experimental situations located in the same area of the graph.

To analyze whether the proteins were ubiquitously abundant among the different tissue seed parts, a Venn diagram was plotted by using Venny (http://bioinfogp.cnb.csic.es/tools/venny/index.html). The proteins identified were subjected to heat mapping using Pearson's distance [90] with the average linkage algorithm of the Genesis software package allowing a definition of some expression groups; the presence or absence of some of them is a characteristic of a particular group [91].

Supplementary Materials: Supplementary materials can be found at http://www.mdpi.com/1422-0067/21/14/4870/s1.

Author Contributions: Conceptualization, B.S.-H.; Data curation, S.B.M.H. and C.G.-D.; Formal analysis, S.B.M.H.; Methodology, B.S.-H.; Software, C.G.-D.; Supervision, J.V.J.-N.; Validation, N.B.; Visualization, N.B.; Writing—original draft, B.S.-H. and J.V.J.-N.; Writing—review & editing, B.S.-H. and J.V.J.-N. All authors have read and agreed to the published version of the manuscript.

Funding: All authors are funded through the Small Research group project from the Deanship of Scientific Research at King Khalid University under research grant number (R.G.P.1/195/41). This work was supported also by a grant from the Spanish "La Agencia Española para la Cooperación Internacional" and project from Tunisian Ministry of Higher Education, Scientific Research and Technology (LR15CBBC02, 19PEJC07-17).

Acknowledgments: The authors extend their appreciation to the Deanship of Scientific Research at King Khalid University for funding this work through the Small Research group project under grant number (R.G.P.1/195/41). We gratefully acknowledge the Proteomics Service SCAI (Cordoba, Spain) for mass spectrometry facilities and technical assistance. Thanks to Diana Badder for English proof editing.

Conflicts of Interest: All authors declare no conflict of interest. The funders had no role in the design of the study; in the collection, analyses, or interpretation of data; in the writing of the manuscript, or in the decision to publish the results.

Abbreviations

1-DE	one dimensional gel electrophoresis
2-DE	two-dimensional gel electrophoresis
WC	water content
SDS-PAGE	sodium dodecyl sulfate polyacrylamide gel electrophoresis
IPG	immobilized pH gradient
IEF	isoelectric focusing
ROS	Reactive oxygen species
EA	embryo axis
Cot	cotyledon
NG	non-germinated acorns
MALDI	Matrix-assisted laser desorption ionization
TOF	Time of flight

References

1. Rey, M.D.; Castillejo, M.Á.; Sánchez-Lucas, R.; Guerrero-Sanchez, V.M.; López-Hidalgo, C.; Romero-Rodríguez, C.; Valero-Galván, J.; Sghaier-Hammami, B.; Simova-Stoilova, L.; Echevarría-Zomeño, S.; et al. Proteomics, holm oak (Quercus ilex L.) and other recalcitrant and orphan forest tree species: How do they see each other? *Int. J. Mol. Sci.* **2019**, *20*, 692. [CrossRef] [PubMed]
2. Bareke, T. Biology of seed development and germination physiology. *Adv. Plants Agric. Res.* **2018**, *8*, 336–346. [CrossRef]
3. Walters, C. Orthodoxy, recalcitrance and in-between: Describing variation in seed storage characteristics using threshold responses to water loss. *Planta* **2015**, *242*, 397–406. [CrossRef] [PubMed]
4. Berjak, P.; Pammenter, N.W. From avicennia to zizania: Seed recalcitrance in perspective. *Ann. Bot.* **2008**, *101*, 213–228. [CrossRef]
5. Pammenter, N.W.; Berjak, P. A review of recalcitrant seed physiology in relation to desiccation-tolerance mechanisms. *Seed Sci. Res.* **1999**, *9*, 13–37. [CrossRef]
6. Bai, F.; Settles, A.M. Imprinting in plants as a mechanism to generate seed phenotypic diversity. *Front. Plant Sci.* **2015**, *5*, 1–10. [CrossRef] [PubMed]
7. Leprince, O.; Pellizzaro, A.; Berriri, S.; Buitink, J. Late seed maturation: Drying without dying. *J. Exp. Bot.* **2017**, *68*, 827–841. [CrossRef] [PubMed]
8. Oracz, K.; Karpiński, S. Phytohormones signaling pathways and ROS involvement in seed germination. *Front. Plant Sci.* **2016**, *7*, 864. [CrossRef] [PubMed]
9. Han, Q.; Bartels, A.; Cheng, X.; Meyer, A.; Charles An, Y.Q.; Hsieh, T.F.; Xiao, W. Epigenetics regulates reproductive development in plants. *Plants* **2019**, *8*, 564. [CrossRef] [PubMed]
10. Rajjou, L.; Duval, M.; Gallardo, K.; Catusse, J.; Bally, J.; Job, C.; Job, D. Seed Germination and Vigor. *Annu. Rev. Plant Biol.* **2012**, *63*, 507–533. [CrossRef] [PubMed]
11. Finch-Savage, W.E.; Footitt, S. Seed dormancy cycling and the regulation of dormancy mechanisms to time germination in variable field environments. *J. Exp. Bot.* **2017**, *68*, 843–856. [CrossRef]
12. Angelovici, R.; Galili, G.; Fernie, A.R.; Fait, A. Seed desiccation: A bridge between maturation and germination. *Trends Plant Sci.* **2010**, *15*, 211–218. [CrossRef]
13. Sghaier-Hammami, B.; Redondo-López, I.; Valero-Galvàn, J.; Jorrín-Novo, J.V. Protein profile of cotyledon, tegument, and embryonic axis of mature acorns from a non-orthodox plant species: Quercus ilex. *Planta* **2016**, *243*, 369–396. [CrossRef]
14. Romero-Rodríguez, M.C.; Archidona-Yuste, A.; Abril, N.; Gil-Serrano, A.M.; Meijón, M.; Jorrín-Novo, J.V. Germination and early seedling development in quercus ilex recalcitrant and non-dormant seeds: Targeted transcriptional, hormonal, and sugar analysis. *Front. Plant Sci.* **2018**, *9*, 1508. [CrossRef]
15. Galván, J.V.; Valledor, L.; Cerrillo, R.M.N.; Gil-Pelegrin, E.; Jorrín-Novo, J.V. Studies of variability in Holm oak (Quercus ilex subsp. ballota [Desf.] Samp.) through acorn protein profile analysis. *J. Proteom.* **2011**, *74*, 1244–1255. [CrossRef] [PubMed]
16. Miernyk, J.A.; Hajduch, M. Seed proteomics. *J. Proteom.* **2011**, *74*, 389–400. [CrossRef]

17. Finch-Savage, W.E.; Blake, P.S. Indeterminate development in desiccation-sensitive seeds of Quercus robur L. *Seed Sci. Res.* **1994**, *4*, 127–133. [CrossRef]

18. Xia, K.; Hill, L.M.; Li, D.Z.; Walters, C. Factors affecting stress tolerance in recalcitrant embryonic axes from seeds of four Quercus (Fagaceae) species native to the USA or China. *Ann. Bot.* **2014**, *114*, 1747–1759. [CrossRef]

19. Sghaier-Hammami, B.; Drira, N.; Jorrín-Novo, J.V. Comparative 2-DE proteomic analysis of date palm (Phoenix dactylifera L.) somatic and zygotic embryos. *J. Proteom.* **2009**, *73*, 161–177. [CrossRef]

20. Miguel, A.; de Vega-Bartol, J.; Marum, L.; Chaves, I.; Santo, T.; Leitão, J.; Varela, M.C.; Miguel, C.M. Characterization of the cork oak transcriptome dynamics during acorn development. *BMC Plant Biol.* **2015**, *15*, 158. [CrossRef] [PubMed]

21. Sobrino-Vesperinas, E.; Viviani, A.B. Pericarp micromorphology and dehydration characteristics of Quercus suber L. acorns. *Seed Sci. Res.* **2000**, *10*, 401–407. [CrossRef]

22. Romero-Rodríguez, M.C.; Jorrín-novo, J.V.; Castillejo, M.A.; Angeles, M. Toward characterizing germination and early growth in the non-orthodox forest tree species Quercus ilex through complementary gel and gel-free proteomic analysis of embryo and seedlings. *J. Proteom.* **2019**, *197*, 60–70. [CrossRef] [PubMed]

23. Xu, N.; Coulter, K.M.; Krochko, J.E.; Bewley, J.D. Morphological stages and storage protein accumulation in developing alfalfa (Medicago sativa L.) seeds. *Seed Sci. Res.* **1991**, *1*, 119–125. [CrossRef]

24. Mouzo, D.; Bernal, J.; López-Pedrouso, M.; Franco, D.; Zapata, C. Advances in the biology of seed and vegetative storage proteins based on two-dimensional electrophoresis coupled to mass spectrometry. *Molecules* **2018**, *23*, 2462. [CrossRef]

25. Wang, W.Q.; Ye, J.Q.; Rogowska-Wrzesinska, A.; Wojdyla, K.I.; Jensen, O.N.; Møller, I.M.; Song, S.Q. Proteomic comparison between maturation drying and prematurely imposed drying of zea mays seeds reveals a potential role of maturation drying in preparing proteins for seed germination, seedling vigor, and pathogen resistance. *J. Proteome Res.* **2014**, *13*, 606–626. [CrossRef] [PubMed]

26. Gallardo, K.; Le Signor, C.; Vandekerckhove, J.; Thompson, R.D.; Burstin, J. Proteomics of medicago truncatula seed development establishes the time frame of diverse metabolic processes related to reserve accumulation. *Plant Physiol.* **2003**, *133*, 664–682. [CrossRef]

27. Erbaş, S.; Tonguç, M.; Karakurt, Y.; Şanli, A. Mobilization of seed reserves during germination and early seedling growth of two sunflower cultivars. *J. Appl. Bot. Food Qual.* **2016**, 89. [CrossRef]

28. Zhao, M.; Zhang, H.; Yan, H.; Qiu, L.; Baskin, C.C. Mobilization and role of starch, protein, and fat reserves during seed germination of six wild grassland species. *Front. Plant Sci.* **2018**, *9*, 234. [CrossRef] [PubMed]

29. Walker, R.P.; Battistelli, A.; Moscatello, S.; Chen, Z.H.; Leegood, R.C.; Famiani, F. Metabolism of the seed and endocarp of cherry (Prunus avium L.) during development. *Plant Physiol. Biochem.* **2011**, *49*, 923–930. [CrossRef]

30. Kabeya, D.; Sakai, S. The role of roots and cotyledons as storage organs in early stages of establishment in Quercus crispula: A quantitative analysis of the nonstructural carbohydrate in cotyledons and roots. *Ann. Bot.* **2003**, *92*, 537–545. [CrossRef] [PubMed]

31. Fahy, B.; Siddiqui, H.; David, L.C.; Powers, S.J.; Borrill, P.; Uauy, C.; Smith, A.M. Final grain weight is not limited by the activity of key starch-synthesising enzymes during grain filling in wheat. *J. Exp. Bot.* **2018**, *69*, 5461–5475. [CrossRef] [PubMed]

32. Li, Q.; Wang, B.C.; Xu, Y.; Zhu, Y.X. Systematic studies of 12S seed storage protein accumulation and degradation patterns during Arabidopsis seed maturation and early seedling germination stages. *J. Biochem. Mol. Biol.* **2007**, *40*, 373–381. [CrossRef] [PubMed]

33. Gábrišová, D.; Klubicová, K.; Danchenko, M.; Gömöry, D.; Berezhna, V.V.; Skultety, L.; Miernyk, J.A.; Rashydov, N.; Hajduch, M. Do cupins have a function beyond being seed storage proteins? *Front. Plant Sci.* **2016**, *6*, 1215. [CrossRef]

34. Uwer, U.; Willmitzer, L.; Altmann, T. Inactivation of a glycyl-tRNA synthetase leads to an arrest in plant embryo development. *Plant Cell* **1998**, *10*, 1277–1294. [CrossRef]

35. Tian, S.; Wu, J.; Liu, Y.; Huang, X.; Li, F.; Wang, Z.; Sun, M.X. Ribosomal protein NtRPL17 interacts with kinesin-12 family protein NtKRP and functions in the regulation of embryo/seed size and radicle growth. *J. Exp. Bot.* **2017**, *68*, 5553–5564. [CrossRef]

36. Oracz, K.; Stawska, M. Cellular recycling of proteins in seed dormancy alleviation and germination. *Front. Plant Sci.* **2016**, *7*, 1–8. [CrossRef]

37. Shen, W.; Yao, X.; Ye, T.; Ma, S.; Liu, X.; Yin, X.; Wu, Y. Arabidopsis aspartic protease ASPG1 affects seed dormancy, seed longevity and seed germination. *Plant Cell Physiol.* **2018**, *59*, 1415–1431. [CrossRef] [PubMed]

38. Cuesta-Seijo, J.A.; Ruzanski, C.; Krucewicz, K.; Meier, S.; Hägglund, P.; Svensson, B.; Palcic, M.M. Functional and structural characterization of plastidic starch phosphorylase during barley endosperm development. *PLoS ONE* **2017**, *12*, e0175488. [CrossRef] [PubMed]

39. Wilson, A.E.; Tian, L. Phylogenomic analysis of UDP-dependent glycosyltransferases provides insights into the evolutionary landscape of glycosylation in plant metabolism. *Plant J.* **2019**, *100*, 1273–1288. [CrossRef]

40. Woo, M.O.; Ham, T.H.; Ji, H.S.; Choi, M.S.; Jiang, W.; Chu, S.H.; Piao, R.; Chin, J.H.; Kim, J.A.; Park, B.S.; et al. Inactivation of the UGPase1 gene causes genic male sterility and endosperm chalkiness in rice (Oryza sativa L.). *Plant J.* **2008**, *54*, 190–204. [CrossRef] [PubMed]

41. Ahn, J.; Verma, R.; Kim, M.; Lee, J.; Kim, Y.; Bang, J.; Reiter, W.; Pai, H. Depletion of UDP-D-apiose/UDP-D-xylose synthases results in rhamnogalacturonan-II deficiency, cell wall thickening, and cell death in higher plants. *J. Biol. Chem.* **2006**, *281*, 13708–13716. [CrossRef] [PubMed]

42. Park, M.R.; Hasenstein, K.H. Oxygen dependency of germinating Brassica seeds. *Life Sci. Sp. Res.* **2016**, *8*, 30–37. [CrossRef] [PubMed]

43. Niu, J.; Cao, D.; Li, H.; Xue, H.; Chen, L.; Liu, B.; Cao, S. Quantitative proteomics of pomegranate varieties with contrasting seed hardness during seed development stages. *Tree Genet. Genomes* **2018**, *14*, 14. [CrossRef]

44. Deng, Y.; Wang, J.; Zhang, Z.; Wu, Y. Transactivation of Sus1 and Sus2 by Opaque2 is an essential supplement to sucrose synthase-mediated endosperm filling in maize. *Plant Biotechnol. J.* **2020**, *18*, 1–11. [CrossRef] [PubMed]

45. Gilbert, L.; Alhagdow, M.; Nunes-Nesi, A.; Quemener, B.; Guillon, F.; Bouchet, B.; Faurobert, M.; Gouble, B.; Page, D.; Garcia, V.; et al. GDP-d-mannose 3,5-epimerase (GME) plays a key role at the intersection of ascorbate and non-cellulosic cell-wall biosynthesis in tomato. *Plant J.* **2009**, *60*, 499–508. [CrossRef]

46. Amir, R.; Galili, G.; Cohen, H. The metabolic roles of free amino acids during seed development. *Plant Sci.* **2018**, *275*, 11–18. [CrossRef]

47. Kamara, J.S.; Hoshino, M.; Satoh, Y.; Nayar, N.; Takaoka, M.; Sasanuma, T.; Abe, T. Japanese sake-brewing rice cultivars show high levels of globulin-like protein and a chloroplast stromal HSP70. *Crop Sci.* **2009**, *49*, 2198–2206. [CrossRef]

48. De Ruiter, H.; Kollöffel, C. Activity of enzymes of arginine metabolism in the cotyledons of developing and germinating pea seeds. *Plant Physiol.* **1982**, *70*, 313–315. [CrossRef]

49. Kollöffel, C.; Verkerk, B.C. Carbamoyl phosphate synthetase activity from the cotyledons of developing and germinating pea seeds. *Plant Physiol.* **1982**, *69*, 143–145. [CrossRef]

50. Guan, M.; Møller, I.S.; Schjoerring, J.K. Two cytosolic glutamine synthetase isoforms play specific roles for seed germination and seed yield structure in Arabidopsis. *J. Exp. Bot.* **2015**, *66*, 203–212. [CrossRef]

51. Bertagnolli, B.L.; Wedding, R.T. Cystine content of legume seed proteins: Estimation by determination of cysteine with 2-vinylquinoline, and relation to protein content and activity of cysteine synthase. *J. Nutr.* **1977**, *107*, 2122–2127. [CrossRef]

52. Cheung, G.P.; Rosenblum, I.Y.; Sallach, H.J. Comparative studies of enzymes related to serine metabolism in higher plants. *Plant Physiol.* **1968**, *43*, 1813–1820. [CrossRef] [PubMed]

53. He, Y.; Mawhinney, T.P.; Preuss, M.L.; Schroeder, A.C.; Chen, B.; Abraham, L.; Jez, J.M.; Chen, S. A redox-active isopropylmalate dehydrogenase functions in the biosynthesis of glucosinolates and leucine in Arabidopsis. *Plant J.* **2009**, *60*, 679–690. [CrossRef]

54. Zhang, Z.Z.; Li, X.X.; Zhu, B.Q.; Wen, Y.Q.; Duan, C.Q.; Pan, Q.H. Molecular characterization and expression analysis on two isogenes encoding 3-deoxy-d-arabino-heptulosonate 7-phosphate synthase in grapes. *Mol. Biol. Rep.* **2011**, *38*, 4739–4747. [CrossRef] [PubMed]

55. Su, P.H.; Li, H.M. Arabidopsis stromal 70-kD heat shock proteins are essential for plant development and important for thermotolerance of germinating seeds. *Plant Physiol.* **2008**, *146*, 1231–1241. [CrossRef]

56. Ricardo, C.P.P.; Martins, I.; Francisco, R.; Sergeant, K.; Pinheiro, C.; Campos, A.; Renaut, J.; Fevereiro, P. Proteins associated with cork formation in Quercus suber L. stem tissues. *J. Proteom.* **2011**, *74*, 1266–1278. [CrossRef] [PubMed]

57. Czolpinska, M.; Rurek, M. Plant glycine-rich proteins in stress response: An emerging, still prospective story. *Front. Plant Sci.* **2018**, *9*, 302. [CrossRef]

58. Šunderlíková, V.; Salaj, J.; Kopecky, D.; Salaj, T.; Wilhem, E.; Matušíková, I. Dehydrin genes and their expression in recalcitrant oak (Quercus robur) embryos. *Plant Cell Rep.* **2009**, *28*, 1011. [CrossRef]

59. Clark, G.B.; Morgan, R.O.; Fernandez, M.P.; Roux, S.J. Evolutionary adaptation of plant annexins has diversified their molecular structures, interactions and functional roles. *New Phytol.* **2012**, *196*, 695–712. [CrossRef]

60. Anil Kumar, S.; Hima Kumari, P.; Shravan Kumar, G.; Mohanalatha, C.; Kavi Kishor, P.B. Osmotin: A plant sentinel and a possible agonist of mammalian adiponectin. *Front. Plant Sci.* **2015**, *6*, 163. [CrossRef]

61. Sinha, M.; Singh, R.P.; Kushwaha, G.S.; Iqbal, N.; Singh, A.; Kaushik, S.; Kaur, P.; Sharma, S.; Singh, T.P. Current overview of allergens of plant pathogenesis related protein families. *Sci. World J.* **2014**, *2014*, 543195. [CrossRef]

62. Catusse, J.; Job, C.; Job, D. Transcriptome- and proteome-wide analyses of seed germination. *C. R.-Biol.* **2008**, *331*, 815–822. [CrossRef] [PubMed]

63. Yin, G.; Xin, X.; Fu, S.; An, M.; Wu, S.; Chen, X.; Zhang, J.; He, J.; Whelan, J.; Lu, X. Proteomic and carbonylation profile analysis at the critical node of seed ageing in Oryza sativa. *Sci. Rep.* **2017**, *7*, 1–12. [CrossRef] [PubMed]

64. Ni, J.; Ma, X.; Feng, Y.; Tian, Q.; Wang, Y.; Xu, N.; Tang, J.; Wang, G. Updating and interaction of polycomb repressive complex 2 components in maize (Zea mays). *Planta* **2019**, *250*, 573–588. [CrossRef]

65. Sarkar Das, S.; Yadav, S.; Singh, A.; Gautam, V.; Sarkar, A.K.; Nandi, A.K.; Karmakar, P.; Majee, M.; Sanan-Mishra, N. Expression dynamics of miRNAs and their targets in seed germination conditions reveals miRNA-ta-siRNA crosstalk as regulator of seed germination. *Sci. Rep.* **2018**, *8*, 1–13. [CrossRef] [PubMed]

66. Jiao, X.; Zhao, X.; Zhou, X.R.; Green, A.G.; Fan, Y.; Wang, L.; Singh, S.P.; Liu, Q. Comparative Transcriptomic Analysis of Developing Cotton Cotyledons and Embryo Axis. *PLoS ONE* **2013**, *8*, e71756. [CrossRef] [PubMed]

67. Schlereth, A.; Becker, C.; Horstmann, C.; Tiedemann, J.; Müntz, K. Comparison of globulin mobilization and cysteine proteinases in embryonic axes and cotyledons during germination and seedling growth of vetch (Vicia sativa L.). *J. Exp. Bot.* **2000**, *51*, 1423–1433. [CrossRef]

68. Collins, D.M.; Wilson, A.T. Metabolism of the axis and cotyledons of phaseolus vulgaris seeds during early germination. *Phytochemistry* **1972**, *11*, 1931–1935. [CrossRef]

69. Roberts, E.H. Predicting the storage life of seeds. *Seed Sci. Technol.* **1973**, *1*, 499–514.

70. Obroucheva, N.; Sinkevich, I.; Lityagina, S. Physiological aspects of seed recalcitrance: A case study on the tree Aesculus hippocastanum. *Tree Physiol.* **2016**, *36*, 1127–1150. [CrossRef]

71. Angelovici, R.; Lipka, A.E.; Deason, N.; Gonzalez-Jorge, S.; Lin, H.; Cepela, J.; Buell, R.; Gore, M.A.; DellaPenna, D. Genome-wide analysis of branched-chain amino acid levels in Arabidopsis Seeds. *Plant Cell* **2013**, *25*, 4827–4843. [CrossRef] [PubMed]

72. Farrant, J.M.; Berjak, P.; Pammenter, N.W. Proteins in development and germination of a desiccation sensitive (recalcitrant) seed species. *Plant Growth Regul.* **1992**, *11*, 257–265. [CrossRef]

73. Moothoo-Padayachie, A.; Macdonald, A.; Varghese, B.; Pammenter, N.W.; Govender, P. Uncovering the basis of viability loss in desiccation sensitive Trichilia dregeana seeds using differential quantitative protein expression profiling by iTRAQ. *J. Plant Physiol.* **2018**, *221*, 119–131. [CrossRef]

74. Pukacka, S.; Malec, M.; Ratajczak, E. ROS production and antioxidative system activity in embryonic axes of Quercus robur seeds under different desiccation rate conditions. *Acta Physiol. Plant.* **2011**, *33*, 2219. [CrossRef]

75. Vishal, B.; Kumar, P.P. Regulation of seed germination and abiotic stresses by gibberellins and abscisic acid. *Front. Plant Sci.* **2018**, *9*, 838. [CrossRef] [PubMed]

76. Köhler, C.; Makarevich, G. Epigenetic mechanisms governing seed development in plants. *EMBO Rep.* **2006**, *7*, 1223–1227. [CrossRef] [PubMed]

77. Michalak, M.; Plitta-Michalak, B.P.; Naskręt-Barciszewska, M.; Barciszewski, J.; Bujarska-Borkowska, B.; Chmielarz, P. Global 5-methylcytosine alterations in DNA during ageing of Quercus robur seeds. *Ann. Bot.* **2015**, *116*, 369–376. [CrossRef]

78. Plitta-Michalak, B.P.; Naskret-Barciszewska, M.Z.; Kotlarski, S.; Tomaszewski, D.; Tylkowski, T.; Barciszewski, J.; Chmielarz, P.; Michalak, M. Changes in genomic 5-methylcytosine level mirror the response of orthodox (Acer platanoides L.) and recalcitrant (Acer pseudoplatanus L.) seeds to severe desiccation. *Tree Physiol.* **2018**, *38*, 617–629. [CrossRef]

79. Lin, J.Y.; Le, B.H.; Chen, M.; Henry, K.F.; Hur, J.; Hsieh, T.F.; Chen, P.Y.; Pelletier, J.M.; Pellegrini, M.; Fischer, R.L.; et al. Similarity between soybean and Arabidopsis seed methylomes and loss of non-CG methylation does not affect seed development. *Proc. Natl. Acad. Sci. USA* **2017**, *114*, E9730–E9739. [CrossRef]

80. Maldonado, A.M.; Echevarría-Zomeño, S.; Jean-Baptiste, S.; Hernández, M.; Jorrín-Novo, J.V. Evaluation of three different protocols of protein extraction for Arabidopsis thaliana leaf proteome analysis by two-dimensional electrophoresis. *J. Proteom.* **2008**, *71*, 461–472. [CrossRef]

81. Wang, W.; Vignani, R.; Scali, M.; Cresti, M. A universal and rapid protocol for protein extraction from recalcitrant plant tissues for proteomic analysis. *Electrophoresis* **2006**, *27*, 2782–2786. [CrossRef]

82. Bradford, M.M. A rapid and sensitive method for the quantitation of microgram quantities of protein utilizing the principle of protein-dye binding. *Anal. Biochem.* **1976**, *72*, 248–254. [CrossRef]

83. Laemmli, U.K. Cleavage of structural proteins during the assembly of the head of bacteriophage T4. *Nature* **1970**, *227*, 680–685. [CrossRef] [PubMed]

84. Neuhoff, V.; Arold, N.; Taube, D.; Ehrhardt, W. Improved staining of proteins in polyacrylamide gels including isoelectric focusing gels with clear background at nanogram sensitivity using Coomassie Brilliant Blue G-250 and R-250. *Electrophoresis* **1988**, *9*, 255–262. [CrossRef] [PubMed]

85. Steinberg, T.H.; Lauber, W.M.; Berggren, K.; Kemper, C.; Yue, S.; Patton, W.F. Fluorescence detection of proteins in sodium dodecyl sulfate-polyacrylamide gels using environmentally benign, nonfixative, saline solution. *Electrophoresis* **2000**, *21*, 497–508. [CrossRef]

86. Chich, J.F.; David, O.; Villers, F.; Schaeffer, B.; Lutomski, D.; Huet, S. Statistics for proteomics: Experimental design and 2-DE differential analysis. *J. Chromatogr. B Anal. Technol. Biomed. Life Sci.* **2007**, *849*, 261–272. [CrossRef]

87. Shevchenko, A.; Wilm, M.; Vorm, O.; Mann, M. Mass spectrometric sequencing of proteins from silver-stained polyacrylamide gels. *Anal. Chem.* **1996**, *68*, 850–858. [CrossRef]

88. Valledor, L.; Romero-Rodríguez, M.C.; Jorrin-Novo, J.V. Standardization of data processing and statistical analysis in comparative plant proteomics experiment. *Methods Mol. Biol.* **2014**, *1072*, 51–60. [CrossRef]

89. Sharov, A.A.; Dudekula, D.B.; Ko, M.S.H. A web-based tool for principal component and significance analysis of microarray data. *Bioinformatics* **2005**, *21*, 2548–2549. [CrossRef]

90. Eisen, M.B.; Spellman, P.T.; Brown, P.O.; Botstein, D. Cluster analysis and display of genome-wide expression patterns. *Proc. Natl. Acad. Sci. USA* **1998**. [CrossRef]

91. Sturn, A.; Quackenbush, J.; Trajanoski, Z. Genesis: Cluster analysis of microarray data. *Bioinformatics* **2002**, *18*, 207–208. [CrossRef] [PubMed]

International Journal of
Molecular Sciences

Article

Comprehensive Comparison of Clinically Relevant Grain Proteins in Modern and Traditional Bread Wheat Cultivars

Olha Lakhneko [1,2], Maksym Danchenko [1,3,*], Bogdan Morgun [2], Andrej Kováč [4], Petra Majerová [4] and Ľudovit Škultéty [1,5]

1   Institute of Virology, Biomedical Research Center, Slovak Academy of Sciences, Dubravska 9, 84505 Bratislava, Slovak Republic; olakhneko@icbge.org.ua (O.L.); viruludo@savba.sk (Ľ.Š.)
2   Institute of Cell Biology and Genetic Engineering, National Academy of Sciences of Ukraine, Akademika Zabolotnoho 148, 03143 Kyiv, Ukraine; bmorgun@icbge.org.ua
3   Institute of Plant Genetics and Biotechnology, Plant Science and Biodiversity Center, Slovak Academy of Sciences, Akademicka 2, 95007 Nitra, Slovak Republic
4   Institute of Neuroimmunology, Slovak Academy of Sciences, Dubravska 9, 84510 Bratislava, Slovak Republic; andrej.kovac@savba.sk (A.K.); petra.majerova@savba.sk (P.M.)
5   Institute of Microbiology, Czech Academy of Sciences, Videnska 1083, 14220 Prague, Czech Republic
*   Correspondence: maksym.danchenko@savba.sk; Tel.: +421-37 6943-346

Received: 25 April 2020; Accepted: 11 May 2020; Published: 13 May 2020

Abstract: Bread wheat (*Triticum aestivum* L.) is one of the most valuable cereal crops for human consumption. Its grain storage proteins define bread quality, though they may cause food intolerances or allergies in susceptible individuals. Herein, we discovered a diversity of grain proteins in three Ukrainian wheat cultivars: Sotnytsia, Panna (both modern selection), and Ukrainka (landrace). Firstly, proteins were isolated with a detergent-containing buffer that allowed extraction of various groups of storage proteins (glutenins, gliadins, globulins, and albumins); secondly, the proteome was profiled by the two-dimensional gel electrophoresis. Using multi-enzymatic digestion, we identified 49 differentially accumulated proteins. Parallel ultrahigh-performance liquid chromatography separation followed by direct mass spectrometry quantification complemented the results. Principal component analysis confirmed that differences among genotypes were a major source of variation. Non-gluten fraction better discriminated bread wheat cultivars. Various accumulation of clinically relevant plant proteins highlighted one of the modern genotypes as a promising donor for the breeding of hypoallergenic cereals.

Keywords: *Triticum aestivum* L.; food quality; cereal allergens; discovery proteomics; gluten; celiac disease

---

1. Introduction

Bread wheat (*Triticum aestivum* L.) is a valuable cereal widely used in the human diet or livestock feed, and the dominant crop in temperate countries. It is an essential source of nutrients and other beneficial components. World production of this crop reaches 725 million tons annually, which is 30% of all harvested cereals (http://www.fao.org/3/a-I8080e.pdf). Altogether, more wheat proteins are consumed by humanity than from any other plant or animal. This crop is traditionally vital for European nations, though it has broad geographic distribution. The success of wheat largely depends on its adaptability to a wide range of environments, high yield potential, and relevance to the human culture [1].

The wheat grain contains about 16% of proteins, which are classified according to their solubility: In water—albumins, in salt—globulins, in alcohol—gliadins, or in alkali—glutenins. Typically, wheat flour proteome consists of 35% glutenins, 45% gliadins, and only 20% other proteins. Glutenins and gliadins are related and defined as gluten; multiple genes encode them at complex loci [2]. Glutenin fraction represents a complex polymer, stabilized by inter-chain disulfide bonds. Glutenins are classified into high molecular weight (HMW) and low molecular weight (LMW) subunits [2]. A combination of different HMW alleles of x- and y-type subunits defines the elasticity and strength of the dough [3]. Likewise, LMW subunits are determinants of dough extensibility in bread wheat [4]. However, the exact role of each specific LMW glutenins remains largely mysterious. For instance, Lee group found that a single genetic locus played only a minor role in quality variation, although it was the most diverse [5]. Monomeric gliadins are another dominant part of storage proteins. They are divided into α/β-, γ-, and ω-classes according to differences in the primary structure and the number of conserved cysteine residues [6,7]. Gliadin genetic regions are characterized by the complex structure and may cover over 50 alleles, a lot of which are actually expressed, but also a number of them are pseudogenes [8,9]. Gliadins contribute to bread-making quality through covalent and non-covalent bonds with other polymeric gluten components, forming the fine gluten film network and improving gas retention, viscosity, and cohesiveness of dough. Some studies demonstrated the importance of the balance between glutenin and gliadin fractions for boosting bread-making quality [10,11]. Globulins and albumins, collectively referred as metabolic proteins, compose a minor part of grain proteome. They are marginally connected to the technological quality by defining milling properties, but are indispensable for the plant physiology [6].

Modern plant breeding has led to the development of multiple wheat cultivars with superior bread-making quality. Albeit, storage proteins can cause food intolerance or allergy in susceptible individuals. People are exposed to wheat-derived products through ingestion, inhalation, or skin contact. Wheat sensitivities are classified in autoimmune conditions (having T-cell or IgA nature): Celiac disease, gluten ataxia, gluten neuropathy, dermatitis herpetiformis; and allergic disorders (mediated by IgE): Respiratory allergy, food allergy, wheat-dependent exercise-induced anaphylaxis, contact urticaria [12–14]. Etiology of wheat intolerances grounds in inefficient digestion of the consumed gluten-containing food. This may happen because glutenins and gliadins are enriched with glutamine and proline, leading to restricted cleavage by gastric enzymes [14]. Notably, a thorough study reported considerable variation in the T-cell responses of 14 celiac patients, indicating the existence of numerous active epitopes [15]. Proteomics greatly contributed to the understanding of allergy and intolerance to wheat products, through qualitative and structural characterization of the allergenic and toxic peptides [16]. Of note, researchers proved that besides gluten, metabolic proteins are also of medical concern. Celiac disease patients showed antibody reactivity to non-gluten proteins: Serpins (the most frequently), purinins, α-amylase/protease inhibitors, globulins, and farinins. Recombinant proteins confirmed a robust humoral immune response [17,18].

Genetic and environmental factors affect the technological properties of wheat in a rather unpredictable way. One route for safe food is biotechnological creation of transgenic lines; another option is through exploiting rich traditional genetic resources to lower the amount of harmful epitopes [8]. There is a serious public perception issue with genetically modified organisms, yet it has no reliable scientific arguments. An effective approach to reduce allergenicity/toxicity is the silencing of target genes, like ω-gliadins. However, thorough proteomic evaluation of transgenic lines before mass production is recommended because even the same construct can have different effects on grain proteome [19]. Piston group reported that suppressing γ-gliadins resulted in counterbalance of ω- and α/β-classes [20]. On the other hand, despite its recent origin, there is immense diversity among *T. aestivum* cultivars. Experts estimated the existence of at least 25,000 genotypes. A comprehensive study, using immunologic assays, demonstrated a large variation in the amount of reactive peptides among wheat cultivars [21]. The authors concluded that sufficient genetic variation enables the selection of cereals with reduced risk of causing disorders. Such genotypes would allow patients to enjoy a more balanced diet. Intuitively, landraces were believed to synthesize lower amounts of clinically relevant proteins. However, a recent study contradicted this theory [22] and the Simsek group failed to discover any association between release year of wheat cultivars and quantity of immunogenic epitopes in α-gliadins [23]. Finally, it was suggested that more thorough investigations are necessary to clarify potential lower toxicity of old landraces because available data are highly heterogeneous and quite often, growth conditions are unaccounted [24]. Differences in metabolic proteins were shown to be more informative for the discrimination of wheat cultivars [25,26].

Proteomics allows a realistic estimation of the amount of specific grain storage proteins, overcoming the limitations of molecular tools detecting genes/transcripts. Accurate quantification of grain proteins, particularly gluten, is far from a trivial task. The first challenge is the extraction of physicochemically diverse subgroups of polypeptides. Often, a sequential extraction protocol is used, but this introduces inherent analytical variation [27]. Anionic detergent SDS-assisted extraction seems superior, allowing to obtain a more complex mixture in a single step [19]. Wheat grain proteins have high sequence homology and content of hydrophobic amino acids; moreover, immunogenic/toxic peptides are usually resistant to proteolysis. Gluten proteins are rich in proline/glutamine; thus, thorough precise analysis requires alternative proteases [28]. Gel-based methods remain commonly used due to the visualization of multiple isoforms. Long stretches of repetitive sequences complicate unique peptide assignment when using mass spectrometry. Nevertheless, ultrahigh-performance liquid chromatography (UHPLC) with targeted mass spectrometry offers accurate multiplex quantification options, emerging as a viable alternative to standard enzyme-linked immunosorbent assays for allergenicity/toxicity assessment of food products [29–31].

Herein, a wide-range of bread wheat proteins was extracted by SDS-containing buffer and separated using two-dimensional gel electrophoresis (2-DE). Then, we identified differentially abundant grain proteins revealing polymorphism among cultivars. UHPLC of chymotryptically digested extracts with ion mobility-enhanced direct mass spectrometry quantification complemented the dataset. Finally, using bioinformatics, we assessed the clinical relevance of variably accumulated proteins.

2. Results and Discussion

2.1. Separation of Grain Proteins by Two-Dimensional Gel Electrophoresis

Wide-range IPG strips (3–10) covered diverse proteins accumulated in wheat grain. Triplicate gel images are presented in the supporting information, confirming the excellent quality and reproducibility of the analysis (Figure S1). Detected spots had the pI range 4.0–9.5 and molecular mass interval 11–129 kDa. This analytical approach detected 810 protein spots among two modern and one historical genotype. Strict statistics and effect size criteria revealed 66 differently abundant gel spots annotated on the reference image (Figure 1). The quantitative values are presented in the supporting information as means and standard deviations of three biological replicates, together with identification details for differentially abundant proteins (Table S1). Notably, unsupervised statistics—principal component analysis—confirmed the clustering of differentially abundant gel spots according to the genotype (Figure 2A). Moreover, it suggested that cultivar Sotnytsia is distant relative to the others. In order to identify differentially accumulated wheat grain proteins, we applied parallel in-gel digestion with multiple enzymes and identified 49 gel spots (Table 1) with a sufficient number of matched peptides (Table S2). Major gluten and non-gluten fractions were well represented: 8 glutenins, 18 gliadins, and 23 metabolic proteins.

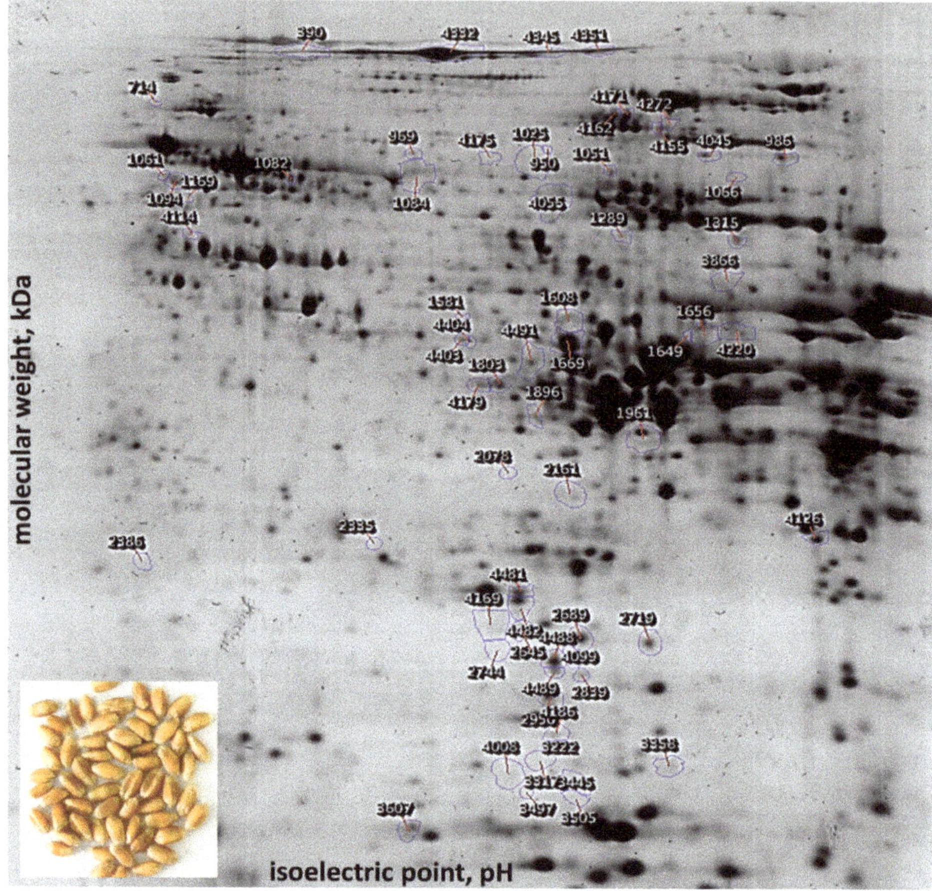

Figure 1. Annotated gel with 66 differentially abundant protein spots among modern and traditional cultivars marked on the image of alignment reference using the SameSpots 5.1 program. In total, 810 analytes extracted from the grain of bread wheat were separated in the 4.0–9.5 pI range and 11–129 kDa molecular mass interval.

Using comprehensive databases, we annotated 22 protein spots as allergenic or toxic. Importantly, all differentially accumulated glutenin subunits are causing disorders. The first group includes food allergen Tri a 26 [32]. HMW-PW212 glutenin (spot 390) was over 10 times more abundant in Sotnytsia compared to Ukrainka and Panna. HMW-12 glutenin (*HMW-GS DY*) was identified in two gel spots 1066 and 4114. The former accumulated in Sotnytsia compared to Ukrainka, and the latter showed higher content in Ukrainka than in Panna. It is noteworthy that these spots had a very distinct isoelectric point (Figure 1). We identified HMW-DX5 glutenin (*Glu-1D-1D*) in four gel spots: (i) 4045 and 4332 accumulated in Panna compared to Sotnytsia, (ii) 4345 was more abundant in Ukrainka versus Sotnytsia, and (iii) 4351 showed a higher amount in both Panna and Ukrainka respective to Sotnytsia (Table 1). The second group, comprising food allergen Tri a 36 [32], included only LMW-B2 glutenin (*Glu-B3-2*) in a single gel spot 4126 that was more abundant in Panna compared to Ukrainka. This protein showed a different trend in LC-MS based quantification.

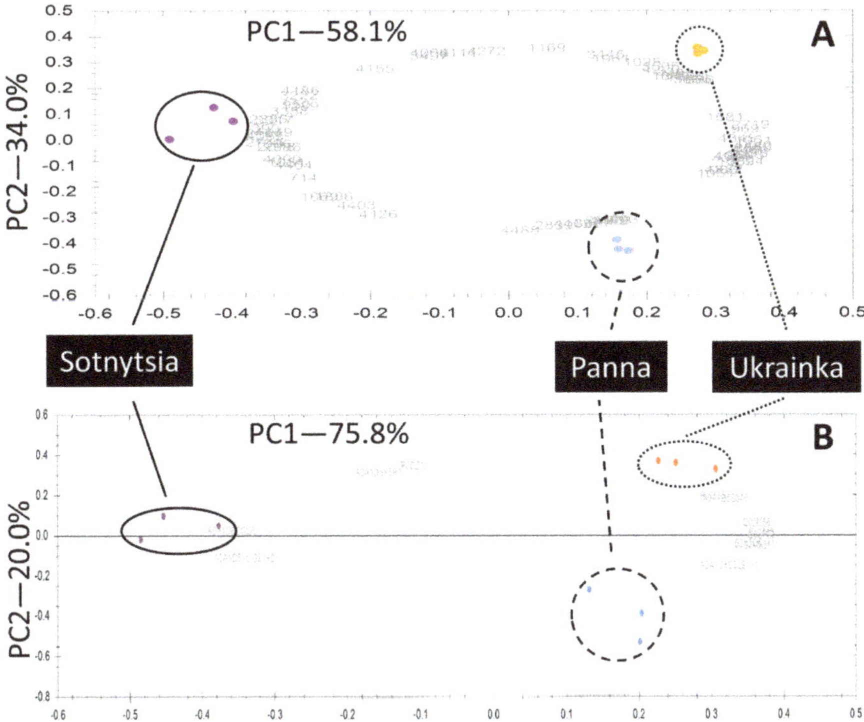

Figure 2. Principal component analysis of differentially abundant proteins showed excellent reproducibility among biological replicates for both analytical approaches and suggested that modern cultivar Sotnytsia is distant: (**A**) 66 gel spots; (**B**) 12 proteins discovered upon direct mass spectrometry quantification. Multidimensional statistics were calculated and visualized using functions integrated into SameSpots 5.1 and Progenesis QI.

The vast majority of differentially accumulated medically relevant gliadins belong to the food allergen Tri a 20 class. This group unites γ-gliadins causing wheat-dependent exercise-induced anaphylaxis [33], perhaps because of homology to ω-gliadins. We detected four different γ-gliadins, each in multiple gel spots. The γ-gliadin P21292 was more abundant in Panna relative to Sotnytsia and Ukrainka in 4179 and more than 5-fold in 1803. The γ-gliadin D4 (*Gli1*) was more abundant in Sotnytsia compared to other genotypes (2161) or only to Ukrainka (2078). The γ-gliadin D3 (*Gli1*) from 1896 accumulated in Sotnytsia relative to Ukrainka, and from 1961 was more abundant in landrace compared to modern bread wheat cultivars. The former is likely a minor post-translational modification (PTM) with a shifted isoelectric point of the latter, which is one order more abundant in all samples. We detected γ-gliadin A1 (*Gli1*) in three gel spots 1649, 1656, and 4220 accumulated more than five times in Ukrainka compared to Panna and Sotnytsia (Table 1). The same pattern characterized α/β-gliadin I0IT62 (spot 3866) not associated with any Allergome identifier. Another α/β-gliadin AII, without this database annotation, accumulated in both modern cultivars compared to landrace (spot 1669) or only in Sotnytsia versus Ukrainka (spot 4403). Gliadins identified in a smaller proportion of differentially accumulated gel spots do not have an association with any human disorder. A protein belonging to the UniRef cluster, which includes avenin-like proteins, was differentially abundant in multiple gel spots 2645, 2689, 4169, 4481, and 4482. The majority of them accumulated in Panna and Ukrainka compared to Sotnytsia, particularly 2689 more than 5-fold. All these spots shared the same area on the gel, but one 4169, having slightly lower pI and opposite accumulation (Figure 1). A protein similar to δ-gliadin D1 (spot 4491) accumulated over 5-fold in landrace relative to newer cultivars.

Table 1. Identified differentially abundant proteins from *Triticum aestivum* L. cultivars discovered by two-dimensional gel electrophoresis. Bold ratios indicate significant differences in the respective comparison (ratio $\geq$ 2.5 and Tukey's test $p \leq 0.01$). Positive value of proportion—protein is more abundant, negative value—protein is less abundant, NA—not available.

Spot #	UniProt Accession	Protein Name (Genetic Locus)	Function	Allergen or Toxin	Protein Group	Panna/Sotnytsia	Panna/Ukrainka	Sotnytsia/Ukrainka
4126	B2Y2Q6	LMW-B2 glutenin (*Glu-B3-2*)	Nutrient reservoir activity	√	Glutenin	1.07	**2.51**	2.34
1066	P08488	HMW-12 glutenin (*HMW-GS DY*)	Nutrient reservoir activity	√	Glutenin	−2.12	1.19	**2.53**
4114	P08488	HMW-12 glutenin (*HMW-GS DY*)	Nutrient reservoir activity	√	Glutenin	−2.37	**−2.82**	−1.19
390	P08489	HMW-PW212 glutenin	Nutrient reservoir activity	√	Glutenin	**−11.01**	1.08	**11.84**
4045	P10388	HMW-DX5 glutenin (*Glu-1D-1D*)	Nutrient reservoir and starch binding activity	√	Glutenin	**2.65**	1.09	−2.43
4332	P10388	HMW-DX5 glutenin (*Glu-1D-1D*)	Nutrient reservoir and starch binding activity	√	Glutenin	**3.00**	1.34	**−2.24**
4345	P10388	HMW-DX5 glutenin (*Glu-1D-1D*)	Nutrient reservoir and starch binding activity	√	Glutenin	2.47	−1.18	**−2.91**
4351	P10388	HMW-DX5 glutenin (*Glu-1D-1D*)	Nutrient reservoir and starch binding activity	√	Glutenin	**4.14**	1.09	**−3.79**
1896	A1EHE7	γ-gliadin D3 (*Gli1*)	Nutrient reservoir activity	√	Gliadin	−1.32	2.61	**3.45**
1961	A1EHE7	γ-gliadin D3 (*Gli1*)	Nutrient reservoir activity	√	Gliadin	−1.36	**−3.97**	**−2.92**
3866	I0IT62	α/β-gliadin	Nutrient reservoir activity	√	Gliadin	1.40	**−3.72**	**−5.20**
1649	M9TG60	γ-gliadin A1 (*Gli1*)	Nutrient reservoir activity	√	Gliadin	1.24	**−4.40**	**−5.47**
1656	M9TG60	γ-gliadin A1 (*Gli1*)	Nutrient reservoir activity	√	Gliadin	1.37	**−4.09**	**−5.59**
4220	M9TG60	γ-gliadin A1 (*Gli1*)	Nutrient reservoir activity	√	Gliadin	−1.04	**−3.67**	**−3.53**
1669	P04722	α/β-gliadin AII	Nutrient reservoir activity	√	Gliadin	−1.32	**2.57**	**3.37**
4403	P04722	α/β-gliadin AII	Nutrient reservoir activity	√	Gliadin	−1.03	2.49	**2.57**
1803	P21292	γ−gliadin	Nutrient reservoir activity	√	Gliadin	**5.59**	**4.47**	−1.25
4179	P21292	γ−gliadin	Nutrient reservoir activity	√	Gliadin	**3.09**	2.78	−1.11
2078	Q94G92	γ-gliadin D4 (*Gli1*)	Nutrient reservoir activity	√	Gliadin	**−2.60**	1.35	**3.50**
2161	Q94G92	γ-gliadin D4 (*Gli1*)	Nutrient reservoir activity	√	Gliadin	**−3.40**	1.11	**3.78**
4491	A0A1D5T3T7	Similar to δ-gliadin D1, obsolete	Nutrient reservoir activity		Gliadin	1.18	**−4.30**	**−5.08**
2645	A0A3B5YPZ7	Similar to avenin-like protein	Nutrient reservoir activity		Gliadin	**4.00**	1.04	**−3.86**
2689	A0A3B5YPZ7	Similar to avenin-like protein	Nutrient reservoir activity		Gliadin	**6.03**	−1.14	**−6.86**
4169	A0A3B5YPZ7	Similar to avenin-like protein	Nutrient reservoir activity		Gliadin	**−3.16**	1.07	**3.37**
4481	A0A3B5YPZ7	Similar to avenin-like protein	Nutrient reservoir activity		Gliadin	**3.90**	−1.03	**−4.03**
4482	A0A3B5YPZ7	Similar to avenin-like protein	Nutrient reservoir activity		Gliadin	**4.33**	−1.10	**−4.77**
1581	Q41593	Serpin-Z1A (*WZCI*)	Serine protease inhibitor, extracellular	√	Metabolic	1.88	−1.57	**−2.95**

Table 1. *Cont.*

Spot #	UniProt Accession	Protein Name (Genetic Locus)	Function	Allergen or Toxin	Protein Group	Panna/Sotnytsia	Panna/Ukrainka	Sotnytsia/Ukrainka
3607	Q43691	Trypsin/α-amylase inhibitor CMX2	Serine protease inhibitor, secreted	√	Metabolic	**2.63**	**2.70**	1.03
4404	A0A1D5US94	Methyltransferase	Protein dimerization activity		Metabolic	−1.87	1.61	**3.00**
1084	A0A1D5YFA7	Similar to β-amylase, obsolete	Hydrolysis of (1->4)-α-D-glucosidic linkages in polysaccharides		Metabolic	−2.55	1.82	**4.64**
2386	A0A1D5ZTV0	Similar to serpin-N3.2, obsolete	Serine protease inhibitor, extracellular		Metabolic	−2.38	1.24	**2.94**
1025	A0A1D6A827	Dehydrin, obsolete	Stress response		Metabolic	−1.21	**−7.08**	**−5.86**
969	A0A1D6D1Q3	Pyrophosphate—fructose 6-phosphate 1-phosphotransferase subunit β (*PFP-β*)	Glycolysis, cytoplasm		Metabolic	2.32	−1.33	**−3.10**
2335	A0A1D6RH21	Similar to dehydroascorbate reductase (*DHAR*), obsolete	Glutathione S-transferase domain		Metabolic	**−5.31**	−1.14	**4.65**
1169	A0A1D6S518	UTP—glucose-1-phosphate uridylyltransferase	Biosynthesis of saccharides, cytoplasm		Metabolic	−1.72	**−2.94**	−1.71
2719	A0A3B5XV32	Cupin domain protein	Nutrient reservoir activity		Metabolic	**2.60**	−1.51	**−3.94**
4272	A0A3B6ILV9	Similar to globulin 3	Nutrient reservoir activity		Metabolic	−2.10	**−2.73**	−1.30
3358	A0A3B6IN56	Similar to alanine-tRNA ligase	Protein synthesis		Metabolic	**−2.93**	−1.22	2.41
1082	A0A3B6KSH4	Similar to β-amylase	Hydrolysis of (1->4)-α-D-glucosidic linkages in polysaccharides		Metabolic	**2.60**	1.10	−2.36
4055	A0A3B6MYZ0	Similar to globulin-1 S allele	Nutrient reservoir activity		Metabolic	**−4.81**	1.19	**5.73**
1051	I6QQ39	Globulin-3A (*Glo−3A*)	Nutrient reservoir activity		Metabolic	**2.66**	1.35	−1.97
4155	I6QQ39	Globulin-3A (*Glo-3A*)	Nutrient reservoir activity		Metabolic	**−3.09**	−2.27	1.37
1315	A0A1D5VMG1	Uncharacterized protein, obsolete	NA		Metabolic	**3.19**	1.03	**−3.11**
950	A0A3B5Z536	Uncharacterized protein	NA		Metabolic	**−5.44**	−1.58	3.44
1289	A0A3B6G0N3	Uncharacterized protein	NA		Metabolic	**−2.63**	−1.28	2.06
4162	A0A3B6JER7	Uncharacterized protein	NA		Metabolic	**4.48**	**5.18**	1.16
4171	A0A3B6JER7	Uncharacterized protein	NA		Metabolic	**3.51**	**3.74**	1.06
4489	A0A3B6LGL1	Uncharacterized protein	NA		Metabolic	**5.72**	**5.05**	−1.13
4008	A0A3B6PFQ8	Uncharacterized protein	NA		Metabolic	**−4.11**	**−4.38**	−1.06

In the non-gluten grain protein fraction, we discovered a trypsin/α-amylase inhibitor CMX2, which belongs to the Tri a CMX group, and predominantly causes a non-celiac gluten sensitivity [18,34]. This protein from spot 3607 accumulated in Panna compared to Sotnytsia and Ukrainka. Serpin-Z1A (*WZCI*) in spot 1581 was more abundant in Ukrainka compared to Sotnytsia. It belongs to the Tri a 33 group and elicits a broad spectrum of sensitizations [35]. The largest proportion of differentially abundant gel spots comprised metabolic proteins, which are neither allergens nor toxins. We identified several storage proteins as globulin-3A (*Glo-3A*) accumulated in Sotnytsia compared to Panna (spot 4155), or vice versa (spot 1051), which showed very distinct pI. A protein similar to globulin-1 S allele (spot 4055) accumulated more than 5-fold in Sotnytsia compared to other genotypes. Protein similar to globulin 3 (spot 4272) was more abundant in Ukrainka versus Panna. Cupin domain protein in spot 2719 (also revealed in complementary LC-MS analysis) accumulated in Panna and Ukrainka compared to Sotnytsia. Among the proteins involved in primary metabolism was similar to β-amylase (spots 1082 and 1084) more abundant in Panna versus Sotnytsia and Sotnytsia versus Ukrainka, respectively. The former seems to be a minor modification of the latter. Cytoplasmic glycolytic enzyme pyrophosphate—fructose 6-phosphate 1-phosphotransferase subunit β (*PFP-β*) from spot 969 accumulated in Ukrainka compared to Sotnytsia. The enzyme involved in the biosynthesis of saccharides (spot 1169) was more abundant in Ukrainka versus Panna. Alanine-tRNA ligase (spot 3358) accumulated in Sotnytsia versus Panna and methyltransferase in spot 4404 showed a higher amount in Sotnytsia compared to Ukrainka. Besides, we discovered several stress proteins, like dehydrin (spot 1025) or a protein similar to dehydroascorbate reductase (*DHAR*) in spot 2335 that more than 5-fold accumulated in landrace compared to the newer genotypes or in Sotnytsia versus Panna and Ukrainka, respectively. A protein similar to serpin-N3.2 (spot 2386) was more abundant in Sotnytsia compared to Ukrainka. The last seven spots, which varied across investigated bread wheat genotypes, are not annotated.

2.2. Label-Free Quantification Complemented the Comparison Between Bread Wheat Cultivars

To complement the gel-based dataset, we employed a comprehensive UHPLC profiling of peptides, followed by ion mobility-enhanced label-free quantification using an accurate peak area principle. Ion mobility allows another separation dimension, complementing reverse phase chromatography. As a result, we reproducibly quantified 127 proteins among samples. We presented the values as means and standard deviations of three biological replicates together with identification and annotation details in the supporting information (Table S3). The whole list of identified peptides is a valuable reference for the future development of targeted quantification methods (Table S4). Criteria for selecting differentially abundant proteins were the same as with gel-based analysis in terms of effect size and statistical significance for consistency. Thereby, 12 proteins satisfied these parameters (Table 2). Of note, the principal component analysis indicated sample clustering according to the genotype and closer relation between cultivars Panna and Ukrainka, concordant to 2-DE (Figure 2B). Principal fractions were six glutenins, two gliadins, and four metabolic proteins.

Table 2. Differentially abundant proteins among wheat cultivars revealed by direct label-free mass spectrometry quantification. Bold values indicate significant differences in the respective comparison (ratio $\geq$ 2.5 and Tukey's test $p \leq 0.01$). Positive value of proportion—protein is more abundant, negative value—protein is less abundant.

UniProt Accession	Protein Name (Genetic Locus)	Function	Allergen or Toxin	Protein Group	Panna/Sotnytsia	Panna/Ukrainka	Sotnytsia/Ukrainka
Q6SPZ3	LMW-A2 glutenin (*Glu-A3-11*)	Nutrient reservoir activity	√	Glutenin	**6.39**	−1.25	**−7.99**
Q00M56	LMW-D1 glutenin (*Glu-D3-3*)	Nutrient reservoir activity	√	Glutenin	2.31	−1.39	**−3.21**
P10386	LMW-1D1 glutenin (*Glu-D3-2*)	Nutrient reservoir activity	√	Glutenin	1.45	−1.92	**−2.78**
D2DII3	LMW glutenin subunit (*Glu-A3-16*)	Nutrient reservoir activity	√	Glutenin	**4.52**	−1.30	**−5.88**
B2Y2Q6	LMW-B2 glutenin (*Glu-B3-2*)	Nutrient reservoir activity	√	Glutenin	**11.01**	1.07	**−10.24**
B2Y2Q1	LMW glutenin subunit (*Glu-B3-1*)	Nutrient reservoir activity	√	Glutenin	**3.06**	1.07	**−2.86**
A0A1D5S346	Similar to γ-gliadin, obsolete	Nutrient reservoir activity	√	Gliadin	1.48	−1.81	**−2.68**
P0CZ10	Avenin-like a6	Nutrient reservoir activity		Gliadin	−3.06	**−3.38**	−1.11
A0A3B6K9J4	Thaumatin family protein	Multiple disulfide bonds		Metabolic	−1.78	1.44	**2.56**
A0A3B5XV32	Cupin domain protein	Nutrient reservoir activity		Metabolic	**2.91**	1.44	−2.02
A0A1D5Y5R1	Cupin domain protein, obsolete	Nutrient reservoir activity		Metabolic	**−2.74**	**−2.57**	1.07
A0A3B6SDE7	Uncharacterized membrane protein	Integral component of membrane		Metabolic	**−4.88**	1.08	**5.25**

Seven of the discovered differentially abundant proteins are known to contain allergenic/toxic epitopes according to the database sources. As in the case of 2-DE quantification, all variable glutenin subunits can affect human health. They belong to the Tri a 36 group [32], according to Allergome classification. Glutenin LMW-D1 (*Glu-D3-3*) and glutenin LMW-1D1 (*Glu-D3-2*) showed higher abundance in Ukrainka compared to Sotnytsia. LMW glutenin subunit (*Glu-A3-16*), LMW glutenin subunit (*Glu-B3-1*), and glutenin LMW-A2 (*Glu-A3-11*, over 5-fold) were more abundant in the grain of cultivars Panna and Ukrainka versus Sotnytsia. Glutenin LMW-B2 (*Glu-B3-2*) was more than 10-fold abundant in Panna and Ukrainka respective to Sotnytsia. Single toxic protein from the gliadin group, belonging to the food allergen class Tri a 20 [33], similar to γ-gliadin accumulated in Ukrainka contrasted to Sotnytsia. Another gliadin related protein, avenin-like a6, was more abundant in Ukrainka compared to Panna. Avenins are not directly integrated into the gluten polymer through disulfide bonds unless incorporated by reduction and reoxidation during dough making [36]. Among variable proteins belonging to a non-gluten group, we discovered an uncharacterized membrane protein having a dentin matrix domain as more abundant in Sotnytsia compared to other cultivars. The thaumatin family (pathogenesis-related group five) protein accumulated in Sotnytsia relative to Ukrainka. A storage protein having a cupin domain A0A3B5XV32 was more abundant in Panna compared only to Sotnytsia, and another cupin domain protein A0A1D5Y5R1 accumulated in Sotnytsia and Ukrainka versus Panna.

Contrasting differentially abundant proteins from complementary approaches, we revealed only two common accessions and just one of them—LMW-B2 glutenin (*Glu-B3-2*)—is a known food allergen. In 2-DE, we detected about 2-fold higher amount of this protein in Panna (significantly) and Sotnytsia (below effect size threshold) relative to Ukrainka, but somehow contradictory, the LC-MS approach pointed to more than 10-fold higher abundance in Panna and Ukrainka compared to Sotnytsia. Likely, this discrepancy reflects the specific characteristics of the particular analytical method. On the other hand, storage protein with cupin domain protein accumulated in Panna and Ukrainka compared to Sotnytsia according to both 2-DE evaluation, and direct mass spectrometry quantification (below effect size threshold for Ukrainka).

A small overlap between LC-MS and 2-DE proteomic datasets is puzzling. However, such phenomenon is rather common [37]. We assume that some discrepancy is explained by PTMs, which were separated on gels, but not distinguished by direct LC-MS. Such as HMW-12 glutenin differentially accumulated in two gel spots (Table 1) but showed the same amount among cultivars according to the label-free quantification (Table S3). On the other hand, some proteins highlighted as varied in the LC-MS dataset, for example, LMW-A2 glutenin (Table 2), could be among a few protein spots, which we failed to identify.

2.3. Effective Workflow to Overcome the Analytical Challenges of Grain Proteome Profiling

Proteomic tools have been exhaustively used to evaluate the genetic diversity of wheat germplasm from different continents at the level of allelic polymorphism of glutenins and gliadins—the two main components of gluten. More recently, proteomics became fundamental to understand the impact of specific gluten proteins on wheat quality [38]. We used anionic detergent SDS to extract physicochemically diverse subgroups of wheat storage proteins [19]. Consequently, we obtained a complex mixture with both gluten and other proteins in a single step for better analytical reproducibility. Since gluten proteins are rich in proline/glutamine, alternative proteases might be necessary for efficient analysis. Multi-enzymatic digestion is preferable to generate unique peptides because thermolytic, chymotryptic, and tryptic peptides match different parts of protein sequences [28]. Considering the fact that more hits could be identified after chymotryptic digestion of gluten proteins than using other enzymes [39], we used chymotrypsin as the first choice protease, even though scoring algorithms are optimized for classic trypsin digestion. Our data confirmed that processing of gel spots, unassigned after chymotryptic digestion, with alternative enzymes, added several more hits to the cumulative list.

The comparative advantage of 2-DE is a visualization of multiple isoforms. Nevertheless, direct mass spectrometry quantification is gaining momentum as a preferred analytical tool for the discovery proteomics since it is fast, robust, deep, and with mature bioinformatic algorithms for reliable quantification.

However, peptide centered analysis—LC-MS—has intrinsic limitations, such as the loss of intact protein information, consequently the inability to grasp combinations of PTMs. Repetitive sequence motifs, frequent in gluten proteins, enormously complicate unique peptide assignment with mass spectrometry. Therefore, using data-independent mass spectrometry, Hajduch laboratory quantified only 34 gliadins and 22 glutenins in wheat grain, concluding that their approach is a reproducible quantitative method for the determination of gluten protein content in the highly complex matrix [40]. Quadrupole time-of-flight mass spectrometer with ion mobility worked very well for sequencing longer peptides, crucial to distinguish individual gluten proteins [41]. Using the same type of instrument, we also confirmed multiple glutamine deamidations, as reported earlier [41]. Nevertheless, due to the inherent challenges, like the dominance of only a few related (sharing long sequence stretches) storage protein classes, direct mass spectrometry in the discovery mode was not so effective. Despite effort into developing an optimal analytical workflow, we quantified a modest number of proteins.

2.4. Both Gluten and Non-Gluten Proteins Contain Allergenic or Toxic Motifs

Consolidating both 2-DE and label-free datasets, we revealed that the majority of proteins showing variable amounts in studied genotypes belong to a metabolic group—27 hits, 44%; the next are gliadins—20 proteins, 33%; and the least glutenins—14 hits, 23% (Figure 3). Contrastingly, among allergenic/toxic proteins, the highest diversity showed glutenins with 14 proteins, 48%; gliadins included 13 hits, 45%; and the non-gluten group consisted of only two proteins, 7%. Traditional landrace Ukrainka showed the highest cumulative coefficient for proteins with allergenic/toxic epitopes (Figure 4). Next, we extracted information on known epitopes in sequences of discovered allergenic/toxic proteins (Table 3). The number, as well as density (number of toxic sequences divided by the quantity of all amino acids) of unique celiac motifs according to GluPro and the number of medically relevant epitopes according to ProPepper, characterized the detected polypeptides. The highest number of such epitopes 127 occurred in α/β-gliadin AII, yet the highest density of the toxic celiac motifs occurred in γ-gliadin P21292; the former was particularly notable in Sotnytsia, accumulating in two gel spots, the latter also in two gel spots was more abundant in Panna. Contrary, HMW-12 glutenin (*HMW-GS DY*), differentially abundant in two gel spots, had only six epitopes and minimal density of unique celiac motifs.

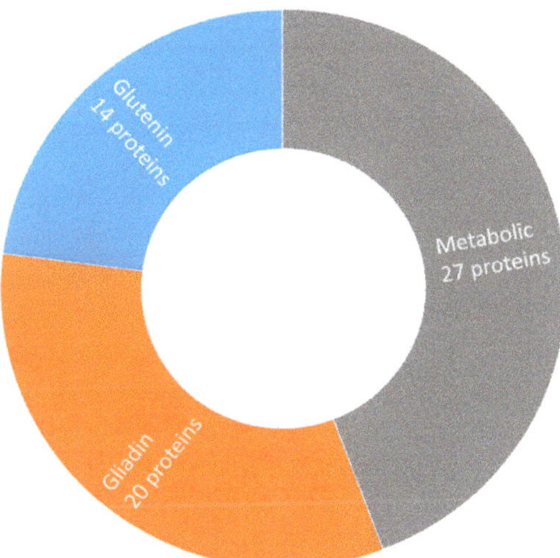

Figure 3. The proportion of grain protein groups differentially accumulated among investigated *Triticum aestivum* L. cultivars revealed the most diverse non-gluten fraction. Information about the functional role of proteins was extracted primarily from UniProt.

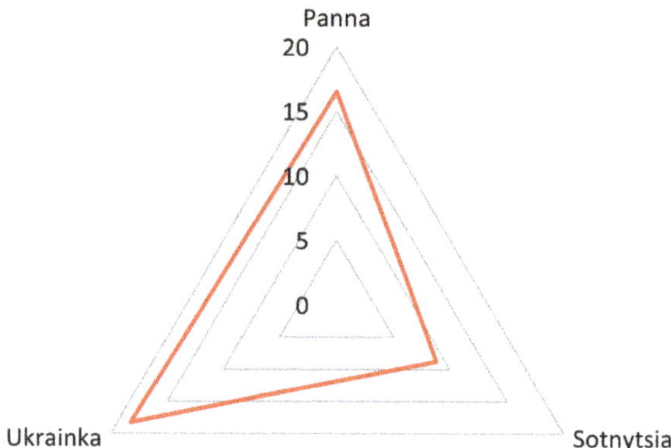

Figure 4. Cumulative coefficient of contrastingly accumulated proteins with harmful epitopes discovered with both experimental approaches among bread wheat genotypes; from a total of 61 identified polypeptides, 29 are of medical concern. Modern cultivar Sotnytsia scored lowest, while landrace Ukrainka accumulated the most allergenic/toxic proteins in grain.

Table 3. Detailed characteristics of discovered allergenic or toxic proteins differentially accumulated among bread wheat cultivars. Information about reactive motifs was extracted from dedicated databases—Allergome, ProPepper, and GluPro, NA—not available.

UniProt Accession	Protein Name (Genetic Locus)	Allergome Identifier	Protein Group	# of Epitopes ProPepper	# of Celiac Motifs GluPro	Density of Celiac Motifs GluPro
P08489	HMW-PW212 glutenin	Tri a 26	Glutenin	48	18	0.02
P08488	HMW-12 glutenin (*HMW-GS DY*)	Tri a 26	Glutenin	6	3	< 0.01
P10388	HMW-DX5 glutenin (*Glu-1D-1D*)	Tri a 26	Glutenin	49	5	0.01
B2Y2Q6	LMW-B2 glutenin (*Glu-B3-2*)	Tri a 36	Glutenin	9	5	0.01
Q6SPZ3	LMW-A2 glutenin (*Glu-A3-11*)	Tri a 36	Glutenin	NA	11	0.03
Q00M56	LMW-D1 glutenin (*Glu-D3-3*)	Tri a 36	Glutenin	7	5	0.01
D2DII3	LMW glutenin subunit (*Glu-A3-16*)	Tri a 36	Glutenin	8	NA	NA
B2Y2Q1	LMW glutenin subunit (*Glu-B3-1*)	Tri a 36	Glutenin	12	4	0.01
P10386	LMW-1D1 glutenin (*Glu-D3-2*)	Tri a 36	Glutenin	6	4	0.01
I0IT62	α/β-gliadin	NA	Gliadin	43	13	0.04
P04722	α/β-gliadin AII	NA	Gliadin	127	22	0.08
P21292	γ-gliadin	Tri a 20	Gliadin	59	46	0.15
A1EHE7	γ-gliadin D3 (*Gli1*)	Tri a 20	Gliadin	49	37	0.13
Q94G92	γ-gliadin D4 (*Gli1*)	Tri a 20	Gliadin	40	23	0.08
A0A1D5S346	Similar to γ-gliadin, obsolete	Tri a 20	Gliadin	NA	NA	NA
M9TG60	γ-gliadin A1 (*Gli1*)	Tri a 20	Gliadin	44	31	0.09
Q41593	Serpin-Z1A (*WZCI*)	Tri a 33	Metabolic	NA	NA	NA
Q43691	Trypsin/α-amylase inhibitor CMX2	Tri a CMX	Metabolic	NA	NA	NA

Herein, we evaluated the variability of all proteome fractions, including metabolic proteins and globulins, because non-gluten proteins better discriminated wheat cultivars and lines [25,26,42]. Structure modeling and epitope prediction confirmed the presence of linear homologs of celiac disease epitopes in seed storage globulins, highlighting that reactive response may be developed not exclusively to prolamins [43]. We confirmed non-gluten proteins as the primary variable group (Figure 3), yet discovered little diversity among allergens/toxins within metabolic proteins (Table 3). Recently, researchers analyzed the composition of epitopes relevant to celiac disease and wheat-dependent exercise-induced anaphylaxis in the gliadins from the flour of cultivar Keumkang using 2-DE. They confirmed a strong immunogenic potential of specific α/β- and γ-gliadins [44]. Our study highlighted five unique γ-gliadins and two α/β-gliadins among bread wheat genotypes (Table 3). High-resolution 2-DE discovered nine subunits of LMW glutenins as predominant IgE-binding antigens causing food allergy [45]. In line with this report, our analysis also showed multiple potentially toxic glutenins differentially accumulated among Ukrainian bread wheat cultivars (Tables 1 and 2).

Multiple databases contain useful curated and annotated information about the role of wheat proteins in different pathologies. However, the heterogeneity of these databases prevents direct connectivity. Developers of one useful resource Allergome have the ambition to make it a common platform where experimental and clinical data will be merged [46]. ProPepper indexes linear epitopes with proven T- or B-cell specific activity [47] and GluPro database associates sequences of gluten proteins with celiac disease by annotating number and density of unique medically relevant motifs [48]. We received largely correlated indices from ProPepper and GluPro analysis of clinically relevant proteins in *T. aestivum* genotypes (Table 3).

2.5. Modern Breeding, Traditional Landraces, and Biotechnology for Safe Wheat

Wheat breeding primarily targeted specific traits such as high yield, disease resistance, and drought tolerance. There is sufficient genetic variation to breed wheat with reduced toxicity, which is beneficial for people with sensitivities [21]. Using accurate targeted mass spectrometry, researchers showed that a modern *T. aestivum* cultivar, Toronto, contained the highest amounts of immunogenic celiac peptides compared with the older one, Minaret, and the tetraploid wheat cultivar [49]. Contrary, a large screening study rebutted the idea that landraces synthesize a lower amount of clinically relevant proteins, showing no association between the age of cultivars and immunogenicity of gliadins [23]. Standardized growth conditions should clarify the genetic potential of lower toxicity for older landraces [24]. In our experiment, grains were harvested from neighboring plots agronomically treated in the same manner, avoiding heterogeneity of growth conditions. Another survey detected immunogenic epitopes having an extremely high quantitative range in both historical and modern spring wheat genotypes. Therein, researchers proposed cultivar Russ with the least amount of immunogenic epitopes, as a suitable starting point for breeding safer wheat for celiac disease patients [23]. Our study also highlighted one of the modern cultivars of Ukrainian selection Sotnytsia, as the genotype with the lowest accumulation of allergenic/toxic proteins in grain (Figure 4). The Igrejas group showed that *T. aestivum* landraces presented higher amounts of immunostimulatory epitopes and concluded that there is a good potential for selection of cultivars with a low content of toxic epitopes via conventional breeding practices [22].

A significant positive correlation was found between release year of cultivars and dough quality characteristics, which could be associated with quantitative variations in glutenin polymeric proteins, and certain subfractions of ω-gliadins [50]. Farmers increasingly replace landraces with modern cultivars, which are less resilient to pests, diseases, and abiotic stresses. Thereby, a valuable source of germplasm may be lost for meeting the future needs of sustainable agriculture in the context of climate change. Landraces are a proven indispensable source of resistance to pathogens, which should be preserved and exploited [51]. Superior stress resilience is due to the natural heterogeneity of the landraces in contrast to modern, more homogeneous cultivars [52].

An alternative approach to make wheat products safer is using powerful genetic engineering. A large survey demonstrated that a wide variation exists in the amount of allergenic polypeptides among wheat cultivars, and the differences detected between genetically engineered lines are within the range of conventional cultivars. Advanced statistics showed that patient sera are yet another key variable [53]. Recently, Barro laboratory reported the first application of genome editing for subtracting immunodominant celiac peptides. Up to 75% of different α/β-gliadin genes were mutated in engineered lines, while immunoreactivity of grain extracts was dramatically reduced. Thus, the authors speculated that such 'transgene-free' wheat genotypes could be used to produce less toxic food products and serve as source material to incorporate new traits into elite wheat cultivars [54]. Currently, the most frequent biotechnological approach—RNA interference—was used to produce low-gliadin wheat lines. Separate analyses of gliadins, glutenin subunits, metabolic, and chloroform/methanol-like proteins by a classical 2-DE allowed thorough safety evaluation of the transgenes. Researchers discovered a compensation of total protein, as those lines showed significant accumulation of HMW glutenins, and non-gluten albumins and globulins [27].

3. Materials and Methods

3.1. Plant Material and Protein Extraction

Three Ukrainian winter wheat cultivars: Sotnytsia, Panna (both modern selection), and Ukrainka (archaic landrace) were selected for this study, being either essential for the contemporary agriculture or older cultivars not widely used in mass production. We hypothesized that they accumulate a different amount of clinically relevant proteins. The crop was grown in the experimental field using standard agrotechnical practice with a random plot design in the Kyiv region. Grain was harvested in the 2017 season. No considerable environmental stress (drought, cold, or pathogens) was recorded during bread wheat growth. Experimental plots were black soil with the appropriate addition of mineral fertilizers.

Panna and Sotnytsia are modern Ukrainian cultivars promising for agricultural mass production. Wheat cultivar Panna, with a unique protein gluten complex, was registered at the beginning of the XXI century. The protein content in its grain is approximately 15.3%, and the technological quality of the flour is above average. Cultivar Panna carries the glutenin allele *Glu-B1-5*, which provides good baking properties, positively influencing elasticity and viscosity of the dough. However, because of lower productivity and the propensity to lay in unfavorable conditions, Panna is not widely planted for grain production. Nevertheless, it is included in breeding programs as a genetic source for improving the quality of winter wheat. On the other hand, cultivar Sotnytsia has a high genetic potential for productivity; moreover, it has superior technological properties for milling and bread making. This cultivar tolerates major pathogens and drought well. Notably, the average productivity of Sotnytsia is about 10% higher than the Ukrainian national standard [55]. Breeding for maximum yield, especially in grain crops, can cause deterioration of protein content. Thus, the modern cultivars of wheat are plausibly inferior to the landrace Ukrainka, comparing the amount and composition of grain storage proteins, particularly gluten. Genotype Ukrainka was developed a century ago, showing decent yield and excellent bread-making qualities. For dozens of years, it was widely grown and even had a reputation of world quality standard. Genetic heterogeneity of glutenin and gliadin genetic loci in historical landrace Ukrainka was reported previously [52,56].

All reagents and solvents of the highest available analytical grade were purchased from Sigma-Aldrich or Merck Millipore, respectively, unless stated otherwise. Grain proteins were extracted according to the developed protocol [57]. Following grinding of 1 g of seeds to fine flour with liquid nitrogen in a mortar, 10 mL of extraction buffer (2% SDS, 10% glycerol, 50 mM dithiothreitol, and 50 mM Tris pH 6.8) was added. The mixture was incubated with vigorous shaking for 1 h, followed by centrifugation at $10,000\times g$ for 15 min. Proteins were precipitated from the supernatant with four volumes of cold acetone and kept overnight at $-20\ °C$. The precipitate was collected by centrifugation at $4000\times g$, $4\ °C$ for 15 min, followed by a single acetone wash. The protein precipitate was dried under vacuum and stored at $-80\ °C$.

3.2. Two-Dimensional Gel Electrophoresis

Protein extracts from different cultivars were resuspended in the solubilization buffer (8 M urea, 2 M thiourea, 2% CHAPS, and 2% Triton X-100). Then, concentration was measured by Pierce detergent compatible Bradford assay (ThermoFisher Scientific). Protein aliquots of 500 µg were adjusted to 340 µL with isoelectric focusing buffer (8 M urea, 2 M thiourea, 2% CHAPS, 1% amidosulfobetaine-14, 1% Triton X-100, 2% ampholytes, and 1% DeStreak). Upon centrifugation, the supernatant was transferred to 18 cm immobilized pH 3-10 non-linear gradient strips (GE Healthcare). Following overnight passive rehydration, the first dimension separation, according to the isoelectric points, was done in the Ettan IPGphor 3 unit (GE Healthcare). After isoelectric focusing, strips were reduced in 4 mL of equilibration buffer (EQB; 0.1 M Tris pH 6.8, 30% glycerol, 6 M urea, and 3% SDS) with 2% dithiothreitol for 15 min, alkylated in 4 mL of EQB with 2.5% iodoacetamide in the dark for 15 min, and washed in 4 mL of running buffer (25 mM Tris, 192 mM glycine, and 0.1% SDS). Next, they were placed on top of 12%

polyacrylamide gels and sealed by 0.5% agarose with 0.002% bromophenol blue in running buffer. The second dimension separation, according to molecular weights, was performed in Protean II xi Cell (Bio-Rad).

Gels were stained by sensitive colloidal Coomassie G-250. Images were digitalized with resolution 300 dpi and 16-bit grayscale pixel depth on Umax ImageScanner (GE Healthcare). Quantitative software-assisted gel analysis was done in SameSpots 5.1 (TotalLab) to reveal protein spots differentially accumulated between cultivars. Firstly, software aligned the images, then dust/stain particles, obvious streaks, and damaged gel areas were filtered. Relative volumes were normalized to the median distribution of reference gel to compensate for minor differences in sample loading. All spots were reviewed and manually edited if necessary. Before statistical analysis, normalized spot volumes were transformed using the inverse hyperbolic sine function. Differentially accumulated proteins were chosen based on ANOVA $p \leq 0.01$ and ratio ≥ 2.5. Additionally, we applied post hoc Tukey's honestly significant difference test to assess changes between specific genotypes.

3.3. In-Gel Digestion and Filter-Aided Sample Preparation

Differentially abundant protein spots were excised from the gels, washed with 300 µL of 50 mM ammonium bicarbonate in 50% acetonitrile, and dehydrated with 300 µL acetonitrile. Proteins were reduced with 100 µL of 10 mM dithiothreitol in 100 mM ammonium bicarbonate at 50 °C for 30 min. Alkylation was performed with 100 µL of 50 mM iodoacetamide in 100 mM ammonium bicarbonate in the dark for 30 min. Protein spots were digested with 20 µL of 10 ng/µL enzyme (Promega) in 10 mM ammonium bicarbonate and 10% acetonitrile at 25 °C (chymotrypsin) or 37 °C (trypsin), in 50 mM Tris pH 8.0 and 0.5 mM CaCl2 at 60 °C (thermolysin). Peptides were extracted twice with 50 µL of 70% acetonitrile and 1% trifluoroacetic acid.

Total protein extracts were digested according to the filter-aided sample preparation (FASP) protocol [58]. Following activation of centrifugal filter units Microcon Ultracel YM-10 (Merck Millipore) by 200 µL of 1% formic acid and centrifugation at 10,000× g for 40 min, 100 µg protein aliquots, adjusted to 200 µL with urea buffer (8 M urea and 100 mM Tris pH 8.5), were loaded on the filter and centrifuged. Subsequently, proteins were washed with 200 µL of urea buffer. The reduction was performed with 200 µL of 10 mM dithiothreitol in the urea buffer at 50 °C for 15 min. Proteins were alkylated with 200 µL of 50 mM iodoacetamide in urea buffer in the dark for 15 min. The excess of iodoacetamide was quenched by dithiothreitol. Final washing was done with 200 µL of 20 mM ammonium bicarbonate, followed by centrifugation of filter units. Then, collection tubes were exchanged, and 75 µL of chymotrypsin 100 ng/µL in 50 mM ammonium bicarbonate was added to the filter. Digestion was carried out at 25 °C overnight. Afterwards, peptides were collected by centrifugation at 10,000× g for 20 min. Another 75 µL of 50 mM ammonium bicarbonate was added on the filter and again collected by centrifugation. Finally, 100 µL of 0.1% trifluoroacetic acid was pipetted to acidify the combined flow-throughs. For peptide purification, after FASP, we used Sep-Pak Light C18 cartridges (Waters). Subsequently, the concentration of the peptides was measured by NanoDrop 2000 spectrophotometer (ThermoFisher Scientific).

3.4. Tandem Mass Spectrometry for Protein Identification from Gel Spots

The peptides were analyzed by liquid chromatography-tandem mass spectrometry (LC-MS/MS), using nanoAcquity UHPLC (Waters) and Q-TOF Premier (Waters) as described earlier [40] with minor modifications. Samples were separated by BEH130 C18 analytical column (200 mm length, 75 µm diameter, 1.7 µm particle size), using a fast 20 min gradient of 5%–40% acetonitrile with 0.1% formic acid at a flow rate 300 nL/min. The data were recorded in the MSE mode (parallel high and low energy traces without precursor ion selection) and processed using ProteinLynx Global Server 3.0 (Waters). Spectra were searched against wheat proteome sequences downloaded from UniProt in April 2018 (136,892 entries, uniprot.org). Search parameters were as specified in the following chapter for chymotrypsin, but one allowed miscleavage for trypsin and thermolysin. Thermolysin was defined as cutting on N-terminus after alanine, phenylalanine, isoleucine, leucine, methionine, and valine, but not before proline. Identities were accepted if two or more different peptides with a score higher than 95% reliability threshold were matched. Reliability scores were adjusted based on the distribution of target/decoy queries. In the cases when several sequences matched spectra from a single gel spot, we reported accession with the highest number of reliable peptides.

3.5. Relative Quantification by Direct Mass Spectrometry and Bioinformatics

Aliquots of purified complex peptide mixtures of 300 ng were separated in biological triplicate, using Acquity M-Class UHPLC (Waters) as described earlier [59]. Samples were loaded onto the nanoEase Symmetry C18 trap column (20 mm length, 180 μm diameter, 5 μm particles size). After 2 min of desalting/concentration by 1% acetonitrile containing 0.1% formic acid at a flow rate 8 μL/min, peptides were introduced to the nanoEase HSS T3 C18 analytical column (100 mm length, 75 μm diameter, 1.8 μm particle size). For the thorough separation, a 90 min gradient of 5%–35% acetonitrile with 0.1% formic acid was applied at a flow rate of 300 nL/min. The samples were nanosprayed (3.1 kV capillary voltage) to the quadrupole time-of-flight mass spectrometer Synapt G2-Si with ion mobility option (Waters). Spectra were recorded in a data-independent manner in high definition MSE mode. Ions with 50–2000 m/z were detected in both channels, with a 1 s spectral acquisition scan rate.

Spectra were preprocessed with the Compression and Archival Tool 1.0 (Waters) to reduce noise, removing ion counts below 15. Data processing was done in Progenesis QI 4.0 (Waters) using the workflow outlined in the literature [59]. For peak picking, the following thresholds were applied: Low energy 320 counts and high energy 40 counts. Precursors and fragment ions were coupled, using correlations of chromatographic elution profiles in low/high energy traces. Then, peak retention times were aligned across all chromatograms. Peak intensities were normalized to the median distribution of all ions, assuming the majority of signals are unaffected by experimental conditions. The label-free quantification relied on measured peak areas of the three most intense precursor peptides, preferentially unique. Before statistical analysis, data were transformed using inverse hyperbolic sine function. For the protein identification, the Ion Accounting 4.0 (Waters) search algorithm was applied. The reference sequence file was as mentioned above. Workflow parameters for the protein identification searches were: Maximum two possible chymotrypsin miscleavages, a fixed carbamidomethyl cysteine, variable oxidized methionine, and deamidated glutamine. Chymotrypsin was defined as cutting on C-terminus after tyrosine, phenylalanine, tryptophan, leucine, and methionine, but not before proline. The software automatically determined the precursor and peptide fragment mass tolerances. Peptide matching was limited to less than 4% false discovery rate against the randomized database. Identifications were accepted if at least two distinct reliable peptides (score $\geq$ 5.5, mass accuracy $\leq$ 15 ppm) matched the protein sequence. The protein grouping feature was then applied to show only hits with unique peptides. We considered as differentially abundant proteins, which satisfied the same strict statistic and effect size criteria as for 2-DE.

To assess the clinical relevance of identified proteins, which differentially accumulated in grain from studied cultivars, we used Allergome (allergome.org), ProPepper (propepper.net), and GluPro [48]. Biological functions and technological quality features of polypeptides were taken primarily from UniProt. Since recently, there was a systematic update of the wheat proteome in UniProt; obsolete accessions were replaced with analogous sequences actual for February 2019 if available. The cumulative coefficient of contrastingly accumulated clinically relevant proteins was calculated by sharing 1.5 points per hit among cultivars. The genotype showing the highest accumulation of a particular allergenic/toxic protein received maximum count while the cultivar with the lowest abundance of the same hit acquired a minimum point.

3.6. Data Availability

All data generated or analyzed during this study are included in this published article and its supplementary information files. The mass spectrometry proteomics data have been deposited to the ProteomeXchange Consortium via the PRIDE partner repository with dataset identifier PXD012940 [60].

4. Conclusions

Human pathologies associated with grain proteins are on the rise, and the only effective treatment is a lifelong gluten-free diet, which is complicated to follow and detrimental to gut health. Therefore, there is a need to improve the quality of life for millions of gluten intolerant patients around the world and prevent new cases. Herein, we showed that the majority of grain proteins, variable among genotypes, belonged to the metabolic group, while detected gliadins were the most health-threatening, due to the highest number/density of epitopes. This group has a somehow lower influence on bread-making quality. Thus, it is promising to investigate promoter polymorphism of gliadin genes further to select germplasm for the breeding of bread wheat cultivars with satisfactory technological properties, yet reduced allergenicity. We discovered the highest accumulation coefficient of allergenic/toxic proteins in landrace Ukrainka. This can be explained by its genetic heterogeneity in contrast to modern genotypes. Among them, particularly Sotnytsia stands out as suitable germplasm for novel less toxic cultivars. Marker peptides from variable proteins can be used in method development for targeted quantification by a triple quadrupole mass spectrometer, offering an accurate high-throughput alternative in both analyte/genotype dimensions, enabling fast and efficient assessment of the medical safety of multiple wheat cultivars.

Supplementary Materials: Supplementary materials can be found at http://www.mdpi.com/1422-0067/21/10/3445/s1. Figure S1: All analytical quality gels, stained by colloidal Coomassie and analyzed with SameSpots from the grain of three bread wheat cultivars; Table S1: List of all 810 quantified gel spots; Table S2: Protein and peptide identification details for specific gel spots; Table S3: List of all 127 reliable protein hits directly quantified by mass spectrometry; Table S4: Peptide identification details for direct liquid chromatography-mass spectrometry quantification.

Author Contributions: Conceptualization, B.M.; methodology, O.L. and P.M.; software, M.D. and A.K.; validation, B.M. and Ľ.Š.; formal analysis, O.L. and M.D.; investigation, O.L., M.D. and P.M.; resources, B.M., A.K. and Ľ.Š.; data curation, M.D.; writing—original draft preparation, O.L. and M.D.; writing—review and editing B.M., A.K. and Ľ.Š.; visualization, O.L. and M.D.; supervision, M.D. and Ľ.Š.; project administration, Ľ.Š.; funding acquisition O.L., B.M., A.K. and Ľ.Š. All authors have read and agreed to the published version of the manuscript.

Funding: This study was supported by the National Scholarship Programme of the Slovak Republic (fellowship to OL), the National Academy of Sciences of Ukraine project 0117U000385 (granted to BM), Slovak Research and Development Agency project APVV-18-0302 (awarded to AK), and the European Regional Development Fund through Operational Programme Research and Development project 26240220096 (awarded to ĽŠ).

Acknowledgments: The authors are grateful to Raphael R. Wood from the University of South Alabama for improving the language quality of the manuscript.

Abbreviations

HMW	high molecular weight
LMW	low molecular weight
UHPLC	ultrahigh-performance liquid chromatography
EQB	equilibration buffer
FASP	filter-aided sample preparation
NA	not available
PTM	post-translational modification
LC-MS/MS	liquid chromatography-tandem mass spectrometry
2-DE	two-dimensional gel electrophoresis

References

1. Kuźniar, A.; Włodarczyk, K.; Grządziel, J.; Goraj, W.; Gałązka, A.; Wolińska, A. Culture-independent analysis of an endophytic core microbiome in two species of wheat: *Triticum aestivum* L. (cv. 'Hondia') and the first report of microbiota in *Triticum spelta* L. (cv. 'Rokosz'). *Syst. Appl. Microbiol.* **2020**, *43*, 126025. [CrossRef] [PubMed]

2. Shewry, P.R.; Halford, N.G.; Lafiandra, D. Genetics of Wheat Gluten Proteins. In *Advances in Genetics*; Hall, J.C., Dunlap, J.C., Friedmann, T., Eds.; Elsevier: Amsterdam, The Netherlands; ISBN 00652660.

3. Payne, P.I. Genetics of wheat storage proteins and the effect of allelic variation on bread-making quality. *Annu. Rev. Plant. Physiol.* **1987**, *38*, 141–153. [CrossRef]

4. Rasheed, A.; Xia, X.; Yan, Y.; Appels, R.; Mahmood, T.; He, Z. Wheat seed storage proteins: Advances in molecular genetics, diversity and breeding applications. *J. Cereal Sci.* **2014**, *60*, 11–24. [CrossRef]

5. Lee, J.-Y.; Beom, H.-R.; Altenbach, S.B.; Lim, S.-H.; Kim, Y.-T.; Kang, C.-S.; Yoon, U.-H.; Gupta, R.; Kim, S.-T.; Ahn, S.-N.; et al. Comprehensive identification of LMW-GS genes and their protein products in a common wheat variety. *Funct. Integr. Genom.* **2016**, *16*, 269–279. [CrossRef] [PubMed]

6. Juhász, A.; Békés, F.; Wrigley, C.W. Wheat Proteins. In *Applied Food Protein Chemistry*; Ustunol, Z., Ed.; Wiley Blackwell: Hoboken, NJ, USA, 2014; Volume 303, p. 219. ISBN 9781118860588.

7. Gil-Humanes, J.; Pistón, F.; Rosell, C.M.; Barro, F. Significant down-regulation of γ-gliadins has minor effect on gluten and starch properties of bread wheat. *J. Cereal Sci.* **2012**, *56*, 161–170. [CrossRef]

8. Shewry, P.R.; Tatham, A.S. Improving wheat to remove coeliac epitopes but retain functionality. *J. Cereal Sci.* **2016**, *67*, 12–21. [CrossRef]

9. Wang, D.-W.; Li, D.; Wang, J.; Zhao, Y.; Wang, Z.; Yue, G.; Liu, X.; Qin, H.; Zhang, K.; Dong, L.; et al. Genome-wide analysis of complex wheat gliadins, the dominant carriers of celiac disease epitopes. *Sci. Rep.* **2017**, *7*, 44609. [CrossRef]

10. Barak, S.; Mudgil, D.; Khatkar, B.S. Relationship of gliadin and glutenin proteins with dough rheology, flour pasting and bread making performance of wheat varieties. *LWT Food Sci. Technol.* **2013**, *51*, 211–217. [CrossRef]

11. Peña, E.; Bernardo, A.; Soler, C.; Jouve, N. Relationship between common wheat (*Triticum aestivum* L.) gluten proteins and dough rheological properties: Gluten proteins and rheological properties in wheat. *Euphytica* **2005**, *143*, 169–177. [CrossRef]

12. Hadjivassiliou, M.; Sanders, D.S.; Grünewald, R.A.; Woodroofe, N.; Boscolo, S.; Aeschlimann, D. Gluten sensitivity: From gut to brain. *Lancet Neurol.* **2010**, *9*, 318–330. [CrossRef]

13. Sapone, A.; Bai, J.C.; Ciacci, C.; Dolinsek, J.; Green, P.H.R.; Hadjivassiliou, M.; Kaukinen, K.; Rostami, K.; Sanders, D.S.; Schumann, M.; et al. Spectrum of gluten-related disorders: Consensus on new nomenclature and classification. *BMC Med.* **2012**, *10*, 13. [CrossRef] [PubMed]

14. Scherf, K.A.; Koehler, P.; Wieser, H. Gluten and wheat sensitivities—An overview. *J. Cereal Sci.* **2016**, *67*, 2–11. [CrossRef]

15. Camarca, A.; Anderson, R.P.; Mamone, G.; Fierro, O.; Facchiano, A.; Costantini, S.; Zanzi, D.; Sidney, J.; Auricchio, S.; Sette, A.; et al. Intestinal T cell responses to gluten peptides are largely heterogeneous: Implications for a peptide-based therapy in celiac disease. *J. Immunol.* **2009**, *182*, 4158–4166. [CrossRef] [PubMed]

16. Mamone, G.; Picariello, G.; Addeo, F.; Ferranti, P. Proteomic analysis in allergy and intolerance to wheat products. *Expert Rev. Proteom.* **2011**, *8*, 95–115. [CrossRef] [PubMed]

17. Huebener, S.; Tanaka, C.K.; Uhde, M.; Zone, J.J.; Vensel, W.H.; Kasarda, D.D.; Beams, L.; Briani, C.; Green, P.H.R.; Altenbach, S.B.; et al. Specific nongluten proteins of wheat are novel target antigens in celiac disease humoral response. *J. Proteome Res.* **2015**, *14*, 503–511. [CrossRef] [PubMed]

18. Junker, Y.; Zeissig, S.; Kim, S.-J.; Barisani, D.; Wieser, H.; Leffler, D.A.; Zevallos, V.; Libermann, T.A.; Dillon, S.; Freitag, T.L.; et al. Wheat amylase trypsin inhibitors drive intestinal inflammation via activation of toll-like receptor 4. *J. Exp. Med.* **2012**, *209*, 2395–2408. [CrossRef] [PubMed]

19. Altenbach, S.B.; Tanaka, C.K.; Allen, P.V. Quantitative proteomic analysis of wheat grain proteins reveals differential effects of silencing of omega-5 gliadin genes in transgenic lines. *J. Cereal Sci.* **2014**, *59*, 118–125. [CrossRef]

20. Pistón, F.; Gil-Humanes, J.; Rodríguez-Quijano, M.; Barro, F. Down-regulating γ-gliadins in bread wheat leads to non-specific increases in other gluten proteins and has no major effect on dough gluten strength. *PLoS ONE* **2011**, *6*, e24754. [CrossRef] [PubMed]

21. Spaenij-Dekking, L.; Kooy-Winkelaar, Y.; Van Veelen, P.; Drijfhout, J.W.; Jonker, H.; Van Soest, L.; Smulders, M.J.M.; Bosch, D.; Gilissen, L.J.W.J.; Koning, F. Natural variation in toxicity of wheat: Potential for selection of nontoxic varieties for celiac disease patients. *Gastroenterology* **2005**, *129*, 797–806. [CrossRef] [PubMed]

22. Ribeiro, M.; Rodriguez-Quijano, M.; Nunes, F.M.; Carrillo, J.M.; Branlard, G.; Igrejas, G. New insights into wheat toxicity: Breeding did not seem to contribute to a prevalence of potential celiac disease's immunostimulatory epitopes. *Food Chem.* **2016**, *213*, 8–18. [CrossRef]

23. Malalgoda, M.; Meinhardt, S.W.; Simsek, S. Detection and quantitation of immunogenic epitopes related to celiac disease in historical and modern hard red spring wheat cultivars. *Food Chem.* **2018**, *264*, 101–107. [CrossRef] [PubMed]

24. Shewry, P.R. Do ancient types of wheat have health benefits compared with modern bread wheat? *J. Cereal Sci.* **2018**, *79*, 469–476. [CrossRef] [PubMed]

25. Pompa, M.; Giuliani, M.M.; Palermo, C.; Agriesti, F.; Centonze, D.; Flagella, Z. Comparative analysis of gluten proteins in three durum wheat cultivars by a proteomic approach. *J. Agric. Food Chem.* **2013**, *61*, 2606–2617. [CrossRef]

26. Yahata, E.; Maruyama-Funatsuki, W.; Nishio, Z.; Tabiki, T.; Takata, K.; Yamamoto, Y.; Tanida, M.; Saruyama, H. Wheat cultivar-specific proteins in grain revealed by 2-DE and their application to cultivar identification of flour. *Proteomics* **2005**, *5*, 3942–3953. [CrossRef] [PubMed]

27. García-Molina, M.D.; Muccilli, V.; Saletti, R.; Foti, S.; Masci, S.; Barro, F. Comparative proteomic analysis of two transgenic low-gliadin wheat lines and non-transgenic wheat control. *J. Proteom.* **2017**, *165*, 102–112. [CrossRef]

28. Vensel, W.H.; Dupont, F.M.; Sloane, S.; Altenbach, S.B. Effect of cleavage enzyme, search algorithm and decoy database on mass spectrometric identification of wheat gluten proteins. *Phytochemistry* **2011**, *72*, 1154–1161. [CrossRef]

29. Martínez-Esteso, M.J.; Nørgaard, J.; Brohée, M.; Haraszi, R.; Maquet, A.; O'Connor, G. Defining the wheat gluten peptide fingerprint via a discovery and targeted proteomics approach. *J. Proteom.* **2016**, *147*, 156–168. [CrossRef]

30. Scherf, K.A.; Poms, R.E. Recent developments in analytical methods for tracing gluten. *J. Cereal Sci.* **2016**, *67*, 112–122. [CrossRef]

31. Nakamura, R.; Teshima, R. Proteomics-based allergen analysis in plants. *J. Proteom.* **2013**, *93*, 40–49. [CrossRef]

32. Lollier, V.; Denery-Papini, S.; Brossard, C.; Tessier, D. Meta-analysis of IgE-binding allergen epitopes. *Clin. Immunol.* **2014**, *153*, 31–39. [CrossRef]

33. Yokooji, T.; Kurihara, S.; Murakami, T.; Chinuki, Y.; Takahashi, H.; Morita, E.; Harada, S.; Ishii, K.; Hiragun, M.; Hide, M.; et al. Characterization of causative allergens for wheat-dependent exercise-induced anaphylaxis sensitized with hydrolyzed wheat proteins in facial soap. *Allergol. Int.* **2013**, *62*, 435–445. [CrossRef] [PubMed]

34. Reig-Otero, Y.; Mañes, J.; Manyes, L. Amylase-trypsin inhibitors in wheat and other cereals as potential activators of the effects of nonceliac gluten sensitivity. *J. Med. Food* **2018**, *21*, 207–214. [CrossRef] [PubMed]

35. Mameri, H.; Denery-Papini, S.; Pietri, M.; Tranquet, O.; Larré, C.; Drouet, M.; Paty, E.; Jonathan, A.-M.; Beaudouin, E.; Moneret-Vautrin, D.-A.; et al. Molecular and immunological characterization of wheat Serpin (Tri a 33). *Mol. Nutr. Food Res.* **2012**, *56*, 1874–1883. [CrossRef] [PubMed]

36. Ma, F.; Li, M.; Yu, L.; Li, Y.; Liu, Y.; Li, T.; Liu, W.; Wang, H.; Zheng, Q.; Li, K.; et al. Transformation of common wheat (*Triticum aestivum* L.) with avenin-like b gene improves flour mixing properties. *Mol. Breed.* **2013**, *32*, 853–865. [CrossRef]

37. Barrachina, M.N.; Sueiro, A.M.; Casas, V.; Izquierdo, I.; Hermida-Nogueira, L.; Guitián, E.; Casanueva, F.F.; Abián, J.; Carrascal, M.; Pardo, M.; et al. A combination of proteomic approaches identifies a panel of circulating extracellular vesicle proteins related to the risk of suffering cardiovascular disease in obese patients. *Proteomics* **2019**, *19*, 1800248. [CrossRef]

38. Ribeiro, M.; Nunes-Miranda, J.D.; Branlard, G.; Carrillo, J.M.; Rodriguez-Quijano, M.; Igrejas, G. One hundred years of grain omics: Identifying the glutens that feed the world. *J. Proteome Res.* **2013**, *12*, 4702–4716. [CrossRef]

39. Rombouts, I.; Lagrain, B.; Brunnbauer, M.; Delcour, J.A.; Koehler, P. Improved identification of wheat gluten proteins through alkylation of cysteine residues and peptide-based mass spectrometry. *Sci. Rep.* **2013**, *3*, 2279. [CrossRef]

40. Uvackova, L.; Skultety, L.; Bekesova, S.; McClain, S.; Hajduch, M. The MSE-proteomic analysis of gliadins and glutenins in wheat grain identifies and quantifies proteins associated with celiac disease and baker's asthma. *J. Proteom.* **2013**, *93*, 65–73. [CrossRef]

41. Bromilow, S.; Gethings, L.A.; Langridge, J.I.; Shewry, P.R.; Buckley, M.; Bromley, M.J.; Mills, E.N.C. Comprehensive proteomic profiling of wheat gluten using a combination of data-independent and data-dependent acquisition. *Front. Plant Sci.* **2017**, *7*, 2020. [CrossRef]

42. Ma, C.-Y.; Gao, L.-Y.; Li, N.; Li, X.-H.; Ma, W.-J.; Appels, R.; Yan, Y.-M. Proteomic analysis of albumins and globulins from wheat variety Chinese spring and its fine deletion line 3BS-8. *Int. J. Mol. Sci.* **2012**, *13*, 13398–13413. [CrossRef]

43. Gell, G.; Kovács, K.; Veres, G.; Korponay-Szabó, I.R.; Juhász, A. Characterization of globulin storage proteins of a low prolamin cereal species in relation to celiac disease. *Sci. Rep.* **2017**, *7*, 39876. [CrossRef] [PubMed]

44. Cho, K.; Beom, H.-R.; Jang, Y.-R.; Altenbach, S.B.; Vensel, W.H.; Simon-Buss, A.; Lim, S.-H.; Kim, M.G.; Lee, J.-Y. Proteomic profiling and epitope analysis of the complex α-, γ-, and ω-gliadin families in a commercial bread wheat. *Front. Plant Sci.* **2018**, *9*, 818. [CrossRef] [PubMed]

45. Akagawa, M.; Handoyo, T.; Ishii, T.; Kumazawa, S.; Morita, N.; Suyama, K. Proteomic analysis of wheat flour allergens. *J. Agric. Food Chem.* **2007**, *55*, 6863–6870. [CrossRef]

46. Mari, A.; Rasi, C.; Palazzo, P.; Scala, E. Allergen databases: Current status and perspectives. *Curr. Allergy Asthma Rep.* **2009**, *9*, 376–383. [CrossRef] [PubMed]

47. Juhász, A.; Haraszi, R.; Maulis, C. ProPepper: A curated database for identification and analysis of peptide and immune-responsive epitope composition of cereal grain protein families. *Database* **2015**, *2015*, bav100. [CrossRef]

48. Bromilow, S.; Gethings, L.A.; Buckley, M.; Bromley, M.; Shewry, P.R.; Langridge, J.I.; Mills, E.N.C. A curated gluten protein sequence database to support development of proteomics methods for determination of gluten in gluten-free foods. *J. Proteom.* **2017**, *163*, 67–75. [CrossRef]

49. van Den Broeck, H.C.; Cordewener, J.H.G.; Nessen, M.A.; America, A.H.P.; van der Meer, I.M. Label free targeted detection and quantification of celiac disease immunogenic epitopes by mass spectrometry. *J. Chromatogr. A* **2015**, *1391*, 60–71. [CrossRef]

50. Malalgoda, M.; Ohm, J.-B.; Meinhardt, S.; Simsek, S. Association between gluten protein composition and breadmaking quality characteristics in historical and modern spring wheat. *Cereal Chem.* **2018**, *95*, 226–238. [CrossRef]

51. Newton, A.C.; Akar, T.; Baresel, J.P.; Bebeli, P.J.; Bettencourt, E.; Bladenopoulos, K.V.; Czembor, J.H.; Fasoula, D.A.; Katsiotis, A.; Koutis, K.; et al. Cereal landraces for sustainable agriculture. A review. *Agron. Sustain. Dev.* **2010**, *30*, 237–269. [CrossRef]

52. Gregová, E.; Hermuth, J.; Kraic, J.; Dotlačil, L. Protein heterogeneity in European wheat landraces and obsolete cultivars. *Genet. Resour. Crop Evol.* **1999**, *46*, 521–528. [CrossRef]

53. Lupi, R.; Masci, S.; Rogniaux, H.; Tranquet, O.; Brossard, C.; Lafiandra, D.; Moneret-Vautrin, D.A.; Denery-Papini, S.; Larré, C. Assessment of the allergenicity of soluble fractions from GM and commercial genotypes of wheats. *J. Cereal Sci.* **2014**, *60*, 179–186. [CrossRef]

54. Sánchez-León, S.; Gil-Humanes, J.; Ozuna, C.V.; Giménez, M.J.; Sousa, C.; Voytas, D.F.; Barro, F. Low-gluten, nontransgenic wheat engineered with CRISPR/Cas9. *Plant Biotechnol. J.* **2018**, *16*, 902–910. [CrossRef]

55. Рибалка, О.; Моргун, Б.; Починок, В. Сучасні дослідження якості зерна пшениці у світі: Генетика, біотехнологія та харчова цінність запасних білків. *физиология и биохимия культурных растений* **2012**, *44*, 3–22.

56. Metakovsky, E.; Melnik, V.; Rodriguez-Quijano, M.; Upelniek, V.; Carrillo, J.M. A catalog of gliadin alleles: Polymorphism of 20th-century common wheat germplasm. *Crop J.* **2018**, *6*, 628–641. [CrossRef]

57. Dupont, F.M.; Vensel, W.H.; Tanaka, C.K.; Hurkman, W.J.; Altenbach, S.B. Deciphering the complexities of the wheat flour proteome using quantitative two-dimensional electrophoresis, three proteases and tandem mass spectrometry. *Proteome Sci.* **2011**, *9*, 101. [CrossRef]
58. Distler, U.; Kuharev, J.; Navarro, P.; Tenzer, S. Label-free quantification in ion mobility-enhanced data-independent acquisition proteomics. *Nat. Protoc.* **2016**, *11*, 795–812. [CrossRef]
59. Nováková, S.; Šubr, Z.; Kováč, A.; Fialová, I.; Beke, G.; Danchenko, M. *Cucumber mosaic virus* resistance: Comparative proteomics of contrasting *Cucumis sativus* cultivars after long-term infection. *J. Proteom.* **2020**, *214*, 103626. [CrossRef]
60. Perez-Riverol, Y.; Csordas, A.; Bai, J.; Bernal-Llinares, M.; Hewapathirana, S.; Kundu, D.J.; Inuganti, A.; Griss, J.; Mayer, G.; Eisenacher, M.; et al. The PRIDE database and related tools and resources in 2019: Improving support for quantification data. *Nucleic Acids Res.* **2019**, *47*, D442–D450. [CrossRef]

International Journal of
Molecular Sciences

Article

Identification and Characterization of Glycoproteins and Their Responsive Patterns upon Ethylene Stimulation in the Rubber Latex

Li Yu [1,2,†], Boxuan Yuan [1,2,†], Lingling Wang [2], Yong Sun [3], Guohua Ding [2], Ousmane Ahmat Souleymane [2], Xueyan Zhang [2], Quanliang Xie [1,*] and Xuchu Wang [1,2,*]

[1] College of Life Sciences, Shihezi University, Shihezi 832003, China; yulixjnu@163.com (L.Y.); yuanboxuan111@163.com (B.Y.)
[2] Key Laboratory for Ecology of Tropical Islands, Ministry of Education, College of Life Sciences, Hainan Normal University, Haikou 571158, China; wll_198927@126.com (L.W.); dingguohuasw@163.com (G.D.); ousmaneben062@gmail.com (O.A.S.); zhangxueyan_caas@126.com (X.Z.)
[3] Rubber Research Institute, Chinese Academy of Tropical Agricultural Sciences, Danzhou 571737, China; sunyong_03119308@126.com
[*] Correspondence: xiequanliang001@163.com (Q.X.); xchwang@hainnu.edu.cn (X.W.); Tel.: +86-898-65891065 (X.W.)
[†] These authors contributed equally to this work.

Received: 10 July 2020; Accepted: 23 July 2020; Published: 25 July 2020

Abstract: Natural rubber is an important industrial material, which is obtained from the only commercially cultivated rubber tree, *Hevea brasiliensis*. In rubber latex production, ethylene has been extensively used as a stimulant. Recent research showed that post-translational modifications (PTMs) of latex proteins, such as phosphorylation, glycosylation and ubiquitination, are crucial in natural rubber biosynthesis. In this study, comparative proteomics was performed to identify the glycosylated proteins in rubber latex treated with ethylene for different days. Combined with Pro-Q Glycoprotein gel staining and mass spectrometry techniques, we provided the first visual profiling of glycoproteomics of rubber latex and finally identified 144 glycosylated protein species, including 65 differentially accumulated proteins (DAPs) after treating with ethylene for three and/or five days. Gene Ontology (GO) functional annotation showed that these ethylene-responsive glycoproteins are mainly involved in cell parts, membrane components and metabolism. Pathway analysis demonstrated that these glycosylated rubber latex proteins are mainly involved in carbohydrate metabolism, energy metabolism, degradation function and cellular processes in rubber latex metabolism. Protein–protein interaction analysis revealed that these DAPs are mainly centered on acetyl-CoA acetyltransferase and hydroxymethylglutaryl-CoA synthase (HMGS) in the mevalonate pathway for natural rubber biosynthesis. In our glycoproteomics, three protein isoforms of HMGS2 were identified from rubber latex, and only one HMGS2 isoform was sharply increased in rubber latex by ethylene treatment for five days. Furthermore, the *HbHMGS2* gene was over-expressed in a model rubber-producing grass *Taraxacum Kok-saghyz* and rubber content in the roots of transgenic rubber grass was significantly increased over that in the wild type plant, indicating HMGS2 is the key component for natural rubber production.

Keywords: comparative proteomics; *Hevea brasiliensis*; Hydroxymethylglutaryl-CoA synthase; glycosylated proteins; natural rubber biosynthesis; rubber latex

---

1. Introduction

Natural rubber (NR), as a kind of high-molecular polymer with unique properties, is produced from laticifer cells in the bark of the rubber tree (*Hevea brasiliensis*), which is the only tree species that is widely planted as a commercially available source of NR used for many products, varying from tires to medical products [1]. In NR production, rubber latex is usually obtained by regular tapping of the trunk bark with a two- or three-day interval with the application of ethylene as a stimulant [2]. As one of the most important phytohormones, ethylene has been widely used to accelerate the processes of plant growth and development, including the production of NR [3]. When applying or spraying the ethylene generator, known as ethephon or ethrel (chloro-2-ethyl phosphonic acid), on the trunk bark of the tapped rubber tree, both the fresh latex production and latex regeneration between different tapping events can be significantly increased [4]. The ^{14}C-labeled experiment demonstrated that ethylene was released from the treated bark to the upper leaf shortly after supplying with ethephon. This stimulation effect is associated with marked changes in both physiology and metabolism in laticifer cells [5,6]. Our previous results also demonstrated that ethylene stimulation can sharply improve the yield of fresh latex and dry matter, prolonging the latex flow time [7], as well as markedly accelerating the generation of small rubber particles (SRPs) in rubber latex [8].

However, it is still a mystery that, although the ethephon-stimulated rubber production has been widely used for several decades in commercial NR production, many genes involved in NR biosynthesis (NRB) were found to be significantly inhibited after ethylene application [4,9]. Our recently published results also proved that the expression levels of the 27 selected genes were inhibited upon ethylene treatment. The other 25 transcripts either did not change or changed less than the level of their encoded proteins. Among the down-regulated ones, the genes encoding several enzymes known to be the key factors in rubber biosynthesis have been determined [5]. More recently published results also revealed that proteins might be the key regulators for NRB [10–14], and post-translational modifications (PTMs) of different protein isoforms may play crucial roles in NRB upon ethylene treatment [7,15,16], especially in small rubber particles [8]. There are more than three hundred kinds of reported PTMs in proteins [17]. Among them, phosphorylation, glycosylation and ubiquitination are considered as the most three important kinds of PTMs in controlling the final function of enzymes involved in NRB [7,8,14]. Our proteomics results have demonstrated that isoform-specific phosphorylation of proteins in rubber particles (RPs) is important for ethylene-stimulated latex production [7,8].

Almost all proteins can produce different kinds of PTMs, and these PTMs are important for responding to the external environmental changes in different plant cells [18,19]. As one of the most important PTMs, protein glycosylation is an essential co- and post-translational modification in secretory and membrane proteins in many eukaryotes, and it is widely observed in many plant species [20]. Recently, the roles of protein glycosylation have received more considerable attention by researchers for their regulation mechanism of the cell biology field in eukaryotic cells [19,20]. In many plants, it was reported that protein glycosylation can control tissue and organ formation, signal recognition, protein activity promotion, protein structure and protein stability [18–20]. Protein-glycosylated modifications result from adding sugar chains to the amino acid residues of proteins and maintain the folding and stability of proteins to perform their biological functions [21]. Two different kinds of sugar chains, named O- and N-sugar chains, are usually examined in glycosylated proteins. Among them, O-glycosylation is a common PTM in many proteins, and these O-sugar chains have different roles in development, cell differentiation, pathogenesis and proteolysis in plant cells [18]. On the other side, N-sugar chains can mainly affect the folding and structure of proteins, and thus are traditionally found to regulate the subcellular localization and protein secretion in different plant cells [21,22]. It was reported that N-sugar chains play regulatory roles in both the plant–pathogen interaction [23] and receptor recognition processes to mediate plant immunity [24].

With the development of proteomics, different kinds of technologies have been used to identify, enrich and separate the glycoproteins. Among them, concanavalin A lectin affinity chromatography has been widely used to separate glycoproteins. With mass spectrometry technology, Catala and coworkers determined many glycoproteins involved in the secretion pathway in ripe tomato fruit [25]. Meanwhile, this technique had also been used to enrich the glycoproteins in the xylem juice of cabbage, and the results demonstrated that most of the enriched proteins are involved in regulating the secretion process [26]. Besides, proteomics analyses also revealed that glycosylated proteins might play important roles in plant defense against fungal invasion [27] and the course of transforming enzymes in white wine as sweet molecular renaissance [28]. Another classic method for detecting glycoproteins is Pro-Q Emerald 488 glycoprotein stain, which can provide direct detection of glycoproteins in gels and on blots rapidly and sensitively. It reacts with periodate-oxidized carbohydrate groups, creating a bright green fluorescent signal on glycoproteins [29]. This stain has been widely used in detecting glycoproteins from human idiopathic pulmonary fibrosis [30], human total serum and mouse liver [31] and skeletal muscle [32]. These above results indicate that the proteomic technique is of great significance for the identification of glycoproteins and the study of plant glycoproteins in pathological regulation and protein catalytic mechanisms.

In this study, we used gel-based experiments to determine the glycoproteins in the total proteins from the collected rubber latex after ethylene stimulation and identified more than one hundred glycoproteins in the stained gel spots with mass spectrometry. To our best knowledge, this is the first proteomics report on ethylene-stimulated glycosylation proteins, which may provide new insights for revealing the regulation mechanism of NRB from the aspects of glycosylation proteins in the rubber latex.

2. Results

2.1. Profiles of the Glycosylated Proteins from Rubber Latex in One-Dimensional Gel

Pro-Q Emerald 488 staining is a sensitive method to detect glycoproteins in both gels and blots. In this study, total proteins from rubber latex treated with double-distilled water (ddH$_2$O) and ethylene for three days (D3, E3) and five days (D5, E5) were respectively extracted to perform one-dimensional electrophoresis (1-DE) analysis and the glycosylated proteins were then stained by using Pro-Q Emerald 488 stain to a red color. More than thirty red bands, ranging from 14 to 180 kDa, were detected in the lines of each 1-DE gel (Figure 1), indicating that most proteins were glycosylated in the rubber latex. Although several differential bands were observed, the main red bands showed similar change patterns after different time length treatments. We excised the most abundant bands from the 1-DE gel and then performed in-gel digestion, and finally identified five proteins by matrix-assisted laser desorption ionization tandem time-of-flight (MALDI TOF/TOF) mass spectrometry (MS). The five protein bands were identified as β-1,3-glucanse (bands 1 and 2), small rubber particle protein (SRPP, band 3) and rubber elongation factor (REF, bands 4 and 5). Among them, the most abundant bands were REF, which were respectively approximately 14 and 23 kDa.

The changed patterns of the protein bands in the 1-DE gels for protein samples from different water and ethylene treatments were compared, and the protein abundance of the bands was found to vary with different treated time points. That means more obviously changed bands can be detected from the latex samples collected from different days. Several main bands with high molecular weights (approximately 60 kDa) could only be detected in the samples treated for three days (D3 and E3) but disappeared after the five-day treatment. On the other hand, several abundant bands with approximately 40 kDa could only be observed in the five-day treatment (D5 and E5) latex samples. These results indicated that the changes in protein abundance between different time points were more obvious than that from different ethylene treatments. Therefore, we divided three-day and five-day treatments as the two groups and paid more attention to the differential proteins that were collected from the same time point with different treatments in the following study.

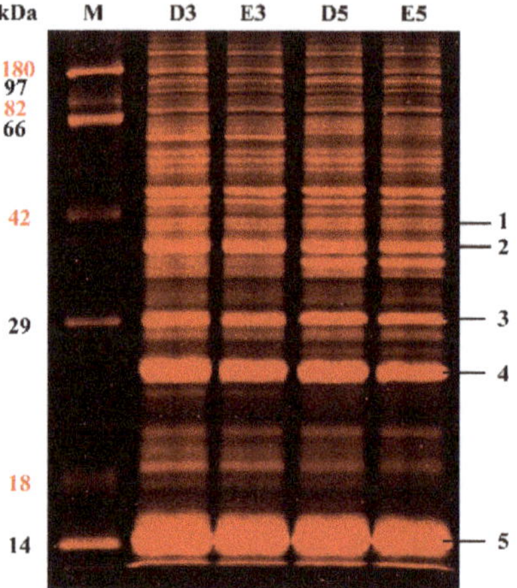

Figure 1. Profiles of glycosylated proteins from the rubber latex in 1-DE gel. Total proteins were extracted from the rubber latex after treating with water and ethylene for three (D3; E3) and five (D5; E5) days, and then the glycosylated protein bands were visualized by Pro-Q Emerald 488 stain. The main bands in the gels were exposed to MALDI TO/TOF MS and five main protein bands were positively identified from these bands. M, protein markers.

2.2. Profiles of Glycosylated Proteins Abundance in the Reference 2-DE Proteome Map for the Rubber Latex

Our results showed that many protein bands in the 1-DE gel (Figure 1) and protein spots in the 2-DE gels (Figure 2) could be dyed to a red color, indicating that these proteins are likely to be glycosylated in total rubber latex. The protein spots with an abundance value (Vol%) > 0.02 in the 2-DE gels of different treatments were selected and used to perform MS, and finally, 144 protein spots were successfully identified from the 2-DE gels for latex proteins from different treatments (Figure 2). These 144 proteins are encoded by 98 genes (Table S1 and Figure S1). The highly abundant proteins identified in this study are REF (spots 34, 35, 63 and 82), SRPP (spot 20), anhydrase (spots 10), protease (spot 74), triosephosphate isomerase (spot 61), enolase (spot51), cis-trans isomerase (spot 106), chitinase (spot 9) and a REF/SRPP-like protein (spot 39). Among them, the most abundant protein spot is REF (spot 35), followed by SRPP (spot 82) and anhydrase (spot 74).

The Kyoto Encyclopedia of Genes and Genomes (KEGG) pathway and Gene Ontology (GO) function analyses were performed to determine the potential metabolism processes and biological functions for the identified 98 unique glycoproteins. KEGG results exhibited that these proteins are involved in 32 global and overview maps. Among them, 20 enzymes are referred to a carbohydrate metabolism pathway, including aldehyde dehydrogenase, phosphoglycerate kinase, enolase, acetyl-CoA acetyltransferase (ACAT), hydroxymethylglutaryl-CoA synthase (HMGS), fructokinase, phospholipase C3, glucosidase and chitinase. Twelve other proteins, including ACAT, HMGS, REF, glutamine synthetase, adenosylhomocysteinase, S-adenosylmethionine synthase, cysteine synthase, malate dehydrogenase, aldehyde dehydrogenase and proline iminopeptidase isoform X1, are involved in amino acid metabolism. Another 12 proteins are annotated to be implicated in an energy metabolism pathway. Moreover, nine members related to folding, sorting and degradation function processes: they are heat shock cognate protein (HSP80, spot 108, only one), enolase members (spots 25 and 51, two) and proteasome subunits or their homologs (six). Additionally, five proteins, named V-type proton ATPase subunit B2 isoform X1 (spot 18), V-type proton ATPase catalytic subunit A (spot 132), tubulin alpha-3 (spots 37, 64 and 137), superoxide dismutase (spot 131) and an early-responsive to dehydration protein (spot 30), were annotated to be involved in the cellular processes, which are important members in transport and catabolism pathways. GO annotation results (biological process term) demonstrated that most proteins are involved in the single-organism process (32 proteins), cellular process (31 proteins), response to stimulus (30 proteins), developmental process (21 proteins), multicellular organismal process (19 proteins), cellular component organization or biogenesis (15 proteins) and metabolic process (Table S1).

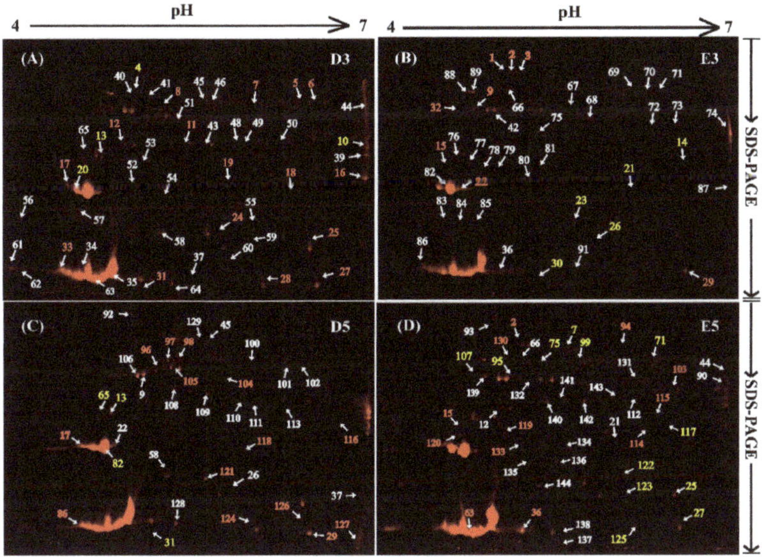

Figure 2. Reference proteome profiles of the glycosylated proteins in 2-DE gels and MS identification of differentially accumulated glycoproteins after ethylene treatments. The glycosylated proteins in total rubber latex from the rubber tree treated with water (**A,C**) and ethylene (**B,D**) for three (D3, E3) and five days (D5, E5) were dyed with Pro-Q Emerald 488 in the 2-DE gels and the protein spots were visualized by a typhoon scanner. The main protein spots were incised, and 144 protein spots were positively identified by MS. The ethylene-induced protein spots are marked with red numbers, and reduced spots are yellow numbers. The white numbers stand for the unchanged protein spots upon stimulation. These typical 2-DE gels were selected from the three biological replicates. Both protein accumulation levels and the fold change patterns are the average value from the three replicates. The glycosylated protein identities are presented in Table S1 and Figure S1.

2.3. Determinnation of Isoforms for the Glycosylated Proteins in 2-DE Gels

It is noteworthy that many spots in the 2-DE gels were identified as the product from the same gene, and these glycoproteins are termed as different protein isoforms or protein species for 2-DE gel-based proteomics. In this study, we found that 23 proteins were identified from 65 protein spots, and at least two isoforms were observed for these proteins (Table S1). Among them, REF contains 12 isoforms and these isoforms are the products from two REF genes. Seven spots (spots 14, 55, 83, 91, 125, 128 and 138) were identified as pro-hevein, suggesting that at least seven proteins are found to be the different isoforms of hevein in 2-DE gels. Four spots were identified as two kinds of SRPP, and three spots (spots 101, 111 and 116) were HMGS isoforms.

Bubble distribution diagram analysis showed that most of these protein isoforms have a pH value varying from 5.0 to 6.0, and their molecular weight ranges from 20 to 60 kDa (Figure 3A). A total of 65 protein spots were identified as protein isoforms, and these isoforms were classified into 23 bubbles, suggesting that these isoforms were the products of 23 genes in the rubber tree genome. Among these bubbles, each of the 13 green bubbles contains two isoforms: they were identified as CAMT, EIF4 (eukaryotic initiation factor 4A-14), SRPP, TPI (triosephosphate isomerase), enolase 1, adenosylmethionine synthase, protease, proteasome, etc. The seven blue bubbles, each of which contains three protein isoforms, in the diagram were known as ACAT (acetyl-CoA acetyltransferase), EF2, HMGS, LCAP (leucine aminopeptidase-1), non-specific phospholipase C3, metacaspase-4 and tubulin alpha-3 chain. Two kinds of REF family members were exhibited as the yellow bubbles, with six isoforms for each. The largest purple bubble is pro-hevein, identified from seven different protein spots in the 2-DE gels. We also noticed that these protein isoforms had different abundance values (Vol%) in the same gel and presented disparate changing patterns after treating with water and ethylene for different days (Figure 3B).

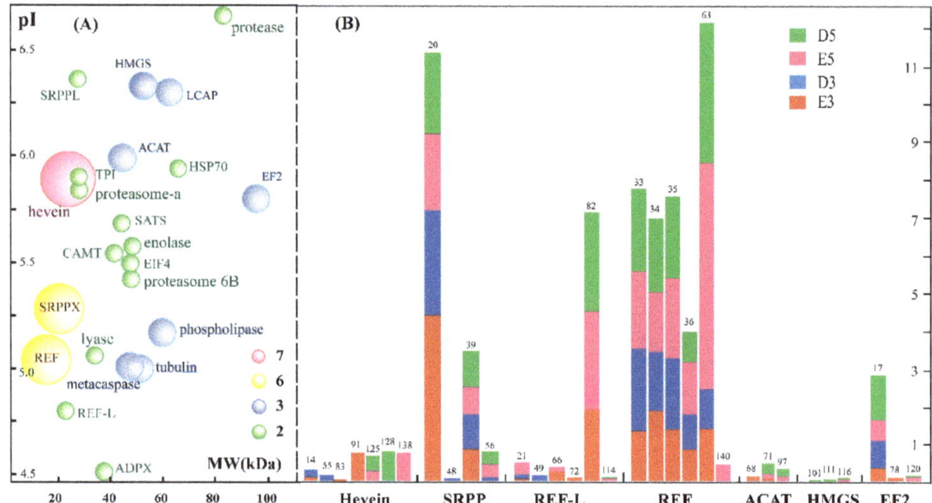

Figure 3. Bubble diagram of all the protein isoforms and comparison of the changed abundance of 2-DE gels upon ethylene application. The glycosylated proteins with at least two isoforms are highlighted as 23 bubbles from 65 protein spots. The bubbles with different sizes and colors stand for different number of protein isoforms identified from different spots (**A**). The accumulation patterns of 7 glycoprotein proteins containing 32 isoforms in the 2DE gels for total rubber latex from the rubber tree treated with water and ethylene for three (D3, E3) and five days (D5, E5) are provided (**B**). The number in the top of the histogram stands for the protein spot number in the 2-DE gels.

2.4. Characterization and Identification of Differentially Accumulated Glycoproteins in the Rubber Latex upon Ethylene Stimulation

Based on the abundance of each protein spot in the 2-DE gels, the changed patterns of the identified 144 protein spots were determined, and the abundance of the 65 protein spots presented more than 1.5-fold changes after supplying with exogenous ethylene for three (E3) and/or five (E5) days (Table S2). The ratios of different treatments demonstrated that half of the differentially accumulated glycoproteins (DAGPs) (33 proteins) were sharply increased with the time elongation of the ethylene (E5/E3) or water (D5/D3) treatment. In this study, our interest is focused on the identification of the ethylene-responsive glycoproteins (ERGPs). After three days of ethylene treatment, the abundance of 12 proteins was increased, while 21 proteins decreased. When elongating the treatment time, more ERGPs were identified, including 14 up-regulated and 27 down-regulated glycoproteins (Figure 4A).

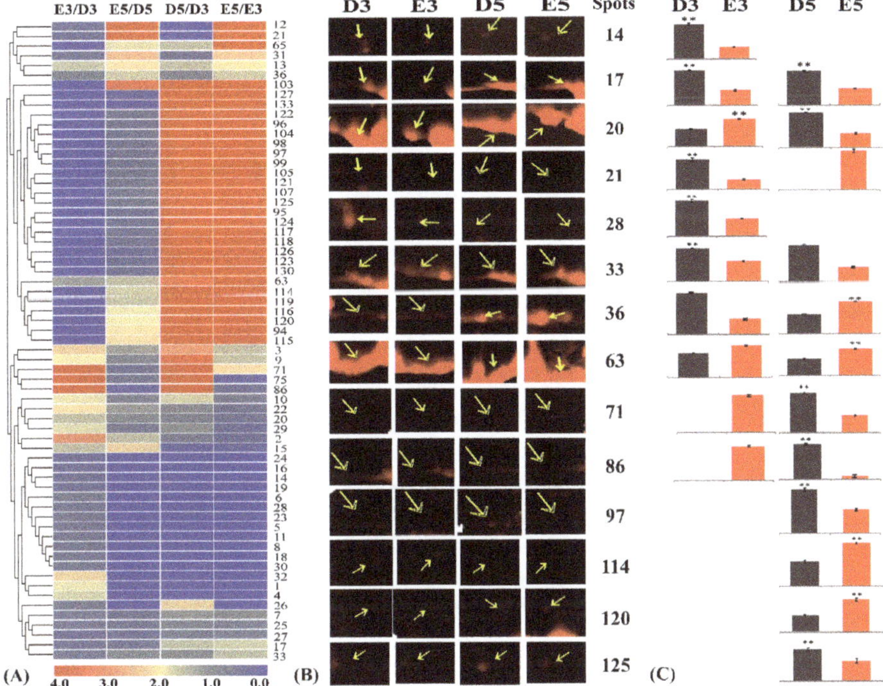

Figure 4. Changing patterns of differentially accumulated glycoproteins (DAGPs) after application of ethylene. The 65 glycosylated proteins with at least 1.5-fold changes were exposed to hierarchical clustering (**A**) to show their changed accumulation patterns after being treated with ethylene or water for three (E3/D3) or five (E5/D5) days. The changed ratios of these proteins after treating with water or ethylene for different times (D5/D3; E5/E3) are also presented. The up- or down-regulated proteins are indicated in red or green, respectively. The intensity of the colors increased with the increased accumulation, as shown in the bar at the bottom. The spot numbers are indicated in the heat plot, which includes five based areas from the top to bottom. The accumulation profiles (**B**) of 14 typical DAGPs in the 2-DE gels and their relatively changed ratios (**C**) are highlighted. The exact position of each protein spot was correspondingly presented in the four 2-DE gel areas.

Among these ERGPs, adenosylhomocysteinase (spot 2), universal stress protein A-like protein (spot 13) and eukaryotic initiation factor 4A-14 (spot 15) were up-regulated at both two time points. Meanwhile, four down-regulated proteins were detected with ethylene stimulation: they are S-adenosylmethionine synthase 1 (spot 7), elongation factor 2 (spot 17), enolase 2 (spot 25) and PITH domain-containing protein (spot 27). Besides, nucleoredoxin 1 (spot 29) was found to be up-regulated at E3, but down-regulated at E5, while another protein (spot 31) presented an opposite trend. Furthermore, eight proteins, including three REFs and one SRPP, were detected at only one time point. We also noticed that a large portion of identified numbers (48) proteins could only be detected in the gel at three or five days, and all of them were either an ethylene-induced protein spot (IPS) or a disappeared protein spot (DPS) (Table S2). We also selected 14 typical DAGPs and highlighted their detail positions in the 2-DE gels (Figure 4B), displaying their relatively changed ratios (Figure 4C) after different treatments. All these results indicated that different protein spots presented different changed patterns in the four treatments, and many protein isoforms were induced or disappeared with the time elongation of the ethylene treatment.

2.5. Functinal Analysis of Differential Glycoproteins in the Rubber Latex upon Ethylene Stimulation

In order to determine related functions of these DAGPs, we used the AgBase software to perform gene ontology (GO) analysis. These proteins were divided into three GO categories based on biological process, cellular component and molecular function (Figure 5A). These GO categories contain 31 subgroups. Among them, 18 subgroups were included in the biological process, among which 18 proteins are annotated to be involved in the single-organism process (single-OP), followed by cellular process (17 proteins) and response to stimulus with 14 proteins. Furthermore, proteins involved in the developmental process (12 proteins), multicellular organismal process (11 proteins) and cellular component organization or biogenesis (9 proteins) were determined. For cellular component terms, 23 proteins were annotated to be related to the cell part components, which is consistent with the fact that rubber latex is a kind of specific cytoplasm of laticifers. Another 19 proteins were annotated to be the component of the plasma membrane, followed by 14 proteins referring to the organelle. In the molecular function category, 12 proteins showed catalytic activity and 10 proteins possessed strong binding ability (Table S2).

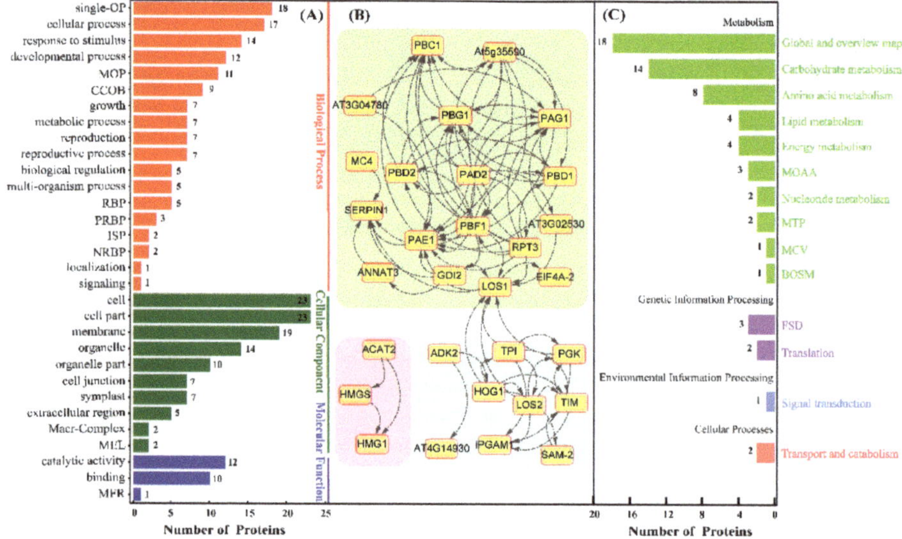

Figure 5. Functional analysis of DAGPs upon ethylene stimulation. The identified 65 DAGPs were exposed to Gene Ontology (GO) (**A**), protein–protein interaction (**B**) and The Kyoto Encyclopedia of Genes and Genomes (KEGG) (**C**) analyses to determine their biological pathways in rubber latex. The abbreviations for these pathways and their detail information are provided in the end section and Table S2.

Before submitting the identified 65 glycosylated proteins to the STRING database to obtain their protein–protein interaction networks, the rubber tree-related protein amino acid sequences were used as queries to obtain their orthologous protein sequences of the *Arabidopsis thaliana* protein sequence, with a high confidence value (0.7) in this study. Excluding the disconnected nodes of the network, 30 proteins showed strong interaction relationships with other proteins. According to the associated degrees between different proteins, the network of the 30 proteins was divided into three sub-clusters (Figure 5B). Cluster 1 is the largest interconnection network containing 18 proteins, including LOS1, PAD2, PAG1, PBD2, PBF1 and PBG1, among others. Among them, LOS1, as a ribosomal protein S5/elongation factor G which catalyzes the GTP-dependent ribosomal translocation step during translation elongation, was simultaneously identified from spots 17 and 120 (Table S2).

PBF1 was identified as a proteasome subunit from spot 19, and it is a member of the N-terminal nucleophile amino hydrolase superfamily. Cluster 2 includes nine proteins. Among them, LOS2, HOG1 and TIM are the three central nodes. LOS2, identified as the enolase 2 in the rubber latex, is a multifunctional enzyme that acts as a metabolism-related enolase and positively regulates the transcription of cold-responsive genes. HOG1 is known as an adenosylhomocysteinase (spot 2), and it was up-regulated after ethylene treatment at both of the time points in the rubber latex. HOG1 is a competitive inhibitor of S- adenosyl-L-methionine-dependent methyl transferase, which may play a key role in controlling the methylation process by regulating the intracellular concentration of adenosylhomocysteine. Besides, TIM is a chloroplastic triosephosphate isomerase, and it presents a narrower expression range limited to roots. Cluster 3 contains three unique proteins: they are ACAT2, HMGS and HMG1. They are three important members involved in the melavonate (MVA) pathway of NR synthesis. Among them, HMGS condenses acetyl-CoA with acetoacetyl-CoA to form HMG-CoA, which is the substrate of HMG-CoA reductase. HMG1 encodes a 3-hydroxy-3-methylglutaryl coenzyme A reductase, which is involved in melavonate biosynthesis and performs the first committed step in isoprenoid biosynthesis (Table S2).

Furthermore, to understand the high-level functions and utilities of the biological system from molecular-level information, the 65 DAGPs were annotated using the online software KOBAS and Blast KOALA (https://www.kegg.jp/blastkoala), and these proteins were found to be involved in 10 main metabolism pathways (Figure 5C). Among them, 18 proteins are involved in global and overview maps, and 14 proteins are known as the main members of the carbohydrate metabolic process. There are also many members involved in amino acid metabolism (eight proteins), lipid metabolism (four proteins) and energy metabolism (four proteins). At the same time, several proteins were observed to take part in genetic information processing, environmental information processing and transport and catabolism (Table S2).

2.6. Functional Verification of the HbHMGS2 Gene

Based on the preliminary mass spectrometry identification results and functional analysis, we noticed that HMGS2 (ref|XP_021666594.1) in the MVA pathway in the NRB pathway is significantly glycosylated, and the changed patterns in the 2-DE gels after ethylene and water treatments for three days (E3, D3) and five days (E5, D5) were highlighted (Figures 2 and 6A). Statistical results showed that HMGS2 could hardly be detected in the 2-DE gels of the rubber latex treated for three days, however, three protein spots (spots 101, 111 and 116) had been identified as three HMGS protein species in the 2-DE gels of the latex samples for five days (D5 and E5). It is noteworthy that only spot 116 was sharply increased in the rubber latex by ethylene treatment for five days (E5), and the other two spots could only be observed in the D5 latex (Table S1). Our results demonstrated that the abundance for spot 116 was increased 1.9-fold after ethylene treatment for five days (E5) in the 2-DE gel (Table S2). These results indicated that some protein species of HMGS might be the important players for ethylene stimulation of NR production. Therefore, in order to determine the detail functions of HMGS2, we obtained its gene accession number in the rubber tree genome (ref|XP_021666594.1) and over-expressed the gene *HbHMGS2* in a model rubber-producing grass *Taraxacum Kok-saghyz* (TKS) (Figure 6B), and compared their phenotypes in the transgenic TKS plants growing for one (1M), three (3M) and six (6M) months. The results demonstrated that these transgenic grasses and the wild type plants are similar to each other (Figure 6C). However, the main root in the transgenic grasses is bigger than the wild type, and photosynthetic pigment contents, including chlorophyll b and total chlorophyll, are significantly increased over that in the wild type plants (Figure 6D). Furthermore, the NR content in the 6M roots of transgenic plants and wild type was measured. The results demonstrated that the NR content was sharply enhanced and reached about a 3-fold increase in the transgenic TKS roots over the wild type (Figure 6E).

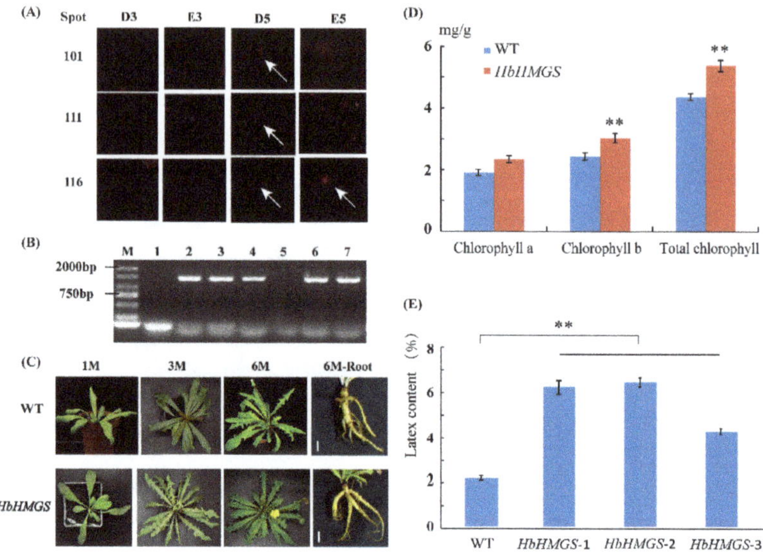

Figure 6. Functional analysis of *HbHMGS2* in the rubber grass *Taraxacum Kok-saghyz*. The protein spots identified as HMGS2 in the 2-DE gels are highlighted, the white arrows indicate the detail location of the corresponding spots that identified as HMGS2 (**A**). The target gene in different transgenic rubber grass lines was determined (**B**) and phenotypes of the transgenic and wild type (WT) plants grown for 1 (1M), 3 (3M) and 6 (6M) months in a greenhouse are presented (**C**). The photosynthetic pigment content in the 6M-old leaves of the transgenic rubber grass and WT plants was determined, and the results proved that the over-expressing of *HbHMGS* can significantly improve the chlorophyll content in the leaves of rubber grass (**D**). The natural rubber content in the 6M roots of wild type and transgenic plants was determined from three independent lines (**E**). Double asterisks indicate the significant difference among different samples.

Based on the analysis of the above proteomics data, a schematic diagram of the rubber latex glycosylation proteins involved in NRB was put forward (Figure 7). In the positively identified 98 unique glycosylated proteins, eight kinds of proteins were found to be related to NRB and/or NR metabolism. They are ACAT (spots 68, 71 and 97), HMGS, REF, SRPP, 14-3-3 protein, pro-hevein, chitinase and beta-glucosidase (Table S1). Among them, ACAT, HMGS, REF and SRPP are four important components in the MVA pathway for NR biosynthesis (Figure 7).

It is noteworthy that ACAT was identified from three independent spots named S68, S71 and S97, and only spot 68 was induced by ethylene in D3 latex, while the other two spots showed a decreased abundance after ethylene stimulation. Several isoforms of REF and SRPP had been determined as glycosylated proteins. Seven protein species from different spots (S17, S21, S33, S36, S63, S114 and S120) were identified as REF. Among them, four REF species (S36, S63, S114 and S120) were increased, but two members (S21 and S33) were decreased after ethylene stimulation for three and/or five days. For SRPP, only one spot (S20) was sharply induced by ethylene for three and five days, the other two spots (S48 and S86) changed not significantly upon ethylene application. Except for chitinase (spot S9), the other three kinds of proteins (including 14-3-3 protein, pro-hevein and beta-glucosidase), which are known to take part in the regulation of NR production, showed a decreased accumulation pattern upon ethylene stimulation (Figure 7). These results indicated that glycosylated proteins might have an important influence on the regulation of NR biosynthesis.

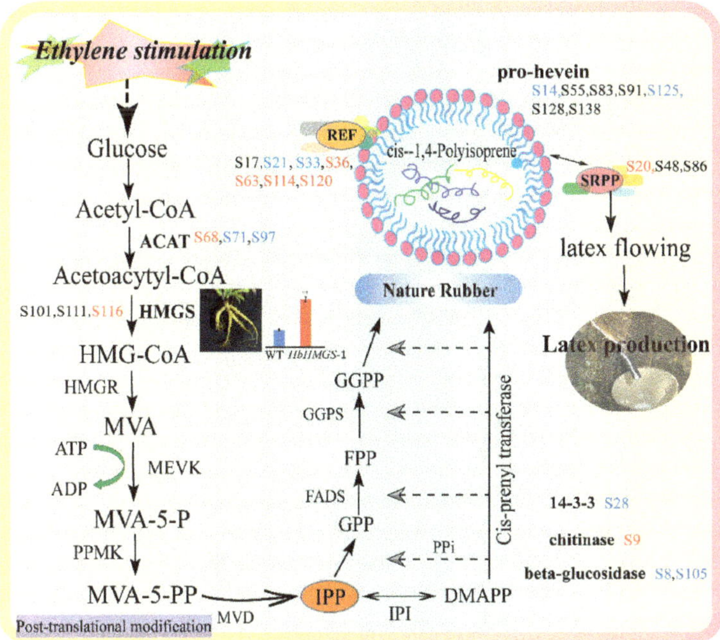

Figure 7. Schematic diagram of the melavonate (MVA) pathway regulated by natural rubber biosynthesis. The identified eight glycosylation proteins in the natural rubber (NR) biosynthesis pathway are highlighted. Among them, seven proteins were identified from different spots (S-) in the 2-DE gels for rubber latex collected from the water- and ethylene-treated rubber trees on the third and fifth day. The red number refers to the up-regulated protein spots, and the blue number stands for the down-regulated protein spots after ethylene stimulation (Table S2).

3. Discussion

3.1. The First Visual Profiling of Glycoproteomics Based on 2-DE Gel Determined Lots of Proteins Involved in Carbohydrate Metabolism in Rubber Latex

Ethylene has been widely applied in agriculture and horticulture as a plant hormone. Ethylene stimulation is considered of great importance for rubber exploitation systems since its effects are favorable to the maintenance of high production levels, and this technology was tested as early as the 1970s, and now it has been a widely used technology in NR production [33]. Surprisingly, the expression levels of several known genes related to rubber biosynthesis did not significantly increase after ethylene stimulation [7], such as farnesyl diphosphate synthase (FADS) [34], cis-isoprene transferase (CPT) [35] and hydroxymethylglutaryl coenzyme A reductase (HMGR) [36]. These results proved that many NRB-related genes may not be induced or even be inhibited by exogenous ethylene stimulation. Therefore, several researchers suggested that ethylene has little direct effect on NRB and the increased latex yield produced by ethylene stimulation might be attributed to the prolongation of latex flow [4,37]. However, at the protein level, some new evidence supports the accelerative effect of ethylene stimulation on NR production [38,39]. Our recently published proteomics results revealed that a large amount of ethylene-responsive latex proteins (ERLPs) are determined in total rubber latex [7] and small rubber particles [8], and many isoforms of NRB-related proteins are sharply induced upon ethylene stimulation. These results indicated that the regulation of rubber latex production by ethylene stimulation might occur not solely at the gene level but also at the protein level, and post-translational modifications (PTMs) of proteins might play crucial roles in controlling the final function of enzymes involved in rubber biosynthesis [7,8]. Among these PTMs, phosphorylation of REF and SRPP isoforms might be crucial for NRB, and small rubber particles may act as a complex natural rubber biosynthetic machine in rubber latex [8].

Protein glycosylation, which is similar to protein phosphorylation [7], is also one important PTM in plants. Glycosylated proteins have been reported to be involved in regulating plant growth and development at the cell level. Arabinogalactan proteins, which are widespread in the plant kingdom, are a class of hydroxyproline-rich glycoproteins that can act on different development aspects of plants, including cell division and expansion, leaf development and reproduction [40]. Hundreds of glycosylated proteins have been identified from the mature stems trapped on Concanavalin A [41] and etiolated hypocotyls [42] in model plant *Arabidopsis thaliana* by 2-DE gel-based glycoproteomics methods, and respectively identified 102 and 126 glycosylated proteins. Among them, many glycosylated proteins are localized in the plant cell wall [41,42]. Meanwhile, cell wall glycoproteomics has been performed to explain how the cell wall peroxidases in response to the changed external environment conditions in the ascorbate-deficient mutant *A. thaliana*, and reveals that the cell wall glycoprotein group is affected by the lack of ascorbic acid, which finally affects the expression of cell wall proteins involved in the pathogen response [43].

It is known that rubber latex contains many macromolecular substances, such as proteins, sugars and alkaloids, and the NRB is a complex process [7,44]. Many proteins involved in the NRB process have been identified by proteomics-based technologies from different latex components [2,7,8,10–13,15,16,45]. These proteomics results revealed that rubber latex has a high protein content, and different kinds of PTMs have been determined in rubber latex to efficiently regulate NRB [7,8]. Among them, glycosylation is one of the most important kinds of protein PTMs, which plays crucial roles in plant growth and development processes, as well as NRB and metabolism. Although some pioneering research on latex glycoproteins have done, a systematic study on the latex glycoproteins upon ethylene is still limited. In this study, the first visual profiling of glycoproteomics was performed by a combination of 2-DE and mass spectrometry technologies to identify the glycosylated proteins in rubber latex after ethylene treatment for three and five days. Finally, 144 species were identified as glycosylated proteins from rubber latex and they were encoded by 98 genes in the rubber tree genome. Our proteomics results demonstrated that these glycosylated rubber latex proteins are mainly involved in carbohydrate metabolism, energy metabolism, degradation function and cellular processes, and might be important for rubber metabolism (Table S1).

3.2. Glycoproteomics of Rubber Latex Revealed Many Ethylene-Responsive Glycoproteins are Important for Nature Rubber Biosynthesis

Among the identified 98 gene products, 65 protein species were found to be ERGPs. KEGG pathway analysis revealed that 14 of these ERGPs are involved in carbohydrate metabolism: they are lactoylglutathione lyase, phosphoglycerate mutase, phosphoglycerate kinase, enolase, phosphate uridylyltransferase, fructokinase, aldehyde dehydrogenase, phospholipase C3 and glutamine synthetase. Lactoylglutathione lyase, also known as glyoxalase, is an enzyme that catalyzes the isomerization of hemithioacetal adducts. It was identified from spots 98 and 126 (Table S2) of rubber latex in this glycoproteomics study, and this enzyme was also detected from the 2-DE gel of milky sap from *Chelidonium majus* [45] and total latex of the rubber tree [7]. Its expression in rubber latex serum was obviously changed when triggered by ethephon treatment to maintain and regulate the plant redox homeostasis process and defense system in the rubber tree [46].

Phosphoglycerate mutase is known to exert a certain physiological impact in plant cells. It was identified as a metabolic protein from lettuce latex sap [47], rubber latex [13] and the washed solutions from rubber particles [48]. In this proteomics study, we found it was glycosylated (spot 3) in the 2-DE gels (Figure 2) and increased by 2.3-fold after ethylene stimulation for three days (Table S2). Phosphoglycerate kinase, which catalyzes the formation of ATP from ADP and 1,3-diphosphoglycerate for cell metabolism, was determined as an ERGP from spot 12 (Figure 2) in this proteomics study.

It was also identified from the latex of *C. majus* in different phases of plant development [45], lettuce latex [47], rubber tree latex [13,49], different washed solutions from rubber particles [8,48] and different rubber latices after ethylene application, in which, phosphoglycerate mutase was determined as a phosphorylation protein [7]. These recently published results indicate that phosphoglycerate mutase and phosphoglycerate kinase are two important components in ethylene stimulation of rubber latex metabolism.

Enolase, typically localized in the cytosol, is widely known as one of the glycolytic enzymes and catalyzes the conversion of 2-phosphoglycerate and phosphoenolpyruvate in glycolysis. This enzyme is ubiquitous in rubber-producing plants and many living organisms [7], and it has also been identified by several proteomics researches from the latex of rubber tree as a new allergen named Hev b 9 [50–52], an ethylene-responsive protein [7,15,46], or a high-abundance protein in rubber particles [13,48]. It was identified by mass spectrometry from the latex of *Opium poppy* [49] and the milky sap of *C. majus* [45]. Phosphate uridylyltransferase was determined as an ERGP in rubber latex in this study (Table S2), and it was also identified from the washed solutions of rubber particles [53] and lettuce latex in a proteomics study [47].

Fructokinase is a kinase that catalyzes the transfer of phosphate groups to fructose and can phosphorylate the sugar at the C-1 position, and thus is considered as one of the rate-limiting enzymes of glycolysis [54]. Recent rubber latex proteomics data revealed that fructokinase can be induced by ethylene treatment [55], but in this study, we found this enzyme can be glycosylated and its abundance was decreased after ethylene stimulation for five days (Table S2). Aldehyde dehydrogenase is a kind of oxidizing enzyme that is involved in detoxification of both exogenous and endogenous aldehyde substrates through $NAD(P)^+$-dependent oxidation. In a proteomics study, it was identified from *Colletotrichum gloeosporioides*, which is a hemibiotrophic fungi that could cause anthracnose in *Hevea brasiliensis* [56].

Phospholipase C is an enzyme that hydrolyzes plasma membrane phospholipids at the third position of the glycerol backbone. Its gene expression was up-regulated in the latex of a tapping panel dryness [57] and a high-yielding rubber tree [54]. Proteomics analyses revealed that this protein is highly abundant in the washed solutions of rubber particles [48] and was significantly induced in rubber latex after ethylene application [55]. Another enzyme, glutamine synthetase, plays an important role in metabolizing nitrogen by catalyzing the reaction of condensation of glutamate and ammonia to form glutamine, and the glutamine–glutamate synthase cycle might be the major pathway for the amino acid and protein synthesis required for natural latex regeneration [4]. Previous proteomics studies demonstrated that glutamine synthetase is a high-abundance protein in the latex of lettuce [58], *O. poppy* [49], the washed solutions from rubber particles [48] and rubber latex serum [59]. The activity and mRNA level of glutamine synthetase are observed to be increased after ethylene stimulation in latex cells [60]. Comparative proteomics analyses proved that accumulation of this enzyme is significantly up-regulated in rubber latex upon ethylene treatment [7,15]. These published results and our glycoproteomics of rubber latex in this study revealed that ethylene-responsive glycoproteins are important for NR production.

3.3. Glycosylation of Different Isforms of NRB-Ralated Proteins Mgiht Play Different Roles and HbHMGS2 Gene Is Crucial for Nature Rubber Production

Several proteomics studies have been conducted and more NRB-related enzymes were identified by mass spectrometry from natural rubber latex and rubber particles. These NRB-related proteins are AACT, HMGS, MEVK, FADS, GGPS, CPT, REF, SRPP, etc., and most of them were identified as membrane-attached proteins from purified rubber particles [7,8,10–13,15,16]. A pioneering proteomics study of rubber particles identified 186 proteins, including REF, SRPP and cis-prenyl transferase [11]. Comparative proteomics of large and small rubber particles revealed 22 gene products, including SRPP, REF, HMGS, HSP70 and phospholipase D, are differentially accumulated [10]. NRB, beginning with isopentenyl pyrophosphate (IPP) synthesis in the MVA pathway, is a typical isoprenoid metabolic process. In the early steps, acetyl-CoA C-acetyltransferase (ACAT) is important for generating acetoacytyl-CoA, where it catalyzes acetyl-CoA to form acetoacetyl-CoA, which is the first step in the MVA pathway [61,62]. Then, HMGS and HMGR activate the supply of mevalonate substrates [7,62]. In this proteomics study, eight kinds of NRB-related proteins in the MVA pathway were positively identified from rubber latex, and some isoforms of these glycosylation proteins were detected to change significantly after ethylene stimulation for three and/or five days (Table S2). Among them, three ACAT isoforms were detected from spots 68, 71 and 97, respectively (Figure 7). These protein isoforms may generate by alternative splicing and different kinds of post-translational modifications [55]. Three ACAT genes were determined in the rubber tree genome [63]. Our previous proteomics results of total rubber latex revealed that both the gene and protein accumulation of ACAT are depressed upon ethylene stimulation, but one protein isoform is induced [7]. In another proteomics study on small rubber particles, one isoform was identified as ACAT from SRPs, and its abundance was induced upon ethylene treatment [8], which is consistent with our previously published latex proteomics results [7] and the new observation in this study (Figure 7).

During the rubber elongation process, REF and SRPP are two key members [62], which are closely combined with the membrane of rubber particles [43]. SRPP, a rubber biosynthesis-related protein that is expressed mainly in small rubber particles [64], has been reported to play more important roles than REF for NBR [10,65]. Comparative localization results of SRPP and REF reveal that the rubber biosynthesis capability in rubber laticifer is mostly concentrated in small rubber particles by SRPP [61,66]. SRPP can recruit CPT to the endoplasmic reticulum to interact with CTP1-REF bridging protein [67]. Interaction network analysis of rubber particles demonstrates that CTP1, REF and CTP1-REF bridging protein (HRBP) may play crucial roles in NRB [12], and they are associated with the endoplasmic reticulum [67]. In a previously published proteomics study, we have detected 18 REF/SRPP gene family members in the rubber tree, and their genes exhibit distinct expression patterns in different tissues in a single 205-kb genome site [16]. In 2-DE gels, we also have determined 28 protein isoforms from five REF/SRPP members, and multiple protein isoforms have been identified [16]. Our proteomics data revealed that the gene expression and protein accumulation for most of these REF/SRPP members are decreased or not changed after ethylene treatment. However, individual isoform members display different changed patterns, and some family members or protein isoforms are sharply increased by ethylene stimulation in rubber latex [7,16] and small rubber particles [8]. Among them, 14 REF isoforms are significantly changed after ethylene treatments, including 10 sharply induced ones. For SRPP, five protein isoforms are increased upon ethylene stimulation [7]. However, in this glycoproteomics of rubber latex, three glycosylated SRPP isoforms were respectively identified from spots 20, 48 and 86 in the 2-DE gels, and only one member (spot 20) was detected as the induced protein species after ethylene treatment for three days (Figure 7). Compared with SRPP, many more REF isoforms were detected to be the induced glycosylation proteins. Among the identified seven REF isoforms, four members (spots 36, 63, 114 and 120) were determined as ethylene-induced proteins (Table S2). These results demonstrated more REF isoforms have been glycosylated in rubber latex and glycosylation modification of REF might play more important roles than SRPP after ethylene stimulation for different times.

Beta-glucosidase is a key enzyme component present in the cellulase enzyme complex and completes the final step during cellulose hydrolysis, which is essential for complete hydrolysis of cellulose into glucose. This reaction is always under control as it gets inhibited by its product glucose, thus is a major bottleneck in the efficient biomass conversion by cellulase. Proteomics analysis of the vegetative vacuole from *A. thaliana* results in the identification of three identified beta-glucosidases containing endoplasmic reticulum retention signals at their respective C termini, which is responsible for the vacuolar localization of target proteins under certain conditions [68]. In this proteomics study, we noticed two isoforms of this enzyme were also detected as glycosylated proteins, and their accumulation was significantly reduced after ethylene application (Figure 7). It is known as a major protein from proteomics analyses of the latex from *H. brasiliensis* [13,48], *C. majus* [45] and *O. poppy* [49], and it is an effect factor on the inhibition of lectin-like protein-induced rubber particle aggregation [69].

Chitinase and hevein, as well as glucosidase, are key activators of lutoid-mediated rubber particle aggregation and sequential latex coagulation in rubber latex [2,65]. Hevein, or its precursor pro-hevein, is highly abundant in lutoids protein [70]. Both hevein and pro-hevein showed a strong chitin binding ability [65]. Immuno-blotting analysis of hevein revealed both mature hevein and a 20-kDa protein recognized by an N domain-specific antibody [70]. Our comparative proteomics of primary and secondary lutoids has revealed that the decreased accumulation of glucosidase and hevein in ethylene-treated latex may inhibit rubber particle aggregation, thus maintaining the latex flow and resulting in an enhanced rubber latex yield [2]. In this proteomics study, we further proved that several isoforms of pro-hevein and beta-glucosidase can be glycosylated and much of their abundance is sharply decreased after ethylene application (Table S2), thus helping to inhibit rubber particle aggregation in the rubber latex production process.

In an early step of the MVA pathway, HMGS activates the supply of mevalonate substrates and demonstrates a positive response to IPP substrate. Therefore, it is considered as a key enzyme for NRB in the rubber tree [7,39,62]. In the recently published rubber tree genome, two gene family members, named *HMGS1* and *HMGS2*, are characterized [63]. It was reported that HMGS gene expression and enzyme activity are significantly enhanced upon the addition of ethylene [39], and the *HbHMGS1* promoter is found to play important roles in regulating ethylene-mediated gene expression [71]. Over-expression of *BjHMGS1* can up-regulate the genes in sterol biosynthesis and enhance the total sterol production and stress tolerance ability in Arabidopsis [72]. At the enzyme level, the rubber biosynthetic pathway is coordinately regulated by both the activities of HbHMGS and HbHMGR [71], and the isoenzyme derived from the HbHMGS1 transcript is likely involved in the biosynthesis of rubber in laticiferous cells [39]. Therefore, HMGS may play a significant role in controlling isoprenoid and rubber biosynthesis in the rubber tree.

However, our recently published proteomics data showed a different changed pattern in that the gene expression and protein accumulation levels of HMGS are down-regulated or not significantly improved upon treatment with exogenous ethylene [7]. Proteomics analysis of rubber particles revealed that HMGS is expressed more predominantly in small rubber particles than large rubber particles [10], which is consistent with the fact that small rubber particles have higher activity of rubber biosynthesis than the large ones. In this glycoproteomics study, three protein isoforms (spots 101, 111 and 116) of HMGS2 (ref|XP_021666594.1) were identified from the 2-DE gels, but only one member (spot 116) was significantly increased after ethylene treatment for five days (Table S2). Our transgenic results proved that the natural rubber content in the mature roots of the over-expressed *HMGS2* gene rubber-producing grass TKS is sharply improved (Figure 6), and these new data showed that the gene *HbHMGS2* is really a key member for NBR in both the *Hevea* rubber tree and rubber grass TKS. The combination of the recently published results with our new proteomics data demonstrated that glycosylation modification of some protein isoforms is important for NRB and HMGS2 might positively correlate with the natural rubber product in the rubber tree and rubber grass.

In conclusion, in this study, comparative proteomics of rubber latex treated with ethylene for three and five days was performed, and finally resulted in the identification of 65 differentially accumulated proteins from the 144 glycosylated proteins. Both GO functional annotation and KEGG pathway analysis results demonstrated that these DAGPs are mainly involved in the functions and pathways related to the biosynthesis of rubber latex, including cell parts, and membrane components. Our results also indicated that glycosylation of HMGS2 protein might play important roles in the biosynthesis of rubber latex. The gene function verification results showed that the latex content of *HbHMGS2* transgenic plants was significantly higher than that of wild type plants, indicating that the *HbHMGS2* gene is a key member in the regulation of the ethylene-stimulated NBR by post-translational glycosylation of some HMGS2 isoforms. This glycoproteomics study of rubber latex may give some new insights on the regulation mechanism of rubber latex biosynthesis under ethylene stimulation and provide more theoretical support for the further usage of ethylene as a stimulant in natural rubber production.

4. Materials and Methods

4.1. Plant Material and Ethylene Treatments

Total latex protein samples were obtained from 90 newly tapped mature rubber plants (8-year-old *H. Brasiliensis* Mull. Arg., clone RY 7-33-97) which were grown at an experimental farm of the Chinese Academy of Tropical Agricultural Sciences in Danzhou City, Hainan Province, China. The rubber trees never treated with ethylene were selected and randomly divided into four groups. The tree cuts were respectively treated with ethephon (3%, *v/v*) and double-distilled water (ddH$_2$O, the control) as described [7], and the latex samples were respectively collected from each rubber tree at two time points: three days after treatment (D3, E3) and the fifth (D5, E5) day after ethylene or ddH$_2$O treatments. For each treatment, the mixed latex samples were obtained from the corresponding five rubber trees. The latex droplets were collected in an iced glass beaker, then immediately frozen in liquid nitrogen and stored for further use at −80 °C.

4.2. Protein Extraction and Electrophoresis

The extraction of the total latex protein for the rubber was performed using the BPP method as described previously [7,8]. Then, the protein concentration of the total latex was determined following the Bradford method by a spectrophotometer (Shimadzu UV-160, Kyoto, Japan) and bovine serum albumin (BSA) was used as a standard. An amount of 20 μL of protein lysate containing 30 μg of total protein for each sample was loaded with 10 μL of 3 × loading buffer for the one-dimensional discontinuous SDS-PAGE electrophoresis, and the electrophoresis conditions were set as: 16 °C, 6 W, for 60 min, and then 16 °C, 8 W, for 4 h. The Image Scanner was used for scanning after Coomassie blue staining, then the image was analyzed with the Image Master 2D Platinum software (GE Healthcare, Uppsala, Sweden).

In the two-dimensional electrophoresis system, 455 μL sample containing about 1000 μg protein was loaded. The first isoelectric focusing was performed using an IEF-100 focusing instrument from Hoefer, a linear IPG gel strip (pH 4–7, 24 cm in length). Then, the gel was equilibrated with an equilibration solution containing 1% DTT (50 mmol/L Tris-HCl, PH 8.8, 6 mol/L urea, 30% glycerol, 2% SDS, 0.02% bromophenol blue) and 4% iodoacetamide liquid for 15 min. Following this, the remaining equilibration buffer was washed with ddH$_2$O, and second-direction vertical plate gel electrophoresis was performed in an Etan Daltsix electrophoresis apparatus (GE Healthcare), with the program set to: 1.5 W/gel, 1 h; 8 W/gel, for 6 h, and three biological replicates were conducted for each sample. Finally, the gels were stained with Coomassie blue and the results were observed as mentioned above.

4.3. Glycoproteomics Analysis

Glycosylated proteome staining was carried out with Pro-Q® Glycoprotein Blot Stain Kit (Molecular Probes, Eugene, WI, USA). Firstly, 500 mL stationary liquid (50% methanol, 5% acetic acid) was used to fix the gel with soft shaking at room temperature overnight; then, 500 mL solution containing 3% glacial acetic acid was used as the rinsing solution, and was gently shaken for 20 min twice. Then, we gently shook 500 mL oxidizing solution (periodic acid in 3% acetic acid) to immerse and oxidize for 1 h, followed by rinsing with 500 mL rinsing solution for 15 min, repeated three times. After staining with 250 mL of Pro-Q Emerald 300 for 2.5 h and 15 min of rinse (repeated once for gel imaging), the result could be observed with a 300 nm UV transmission.

4.4. Protein Identification via Mass Spectrometry

After the target protein spots were cut out, the in-gel digestion was performed as described [8]. After enzymatic digestion, these peptides were collected and identified by the AB 5800 MALDI-TOF/TOF mass spectrometry (MS) instrument (AB SCIEX, Foster City, CA, USA). The ProteinPilot Software (Version 4.5) and a Mascot Algorithm (version 2.3) were used to search against the *Hevea* genome scaffolds (BioProject ID: PRJNA80191, www.ncbi.nlm.nih.gov/nuccore/448814761) and the draft genome (GenBank: AJJZ01000000) with 46,718 sequences and 17,435,757 residues [63]. The parameters were set as: precursor tolerance 300 ppm, 0.3 Da tolerance as MS/MS fragment, trypsin as the enzyme, carbamidomethylation as the fixed modification (C) and oxidation (M) as the variable modification. Detailed information on all identified protein spot mass searches can be found in Supplemental Figure S1 and Table S1.

4.5. GO and KEGG Annotations for the Identified Glycosylated Proteins

In order to clarify the specific functions of these identified glycosylated proteins, we performed Gene Ontology (GO) function and Kyoto Encyclopedia of Genes and Genomes (KEGG) pathway annotation for the identified glycosylated proteins. The related protein sequences were submitted to AgBase (Version 2.0) (https://agbase.arizona.edu) for the GO function annotation, and the annotated results were visualized by Omicshare analysis (https://www.omicshare.com). Meanwhile, the KEGG pathway was annotated with the KOBAS 3.0 database (http://kobas.cbi.pku.edu.cn/kobas3). The detailed information is shown in Supplementary Table S2.

4.6. Protein-Protein Interaction (PPI) Analysis

In order to further identify the correlation between these proteins, a protein–protein interaction analysis was performed with the web tool STRING 11.0 (http://stringdb.org). Before submitting the data to the STRING database, the amino acid sequences of the identified proteins were aligned with the *Arabidopsis thaliana* database to obtain the homologous protein sequences. Finally, the PPI network was visualized by the Cytoscape 3.7.2 software. All the details can be found in Supplementary Table S2.

4.7. Phenotypic Analysis of the Transgenic Plants and Determination of Photosynthetic Pigment and Latex Content

In this study, the rubber grass *Taraxacum Kok-saghyz* (TKS) was selected as the model plant as a transgenic receptor material for further function analysis of the target protein HMGS as described [73]. The phenotypes of the wild type rubber grass and the *HbHMGS* over-expressed transgenic plants at the age of 1 month, 3 months and 6 months were compared. The contents of photosynthetic pigment and latex content were measured when the plant was 6 months old. The photosynthetic pigment content was measured with the method of 90% acetone extraction as described [74], and then infrared spectroscopy was used to measure the content of the natural rubber in the 6M roots of the rubber grasses as described [75].

4.8. Statistical Analysis

The data of the protein expression levels and the fold changes in the 2-DE gels are the average of three biological replicates. Student's *t*-test and one-way analysis of variance (ANOVA) followed by Tukey's multiple comparison test ($p < 0.05$) were performed using SPSS18.0. The mean differences were significant by *t*-test at $p < 0.05$, whilst the least significant difference at the 5% level was considered statistically significant among different treatments. The following asterisks indicate the results of significance testing: * $p < 0.05$ and ** $p < 0.01$. Different colors in the graphs indicate differences. Data represent mean values and error bars are means ± standard deviation (SD).

Supplementary Materials: The following are available online at http://www.mdpi.com/1422-0067/21/15/5282/s1. Figure S1: Detail information for MS identification of proteins in the 2-DE gel of the glycosylation spots; Table S1: All proteins identified from 2-DE gels and the potential functions; Table S2: Determination and functional analysis of the DAGPs in rubber latex upon ethylene stimulation.

Author Contributions: X.W. and Q.X.; conceived and designed the experiments; L.Y.; B.Y. and Y.S.; performed the experiments; L.Y. wrote the original manuscript, X.W. and L.W.; revised this manuscript. L.Y.; B.Y.; Q.X.; G.D.; O.A.S.; and X.Z.; analyzed the data and polished the manuscript. All authors have read and agreed to the published version of the manuscript.

Funding: Thanks for the funding support provided by the National Key Research and Development Program of China (No. 2018YFD1000502), the National Natural Science Foundation of China (No. 31860224) and the International scientific and technological cooperation to advance the project by the Shihezi University (No. GJHZ201708).

Acknowledgments: The author wants to thank, in particular, Hongbin Li for his kind help in this study.

Conflicts of Interest: The authors declare that they have no conflict of interest.

Abbreviations

ACAT	acetyl-CoA acetyltransferase
DMAPP	elongation factor 2-like protein
EF2	Three letter acronym
FPP	farnesyl pyrophosphate
FPS	farnesyl pyrophosphate synthase
GGPP	geranylgeranyl pyrophosphate
GPP	geranyl pyrophosphate
GPS	geranyl pyrophosphate synthase
Hevein	pro-hevein
HMG-CoA	hydroxymethylglutaryl-CoA
HMGR	HMG-CoA reductase
HMGS	hydroxymethylglutaryl-CoA synthase
IPP	isopentenyl pyrophosphate
IPI	IPP isomerase
MEVK	mevalonate kinase
MVA	mevalonate acid
MVA-5-P	mevalonate-5-phosphate
MVA-5-PP	mevalonate pyrophosphate
MVD	mevalonate pyrophosphate decarboxylase
PPMK	MVA-5-P kinase
SRPP	small rubber particle protein
REF	rubber elongation factor
REF-L	rubber elongation factor-like protein
14-3-3	14-3-3 protein

References

1. Cornish, K. Biochemistry of natural rubber, a vital raw material, emphasizing biosynthetic rate, molecular weight and compartmentalization, in evolutionarily divergent plant species. *Nat. Prod. Rep.* **2001**, *18*, 182–189. [CrossRef] [PubMed]

2. Wang, X.C.; Shi, M.J.; Wang, D.; Chen, Y.Y.; Cai, F.G.; Zhang, S.X.; Wang, L.M.; Tong, Z.; Tian, W.M. Comparative proteomics of primary and secondary lutoids reveals that chitinase and glucanase play a crucial combined role in rubber particle aggregation in Hevea brasiliensis. *J. Proteome Res.* **2013**, *12*, 5146–5159. [CrossRef] [PubMed]

3. Liu, J.P.; Zhuang, Y.F.; Guo, X.L.; Li, Y.J. Molecular mechanism of ethylene stimulation of latex yield in rubber tree (Hevea brasiliensis) revealed by de novo sequencing and transcriptome analysis. *BMC Genomics.* **2016**, *17*, 257. [CrossRef] [PubMed]

4. Zhu, J.H.; Zhang, Z.L. Ethylene stimulation of latex production in Hevea brasiliensis. *Plant Signal. Behav.* **2009**, *4*, 1072–1074. [CrossRef] [PubMed]

5. Coupe, M.; Chrestin, H. Physico-chemical and biochemical mechanisms of hormonal (ethylene) stimulation. *Physiol. Rubber Tree Latex* **1989**, 295–319. [CrossRef]
6. Lestari, R.; Rio, M.; Martin, F.; Leclercq, J.; Woraathasin, N.; Roques, S.; Dessailly, F.; Clément-Vidal, A.; Sanier, C.; Fabre, D.; et al. Overexpression of Hevea brasiliensis ethylene response factor HbERF-IXc5 enhances growth, tolerance to abiotic stress and affects laticifer differentiation. *Plant Biotechnol. J.* **2017**, *15*, 1–15.
7. Wang, X.C.; Wang, D.; Sun, Y.; Yang, Q.; Chang, L.L.; Wang, L.M.; Meng, X.R.; Huang, Q.X.; Jin, X.; Tong, Z. Comprehensive proteomics analysis of laticifer latex reveals new insights into ethylene stimulation of natural rubber production. *Sci. Rep.* **2015**, *5*, 13778. [CrossRef]
8. Wang, D.; Xie, Q.L.; Sun, Y.; Tong, Z.; Chang, L.L.; Yu, L.; Zhang, X.Y.; Yuan, B.X.; He, P.; Jin, X.; et al. Proteomic landscape has revealed small rubber particles are crucial rubber biosynthetic machines for ethylene-stimulation in natural rubber production. *Int. J. Mol. Sci.* **2019**, *20*, 5082. [CrossRef]
9. Yeang, H.Y.; Arif, S.M.; Yusof, F.; Sunderasan, E. Allergenic proteins of natural rubber latex. *Methods* **2002**, *27*, 32–45. [CrossRef]
10. Xiang, Q.L.; Xia, K.C.; Dai, L.J.; Kang, G.J.; Li, Y.; Nie, Z.Y.; Duan, C.F.; Zeng, R.Z. Proteome analysis of the large and the small rubber particles of Hevea brasiliensis using 2D-DIGE. *Plant Phys. Biochem.* **2012**, *60*, 207–213. [CrossRef]
11. Dai, L.J.; Kang, G.J.; Li, Y.; Nie, Z.Y.; Duan, C.F.; Zeng, R.Z. In-depth proteome analysis of the rubber particle of Hevea brasiliensis (para rubber tree). *Plant Mol. Biol.* **2013**, *82*, 155–168. [CrossRef] [PubMed]
12. Yamashita, S.; Yamaguchi, H.; Waki, T.; Aoki, Y.; Mizuno, M.; Yanbe, F.; Ishii, T.; Funaki, A.; Tozawa, Y.; Miyagi-Inoue, Y.; et al. Identification and reconstitution of the rubber biosynthetic machinery on rubber particles from Hevea brasiliensis. *eLife* **2016**, *5*, 19022. [CrossRef] [PubMed]
13. Habib, M.H.; Yuen, G.C.; Othman, F.; Zainudin, N.N.; Latiff, A.A.; Ismail, M.N. Proteomics analysis of latex from Hevea brasiliensis (clone RRIM 600). *Biochem. Cell Biol.* **2017**, *95*, 232–242. [CrossRef] [PubMed]
14. Men, X.; Wang, F.; Chen, G.Q.; Zhang, H.B.; Xian, M. Biosynthesis of natural rubber: Current state and perspectives. *Int. J. Mol. Sci.* **2019**, *20*, 50. [CrossRef] [PubMed]
15. Dai, L.J.; Kang, G.J.; Nie, Z.Y.; Li, Y.; Zeng, R.Z. Comparative proteomic analysis of latex from Hevea brasiliensis treated with ethrel and methyl jasmonate using iTRAQ-coupled two-dimensional LC-MS/MS. *J. Proteom.* **2016**, *132*, 167–175. [CrossRef] [PubMed]
16. Tong, Z.; Wang, D.; Sun, Y.; Yang, Q.; Meng, X.R.; Wang, L.M.; Feng, W.Q.; Li, L.; Wurtele, E.S.; Wang, X.C. Comparative proteomics of rubber latex revealed multiple protein species of REF/SRPP family respond diversely to ethylene stimulation among different rubber tree clones. *Int. J. Mol. Sci.* **2017**, *18*, 958. [CrossRef]
17. Linding, R.; Jensen, L.J.; Ostheimer, G.J.; van Vugt, M.A.; Jorgensen, C.; Miron, I.M.; Diella, F.; Colwill, K.; Taylor, L.; Elder, K.; et al. Systematic discovery of in vivo phosphorylation networks. *Cell* **2007**, *129*, 1415–1426. [CrossRef]
18. Zielinska, D.F.; Gnad, F.; Schropp, K.; Wisniewski, J.R.; Mann, M. Mapping N-glycosylation sites across seven evolutionarily distant species reveals a divergent substrate proteome despite a common core machinery. *Mol. Cell* **2012**, *46*, 542–548. [CrossRef]
19. Ruiz-May, E.; Hucko, S.; Howe, K.J.; Zhang, S.; Sherwood, R.W.; Thannhauser, T.W.; Rose, J.K. A comparative study of lectin affinity-based plant N-glycoproteome profiling using tomato fruit as a model. *Mol. Cell. Proteom.* **2014**, *13*, 566–579. [CrossRef]
20. Lige, B.; Ma, S.; Van-huystee, R.B. The effects of the site-directed removal of N-glycosylation from cationic peanut peroxidase on its function. *Arch. Biochem. Biophys.* **2001**, *386*, 17–24. [CrossRef]
21. Strasser, A. Plant protein glycosylation. *Glycobiology* **2016**, *26*, 926–939. [CrossRef] [PubMed]
22. Ceriotti, A.; Duranti, M.; Bollini, R. Effects of N-glycosylation on the folding and structure of plant proteins. *J. Exp. Bot.* **1998**, *49*, 1091–1103. [CrossRef]
23. Pattison, R.J.; Amtmann, A. N-glycan production in the endoplasmic reticulum of plants. *Trends Plant Sci.* **2009**, *14*, 92–99. [CrossRef] [PubMed]
24. Haweker, H.; Rips, S.; Koiwa, H.; Salomon, S.; Saijo, Y.; Chinchilla, D.; Robatzek, S.; Schaewen, A.V. Pattern recognition receptors require N-glycosylation to mediate plant immunity. *J. Biol. Chem.* **2010**, *285*, 4629–4636. [CrossRef]

25. Catala, C.; Howe, K.J.; Hucko, S.; Rose, J.C.; Thannhauser, T.W. Towards characterization of the glycoproteome of tomato (Solanum lycopersicum) fruit using Concanavalin A lectin affinity chromatography and LC-MALDI-MS/MS analysis. *Proteomics* **2011**, *11*, 1530–1544. [CrossRef]

26. Ligat, L.; Lauber, E.; Albenne, C.; Clemente, H.S.; Valot, B.; Zivy, M.; Pont-Lezica, R.; Arlat, M.; Jamet, E. Analysis of the xylem sap proteome of Brassica oleracea reveals a high content in secreted proteins. *Proteomics* **2011**, *11*, 1798–1833. [CrossRef]

27. Melo-braga, M.N.; Verano-braga, T.; Leon, I.R.; Antonacci, D.; Nogueira, F.C.; Thelen, J.J.; Larsen, M.R.; Palmisano, G. Modulation of protein phosphorylation, N-glycosylation and lys-acetylation in grape mesocarp and exocarp owing to Lobesia botrana infection. *Mol. Cell. Proteom.* **2012**, *11*, 945–956. [CrossRef]

28. Palmisano, G.; Antonacci, D.; Larsen, M.R. Glycoproteomic profile in wine: A sweet molecular renaissance. *J. Proteome Res.* **2010**, *9*, 6148–6159. [CrossRef]

29. Hart, C.; Schulenberg, B.; Steinberg, T.H.; Leung, W.Y.; Patton, W.F. Detection of glycoproteins in polyacrylamide gels and on electroblots using Pro-Q Emerald 488 dye, a fluorescent periodate Schiff-base stain. *Electrophoresis* **2003**, *24*, 588–598. [CrossRef]

30. Ohlmeier, S.; Mazur, W.; Salmenkivi, K.; Myllärniemi, M.; Bergmann, U.; Kinnula, V.L. Proteomic studies on receptor for advanced glycation end product variants in idiopathic pulmonary fibrosis and chronic obstructive pulmonary disease. *Proteomics* **2010**, *4*, 97–105. [CrossRef]

31. Cong, W.T.; Zhou, A.; Liu, Z.G.; Shen, J.; Zhou, X.; Ye, W.; Zhu, Z.; Zhu, X.; Lin, J.J.; Jin, L.T. Highly sensitive method for specific, brief, and economical detection of glycoproteins in sodium dodecyl sulfate-polyacrylamide gel electrophoresis by the synthesis of a new hydrazide derivative. *Anal. Chem.* **2015**, *87*, 1462–1465. [CrossRef] [PubMed]

32. Wang, X.; Shi, M.; Lu, X.; Ma, R.; Wu, C.; Guo, A.; Peng, M.; Tian, W. A method for protein extraction from different subcellular fractions of laticifer latex in Hevea brasiliensis compatible with 2-DE and MS. *Proteome Sci.* **2010**, *8*, 35. [CrossRef] [PubMed]

33. Putranto, R.A.; Herlinawati, E.; Rio, M.; Leclercq, J.; Piyatrakul, P.; Gohet, E.; Sanier, C.; Oktavia, F.; Pirrello, J.; Montoro, P. Involvement of ethylene in the latex metabolism and tapping panel dryness of Hevea brasiliensis. *Int. J. Mol. Sci.* **2015**, *16*, 17885–17908. [CrossRef] [PubMed]

34. Adiwilaga, K.; Kush, A. Cloning and characterization of cDNA encoding farnesyl diphosphate synthase from rubber tree (*Hevea brasiliensis*). *Plant Mol. Biol.* **1996**, *30*, 935–946. [CrossRef]

35. Luo, M.W.; Deng, L.H.; Yi, X.P.; Zeng, H.C.; Xiao, S.H. Cloning and sequence analysis of a novel cis-prenyltransferases gene from Hevea brasiliensis. *J. Trop. Subtrop. Bot.* **2009**, *17*, 223–228.

36. Chye, M.L.; Tan, C.T.; Chua, N.H. Three genes encode 3-hydroxy-3-methylglutaryl-coenzyme A reductase in Hevea brasiliensis: Hmg1 and hmg3 are differentially expressed. *Plant Mol. Biol.* **1992**, *19*, 473–484. [CrossRef]

37. Tungngoen, K.; Kongsawadworakul, P.; Viboonjun, U.; Katsuhara, M.; Brunel, N.; Sakr, S.; Narangajavana, J.; Chrestin, H. Involvement of HbPIP2;1 and HbTIP1;1 aquaporins in ethylene stimulation of latex yield through regulation of water exchanges between inner liber and latex cells in Hevea brasiliensis. *Plant Physiol.* **2009**, *151*, 843–856. [CrossRef]

38. Pluang, S.; Nualpun, S.; Wallie, S. Regulation of the expression of 3-hydroxy-3-methylglutaryl-CoA synthase gene in Hevea brasiliensis (B.H.K.) Mull. Arg. *Plant Sci.* **2004**, *166*, 531–537.

39. Sirinupong, N.; Suwanmanee, P.; Doolittle, R.F.; Suvachitanont, W. Molecular cloning of a new cDNA and expression of 3-hydroxy-3-methylglutaryl-CoA synthase gene from Hevea brasiliensis. *Planta* **2005**, *221*, 502–512. [CrossRef]

40. Yang, J.; Sardar, H.S.; McGovern, K.R.; Zhang, Y.; Showalter, A.M. A lysine-rich arabinogalactan protein in Arabidopsis is essential for plant growth and development, including cell division and expansion. *Plant J.* **2007**, *49*, 629–640. [CrossRef]

41. Minic, Z.; Jamet, E.; Négroni, L.; Arsene der Garabedian, P.; Zivy, M.; Jouanin, L. A sub-proteome of Arabidopsis thaliana mature stems trapped on Concanavalin A is enriched in cell wall glycoside hydrolases. *J. Exp. Bot.* **2007**, *58*, 2503–2512. [CrossRef] [PubMed]

42. Zhang, Y.; Giboulot, A.; Zivy, M.; Valot, B.; Jamet, E.; Albenne, C. Combining various strategies to increase the coverage of the plant cell wall glycoproteome. *Phytochemistry* **2011**, *72*, 1109–1123. [CrossRef] [PubMed]

43. Sultana, N.; Florance, H.V.; Johns, A.; Smirnoff, N. Ascorbate deficiency influences the leaf cell wall glycoproteome in Arabidopsis thaliana. *Plant Cell Environ.* **2015**, *38*, 375–384. [CrossRef] [PubMed]

44. Hunter, J.R. Reconsidering the functions of latex. *Trees* **1994**, *9*, 1–5. [CrossRef]

45. Nawrot, R.; Kalinowski, A.; Gozdzicka-Jozefiak, A. Proteomic analysis of Chelidonium majus milky sap using two-dimensional gel electrophoresis and tandem mass spectrometry. *Phytochemistry* **2007**, *68*, 1612–1622. [CrossRef]

46. Abd-Rahman, N.; Kamarrudin, M.F. Proteomic profiling of Hevea latex serum induced by ethephon stimulation. *J. Trop. Plant Physiol.* **2018**, *10*, 11–22.

47. Cho, W.K.; Chen, X.Y.; Rim, Y.; Chu, H.; Jo, Y.; Kim, S.; Park, Z.Y.; Kim, J.Y. Extended latex proteome analysis deciphers additional roles of the lettuce laticifer. *Plant Biotechnol. Rep.* **2010**, *4*, 311–319. [CrossRef]

48. Wang, D.; Sun, Y.; Chang, L.L.; Tong, Z.; Xie, Q.L.; Jin, X.; Zhu, L.P.; He, P.; Li, H.B.; Wang, X.C. Subcellular proteome profiles of different latex fractions revealed washed solutions from rubber particles contain crucial enzymes for natural rubber biosynthesis. *J. Proteom.* **2018**, *182*, 53–64. [CrossRef]

49. Cho, W.K.; Jo, Y.; Chu, H.; Park, S.H.; Kim, K.H. Integration of latex protein sequence data provides comprehensive functional overview of latex proteins. *Mol. Biol. Rep.* **2014**, *41*, 1469–1481. [CrossRef]

50. Gonzalez-Buitrago, J.M.; Ferreira, L.; Isidoro-Garcia, M.; Sanz, C.; Lorente, F.; Davila, I. Proteomic approaches for identifying new allergens and diagnosing allergic diseases. *Clinica Chim. Acta* **2007**, *385*, 21–27. [CrossRef]

51. D'Amato, A.; Bachi, A.; Fasoli, E.; Boschetti, E.; Peltre, G.; Sénéchal, H.; Sutra, J.P.; Citterio, A.; Righetti, P.G. In-depth exploration of Hevea brasiliensis latex proteome and "hidden allergens" via combinatorial peptide ligand libraries. *J. Proteom.* **2010**, *73*, 1368–1380. [CrossRef] [PubMed]

52. Yagami, T.; Haishima, Y.; Tsuchiya, T.; Tomitaka-Yagami, A.; Kano, H.; Matsunaga, K. Proteomic analysis of putative latex allergens. *Int. Arch. Allergy Immunol.* **2004**, *135*, 3–11. [CrossRef] [PubMed]

53. Wang, D.; Sun, Y.; Tong, Z.; Yang, Q.; Chang, L.L.; Meng, X.R.; Wang, L.M.; Tian, W.M.; Wang, X.C. A protein extraction method for low protein concentration solutions compatible with the proteomic analysis of rubber particles. *Electrophoresis* **2016**, *37*, 2930–2939. [CrossRef]

54. Tang, C.R.; Xiao, X.H.; Li, H.P.; Fan, Y.J.; Yang, J.H.; Qi, J.Y.; Li, H.B. Comparative analysis of latex transcriptome reveals putative molecular mechanisms underlying super productivity of Hevea brasiliensis. *PLoS ONE* **2013**, *8*, 1–18. [CrossRef] [PubMed]

55. Gao, L.; Sun, Y.; Wu, M.; Wang, D.; Wei, J.S.; Wu, B.S.; Wang, G.H.; Wu, W.G.; Jin, X.; Wang, X.C.; et al. Physiological and proteomic analyses of molybdenum- and ethylene-responsive mechanisms in rubber latex. *Front. Plant Sci.* **2018**, *9*, 621. [CrossRef] [PubMed]

56. Wang, Q.; An, B.; Hou, X.; Guo, Y.; Luo, H.L.; He, C.Z. Dicer-like proteins regulate the growth, conidiation, and pathogenicity of Colletotrichum gloeosporioides from Hevea brasiliensis. *Front. Microbiol.* **2018**, *8*, 2621. [CrossRef]

57. Li, D.J.; Deng, Z.; Chen, C.L.; Xia, Z.H.; Wu, M.; He, P.; Chen, S.C. Identification and characterization of genes associated with tapping panel dryness from Hevea brasiliensis latex using suppression subtractive hybridization. *BMC Plant Biol.* **2010**, *10*, 140. [CrossRef]

58. Cho, W.K.; Chen, X.Y.; Uddin, N.M.; Rim, Y.; Moon, J.; Jung, J.H.; Shi, C.; Chu, H.; Kim, S.; Kim, S.W.; et al. Comprehensive proteome analysis of lettuce latex using multidimensional protein-identification technology. *Phytochemistry* **2009**, *70*, 570–578. [CrossRef]

59. Havanapan, P.; Bourchookarn, P.; Ketterman, A.J.; Krittanai, C. Comparative proteome analysis of rubber latex serum from pathogenic fungi tolerant and susceptible rubber tree (Hevea brasiliensis). *J. Proteom.* **2016**, *131*, 82–92. [CrossRef]

60. Pujade-Renaud, V.; Clement, A.; Perrot-Rechenmann, C.; Prevot, J.C.; Chrestin, H.; Jacob, J.L.; Guern, J. Ethylene-induced increase in glutamine synthetase activity and mRNA levels in Hevea brasiliensis latex cells. *Plant Physiol.* **1994**, *105*, 127–132. [CrossRef]

61. Sando, T.; Takaoka, C.; Mukai, Y.; Yamashita, A.; Hattori, M.; Ogasawara, N.; Fukusaki, E.; Kobayashi, A. Cloning and characterization of mevalonate pathway genes in a natural rubber producing plant, Hevea brasiliensis. *Biosci. Biotechnol. Biochem.* **2008**, *72*, 2049–2060. [CrossRef] [PubMed]

62. Puskas, J.E.; Gautriaud, E.; Deffieux, A.; Kennedy, J.P. Natural rubber biosynthesis-A living carbocationic polymerization. *Prog. Polym. Sci.* **2006**, *31*, 533–548. [CrossRef]

63. Tang, C.R.; Yang, M.; Fang, Y.J.; Luo, Y.F.; Gao, S.H.; Xiao, X.H.; An, Z.W.; Zhou, B.H.; Zhang, B.; Tan, X.Y.; et al. The rubber tree genome reveals new insights into rubber production and species adaptation. *Nature Plants.* **2016**, *6*, 16073. [CrossRef] [PubMed]

64. Oh, S.K.; Kang, H.; Shin, D.H.; Yang, J.; Chow, K.S.; Yeang, H.Y.; Wagner, B.; Breiteneder, H.; Han, K.H. Isolation, characterization, and functional analysis of a novel cDNA clone encoding a small rubber particle protein from Hevea brasiliensis. *J. Biol. Chem.* **1999**, *274*, 17132–17138. [CrossRef] [PubMed]

65. Rojruthai, P.; Sakdapipanich, J.T.; Takahashi, S.; Hyegin, L.; Noike, M.; Koyama, T.; Tanaka, Y. In vitro synthesis of high molecular weight rubber by Hevea small rubber particles. *J. Biosci. Bioeng.* **2010**, *109*, 107–114. [CrossRef] [PubMed]

66. Dai, L.J.; Nie, Z.Y.; Kang, G.J.; Li, Y.; Zeng, R.Z. Identification and subcellular localization analysis of two rubber elongation factor isoforms on Hevea brasiliensis rubber particles. *Plant Physiol. Biochem.* **2017**, *111*, 97–106. [CrossRef]

67. Brown, D.; Feeney, M.; Ahmadi, M.; Lonoce, C.; Sajari, R.; Di Cola, A.; Frigerio, L. Subcellular localization and interactions among rubber particle proteins from Hevea brasiliensis. *J. Exp. Bot.* **2017**, *68*, 5045–5055. [CrossRef]

68. Carter, C.; Pan, S.; Zouhar, J.; Avila, E.L.; Girke, T.; Raikhel, N.V. The vegetative vacuole proteome of Arabidopsis thaliana reveals predicted and unexpected proteins. *Plant Cell* **2004**, *16*, 3285–3303. [CrossRef]

69. Wititsuwannakul, R.; Rukseree, K.; Kanokwiroon, K.; Wititsuwannakul, D. A rubber particle protein specific for Hevea latex lectin binding involved in latex coagulation. *Phytochemistry* **2008**, *69*, 1111–1118. [CrossRef]

70. Lee, H.; Broekaert, W.F.; Raikhel, N.V. Co-and post-translational processing of the hevein preproprotein of latex of the rubber tree (Hevea brasiliensis). *J. Biol. Chem.* **1991**, *266*, 15944–15946.

71. Gong, X.X.; Yan, B.Y.; Tan, Y.R.; Gao, X.; Wang, D.; Zhang, H.; Wang, P.; Li, S.J.; Wang, Y.; Zhou, L.Y.; et al. Identification of cis-regulatory regions responsible for developmental and hormonal regulation of *HbHMGS1* in transgenic Arabidopsis thaliana. *Biotechnol. Lett.* **2019**, *41*, 1077–1091. [CrossRef] [PubMed]

72. Wang, H.; Nagegowda, D.A.; Rawat, R.; Bouvier-Nave, P.; Guo, D.; Bach, T.J.; Chye, M.L. Overexpression of Brassica juncea wild-type and mutant HMG-CoA synthase 1 in Arabidopsis up-regulates genes in sterol biosynthesis and enhances sterol production and stress tolerance. *Plant Biotechnol. J.* **2012**, *10*, 31–42. [CrossRef] [PubMed]

73. Putter, K.M.; Deenen, N.V.; Unland, K.; Prufer, D.; Gronover, C.S. Isoprenoid biosynthesis in dandelion latex is enhanced by the overexpression of three key enzymes involved in the mevalonate pathway. *BMC Plant Biol.* **2017**, *17*, 88. [CrossRef] [PubMed]

74. Yang, M.W. Study on rapid determination of chlorophyll content of leaves. *Chin. J. Spect. Lab.* **2002**, *4*, 478–481.

75. Zhang, X.; Guo, T.; Xiang, T.; Dong, Y.Y.; Zhang, J.C.; Zhang, L.Q. Quantitation of isoprenoids for natural rubber biosynthesis in natural rubber latex by liquid chromatography with tandem mass spectrometry. *J. Chromatogr. A* **2018**, *1558*, 115–119. [CrossRef]

International Journal of
Molecular Sciences

Article

Hypoxia-Responsive Class III Peroxidases in Maize Roots: Soluble and Membrane-Bound Isoenzymes

Anne Hofmann [1], Stefanie Wienkoop [2], Sönke Harder [3], Fabian Bartlog [1] and Sabine Lüthje [1,*]

[1] Oxidative Stress and Plant Proteomics Group, Institute of Plant Science and Microbiology, Universität Hamburg, Ohnhorststrasse 18, 22609 Hamburg, Germany; anne.hofmann@uni-hamburg.de (A.H.); Fabian.Bartlog@studium.uni-hamburg.de (F.B.)

[2] Department of Ecogenomics and Systems Biology, University of Vienna, Althanstrasse 14, 1090 Vienna, Austria; stefanie.wienkoop@univie.ac.at

[3] Center for Diagnostics Clinical Chemistry and Laboratory Medicine, Core Facility of Mass Spectrometric Proteomic, Campus Forschung N27, University Hospital Hamburg-Eppendorf (UKE), Martinistrasse 52, 20246 Hamburg, Germany; harder@uke.de

* Correspondence: sabine.luethje@uni-hamburg.de

Received: 14 October 2020; Accepted: 20 November 2020; Published: 23 November 2020

Abstract: Flooding induces low-oxygen environments (hypoxia or anoxia) that lead to energy disruption and an imbalance of reactive oxygen species (ROS) production and scavenging enzymes in plants. The influence of hypoxia on roots of hydroponically grown maize (*Zea mays* L.) plants was investigated. Gene expression (RNA Seq and RT-qPCR) and proteome (LC–MS/MS and 2D-PAGE) analyses were used to determine the alterations in soluble and membrane-bound class III peroxidases under hypoxia. Gel-free peroxidase analyses of plasma membrane-bound proteins showed an increased abundance of *Zm*Prx03, *Zm*Prx24, *Zm*Prx81, and *Zm*Pr85 in stressed samples. Furthermore, RT-qPCR analyses of the corresponding peroxidase genes revealed an increased expression. These peroxidases could be separated with 2D-PAGE and identified by mass spectrometry. An increased abundance of *Zm*Prx03 and *Zm*Prx85 was determined. Further peroxidases were identified in detergent-insoluble membranes. Co-regulation with a respiratory burst oxidase homolog (Rboh) and key enzymes of the phenylpropanoid pathway indicates a function of the peroxidases in membrane protection, aerenchyma formation, and cell wall remodeling under hypoxia. This hypothesis was supported by the following: (i) an elevated level of hydrogen peroxide and aerenchyma formation; (ii) an increased guaiacol peroxidase activity in membrane fractions of stressed samples, whereas a decrease was observed in soluble fractions; and (iii) alterations in lignified cells, cellulose, and suberin in root cross-sections.

Keywords: aerenchyma; cell wall remodeling; class III peroxidases; hypoxia; maize roots; plasma membrane; respiratory burst oxidase homolog; *Zea mays* L.

1. Introduction

Plants worldwide have to cope with flooding events, and humanity has to manage the resulting agricultural yield loss of crop plants. The main reason for the dramatic effect of flooding seems to be the energy disruption, caused by the lack of oxygen needed in respiratory metabolisms. Fortunately, some plants can tolerate or adapt to this abiotic stress with the "low-oxygen escape strategy" or the "low-oxygen quiescence strategy" [1,2]. These adaptations to flooding-induced low-oxygen stress (hypoxia or anoxia) have been well studied so far [3–5]. Besides the energy disruption, the production of reactive oxygen species (ROS) and the imbalance of ROS-scavenging enzymes might be another factor for cell-damaging effects [6].

Reactive oxygen species (superoxide anion radical, hydroxyl radical, hydrogen peroxide, etc.) are produced even under physiological conditions via the aerobic pathway. They function as signaling molecules in plant growth, cell development, and programmed cell death [7,8]. Hypoxia has been shown to be responsible for ROS-induced oxidative stress [9,10]. The produced oxygen radicals cause lipid, nucleic acid, and protein oxidation, as well as total cell damage [6]. Increased ROS levels can be reduced by ROS-scavenging molecules (ascorbate and glutathione) or enzymes (superoxide dismutase, catalase, ascorbate peroxidases, glutathione peroxidases, and class III peroxidases).

Maize (*Zea mays* L.) belongs to the waterlogging-tolerant plant species that adapt to hypoxia by developing aerenchyma in roots for ventilation. It has been demonstrated that hypoxia-induced development of aerenchyma occurred in maize roots after 12 to 60 h [11]. Reactive oxygen species and cell wall degradation play a crucial role in the formation of aerenchyma by several stressors. A function of respiratory burst oxidase homologs (Rboh) in programmed cell death during aerenchyma formation has been suggested by a strong upregulation of those genes [12–14]. Additionally, some wetland plant species form a suberin barrier at the outer cell layers of roots to reduce radial oxygen loss (ROL) from aerenchyma [15]. Accumulation of suberin at the hypodermal/exodermal cell layers and deposition of lignin was observed in adventitious roots under waterlogged soil conditions [16]. It was concluded that ROL barrier formation contributes to higher waterlogging tolerance in plants. Class III peroxidases were upregulated by hypoxia [11,14]. Although guaiacol peroxidase activity of plant extracts has been used as a general stress marker for a long time [17], results might be not clear, because of the high amount of isoenzymes that could be differentially regulated. To distinguish between several isoenzymes and to identify low-abundant peroxidases involved in a specific stress response, proteomic approaches are state of the art [18,19]. Although peroxidases involved in biotic and abiotic stress are well studied, the impact of hypoxia, e.g., induced by flooding, on class III peroxidases is still rarely investigated [12,20,21].

The class III peroxidases are heme-containing proteins of the secretory pathway in higher plants with a high number of isoforms. The secretory pathway delivered the glycosylated peroxidases into the apoplast and cell wall, but also to the vacuole, plasma membrane (PM), and thylakoid [22,23]. Due to their peroxidative or hydroxylic reaction cycles, peroxidases are involved in cell wall-related reactions, metabolic pathways, and stress-related processes [23–25]. Besides the ROS scavenging, peroxidases take part in the final steps of lignin and suberin synthesis [26,27]. Sinapyl and cinnamyl alcohols are precursors of lignin monomers [28], whereas ferulic, caffeic, sinapinic, and *p*-coumaric acid are precursors of suberin [27]. Peroxidases mediate the crosslinking of cell wall compounds in response to different stimuli [29]. In grasses, arabinose and arabinoxylans are crosslinking by peroxidase-generated diferulates [30,31]. So far, 158 peroxidases have been identified in the maize (*Zea mays* L.) genome (RedoxiBase, as of 6 October 2020). A majority of maize peroxidase genes appear expressed in root tissues [32,33]. However, evidence has been presented for localization of about 25% of the class III peroxidases in the PM [23]. To date, four peroxidases have been purified from PM of maize roots and were characterized biochemically [34,35]. Evidence for a function of these peroxidases in biotic and abiotic stress has been given by a proteomic approach [19].

In the present study, we used a systems biological approach to identify hypoxia-responsive class III peroxidases with a special focus on PM-bound isoenzymes. We also discuss their possible functions in maize roots.

2. Results

2.1. RNA Sequence Analyses

In the RNA Sequence (RNA Seq) analyses of control and stressed maize root samples, 152 Prx transcripts (135 genes) and 20 Rboh transcripts (11 genes) were identified. Pseudogenes were excluded from the analyses. Some of the peroxidases, catalogued by RedoxiBase database, could not be detected. This contains *zmprx12*, *zmprx138*, *zmprx36*, *zmprx40*, *zmprx41*, *zmprx66*, *zmprx75*, *zmprx 01_W64A*, *zmprx100_35A19*, *zmprx108_35A19*, *zmprx03_Du101*, *zmprx03_F66*, *zmprx03_Lan496*, *zmprx03_W64A*, and *zmprx03_Wis93*. The statistical analysis of the RNA Seq data revealed a significant upregulation of *zmrboh10* and 17 differentially expressed class III peroxidase genes (DEGs), of which three were upregulated and 13 were downregulated based on 2-fold change and *p*-value of $p < 0.05$ of comparison pair (hypoxia versus control) (Figure 1A and Supplementary Materials Table S1A,B). In silico prediction of the 17 class III peroxidases showed localization in the endoplasmic reticulum (*zmprx35*, *zmprx46*, and *zmprx135*), the PM (*zmprx109* and *zmprx140*), and "outside", i.e., apoplast or cell wall (Figure 1B). The two PM predicted peroxidases, *zmprx109* and *zmprx140*, were 2–5-fold downregulated in stressed maize plants (*zmprx109* with mean of 2.16 ± 0.16 and *zmprx140* with mean of 0.47 ± 0.15) compared to controls (*zmprx109* with mean of 4.44 ± 0.24 and *zmprx140* with mean of 1.56 ± 0.43) (Figure 1A,C).

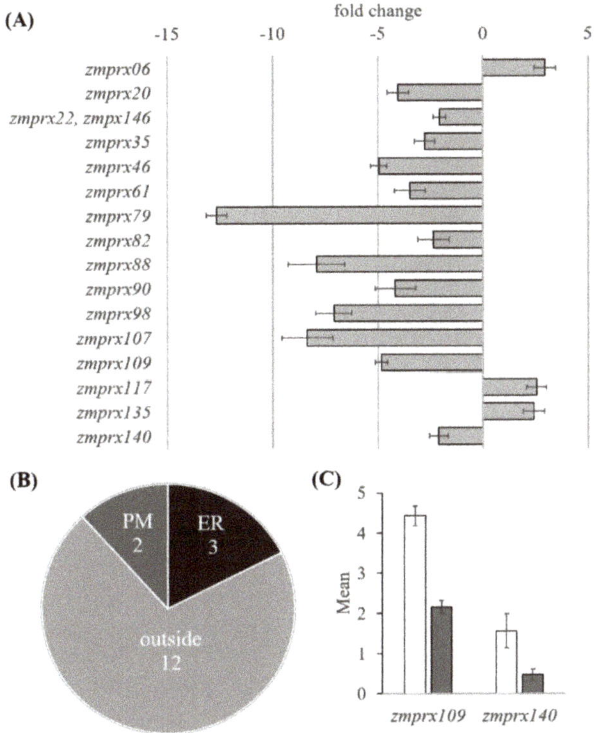

Figure 1. Differential regulation and putative localization of class III peroxidases under hypoxia. From three non-stressed and 24 h hypoxia-stressed root samples, expression profiles were calculated as Fragments Per Kilobase Million (FPKMs). Differentially expressed gene (DEG) results were performed on a comparison pair (stressed versus controls), using FPKM with a fold change >2 and Student's *t*-test $p < 0.05$. (**A**) Differential expressed peroxidases (excluding pseudogenes) with their fold change expression and (**B**) their predicted localization, using PSORT (http://psort1.hgc.jp/form.html) at either the plasma membrane (PM), the endoplasmic reticulum (ER) or "outside" (apoplast, cell wall). (**C**) Gene expression (mean = mean of normalized signal for each sample or group) of the two putative plasma membrane-bound class III peroxidases that were differentially expressed in controls (white columns) and hypoxia-stressed samples (gray columns). Error bars indicate standard deviation (three biological replicates).

Besides these peroxidases, genes of key enzymes of cell wall synthesis, degradation, and reinforcement were found to be differentially regulated by hypoxia (Supplementary Materials

Table S1C). The phenylpropanoid pathway was presented by caffeoylshikimate esterase (*zmcse*, A0A1D6N7M7) and several cinnamyl alcohol dehydrogenases (*zmcad*). The expression level of *zmcse* increased 17-fold, whereas that of *zmcad1* (B4FAJ0) was moderately upregulated. Expression levels of *zmcad6* (B4FR97) and *zmcad* (O24562) decreased significantly, 2.5-fold and 2.9-fold, respectively. Most of the *zmcad* genes were downregulated. The expression level of cellulose synthase 5 (*zmcesa*, A0A1D6L8J3) increased 2.8-fold, whereas two other *zmcesa* genes (B4FJI1 and B6TTA1) were significantly downregulated. The expression of further *zmcesa* genes decreased. Omega-hydroxypalmitate O-feruloyl transferase (*zmhht*, B4FV3), a key enzyme of suberin biosynthesis, showed a 2.9-fold downregulation. For cell wall degradation, two expansins (*zmexpl3*, B4FL59, and *zmexp3A*, A0A1D6EVK8) were significantly upregulated by hypoxia, whereas 20 other expansins were downregulated. Finally, caffeoyl-CoA O-methyltransferase 1 (*zmccoaomt*, B6UF45), involved in cell wall reinforcement, was downregulated.

Among the 15 dirigent (*zmdir*) genes, one transcript (B6U4X5) was 2.1-fold upregulated, six were downregulated and eight were not differentially regulated by hypoxia (Supplementary Materials Table S1C). Expression levels of eighteen fasciclin-like arabinogalactan transcripts (*zmfla*) were either downregulated (*zmfla2*, B6SZA0; *zmfla7*, B6SHU5; *zmfla10*, A0A096SZF8; *zmfla7*, B4FB81; *zmfla6*, B4F7Z4) or not significantly affected. A glycerophosphodiesterase (*zmgdpd3*, A0A1D6KIK2) was 2.7-fold upregulated.

2.2. Hydrogen Peroxide Determination, Total Guaiacol Peroxidase Activity and Abundance

Control and 24 h hypoxia-stressed root samples were divided into total extracts and the sub-proteomes soluble proteins, microsomes, and PM. The PM fraction showed an enrichment of the H^+-ATPase and a lower amount of V-PPase and Cox2 signals compared to the corresponding microsomal proteins (Supplementary Materials Figure S1). Total extracts of 24 h hypoxia-stressed root tissue showed a significant ($p < 0.01$) 1.5-fold increased level of hydrogen peroxide compared to controls. The quantification of hydrogen peroxide in those samples revealed amounts of 227 ± 12 µM in control and 365 ± 19 µM in stressed samples. The guaiacol peroxidase activities decreased by about 25% in total fractions of stressed samples (5.95 ± 4.73 µmol min^{-1} mg^{-1}) compared to controls (7.82 ± 5.51 µmol min^{-1} mg^{-1}) as well as by about 30% in soluble fractions of stressed samples (30.68 ± 23.39 µmol min^{-1} mg^{-1}) compared to controls (45.55 ± 25.04 µmol min^{-1} mg^{-1}). Contrary to this, membrane fractions showed a about 33% increased activity of 0.97 ± 1.04 µmol min^{-1} mg^{-1} (stressed) to 0.64 ± 0.45 µmol min^{-1} mg^{-1} (controls) in microsomes as well as an about 44% increase of 0.71 ± 0.46 µmol min^{-1} mg^{-1} (stressed) to 0.54 ± 0.23 µmol min^{-1} mg^{-1} (controls) in PM (Figure 2A). Separation of those fractions by modified SDS-PAGE revealed a higher abundance of class III peroxidases in total and soluble fractions compared to membrane fractions (Figure 2B). Several peroxidase isoforms were detected with different protein masses or complexes of proteins. Six bands (nos. 1–6) in total and soluble fraction (148, 103, 70, 55, 43, and 31 kDa) could be determined that showed only a slight increase in abundance in stressed samples compared to controls. Seven bands (nos. 1–7) in microsomal fractions (147, 84, 77, 65, 59, 41, and 29 kDa) were visible, of which the 84, 77, and 29 kDa bands showed an increased abundance of more than 2.0-fold contrary to the 59 kDa band that showed a decrease of 0.4-fold compared to controls. Only two bands (nos. 1 and 2) in PM fractions (148 and 65 kDa) could be seen with no significant difference in abundance between control and stressed samples.

2.3. Gel-Free Peroxidase Analyses

The gel-free approach revealed four class III peroxidases in PM fractions of control and stressed roots. Here, *Zm*Prx03 (A0A1D6LYW3, two unique peptides), *Zm*Prx24 (B4FHG3, five unique peptides), *Zm*Prx81 (B4FG39, seven unique peptides), and *Zm*Prx85 (A0A1D6E530, three unique peptides) were identified (Supplementary Materials Table S2A). Abundance of *Zm*Prx03, *Zm*Prx81 and *Zm*Prx24 significantly increased in stressed PM samples, whereas *Zm*Prx85 abundance shows a non-significant tendency to increase (Figure 3).

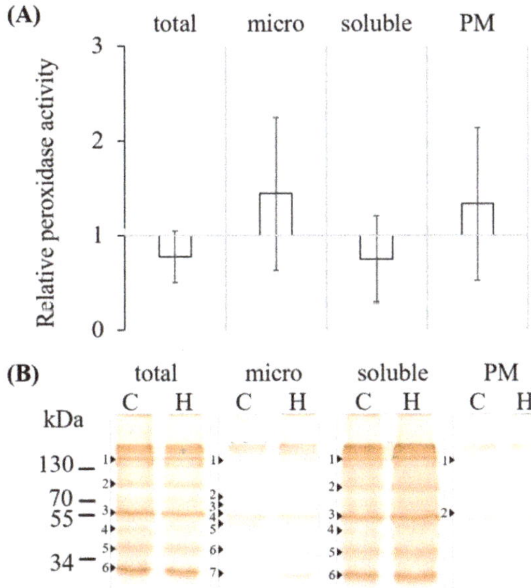

Figure 2. Total guaiacol peroxidase activity and abundance of control and hypoxia-stressed maize roots. Non-stressed (C) and 24 h hypoxia-stressed root samples (H) were divided into total extracts (total) and the sub-proteomes soluble proteins, microsomes (micro), and plasma membranes (PM). (**A**) Using the substrate guaiacol, the total activity was detected spectrophotometrically of the cellular fractions. Data of the stressed samples were related to the controls (three biological and three technical replicates). (**B**) The abundances of guaiacol peroxidases were detected after separation by 11% polyacrylamide gels of the cellular fractions. Peroxidase bands (nos. 1-7) were marked by arrows. Shown is one representative replicate.

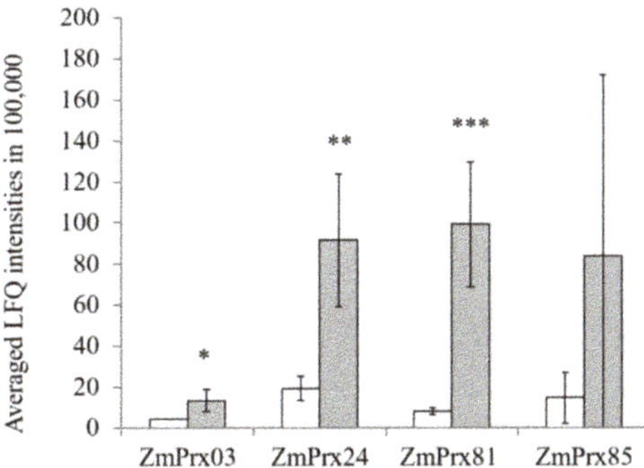

Figure 3. Mass spectrometry analyses of the plasma membrane for class III peroxidases in control and hypoxia-stressed maize roots. Plasma membranes of controls (white columns) and 24 h hypoxia-stressed maize roots (gray columns) were used for mass spectrometry analyses. Label-free quantifications (LFQ) data were determined with MaxQuant software and used for further statistical analyses. Error bars indicate standard deviation. Asterisks indicate significances ($p < 0.05$ *, $p < 0.01$ **, and $p < 0.001$ ***) determined with Student's *t*-test for three biological and two technical replicates per treatment.

Besides these peroxidases, increased abundances were found for several cell wall-related proteins (Supplementary Materials Table S2B). Among these were two dirigent proteins (B4FV87 and B6T6D2),

FLA10 (C0PD01), and two glycerophosphodiester phosphodiesterases (GDPDL3, A0A1D6HBU2, and C0PGU8) that showed a weak increase in hypoxia-stressed samples.

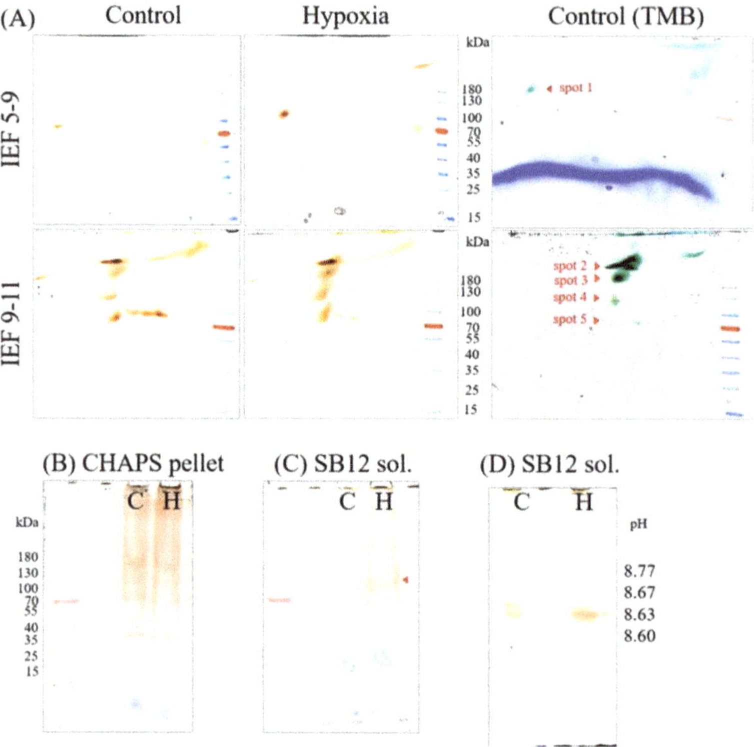

Figure 4. Abundance of guaiacol peroxidases under hypoxia-stress. Plasma membranes (250 µg total protein) of control (C) and 24 h hypoxia-stressed (H) maize roots were solubilized with 8% CHAPS for 1 h. The supernatant was separated on (**A**) native IEF gels pH 5–8 and pH 9–11 followed by 4–18% non-denaturing polyacrylamide gels in second dimensions. The remaining pellet was either separated on 4–18% non-denaturing gels (**B**) or solubilized with SB12 (protein-detergent ratio of 1:7) for 1 h. The SB12 supernatant (sol.) was loaded on 4–18% non-denaturing gels (**C**) and IEF gels pH 9–11 (**D**). Class III peroxidases were visualized by staining with hydrogen peroxide and guaiacol (orange color). No or only weak signal could be determined in the pellet after SB12 solubilization (data not shown). For MS analyses, gels were stained with TMB (blue color). The protein spots and bands, indicated with red arrows, were identified by mass spectrometry. Molecular weight in kDa was determined using a protein standard (PageRuler Prestained Protein Ladder, Thermo Scientific, Waltham, MA, USA). For further details, see the text. SB12, n-dodecyl-*N*, *N*-dimethyl-3-ammonio1-propanesulfonate; CHAPS, 3-[(3-Cholamidopropyl)dimethylammonio]1-propanesulfonate; TMB, 3,3′,5,5′-Tetramethylbenzidine.

2.4. Gel-Based Peroxidase Analyses

Plasma membrane that was solubilized with 3-[(3-Cholamidopropyl)dimethylammonio]-1-propanesulfonate (CHAPS) and separated on native isoelectric focusing (IEF) gels with a pH range from 5 to 9 revealed one recurring spot at acidic pH (Figure 4A). This spot was identified by MS as *Zm*Prx85 with a coverage of 27%. The isoelectric point (pI) and molecular weight (MW in kDa) of this peroxidase (spot no. 1) ranged from pI 4.73 ± 0.11 to 109 ± 11 kDa in control samples and not significantly different from pI 4.82 ± 0.14 to 106 ± 11 kDa in stressed samples. The analysis of the spot intensity revealed a

2.2-fold increased abundance of *Zm*Prx85 in stressed samples compared to controls (four biological and two technical replicates; Figures S3 and S4). Besides, four main spots at pH > 9 were determined in these gels but were furthermore separated on IEF gels with pH range from 9 to 11 (Figure 4A). These spots were identified by MS as *Zm*Prx101 (B4FU88) with pI 8.8 ± 0.1 and 290 ± 37 kDa (spot no. 2), as *Zm*Prx03 with pI 8.7 ± 0.5 and 227 ± 27 kDa (spot no. 3), as *Zm*Prx01 (A5H8G4) and *Zm*Prx03 with pI 8.6 ± 0.4 and 131 ± 9 kDa (spot no. 4), and as *Zm*Prx01 with pI 8.7 ± 0.1 and 82 ± 5 kDa (spot no. 5), respectively. The pI and MW of these spots did not differ significantly between control and stressed samples. The analyses of the abundances showed an increase of *Zm*Prx101 (spot no. 2, 1.3-fold), *Zm*Prx03 (spot no. 3, 2.3-fold), *Zm*Prx01 and *Zm*Prx03 (spot no 4, 3.6-fold), and *Zm*Prx01 (spot no. 5, 1.2-fold) of the stressed samples compared to the control samples, respectively. Spot no. 4 contains the two class III peroxidases *Zm*Prx01 and *Zm*Prx03. In the spot of the control sample, *Zm*Prx01 was the dominant peroxidase (with six peptides) compared to the spot of the stressed sample (with three peptides). Contrary to this, *Zm*Prx03 was the dominant peroxidase in the spot of the stressed sample (with six peptides) compared to the spot of the control sample (with three peptides). Besides these five representative spots, that were found in all biological and technical replicates, some additional spots (spots nos. 6–8) appeared in only some biological samples (Supplementary Materials Figure S2). Four peroxidases (*Zm*Prx24, *Zm*Prx87 (B4FSW5), *Zm*Prx118 (B4FK72), and *Zm*Prx85) were identified in spot no. 6 with pI 8.03–8.3 and 31 ± 2 kDa. Spot no. 7 contained *Zm*Prx85 with pI 8.0–8.4 and 123 ± 9 kDa, and spot no. 8 revealed *Zm*Prx01 with pI 8.7–8.8 and 91 ± 12 kDa. The abundance of these spots was higher in stressed samples compared to control samples (Supplementary Materials Figure S2). Co-separation of these peroxidases with *Zm*RbohB (A0A1D6MT17) and/or *Zm*Rboh04 (A0A1D6QI90) was found in spots nos. 2–5 of the 2D-PAGE (Supplementary Materials Table S3).

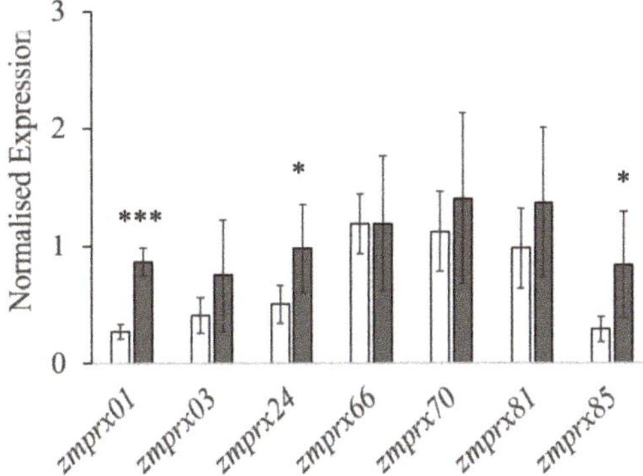

Figure 5. Expression profiles of class III peroxidases identified in the plasma membrane of maize roots. Total RNAs of non-stressed (white columns) and 24 h hypoxia-stressed roots (gray columns) were extracted for gene expression analyses (RT-qPCR) with SYBRGreen. Expression was normalized to *zmtufM* as housekeeping gene. Significancies, calculated with Student's *t*-test, are marked with ($p < 0.05$ * and $p < 0.001$ ***) for three biological and two technical replicates per treatment.

Class III peroxidases were detected in the remaining PM pellets of controls and stressed samples (Figure 4B). After solubilization of these pellets with n-dodecyl-*N*,*N*-dimethyl-3-ammonio-1-propanesulfonate (SB12), *Zm*Prx81 (B4FG39) and *Zm*Prx85 were found in controls and, additionally, *Zm*Prx01 and *Zm*Prx70 (A5H452) were found in stressed samples (Figure 4C). In IEF gels with pH 9–11, the SB12 solubilized supernatant showed guaiacol positive proteins at pH 8.6 (Figure 4D). All identified class III peroxidases with the corresponding peptides can be found in Supplementary Materials Table S3.

2.5. RT-qPCR Analyses

The regulation of the six PM-bound class III peroxidases, identified by gel-free and gel-based analyses, was investigated by real-time quantitative polymerase chain reaction (RT-qPCR) (Figure 5). Although all genes were upregulated, a significant increase was found for *zmprx01* (3.2-fold), *zmprx24* (1.9-fold), and *zmprx85* (2.9-fold), whereas expression of *zmprx03* (1.8-fold), *zmprx70* (1.3-fold), and *zmprx81* (1.4-fold) showed higher levels compared to controls. Additionally, the expression of *zmprx66* was not increased in comparison to controls. The housekeeping gene, used in RT-qPCR (*zmtufM*, Q9FUZ6), could not be found in the RNA Seq data. The NCBI blast with the corresponding sequence led to the protein (NP_001141314) with less than 100% coverage. This protein had a slight non-significant decrease in expression (−1.4-fold change) in RNA Seq analyses, as well as another elongation factor (gene entry 542581, −1.5-fold change).

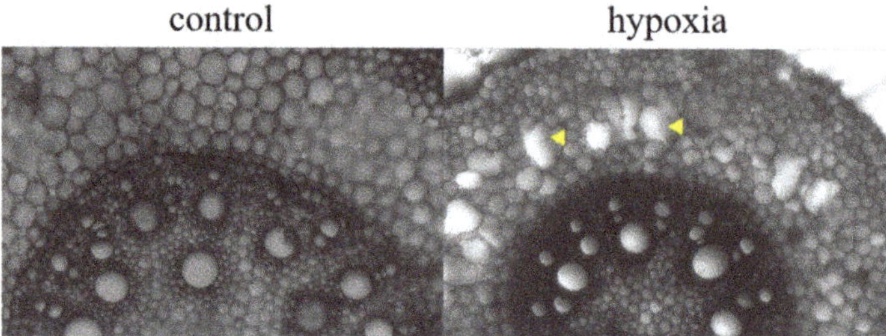

Figure 6. Aerenchyma formation in the mature zone of maize primary root cross-sections. Control and 24 h hypoxia-stressed maize roots were cross-sectioned, by hand, for observation of aerenchyma (indicated with arrows). Images taken with 10x magnification using a Leica DM500 binocular microscope.

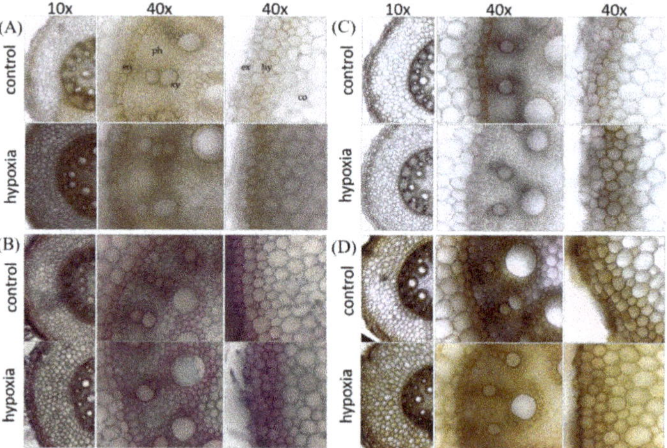

Figure 7. Visualization of lignin, its precursors, and cellulose in the mature zone of primary root cross-sections from control and hypoxia-stressed maize plants. Control and 24 h hypoxia-stressed maize roots were cross-sectioned by hand and analyzed with (**A**) Mäule staining for syringyl-rich polyphenols (deep-red color), (**B**) fuchsin, chrysoidin, and Astra blue (FCA) staining for lignified cells (red color) and non-ligneous cells (blue color), (**C**) Phloroglucinol staining for lignified cells (pink color), and (**D**) chlorine–zinc–iodine staining for cellulose (violet color). For detailed explanation of the colours, see methods. Co, cortex; en, endodermis; ex, exodermis; hy, hypodermis; ph, phloem; xy, xylem.

2.6. Root Cross-Sections and In Vivo Root Staining

During cross-sectioning, aerenchyma formation in the root cortex was visible in some but not all of the stressed plants, while there were no such signs in control plants (Figure 6).

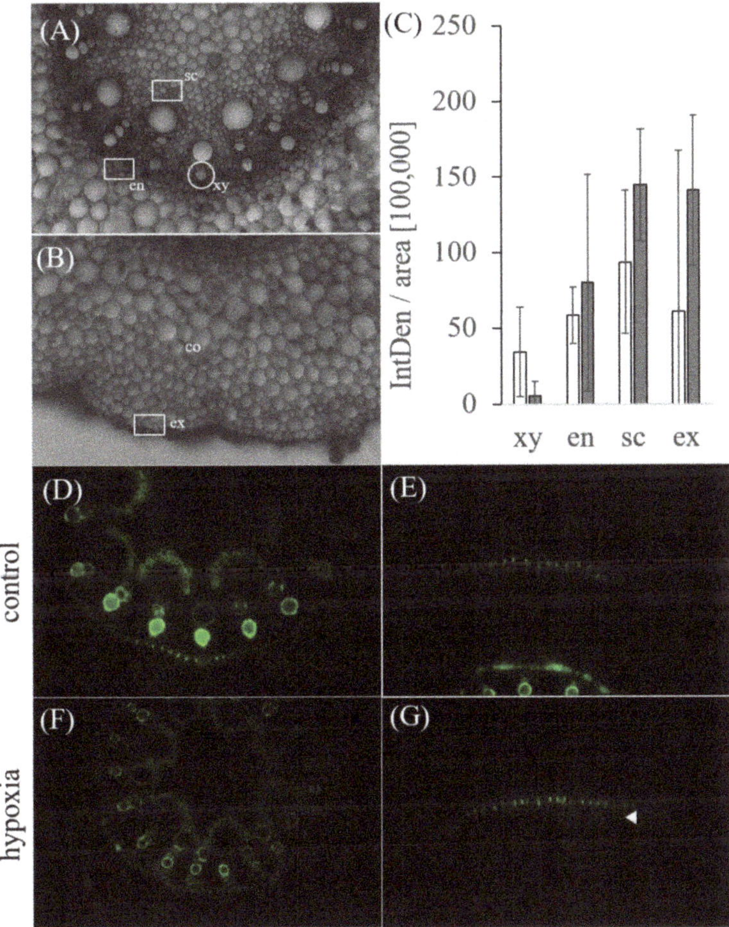

Figure 8. Visualization of suberin in the mature zone of primary root cross-sections from control and hypoxia-stressed maize plants. Control and 24 h hypoxia-stressed maize roots were cross-sectioned by hand and stained with the fluorescent berberine–aniline blue to detect suberin. Images were taken with 10x magnification using an Olympus BHS fluorescent microscope. (**A,D,F**) Shown is an overview of the central cylinder with vascular bundles and (**B,E,F**) the cortex and exodermis of maize root transversal section in brightfield image (**A,B**) and fluorescent image (**D–G**). (G) White arrow indicates an additional suberin layer beneath the exodermis in stressed samples. (**C**) Intensity measurements of endodermis (en), exodermis (ex), vascular sclerenchyma cells (sc) and xylem vessels (xy) were calculated as intensity per area (IntDen/area) for controls (white columns) and stressed samples (grey columns). Error bars indicate standard deviation for three to five biological replicates.

Mäule staining revealed overall darker staining of stressed samples compared to controls including hypodermis, cortical cell region, and vascular sclerenchyma cells (Figure 7A). The yellow-brownish color indicates the presence of guaiacyl lignin monomers in both, controls and stressed samples. No red staining was observed that would show the presence of syringyl lignin monomers. Etzold staining with fuchsin, chrysoidin, and Astra blue (FCA) revealed overall stronger staining of stressed samples compared to controls (Figure 7B). Phloem and pith of the vascular sclerenchyma cells seem non-lignified (blue appearance), hypodermis, endodermis, and xylem cells are lignified (pink appearance). In stressed cells, more hypodermic layers are lignified compared to controls. Phloroglucinol revealed strong staining of hypodermis, endodermis, and xylem vessels. All of those cells were less stained in stressed samples compared to controls (Figure 7C). Controls show strong yellowish staining of the Casparian band in endodermal cell walls. The exodermis and xylem vessels of controls have a positive pink color, indicating lignified cells, which is not observed in stressed samples. There are more yellow-brownish stained hypodermic cell layers in stressed samples compared to controls. With chlorine–zinc–iodine staining, a positive cellulose reaction in cortical and vascular sclerenchyma cells (phloem and pith) of controls was observed, that did not appear in stressed samples. Hypodermis, endodermis, and xylem cells are stained in a brownish color with less intensity in stressed samples (Figure 7D).

Berberin–aniline staining, quantified as intensity per area (IntDen/area), revealed a decreased intensity of suberin in xylem vessels (controls with 34.52 ± 29.71 and stressed with 5.60 ± 9.29 IntDen/area in 100,000). Contrary to this, there is a slightly, but not significant, increased intensity in vascular sclerenchyma cells (controls with 93.57 ± 47.36 and stressed with 144.87 ± 37.06 IntDen/area in 100,000), endodermis (controls with 58.60 ± 18.89 and stressed with 80.10 ± 71.88 IntDen/area in 100,000), and exodermis cells (controls with 61.33 ± 105.91 and stressed with 141.03 ± 49.52 IntDen/area in 100,000) (Figure 8A). A closer look at the exodermis showed an additional suberin layer beneath the exodermis in stressed samples (Figure 8B).

3. Discussion

For the first time, this study shows the induction of PM-bound class III peroxidases in maize roots by hypoxia. Furthermore, we show that the differential regulation of specific class III peroxidases caused cell wall modifications in response to hypoxia and present evidence for the interaction of PM bound class III peroxidases with ZmRbohB and ZmRboh4.

3.1. Hypoxia-Responsive Class III Peroxidases

The transcriptome analyses of maize roots under hypoxia revealed that most of the differentially expressed peroxidase genes were soluble proteins (70%), localized either in the apoplast or bound to cell walls (Figure 1). Only two of these soluble peroxidases (*zmprx06* and *zmprx117*) were upregulated. Although gene regulation may be different in root and leaf, ZmPrx06 was found in a leaf soluble fraction together with other peroxidases [20]. For this fraction, guaiacol peroxidase activity increased by hypoxia. In roots, the majority of soluble peroxidases (n = 10) was downregulated. For membrane-bound peroxidases, the transcript of a putative peroxidase (*zmprx135*) of the endoplasmic reticulum was upregulated, whereas putative PM peroxidase transcripts were either not differentially regulated or downregulated (*zmprx109* and *zmprx140*) (Figure 1C). These peroxidases have not been investigated on the protein level. In silico analyses of *zmprx135* predicted a function related to abscisic acid and heat stress [36].

Fractionation of the root samples showed an increase in membrane-bound peroxidase activities and a decrease in soluble peroxidase activity by hypoxia (Figure 2A). This result fits well with the downregulation of several soluble peroxidase genes (Figure 1A). The increase of peroxidase activity in microsomal fractions (Figure 2A) could be partially explained by the upregulation of *zmprx135*. This assumption will need further proof for the gene product.

However, peroxidase abundance appeared to be higher for the soluble fraction compared to membrane fractions (Figure 2B). This observation may partially depend on the detergent used for solubilization of membrane proteins. Solubilization of PM by CHAPS revealed the maximal peroxidase activity of non-stressed samples for maize seedlings [35]. CHAPS-insoluble membranes still showed guaiacol peroxidase bands with ZmPrx81 and ZmPrx85 that were more intense in stressed samples.

Additionally, *Zm*Prx01 and *Zm*Prx70 were found only in stressed samples (Figure 4C,D). It has been shown that CHAPS and Triton X-100 selectively extract glycerophospholipids and some proteins, whereas the resulting insoluble membranes are strongly enriched in sphingolipids and cholesterol [37]. Identification of these peroxidases in CHAPS-insoluble membranes supports not only a strong interaction of the proteins with the PM [34] but also a localization of the enzymes in microdomains [23].

Both, gel-free and gel-based peroxidase analyses of PM revealed significant increases in the abundance of *Zm*Prx03, *Zm*Prx24, *Zm*Prx81, and *Zm*Prx85 (Figures 3 and 4). None of these peroxidases appeared differentially regulated on the transcriptional level, using RNA Seq analyses (Supplementary Materials Table S1A). The observed downregulation of *zmprx109* and *zmprx140* (Figure 1C) was in agreement with the fact that these putative PM peroxidases could not be detected on the protein level.

To verify the gene regulation of the identified peroxidases as well as PM peroxidases that were already identified in primary maize roots [34], RT-qPCR was performed. The results revealed a significant upregulation of *zmprx01*, *zmprx24*, and *zmprx85* and a tendency of higher expression levels for *zmprx03*, *zmprx70*, and *zmprx81* in stressed samples whereas *zmprx66* was not differentially regulated by hypoxia (Figure 5). A difference between RNA Seq and RT-qPCR was observed for smaller genes with fewer exons and lower expression [38]. Hypoxia-induced upregulation of *zmprx01* confirmed results of chip-based analyses [21]. The higher expression of *zmprx01* did not correlate with changes in abundance of *Zm*Prx01 in gel-free or gel-based proteome analyses (Figures 3 and 4). The protein was not found in LC–MS/MS in contrast to *Zm*Prx03, *Zm*Prx24, *Zm*Prx81, and *Zm*Prx85. A possible reason might be the higher amount of protein (250 µg) used for 2D-PAGE compared to LC–MS/MS (100 µg). In gel-based analyses, *Zm*Prx01 was not only detected in spots 4 and 5 but also in CHAPS-insoluble membranes (Figure 4). Additionally, *Zm*Prx01 revealed a stronger interaction with the PM in the hypoxia-stressed samples compared to the control. These results complicate a quantitative analysis of *Zm*Prx01 compared to the other peroxidases.

3.2. Membrane Protection and Aerenchyma Formation

The observed higher level of hydrogen peroxide in hypoxia-stressed root samples confirmed a production of ROS that lead to oxidative stress and lipid peroxidation [6]. A 5.9-fold upregulation of *zmrboh10* was found (Supplementary Materials Table S1B). This isoform has a function in ROS production and aerenchyma formation in maize roots [13] which fits nicely with the observed aerenchyma in hypoxia-stressed root cross-sections (Figure 6). The higher expression levels of expansins and the downregulation of *zmccoaomt* correlate with the cell wall loosening and degradation during aerenchyma formation (Supplementary Materials Table S1C). The consumption of hydrogen peroxide by hypoxia-induced PM peroxidases (*Zm*Prx01, *Zm*Prx03, *Zm*Prx24, *Zm*Prx70, *Zm*Prx81, and *Zm*Prx85) regulates not only the level of ROS but also protects the membranes of neighbor cells against oxidative stress and lipid peroxidation. Additionally, cell lysis will release NADH that can react with apoplastic and cell wall class III peroxidases and thereby increase ROS production and cell wall loosening [6,39].

3.3. Peroxidase–Rboh Interaction and Cell Wall-Remodeling

Protein assemblies may not be destructed by native IEF or non-denaturing SDS-PAGE as used in the present study and could explain the high molecular masses (100–290 kDa) of the peroxidase spots in 2D-PAGE (Figure 4). The observed co-separation of *Zm*RbohB and *Zm*Rboh04 isoforms with at least *Zm*Prx01, *Zm*Prx03, and *Zm*Prx24 in 2D-PAGE (Supplementary Materials Table S3) suggests an interaction between these proteins.

Transmembrane helices were predicted for *Zm*Prx01, *Zm*Prx70, *Zm*Prx81, and *Zm*Prx85 (Supplementary Materials Table S4). The co-localization of peroxidases with Rboh in functional microdomains could present a mechanism for the fine-tuning of ROS levels and may prevent lipid peroxidation not only under stress conditions [34]. In contrast to the peroxidases, none of the putative interaction partners (*zmrbohB* isoforms, *zmrboh04*) were differentially regulated by hypoxia (Supplementary Materials Table S1B). Opposite results have been found for tomato (*Solanum lycopersicum* L.) roots that showed a strong upregulation of *slrbohB* [14]. In contrast to maize, tomato is a flooding-sensitive species, which may explain this observation.

Protein assemblies may also explain the occurrence of *Zm*Prx03 and *Zm*Prx24 at the PM. These peroxidases appeared to be soluble because transmembrane helices were not predicted for these proteins (Supplementary Materials Table S4). A soluble class III peroxidase in *Arabidopsis thaliana* (L.) HEYNH. (*At*Prx64) was shown to interact with *At*RbohF via a Casparian strip protein and a dirigent-like protein [40]. This protein assembly has a function in lignification during Casparian strip formation in the endodermis. Although homologs of *At*Prx64 were not found in maize, the soluble peroxidases (*Zm*Prx03, *Zm*Prx24) identified in PM may have comparable functions in cell wall processes during hypoxia. Further, higher abundances of dirigent proteins in hypoxia-stressed samples support an interaction between the upregulated peroxidases and *Zm*Rboh (Supplementary Materials Table S2B). To confirm this assumption, a detailed biochemical characterization will be necessary for the future.

Although the data at hand did not allow a detailed discussion of the specific functions of hypoxia-induced peroxidases yet, the higher abundances of *Zm*Prx01, *Zm*Prx03, *Zm*Prx24, *Zm*Prx70, *Zm*Prx81, and *Zm*Prx85 in PM of stressed samples (Figures 3 and 4) point to a function of these peroxidases in cell wall remodeling. Other class III peroxidases (*zmprx06*, *zmprx117*, and *zmprx135*), that were found to be differentially regulated by hypoxia, indicate participation in these processes as well.

Deposition of lignin was observed in hypoxia-stressed maize roots by specific stains (Figure 7A–C). Due to a lack of biochemical characterization of hypoxia-responsive peroxidases, a statement on specific peroxidases involved in these processes may be incomplete. Although the higher expression level of *zmprx70* was not significant (Figure 5), its gene product was detected with higher abundance in microdomains of stressed samples (Figure 4). Biochemical characterization of the partially purified *Zm*Prx01 and *Zm*Prx70 showed a significant increase of guaiacol peroxidase activity in the presence of ferulic acid [35]. For both peroxidases, the preferred substrate—in the absence of guaiacol—was the lignin precursor coniferyl alcohol. Thus, at least *Zm*Prx01 and *Zm*Prx70 should be involved in the observed lignification of hypoxia-stressed roots. This hypothesis was further supported by the upregulation of key enzymes (*zmces* and *zmcad*) of the phenylpropanoid pathway (Supplementary Materials Table S1C) and the higher abundances of dirigent proteins and *Zm*FLA10 (Supplementary Materials Table S2B). Fasciclin-like arabinogalactan proteins precede lignification [41]. Association of *Zm*Prx01 and *Zm*Prx70 with *Zm*CAD1 was suggested by the Search Tool for the Retrieval of Interacting Genes/Proteins (STRING) database [42]. Cinnamyl alcohol dehydrogenase facilitates the substrates for the final peroxidase dependent step of lignin-monomer formation [26]. A function of *Zm*Prx01 and *Zm*Prx70 in monolignol biosynthesis fits nicely with the increase of their abundances observed by fungal elicitors [19]. The function of monolignol biosynthesis plays a crucial role in cell wall apposition-mediated defense against pathogens [43].

Chlorine–zinc–iodine staining revealed lower amounts of cellulose in cross-sections of stressed samples (Figure 7D). This observation matches the downregulation of several cellulose synthases (Supplementary Materials Table S1C), but appear to disagree with the upregulation of *zmcesa5*. Cellulases were also downregulated after 24 h. Although *zmgdpdl3* (A0A1D6KIK2, C0PL13) was significantly upregulated in the RNA Seq experiment, its gene product was not detected in the PM. Abundances of the two *Zm*GDPDL (A0A1D6HBU2, C0PGU8), identified in PM, showed a weak increase compared to controls. Transcripts of these *Zm*GDPDL (C0PGU8, A0A1D6HBU2) were either not differentially regulated or downregulated. Thus, the lower expression of several *zmgdpdl* further supports the decrease in cellulose. Knockouts of this enzyme revealed a lower content of crystalline cellulose [44].

Besides the cellulose, cell walls of grasses contain high amounts of hemicellulose (55%) consisting of arabinoxylans, xyloglucan, and mixed-linked glucans [45]. The higher levels of hydrogen peroxide in hypoxia-stressed samples support a generation of diferulates by *Zm*Prx03, *Zm*Prx24, *Zm*Prx81, or *Zm*Prx85 and thereby a crosslinking of arabinoxylans and lignin by phenolics [31,46]. In maize suspension-cultured cells, it has been demonstrated that a decrease in cellulose content can be compensated by the deposition of lignin-like polymers and a network of highly crosslinked feruloylated arabinoxylans [31].

As shown in Figure 8, quantitative analyses of the berberin–aniline staining revealed a weak increase of suberin in vascular sclerenchyma cells, endodermis and exodermis cells. Peroxidases catalyze the oxidation of cinnamyl alcohols before their polymerization by a peroxidase/hydrogen peroxide-mediated process during suberin formation [27,47,48]. However, *zmhht* that is related to suberin synthesis was downregulated by hypoxia. Enstone and Peterson [49] showed that maize roots grown in hydroponics had significantly fewer suberin lamellae in endodermis and exodermis compared to plants grown in other substrates (e.g., vermiculite). Additionally, induction of exodermal Casparian bands or suberin lamellae failed in the lateral roots of maize grown in hydroponics [50]. Although the formation of a ROL barrier by peroxidases identified was supported by the deposition of lignin and suberin in hypodermal/exodermal cell layers in hypoxia-stressed samples (Figures 7 and 8) and a possible crosslinking of arabinoxylans and lignin, the function of the peroxidases in these processes will need further investigations.

4. Materials and Methods

4.1. Plant Material and Growth Conditions

Maize caryopses (*Zea mays* L. cv. Gelber Badischer Landmais, Saatenunion, Hannover, Germany) were soaked in fully desalted water for 4–6 h and sterilized with 3% H_2O_2 for 10 min. In trays, sterilized with 70% ethanol, the kernels were placed onto and covered with wetted germination tissue. The trays were covered with aluminum foil and stored in dark for four days at 26 °C. The seedlings were transferred into 9 L boxes filled with hydroponic culture medium (5.25 mM KNO_3, 7.75 mM $Ca(NO_3)_2$ $4H_2O$, 4.06 mM $MgSO_4$ $7H_2O$, 1.0 mM KH_2PO_4, 100 µM Fe(III)-EDTA, 46 µM H_3BO_4, 9.18 µM $MnSO_4$ H_2O, 5.4 µM $ZnSO_4$ $7H_2O$, 9.0 µM $CuSO_4$ $7H_2O$, 2.0 µM Na_2MoO_4 $2H_2O$, pH 5.5) which was changed once after ten days. Culturing was performed in a climate chamber (light source: Philips SGR 140 with Philips SON-T Agro 400 W sodium vapor lamp, about 400–500 µmol m^{-2} s^{-1}, 12 h day/night, temperature: 22 °C day/18 °C night), the medium was oxygenated by KOH washed air (compressor type LK60, OSAGA, Glandorf, Germany). After 14 days of culturing, oxygenation was stopped and hypoxia stress was induced by preventing the oxygen supply with 500 mL commercially available rape oil that led to a reduction of oxygen from 21% to 3.5 ± 0.5% after 24 h of stress. The pH thereby stayed stable at 6.8 ± 0.5 to 6.3 ± 0.7 within 24 h of stress induction. Contrary to these stressed plants, control plants were continuously supplied with air (21% oxygen). After 24 h, the roots were harvested between 9 and 10 a.m. (CET). Adhered oil was removed from the roots by washing with 0.1% Triton X-100 for 15–30 s.

4.2. Preparation of Subcellular Fractions

Maize roots were washed (3 mM KCl, 0.5 mM $CaCl_2$, 0.125 mM $MgSO_4$) and homogenized (0.25 M sucrose, 50 mM HEPES, 5 mM Na_2-EDTA, pH 7.5, supplied with 1 mM dithiothreitol and 1% polyvinylpolypyrrolidone), using a Waring blender 7011HS (Co. Waring, Stamford, CT, USA). The homogenate was filtered through a nylon net (125 µm mesh, Co. Hydro-Bios, Kiel, Germany) and 1 mM phenylmethylsulfonyl fluoride was added (=total fraction). After the first centrifugation at 10,000× *g* for 10 min at 4 °C (Avanti J-E centrifuge, rotor type JA-14, Beckman Coulter, Krefeld, Germany), the supernatant was centrifuged at 48,000× *g* (Avanti J-E centrifuge, rotor type JA-25.50, Beckman Coulter, Krefeld, Germany) for 30 min at 4 °C, which resulted in a supernatant with mainly cytosolic, soluble components, and a microsomal pellet. The proteins of the soluble fraction were precipitated with 90% saturated (662 g/L) ammonium sulfate overnight at 4 °C, pelleted at 15,000× *g* (Avanti J-E centrifuge, rotor type JA-25.50, Beckman Coulter, Krefeld, Germany) for 20 min at 4 °C and resolved (0.25 M sucrose, 50 mM HEPES, pH 7.0). The microsomal pellet was resolved in phase buffer (0.25 M sucrose, 5 mM KCl, 5 mM phosphate buffer, pH 7.8) and were used either directly for the following aqueous polymer two-phase partitioning or stored at −76 °C until further use. Plasma membranes were isolated from the microsomal fractions by 36 g phase systems (0.25 M sucrose, 5 mM phosphate buffer, pH 7.8, 5 mM KCl, and 6.5% Dextran T500, 6.5% polyethylene glycol 3350) [51]. Proteins were quantified by using Pierce™ Bovine Serum Albumin Standard (BSA, Co. ThermoFisher Scientific, Waltham, MA, USA, from 2 to 20 µg) for calibration [52].

4.3. Hydrogen Peroxide Assay

For H_2O_2 determination, root tissue of three controls and three 24 h hypoxia-stressed maize plants was ground with liquid nitrogen (fresh weight about 1.3 g) and homogenized in 4 mL buffer (250 mM sucrose, 50 mM HEPES pH 6.8, 1 mM dithiothreitol, and 1% polyvinylpolypyrrolidone) per g fresh weight and centrifuged at 16.000× g for 10 min at 4 °C (rotor Sorvall #3325B, Heraeus Biofuge fresco, ThermoScientific). Hydrogen peroxide was estimated by the ferric-xylenol orange assay [53,54]. For short, two reagents (reagent A with 25 mM $(NH_4)_2Fe(SO_4)_2$, 110 mM $HClO_4$, and reagent B with 125 mM xylenol orange, 100 mM sorbitol) were mixed at a ratio of 1:100. The supernatant of the samples (200 µL) was mixed with xylenol orange solution (1 mL) and incubated in the dark for 30 min. The absorbance was measured with a dual-beam UV/Vis-Spectrophotometer (Type UV-1800, Co. Shimadzu, Hamburg, Germany) at 560 nm. For quantification of H_2O_2, a dilution series of H_2O_2 was produced in oxygen-depleted water which resulted in a calibration curve with a linear range up to 625 µmol H_2O_2.

4.4. Peroxidase Activity

The activity of class III peroxidases was determined in different fractions (total, microsomes, soluble, and PM). The assay contained 775 µL 25 mM sodium acetate buffer pH 5.0, 100 µL 0.3% H_2O_2 (Co. AppliChem, Darmstadt, Germany), 100 µL 89 mM guaiacol (Co. Merck KGaA, Darmstadt, Germany), and 25 µL protein sample with different protein amount (0.02–50 µg). The turnover of guaiacol to tetraguaiacol ($\varepsilon_{470\,nm} = 26.6$ mM^{-1}·cm^{-1}) was measured for 2 min at 470 nm with the UV-1800 spectrophotometer. Values given were from six biological and three technical replicates per sample. The buffer and the two substrates served as a reference.

4.5. Gel-Based Analyses and Mass Spectrometry

Modified SDS-PAGE with 11% polyacrylamid gels was used for the separation of subcellular fractions. Therefore, samples were mixed with 4 x non-reducing loading buffer (500 mM Tris-HCl pH 6.8, 80% (*w/v*) glycerol, 0.08% (*w/v*) SDS, bromophenol blue; [51]) and separated 10 min at 80 V and about 120 min at 120 V.

For two-dimensional polyacrylamide gel electrophoresis (2D-PAGE), washed PM (250 µg total protein content) were pelleted at 105,000 g for 30 min at 4 °C, solubilized in 3x IEF loading buffer (8% ampholytes, 8% CHAPS, 40% glycerol, 3 M urea) on ice for 1 h and centrifuged again. Afterwards, the remaining pellet was solubilized with 7:1 SB12 on ice for 2 h and centrifuged at 105,000× g for 45 min or 13,000× g for 60 min at 4 °C [23]. The first dimension was performed with native IEF gels (2% CHAPS, 3 M urea, 7.5% acrylamide, 2% ampholytes pH 5–8 and 8–11) with 20 mM NaOH (cathode buffer) and 10 mM phosphoric acid (anode buffer) in an electric gradient (12 h 30 V, 2 h 100 V, 1.5 h 250 V, 1 h 300 V) at 4 °C. The pH of the 0.5 cm thick gel pieces was measured after the run. The sample-loaded gel lanes were equilibrated (125 mM Tris, 1% SDS, 10% glycerol, pH 8.8) at 4 °C for 1 h and transferred to a 4–18% polyacrylamide gradient gel for the second dimension under non-reducing conditions. This gel electrophoresis was performed at 4 °C with 30 mA per gel for 10 min at 80 V and about 120 min at 150–200 V. After the run, gels were stained with guaiacol (0.5% (*v/v*) in 50 mM sodium acetate buffer pH 5.0 and 0.5% H_2O_2) or 3,3′,5,5′-tetramethylbenzidine (4.7 mM TMB, 30% methanol in 50 mM sodium acetate buffer pH 5.0 and 0.1% H_2O_2). The intensity of single spots was quantified by using Image J

(Image J software version 1.53a, Bethesda, MD, USA). The determination of pI and MW was performed by using pH and protein standard [51]. TMB stained protein spots were cut out and in-gel digestion of proteins and liquid chromatography MS (LC–MS/MS) was done [55]. LC–MS/MS data were processed with Proteome Discoverer 2.0 (Thermo Scientific, Bremen, Germany). Identification of the proteins from the MS/MS spectra was performed with the search engine Sequest HT, using the MaizeGeneDatabase (https://www.maizegdb.org/) and the Peroxibase (http://peroxibase.toulouse.inra.fr/). For the searches, the following parameters were applied: precursor mass tolerance: 10 ppm and fragment mass tolerance: 0.2 Da. Two missed cleavages were allowed. Carbamidomethylation on cysteine residues as a fixed modification and oxidation of methionine residues as a variable modification was used for the search. Peptides with a false discovery rate of 1%, using Percolator, were identified. At least two unique peptides per protein were used as a condition for reliable identification. Peroxidase nomenclature was used in accordance with Peroxibase.

4.6. Gel-Free Peroxidase Analyses

Plasma membrane preparation and further MS analyses were done according to previous studies [56]. PM of three stressed and three control plants were enriched and prepared before MS analyses as described as followed: 100 µg total PM protein were washed (250 mM sucrose, 50 mM HEPES, 150 mM KCl, 0.01% Triton X-100) for 30 min, then pelleted for 1 h at 13,000× g at 4 °C. The pellet was incubated in 200 µL of solubilization buffer (125 mM Tris-HCl pH 6.5, 2% SDS, 5% mercaptoethanol, 6 M urea) for 1 h at room temperature and centrifuged again at 13,000× g for 60 min. Proteins in the resulting supernatant were precipitated with 1.8 mL methanol:chloroform (4:1) at −20 °C overnight, centrifuged, and washed three times in 0.5 mL pure methanol by centrifugation at 13,000× g for 20 min. The washed pellet was dried for 30 min, resuspended in 50 µL digestion buffer (200 mM NH_4CO_3 pH 8.5, 8 M urea, 10% acetonitrile (ACN)) and incubated in addition of 0.1–0.5 µg lysin C at 37 °C for 16–18 h. Then, the sample was diluted 1:3 in 10% ACN, 10 µL of trypsin beads were added and incubated at 37 °C for 16–24 h for hybridization. This digestion was stopped by adding three times volume of 0.3% heptafluorobutyric acid and trypsin beads were removed by centrifugation. Peptides were washed and dried, using ZipTips (Co. Agilent Technologies, Santa Clara, CA, USA) according to the manufacturer's protocol. Peptides were dissolved in 2% ACN, 0.1% formic acid. In random order 1 µg was applied on a C18 column (15 cm, 50 mm column, PepMapR RSLC, Thermo Scientific, 2 mm particle size) for separation during a 90 min gradient at a flow rate of 300 nL min^{-1}. Measurement was done on an LTQ-Orbitrap Elite (Thermo Fisher Scientific, Bremen, Germany) with the following settings: full scan range 350–1800 m/z, max 20 MS2 scans (activation type CID), repeat count 1, repeat duration 30 s, exclusion list size 500, exclusion duration 60 s, charge state screening enabled with a rejection of unassigned and +1 charge states, minimum signal threshold 500. Proteins were identified and quantified as described earlier [57], using a UniprotKB FASTA download for *Zea mays* (UP000007305) and the software MaxQuant v1.6.5.0 with the following parameters: first search peptide tolerance 20 ppm, main search tolerance 4.5 ppm, ITMS MS/MS match tolerance 0.6 Da. A maximum of 3 of the following variable modifications were allowed per peptide: oxidation of methionine and acetylation of the N-term. A maximum of two missed cleavages were tolerated. The best retention time alignment function was determined in a 20 min window. Identifications were matched between runs in a 0.7 min window. An FDR cutoff at 0.01 (at Peptide Spectrum Match and protein level) was set with a reversed decoy database. A minimum of seven amino acids was required for the identification of peptides and at least two peptides were required for protein identification. The resulting data matrix was filtered so that there are label-free quantifications (LFQ) in at least one of the treatments (control and stressed) and more than four replicates (biological and/or technical replicates). Missing values that appear due to low abundant proteins or an oversupply of peptides during MS run were corrected with COVAIN [58]. The LFQ intensities (the normalized intensities) of the control and stressed samples were averaged of three biological and two technical replicates. Standard deviation and Student's t-test were used to determine significant changes. The stressed samples were normalized to the controls. The obtained ratios show either increase (ratio > 1.05) or decrease (ratio < 0.95) of the proteins on a comparison pair (stressed versus control).

4.7. Isolation of Total RNA

Stressed and control maize roots of three biological replicates were harvested after 24 h. For each biological replicate, at least five plants were pooled. The roots were ground with a mortar and pestle, using liquid nitrogen to get a very fine powder (about 0.3 g fresh weight). Total RNA isolation from this powder was done with the NucleoSpin® RNA Plant and Fungi Kit (Co. Macherey-Nagel, Düren, Germany). For RNA Seq analyses, a final step of ethanol precipitation with 1/10th volume of 3 M sodium acetate pH 5.2 and three volumes of 100% ethanol absolute was added before delivering the samples to Macrogen Inc. (Seoul, South Korea) for further analyses.

4.8. Quality Control and RNA Sequencing (RNA Seq)

Quality control (QC) and analyses of the total RNA samples were done by Macrogen Inc. (Seoul, South Korea). QC analyses for verifying the quantity and quality of the RNA samples were performed by using agarose gel electrophoresis and an Agilent Technologies 2100 Bioanalyzer (Agilent Technologies, Santa Clara, CA, USA). Six high-quality RNA samples with an RNA Integrity Number (RIN) value greater than or equal to seven were used for cDNA library construction. Sequencing was done, using Illumina Sequencing. RefGen_v4 of maize was used as a reference gene (ftp://ftp.ncbi.nlm.nih.gov/genomes/all/GCF/000/005/005/GCF_000005005.2_B73_RefGen_v4/)

4.9. Quantitative Reverse-Transcription Polymerase Chain Reaction (RT-qPCR)

Expression levels of proteins, identified by MS, were verified by RT-qPCR. The concentration and purity of the isolated total RNA were determined with a Nanodrop spectrophotometer (Fisher Scientific GmbH, Schwerte, Germany) and agarose gel electrophoresis (1.5% agarose in 1x TAE (40 mM Tris, 20 mM acetic acid, 1 mM EDTA), run 60 V 3 h). The cDNA was prepared from 100 ng of total RNA with the First Strand cDNA Synthesis Kit (Co. Fisher Scientific GmbH, Schwerte, Germany) according to the manufacturer´s protocol. Efficiencies of the primers were checked first with PCR (7 min 95 °C, 30–35 cycles of 20 s 95 °C, 30 s 60 °C, 30 s 72 °C, and finally 7 min 72 °C), using the Maxima Hot Start Kit (Co. Thermo Scientific, Massachusetts, USA) followed by agarose gel electrophoresis (1% agarose in 1x TAE, run 60 V) and second with RT-qPCR (5 min 95 °C, 40 cycles of 10 s 95 °C and 30 s 60 °C terminating in a melting curve from 65 to 95 °C with $0.5\ °C\ s^{-1}$ steps), using Quantifast SYBR green PCR kit (Qiagen GmbH, Hilden, Germany) and the CFX 96 Cycler (CFX96 Touch system, Bio-Rad, Munich, Germany). To analyse specific maize peroxidases (ZmPrx), a set of primers were designed (Eurofins Genomics Germany GmbH, Ebersberg, Germany, Table 1). As a housekeeping gene, *Zea mays* translational elongation factor EF-Tu (*zmtufM*, AF264877.1, Q9FUZ6) was used. For statistical analysis, RT-qPCR was performed twice for three biological replicates of each treatment and compared to the housekeeping gene, using the CFX manager software version 3.1 (Co. Bio-Rad, Hercules, USA).

Table 1. Primer sequences in 5′-3′-orientation for RT-qPCR.

Name of Peroxidase	Forward Primer	Reverse Primer
zmprx01	ACTTGTTCAAGGCCAAGGAG	TTCGTGCTTGTGTTCCAGAC
zmprx03	TCAAGATGGGGCAGATCGAG	ACTCCAGTGAATCCTGATGGG
zmprx24	GGCTCATCCGCATCTTCTT	TGGTTGGGTACCTCGATCT
zmprx66	CGACATGGTTGCACTCTCAG	CGAAGGCGGAGTTGATGTTG
zmprx70	CCACCTCCATGACTGCTTTG	TTCGGATTAGCGGTCTGCTC
zmprx81	CAGGAGGATGACTTCGCCAG	CCGTTGTAGGGTCCCTGATG
zmprx85	GACGCTGAGGAAGAACAAGG	CTGGTCGAAGAACCACCAG
zmtufM	CGCAGTTGATGAGTACATCC	AACACGCCCAGTAACAACAG

4.10. In Vivo Cell Wall Staining

Handmade cross-sections of maize primary roots (24 h hypoxia-stressed and controls; mature differentiation zone) were prepared with a razor blade, stained with different methods, and imaged. A Leica DM500 binocular microscope (10x objective #13613241 and 40x objective #13613242, Leica, Wetzlar, Germany) and an Olympus BHS fluorescent microscope (10x SPLAN Apo objective, Olympus "B" dichroic mirror (DM500) and EY455 excitation filter) with excitation from 455 to 490 nm and emission LP at about 515 nm) were used. For Mäule staining, sections were placed in 1% $KMnO_4$ solution for 5 min, and then washed 3x with water. After 30 min incubation in fresh prepared 1 N HCl solution, sections were washed again, and 1 M Tris-HCl pH 8.0 was added [59]. Mäule staining results in a positive deep-red colored reaction produced by 3-methoxy-o-quinone structures generated from syringyl lignin monomers (derived from sinapic acid) or a negative yellow reaction. The latter indicates the localization of guaicyl lignin monomers (derived from ferulic acid) [59–61]. For Wiesner stain, sections were incubated in 3% phloroglucinol-in-ethanol solution for 20 min. After adding 37% HCl the sections were directly imaged. "Using the phloroglucinol reagent, a distinction can be made between (I) aldehydes (intense orange-red colour), (II) anethole, asarones, isosafrole (no colour) and (III) the group of eugenol, methyleugenol, myristicin and safrole (pink)" [62]. Yellow-to-red colors develop with certain compounds containing aldehydes. A strong orange staining is determined when a reaction with cinnamic aldehydes or cinnamic alcohols occurs [63,64]. Phloroglucinol-HCl (Wiesner reagent) reacts with the cinnamaldehyde groups in lignin, resulting in a pink color of lignified cell walls [61]. Etzold staining is a simultaneous staining with fuchsin, chrysoidin, and Astra blue (FCA). A ready-to-use staining solution is commercially available (Co. Morphisto GmbH, Frankfurt am Main, Germany) and added directly onto the sections. Non-ligneous cells, cell walls, and phloem show blue, ligneous cell walls, and xylem show red color. A ready-to-use solution of chlorine–zinc–iodine according to Behrens (Co. Morphisto GmbH, Frankfurt am Main, Germany) was used for the detection of cellulose. A positive reaction results in blue to violet color. To detect suberines [65,66], sections were incubated in dark for 1 h in 0.1% berberine hemisulphate, then washed and incubated 30 min in 0.5% anilin blue, then washed again and imaged. Shading correction of the fluorescent images was done with Image J plugin BaSiC [67], using 2.5% Lucifer Yellow as a flat field image.

5. Conclusions

For the first time, regulation, abundance, and activity of hypoxia-responsive class III peroxidases of the PM were studied. The data at hand revealed functions in (i) cell-wall loosening and membrane protection during aerenchyma formation; and (ii) lignification (*ZmPrx01*, *ZmPrx70*), suberization, and cell wall crosslinking during hypoxia-induced cell wall remodeling. To clarify specific functions of hypoxia-responsive peroxidases (*ZmPrx01*, *ZmPrx03*, *ZmPrx24*, *ZmPrx70*, *ZmPrx81*, and *ZmPrx85*), future research needs to be focused on peroxidase–Rboh interaction and biochemical characterization of these peroxidases. Due to the significant upregulation of *zmpr01*, *zmprx24*, and *zmprx85* by hypoxia, these peroxidases are suitable hypoxia-specific stress marker candidates.

Supplementary Materials: Supplementary Materials can be found at http://www.mdpi.com/1422-0067/21/22/8872/s1. Figure S1: Western blot. Figure S2: Abundance of guaiacol peroxidases under hypoxia stress. Figure S3: Biological and technical replicates of 2D-PAGE pH 9–11. Figure S4: Biological and technical replicates of 2D-PAGE pH 5–9. Table S1A: RNA Seq analyses of class III peroxidases. Table S1B: RNA Seq analyses of RBOH. Table S1C: RNA Seq analyses of cell wall-related genes. Table S2A: Mass spectrometry analyses of class III peroxidases. Table S2B: Mass spectrometry analyses of cell wall-related proteins. Table S3: Mass spectrometry analyses of the peroxidases. Table S4: In silico prediction of the identified class III peroxidases.

Author Contributions: A.H. and S.L. designed the experiments; A.H. performed the experiments and analyzed the data; S.W. and S.H. executed MS analyses. A.H. and F.B. performed the Xylenol assay. A.H. and S.L. wrote the manuscript with the collaboration of all co-authors and discussed the results. All authors have read and agreed to the published version of the manuscript.

Funding: This research was funded by Elisabeth-Appuhn-Foundation.

Acknowledgments: Thanks go to Company Macrogen for RNA Sequence analyses and PhD Teresa Martínez-Cortés (Universidade da Coruña, Coruña, Spain) for scientific support of RT-qPCR.

Conflicts of Interest: The authors declare no conflict of interest. The funders had no role in the design of the study; in the collection, analyses, or interpretation of data; in the writing of the manuscript, or in the decision to publish the results.

Abbreviations

2D	two-dimensional
ACN	acetonitrile
At	*Arabidopsis thaliana*
BSA	bovine serum albumin
C	control sample
CAD	cinnamyl alcohol dehydrogenase
CCoAOMT	caffeoyl-CoA *O*-methyltransferase 1
cDNA	copy desoxyribonucleic acid
CESA	cellulose synthase
CET	Central European Time
CHAPS	3-[(3-cholamidopropyl)dimethylammonio]-1-propanesulfonate
co	Cortex
Cox2	cytochrome c oxidase
CSE	caffeoylshikimate esterase
Cv	Cultivar
DEGs	differentially expressed genes
DIR	Dirigent
$\varepsilon_{470\,nm}$	extinction coefficient at 470 nm
ECL	enhanced chemiluminescence
EDTA	ethylenediaminetetraacetic acid
EF	elongation factor
en	Endodermis
ER	endoplasmic reticulum
ex	Exodermis
EXP	Expansin
EXPL	expansin-like
FCA	fuchsin, chrysoidin, and Astra blue
FDR	false discovery rate
FLA	fasciclin-like arabinogalactan
FPKM	fragments per kilobase million
GDPD	glycerophosphodiesterase
GDPDL	glycerophosphodiester phosphodiesterase-like
H	hypoxia-stressed sample
H^{+}-ATPase	PM specific H^{+}ATPase
HEPES	4-(2-hydroxyethyl)-1-piperazineethanesulfonic acid
HHT	hydroxycinnamoyl-CoA:ω-hydroxyacid *O*-hydroxycinnamoyltransferase
hy	hypodermis
IEF	isoelectric focusing
IntDen/area	Integrated Density per area
kDa	kilodalton
LC–MS/MS	liquid chromatography mass spectrometry
LFQ	label-free quantifications
micro	microsomes
MS	mass spectrometry
MW	molecular weight

NADH	nicotinamide adenine dinucleotide
NCBI	National Center for Biotechnology Information
PAGE	polyacrylamid gel electrophoresis
ph	phloem
pI	point isoelectric
PM	plasma membrane
ppm	parts per million
Prx	class III peroxidases
QC	quality control
Rboh	respiratory burst oxidase homologs
RIN	RNA integrity number
RNA	ribonucleic acid
RNA Seq	RNA sequence analyses
ROL	radial oxygen loss
ROS	reactive oxygen species
RT	room temperature
RT-qPCR	real-time quantitative polymerase chain reaction; quantitative reverse-transcription polymerase chain reaction
SB12	n-dodecyl-*N*, *N*-dimethyl-3-ammonio-1-propanesulfonate
sc	vascular sclerenchyma cells
SDS	sodiumdodecylsulfate
STRING	Search Tool for the Retrieval of Interacting Genes/Proteins
TAE	tris-acetate-EDTA
TMB	3,3′,5,5′-Tetramethylbenzidine
Tris	tris(hydroxymethyl)aminomethane
Triton X-100	2-[4-(2,4,4-trimethylpentan-2-yl)phenoxyl] ethanol
tufM	thermo unstable translation elongation factor, mitochondrial
V-PPase	pyrophosphate-energized vacuolar membrane proton pump 1
xy	xylem
Zm	*Zea mays*

References

1. Bailey-Serres, J.; Lee, S.C.; Brinton, E. Waterproofing crops: Effective flooding survival strategies. *Plant. Physiol.* **2012**, *160*, 1698–1709. [CrossRef]
2. Nishiuchi, S.; Yamauchi, T.; Takahashi, H.; Kotula, L.; Nakazono, M. Mechanisms for coping with submergence and waterlogging in rice. *Rice* **2012**, *5*, 2. [CrossRef]
3. Komatsu, S.; Shirasaka, N.; Sakata, K.J. 'Omics' techniques for identifying flooding-response mechanisms in soybean. *J. Proteom.* **2013**, *93*, 169–178. [CrossRef]
4. Yamauchi, T.; Shimamura, S.; Nakazono, M.; Mochizuki, T. Aerenchyma formation in crop species: A review. *Field Crops Res.* **2013**, *152*, 8–16. [CrossRef]
5. Yordanova, R.Y.; Popova, L.P. Flooding-induced changes in photosynthesis and oxidative status in maize plants. *Acta Physiol. Plant.* **2007**, *29*, 535–541. [CrossRef]
6. Blokhina, O.; Virolainen, E.; Fagerstedt, K.V. Antioxidants, oxidative damage and oxygen deprivation stress: A review. *Ann. Bot.* **2003**, *91*, 179–194. [CrossRef]
7. Noctor, G.; Reichheld, J.-P.; Foyer, C.H. ROS-related redox regulation and signaling in plants. *Semin. Cell Dev. Biol.* **2018**, *80*, 3–12. [CrossRef] [PubMed]
8. Smirnoff, N.; Arnaud, D. Hydrogen peroxide metabolism and functions in plants. *New Phytol.* **2019**, *221*, 1197–1214. [CrossRef] [PubMed]
9. Monk, L.S.; Fagerstedt, K.V.; Crawford, R.M.M. Oxygen toxicity and superoxide dismutase as an antioxidant in physiological stress. *Physiol. Plant* **1989**, *76*, 456–459. [CrossRef]
10. Yu, Q.; Rengel, Z. Drought and salinity differentially influence activities of superoxide dismutase in narrow-leafed lupins. *Plant Sci.* **1999**, *142*, 1–11. [CrossRef]

11. Gunawardena, A.H.L.A.N.; Pearce, D.M.E.; Jackson, M.B.; Hawes, C.R.; Evans, D.E. Rapid changes in cell wall pectic polysaccharides are closely associated with early stages of aerenchyma formation, a spatially localized form of programmed cell death in roots of maize (*Zea mays* L.) promoted by ethylene. *Plant Cell Environ.* **2001**, *24*, 1369–1375. [CrossRef]

12. Arora, K.; Panda, K.K.; Mittal, S.; Mallikarjuna, M.G.; Rao, A.R.; Dash, P.K.; Thirunavukkarasu, N. RNAseq revealed the important gene pathways controlling adaptive mechanisms under waterlogged stress in maize. *Sci. Rep.* **2017**, *7*, 10950. [CrossRef] [PubMed]

13. Rajhi, I.; Yamauchi, T.; Takahashi, H.; Nishiuchi, S.; Shiono, K.; Watanabe, R.; Mliki, A.; Nagamura, Y.; Tsutsumi, N.; Nishizawa, N.K.; et al. Identification of genes expressed in maize root cortical cells during lysigenous aerenchyma formation using laser microdissection and microarray analyses. *New Phytol.* **2011**, *190*, 351–368. [CrossRef] [PubMed]

14. Safavi-Rizi, V.; Herde, M.; Stöhr, C. RNA-Seq reveals novel genes and pathways associated with hypoxia duration and tolerance in tomato root. *Sci. Rep.* **2020**, *10*, 1692. [CrossRef] [PubMed]

15. Watanabe, K.; Nishiuchi, S.; Kulichikhin, K.; Nakazono, M. Does suberin accumulation in plant roots contribute to waterlogging tolerance? *Front. Plant. Sci.* **2013**, *4*, 178. [CrossRef] [PubMed]

16. Abiko, T.; Kotula, L.; Shiono, K.; Malik, A.I.; Colmer, T.D.; Nakazono, M. Enhanced formation of aerenchyma and induction of a barrier to radial oxygen loss in adventitious roots of *Zea nicaraguensis* contribute to its waterlogging tolerance as compared with maize (*Zea mays* ssp. mays). *Plant. Cell Environ.* **2012**, *35*, 1618–1630. [CrossRef] [PubMed]

17. Castillo, F. Extracellular peroxidases as markers of stress. In *Molecular and Physiological Aspects of Plant. Peroxidases*; Greppin, H., Penel, C., Gaspar, T., Eds.; Université de Genève, Centre de botanique: Geneva, Switzerland, 1986; pp. 419–426.

18. Komatsu, S.; Hiraga, S.; Yanagawa, Y. Proteomics techniques for the development of flood tolerant crops. *J. Proteome Res.* **2012**, *11*, 68–78. [CrossRef]

19. Mika, A.; Boenisch, M.J.; Hopff, D.; Lüthje, S. Membrane-bound guaiacol peroxidases from maize (*Zea mays* L.) roots are regulated by methyl jasmonate, salicylic acid, and pathogen elicitors. *J. Exp. Bot.* **2010**, *61*, 831–841. [CrossRef]

20. Meisrimler, C.N.; Buck, F.; Lüthje, S. Alterations in soluble class III peroxidases of maize shoots by flooding stress. *Proteomes* **2014**, *2*, 303–322. [CrossRef]

21. Thirunavukkarasu, N.; Hossain, F.; Mohan, S.; Shiriga, K.; Mittal, S.; Sharma, R.; Singh, R.K.; Gupta, H.S. Genome-wide expression of transcriptomes and their co-expression pattern in subtropical maize (*Zea mays* L.) under waterlogging stress. *PLoS ONE* **2013**, *8*, e70433. [CrossRef]

22. De Gara, L. Class III peroxidases and ascorbate metabolism in plants. *Phytochem. Rev.* **2004**, *3*, 195–205. [CrossRef]

23. Lüthje, S.; Meisrimler, C.N.; Hopff, D.; Möller, B. Phylogeny, topology, structure and functions of membrane-bound class III peroxidases in vascular plants. *Phytochemistry* **2011**, *72*, 1124–1135. [CrossRef] [PubMed]

24. Cosio, C.; Dunand, C. Specific functions of individual class III peroxidase genes. *J. Exp. Bot.* **2009**, *60*, 391–408. [CrossRef] [PubMed]

25. Passardi, F.; Cosio, C.; Penel, C.; Dunand, C. Peroxidases have more functions than a Swiss army knife. *Plant. Cell Rep.* **2005**, *24*, 255–265. [CrossRef] [PubMed]

26. Liu, Q.; Luo, L.; Zheng, L. Lignins: Biosynthesis and biological functions in plants. *Int. J. Mol. Sci.* **2018**, *19*, 335. [CrossRef]

27. Vishwanath, S.J.; Delude, C.; Domergue, F.; Rowland, O. Suberin: Biosynthesis, regulation, and polymer assembly of a protective extracellular barrier. *Plant. Cell Rep.* **2015**, *34*, 573–586. [CrossRef]

28. Novo-Uzal, E.; Fernandez-Perez, F.; Herrero, J.; Gutierrez, J.; Gomez-Ros, L.V.; Bernal, M.A.; Diaz, J.; Cuello, J.; Pomar, F.; Pedreno, M.A. From *Zinnia* to *Arabidopsis*: Approaching the involvement of peroxidases in lignification. *J. Exper. Bot.* **2013**, *64*, 3499–3518. [CrossRef]

29. Almagro, L.; Gomez Ros, L.V.; Belchi-Navarro, S.; Bru, R.; Ros Barcelo, A.; Pedreno, M.A. Class III peroxidases in plant defence reactions. *J. Exp. Bot.* **2009**, *60*, 377–390. [CrossRef]

30. Fry, S.C. Oxidative coupling of tyrosine and ferulic acid residues: Intra- andextra-protoplasmic occurrence, predominance of trimers and larger products, and possible role in inter-polymeric cross-linking. *Phytochem. Rev.* **2004**, *3*, 97–111. [CrossRef]

31. Martinez-Rubio, R.; Acebes, J.L.; Encina, A.; Karkonen, A. Class III peroxidases in cellulose deficient cultured maize cells during cell wall remodeling. *Physiol. Plant.* **2018**, *164*, 45–55. [CrossRef]

32. Sekhon, R.S.; Lin, H.; Childs, K.L.; Hansey, C.N.; Buell, C.R.; de Leon, N.; Kaeppler, S.M. Genome-wide atlas of transcription during maize development. *Plant. J.* **2011**, *66*, 553–563. [CrossRef] [PubMed]

33. Wang, Y.; Wang, Q.; Zhao, Y.; Han, G.; Zhu, S. Systematic analysis of maize class III peroxidase gene family reveals a conserved subfamily involved in abiotic stress response. *Gene* **2015**, *566*, 95–108. [CrossRef]

34. Mika, A.; Buck, F.; Lüthje, S. Membrane-bound class III peroxidases: Identification, biochemical properties and sequence analysis of isoenzymes purified from maize (*Zea mays* L.) roots. *J. Proteom.* **2008**, *71*, 412–424. [CrossRef] [PubMed]

35. Mika, A.; Lüthje, S. Properties of guaiacol peroxidase activities isolated from corn root plasma membranes. *Plant. Physiol.* **2003**, *132*, 1489–1498. [CrossRef] [PubMed]

36. Lüthje, S.; Martinez-Cortes, T. Membrane-bound class III peroxidases: Unexpected enzymes with exciting functions. *Int. J. Mol. Sci.* **2018**, *19*, 2876. [CrossRef] [PubMed]

37. Pike, L.J. Lipid rafts heterogeneity on the high seas. *Biochem. J.* **2004**, *378*, 281–292. [CrossRef] [PubMed]

38. Everaert, J.; Podina, I.R.; Koster, E.H.W. A comprehensive meta-analysis of interpretation biases in depression. *Clin. Psychol. Rev.* **2017**, *58*, 33–48. [CrossRef]

39. Liszkay, A.; Kenk, B.; Schopfer, P. Evidence for the involvement of cell wall peroxidase in the generation of hydroxyl radicals mediating extension growth. *Planta* **2003**, *217*, 658–667. [CrossRef]

40. Lee, Y.; Rubio, M.C.; Alassimone, J.; Geldner, N. A mechanism for localized lignin deposition in the endodermis. *Cell* **2013**, *153*, 402–412. [CrossRef]

41. Ito, S.; Suzuki, Y.; Miyamoto, K.; Ueda, J.; Yamaguchi, I. AtFLA11, a fasciclin-like arabinogalactan-protein, specifically localized in sclerenchyma cells. *Biosci. Biotechnol. Biochem.* **2005**, *69*, 1963–1969. [CrossRef]

42. Von Mering, C.; Jensen, L.J.; Snel, B.; Hooper, S.D.; Krupp, M.; Foglierini, M.; Jouffre, N.; Huynen, M.A.; Bork, P. STRING: Known and predicted protein-protein associations, integrated and transferred across organisms. *Nucleic Acids Res.* **2005**, *33*, D433–D437. [CrossRef] [PubMed]

43. Bhuiyan, N.H.; Selvaraj, G.; Wei, Y.; King, J. Gene expression profiling and silencing reveal that monolignol biosynthesis plays a critical role in penetration defence in wheat against powdery mildew invasion. *J. Exp. Bot.* **2009**, *60*, 509–521. [CrossRef] [PubMed]

44. Hayashi, S.; Ishii, T.; Matsunaga, T.; Tominaga, R.; Kuromori, T.; Wada, T.; Shinozaki, K.; Hirayama, T. The glycerophosphoryl diester phosphodiesterase-like proteins SHV3 and its homologs play important roles in cell wall organization. *Plant Cell Physiol.* **2008**, *49*, 1522–1535. [CrossRef] [PubMed]

45. Santiago, R.; Barros-Rios, J.; Malvar, R.A. Impact of cell wall composition on maize resistance to pests and diseases. *Int. J. Mol. Sci.* **2013**, *14*, 6960–6980. [CrossRef]

46. Encina, A.; Fry, S.C. Oxidative coupling of a feruloyl-arabinoxylan trisaccharide (FAXX) in the walls of living maize cells requires endogenous hydrogen peroxide and is controlled by a low-Mr apoplastic inhibitor. *Planta* **2005**, *223*, 77–89. [CrossRef]

47. Roberts, E.; Kutchan, T.; Kolattukudy, P.E. Cloning and sequencing of cDNA for a highly anionic peroxidase from potato and the induction of its mRNA in suberizing potato tubers and tomato fruits. *Plant. Mol. Biol.* **1988**, *11*, 15–26. [CrossRef]

48. Whetten, R.W.; MacKay, J.J.; Sederoff, R.R. Recent advantages in understanding lignin biosynthesis. *Ann. Rev. Plant. Physiol. Plant. Mol. Biol.* **1998**, *49*, 585–609. [CrossRef]

49. Enstone, D.; Peterson, C.A. Suberin lamella development in maize seedling roots grown in aerated and stagnant conditions. *Plant Cell. Environ.* **2006**, *28*, 444–455. [CrossRef]

50. Tylova, E.; Peckova, E.; Blascheova, Z.; Soukup, A. Casparian bands and suberin lamellae in exodermis of lateral roots: An important trait of roots system response to abiotic stress factors. *Ann. Bot.* **2017**, *120*, 71–85. [CrossRef]

51. Luthje, S.; Meisrimler, C.N.; Hopff, D.; Schutze, T.; Koppe, J.; Heino, K. Class III peroxidases. *Methods Mol. Biol.* **2014**, *1072*, 687–706. [CrossRef]

52. Bradford, M.M. Rapid and sensitive method for the quantitation of microgram quantities of protein utilizing the principle of protein-dye binding. *Anal. Biochem.* **1976**, *72*, 248–254. [CrossRef]

53. Gay, C.; Collins, J.; Gebicki, J.M. Hydroperoxide assay with the ferric-xylenol orange complex. *Anal. Biochem.* **1999**, *273*, 149–155. [CrossRef] [PubMed]

Int. J. Mol. Sci. **2020**, *21*, 8872

54. Gay, C.A.; Gebicki, J.M. Perchloric acid enhances sensitivity and reproducibility of the ferric-xylenol orange peroxide assay. *Anal. Biochem.* **2002**, *304*, 42–46. [CrossRef] [PubMed]
55. Sturmer, L.R.; Dodd, D.; Chao, C.S.; Shi, R.Z. Clinical utility of an ultrasensitive late night salivary cortisol assay by tandem mass spectrometry. *Steroids* **2018**, *129*, 35–40. [CrossRef] [PubMed]
56. Hopff, D.; Wienkoop, S.; Luthje, S. The plasma membrane proteome of maize roots grown under low and high iron conditions. *J. Proteom.* **2013**, *91*, 605–618. [CrossRef] [PubMed]
57. Turetschek, R.; Desalegn, G.; Epple, T.; Kaul, H.P.; Wienkoop, S. Key metabolic traits of *Pisum sativum* maintain cell vitality during Didymella pinodes infection: Cultivar resistance and the microsymbionts' influence. *J. Proteom.* **2017**, *169*, 189–201. [CrossRef] [PubMed]
58. Sun, X.; Weckwerth, W. Covain: A toolbox for uni- and multivariate statistics, time-series and correlation network analysis and inverse estimation of the differential Jacobian from metabolomics covariance data. *Metabolomics* **2012**, *8*, 81–93. [CrossRef]
59. Yamashita, D.; Kimura, S.; Wada, M.; Takabe, K. Improved Mäule color reaction provides more detailed information on syringyl lignin distribution in hardwood. *J. Wood Sci.* **2016**, *62*, 131–137. [CrossRef]
60. Chapple, C.C.S.; Vogt, T.; Ellis, B.E.; Somerville, C.R. An Arabidopsis mutant defective in the general phenylpropanoid pathway. *Plant. Cell* **1992**, *4*, 1413–1424.
61. Franke, R.; McMichael, C.M.; Meyer, K.; Shirley, A.M.; Cusumano, J.C.; Chapple, C. Modified lignin in tobacco and poplar plants over-expressing the Arabidopsis gene encoding ferulate 5-hydroxylase. *Plant J.* **2000**, *22*, 223–234. [CrossRef]
62. Leitner, A.; Lechner, H.; Peter, K. *Colour Tests for Precursor Chemicals of Amphetamine-Type Substances*; United Nations, Office of Drugs and Crime: Vienna, Austria, 2007.
63. Crocker, E.C. An experimental study of the significance of "lignin" color reactions. *J. Industr. Engineer. Chem.* **1921**, *13*, 625–627. [CrossRef]
64. Turrell, F.M.; Fisher, P.L. The proximate chemical constituents of citrus woods, with special reference to lignin. *Plant Physiol.* **1942**, *17*, 558–581. [CrossRef] [PubMed]
65. Brundrett, M.C.; Enstone, D.E.; Peterson, C.A. A berberine-aniline blue fluorescent stainingprocedure for suberin, lignin and callose in plant tissue. *Protoplasma* **1988**, *146*, 133–142. [CrossRef]
66. Ranathunge, K.; Kim, Y.X.; Wassmann, F.; Kreszies, T.; Zeisler, V.; Schreiber, L. The composite water and solute transport of barley (*Hordeum vulgare*) roots: Effect of suberized barriers. *Ann. Bot.* **2017**, *119*, 629–643. [CrossRef]
67. Peng, T.; Thorn, K.; Schroeder, T.; Wang, L.; Theis, F.J.; Marr, C.; Navab, N. A BaSiC tool for background and shading correction of optical microscopy images. *Nat. Commun.* **2017**, *8*, 14836. [CrossRef]

Publisher's Note: MDPI stays neutral with regard to jurisdictional claims in published maps and institutional affiliations.

International Journal of
Molecular Sciences

Article

Proteomic and Transcriptomic Patterns during Lipid Remodeling in *Nannochloropsis gaditana*

Chris J. Hulatt [1,*], Irina Smolina [1], Adam Dowle [2], Martina Kopp [1], Ghana K. Vasanth [1], Galice G. Hoarau [1], René H. Wijffels [3] and Viswanath Kiron [1]

[1] Faculty of Biosciences and Aquaculture, Nord University, PB 1490, 8049 Bodø, Norway; irina.smolina@nord.no (I.S.); martina.kopp@nord.no (M.K.); ghana.k.vasanth@nord.no (G.K.V.); galice.g.hoarau@nord.no (G.G.H.); kiron.viswanath@nord.no (V.K.)

[2] Department of Biology, Bioscience Technology Facility, University of York, York YO10 5DD, UK; adam.dowle@york.ac.uk

[3] Bioprocess Engineering, AlgaePARC, Wageningen University, 6700 AA Wageningen, The Netherlands; rene.wijffels@wur.nl

* Correspondence: christopher.j.hulatt@nord.no; Tel.: +47-9009-8564

Received: 6 August 2020; Accepted: 17 September 2020; Published: 22 September 2020

Abstract: Nutrient limited conditions are common in natural phytoplankton communities and are often used to increase the yield of lipids from industrial microalgae cultivations. Here we studied the effects of bioavailable nitrogen (N) and phosphorus (P) deprivation on the proteome and transcriptome of the oleaginous marine microalga *Nannochloropsis gaditana*. Turbidostat cultures were used to selectively apply either N or P deprivation, controlling for variables including the light intensity. Global (cell-wide) changes in the proteome were measured using Tandem Mass Tag (TMT) and LC-MS/MS, whilst gene transcript expression of the same samples was quantified by Illumina RNA-sequencing. We detected 3423 proteins, where 1543 and 113 proteins showed significant changes in abundance in N and P treatments, respectively. The analysis includes the global correlation between proteomic and transcriptomic data, the regulation of subcellular proteomes in different compartments, gene/protein functional groups, and metabolic pathways. The results show that triacylglycerol (TAG) accumulation under nitrogen deprivation was associated with substantial downregulation of protein synthesis and photosynthetic activity. Oil accumulation was also accompanied by a diverse set of responses including the upregulation of diacylglycerol acyltransferase (DGAT), lipase, and lipid body associated proteins. Deprivation of phosphorus had comparatively fewer, weaker effects, some of which were linked to the remodeling of respiratory metabolism.

Keywords: proteomics; transcriptomics; *Nannochloropsis*; EPA; TAG; phosphorus; nitrogen; bioreactor

---

1. Introduction

Bioavailable nitrogen and phosphorus are essential macronutrients required by microalgae for optimal, balanced growth. In the oceans, the effects of nitrogen (N) and phosphorus (P) supply on phytoplankton physiology and elemental stoichiometry are well recognized [1,2], where nutrient abundance often controls primary production, community structure, and ultimately the flux of matter and energy through ecosystems [3,4]. Many species of microalgae also have applications in biotechnology, where modulating the nutrient supply to intensive cell cultures is a common technique used to induce the accumulation of triacylglycerol (TAG) and secondary carotenoids [5,6]. Understanding how microalgae respond to changes in nutrient availability, especially the supply of N and P, is therefore valuable for characterizing their behavior in natural and industrial settings.

Protein accounts for a large share of cellular N, but nitrogen is also a component of nucleic acids (RNA and DNA) and chlorophyll. Phosphorus is required in lower amounts, but is nevertheless embodied in nucleic acids, phospholipids, post-translational modifications (e.g., phosphoproteins), and ATP [7–9]. Though N and P are often found in the same molecules, the effects of their abundance on microalgae physiology can be profoundly different. Nitrogen deprivation typically leads to substantial reductions in growth, protein and chlorophyll content, concomitant with increased neutral lipids, carbohydrates, or secondary carotenoids, depending on the species. The effects of P- deprivation are often more subtle, but have been consistently linked to remodeling of the lipid profile [10,11], where phosphorus-containing lipid classes are substituted for nonphosphorus lipids [9]. The active remodeling of the microalgae cell under N and P stress implicates the roles of a large number of regulatory pathways, but we still lack a deep understanding of the molecular mechanisms at work.

Transcriptome-based studies have identified patterns of gene expression during nutrient stress response and product formation [12,13]. However, eukaryotic microalgae have evolved through diverse endosymbiotic routes, and different families, genera, and species may respond differently to similar treatments. Quantitative transcript sequencing can imply that gene expression directly regulates the abundance of proteins, yet there is often only moderate association between mRNA and protein expression [14,15]. For example, studies on human cell lines have found low correlation (R^2 = 0.22–0.29) between mRNA and protein measurements [16,17], although stronger relationships have been reported from mouse cells (R^2 = 0.41), bacteria (R^2 = 0.47), and yeast (R^2 = 0.58) [16,18]. One explanation for this is the variable role of post-transcriptional mechanisms in different organisms and conditions [15,18,19]. Compared to transcriptomics, then, proteomics should provide more direct measurement of metabolic activity inside the cell, but such studies in microalgae are relatively few. Key questions include, how does macronutrient supply reshape the algal proteome, and do proteomic and transcriptomic methods describe similar metabolic patterns?

The marine eustigmatophyte *Nannochloropsis* is one of a handful of industrially tractable oleaginous microalgae. Its ~30 Mbp haploid nuclear genome is compact, containing around ten-and-a-half thousand protein coding genes, varying slightly amongst the assemblies of different strains [20,21]. Despite its modest size, the *Nannochloropsis* genome encodes a disproportionately large number of genes involved in lipid synthesis, including 11 or more copies of diacylyglycerol acyltransferase-2 (DGAT2), which performs the terminal step in TAG synthesis via the Kennedy pathway [20,22]. Under adverse conditions, especially N starvation, *Nannochloropsis* can accumulate substantial quantities of TAG in oil bodies, reaching 50% or more of the cell dry mass [23]. *Nannochloropsis* is also remarkable as a genus that can synthesize large amounts of the long-chain polyunsaturated fatty-acid C20:5*n*-3 (eicosapentanoic acid or EPA), which is highly valued in human and animal diets [24,25].

Here we used flat-plate photobioreactors operated as turbidostats to selectively apply nitrogen and phosphorus deprivation to *Nannochloropsis gaditana*. The molecular patterns emerging under N and P deficient conditions were characterized using Tandem Mass Tag (TMT) based quantitative proteomics and are supported by transcriptome (mRNA) sequencing of the same samples. Our analysis first examines the global (cell-wide) patterns of protein and transcript abundance, before exploring the primary effects of N and P starvation on the subcellular proteomes, gene clusters, and metabolic pathways. Individual pathways and proteins that were either highly impacted, or relevant to biotechnology applications, are investigated and discussed.

2. Results

2.1. Turbidostat Cultivation Dynamics, Lipids, and Fatty-Acids

Control cultures were maintained in nutrient-replete, steady-state conditions throughout the experiments with a specific growth rate of 0.55 ± 0.07 d^{-1} and a cell density of 2.6 ± 0.3 g L^{-1}. In nitrogen (N-) and phosphorus (P-) deprived treatments the growth rates declined, but other variables inside the bioreactor including the average light intensity, were largely maintained (Figure 1a). In the N- and P- cultures, either nitrate or phosphate was exhausted within 28 h due to rapid nutrient uptake coupled with high biomass turnover and dilution with fresh medium (Figure 1a). Nitrate-starved cultures showed a gradual increase in cell density toward the end of the experiment, a result of maintaining constant turbidity whilst the cells experienced chlorosis (loss of pigmentation). The N- cultures experienced an immediate reduction in growth rate to 0.11 ± 0.02 d^{-1} at day 3 and 0.05 ± 0.02 d^{-1} at day 5. In comparison the onset of P- conditions was more dampened with the growth rate 0.49 ± 0.06 d^{-1} at day 3 and 0.44 ± 0.07 d^{-1} at day 5. Analysis of fatty-acids showed a substantial increase in TAG comprised primarily of C16:0 and C16:1 fatty-acids in the N- treatments (Figure 1b). After 5 days in N- conditions, fatty-acids in TAG comprised 21.4% of the cell dry weight but remained at only 1.0% and 2.2% of the dry weight in the control (C) and P- treatments, respectively. The long-chain PUFAs eicosapentanoic acid (EPA, C20:5n-3) and arachidonic acid (ARA, C20:4n-6) were mostly present in the polar lipids. At day 5 the EPA accounted for 26.5% and ARA for 2.5% of total fatty acids (TFA) in control cultures. In N- cultures the EPA content was reduced substantially to 6.3% TFA after 5 days, due to the reduction of polar lipids and the accumulation of fatty acids in TAG.

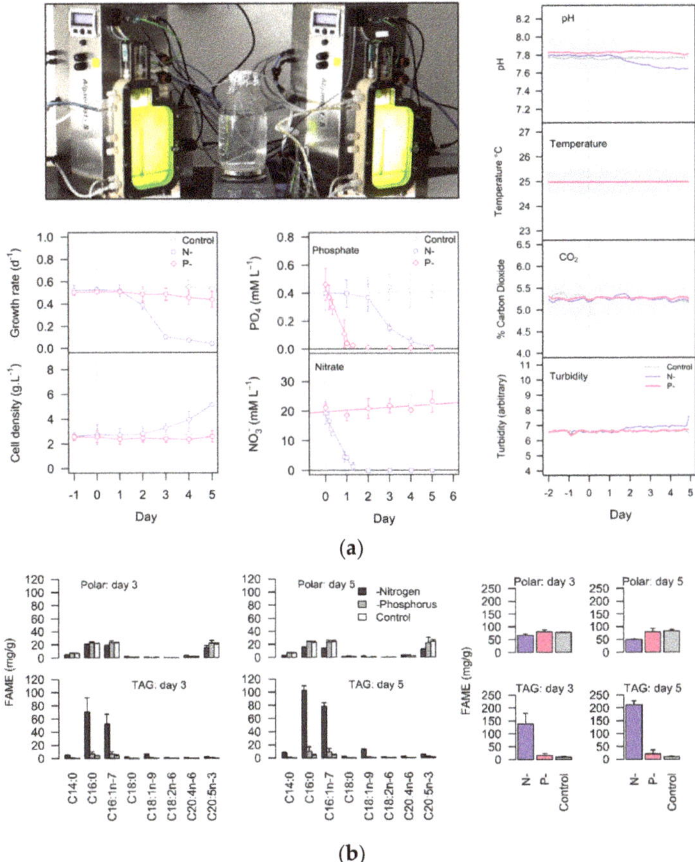

Figure 1. (**a**) Image of the flat-plate photobioreactors operated as turbidostats including measurement of pH, temperature, CO$_2$ concentration in the sparging gas, and turbidity. The growth rate (d^{-1}) and the cell density (g L^{-1}) are shown with the changes in the dissolved extracellular nitrate (NO$_3^-$) and phosphate (PO$_4^{3-}$) concentrations (mean $\pm$ sd, $n = 4$). (**b**) Lipid analysis including the fatty-acid profiles (left) of polar and neutral lipids (TAG) in control, N-, and P- treatments after 3 and 5 days of the experiment, as fatty-acid methyl-esters—FAME (mg/g dry weight). The total FAMEs in control, N-, and P- treatments after 3 and 5 days of the experiment (right). Data are the mean $\pm$ sd of $n = 4$ experimental replicates (except $n = 3$ for N- treatments at day 5).

2.2. Identification and Differential Expression of Proteins and Their Transcripts

In total 3423 proteins were identified across all of the tested conditions. After 3 days of N- deprivation 1543 of these proteins were significantly differentially regulated, whilst in P- treatments only 113 proteins were significantly differentially regulated (Figure 2). Transcriptome analysis showed that after 3 days of N- treatment, 1448 of the 10,496 genes in the B31 genome were differentially expressed, where 528 transcripts were upregulated and 920 were downregulated. After 5 days of N- treatment, the number of differentially expressed genes (DEGs) increased to 2371, where 859 were upregulated and 1512 were downregulated. Phosphorus depletion resulted in far fewer DEGs, where only 52 genes were upregulated and two were downregulated after 3 days, increasing to a total of 122 DEGs after 5 days. Principal components analysis showed that in the protein dataset there was distinct clustering of N- samples, but much weaker demarcation between P- and control treatments (Figure S3). Principal components analysis of the transcriptomic data indicated clear divergence between each of the treatments after 3 days, strengthening further after 5 days.

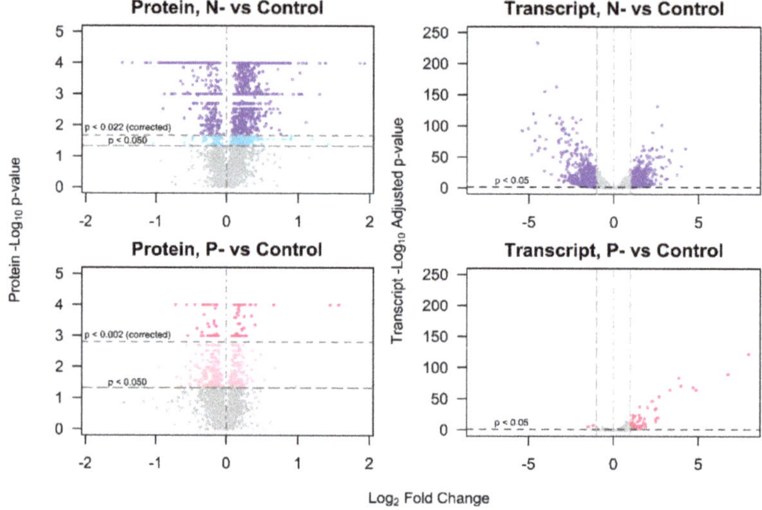

Figure 2. Volcano plots showing the differential expression of proteins and transcripts in the nitrogen starved (N-) and phosphorus starved (P-) treatments, vs. controls. The *x*-axis displays the $\log_2$ fold change (L_2fc) of protein or transcript expression, where positive values indicate upregulated proteins and negative values correspond to downregulated proteins. The *p*-values are presented on -$\log_{10}$ scale on the *y*-axis, and for transcripts these are the adjusted *p*-values from the DESeq2 methodology. Proteins determined significantly differently regulated at corrected thresholds $p < 0.022$ (N-/C treatments) or $p < 0.002$ (P-/C treatments) are indicated in the uppermost segment. Proteins differentially expressed at $p < 0.050$, but not reaching the adjusted threshold, are indicated in the central segment.

2.3. Correlation between the Nannochloropsis Proteome and Transcriptome

The global patterns in protein and mRNA abundance were examined using three complimentary approaches. First, the correlation between the $\log_2$ fold changes (L_2fc) of mRNA transcripts and their corresponding proteins was performed (Figure 3a). The N-/C treatment yielded moderate correlation ($R^2 = 0.25$), whilst the correlation in P-/C treatments was much weaker ($R^2 = 0.08$). Our second method combined data for all observations (C, N-, and P- treatments) together, and a linear mixed-effects model was used to describe the relationship between mRNA abundance (log (RPKM)) and protein abundance (log (Mol%)) across all gene/protein accessions (Figure 3b). For comparative purposes, a conventional Pearson's R^2 of 0.31 was also calculated for the same data, indicating moderate positive correlation between transcript and protein abundance. Our third method fitted individual linear regression models to each gene/protein pair, yielding 2576 regression models. The distribution of R^2 values from these linear models are presented in Figure 3c (upper panel), and for only the subset of proteins which showed significant differential expression (Figure 3c lower panel). The median R^2 for all accessions was 0.29, but increased substantially to $R^2 = 0.58$, with a shoulder at $R^2 \sim 0.8$, when only the significantly differentially expressed proteins were included. For those significantly differentially expressed proteins, 79% of the gene/protein correlation slopes were positive, the remaining 21% were negative (Figure 3d). Together, these three alternative approaches characterize a moderate but detectible cell-wide association between mRNA and protein expression in these data.

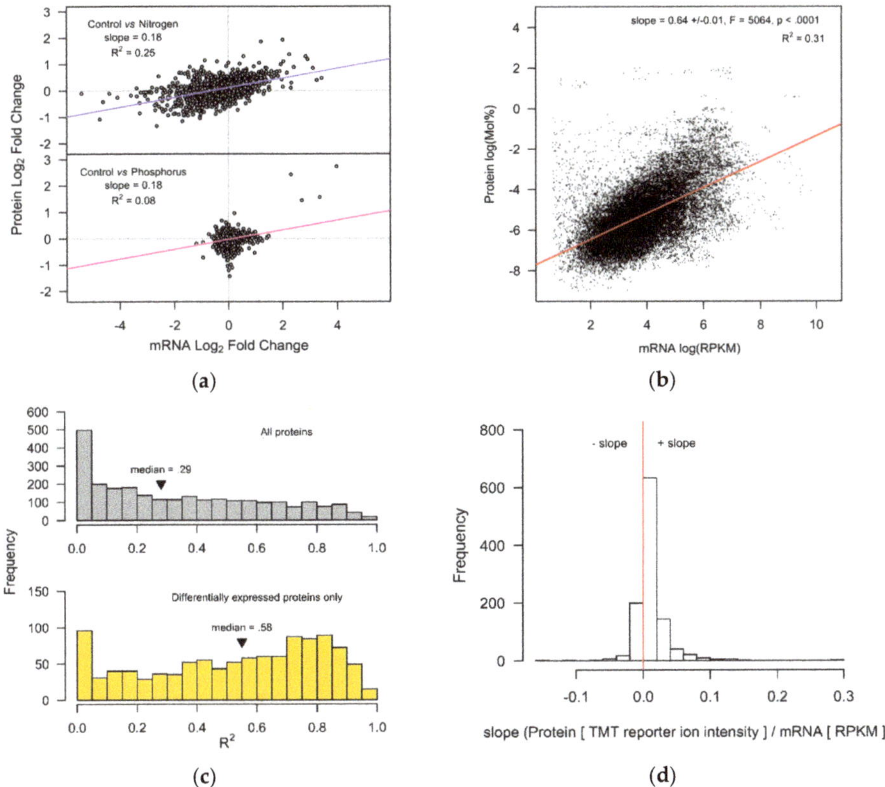

Figure 3. Global patterns in protein and mRNA abundance in *Nannochloropsis gaditana*. (**a**) The L_2fc mRNA abundance vs. the L_2fc protein abundance for N- and P- treatments, vs. controls ($n = 2578$ each). (**b**) Protein abundance in log (Mol%) vs. mRNA transcript abundance (RPKM) for all samples. The regression line was fitted with a linear mixed-effects model with random slopes and random intercepts fitted for each experimental unit ($n = 10$). Very low abundance transcripts < 2.0 RPKM were excluded. (**c**) Histograms showing the population of R^2 values that describe the relationship between mRNA abundance (RPKM) and protein abundance (normalized TMT reporter ion intensities) for each gene/protein set. The R^2 values are collected from $n = 2576$ linear regression models fitted separately to each gene/protein pair from the B31 genome assembly (Figure S5). The upper panel contains all of the correlations, whilst the lower panel shows only those where the proteins were significantly differently regulated ($n = 1083$), as determined by the Benjamini–Hochberg adjusted *p*-values. (**d**) The slopes showing positive or negative correlations for the same 1083 linear regression fits.

2.4. The Effect of Nitrogen and Phosphorus Stress on Subcellular Proteome Remodeling

To investigate large-scale changes in subcellular proteomes under N- and P- conditions, we examined the overall fold changes of proteins after grouping them into their respective cellular locations. For most compartments, N- treatments exhibited greater variance in protein abundance than P- treatments (Figure 4). Proteins associated with the plastid were mostly downregulated under nitrogen deprivation, with a median L_2fc of -0.42. Proteins localized to the mitochondrion, membranes and the endoplasmic reticulum (ER) also displayed variation in L_2fc, but their median fold changes each remained around zero (L_2fc 0.00, 0.02, and -0.08, respectively). The data indicate that under N- conditions the plastid proteome shrank, whilst the ER, mitochondrial and membrane proteins were remodeled but did not substantially change overall size. In P- treatments there were no substantial shifts in expression of any of the subcellular proteomes, and variation in L_2fc was much lower than those in N- treatments, indicating only limited remodeling.

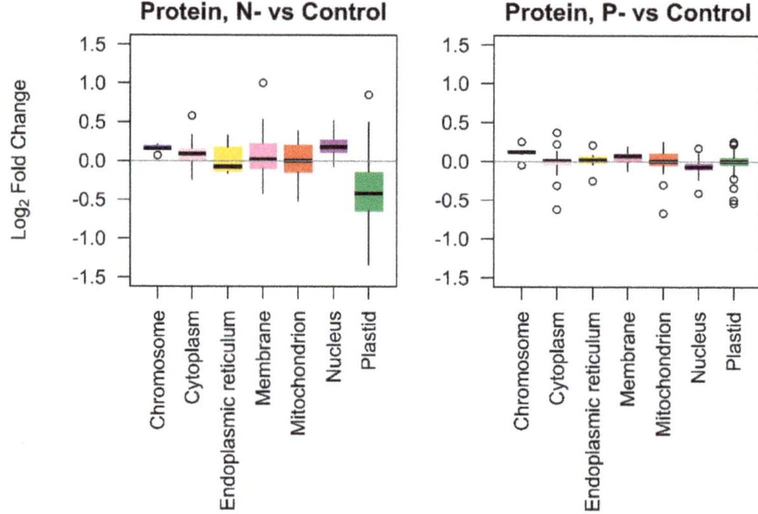

Figure 4. The L_2fc of proteins localized in different subcellular compartments. Panels represent N- ($n = 4$) or P- conditions ($n = 2$), relative to the control group ($n = 4$). Annotation of locations was provided by the UniProtKB database.

2.5. Functional Enrichment Analysis of Differentially Expressed Proteins and Transcripts

To capture the main patterns in gene expression and protein abundance, gene ontology (GO) and KEGG pathway ontology (KO) terms were examined (Figures 5 and 6). Under N- conditions changes in the proteome and transcriptome were mostly concordant, where downregulation of proteins and mRNA transcripts was observed in protein translation processes (GO:0006412), protein-chromophore linkage (GO:0018298), and light-independent chlorophyll biosynthesis (GO:0036068), together with photosynthesis (GO:0015979) and its light-dependent (GO:0009765) and light-independent reactions (GO:0019685). Fewer gene and protein GO categories were significantly upregulated in N- treatments, but genes and proteins with roles in amine metabolism (GO:0009308), the tricarboxylic acid cycle (GO:0006099), and nucleotide catabolism (GO:0009166) were increased.

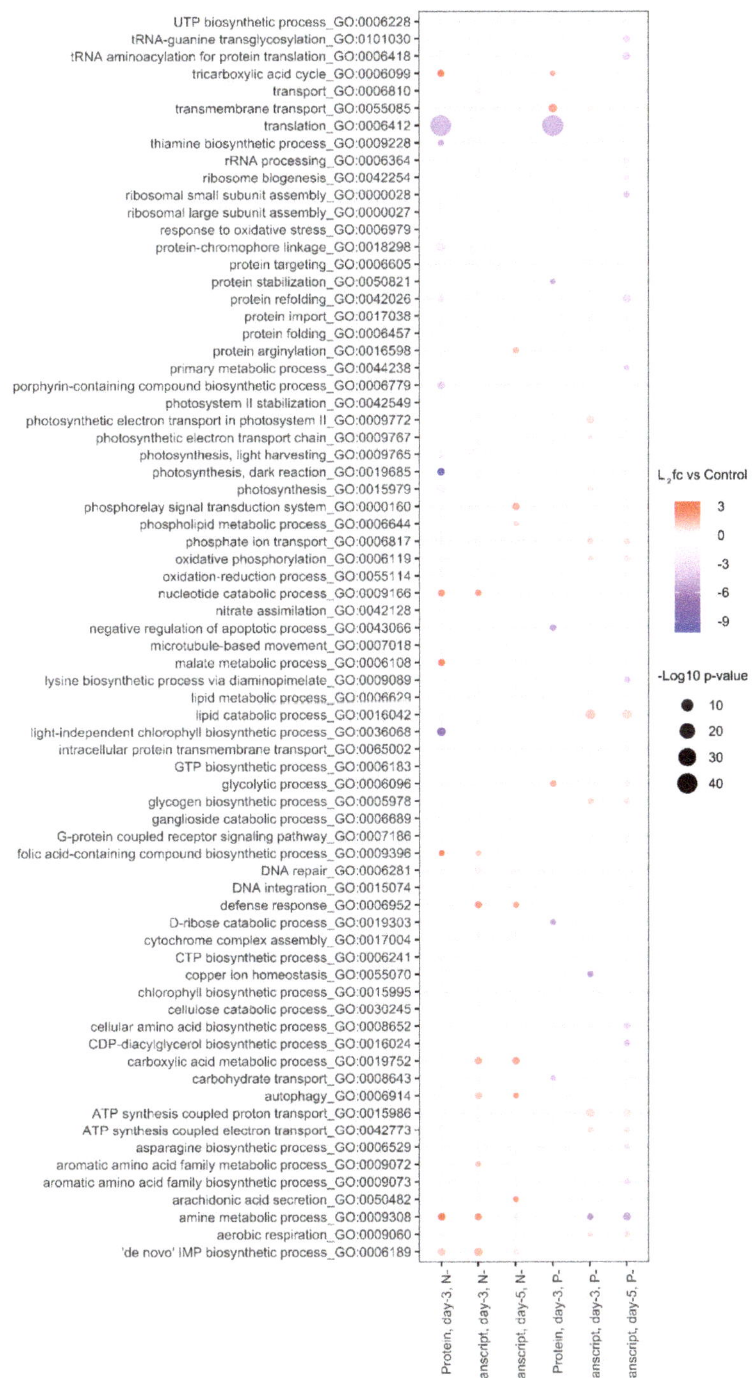

Figure 5. Gene set enrichment. The gene ontology classifications (GO: biological processes) of proteins and transcripts differentially expressed under nitrogen and phosphorus deprivation.

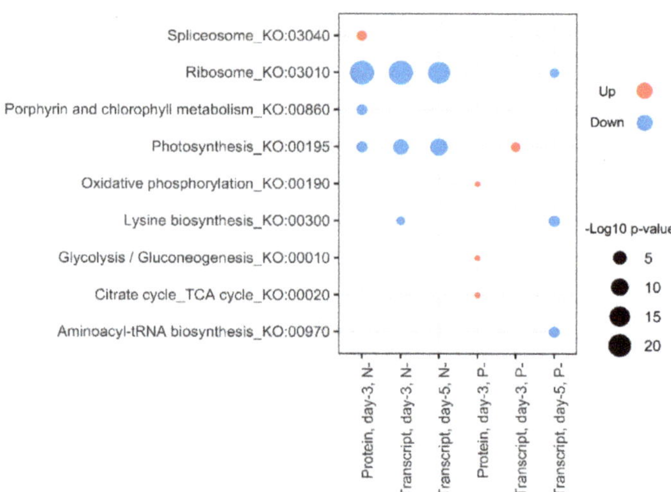

Figure 6. Changes in metabolic pathways. The most perturbed KEGG (KO:) metabolic pathways in the proteome and the transcriptome.

In P- treatments, over-represented GO terms for proteomic and transcriptomic data were less concordant. The downregulation of proteins involved in translation (GO:0006412), protein stabilization (GO:0050821), D-ribose catabolic process (GO:0019303), and carbohydrate transport (GO:0008643), together with the upregulation of tricarboxylic acid cycle (GO:0006099) and glycolytic process (GO:0006096), was not echoed by the transcriptome (Figure 5). After 5 days of phosphorus starvation, gene expression associated with tRNA (GO:0006418) and rRNA processing (GO:0006364) were also lowered, together with reductions in ribosome biogenesis (GO:0042254), ribosome assembly (GO:0000028), protein refolding (GO:0042026), and amino-acid biosynthesis (GO:0008652). Transcripts associated with amine metabolism (GO:0009308) were also downregulated after 5 days of P deprivation, contrasting with the upregulation of the same group during N deprivation. Upregulated gene clusters in P- treatments included increases in phosphate-ion transport (GO:0006817) and increases in transcripts associated with lipid catabolism (GO:0016042), ATP synthesis (GO:0015986, GO:0042773), and oxidative phosphorylation (GO:0006119).

In nitrogen-starved cells, KEGG pathways related to photosynthesis (KO:00195) and ribosomes (KO:03010) were downregulated in both proteomic and transcriptomic data. (Figure 6). Under P- conditions proteins in the KEGG pathways glycolysis/gluconeogenesis (KO:00010), the TCA cycle (KO:00020), and oxidative phosphorylation (KO:00190) were upregulated. However, these increases in respiration-associated protein groups were not mirrored by the transcriptome. Instead, after 5 days transcriptome data indicated downregulation of several pathways linked to lysine biosynthesis (KO:00300) and aminoacyl-tRNA biosynthesis (KO:00970), implying reduced translation activity under protracted P-deprivation.

2.6. Translation, Nitrogen Acquisition, and Metabolism

Under N- conditions, 12 of the 30 most downregulated proteins were ribosomal (Table 1), mostly 30S and 50S that are plastid-associated. The L_2fc of all ribosomal proteins were examined, and we found that both plastidic ribosomes and ribosomal proteins of eukaryotic origin (40S and 60S) were downregulated after 3 days of N- conditions (Figure S6). In P- treatments the expression of ribosomal proteins and their transcripts was not substantially changed. Both nitrate and nitrite reductase were among the most downregulated proteins in the N- treatments, highlighting the reduced investments in N acquisition from the extracellular environment.

Table 1. The 30 proteins with largest fold increase and 30 proteins with the largest fold decrease in the N- treatments ($n = 4$), relative to the controls ($n = 4$). Proteins annotated as "uncharacterized" were omitted and the *p*-values are from permutation tests. The suffix string of the Accession Number "9STRA" or "NANGC" refers to the B31 or CCMP526 *N. gaditana* reference proteomes, respectively.

Rank	Identified Proteins	Accession Number	kDa	L_2fc	*p*-Value
Upregulated					
1	Lipid droplet surface protein	W7TWF7_9STRA	18	1.93	0.0001
2	Amine oxidase	W7TFN3_9STRA	75	1.38	0.0001
3	Methylenetetrahydrofolate dehydrogenase	W7T6I6_9STRA	39	1.3	0.0001
4	Acid sphingomyelinase-like phosphodiesterase 3b	W7TQ09_9STRA	76	1.3	0.001
5	EF-Hand 1, calcium-binding site	W7TRW6_9STRA	64	1.11	0.0001
6	Lipase family protein	W7TUB0_9STRA	54	1.06	0.0001
7	Two component regulator propeller domain-containing protein	K8Z0G9_NANGC	27	1.03	0.001
8	Lipocalin protein	W7TQX7_9STRA	29	1.02	0.00021
9	Ammonium transporter	W7U477_9STRA	58	1	0.0001
10	Carbonic anhydrase, alpha-class	W7T0A1_9STRA	37	0.9	0.028
11	Cathepsin a	W7TYE0_9STRA	60	0.87	0.0001
12	Nadp-dependent glyceraldehyde-3-phosphate dehydrogenase	W7U8W3_9STRA	66	0.86	0.0001
13	Cluster of Sodium hydrogen exchanger 8	W7TNK5_9STRA	72	0.86	0.0001
14	Light harvesting complex protein	K8YPR7_NANGC	19	0.85	0.0001
15	Subfamily member 9	W7TPA4_9STRA	41	0.82	0.028
16	Plasma membrane ATPase	K8YQB4_NANGC	107	0.77	0.0001
17	Manganese lipoxygenase	W7TYD4_9STRA	73	0.77	0.0001
18	Quinoprotein amine dehydrogenase, beta chain	W7TI92_9STRA	66	0.77	0.0001
19	4-hydroxyphenylpyruvate dioxygenase	W7TNB7_9STRA	50	0.77	0.001
20	Malate cytoplasmic isoform 2	W7TPM0_9STRA	37	0.76	0.0001
21	Cluster of Violaxanthin de-epoxidase	K8YTT8_NANGC	35	0.75	0.019
22	Had-superfamily subfamily iia hydrolase	W7U270_9STRA	43	0.74	0.0001
23	Glutaryl-mitochondrial	W7TTQ4_9STRA	48	0.74	0.0001
24	Pyruvate dehydrogenase	W7TN62_9STRA	55	0.74	0.0001
25	Myotubularin-related protein 2	W7TSB4_9STRA	109	0.74	0.004
26	Cdgsh iron sulfur domain-containing protein 1	W7TPN8_9STRA	23	0.72	0.001
27	Arachidonate 5-lipoxygenase	K8Z8I5_NANGC	60	0.71	0.0001
28	Cluster of Purple acid phosphatase	W7TLQ2_9STRA	56	0.71	0.0001
29	Cluster of Expulsion defective family member (Exp-2)	K8YVZ3_NANGC	62	0.71	0.049
30	V-type proton ATPase subunit F	W7TU11_9STRA	13	0.7	0.0001
Downregulated					
30	Cytochrome p450	W7UBA8_9STRA	70	−0.77	0.0001
29	30s ribosomal protein s15	W7TEF2_9STRA	34	−0.77	0.0001
28	RNA binding s1 domain protein	W7U882_9STRA	45	−0.77	0.0001
27	Cluster of Solute carrier family 35 member b1	W7TCR9_9STRA	43	−0.77	0.7
26	Cytochrome P450 enzyme	I2CNY8_NANGC	67	−0.78	0.001
25	Heat shock protein DNAJ, cysteine-rich domain protein	W7TJ91_9STRA	13	−0.78	0.001
24	Geranylgeranyl reductase	W7THD6_9STRA	57	−0.79	0.0001
23	Coproporphyrinogen iii oxidase chloroplast	W7TZ92_9STRA	46	−0.79	0.0001
22	50S ribosomal protein L18, chloroplastic	K9ZX62_9STRA	12	−0.8	0.0001
21	50S ribosomal protein L19	K9ZV73_9STRA	14	−0.81	0.0001
20	30S ribosomal protein S9, chloroplastic	A0A023PLK7_9STRA	15	−0.82	0.0001
19	30S ribosomal protein S2, chloroplastic	K9ZWC8_9STRA	29	−0.83	0.0001
18	Nitrite reductase	W7T0E9_9STRA	46	−0.85	0.0001
17	30S ribosomal protein S8, chloroplastic	K9ZV68_9STRA	15	−0.86	0.0001
16	Cluster of H+-transporting ATPase	K8YQ29_NANGC	152	−0.87	0.0001
15	30S ribosomal protein S12, chloroplastic	K9ZVC5_9STRA	14	−0.88	0.0001
14	50S ribosomal protein L36, chloroplastic	K9ZXS5_9STRA	4	−0.88	0.001
13	Magnesium chelatase ATPase subunit I	K9ZV21_9STRA	47	−0.9	0.0001
12	50S ribosomal protein L16, chloroplastic	K9ZWF3_9STRA	16	−0.9	0.0001
11	Ribosomal protein s21	W7TSY1_9STRA	14	−0.9	0.003
10	Cluster of Mfs transporter	W7U968_9STRA	66	−0.93	0.14
9	30S ribosomal protein S17, chloroplastic	K9ZVE6_9STRA	10	−0.94	0.0001
8	30S ribosomal protein S20, chloroplastic	K9ZX69_9STRA	11	−0.94	0.0001
7	Delta 5 fatty acid desaturase	K8YSX2_NANGC	54	−0.95	0.0001
6	30S ribosomal protein S18, chloroplastic	K9ZV97_9STRA	8	−0.97	0.0001
5	Nitrate reductase	W7TAR6_9STRA	70	−1.08	0.0001
4	Ferredoxin nitrite reductase	K8YST4_NANGC	40	−1.13	0.0001
3	Light-independent protochlorophyllide reductase subunit N	K9ZV79_9STRA	50	−1.15	0.0001
2	Light-independent protochlorophyllide reductase iron-sulfur ATP-binding protein	K9ZV32_9STRA	32	−1.34	0.0001
1	NAD(P)H nitrate reductase	K8YSU6_NANGC	63	−1.48	0.0001

The reduced plastid proteome and diminished photosynthetic capacity associated with N starvation led us to hypothesize that enzymes involved with protein/amino-acid catabolism, nitrogen recycling, and recovery could be upregulated. Consistent with increases in amine metabolic processes (GO:0009308, Figure 5), an amine oxidase (W7TFN3_9STRA) was the second-most upregulated protein under N- conditions with an L_2fc of +1.38 (Table 1). In P- treatments, the same protein was significantly downregulated (L_2fc −0.32, $p < 0.001$). Further searching through the proteome revealed an additional six proteins annotated as amine oxidases, and of these a further two were significantly upregulated under N- conditions (Table S5). Additional proteins associated with amine metabolism were also significantly upregulated in N- treatments, including an amine dehydrogenase (W7TI92_9STRA) with an L$_2$fc of +0.77.

2.7. Tricarboxylic Acid (TCA) Cycle, Glycolytic Processes, and Oxidative Phosphorylation

Evidence from Figures 4–6 indicated that remodeling of mitochondrial or respiratory activity took place under both N- and P- conditions. To establish which proteins and transcripts were differentially expressed, and how regulatory activity potentially differed under N- and P- conditions, the L$_2$fc of respiratory-associated proteins were examined together with their transcripts (Figure 7). In N- conditions, most proteins and transcripts associated with the TCA cycle were upregulated, but those associated with glycolytic processes were both up- and downregulated. Two glycolytic enzymes, glyceraldehyde-3-phosphate dehydrogenase and phosphoglycerate kinase included multiple copies that were not coregulated with one another, with different accessions showing divergent patterns of regulation (e.g., W7U208_9STRA vs. W7T2R0_9STRA). In P- conditions, most TCA cycle and glycolytic proteins and transcripts were weakly upregulated.

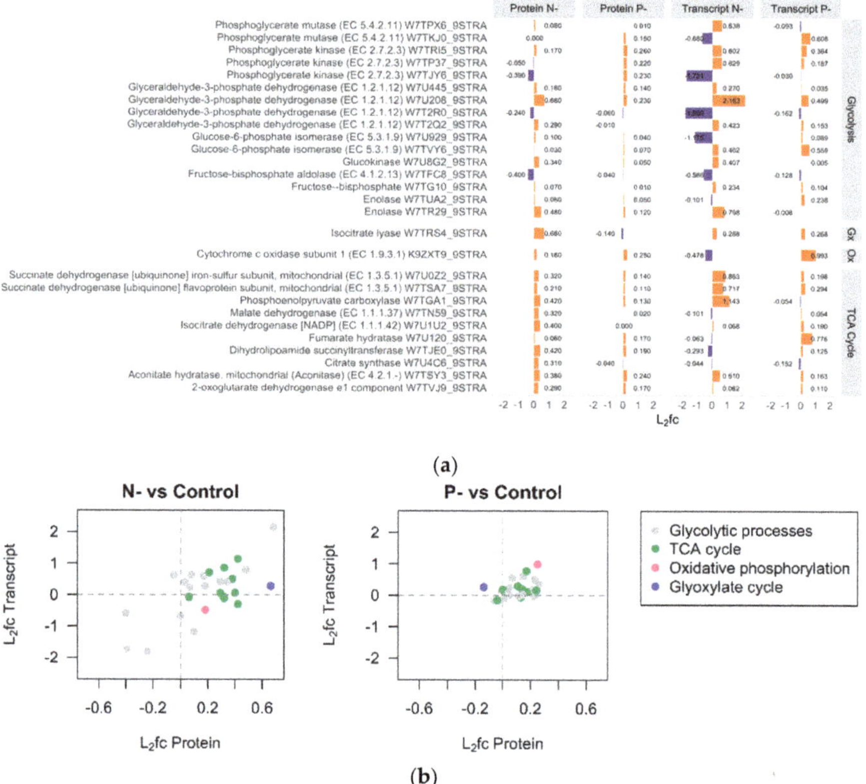

Figure 7. Respiratory activity under N- and P- conditions. (**a**) The L$_2$fc of proteins and genes linked to glycolytic processes, the TCA cycle, glyoxylate cycle (Gx), and oxidative phosphorylation (Ox). Proteins were identified manually using GO terms and by searching for specific accessions. Transcript data were then matched to the proteins using the unique accession number. (**b**) The fold changes of the proteins and transcripts.

2.8. Fatty-Acid and Acyl-CoA Metabolism

An Acetyl-CoA carboxylase protein (I2CQP5_NANGC) was significantly upregulated during P-starvation (L_2fc +0.12, $p < 0.001$), but significantly downregulated under N- conditions (L_2fc −0.50, $p < 0.001$). Two proteins annotated as Acyl CoA synthetase were identified, but only one long-chain Acyl-CoA synthetase (LACS, W7TGG5_9STRA) was significantly upregulated under N- conditions (L_2fc +0.36, $p < 0.001$).

2.9. Polyunsaturated Fatty Acid (PUFA) Metabolism

The primary route to medium and long-chain polyunsaturated fatty-acid biosynthesis in microalgae is via a series of steps involving desaturase and elongase enzymes. A Δ5 desaturase (K8YSX2_NANGC) was amongst the most downregulated proteins in N- treatments (Table 1). Six other desaturase enzymes were also significantly downregulated during N- conditions (Table S6), including a Δ12 ω-6 desaturase (K8YR13_NANGC) and a glycerolipid ω-3 desaturase (I2CR09_NANGC), with L_2fc of −0.37 and −0.53 respectively ($p \leq 0.005$). Under P- conditions the abundance of the same Δ5, Δ12, and glycerolipid desaturases did not significantly change.

2.10. Proteins Associated with TAG Biosynthesis and Storage in Oil Bodies

The most upregulated protein in N- treatments with an L_2fc of +1.93 ($p < 0.001$) was a lipid droplet surface protein (W7TWF7_9STRA), which is concordant with the substantial increases in TAG observed in the same samples (Table 1, Figure 1). Although the *N. gaditana* genome is reported to encode 11 copies of DGAT2, only one diacylglycerol acyltransferase (DGAT) family protein (W7U9S5_9STRA) was identified. This protein was significantly upregulated under N- conditions (L_2fc +0.30, $p = 0.004$), but not under P- conditions (L_2fc −0.14, $p = 0.420$). In comparison, the transcript data quantified the expression of eight different genes annotated as DGAT or DGAT2, where three were significantly upregulated under N- conditions and two were significantly downregulated (Table S7). Further upstream in lipid biosynthesis, Lysophosphatidylglycerol acyltransferase (LPAT) catalyzes the conversion of lysophosphatidic acid to phosphatidic acid. We identified a single LPAT protein (K8YP17_NANGC), that did not respond significantly in either N- or P- conditions.

2.11. Glycerolipid and Phospholipid Biosynthesis

A single protein annotated as monogalactosyldiacylglycerol synthase (MGDG synthase, W7TN13_9STRA) was not significantly differently expressed in either N- or P- conditions (L_2fc < 0.07, $p > 0.130$). A choline/ethanolamine kinase family protein (K8YV04_NANGC) was significantly upregulated (L_2fc +0.28, $p = 0.001$) in P- conditions, but was not significantly changed in N- conditions (L_2fc +0.13, $p = 0.072$). The proteomics data also identified a Udp-sulfoquinovose synthase (W7TMH8_9STRA) that was significantly downregulated in N- conditions (L_2fc −0.2, $p < 0.001$), but significantly upregulated in P- conditions (L_2fc +0.24, $p < 0.001$). In P- conditions an Acid sphingomyelinase-like phosphodiesterase 3b (W7TQ09_9STRA) was amongst the most upregulated proteins with an L_2fc of 0.68 ($p = 0.011$) (Table 2).

Table 2. The 30 proteins with largest fold increase and 30 proteins with the largest fold decrease in P- treatments (n = 2), relative to the controls (n = 4). Proteins annotated as "uncharacterized" were omitted and the p-values are from permutation tests. The suffix string of the Accession Number "9STRA" or "NANGC" refers to the B31 or CCMP526 *N. gaditana* reference proteomes, respectively.

Rank	Identified Proteins	Accession Number	kDa	L$_2$fc	p-Value
Upregulated					
1	Sse2p	W7TMT9_9STRA	32	0.96	0.04
2	Acid sphingomyelinase-like phosphodiesterase 3b	W7TQ09_9STRA	76	0.68	0.011
3	Cluster of Calcium binding protein 39	W7T646_9STRA	51	0.61	0.59
4	Snf7 family protein	W7U1R3_9STRA	22	0.53	0.026
5	Ddi1p	W7U1J9_9STRA	41	0.52	0.97
6	Nad-dependent deacetylase	W7TT51_9STRA	38	0.48	0.004
7	Elongation of fatty acids protein	W7TSM8_9STRA	36	0.48	0.06
8	Lysyl-tRNA synthetase	W7TMK7_9STRA	20	0.45	0.13
9	Aminoglycoside phosphotransferase	W7TK75_9STRA	37	0.43	0.2
10	Pyruvate decarboxylase	K8YS66_NANGC	62	0.41	0.0001
11	Splicing arginine serine-rich 19	W7T8W4_9STRA	34	0.41	0.0001
12	Ribosomal protein	K8Z5W4_NANGC	33	0.39	0.067
13	Cluster of Trypsin family	K8Z6K0_NANGC	65	0.38	0.36
14	Cluster of Methylthioribose kinase	W7TVE0_9STRA	94	0.37	0.98
15	Ferredoxin	K8YW46_NANGC	12	0.36	0.055
16	Otu-like cysteine type protease	W7TUL0_9STRA	102	0.36	0.15
17	Protein-tyrosine low molecular weight	K8YTE7_NANGC	16	0.35	0.00023
18	Threonine aldolase	W7TQZ9_9STRA	47	0.35	0.012
19	Protein phosphatase	W7TA28_9STRA	48	0.35	0.2
20	Pre-mRNA-processing factor 17	K8Z4U6_NANGC	86	0.34	0.0001
21	Beta-ketoacyl-thiolase	W7SYP3_9STRA	8	0.34	0.022
22	Ethylmalonic encephalopathy 1	K8Z7T8_NANGC	47	0.33	0.0001
23	Soluble pyridine nucleotide transhydrogenase	W7T7X5_9STRA	17	0.32	0.14
24	Ring-finger-containing e3 ubiquitin	W7UAK3_9STRA	76	0.32	0.25
25	Glycerol kinase	W7U0M7_9STRA	24	0.3	0.072
26	Ig family protein	W7T9Y3_9STRA	60	0.3	0.082
27	Cluster of Mfs transporter	W7U968_9STRA	66	0.3	0.89
28	Mitochondrial tricarboxylate carrier family	W7TKI7_9STRA	36	0.29	0.009
29	Cdgsh iron sulfur domain-containing protein 1	W7TPN8_9STRA	23	0.29	0.025
30	NAD(P)-binding domain protein	W7TM45_9STRA	40	0.29	0.032
Downregulated					
30	Vacuolar protein-sorting-associated protein 36	W7TG31_9STRA	49	−0.44	0.34
29	Exocyst complex	W7U8I8_9STRA	115	−0.45	0.009
28	Methyltransferase type 11	W7U3Q9_9STRA	34	−0.45	0.027
27	RNA binding protein	W7TAT7_9STRA	20	−0.47	0.29
26	Light harvesting complex protein	K8YPR7_NANGC	19	−0.5	0.007
25	Diaminopimelate decarboxylase	W7TNX0_9STRA	56	−0.5	0.04
24	DNA polymerase subunit Cdc27	W7TMW3_9STRA	62	−0.51	0.013
23	Tubulin-tyrosine ligase-like protein	W7TWY1_9STRA	79	−0.51	0.024
22	Translocase of inner mitochondrial membrane 50-like protein	K8YTV0_NANGC	43	−0.51	0.13
21	Cluster of Protease do-like 9	W7TU24_9STRA	69	−0.52	0.61
20	TatA-like sec-independent protein translocator subunit	W7T3A7_9STRA	22	−0.54	0.001
19	Photosystem II reaction center protein H	K9ZXQ7_9STRA	7	−0.54	0.084
18	Cyclic nucleotide-binding protein	W7TMP7_9STRA	25	−0.56	0.002
17	Ubiquilin	I2CQX3_NANGC	47	−0.6	0.07
16	Ribokinase	W7TXK5_9STRA	34	−0.62	0.21
15	Ankyrin	W7TWU3_9STRA	48	−0.63	0.068
14	Soluble nsf attachment protein receptor	W7TW41_9STRA	32	−0.66	0.27
13	Elongation of fatty acids protein	W7U1Y8_9STRA	37	−0.67	0.098
12	Anamorsin homolog	W7TKP2_9STRA	30	−0.67	0.19
11	Adenylate kinase	K8ZCS9_NANGC	19	−0.68	0.1
10	Mitochondrial carrier domain protein	W7TRC0_9STRA	50	−0.68	0.13
9	Set domain protein	W7TKH2_9STRA	119	−0.73	0.75
8	ATP-dependent RNA helicase DDX23/PRP28	K8YWH1_NANGC	91	−0.77	0.13
7	Pentatricopeptide repeat-containing protein	W7TSL2_9STRA	138	−0.81	0.24
6	Fgd6 protein	K8Z5M8_NANGC	33	−0.84	0.26
5	Polypyrimidine tract binding protein	I2CQY0_NANGC	35	−0.96	0.01
4	Major facilitator superfamily	W7UAL7_9STRA	66	−0.99	0.21
3	U3 small nucleolar RNA-associated	W7UBP4_9STRA	207	−1.1	0.049
2	Phytanoyl-dioxygenase	W7T3Z1_9STRA	24	−1.2	0.047
1	DNA damage-binding protein 1a	I2CQY4_NANGC	41	−1.45	0.14

2.12. Lipase Activity and Lipid Catabolism

In P- conditions a single lipase (W7TUB0_9STRA) was significantly downregulated (L_2fc −0.32, $p = 0.001$). The same accession was substantially upregulated under N- conditions (L_2fc +1.06, $p < 0.001$), in addition to the significant upregulation of five other lipase family proteins, including two lysophospholipases (Table S8).

2.13. Polyketide Synthase, Fatty Acid Synthase, and Lipoxygenase Expression

Six proteins annotated as polyketide synthases (PKS) were detected in the proteomics data, but none responded significantly in either the nitrogen-starved or phosphorus-starved treatments (Table S9). A single fatty acid synthase (FAS1) domain protein (W7TBQ5_9STRA) was significantly downregulated in nitrogen-starved conditions (L_2fc −0.47, $p < 0.001$) but not phosphorus-starved conditions (L_2fc −0.10, $p = 0.091$). An Arachidonate 5-lipoxygenase (K8Z8I5_NANGC) was also amongst the most upregulated proteins with an L_2fc of +0.71 (Table 1), whilst a manganese lipoxygenase protein (W7TYD4_9STRA) was also significantly upregulated under N- conditions, providing evidence for the upregulation of oxylipin pathways during nitrogen starvation.

3. Discussion

The 3423 proteins identified in this study represent a third of the gene models in the *N. gaditana* genome [20,21] providing deep profiling of the *Nannochloropsis* proteome. The data also offers the opportunity to compare the expression of proteins with their mRNA transcripts.

3.1. Global Correlation of Nannochloropsis Protein and Transcript Expression

Integrating different 'omics datasets is a challenge but offers the chance to ask valuable questions. On one hand, transcriptome sequencing provides high-throughput measurements of global responses to physiological stress and has been widely adopted. Nevertheless, the abundance and activity of proteins in cells, which ultimately determines the phenotype, is regulated by numerous mechanisms beyond mRNA expression alone [18]. Our proteomic and transcriptomic data presented here are concordant with studies on other organisms, where generally only weak-moderate associations have been observed at the whole-cell level. Whether the unexplained residual variation is due to post-transcriptional mechanisms or to methodological sensitivity, is not always clear [15].

Correlating the L_2fc (Figure 3a) is a straightforward method of associating transcript and protein data that relies only on relative changes in expression. Here N starvation produced a stronger correlation than P starvation, likely due to larger changes in protein and transcript abundance under N stress. However, our additional correlation methods help to provide a more complete picture. In Figure 3b we used measures of protein and transcript abundance, rather than their relative fold changes, and obtained an $R^2 = 0.31$. This value is comparable to observations in the model plant *Arabidopsis thaliana* ($R^2 = 0.27$–0.46) and bacteria ($R^2 = 0.20$–0.47), but lower than yeasts ($R^2 = 0.34$–0.87) [16]. When individual linear models were fitted separately to data from each protein/transcript, we were able to show the heterogeneity of correlations across different genes (Figure 3c). Proteins that were significantly differentially expressed often exhibited higher correlation with their transcripts, providing support for the role of effect-size in determining the strength of gene–protein correlations. Nevertheless, a proportion of significantly regulated proteins remained only weakly correlated with their transcripts. Like other eukaryotes, microalgae employ a multitude of post-transcriptional systems, but to what extent ncRNAs, splicing, post-translational modifications, and protein turnover [19,26–29] impact transcript/protein/metabolome relations in oleaginous microalgae, is not yet very clear. The effect of N, but not P deprivation, on reducing ribosomal protein abundance illustrates that ribosome density varies with certain stress responses, representing a further layer of regulation between transcription and translation. Lastly, the dynamic nature of gene–protein regulatory circuits may be a critical variable [30]. Our turbidostat cultures controlled for the light intensity, but during the experimental treatments the cultures remained non-steady-state systems, where there may be overshoot in the transcriptional control of protein abundance [30,31]. Future studies can address this aspect by using alternative bioreactor control strategies.

3.2. N and P Deprivation Remodels Organelle Proteomes and Energy Metabolism

Eukaryotic cells are highly compartmentalized and the size, spatial arrangement, and contacting of subcellular compartments is re-optimized under stress conditions. In our data the dampened onset of P- stress contrasted with the rapid reduction in growth and changes in protein/gene expression observed in N- conditions. These differences can be reconciled by the way phosphorus is utilized inside the cell. Under nutrient-replete conditions luxury phosphorus uptake takes place and cells can accumulate excess reserves of intracellular phosphorus which acts as a short-term buffer during P- conditions [32]. Secondarily, certain classes of phosphorus-containing compounds can be functionally replaced by phosphorus-free alternatives (see Section 3.3), which reduces the impact of P- conditions on metabolism.

Our analysis showed that the plastid proteome was downregulated under N- conditions, consistent with the nitrogen-starved phenotypes (chlorosis and reduced polar lipid content) and the downregulation of mRNAs and proteins associated with photosynthesis. Despite the reduced photosynthetic capacity, mitochondrial proteins remained on average at comparable abundance to the control treatments, but there was evidence of reorganization, The changes in TCA cycle and glycolytic proteins under N- and, weakly, under P- conditions, highlights the active role played by respiratory processes during macronutrient stress. Previous research has indicated increased expression of glycolytic enzymes including glyceraldehyde-3-P dehydrogenase during N- conditions [33]. Our data indicates that these proteins, which are present in multiple copies, can show opposing patterns of regulation and therefore more information e.g., on cellular localization and targeting is required before their roles can be fully understood. In plants, phosphate deprivation is associated with regulation of alternative pathways in glycolysis and oxidative phosphorylation [34], and evidence from the proteome of the diatom *Phaeodactylum* [35] also indicates upregulation of TCA cycle activity under N-limited conditions. As mitochondrial activity is central to pathways in energy metabolism and amino acid cycling, alternative configurations of the mitoproteome play a central role in acclimation to protracted macronutrient deficits and further research is needed on mitochondrial metabolic flux under nutrient stress.

3.3. Lipid Metabolism and Remodeling

The regulation of lipid metabolism in oleaginous microalgae has been the subject of substantial scientific and commercial attention, yet the underlying mechanisms are still not completely resolved [36]. Transcriptome sequencing studies have shown that the genes involved in lipid biosynthesis are actively regulated during nutrient-induced stress [12], yet attempts to increase oil yields by overexpression of key genes have yielded mixed results [37], indicating that lipid biosynthetic enzymes are not necessarily rate-limiting. In *Nannochloropsis*, nitrogen starvation is primarily associated with TAG production and lipid storage in oil droplets, but surprisingly our GO and KEGG enrichment analysis (Figures 5 and 6) did not prioritize lipid-related protein or gene families during oil accumulation. However, several lipid-related proteins were strongly upregulated under N starvation, including a lipid droplet surface protein (LDSP) with the highest fold change in the whole dataset. Similar proteins have been characterized from *Nannochloropsis oceanica*, *Chlamydomonas*, and *Phaeodactylum* [38,39]. These proteins play a structural role in oil bodies, and so their abundance scales with neutral lipid accumulation [39].

The *N. gaditana* genome is reported to encode 11 DGAT2 genes, but we were only able to distinguish one diacylglycerol acyltransferase protein, although the expression of eight different DGAT2 genes were counted in the transcript data. The upregulation of DGAT under N- stress, but not under P- stress, indicates a regulatory role in TAG accumulation, and the same accession (corresponding to gene Naga_100006g86) also responds to changing light conditions in this species [40]. We identified a single LPAT protein that was unresponsive to either N- or P- conditions. However, several LPAT orthologs are present in the *Nannochloropsis* genome and their subcellular localization and functional role is not shared equally among them [36]. The protein Acetyl-CoA carboxylase, which drives lipid biosynthesis in the plastid [41], was downregulated under N-deprived conditions indicating reduced de-novo fatty acyl chain biosynthesis.

Macronutrient deprivation not only induces accumulation of TAG, but the remodeling of membrane (polar) lipids. Nitrogen deprivation especially induces the degradation of plastidic glycerolipids, especially phosphatidylglycerol (PG), monogalactosyldiacylglycerol (MGDG), and digalactosyldiacylglycerol (DGDG) that that contain the majority of the EPA [42]. The fate of PUFAs under nutrient stress has important consequences for the lipid and fatty acid composition of the cell, and different processes including de-novo PUFA synthesis, translocation, and degradation/oxidation of fatty acids together contribute to the overall lipid profile. Recent evidence indicates that limited de-novo synthesis of LC-PUFAs does occur during nutrient deprivation [43], but the degradation of polar lipids and the translocation of PUFAs into TAG are significant processes that can affect the nutritional properties of microalgae. We found that PUFA biosynthesis was strongly downregulated in N- conditions, with major reductions in desaturase activity. In *Nannochloropsis*, $\Delta 5$ desaturase activity is associated with ARA and EPA biosynthesis [23], and together the proteomic data and fatty-acid profiles indicate that de-novo LC-PUFA biosynthesis probably plays only a minor role in lipid composition under N starvation.

Lipid-class remodeling has been associated with phosphorus starvation, where specific classes of P-containing membrane lipids are substituted with nonphospholipids [10]. Phospholipid remodeling in plants and microalgae involves acyltransferase and phospholipase activity [44]. Whilst various proteins annotated as phospholipases were identified in our data, none were significantly upregulated under P- conditions. Instead, increased lipase activity was a signature of oil-accumulating cells under N- starvation. However, we found that a choline/ethanolamine kinase was upregulated under P- conditions, which could indicate attempts to maintain phospholipid (phosphatidylcholine, PC and phosphatidylethanolamine, PE) production in these conditions. We also identified a Udp-sulfoquinovose synthase protein that was significantly downregulated in N- conditions, but significantly upregulated in P- treatments. This enzyme is associated with the synthesis of sulfoquinovosyldiacylglycerol (SQDG), a thylakoid lipid that can potentially replace and compensate for loss of phospholipids, especially PG, during phosphorus-scarce conditions [11].

An interesting feature of our data was the upregulation of two putative lipoxygenase (LOX) proteins under N- stress. Lipoxygenases provide the enzymatic route to oxylipin production where PUFAs, primarily C18 and C20 series, are converted to various oxidized lipid derivatives [45]. Oxylipins have roles in cell signaling and stress response and, although LOX activity has not been widely investigated in different microalgae species, oxylipin production has been measured in *Nannochloropsis* [46], and hydroxylated EPA was abundant in the metabolome of the diatom *Phaeodactylum tricornitum* under similar experimental conditions [35].

4. Materials and Methods

4.1. Cultivation

Nannochloropsis gaditana (CCMP 526, National Center for Marine Algae and Microbiota, East Boothbay, ME, USA) was cultivated in 400 mL flat plate photobiorectors (Algaemist-S, Wageningen UR, The Netherlands) using f/2 medium (Guillard and Ryther, 1962). The nutrient concentrations were increased proportionally to support high cell density, equivalent to 3.0 g L^{-1} NaNO$_3$. Cultures were maintained as turbidostats (constant optical density) by automatically adding fresh medium and collecting the overflowing broth. Turbidostat cultures provide a high level of experimental control by eliminating variables such as changes in internal irradiance that typically occur in batch or flask cultures. The temperature (25.0 ± 0.2 °C) was maintained by internal heating/external cooling modules and a constant irradiance of 350 µmol m^{-2} s^{-1} was provided by warm-white light emitting diodes. These conditions ensured high cell density and rapid biomass turnover. Before experimental treatments the cultures were maintained for several days, where they reached a constant growth/dilution rate. Control (C) treatments were subsequently maintained at the same steady-state, whilst nitrogen (N-) and phosphorus (P-) stress treatments were selectively applied by omitting either nitrate or phosphate from the feed medium. The high biomass turnover ensured cells in stress treatments were subjected to a rapid, natural depletion of either N or P. Since there were two photobioreactor units, the cultivation sequence was designed to avoid treatment bias (Table S1), and in total there were $n = 4$ independent replicate cultures for C, N-, and P- conditions. Conditions inside the photobioreactors were recorded by a program written in Python v2.7, running on a Raspberry Pi single-board computer (Raspberry Pi foundation, UK). The maximum duration of our experiment was 5 days, by which time growth in N- treatments had nearly ceased and the limit of turbidity control was reached. Based on the cultivation data in Figure 1 we selected day 3 for proteomics and transcriptomics analysis, because it represented the mid-point in the onset of stress conditions, allowing sufficient time to detect metabolic and molecular changes in the cells.

4.2. Sample Collection

Samples for proteomic and transcriptomic analysis were each collected into 2.0 mL tubes. Cells were immediately pelleted by centrifugation (5000 rcf, 2 min) and quenched in liquid nitrogen, then stored at −80 °C. Samples for metabolite analysis were collected in 2.0 mL tubes and additionally desalted by washing with isotonic ammonium formate, then stored at −20 °C. The sample supernatant was retained for analysis of nitrate and phosphate. The sample time points selected for molecular characterization are shown in Table 3. Our experiment comprised 12 turbidostat cultivations, but only 10 TMT labels were available for proteomic analysis. Thus, control and N- proteome treatments each have four biological replicates, whilst P- treatments have two replicates for the proteome. Statistical analysis accounted for the degrees of freedom and multiple comparisons.

Table 3. Summary of the experimental samples used for proteomic and transcriptomic analysis. The "No. Cultivations" is the total number of replicate turbidostat cultures available for each treatment. Ten proteome samples were obtained after 3 days of C, N-, or P- treatment. Twelve RNA samples were obtained after both 3 and 5 days and are repeated measurements from the same experimental units.

Treatment	No. Cultivations	Day 3 Protein	Day 3 Transcript	Day 5 Protein	Day 5 Transcript
Control (C)	4	4	4	−	4
Nitrogen (N-)	4	4	4	−	4
Phosphorus (P-)	4	2	4	−	4
Total	12	10	12	−	12

4.3. Lipid Analysis

Polar and neutral lipids were separated by solid phase extraction and the fatty acids were analyzed with a Gas Chromatograph and Flame Ionization Detector (GC-FID). Approximately 8 mg lyophilized samples were weighed with a precision balance (Mettler Toledo, Columbus, OH, USA, MX5) and transferred into 2.0 mL tubes containing 300 µL of 0.1 mm glass beads. Cell disruption was performed by adding 1.0 mL chloroform:methanol (2:2.5) spiked with C15:0 TAG (tripentadecanoin) internal standard, before bead-milling. The homogenate was transferred to a 10 mL glass tube with the addition of another 3.0 mL chloroform:methanol. Phase separation was used to recover the chloroform fraction, which was then dried under a stream of N_2 to recover total lipids. Polar and neutral lipid extracts were then prepared using solid-phase columns (Waters Sep-Pak 6cc/1g silica) and derivatized to fatty-acid methyl-esters (FAMEs) by adding 3.0 mL of 12% H_2SO_4 in methanol, then heating at 70 °C for 3 h. FAMEs were separated and quantitated using a Scion 436 GC-FID (Bruker, USA) fitted with a splitless injector and a 30 m CP-WAX column (Agilent Technologies, USA).

Supelco 37-component standards (Sigma-Aldrich, Oslo, Norway) were used for identification and quantitation of the FAMEs with five-point calibrations. Blanks were included throughout extraction and derivatization, to eliminate trace background peaks.

4.4. Nutrient Analysis

The concentration of nitrate in the broth was measured with standard colorimetric reagents using a miniaturized microplate method and NADH:nitrate reductase [25]. The absorbance was measured at 540 nm with a Tecan Sunrise microplate reader. Seven-point calibrations were included in each plate ($R^2 > 0.995$). Phosphate was analyzed with the ammonium molybdate/ascorbic acid method, and the absorbance was measured at 650 nm with a 1.0 cm cell.

4.5. Proteomics

Protein was extracted by resuspending cell pellets in 1.0 mL of extraction buffer (phosphate buffered saline +0.03% Triton X-100 + protease inhibitor cocktail) on ice, and homogenized briefly with a bead mill (Precellys, Bertin Instruments, Montigny-le-Bretonneux, France, 0.1 mm glass beads, 6500 rpm, 15 s). The suspension was centrifuged (20,000 rcf, 15 min, 4 °C) and the supernatant transferred to new tubes. Proteins were then precipitated by adding five volumes of ice-cold acetone, followed by centrifugation (20,000 rcf, 15 min, 4 °C). The supernatant was removed, and the protein pellets were allowed to air dry for 2 min at room temperature. Protein pellets were suspended in Laemlli buffer and the protein concentration of each sample was measured in duplicate with a BCA protein assay kit (Microplate BCA™ Protein Assay Kit—Reducing Agent Compatible, Thermo Scientific, Waltham, MA, USA). A seven-point calibration was used ($R^2 > 0.999$) and samples were blank-corrected using the sample buffer (Figure S1). A standardized 95.1 µg of protein from each sample was loaded to an SDS-PAGE gel and trapped for analysis.

Analysis and database searching was performed by University of York metabolomics and proteomics facility (York, UK) using 10-plex Tandem Mass Tags (Thermo Scientific, TMT10plex™). In-gel tryptic digestion was performed after reduction with dithioerythritol and *S*-carbamidomethylation with iodoacetamide. Digests were incubated overnight at 37 °C, then peptides were extracted with 50% aqueous acetonitrile containing 0.1% trifluoroacetic acid, before drying in a vacuum concentrator and reconstituting in aqueous 0.1% trifluoroacetic acid. Peptides were buffer exchanged into aqueous 50 mM triethylammonium bicarbonate using Strata C_{18}-E cartridges before TMT labelling (Table S2 for label-sample assignments). Labelled samples were combined together, loaded onto a conditioned reversed-phase C_{18} spin column (Pierce) and subject to centrifugation at 5000 rcf for 2 min before washing with 300 µL of LC-MS grade water. Peptides were eluted from columns into eight fractions using increasing concentrations of acetonitrile in aqueous triethylamine. Fractions were dried in a vacuum concentrator before reconstituting in aqueous 0.1% trifluoroacetic acid. Fractions were analyzed over 4 h acquisitions with elution from a 50 cm C_{18} EasyNano PepMap nanocapillary column using an UltiMate 3000 RSLCnano HPLC system (Thermo) interfaced with an Orbitrap Fusion hybrid mass spectrometer (Thermo). Positive ESI-MS, MS^2 and MS^3 spectra were acquired with multi-notch synchronous precursor selection using Xcalibur software (version 4.0, Thermo). Mascot Daemon (version 2.5.1, Matrix Science) was used to search against the *Nannochloropsis gaditana* subset of the UniProt database. To maximize the number of identified proteins, the search was conducted on a database containing concatenated data from the B31 and CCMP526 proteomes (15,363 sequences; 5,747,225 residues). The Mascot 0.dat result file was imported into Scaffold Q+ (version 4.7.5, Proteome Software) and a second search run against the same database using X!Tandem. Protein identifications were filtered to require a maximum protein and peptide false discovery rate of 3% [47] with a minimum of two unique peptide identifications per protein. Protein probabilities were assigned by the Protein Prophet algorithm [48]. Relative quantitation of protein abundance was calculated from the TMT reporter ion intensities using Scaffold Q+. TMT isotope correction factors were applied according to the manufacturer. Differentially expressed proteins were determined by applying

Permutation Tests with significance levels (*p*-values) adjusted with the Benjamini–Hochberg method. TMT labelling provides sensitive measurements of differential expression of individual proteins in multiplexed samples. However, the effect of peptide length and composition means that the reporter ion responses across different proteins are only semi-quantitative estimates of abundance, i.e., different peptides/proteins have different response factors. To more accurately estimate protein quantities, the "protein abundance in multiplexed samples" (PAMUS) method [49] was applied, which is based on the empirical linear relationship between the protein abundance index (PAI) and the logarithm of absolute protein abundance [50]. The exponentially modified PAI (emPAI) for each protein was first obtained from Scaffold Q+ to estimate the relative amount of each protein in the multiplexed sample. Then for each protein, the TMT reporter ion intensities were used to quantify the proportion of emPAI attributed to each individual sample/label. The abundance of the proteins in the individual samples was then expressed in Mol% [50]. The location of mature proteins in the cell was annotated based on the "Subcellular location" field of the UniProtKB database (www.uniprot.org). Complete mass spectrometry data sets are open-access and available to download from MassIVE (MSV000085294) and ProteomeXchange (PXD018605) (doi:10.25345/C5GQ50).

4.6. Transcriptomics

Total RNA was extracted from cell pellets by adding 1.0 mL QIAzol (Qiagen) followed by lysis with a bead-beater (Precellys, Bertin Instruments, Montigny-le-Bretonneux, France, 0.1 mm glass beads, 6500 rpm, 15 s). After adding 0.2 mL chloroform, the sample was centrifuged (20,000 rcf, 15 min, 4 °C) and the aqueous supernatant was added directly to RNA Clean and Concentrator columns (Zymo Research, Irvine, CA, USA) and prepared according to the manufacturer instructions. The cleaned RNA was eluted from the columns using molecular grade water and quality and quantity checked using a 2200 TapeStation instrument (Agilent Genomics, Santa Clara, CA, USA) and Nanodrop Spectrophotometer (Thermo Fisher Scientific). Libraries were prepared using Poly(A) selection to enrich for mRNA and a NEBNext Ultra Directional RNA Library Prep kit for Illumina (New England Biolabs Inc., Ipswich, MA, USA) according the manufacturer protocols. Barcoded sample libraries were pooled in equal amount and sequenced on an Illumina NextSeq 500 platform using High Output Kit v2. A total of 443 million 150 bp paired-end reads were obtained and archived at NCBI web portal under Bioproject PRJNA589063.

The quality of reads was assessed with FastQC (Babraham Bioinformatics, Cambridge, UK) and gentle adapter and quality trimming (Q > 20, L > 50) was applied using cutadapt v1.13 [51]. The annotated reference genomes of *N. gaditana* were downloaded for strains CCMP526 (assembly ASM24072v1) and B31 (assembly NagaB31_1.0) and assessed. Although we used strain CCMP526 in our study (verified genetically, Figure S2), the more recent reference genome for strain B31 provided more unique mapped reads in our data (for reference comparisons see Table S3). Our analysis therefore uses reads that were aligned to the B31 reference genome using the splice-aware aligner STAR 2.5.3a [52], with the annotation aware option. The PCR duplication rate was assessed using the Bioconductor package "dupRADAR" [53] in R v. 3.3.3 and was found to be low (<0.1%). Counts of reads for gene-level quantification were extracted using "featureCounts" [54] supplied with annotation information and strands of reads. Raw counts were imported into the Bioconductor package "DESeq2" v 1.14.1 [55] and differential expression analysis was performed with independent filtering enabled and alpha = 0.05. Genes that had an FDR *p*-adjusted value < 0.05 and $L_2fc > 1.0$ (fold change of > 2) were chosen as the differentially expressed genes. Taking into account our design (four replicates in each group and fold change > 2.0) we reached more than 90% statistical power to detect differentially expressed genes [56].

4.7. Gene Ontology and KEGG Pathway Gene Set Enrichment Analysis

Gene ontology (GO) terms were obtained from the UniProtKB database (http://www.uniprot.org). Annotation of genes for KEGG Orthology (KO) numbers was performed using GhostKOALA [57].

Gene ontology and KEGG pathway enrichment analyses were performed for both transcriptome and proteome data sets. Gene set enrichment analysis implemented in Babelomics 5.0 suite [58], was used to detect GO functional sets of genes and proteins significantly affected by nutrient deprivation. The logistic model using the L_2fc of all genes or proteins was employed with significance cut-off FDR-adjusted *p*-value of 0.01. GOs with log-odds ratio (LOR) <0.0 were taken to be over-represented for downregulated genes/proteins, and LOR > 0.0 were over-represented for upregulated genes/proteins. The gene set approach was also used to identify the most perturbed KEGG pathways with unidirectional changes of gene and protein expression. The analysis was performed using the Bioconductor package GAGE 2.24 [59] with L_2fc values as per gene statistics, q < 0.05 and only pathways with more than five annotated KO numbers. GO enrichment analysis was performed separately for up- and downregulated genes using classic Fisher's exact test in R package topGO v2.26 [60] with FDR correction at 0.05 and pruning the GO hierarchy from terms which have less than five annotated genes. To identify the most perturbated KEGG pathways with unidirectional or bidirectional changes of gene expression the gene set approach was used. The analysis was performed using the Bioconductor package GAGE 2.24 [59] with L_2fc values as per gene statistics and only pathways with more than five annotated KO numbers.

4.8. Data Analysis

The protein and transcript data were associated together using their unique ID (gene, UniProt) numbers. Data was analyzed using the R programming language, and the package "nlme" [61], was used to fit a linear mixed-effects model (Figure 3b, Table S4, Figure S4). The mixed-model fixed effects were (log RPKM~log Mol%) with the random effects formula (~1 + logMol%|replicate) following nlme notation, where "log RPKM" is the natural logarithm of transcript counts in units RPKM and "log Mol%" is the natural logarithm of protein abundance in Mol%. The "replicate" term is the individual turbidostat cultivation ($n = 10$). Correlation coefficients, summary statistics, and linear regression models were implemented in base R.

5. Conclusions

This study provides new insights into global protein and gene expression in the oleaginous microalga *Nannochloropsis gaditana*. Both proteomic and transcript sequencing methods each tended to capture the major patterns in expression, but at the whole-cell level protein and transcript associations were characteristically noisy. In *Nannochloropsis* macronutrient stress is associated with lipid remodeling and oleaginous phenotypes, but lipid metabolic processes were not highly enriched in our GO and KEGG analyses. We did however find major changes in several lipid-related proteins, including increased expression of DGAT and lipid body proteins under N-starved conditions. Pathways in lipid remodeling, fatty-acid oxidation and signaling could be prioritized for future studies, as these are key processes that determine the fate of valuable long-chain polyunsaturated fatty acids. Adjustments in respiratory/mitochondrial activity featured in our data, with shifts in TCA cycle activity and glycolytic processes providing metabolic compensation under stress. The active reshaping of organelle (compartment) proteomes and the control of inter-organelle metabolic flux are therefore important research areas. Finally, our data raises the topic of post-transcriptional mechanisms, which may in part explain the observed patterns of gene/protein/metabolite correlations.

Supplementary Materials: The following are available online at http://www.mdpi.com/1422-0067/21/18/6946/s1. Attached to this submission: < Supplementary.information.pdf > single file containing all additional text and figures < Proteomics.results.data.TMT.xlsx > single Excel file containing prepared proteomics data and statistical tests. The original sequencing and proteomics datasets generated in this study can be found in the following repositories; transcript sequences are open-access and deposited in the NCBI sequence read archive SRA (https://www.ncbi.nlm.nih.gov/) under Bioproject PRJNA589063. Mass spectrometry data sets are open-access and available to download from MassIVE (MSV000085294) and ProteomeXchange (PXD018605) (doi:10.25345/C5GQ50).

Author Contributions: Conceptualization, C.J.H., G.G.H., R.H.W. and V.K.; methodology, C.J.H., I.S., A.D., G.G.H. and V.K.; formal analysis, C.J.H., I.S., A.D., G.K.V. and M.K.; data curation, C.J.H., I.S. and A.D.; writing—original draft preparation, C.J.H., I.S. and V.K.; writing—review and editing, C.J.H., I.S., M.K., G.K.V., G.G.H., R.H.W. and

V.K.; visualization, C.J.H., I.S.; project administration, C.J.H., V.K.; funding acquisition, C.J.H. and V.K. All authors have read and agreed to the published version of the manuscript.

Funding: The research has received funds from Nord University and Nordland County Government. C.J.H. was funded by a grant from the European Union's Horizon 2020 research and innovation programme under Marie Skłodowska-Curie project number (749910). The APC was funded by Nord University.

Acknowledgments: The authors are grateful to Anjana Palihawadana for analytical support in lipid analysis.

Conflicts of Interest: The authors declare no conflict of interest. The funders had no role in the design of the study; in the collection, analyses, or interpretation of data; in the writing of the manuscript, or in the decision to publish the results.

References

1. Quigg, A.; Finkel, Z.V.; Irwin, A.J.; Rosenthal, Y.; Ho, T.-Y.; Reinfelder, J.R.; Schofield, O.; Morel, F.M.; Falkowski, P.G. The evolutionary inheritance of elemental stoichiometry in marine phytoplankton. *Nature* **2003**, *425*, 291–294. [CrossRef] [PubMed]

2. Redfield, A.C. The influence of organisms on the composition of seawater. In *The Sea*; Wiley Interscience: New York, NY, USA, 1963; Volume 2, pp. 26–77.

3. Browning, T.J.; Achterberg, E.P.; Rapp, I.; Engel, A.; Bertrand, E.M.; Tagliabue, A.; Moore, C.M. Nutrient co-limitation at the boundary of an oceanic gyre. *Nature* **2017**, *551*, 242. [CrossRef]

4. Marañón, E.; Lorenzo, M.P.; Cermeño, P.; Mouriño-Carballido, B. Nutrient limitation suppresses the temperature dependence of phytoplankton metabolic rates. *ISME J.* **2018**, *12*, 1836–1845. [CrossRef] [PubMed]

5. Hu, Q.; Sommerfeld, M.; Jarvis, E.; Ghirardi, M.; Posewitz, M.; Seibert, M.; Darzins, A. Microalgal triacylglycerols as feedstocks for biofuel production: Perspectives and advances. *Plant J.* **2008**, *54*, 621–639. [CrossRef]

6. Roth, M.S.; Cokus, S.J.; Gallaher, S.D.; Walter, A.; Lopez, D.; Erickson, E.; Endelman, B.; Westcott, D.; Larabell, C.A.; Merchant, S.S. Chromosome-level genome assembly and transcriptome of the green alga Chromochloris zofingiensis illuminates astaxanthin production. *Proc. Natl. Acad. Sci. USA* **2017**, *114*, E4296–E4305. [CrossRef]

7. Elser, J.; Fagan, W.; Kerkhoff, A.; Swenson, N.; Enquist, B. Biological stoichiometry of plant production: Metabolism, scaling and ecological response to global change. *New Phytol.* **2010**, *186*, 593–608. [CrossRef]

8. Toseland, A.; Daines, S.J.; Clark, J.R.; Kirkham, A.; Strauss, J.; Uhlig, C.; Lenton, T.M.; Valentin, K.; Pearson, G.A.; Moulton, V. The impact of temperature on marine phytoplankton resource allocation and metabolism. *Nat. Clim. Chang.* **2013**, *3*, 979–984. [CrossRef]

9. Van Mooy, B.A.; Fredricks, H.F.; Pedler, B.E.; Dyhrman, S.T.; Karl, D.M.; Koblížek, M.; Lomas, M.W.; Mincer, T.J.; Moore, L.R.; Moutin, T. Phytoplankton in the ocean use non-phosphorus lipids in response to phosphorus scarcity. *Nature* **2009**, *458*, 69–72. [CrossRef]

10. Cañavate, J.P.; Armada, I.; Hachero-Cruzado, I. Interspecific variability in phosphorus-induced lipid remodelling among marine eukaryotic phytoplankton. *New Phytol.* **2017**, *213*, 700–713. [CrossRef]

11. Mühlroth, A.; Winge, P.; El Assimi, A.; Jouhet, J.; Marechal, E.; Hohmann-Marriott, M.F.; Vadstein, O.; Bones, A.M. Mechanisms of phosphorus acquisition and lipid class remodelling under P limitation in a marine microalga. *Plant Physiol.* **2017**, *175*, 1543–1559. [CrossRef]

12. Li, J.; Han, D.; Wang, D.; Ning, K.; Jia, J.; Wei, L.; Jing, X.; Huang, S.; Chen, J.; Li, Y. Choreography of transcriptomes and lipidomes of Nannochloropsis reveals the mechanisms of oil synthesis in microalgae. *Plant Cell* **2014**, *26*, 1645–1665. [CrossRef]

13. Rismani-Yazdi, H.; Haznedaroglu, B.Z.; Hsin, C.; Peccia, J. Transcriptomic analysis of the oleaginous microalga Neochloris oleoabundans reveals metabolic insights into triacylglyceride accumulation. *Biotechnol. Biofuels* **2012**, *5*, 74. [CrossRef]

14. Peng, X.; Qin, Z.; Zhang, G.; Guo, Y.; Huang, J. Integration of the proteome and transcriptome reveals multiple levels of gene regulation in the rice dl2 mutant. *Front. Plant Sci.* **2015**, *6*. [CrossRef]

15. Li, J.J.; Chew, G.-L.; Biggin, M.D. Quantitating translational control: mRNA abundance-dependent and independent contributions and the mRNA sequences that specify them. *Nucleic Acids Res.* **2017**, *45*, 11821–11836. [CrossRef]

16. De Sousa Abreu, R.; Penalva, L.O.; Marcotte, E.M.; Vogel, C. Global signatures of protein and mRNA expression levels. *Molecular BioSystems* **2009**, *5*, 1512–1526. [CrossRef]

17. Vogel, C.; de Sousa Abreu, R.; Ko, D.; Le, S.Y.; Shapiro, B.A.; Burns, S.C.; Sandhu, D.; Boutz, D.R.; Marcotte, E.M.; Penalva, L.O. Sequence signatures and mRNA concentration can explain two-thirds of protein abundance variation in a human cell line. *Mol. Syst. Biol.* **2010**, *6*, 400. [CrossRef]

18. Vogel, C.; Marcotte, E.M. Insights into the regulation of protein abundance from proteomic and transcriptomic analyses. *Nat. Rev. Genet.* **2012**, *13*, 227–232. [CrossRef]

19. Lahtvee, P.-J.; Sánchez, B.J.; Smialowska, A.; Kasvandik, S.; Elsemman, I.E.; Gatto, F.; Nielsen, J. Absolute Quantification of Protein and mRNA Abundances Demonstrate Variability in Gene-Specific Translation Efficiency in Yeast. *Cell Syst.* **2017**, *4*, 495–504. [CrossRef]

20. Radakovits, R.; Jinkerson, R.E.; Fuerstenberg, S.I.; Tae, H.; Settlage, R.E.; Boore, J.L.; Posewitz, M.C. Draft genome sequence and genetic transformation of the oleaginous alga Nannochloropsis gaditana. *Nat. Commun.* **2012**, *3*, 686. [CrossRef]

21. Carpinelli, E.C.; Telatin, A.; Vitulo, N.; Forcato, C.; D'Angelo, M.; Schiavon, R.; Vezzi, A.; Giacometti, G.M.; Morosinotto, T.; Valle, G. Chromosome scale genome assembly and transcriptome profiling of Nannochloropsis gaditana in nitrogen depletion. *Mol. Plant* **2014**, *7*, 323–335. [CrossRef]

22. Zienkiewicz, K.; Zienkiewicz, A.; Poliner, E.; Du, Z.-Y.; Vollheyde, K.; Herrfurth, C.; Marmon, S.; Farré, E.M.; Feussner, I.; Benning, C. Nannochloropsis, a rich source of diacylglycerol acyltransferases for engineering of triacylglycerol content in different hosts. *Biotechnol. Biofuels* **2017**, *10*, 8. [CrossRef]

23. Ma, X.-N.; Chen, T.-P.; Yang, B.; Liu, J.; Chen, F. Lipid production from Nannochloropsis. *Mar. Drugs* **2016**, *14*, 61. [CrossRef]

24. Calder, P.C. Very long chain omega-3 (n-3) fatty acids and human health. *Eur. J. Lipid Sci. Technol.* **2014**, *116*, 1280–1300. [CrossRef]

25. Hulatt, C.J.; Wijffels, R.H.; Bolla, S.; Kiron, V. Production of fatty acids and protein by Nannochloropsis in flat-plate photobioreactors. *PLoS ONE* **2017**, *12*, e0170440. [CrossRef]

26. Nelson, C.J.; Millar, A.H. Protein turnover in plant biology. *Nat. Plants* **2015**, *1*, 15017. [CrossRef]

27. Floor, S.N.; Doudna, J.A. Tunable protein synthesis by transcript isoforms in human cells. *Elife* **2016**, *5*, e10921. [CrossRef]

28. Calixto, C.P.; Guo, W.; James, A.B.; Tzioutziou, N.A.; Entizne, J.C.; Panter, P.E.; Knight, H.; Nimmo, H.; Zhang, R.; Brown, J.W. Rapid and dynamic alternative splicing impacts the Arabidopsis cold response transcriptome. *Plant Cell* **2018**, *30*, 1424–1444. [CrossRef]

29. Rastogi, A.; Maheswari, U.; Dorrell, R.G.; Vieira, F.R.J.; Maumus, F.; Kustka, A.; McCarthy, J.; Allen, A.E.; Kersey, P.; Bowler, C. Integrative analysis of large scale transcriptome data draws a comprehensive landscape of Phaeodactylum tricornutum genome and evolutionary origin of diatoms. *Sci. Rep.* **2018**, *8*, 4834. [CrossRef]

30. La Manno, G.; Soldatov, R.; Zeisel, A.; Braun, E.; Hochgerner, H.; Petukhov, V.; Lidschreiber, K.; Kastriti, M.E.; Lönnerberg, P.; Furlan, A. RNA velocity of single cells. *Nature* **2018**, *560*, 494–498. [CrossRef]

31. Airoldi, E.M.; Miller, D.; Athanasiadou, R.; Brandt, N.; Abdul-Rahman, F.; Neymotin, B.; Hashimoto, T.; Bahmani, T.; Gresham, D. Steady-state and dynamic gene expression programs in Saccharomyces cerevisiae in response to variation in environmental nitrogen. *Mol. Biol. Cell* **2016**, *27*, 1383–1396. [CrossRef]

32. Sforza, E.; Calvaruso, C.; La Rocca, N.; Bertucco, A. Luxury uptake of phosphorus in Nannochloropsis salina: Effect of P concentration and light on P uptake in batch and continuous cultures. *Biochem. Eng. J.* **2018**, *134*, 69–79. [CrossRef]

33. Dong, H.-P.; Williams, E.; Wang, D.-z.; Xie, Z.-X.; Hsia, R.-c.; Jenck, A.; Halden, R.; Li, J.; Chen, F.; Place, A.R. Responses of Nannochloropsis oceanica IMET1 to long-term nitrogen starvation and recovery. *Plant Physiol.* **2013**, *162*, 1110–1126. [CrossRef]

34. Plaxton, W.C.; Tran, H.T. Metabolic adaptations of phosphate-starved plants. *Plant Physiol.* **2011**, *156*, 1006–1015. [CrossRef]

35. Remmers, I.M.; D'Adamo, S.; Martens, D.E.; de Vos, R.C.; Mumm, R.; America, A.H.; Cordewener, J.H.; Bakker, L.V.; Peters, S.A.; Wijffels, R.H. Orchestration of transcriptome, proteome and metabolome in the diatom Phaeodactylum tricornutum during nitrogen limitation. *Algal Res.* **2018**, *35*, 33–49. [CrossRef]

36. Nobusawa, T.; Hori, K.; Mori, H.; Kurokawa, K.; Ohta, H. Differently localized lysophosphatidic acid acyltransferases crucial for triacylglycerol biosynthesis in the oleaginous alga Nannochloropsis. *Plant J.* **2017**, *90*, 547–559. [CrossRef]

37. La Russa, M.; Bogen, C.; Uhmeyer, A.; Doebbe, A.; Filippone, E.; Kruse, O.; Mussgnug, J.H. Functional analysis of three type-2 DGAT homologue genes for triacylglycerol production in the green microalga Chlamydomonas reinhardtii. *J. Biotechnol.* **2012**, *162*, 13–20. [CrossRef]

38. Yoneda, K.; Yoshida, M.; Suzuki, I.; Watanabe, M.M. Identification of a major lipid droplet protein in a marine diatom Phaeodactylum tricornutum. *Plant Cell Physiol.* **2016**, *57*, 397–406. [CrossRef]

39. Vieler, A.; Brubaker, S.B.; Vick, B.; Benning, C. A lipid droplet protein of Nannochloropsis with functions partially analogous to plant oleosins. *Plant Physiol.* **2012**, *158*, 1562–1569. [CrossRef]

40. Alboresi, A.; Perin, G.; Vitulo, N.; Diretto, G.; Block, M.A.; Jouhet, J.; Meneghesso, A.; Valle, G.; Giuliano, G.; Maréchal, E. Light remodels lipid biosynthesis in Nannochloropsis gaditana by modulating carbon partitioning between organelles. *Plant Physiol.* **2016**, *171*, 2468–2482. [CrossRef]

41. McKew, B.A.; Lefebvre, S.C.; Achterberg, E.P.; Metodieva, G.; Raines, C.A.; Metodiev, M.V.; Geider, R.J. Plasticity in the proteome of Emiliania huxleyi CCMP 1516 to extremes of light is highly targeted. *New Phytol.* **2013**, *200*, 61–73. [CrossRef] [PubMed]

42. Han, D.; Jia, J.; Li, J.; Sommerfeld, M.; Xu, J.; Hu, Q. Metabolic remodeling of membrane glycerolipids in the microalga Nannochloropsis oceanica under nitrogen deprivation. *Front. Mar. Sci.* **2017**, *4*, 242. [CrossRef]

43. Janssen, J.H.; Lamers, P.P.; de Vos, R.C.; Wijffels, R.H.; Barbosa, M.J. Translocation and de novo synthesis of eicosapentaenoic acid (EPA) during nitrogen starvation in Nannochloropsis gaditana. *Algal Res.* **2019**, *37*, 138–144. [CrossRef]

44. Yamashita, A.; Sugiura, T.; Waku, K. Acyltransferases and transacylases involved in fatty acid remodeling of phospholipids and metabolism of bioactive lipids in mammalian cells. *J. Biochem.* **1997**, *122*, 1–16. [CrossRef]

45. Barbosa, M.; Valentão, P.; Andrade, P.B. Biologically active oxylipins from enzymatic and nonenzymatic routes in macroalgae. *Mar. Drugs* **2016**, *14*, 23. [CrossRef]

46. De los Reyes, C.; Ávila-Román, J.; Ortega, M.J.; de la Jara, A.; García-Mauriño, S.; Motilva, V.; Zubía, E. Oxylipins from the microalgae Chlamydomonas debaryana and Nannochloropsis gaditana and their activity as TNF-α inhibitors. *Phytochemistry* **2014**, *102*, 152–161. [CrossRef]

47. Longworth, J.; Wu, D.; Huete-Ortega, M.; Wright, P.C.; Vaidyanathan, S. Proteome response of Phaeodactylum tricornutum, during lipid accumulation induced by nitrogen depletion. *Algal Res.* **2016**, *18*, 213–224. [CrossRef]

48. Nesvizhskii, A.I.; Keller, A.; Kolker, E.; Aebersold, R. A statistical model for identifying proteins by tandem mass spectrometry. *Anal. Chem.* **2003**, *75*, 4646–4658. [CrossRef]

49. Adav, S.S.; Chao, L.T.; Sze, S.K. Protein abundance in multiplexed samples (PAMUS) for quantitation of Trichoderma reesei secretome. *J. Proteom.* **2013**, *83*, 180–196. [CrossRef]

50. Ishihama, Y.; Oda, Y.; Tabata, T.; Sato, T.; Nagasu, T.; Rappsilber, J.; Mann, M. Exponentially modified protein abundance index (emPAI) for estimation of absolute protein amount in proteomics by the number of sequenced peptides per protein. *Mol. Cell. Proteom.* **2005**, *4*, 1265–1272. [CrossRef]

51. Martin, M. Cutadapt removes adapter sequences from high-throughput sequencing reads. *EMBnet. J.* **2011**, *17*, 10–12. [CrossRef]

52. Dobin, A.; Davis, C.A.; Schlesinger, F.; Drenkow, J.; Zaleski, C.; Jha, S.; Batut, P.; Chaisson, M.; Gingeras, T.R. STAR: Ultrafast universal RNA-seq aligner. *Bioinformatics* **2013**, *29*, 15–21. [CrossRef]

53. Sayols, S.; Scherzinger, D.; Klein, H. dupRadar: A Bioconductor package for the assessment of PCR artifacts in RNA-Seq data. *BMC Bioinform.* **2016**, *17*, 428. [CrossRef]

54. Liao, Y.; Smyth, G.K.; Shi, W. featureCounts: An efficient general purpose program for assigning sequence reads to genomic features. *Bioinformatics* **2013**, *30*, 923–930. [CrossRef]

55. Love, M.I.; Huber, W.; Anders, S. Moderated estimation of fold change and dispersion for RNA-seq data with DESeq2. *Genome Biol.* **2014**, *15*, 550. [CrossRef]

56. Conesa, A.; Madrigal, P.; Tarazona, S.; Gomez-Cabrero, D.; Cervera, A.; McPherson, A.; Szcześniak, M.W.; Gaffney, D.J.; Elo, L.L.; Zhang, X. A survey of best practices for RNA-seq data analysis. *Genome Biology* **2016**, *17*, 13. [CrossRef]

57. Kanehisa, M.; Sato, Y.; Morishima, K. BlastKOALA and GhostKOALA: KEGG tools for functional characterization of genome and metagenome sequences. *J. Mol. Biol.* **2016**, *428*, 726–731. [CrossRef]

58. Alonso, R.; Salavert, F.; Garcia-Garcia, F.; Carbonell-Caballero, J.; Bleda, M.; Garcia-Alonso, L.; Sanchis-Juan, A.; Perez-Gil, D.; Marin-Garcia, P.; Sanchez, R. Babelomics 5.0: Functional interpretation for new generations of genomic data. *Nucleic Acids Res.* **2015**, *43*, W117–W121. [CrossRef]

59. Luo, W.; Friedman, M.S.; Shedden, K.; Hankenson, K.D.; Woolf, P.J. GAGE: Generally applicable gene set enrichment for pathway analysis. *BMC Bioinform.* **2009**, *10*, 161. [CrossRef]

60. Alexa, A.; Rahnenfuhrer, J. topGO: Enrichment Analysis for Gene Ontology. R Package Version 2. 2010. Available online: https://bioc.ism.ac.jp/packages/2.14/bioc/html/topGO.html (accessed on 20 September 2020).

61. Pinheiro, J.C.; Bates, D.M. *Mixed-Effects Models in S and S-PLUS*; Springer: New York, NY, USA, 2000.

International Journal of
Molecular Sciences

Article

Integrative Transcriptomic and Proteomic Analyses of Molecular Mechanism Responding to Salt Stress during Seed Germination in Hulless Barley

Yong Lai [1,†], Dangquan Zhang [1,†], Jinmin Wang [2], Juncheng Wang [3], Panrong Ren [4,5], Lirong Yao [3], Erjing Si [4], Yuhua Kong [1,*] and Huajun Wang [3,4,*]

1   College of Forestry, Henan Agricultural University, Zhengzhou 450002, China; xlaiyong@163.com (Y.L.); zhangdangquan@163.com (D.Z.)
2   College of Agriculture and Animal Husbandry, Qinghai University, Xining 810016, China; jinminw@163.com
3   Gansu Provincial Key Lab of Aridland Crop Science, Lanzhou 730070, China; 13819288385@163.com (J.W.); ylr0384@163.com (L.Y.)
4   Gansu Key Lab of Crop Improvement and Germplasm Enhancement, Lanzhou 730070, China; prren@genetics.ac.cn (P.R.); sierjing@163.com (E.S.)
5   State Key Laboratory of Plant Genomics, National Centre for Plant Gene Research, Institute of Genetics and Developmental Biology, Chinese Academy of Sciences, Beijing 100101, China
*   Correspondence: y.kong@henau.edu.cn (Y.K.); huajunwang@gsau.edu.cn (H.W.)
†   These authors contributed equally to this work.

Received: 3 December 2019; Accepted: 3 January 2020; Published: 6 January 2020

Abstract: Hulless barley (*Hordeum vulgare* L. var. *nudum*) is one of the most important crops in the Qinghai-Tibet Plateau. Soil salinity seriously affects its cultivation. To investigate the mechanism of salt stress response during seed germination, two contrasting hulless barley genotypes were selected to first investigate the molecular mechanism of seed salinity response during the germination stage using RNA-sequencing and isobaric tags for relative and absolute quantitation technologies. Compared to the salt-sensitive landrace lk621, the salt-tolerant one lk573 germinated normally under salt stress. The changes in hormone contents also differed between lk621 and lk573. In lk573, 1597 differentially expressed genes (DEGs) and 171 differentially expressed proteins (DEPs) were specifically detected at 4 h after salt stress, and correspondingly, 2748 and 328 specifically detected at 16 h. Most specific DEGs in lk573 were involved in response to oxidative stress, biosynthetic process, protein localization, and vesicle-mediated transport, and most specific DEPs were assigned to an oxidation-reduction process, carbohydrate metabolic process, and protein phosphorylation. There were 96 genes specifically differentially expressed at both transcriptomic and proteomic levels in lk573. These results revealed the molecular mechanism of salt tolerance and provided candidate genes for further study and salt-tolerant improvement in hulless barley.

Keywords: salt stress; seed germination; transcriptome; proteome; hulless barley

---

1. Introduction

Barley (*Hordeum vulgare* L.) is one of the most important cereal crops in the world that is widely used in the brewing industry and for healthy food products [1]. Moreover, hulless barley (*H. vulgare* L. var. *nudum*) is well adapted to extreme environmental conditions and has a long cultivation history as a food crop for Tibetans, mainly cultivated in the Qinghai-Tibet Plateau [2]. Soil salinization is one of the most serious environmental issues. More than 6% of the world's total land area and 20% of the total agricultural land are affected by salinity [3]. With the degradation of grassland and increased soil salinization in the Qinghai-Tibet Plateau, crop cultivation has also been seriously affected by salt stress. Barley is a relatively salt-tolerant species compared with wheat, rice, and other cereal crops [3].

A good understanding of the response mechanisms to salt stress is critical for crop improvement in salt tolerance. There are numerous reports on the response of barley leaves and roots to salinity [4–6]. The inhibitory effects of salinity on barley seed germination have been reported [7–9], but there is no report concerning the comprehensive response mechanism to salt stress during the germination stage at the transcriptomic and proteomic levels in hulless barley.

Soil salinity is one of the major factors harmful to agriculture due to its side effects of osmotic stress and ion toxicity on the growth and development of crop plants [10]. Plant salt tolerance at different developmental stages is controlled by different mechanisms, and each stage must be separately studied using special screening procedures [11]. The successful germination of mature seeds in a saline environment is the beginning of salt tolerance in the plant life cycle. Seed germination is considered to be the most critical phase in the life cycle because of its high vulnerability to injury, diseases, and environmental stresses [12]. This process starts with the uptake of water (phase I), followed by a plateau phase (phase II), and terminates with elongation of the embryonic axis (phase III) [13]. Sufficient stores of mRNAs and proteins in mature seeds are imperative for seed germination, showing that germination is prepared during seed maturation, and seedling growth is prepared in the germination stage [12]. Phytohormones are important to seed germination. The balance between abscisic acid (ABA) and gibberellins (GA) critically affects seed dormancy and germination, based on the inhibition of germination by ABA and activation of dormant seed by GA [13]. Auxin is not usually necessary for seed germination but can influence this process when ABA is present [14]. In addition, the universal second messenger calcium (Ca) also regulates seed germination by interacting with the effect of ABA [15].

The inhibition of salinity stress on seed germination results from osmotic stress, oxidative stress, and ion toxicity, shown by decreasing germination rate and extended germination time [16]. Salt stress leads to excessive reactive oxygen species (ROS), which damages proteins, lipids, and nucleic acids or the cellular structure, resulting in oxidative stress [3], and alters the phytohormone balance, with decreases in GA, auxin, and cytokinin and increases in ABA and jasmonates in plant tissues [14]. Under salt stress, the Ca^{2+} concentration increases and triggers the ABA signal and salt overly sensitive (SOS) pathways to decrease the damage from ROS and to regulate sodium ion homeostasis in plants [17–19]. The ABA is central to salt stress responses in plants and triggers ROS signals to alleviate the effect of salinity on seed germination [18–20]. In the SOS pathway, SOS3 and SCaBP8 perceive Ca^{2+} signals and interact with SOS2, which activates the plasma membrane Na^+/H^+ antiporter (SOS1) and vacuolar Na^+/H^+ exchanger (NHX) to limit the Na^+ concentration in cells [17,21]. Moreover, energy production is regulated in response to salt stress [22]. The energy and nutrient substrates are supplied from endosperm for the growth and development of the embryo [12]. Programmed cell death (PCD) occurs in aleurone layer cells to enhance the supply process [12]. Thus, more energy and substrates may be required not only for seed germination but also for resistance to salinity stress.

Recently, the integration of various omics technologies is an effective strategy to promote a better understanding of mechanisms in response to environmental stresses. The molecular mechanism of seed salt response during the germination stage is extremely complex. In the current study, two contrasting hulless barley landraces lk621 and lk573 were used to investigate the molecular mechanism in response to salinity during seed germination by comprehensive transcriptomic and proteomic analyses. The changes in hormone contents under salt stress were also analyzed between lk621 and lk573. Our results provided deeper insights into the molecular mechanisms of salt tolerance during the germination stage and candidate genes for further study and breeding salt-tolerant cultivars.

2. Results

2.1. Differences of Seed Germination between Two Landraces under Salt Stress

It is well known that salt stress significantly influences seed germination, and this effect differs among varieties [7,10]. In this study, two hulless barley genotypes lk621 and lk573 showed different responses to salt stress. The imbibition of lk621 and lk573 seeds was completed at 4 h under distilled water (control, CK) or 200 mM NaCl solution (salt treatment, T). At 16 h, lk573 seeds could complete germination under both treatments; however, lk621 seeds germinated under CK but not T treatments (Figure 1A–J). The germination rate of lk621 was significantly affected by salt stress, and there was no significant difference in germination rate of lk573 between CK and T (Figure 1K). These results illustrated that lk621 was more sensitive to salt stress than lk573 during the seed germination stage.

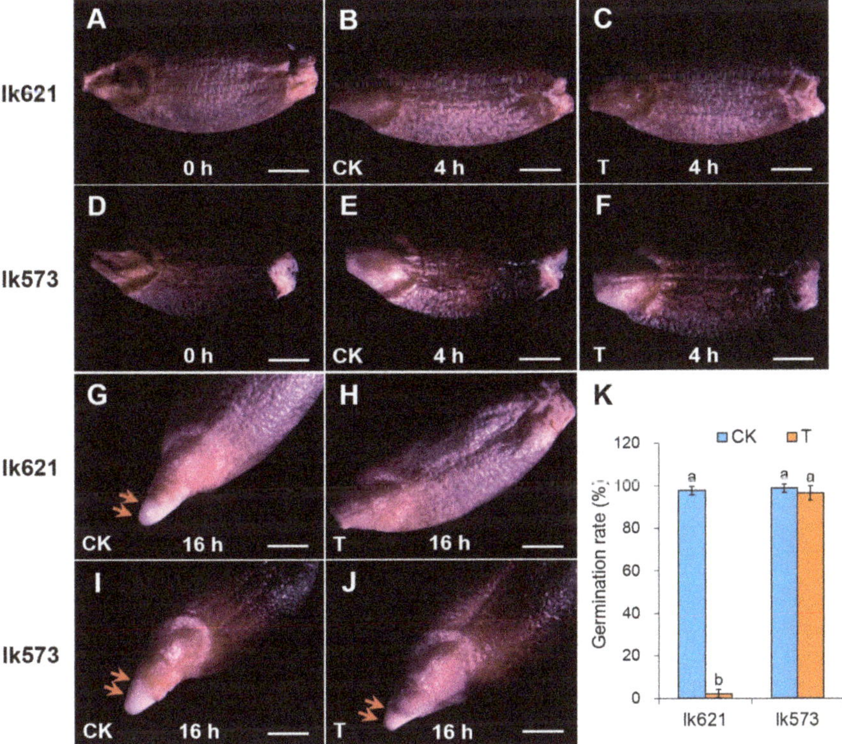

Figure 1. Seed morphology of hulless barley lk621 and lk573 treated with distilled water (control, CK) and 200 mM NaCl solution (salt treatment, T) (**A–J**) and seed germination rate (**K**). (**A,D**) Dormant seed of lk621 and lk573; (**B,C**) lk621 seed under CK and T at 4 h; (**E,F**) lk573 seed under CK and T at 4 h; (**G,H**) lk621 seed under CK and T at 16 h; and (**I,J**) lk573 seed under CK and T at 16 h. Red arrows show the radicles. Bar = 50 µm. Values presented are means of three replicates ± standard error (SE). Different lowercase letters indicate significant difference at $p < 0.05$ as determined by Tukey's Honestly Significant Difference (HSD) test in each sample.

The plant hormone contents in seeds were also affected by salt stress. Compared with CK, ABA contents in lk621 and lk573 were increased by T treatment at 16 h (Figure 2A). The GA content in lk621 was decreased by 26.87% and 46.76% after 4 and 16 h of salt stress, respectively. The GA content in lk573 was reduced by 10.19% at 4 h and by 14.52% at 16 h of salt stress (Figure 2B). The auxin contents in both lk621 and lk573 were significantly decreased by salt stress at 4 h and 16 h (Figure 2C). The ratio of GA/ABA significantly decreased by salt stress in both lk621 and lk573 at 4 and 16 h under salt stress (Figure 2D). These results illustrated the side effect of salt stress on plant hormone in sensitive and tolerant genotypes during the germination stage.

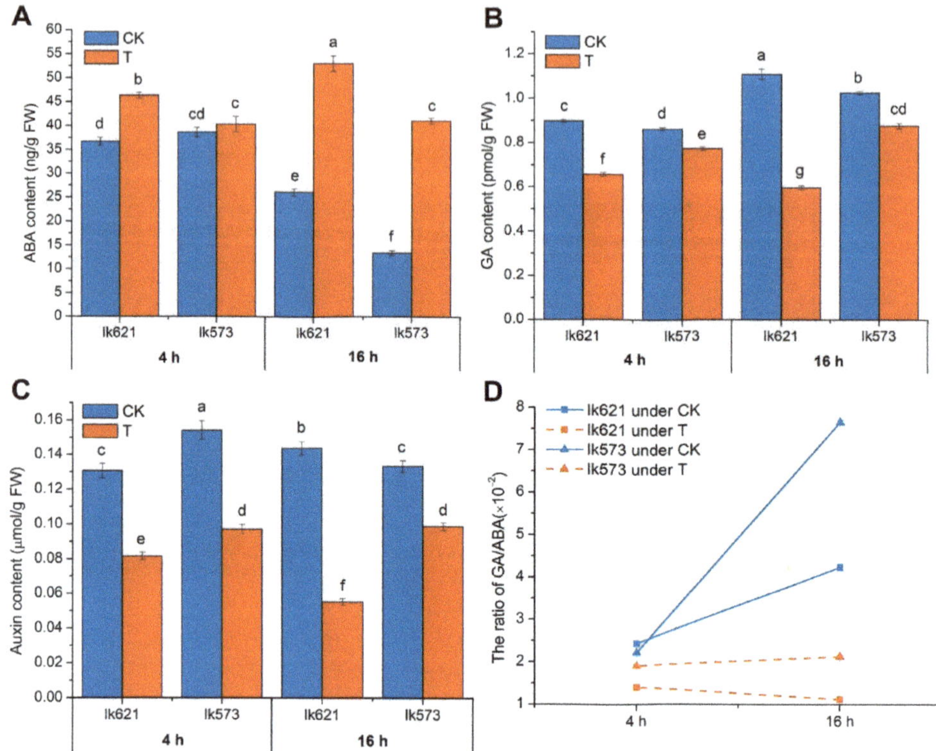

Figure 2. Contents of (**A**) abscisic acid (ABA), (**B**) gibberellins (GA), (**C**) auxin, and (**D**) the ratio of GA/ABA in hulless barley lk621 and lk573 with treatments of distilled water (control, CK) and 200 mM NaCl solution (salt treatment, T) for 4 and 16 h. Values presented are means of three replicates ± standard error (SE). Different lowercase letters indicate significant difference at $p < 0.05$ as determined by Tukey's HSD test in each sample.

2.2. Overview of Transcriptomic and Quantitative Proteomic Analyses

A total of 1238.5 million clean reads were obtained after filtering with a number of 1290.8 million raw reads from the 24 samples of lk621 and lk573 under CK and T treatments at 4 and 16 h. Of clean reads, 79.92–84.72% were successfully mapped to the barley genome, and 74.54–77.89% were uniquely mapped (Table S1A). In the proteomic analysis, a total of 2,139,488 spectra were matched to 10,841 peptides, and 6036 proteins were identified (Table S1B). The number of genes specifically expressed in lk621 and lk573 under T treatment at 4 h was 2233 and 1823, respectively, with 335 genes overlapping (Figure S1A). At 16 h, 1545 and 374 genes were specifically detected in lk621 and lk573 under T treatment, respectively, with 48 overlapping genes (Figure S1B).

Four pairwise comparisons of transcriptomes and proteomes were made to identify differentially expressed genes (DEGs) and differentially expressed proteins (DEPs) between CK and T treatments in lk621 and lk573 at 4 and 16 h. There were 507 and 400 up- and down-regulated DEGs, respectively, identified in lk621 at 4 h after salt stress, and correspondingly 244 and 1461 in lk573 (Figure 3A). With the extension of salt stress time, more DEGs were identified at 16 h: 3972 and 4814 up- and down-regulated DEGs in lk621, respectively, and correspondingly 2279 and 3908 in lk573 (Figure 3A). At the proteomic level, 143 and 212 DEPs were up-and down-regulated in lk621 by salt stress at 4 h, respectively, and correspondingly 123 and 152 in lk573 (Figure 3B). At 16 h, there were more

DEPs identified: 168 and 136 up- and down-regulated in lk621, respectively, and 212 and 269 in lk573 (Figure 3B).

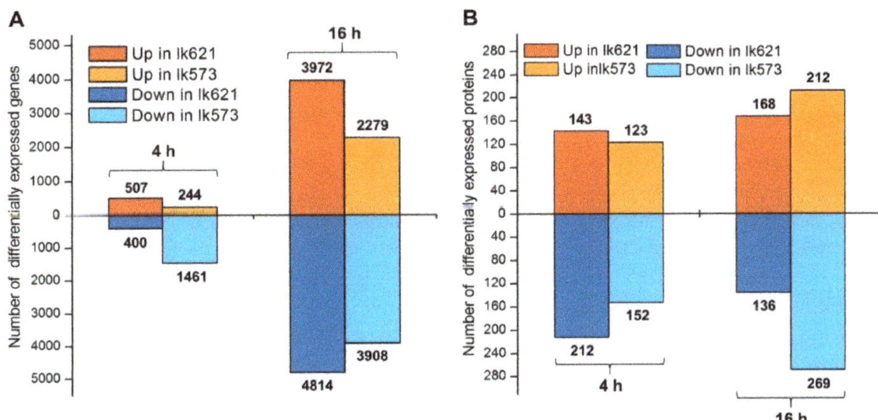

Figure 3. Number of differentially expressed genes (**A**) and differentially expressed proteins (**B**) in hulless barley lk621 and lk573 between treatments of distilled water (control, CK) and 200 mM NaCl solution (salt treatment, T) at 4 and 16 h.

In addition, the specific DEGs and DEPs in lk573 after salt stress were detected, which could be associated with salt tolerance of lk573. Compared with lk621, 1567 DEGs and 171 DEPs were confirmed as specifically expressed in lk573 at 4 h in the T treatment and, correspondingly, 2744 and 328 at 16 h (Table S2). These results suggested the different responses to salt stress between genotypes at transcriptomic and proteomic levels.

2.3. Gene Ontology (GO) and Kyoto Encyclopedia of Genes and Genomes (KEGG) Pathway Analysis of DEGs and DEPs

To gain more insights concerning DEGs and DEPs, GO functional enrichment analysis was also conducted. The GO annotations showed that all enriched DEGs and DEPs were classified into three categories: biological processes, cellular components, and molecular function. In the transcriptomic analysis, 399 and 496 GO terms were searched in lk621 and lk573 after salt stress at 4 h, respectively, among which 32 and 54 corresponding terms were significantly enriched (Table S3). At 16 h, 845 and 820 GO terms were searched, and 117 and 92 were significantly enriched in lk621 and lk573, respectively (Table S3). In the proteomic analysis, 261 and 200 GO terms were searched in lk621 and lk573 after salt stress at 4 h, respectively; and correspondingly at 16 h, 261 and 285 GO terms were searched (Table S4).

In the transcriptomic analysis, various DEGs were enriched in 53 and 107 KEGG pathways in lk621 after salt stress at 4 and 16 h, respectively, and correspondingly, for lk573, in 77 and 105 KEGG pathways (Table S5). In lk621, protein processing in endoplasmic reticulum, arginine and proline metabolism, phenylpropanoid biosynthesis, and carotenoid biosynthesis were the significant pathways after salt stress at 4 h, and most DEGs were significantly enriched in pathways of carbon metabolism, plant hormone signal transduction, and biosynthesis of amino acids. In lk573, most DEGs were significantly enriched in phenylpropanoid biosynthesis and plant hormone signal transduction, pathways after salt stress at 4 h, and correspondingly in phenylpropanoid biosynthesis, MAPK signaling pathway-plant, and plant hormone signal transduction pathways at 16 h. In the proteomic analysis, DEPs were enriched in 52 and 49 KEGG pathways in lk621 at 4 and 16 h, respectively, and correspondingly 53 and 55 pathways in lk573 (Table S6). In lk621, most DEPs were significantly enriched in metabolic pathways at 4 h under salt stress, and correspondingly in terpenoid backbone biosynthesis and fatty acid biosynthesis pathways at 16 h. In lk573, propanoate metabolism and protein export were the significantly enriched pathways at 4 h, and DEPs were significantly enriched in pathways of nitrogen metabolism, pentose and glucuronate interconversions, galactose metabolism, homologous recombination, ascorbate and aldarate metabolism, ABC transporters, folate biosynthesis, and ubiquinone and other terpenoid-quinone biosyntheses. These results revealed that the differences at metabolism levels between salt-sensitive and salt-tolerant genotypes indeed existed in hulless barley.

In addition, GO analysis was also conducted based on the specific DEGs and DEPs in lk573. In the biological process category, most of the specific DEGs were involved in response to stimulus and biosynthetic process at 4 h, and correspondingly response to oxidation-reduction process and metabolic process for most of the specific DEPs (Figure 4A). In the cellular component category at 4 h, most of the specific DEGs were assigned to the cell periphery and cell wall, and most specific DEPs were enriched to membrane and cytoplasm (Figure 4B). For the molecular function category, most specific DEGs were annotated to the GO terms of protein heterodimerization activity and protein dimerization activity at 4 h, and most of the specific DEPs to the GO terms of protein binding and oxidoreductase activity (Figure 4C). The GO terms of specific DEGs and DEPs at 16 h are shown in Figure S2. In the biological process category, most specific DEGs were involved in vesicle-mediated transport and response to oxidative stress, and correspondingly oxidation-reduction process and metabolic process for most specific DEPs. In the cellular component category, most specific DEGs were assigned to macromolecular complex and cytoplasm, and most specific DEPs to nucleolus and membrane. For the molecular function category, most specific DEGs were annotated to the GO terms of heme binding and tetrapyrrole binding, and most specific DEPs to the GO terms of ATP binding and nucleic acid-binding. Finally, the networks of GO terms were obtained using BiNGO to identify the GO terms enriched among these specific DEGs in lk573 (Figure 5 and Figures S3–S5).

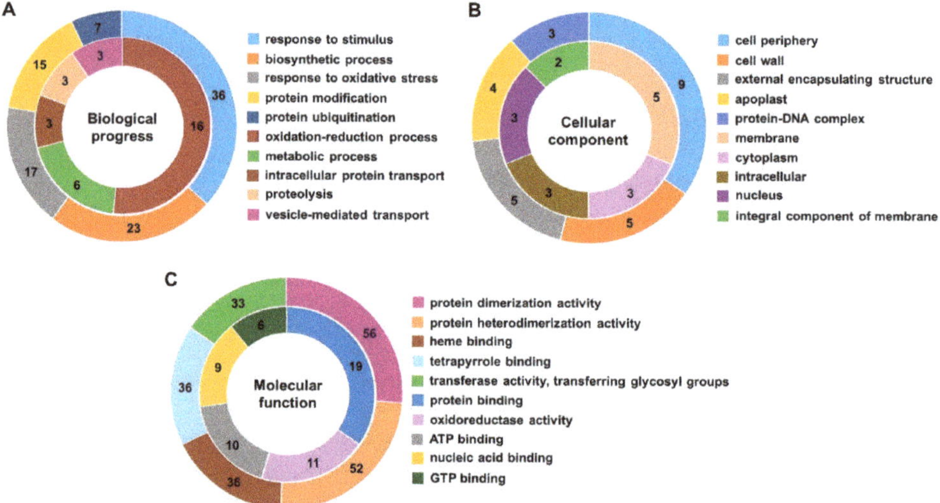

Figure 4. Top five gene ontology (GO) categories: biological process (**A**), cellular component (**B**), and molecular function (**C**) assigned to most of the specific differentially expressed genes (DEGs, outer cycle) and the specific differentially expressed proteins (DEPs, inner cycle) in lk573 at 4 h after salt stress. The numbers represent the number of DEGs or DEPs assigned to each GO term.

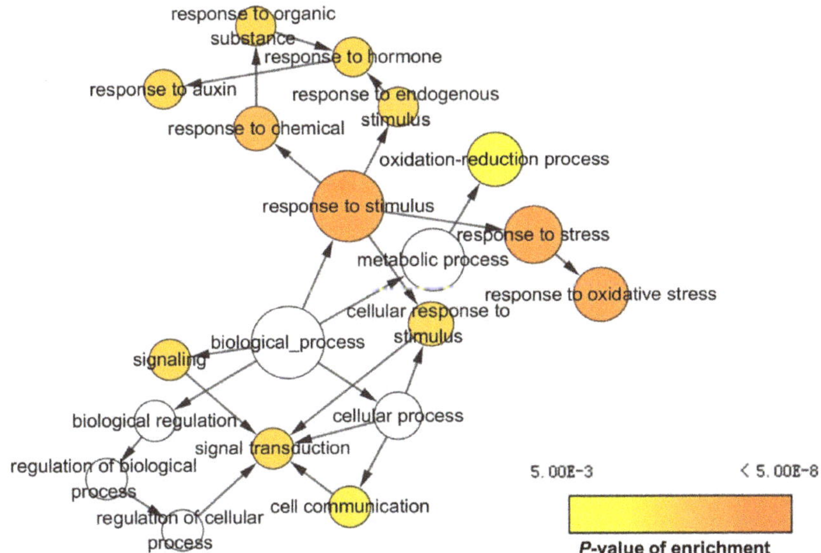

Figure 5. Example of networks representing gene ontology (GO) terms in the biological process category enriched among differentially expressed genes specifically affected by salt stress in hulless barley lk573 at 4 h. Enriched GO terms were identified using BiNGO and visualized with Cytoscape. The GO terms were connected based on their parent-child relationships. Colors of circles indicate the *p*-value of enrichment. Sizes of circles represent the size of GO terms in the background GO annotation. The complete networks of enriched GO terms are presented in Figures S3–S5.

2.4. Correlation between Transcripts and Proteins

The correlations between transcriptome and proteome profiles were assessed using r values. At 4 and 16 h, r values between expressed transcripts and proteins were 0.0387 (p = 0.017) and 0.0316 (p = 0.047) for lk621, respectively, and correspondingly −0.0008 (p = 0.960) and −0.037 (p = 0.023) for lk573 (Figure S6). Between DEGs and DEPs, there was a negative correlation (r = −0.3225, p = 0.1) for lk621, and a positive correlation (r = 0.1634, p = 0.517) for lk573 at 4 h (Figure S7). At 16 h, the corresponding r values were 0.0254 (p = 0.789) and −0.0458 (p = 0.651) (Figure S7). These results indicated that the correlation between transcriptome and proteome under salt stress condition was weak during seed germination. The genes with significant differential expression listed in Table S7 were co-expressed at both transcriptomic and proteomic levels. At 4 h, the expression trend of 11 genes in lk621 at the transcriptomic level was consistent with that at the proteome level, while 13 genes were expressed in an opposite fashion, and, in lk573, the corresponding numbers of genes were 13 and five; in lk621, at 16 h, the numbers were 66 and 58; and, in lk573, at 16 h, there were 52 and 50.

2.5. Genes Related to Salt Tolerance

Based on the different response of molecular mechanisms to salt stress between lk621 and lk573, 96 genes were specifically differentially expressed in lk573 at both transcriptomic and proteomic levels and are listed in Table S8. Among these genes, 16 were specifically expressed at 4 h, 80 genes expressed at 16 h, and the gene HORVU6Hr1G066250 expressed both at 4 h and 16 h. These genes might be responsible for salt tolerance of lk573. The relative expression levels of 10 genes selected from among the 96 DEGs at 4 and 16 h were validated by qRT-PCR, and the expression trends were consistent with the expression patterns measured by RNA-seq (Figure 6). Based on the functions of proteins encoded by 27 selected genes (Table 1), a putative model for salt tolerance was constructed for hulless barley (Figure 7). These genes were involved in the balances of energy and substrates supply, cell wall resistance, ion transport, Ca-dependent regulation, phytohormone pathways, ROS reducing, and vesicular trafficking.

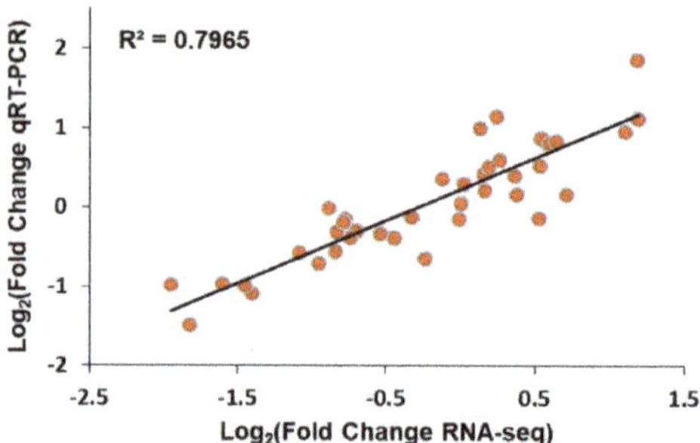

Figure 6. Correlation between RNA-seq and qRT-PCR.

Table 1. List of selected genes related to salt tolerance in lk573.

Gene ID	Tran(log$_2$FC)	Pro(log$_2$FC)	Description
HORVU4Hr1G058810	−1.59814	−0.37389	*Histone H2A 6*
HORVU6Hr1G066250	−0.86759	−0.43371	*Tetratricopeptide repeat (TPR)-like superfamily protein*
HORVU2Hr1G065120	−1.22601	−0.65284	*Pyridoxal 5-phosphate synthase subunit PdxS*
HORVU3Hr1G085130	−0.71579	−0.46692	*Calcium-dependent lipid-binding (CaLB domain) family protein*
HORVU5Hr1G109880	−0.6552	0.28447	*Protein of unknown function, DUF642*
HORVU7Hr1G100810	−0.58782	0.408756	*2-oxoglutarate (2OG) and Fe(II)-dependent oxygenase superfamily protein*
HORVU3Hr1G088130	−1.28183	0.329661	*PRA1 (Prenylated rab acceptor) family protein*
HORVU3Hr1G109590	−0.79428	−0.26857	*Glycosyltransferase family 61 protein*
HORVU5Hr1G098960	−0.83372	−0.38633	*Tubulin alpha-4 chain*
HORVU7Hr1G001030	−1.22656	−0.37861	*Inter-alpha-trypsin inhibitor heavy chain-related*
HORVU5Hr1G087760	−1.60172	−0.48224	*Glutelin type-B-like protein*
HORVU1Hr1G071460	−1.34747	−0.29342	*Serpin 3*
HORVU2Hr1G080100	−0.92463	−0.37249	*Aldose 1-epimerase*
HORVU3Hr1G077000	−0.58028	−0.28372	*Adenine nucleotide alpha hydrolases-like superfamily protein*
HORVU6Hr1G074940	−0.5826	−0.38866	*Lipase/lipooxygenase, PLAT/LH2 family protein*
HORVU5Hr1G067630	−0.82874	−0.90215	*DNA-binding storekeeper protein-related transcriptional regulator*
HORVU3Hr1G100410	−1.02777	0.449302	*Xylanase inhibitor*
HORVU1Hr1G095430	0.561136	0.345589	*UDP-glucose 4-epimerase 1*
HORVU5Hr1G054060	0.639856	0.418968	*Ferrochelatase 1*
HORVU6Hr1G070120	−0.58765	0.266884	*Thioredoxin reductase 2*
HORVU0Hr1G012990	−1.37722	0.861165	*Flavin-containing monooxygenase family protein*
HORVU7Hr1G085130	1.103488	0.418055	*Multiprotein bridging factor 1C*
HORVU3Hr1G067370	0.632487	0.319481	*KRR1 small subunit processome component homolog*
HORVU7Hr1G120030	−0.87435	0.474998	*Delta(24)-sterol reductase*
HORVU5Hr1G025320	0.5301	0.40109	*Tryptophan–tRNA ligase*
HORVU4Hr1G011940	0.657604	0.878678	*Exocyst complex component 1*
HORVU5Hr1G079100	0.495959	0.321981	*TBC1 domain family member 15*

These genes were differentially expressed at transcriptomic and proteomic levels. Tran(log$_2$FC), fold change of transcript regulated by salt stress. Pro(log$_2$FC), fold change of protein regulated by salt stress. Fold change values are color-coded with red and green for up- and down-regulation, respectively.

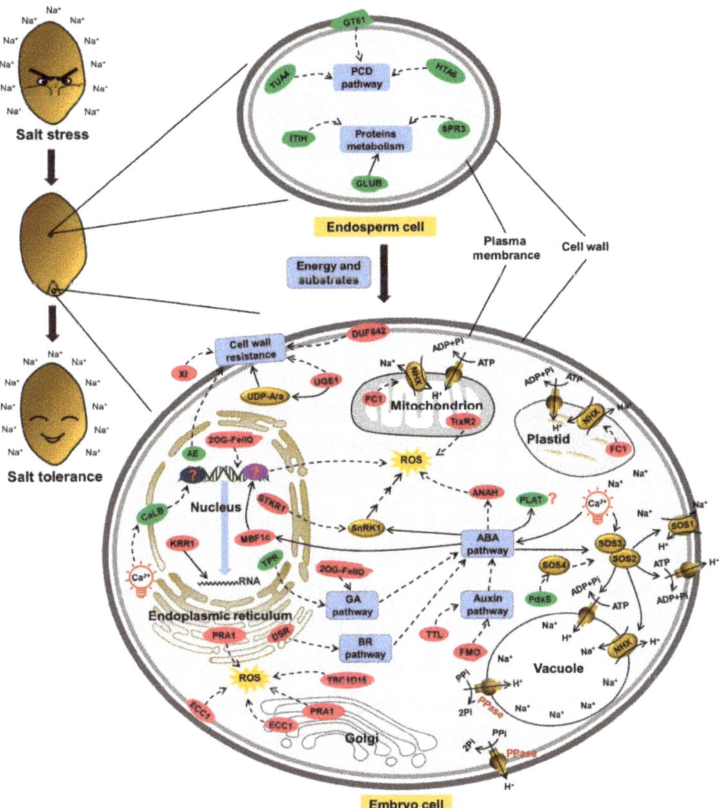

Figure 7. A putative model for salt tolerance of lk573 during seed germination. The enhanced salt tolerance requires energy and substances supplied by endosperm cells, including the critical role of aleurone layer cells. In the embryo, cells need cell wall resistance, reduction in reactive oxygen species (ROS) levels, ionic transport pathway, and different plant hormone pathways, including abscisic acid (ABA), gibberellins (GA), and auxin, along with vesicular trafficking and calcium-dependent regulation. GT61, glycosyltransferase family 61 protein; HTA6, histone H2A 6; TUA4, tubulin alpha-4 chain; ITIH, inter-alpha-trypsin inhibitor heavy chain-related; GLUB2, Glutelin type-B-like protein; SPR3, serpin 3; XI, xylanase inhibitor; DUF642, protein of unknown function DUF642; UGE1, UDP-glucose 4-epimerase 1; AE, aldose 1-epimerase; FC1, ferrochelatase 1; TrxR2, thioredoxin reductase 2; ANAH, adenine nucleotide alpha hydrolases-like superfamily protein; 2OG-FeIIO, 2-oxoglutarate (2OG) and Fe(II)-dependent oxygenase superfamily protein; FMO, flavin-containing monooxygenase family protein; PLAT, lipase/lipooxygenase, PLAT/LH2 family protein; STKR, DNA-binding storekeeper protein-related transcriptional regulator; MBF1c, multiprotein bridging factor 1c; KRR1, KRR1 small subunit processome component homolog; TPR, tetratricopeptide repeat (TPR)-like superfamily protein; DSR, delta(24)-sterol reductase; PRA1, PRA1 (prenylated rab acceptor) family protein; TTL, tryptophan–tRNA ligase; ECC1, exocyst complex component 1; TBC1D15, TBC1 domain family member 15; PdxS, pyridoxal 5-phosphate synthase subunit PdxS; CaLB, calcium-dependent lipid-binding (CaLB domain) family protein; SnRK1, SNF1-related protein kinase 1; UDP-Ara, uridine diphosphate arabinose; PPase, pyrophosphate-energized proton pump; NHX, Na+/H+ exchanger; SOS, salt overly sensitive. The green represents down-regulated differentially expressed proteins (DEPs) by salt stress, and the red represents up-regulated DEPs.

During seed germination, many catabolic reactions occur in endosperm cells to provide energy and substrates for seed germination, a large number of proteins involved in energy and substrates metabolism are synthesized or degraded, and PCD occurs in aleurone layer cells [12]. In the putative model (Figure 7), some genes involved in protein metabolism in endosperm cells were specifically regulated at both transcriptomic and proteomic levels, and the proteins involved in PCD of the aleurone layer cells were down-regulated in lk573 after salt stress, which might provide more energy and substrates for seed germination and response to salt stress. The complex mechanism of salt stress response in the embryo cell is shown in Figure 7. First, cell wall resistance plays an important role in lk573 salt tolerance, and some proteins involved in cell wall synthesis and remodeling are specific DEPs in lk573 after salt stress [23]. Then, the NHX and SOS pathways involved in limiting the Na^+ content in cells and some proteins may influence these pathways to enhance salt tolerance of lk573 [18]. In the process of seed germination during salt stress, Ca^{2+} acts as the messenger, participating in the regulation of the ABA and SOS pathways [24,25]. Moreover, phytohormones play an important role in seed salt tolerance. The ABA pathway is central to salt tolerance by activating ROS reduction, and the GA, auxin, and brassinosteroid (BR) pathways trigger the ABA pathway to influence seed tolerance to salt stress during the germination stage [14,17,26,27]. Finally, some proteins, specifically regulated by salt stress in lk573, that are involved in vesicular trafficking may increase seed salt tolerance by reducing ROS damage [28–30].

3. Discussion

Seed germination is vulnerable to salinity and varies among different genotypes [9,12]. Two genotypes with contrasting performance under salt stress were used in the current study. Being relatively salt-tolerant, the germination rate was significantly higher for lk573 than lk621 under 200 mM NaCl stress (Figure 1K). As effective strategies, transcriptomic and proteomic analyses were widely used in revealing the molecular mechanism, responding to abiotic stresses [31,32]. To further reveal the molecular mechanism in response to salt stress and to obtain a novel understanding of salt tolerance during seed germination, an integrated analysis of transcriptomic and proteomic levels in lk573 and lk621 was conducted using RNA-seq and iTRAQ technologies. It is reported that salt stress delays phases I and II during seed germination [7,33]. Seedling growth is mainly prepared in phase II [12]. Thus, more biological processes occur in phase II, and seeds should be more vulnerable to salinity in phase II than in phase I. In this study, fewer DEGs and DEPs were detected at 4 than at 16 h (Figure 3), suggesting that seeds were more seriously affected in phase II than in phase I by salt stress during the germination stage. Compared with lk621, phases I and II were not delayed in lk573, suggesting that there were some special mechanisms of lk573 in salt tolerance. It is possible to exploit favorable genes for barley breeding in salt resistance through the integration of transcriptomic and proteomic analyses.

Salt stress triggers a series of responses in plants, including signal transduction, ion transport, and energy and substrate metabolism [17–19]. The GO and KEGG pathway analysis showed the complexity of molecular mechanisms in response to salt stress during the seed germination stage. The correlation between the transcriptome and proteome profiles varies with species, response environments, and different stages of growth and development [34–36]. In this study, the r values indicated a weak correlation between the transcriptome and proteome profiles under salt stress. Thus, a comprehensive analysis of transcriptome and proteome is necessary to find out the response mechanisms to salt stress during seed germination. We also found some genes were possibly related to salt tolerance by comparing lk573 with lk621 (Table S8). These genes were involved in energy and substrate metabolism, ion transport, PCD, signal transduction, cell wall stability, phytohormone balance, vesicular trafficking, and ROS reducing. According to co-expression at transcriptomic and proteomic levels, 27 specifically expressed genes in lk573 were focused on, and a putative model for seed salt tolerance was constructed (Figure 7). Thus, some novel insights into salt tolerance during seed germination were obtained.

3.1. Energy and Substrates Supplied by Endosperm under Salt Stress

During seed germination, many catabolic reactions occur in endosperm cells to provide energy and substrates for seed germination, and a large number of enzymes involved in metabolism are synthesized or degraded, and PCD occurs in aleurone layer cells [12]. The glycosyltransferase family 61 protein (GT61) has been confirmed to be involved in the synthesis of xylan, one of the main components of the cell wall [37]. In the putative model (Figure 7), GT61 was down-regulated by salt stress, which could result in decreasing the cell wall formation along with the PCD pathway in aleurone layer cells. At the same time, the histone H2A 6 (HTA6) of lk573 was down-regulated in lk573 to match the decrease in chromatin. One tubulin alpha-4 chain (TUA4), the major constituent of microtubules [38], was down-regulated in lk573 under salt stress, which might also occur in the aleurone layer cells with the emergence of PCD. These results suggested that the increase of PCD under salt stress could supply more energy and substrates for seed salt-resistance and germination, and the detection of PCD and energy metabolism would demonstrate this point.

Protein metabolism occurs in endosperm cells during seed germination. Under salt stress, proteins are hydrolyzed into various amino acids, and thus more energy is supplied (Figure 7). The storage protein glutelin type-B-like protein (GLUB2) was significantly down-regulated in lk573 at 16 h after salt stress in this study, showing that protein metabolism was enhanced in response to salt stress. Moreover, the inter-alpha-trypsin inhibitor heavy chain-related (ITIH) and serpin 3 (SPR3) were specifically down-regulated in lk573 under salt stress. ITIH is one component of the inter-alpha-trypsin inhibitor, which was originally discovered in urine and serum due to its inhibitory activity against trypsin [39]. The serpin family is the largest and the most widespread superfamily of protease inhibitors and plays an important role in the process of development and abiotic stress in plants [40], but the definite mechanisms of stress defense are unknown. In this study, the down-regulation of ITIH and SPR3 meant more proteins were degraded, showing their critical roles in energy and substrates metabolism for a response to salt stress. Further research is needed to classify their functions in response to salt stress during seed germination.

3.2. Cell Wall in Salt Tolerance

The cell wall is important for protecting plants from biotic stress, and its critical role in abiotic stress is widely discussed [18,31]. The UDP-glucose 4-epimerase (UGE) plays an important role in galactose metabolism and the biosynthesis of galactose-containing polysaccharides and can also participate in arabinose metabolism, affecting cell wall resistance [41,42]. The barley *HvUGE1* gene is orthologous to *AtUGE4* and also necessary to control carbohydrate partitioning in the cell wall [43]. The HvUGE1 protein was only significantly up-regulated by salt stress at 16 h, which suggested that lk573 exhibited a strong salt tolerance by regulating the formation and stability of the cell wall. Barley xylanase inhibitor genes encode endoxylanase inhibitors, which are orthologous to *Triticum aestivum* xylanase inhibitors (TAXI) [44]. It has been confirmed that TAXI-type xylanase is important for the defense against pathogens and wounding [45]. The barley HvXI proteins detected in this research, orthologous to TAXI-IV proteins, might improve salt tolerance of lk573 by maintaining the stability of xylan in the cell wall through inhibiting xylanase activity. The DUF642 (domain of unknown function 642) proteins are also involved in cell wall synthesis [46]. In *Arabidopsis thaliana*, the DUF642 proteins encoded by the *At4g32460* gene can improve seed germination by increasing the activity of pectin methylesterase (PME) [47]. The PME acts an important role in the completion of cell division, involved in some biotic and abiotic stress responses, and is negatively regulated by aldose 1-epimerase (AE) proteins [48,49]. In the current study, the DUF642 protein was exactly up-regulated under salt stress, meaning the enhanced activity of PME to promote seed salt tolerance by regulating the formation of cell walls during the germination stage. The down-regulated AE proteins might coordinate DUF642 protein to increase the PME activity and enhance salt tolerance of lk573 during seed germination. These proteins, involved in the formation and stability of cell wall, were specifically regulated by salt stress in tolerant genotype lk573, suggesting their vital roles of the cell wall in salt resistance during the seed germination stage.

3.3. Ion Transport for Salt Tolerance

Various ionic transporters are involved in ion homeostasis in cells by selective uptake and exclusion of ions [18]. With respect to the NHX and SOS pathways, some Na^+/H^+ antiporters, pyrophosphate-energized proton pumps, and H^+-ATPases were identified in lk621 and lk573 during seed germination. Only one plasma membrane H^+-ATPase was up-regulated under salt stress in lk621 and lk573 at 4 h, but just at the proteome level, and was unchanged at 16 h at both transcriptome and proteome level. One F-type ATPase was only up-regulated at 4 h, both in lk621 and lk573. Most H^+-ATPases expressed in lk621 were unchanged and had the same trend as in lk573. The *SOS4* gene encoding a pyridoxal kinase is involved in the biosynthesis of pyridoxal-5-phosphate (PLP), and PLP can modulate the activities of ion transporters to regulate Na^+ and K^+ homeostasis [50]. One pyridoxal 5-phosphate synthase subunit (PdxS) was specifically differentially expressed in lk573 at the transcriptomic and proteomic levels. The PdxS protein might influence the expression of SOS4 to regulate ion homeostasis. Furthermore, a ferrochelatase 1 (FC1) was up-regulated specifically in lk573 under salt stress. It has been demonstrated that FC1 can improve salt tolerance by limiting the accumulation of Na^+ in cells, possibly through the NHX pathway in *Arabidopsis* [51]. FC1 is the terminal enzyme of heme biosynthesis [51]; thus, the heme biosynthesis pathway may coordinate with the NHX pathway to regulate the Na^+ content in cells during the germination stage.

The NHX and SOS systems have been clearly demonstrated to play important roles in response to salt stress in plant roots and leaves [17,18]. During seed germination, many ionic transporters were not differentially expressed under salt stress in our study. The unchanged expression of these ionic transporters in seeds suggested that the response mechanism to salt stress in seeds is more complicated than in roots and leaves. Ion transport might coordinate other resistant mechanisms to maintain a strong tolerance of lk573 to salt stress during seed germination.

3.4. Ca-Dependent Regulation under Salt Stress

The universal second messenger Ca^{2+} triggers the ABA signal and SOS pathways and acts as an important regulator of many processes in plant stress resistance [21,52]. Different abiotic stresses induce Ca^{2+} fluctuations, and the change is decoded by different Ca^{2+}-sensing proteins, which contain Ca^{2+}-dependent lipid-binding (CaLB) domains or C2 domains [25]. The C2 domain is involved in calcium-dependent phospholipid binding and membrane targeting processes, and it is confirmed to be a CaLB domain [53,54]. Two C2 domain proteins, AtBAP1 and AtCLB, have been confirmed to negatively regulate defense responses in *Arabidopsis* [24,25]. The loss of AtCLB protein function, which is localized in the nucleus of cells and promotes the expression of thalianol synthase gene *AtTHAS1*, has enhanced drought and salt tolerance of *Arabidopsis* [25]. There are also some reports on C2 domain proteins up-regulated by salt stress [21,55]. Thus, the responses of C2 domain proteins to Ca^{2+} fluctuations are various. In our study, a CaLB domain protein (HORVU3Hr1G085130.1) was down-regulated in lk573 after salt stress during seed germination at 4 h. After receiving the Ca^{2+} fluctuation, it might act as a transcriptional repressor to negatively regulate salt tolerance of lk573 during seed germination (Figure 7). However, the downstream protein regulated by CaLB is still unknown, and more details are required to reveal this pathway and demonstrate the critical role of CaLB in seed salt tolerance.

3.5. Phytohormones in Salt Stress

It has been well demonstrated that ABA and GA are the primary hormones that antagonistically regulate seed dormancy and germination [13]. Under salt stress, ABA can alleviate the effects of salt stress on seed germination [20]. In this study, the DNA-binding storekeeper protein-related transcriptional regulator, orthologous to the *Arabidopsis* storekeeper-related 1/G-element-binding protein (STKR1), was specifically up-regulated in lk573 at 16 h of salt stress. The STKR1 is involved in the response of plants to environmental stress by interacting with SNF1-related protein kinase 1, the activity of which can be modulated by ABA [56]. So, up-regulated STKR1 may connect to the ABA pathway to regulate seed tolerance of lk573 to salt stress. The lipase/lipoxygenase (PLAT/LH2 family protein) (PLAT) is the downstream target of the ABA signaling pathway and acts as a positive regulator of abiotic stress tolerance [57]. Confusingly, the PLAT protein in our study was down-regulated in lk573 under salt stress. Whether PLAT protein negatively regulates salt tolerance during seed germination needs further investigation.

The role of GA in the regulation of plant responses to abiotic stress has been well discussed, and it is generally concluded that the reduction of GA levels is helpful for restricting plant growth to adapt to several stresses [26]. However, adequate GA levels are necessary for seed germination. Under salt stress, the mechanism of GA balance in seeds to coordinate germination and stress remains unknown. In this research, a tetratricopeptide repeat (TPR)-like (TPR) superfamily protein, homologous to the SPINDLY (SPY) protein, was down-regulated by salt stress in lk573. The SPY has been identified as a negative regulator of abiotic stress, probably by integrating environmental stress signals via GA and cytokinin cross-talk [58]. In the putative model, the down-regulated TPR might involve in GA balance to improve seed germination of lk573 under salt stress. The 2-oxoglutarate (2OG) / Fe(II)-dependent oxygenase (2OG-FeIIO) superfamily proteins catalyze various oxidation reactions of organic substances by using a dioxygen molecule and are involved in a wide range of biological processes in plants, including DNA repair, hormone biosynthesis, and various specialized metabolites [59]. The 2OG-FeIIOs in the putative model might participate in GA biosynthesis or repair DNA damaged by Na^+ to promote seed germination under salt stress (Figure 7).

Auxin is not necessary for seed germination but influences seed germination through its interaction with the ABA pathway [14]. Under salt stress, auxin negatively regulates seed germination [60]. The flavin-containing monooxygenase (FMO) can oxidize a diverse range of substrates and positively regulate the biosynthesis of auxin [61]. In our study, the FMO and tryptophan-tRNA ligase (TTL) were specifically up-regulated in lk573 at 16 h. The reason may be that, when seed germination is completed, auxin is prepared for seedling growth, and FMOs are involved in its content balance. Tryptophan is an important factor affecting auxin synthesis [62]. The up-regulated expression of TTL could reduce the content of tryptophan, and thus auxin synthesis. These results suggested that there were complex regulatory mechanisms to control auxin content in response to salt stress during seed germination (Figure 7).

The BR plays an important role in promoting seed germination by overcoming inhibition of ABA [63] and can increase plant resistance to such environmental stresses as cold, drought, and salinity [27]. Under salt stress, BR also promotes seed germination in *Arabidopsis thaliana* and *Brassica napus* [64]. A delta (24)-sterol reductase (DSR), homologous to *Arabidopsis* DIMINUTO / DWARF1 (DIM) protein involved in BR synthesis [65], was up-regulated in lk573 under salt stress. We suggested that up-regulated DSR facilitated BR synthesis to improve salt tolerance of barley seeds during the germination stage (Figure 7). The results in the present study just suggested that ABA, GA, auxin, and BR could influence seed tolerance to salt stress. Cytokinins are able to enhance seed germination by alleviating stresses of drought, salinity, and heavy metals [14,66,67]. Jasmonic acid has been confirmed to enhance barley seedling salt tolerance [68]. So, other phytohormones may also participate in regulating seed germination under salt stress. In general, the balance of phytohormones not only plays an important regulatory role in seed germination but also coordinates the response to salt stress at the germination stage.

3.6. ROS Reduction for Salt Resistance

Salt stress leads to superfluous production of ROS in plants, and some enzymes can relieve their damage to cells [3]. Thioredoxin reductase (TrxR) catalyzes the NADPH-dependent reduction of oxidized thioredoxin, which could play a key role in protecting cells from ROS damage [69,70]. In this study, the specific up-regulation of TrxR in lk573 might have increased the activity of oxidized thioredoxin to reduce ROS and enhance seed salt tolerance. An adenine nucleotide alpha hydrolases-like (ANAH) superfamily protein was also up-regulated by salt stress in lk573. This is orthologous to the *Arabidopsis* universal stress proteins (USPs) (At3g53990), and overexpression of USPs can confer a strong tolerance to heat shock and oxidative stress [71]. Confusingly, another ANAH protein (HORVU3Hr1G077000.2) was down-regulated at 16 h after salt stress in lk573, suggesting a diversity of ANAH proteins' functions in response to salt stress.

Multiprotein bridging factor 1 (MBF1) is a highly conserved transcriptional coactivator with three members in its family: MBF1a, MBF1b, and MBF1c [72]. The MBF1c is specifically elevated under different abiotic stresses, such as salinity, drought, and heat, and may function as a regulatory component between ABA and stress signal pathways [72,73]. In barley, there is still no report on the characterization of the *HvMBF1c* gene. In the putative model, MBF1c might be affected by ABA and activate a gene that participates in reducing ROS levels (Figure 7). In a word, ROS reduction maintained a strong tolerance of lk573 seeds to salt stress during the germination stage.

3.7. Vesicular Trafficking in Salt Tolerance

The protein TBC1D15 contains a TBC (Tre-2/Bub2/Cdc16) domain, which can stimulate the intrinsic GTPase activity of Rab7 and functions as the key regulator of intracellular vesicular trafficking [28]. In *Arabidopsis*, Rab7 has been demonstrated to be a positive regulator in tolerance to salt and osmotic stresses by reducing the ROS levels in cells [29]. The exocyst acts as a tethering complex and effector of Rho and Rab GTPases to participate in vesicular trafficking [30]. In the current study, one TBC1D15 protein and one exocyst complex component 1 (ECC1) were specifically up-regulated in lk573 under salt stress (Figure 7). This could improve the GTPase activity of Rab7, and so reduce the damage from ROS to cells. Moreover, one prenylated rab acceptor (PRA1, HORVU3Hr1G088130.1) family protein was up-regulated in lk573 at 4 h under salt stress. The PRA1 proteins also act as receptors of Rab GTPases to regulate vesicle trafficking [74]. In *Arabidopsis*, both overexpression and knockdown of the *PRA1.F4* gene have increased sensitivity to high salt stress and lowered vacuolar Na$^+$/K$^+$-ATPase and plasma membrane ATPase activities of plants [75]. Two other PRA1 proteins (HORVU3Hr1G018940.1 and HORVU1Hr1G070360.1) were down-regulated at 16 h. Thus, the functions of PRA1 proteins in response to salt stress during seed germination are complex. These results suggested that vesicular trafficking could improve salt tolerance by triggering ROS regulation during seed germination.

We also found that a small subunit processome component homolog (KRR1) was specifically differentially expressed in lk573 under salt stress (Figure 7). The KRR1 is involved in the assembly of the 40S ribosomal subunit [76], which will influence protein translation. However, the mechanism involved in enhancing salt tolerance with this protein during seed germination is unknown, and further identification is needed. The up-regulated KRR1 may increase the expression of genes involved in regulating seed salt responses.

4. Materials and Methods

4.1. Seed Germination under Salt Stress

Two hulless barley landraces with purple seeds (lk621 and lk573) collected from Menyuan and Huangyuan countries of Qinghai province, China, respectively, were selected as research objects. Each of the 30 grains of lk621 and lk573 was washed three times with sterile deionized water and then placed in 9-cm Petri dishes containing two sheets of filter paper, moistened by distilled water (control, CK) or 200 mM NaCl solution (salt treatment, T) for 4 and 16 h. Seed morphology was surveyed and photographed by a stereomicroscope (Leica-M165 C; Leica, Wetzlar, Germany). There were three biological replications. Seed germination was determined by the protrusion of the radicle.

4.2. Plant Hormone Detection

For content analysis of ABA, GA, and auxin in germinating seeds, 1 g of seeds of each sample were ground in liquid nitrogen. The quantification of these hormones in seeds was performed by indirect ELISA, as previously described [77]. There were three biological replications.

4.3. Transcriptome Sequencing Analysis

Total RNA was isolated from 1 g each of seeds of 24 samples [2 genotypes (lk621 and lk573) × 2 treatments (CK and T) × 2 germination time points (4 and 16 h) × 3 biological replications] using the TRIzol reagent (Invitrogen, Carlsbad, CA, USA) according to the manufacturer's instructions. RNA purity was checked by a NanoPhotometer® spectrophotometer (IMPLEN, Schatzbogen, Munich, Germany). The RNA concentration was measured on a Qubit® 2.0 Fluorimeter (Life Technologies, Carlsbad, CA, USA). The assessment of RNA integrity was performed by the Bioanalyzer 2100 system (Agilent Technologies, Santa Clara, CA, USA). A total amount of 3 µg of RNA per sample was prepared for building sequencing libraries, which were generated using NEBNext® UltraTM RNA Library Prep Kit for Illumina® (NEB, Ipswich, MA, USA), following the manufacturer's recommendations. The PCR was performed with Phusion High-Fidelity DNA polymerase, and library quality was assessed on an Agilent Bioanalyzer 2100 system. The libraries were sequenced on an Illumina Hiseq 2000 platform (NCBI; BioProject ID: PRJNA578897). Barley genome and gene model annotation files were downloaded from the genome website (http://webblast.ipk-gatersleben.de/barley_ibsc/downloads/). Clean reads were obtained by removing reads containing adapters, reads containing poly-N, and low-quality reads from raw data. At the same time, Q20, Q30, and GC content of the clean data were calculated. Paired-end clean reads were aligned to the reference genome using Hisat2 v2.0.5. The featureCounts v1.5.0-p3 was used to count the read numbers mapped to each gene. Then, FPKM (expected number of fragments per kilobase of transcript sequence per million base pairs sequenced) of each gene was calculated based on gene length and reads count aligned to this gene. Genes with FPKM > 1 were defined as being expressed. Differentially expressed genes (DEGs) were screened between the treated and control groups with three replicates, performed using the DESeq2 R package (1.16.1). Genes with an adjusted p-value < 0.05 were assigned as DEGs.

4.4. Proteome Analysis

Samples for protein extraction were prepared as for RNA-seq. The extraction of protein was performed with NitroExtraTM (Cat. PEX-001-250ML, N-Cell Technology, Shenzhen, China) using the manufacturer's instructions. The protein concentration was determined with a Bradford assay. The protein of each sample was digested with trypsin and then desalted with C18 cartridge. Desalted peptides were labeled with iTRAQ reagents (iTRAQ® Reagent-8PLEX Multiplex Kit, Sigma, Foster, CA, USA), as instructed by the manufacturer. The labeled peptide mix was fractionated using a C18 column (Waters BEH C18 4.6 × 250 mm, 5 µm) on a Rigol L3000 HPLC system (Rigol, Beijing, China). The resulting spectra from each fraction were searched against the Hordeum_vulgare_Customer FASTA database containing 81,279 sequences (http://webblast.ipk-gatersleben.de/barley_ibsc/). The raw data were analyzed by Proteome Discoverer software (ver. 2.2, Thermo Fisher Scientific, Waltham, MA, USA) and then probed on the Mascot search engine (ver. 2.3.02; Matrix Science, London, UK). To reduce the probability of false peptide identification, only peptides at a 95% confidence interval ($p < 0.05$), with a false discovery rate estimation ≤5%, were counted as successfully identified. Each positive protein identification contained at least one unique peptide. Proteins containing at least two unique spectra were selected for quantification analysis. Quantitative protein ratios were weighted and normalized by the median ratio in Mascot. Statistical analysis was conducted using Fisher's test. Differentially expressed proteins (DEPs) with $p < 0.05$, and a > 1.2-fold or <0.83-fold cutoff were considered as up- or down-regulated, respectively. The MS-based proteomics data is available via ProteomeXchange with identifier PXD016100.

4.5. Bioinformatics Analysis

Gene ontology (GO) enrichment analysis was implemented by the cluster Profiler R package. The GO terms with corrected *p*-value < 0.05 were considered significantly enriched by DEGs and DEPs. The Kyoto Encyclopedia of Genes and Genomes (KEGG) pathway analysis was performed to investigate high-level functions and utilities of the biological system. The cluster Profiler R package was used to test the statistical enrichment of DEGs and DEPs in KEGG pathways. The significant GO terms enriched among specific DEGs in lk573 were also analyzed via BiNGO to create networks of GO terms [78].

4.6. Correlation between Transcript and Protein

According to the fold change of expressed transcripts and proteins between the treated and control groups in lk621 and lk573, the Pearson correlation coefficient (r) was calculated to evaluate the concordance between transcriptome and the proteome profiles [25].

4.7. Quantitative RT-PCR

Quantitative RT-PCR (qRT-PCR) analysis was used to verify RNA-seq results based on 10 selected genes related to salt stress. The primers of these 10 genes are listed in Table S9. The RNA samples used for qRT-PCR assays were the same as those used for RNA-seq. The qRT-PCR was performed with SYBR® PremixDimerEraser™ (Takara, Dalian, China) according to the manufacturer's specifications. The reaction mixtures were incubated at 95 °C for 3 min, followed by 40 cycles of 95 °C for 5 s, 60 °C for 60 s, and 72 °C for 30 s. Barley ACTIN gene (AY145451) was used as a control to normalize the amount of gene-specific RT-PCR products. Based on the melting curve analysis of PCR amplicons, the results with specific peaks were selected to assess the expression level of selected genes. The relative expression levels of selected genes were calculated with the $2^{-\Delta\Delta Ct}$ method [79]. According to the expression levels assessed by qRT-PCR and RNA-seq, the correlation was estimated with the fold changes regulated by salt stress.

5. Conclusions

Seed germination is sensitive to salt stress, and the mechanism of this process is complex. This is the first report on the research of molecular mechanisms in response to salt stress during the germination stage at transcriptomic and proteomic levels in hulless barley. A large number of genes and proteins associated with salt response were detected. Landrace lk573 was much more tolerant to salt stress than lk621. Moreover, a putative model for expounding salt tolerance of lk573 was constructed (Figure 7), which involved energy and substrate metabolism, cell wall resistance, ion transport, Ca-dependent regulation, phytohormone pathways, reducing ROS levels, and vesicular trafficking. These results deepened our understanding of the mechanism responding to salt stress during seed germination and provided candidate genes for salt-tolerant improvement in hulless barley.

Supplementary Materials: Supplementary materials can be found at http://www.mdpi.com/1422-0067/21/1/359/s1.

Author Contributions: Y.K., D.Z., and H.W. designed the experiments; Y.L., P.R., E.S., L.Y. and J.W. (Juncheng Wang). conducted the experiments; Y.L. wrote the manuscript; Y.L. and J.W. (Jinming Wang) analyzed the data; All authors have read and agreed to the published version of the manuscript.

Funding: This work was supported by the National Natural Science Foundation of China (No. 31660429) and the China Agriculture Research System (CARS-05-03B-03).

Conflicts of Interest: The authors declare no conflict of interest.

References

1. Mascher, M.; Gundlach, H.; Himmelbach, A.; Beier, S.; Twardziok, S.O.; Wicker, T.; Radchuk, V.; Dockter, C.; Hedley, P.E.; Rusell, J.; et al. A chromosome conformation capture ordered sequence of the barley genome. *Nature* **2017**, *544*, 427–433. [CrossRef] [PubMed]

2. Zeng, X.; Long, H.; Wang, Z.; Zhao, S.; Tang, Y.; Huang, Z.; Wang, Y.; Xu, Q.; Mao, L.; Deng, G.; et al. The draft genome of Tibetan hulless barley reveals adaptive patterns to the high stressful Tibetan Plateau. *Proc. Natl. Acad. Sci. USA* **2015**, *112*, 1095–1100. [CrossRef] [PubMed]

3. Munns, R.; Tester, M. Mechanisms of salinity tolerance. *Annu. Rev. Plant Biol.* **2008**, *59*, 651–681. [CrossRef] [PubMed]

4. Hasanuzzaman, M.; Davies, N.W.; Shabala, L.; Zhou, M.; Brodribb, T.J.; Shabala, S. Residual transpiration as a component of salinity stress tolerance mechanism: A case study for barley. *BMC Plant Boil.* **2017**, *107*. [CrossRef] [PubMed]

5. Hazzouri, K.M.; Khraiwesh, B.; Amiri, K.; Pauli, D.; Blake, T.; Shahid, M.; Mullath, S.K.; Nelson, D.; Mansour, A.L.; Salehi-Ashtiani, K.; et al. Mapping of HKT1; 5 gene in barley using GWAS approach and its implication in salt tolerance mechanism. *Front. Plant Sci.* **2018**, *9*, 156. [CrossRef] [PubMed]

6. Witzel, K.; Matros, A.; Møller, A.L.; Ramireddy, E.; Finnie, C.; Peukert, M.; Rutten, T.; Herzog, A.; Kunze, G.; Melzer, M.; et al. Plasma membrane proteome analysis identifies a role of barley membrane steroid binding protein in root architecture response to salinity. *Plant Cell Environ.* **2018**, *41*, 1311–1330. [CrossRef]

7. Zhang, H.; Irving, L.J.; McGill, C.; Matthew, C.; Zhou, D.; Kemp, P. The effects of salinity and osmotic stress on barley germination rate: Sodium as an osmotic regulator. *Ann. Bot.* **2010**, *106*, 1027–1035. [CrossRef]

8. Witzel, K.; Weidner, A.; Surabhi, G.K.; Varshney, R.K.; Kunze, G.; Buck-sorlin, G.H.; Börner, A.; Mock, H.P. Comparative analysis of the grain proteome fraction in barley genotypes with contrasting salinity tolerance during germination. *Plant Cell Environ.* **2010**, *33*, 211–222. [CrossRef]

9. Adem, G.D.; Roy, S.J.; Zhou, M.; Bowman, J.P.; Shabala, S. Evaluating contribution of ionic, osmotic and oxidative stress components towards salinity tolerance in barley. *BMC Plant Biol.* **2014**, *14*, 113. [CrossRef]

10. Ismail, A.M.; Horie, T. Genomics, physiology, and molecular breeding approaches for improving salt tolerance. *Annu. Rev. Plant Biol.* **2017**, *68*, 405–434. [CrossRef]

11. Foolad, M.R.; Lin, G.Y. Absence of a genetic relationship between salt tolerance during seed germination and vegetative growth in tomato. *Plant Breed.* **1997**, *116*, 363–367. [CrossRef]

12. Rajjou, L.; Duval, M.; Gallardo, K.; Catusse, J.; Bally, J.; Job, C.; Job, D. Seed germination and vigor. *Annu. Rev. Plant Biol.* **2012**, *63*, 507–533. [CrossRef] [PubMed]

13. Shu, K.; Liu, X.D.; Xie, Q.; He, Z.H. Two faces of one seed: Hormonal regulation of dormancy and germination. *Mol. Plant* **2016**, *9*, 34–45. [CrossRef] [PubMed]

14. Miransari, M.; Smith, D.L. Plant hormones and seed germination. *Environ. Exp. Bot.* **2014**, *99*, 110–121. [CrossRef]

15. Kong, D.; Ju, C.; Parihar, A.; Kim, S.; Cho, D.; Kwak, J.M. *Arabidopsis* glutamate receptor homolog3.5 modulates cytosolic Ca^{2+} level to counteract effect of abscisic acid in seed germination. *Plant Physiol.* **2015**, *167*, 1630–1642. [CrossRef] [PubMed]

16. Ibrahim, E.A. Seed priming to alleviate salinity stress in germinating seeds. *J. Plant Physiol.* **2016**, *192*, 38–46. [CrossRef] [PubMed]

17. Zhu, J.K. Salt and drought stress signal transduction in plants. *Annu. Rev. Plant Biol.* **2002**, *53*, 247–273. [CrossRef]

18. Zhu, J.K. Abiotic stress signaling and responses in plants. *Cell* **2016**, *167*, 313–324. [CrossRef]

19. Yang, Y.; Guo, Y. Elucidating the molecular mechanisms mediating plant salt-stress responses. *New Phytol.* **2018**, *217*, 523–539. [CrossRef]

20. He, Y.; Yang, B.; He, Y.; Zhan, C.; Cheng, Y.; Zhang, J.; Zhang, H.; Cheng, J.; Wang, Z. A quantitative trait locus, *qSE 3*, promotes seed germination and seedling establishment under salinity stress in rice. *Plant J.* **2019**, *97*, 1089–1104. [CrossRef]

21. Yang, Z.; Wang, C.; Xue, Y.; Liu, X.; Chen, S.; Song, C.; Yang, Y.; Guo, Y. Calcium-activated 14-3-3 proteins as a molecular switch in salt stress tolerance. *Nat. Commun.* **2019**, *10*, 1199. [CrossRef] [PubMed]

22. Zhang, N.; Zhang, H.J.; Sun, Q.Q.; Cao, Y.Y.; Li, X.; Zhao, B.; Wu, P.; Guo, Y. Proteomic analysis reveals a role of melatonin in promoting cucumber seed germination under high salinity by regulating energy production. *Sci. Rep.* **2017**, *7*, 503. [CrossRef] [PubMed]

23. Tenhaken, R. Cell wall remodeling under abiotic stress. *Front. Plant Sci.* **2015**, *5*, 771. [CrossRef] [PubMed]

24. Yang, H.; Li, Y.; Hua, J. The C2 domain protein BAP1 negatively regulates defense responses in *Arabidopsis*. *Plant J.* **2006**, *48*, 238–248. [CrossRef] [PubMed]

25. de Silva, K.; Laska, B.; Brown, C.; Sederoff, H.W.; Khodakovskaya, M. *Arabidopsis thaliana* calcium-dependent lipid-binding protein (AtCLB): A novel repressor of abiotic stress response. *J. Exp. Bot.* **2011**, *62*, 2679–2689. [CrossRef]

26. Colebrook, E.H.; Thomas, S.G.; Phillips, A.L.; Hedden, P. The role of gibberellin signalling in plant responses to abiotic stress. *J. Exp. Boil.* **2014**, *217*, 67–75. [CrossRef]

27. Krishna, P. Brassinosteroid-mediated stress responses. *J. Plant Growth Regul.* **2003**, *22*, 289–297. [CrossRef]

28. Zhang, X.M.; Walsh, B.; Mitchell, C.A.; Rowe, T. TBC domain family, member 15 is a novel mammalian Rab GTPase-activating protein with substrate preference for Rab7. *Biochem. Biophys. Res. Commun.* **2005**, *335*, 154–161. [CrossRef]

29. Mazel, A.; Leshem, Y.; Tiwari, B.S.; Levine, A. Induction of salt and osmotic stress tolerance by overexpression of an intracellular vesicle trafficking protein AtRab7 (AtRabG3e). *Plant Physiol.* **2004**, *134*, 118–128. [CrossRef]

30. Zhang, Y.; Immink, R.; Liu, C.M.; Emons, A.M.; Ketelaar, T. The *Arabidopsis* exocyst subunit SEC3A is essential for embryo development and accumulates in transient puncta at the plasma membrane. *New Phytol.* **2013**, *199*, 74–88. [CrossRef]

31. Geng, G.; Lv, C.; Stevanato, P.; Li, R.; Liu, H.; Yu, L.; Wang, Y. Transcriptome analysis of salt-sensitive and tolerant genotypes reveals salt-tolerance metabolic pathways in Sugar Beet. *Int. J. Mol. Sci.* **2019**, *20*, 5910. [CrossRef] [PubMed]

32. Jiang, Z.; Jin, F.; Shan, X.; Li, Y. iTRAQ-based proteomic analysis reveals several strategies to cope with drought stress in maize seedlings. *Int. J. Mol. Sci.* **2019**, *20*, 5956. [CrossRef] [PubMed]

33. Xu, E.; Chen, M.; He, H.; Zhan, C.; Cheng, Y.; Zhang, H.; Wang, Z. Proteomic analysis reveals proteins involved in seed imbibition under salt stress in rice. *Front. Plant Sci.* **2017**, *7*, 2006. [CrossRef] [PubMed]

34. Dong, T.; Zhu, M.; Yu, J.; Han, R.; Tang, C.; Li, Z. RNA-Seq and iTRAQ reveal multiple pathways involved in storage root formation and development in sweet potato (*Ipomoea batatas* L.). *BMC Plant Boil.* **2019**, *19*, 136. [CrossRef]

35. Lan, P.; Li, W.; Schmidt, W. Complementary proteome and transcriptome profiling in phosphate-deficient *Arabidopsis* roots reveals multiple levels of gene regulation. *Mol. Cell. Proteom.* **2012**, *1*, 1156–1166. [CrossRef]

36. Li, J.M.; San Huang, X.; Li, L.T.; Zheng, D.M.; Xue, C.; Wu, J. Proteome analysis of pear reveals key genes associated with fruit development and quality. *Planta* **2015**, *241*, 1363–1379. [CrossRef]

37. Phan, J.L.; Tucker, M.R.; Khor, S.F.; Shirley, N.; Lahnstein, J.; Beahan, C.; Bacic, A.; Burton, R.A. Differences in glycosyltransferase family 61 accompany variation in seed coat mucilage composition in *Plantago* spp. *J. Exp. Bot.* **2016**, *67*, 6481–6495. [CrossRef]

38. Kopczak, S.D.; Haas, N.A.; Hussey, P.J.; Silflow, C.D.; Snustad, D.P. The small genome of *Arabidopsis* contains at least six expressed alpha-tubulin genes. *Plant Cell* **1992**, *4*, 539–547. [CrossRef]

39. Zhuo, L.; Kimata, K. Structure and function of inter-α-trypsin inhibitor heavy chains. *Connect. Tissue Res.* **2008**, *49*, 311–320. [CrossRef]

40. Fluhr, R.; Lampl, N.; Roberts, T.H. Serpin protease inhibitors in plant biology. *Physiol. Plantarum* **2012**, *145*, 95–102. [CrossRef]

41. Kotake, T.; Takata, R.; Verma, R.; Takaba, M.; Yamaguchi, D.; Orita, T.; Kaneko, S.; Matsuoka, K.; Koyama, T.; Reiter, W.D.; et al. Bifunctional cytosolic UDP-glucose 4-epimerases catalyse the interconversion between UDP-D-xylose and UDP-L-arabinose in plants. *Biochem. J.* **2009**, *424*, 169–177. [CrossRef] [PubMed]

42. Zhao, C.; Zayed, O.; Zeng, F.; Liu, C.; Zhang, L.; Zhu, P.; Hsu, C.C.; Tuncil, Y.E.; Tao, W.A.; Carpita, N.C.; et al. *Arabinose biosynthesis is critical for salt stress tolerance in Arabidopsis.* *New Phytol.* **2019**, *224*, 274–290. [CrossRef] [PubMed]

43. Zhang, Q.; Hrmova, M.; Shirley, N.; Lahnstein, J.; Fincher, G. Gene expression patterns and catalytic properties of UDP-D-glucose 4-epimerases from barley (*Hordeum vulgare* L.). *Biochem. J.* **2006**, *394*, 115–124. [CrossRef] [PubMed]

44. Raedschelders, G.; Debefve, C.; Goesaert, H.; Delcour, J.A.; Volckaert, G.; Van Campenhout, S. Molecular identification and chromosomal localization of genes encoding *Triticum aestivum* xylanase inhibitor I-like proteins in cereals. *Theor. Appl. Genet.* **2004**, *109*, 112–121. [CrossRef] [PubMed]

45. Igawa, T.; Ochiai-Fukuda, T.; Takahashi-Ando, N.; Ohsato, S.; Shibata, T.; Yamaguchi, I.; Kimura, M. New TAXI-type xylanase inhibitor genes are inducible by pathogens and wounding in hexaploid wheat. *Plant Cell Physiol.* **2004**, *45*, 1347–1360. [CrossRef] [PubMed]

46. Vázquez-Lobo, A.; Roujol, D.; Zuñiga-Sánchez, E.; Albenne, C.; Piñero, D.; de Buen, A.G.; Jamet, E. The highly conserved spermatophyte cell wall DUF642 protein family: Phylogeny and first evidence of interaction with cell wall polysaccharides in vitro. *Mol. Phylogenet. Evol.* **2012**, *63*, 510–520. [CrossRef] [PubMed]

47. Zúñiga-Sánchez, E.; Soriano, D.; Martínez-Barajas, E.; Orozco-Segovia, A.; Gamboa-deBuen, A. BIIDXI, the *At4g32460* DUF642 gene, is involved in pectin methyl esterase regulation during *Arabidopsis thaliana* seed germination and plant development. *BMC Plant Biol.* **2014**, *14*, 338. [CrossRef]

48. Sheshukova, E.V.; Komarova, T.V.; Pozdyshev, D.V.; Ershova, N.M.; Shindyapina, A.V.; Tashlitsky, V.N.; Sheval, E.V.; Dorokhov, Y.L. The intergenic interplay between aldose 1-epimerase-like protein and pectin methylesterase in abiotic and biotic stress control. *Front. Plant Sci.* **2017**, *8*, 1646. [CrossRef]

49. Dorokhov, Y.L.; Sheshukova, E.V.; Komarova, T.V. Methanol in plant life. *Front. Plant Sci.* **2018**, *9*, 1623. [CrossRef]

50. Shi, H.; Xiong, L.; Stevenson, B.; Lu, T.; Zhu, J.K. The *Arabidopsis salt overly sensitive 4* mutants uncover a critical role for vitamin B6 in plant salt tolerance. *Plant Cell* **2002**, *14*, 575–588. [CrossRef]

51. Zhao, W.T.; Feng, S.J.; Li, H.; Faust, F.; Kleine, T.; Li, L.N.; Yang, Z. Salt stress-induced FERROCHELATASE 1 improves resistance to salt stress by limiting sodium accumulation in *Arabidopsis thaliana*. *Sci. Rep.* **2017**, *7*, 14737. [CrossRef] [PubMed]

52. Diaz, M.; Sanchez-Barrena, M.J.; Gonzalez-Rubio, J.M.; Rodriguez, L.; Fernandez, D.; Antoni, R.; Yunta, C.; Belda-Palazon, B.; Gonzalez-Guzman, M.; Peirats-Liobet, M.; et al. Calcium-dependent oligomerization of CAR proteins at cell membrane modulates ABA signaling. *Proc. Natl. Acad. Sci. USA* **2016**, *113*, E396–E405. [CrossRef] [PubMed]

53. Nalefski, E.A.; Falke, J.J. The C2 domain calcium-binding motif: Structural and functional diversity. *Protein Sci.* **1996**, *5*, 2375–2390. [CrossRef] [PubMed]

54. Rizo, J.; Südhof, T.C. C2-domains, structure and function of a universal Ca^{2+}-binding domain. *J. Biol. Chem.* **1998**, *273*, 15879–15882. [CrossRef]

55. Rodriguez, L.; Gonzalez-Guzman, M.; Diaz, M.; Rodrigues, A.; Izquierdo-Garcia, A.C.; Peirats-Llobet, M.; Fernandez, M.A.; Antoni, R.; Fernandez, D.; Marquez, J.A.; et al. C_2-domain abscisic acid-related proteins mediate the interaction of PYR/PYL/RCAR abscisic acid receptors with the plasma membrane and regulate abscisic acid sensitivity in *Arabidopsis*. *Plant Cell* **2014**, *26*, 4802–4820. [CrossRef]

56. Nietzsche, M.; Guerra, T.; Alseekh, S.; Wiermer, M.; Sonnewald, S.; Fernie, A.R.; Börnke, F. STOREKEEPER RELATED1/G-element binding protein (STKR1) interacts with protein kinase SnRK1. *Plant Physiol.* **2018**, *176*, 1773–1792. [CrossRef]

57. Hyun, T.K.; van der Graaff, E.; Albacete, A.; Eom, S.H.; Großkinsky, D.K.; Böhm, H.; Janschek, U.; Rim, Y.; Ali, W.W.; Kim, S.Y.; et al. The *Arabidopsis* PLAT domain protein1 is critically involved in abiotic stress tolerance. *PLoS ONE* **2014**, *9*, e112946. [CrossRef]

58. Qin, F.; Kodaira, K.S.; Maruyama, K.; Mizoi, J.; Tran, L.S.P.; Fujita, Y.; Morimoto, K.; Shinozaki, K.; Yamaguchi-Shinozaki, K. SPINDLY, a negative regulator of gibberellic acid signaling, is involved in the plant abiotic stress response. *Plant Physiol.* **2011**, *157*, 1900–1913. [CrossRef]

59. Kawai, Y.; Ono, E.; Mizutani, M. Evolution and diversity of the 2-oxoglutarate-dependent dioxygenase superfamily in plants. *Plant J.* **2014**, *78*, 328–343. [CrossRef]

60. Park, J.; Kim, Y.S.; Kim, S.G.; Jung, J.H.; Woo, J.C.; Park, C.M. Integration of auxin and salt signals by the NAC transcription factor NTM2 during seed germination in *Arabidopsis*. *Plant Physiol.* **2011**, *156*, 537–549. [CrossRef]

61. Schlaich, N.L. Flavin-containing monooxygenases in plants: Looking beyond detox. *Trends Plant Sci.* **2007**, *12*, 412–418. [CrossRef] [PubMed]

62. Cohen, J.D.; Slovin, J.P.; Hendrickson, A.M. Two genetically discrete pathways convert tryptophan to auxin: More redundancy in auxin biosynthesis. *Trends Plant Sci.* **2003**, *8*, 197–199. [CrossRef]

63. Steber, C.M.; McCourt, P. A role for brassinosteroids in germination in *Arabidopsis*. *Plant Physiol.* **2001**, *125*, 763–769. [CrossRef] [PubMed]

64. Kagale, S.; Divi, U.K.; Krochko, J.E.; Keller, W.A.; Krishna, P. Brassinosteroid confers tolerance in *Arabidopsis thaliana* and *Brassica napus* to a range of abiotic stresses. *Planta* **2007**, *225*, 353–364. [CrossRef]

65. Klahre, U.; Noguchi, T.; Fujioka, S.; Takatsuto, S.; Yokota, T.; Nomura, T.; Yoshida, S.; Chua, N.H. The *Arabidopsis DIMINUTO/DWARF1* gene encodes a protein involved in steroid synthesis. *Plant Cell* **1998**, *10*, 1677–1690. [CrossRef]

66. Nikolić, R.; Mitić, N.; Miletić, R.; Nešković, M. Effects of cytokinins on in vitro seed germination and early seedling morphogenesis in *Lotus corniculatus* L. *J. Plant Growth Regul.* **2006**, *25*, 187. [CrossRef]
67. Peleg, Z.; Blumwald, E. Hormone balance and abiotic stress tolerance in crop plants. *Curr. Opin. Plant Boil.* **2011**, *14*, 290–295. [CrossRef]
68. Walia, H.; Wilson, C.; Condamine, P.; Liu, X.; Ismail, A.M.; Close, T.J. Large-scale expression profiling and physiological characterization of jasmonic acid-mediated adaptation of barley to salinity stress. *Plant Cell Environ.* **2007**, *30*, 410–421. [CrossRef]
69. Lennon, B.W.; Williams, C.H.; Ludwig, M.L. Twists in catalysis: Alternating conformations of *Escherichia coli* thioredoxin reductase. *Science* **2000**, *289*, 1190–1194. [CrossRef] [PubMed]
70. Piaz, F.D.; Braca, A.; Belisario, M.A.; De Tommasi, N. Thioredoxin system modulation by plant and fungal secondary metabolites. *Curr. Med. Chem.* **2010**, *17*, 479–494. [CrossRef]
71. Jung, Y.J.; Melencion, S.M.B.; Lee, E.S.; Park, J.H.; Alinapon, C.V.; Oh, H.T.; Yun, D.J.; Chi, Y.; Lee, S.Y. Universal stress protein exhibits a redox-dependent chaperone function in *Arabidopsis* and enhances plant tolerance to heat shock and oxidative stress. *Front. Plant Sci.* **2015**, *6*, 1141. [CrossRef] [PubMed]
72. Suzuki, N.; Rizhsky, L.; Liang, H.; Shuman, J.; Shulaev, V.; Mittler, R. Enhanced tolerance to environmental stress in transgenic plants expressing the transcriptional coactivator multiprotein bridging factor 1c. *Plant Physiol.* **2005**, *139*, 1313–1322. [CrossRef] [PubMed]
73. Arce, D.P.; Godoy, A.V.; Tsuda, K.; Yamazaki, K.I.; Valle, E.M.; Iglesias, M.J.; Di Mauroa, M.; Casalongué, C.A. The analysis of an *Arabidopsis* triple knock-down mutant reveals functions for MBF1 genes under oxidative stress conditions. *J. Plant Physiol.* **2010**, *167*, 194–200. [CrossRef] [PubMed]
74. Kamei, C.L.A.; Boruc, J.; Vandepoele, K.; Van den Daele, H.; Maes, S.; Russinova, E.; Inzé, D.; De Veylder, L. The *PRA1* gene family in *Arabidopsis*. *Plant Physiol.* **2008**, *147*, 1735–1749. [CrossRef]
75. Lee, M.H.; Yoo, Y.J.; Kim, D.H.; Hanh, N.H.; Kwon, Y.; Hwang, I. The prenylated rab GTPase receptor PRA1. F4 contributes to protein exit from the Golgi apparatus. *Plant Physiol.* **2017**, *174*, 1576–1594. [CrossRef] [PubMed]
76. Sturm, M.; Cheng, J.; Baßler, J.; Beckmann, R.; Hurt, E. Interdependent action of KH domain proteins Krr1 and Dim2 drive the 40S platform assembly. *Nat. Commun.* **2017**, *8*, 2213. [CrossRef]
77. Lafuente, M.T.; Martínez-Téllez, M.A.; Zacarias, L. Abscisic acid in the response of 'Fortune' mandarins to chilling. Effect of maturity and high-temperature conditioning. *J. Sci. Food Agric.* **1997**, *73*, 494–502. [CrossRef]
78. Maere, S.; Heymans, K.; Kuiper, M. BiNGO: A Cytoscape plugin to assess overrepresentation of gene ontology categories in biological networks. *Bioinformatics* **2005**, *21*, 3448–3449. [CrossRef]
79. Schmittgen, T.D.; Livak, K.J. Analyzing real-time PCR data by the comparative CT method. *Nat. Protoc.* **2008**, *3*, 1101–1108. [CrossRef]

International Journal of
Molecular Sciences

Article

Integrated Omics Analyses Identify Key Pathways Involved in Petiole Rigidity Formation in Sacred Lotus

Ming Li [1,†], Ishfaq Hameed [2,†], Dingding Cao [3], Dongli He [1] and Pingfang Yang [1,*]

[1]   State Key Laboratory of Biocatalysis and Enzyme Engineering, School of Life Sciences, Hubei University, Wuhan 430062, China; limit@hubu.edu.cn (M.L.); hedongli@hubu.edu.cn (D.H.)

[2]   Departments of Botany, University of Chitral, Chitral 17200, Khyber Pukhtunkhwa, Pakistan; pakoonleo@yahoo.com

[3]   Institue of Oceanography, Minjiang University, Fuzhou 350108, China; caodingding@wbgcas.cn

*   Correspondence: yangpf@hubu.edu.cn

†   These authors contributed equally to this work.

Received: 12 May 2020; Accepted: 15 July 2020; Published: 18 July 2020

Abstract: Sacred lotus (*Nelumbo nucifera* Gaertn.) is a relic aquatic plant with two types of leaves, which have distinct rigidity of petioles. Here we assess the difference from anatomic structure to the expression of genes and proteins in two petioles types, and identify key pathways involved in petiole rigidity formation in sacred lotus. Anatomically, great variation between the petioles of floating and vertical leaves were observed. The number of collenchyma cells and thickness of xylem vessel cell wall was higher in the initial vertical leaves' petiole (IVP) compared to the initial floating leaves' petiole (IFP). Among quantified transcripts and proteins, 1021 and 401 transcripts presented 2-fold expression increment (named DEGs, genes differentially expressed between IFP and IVP) in IFP and IVP, 421 and 483 proteins exhibited 1.5-fold expression increment (named DEPs, proteins differentially expressed between IFP and IVP) in IFP and IVP, respectively. Gene function and pathway enrichment analysis displayed that DEGs and DEPs were significantly enriched in cell wall biosynthesis and lignin biosynthesis. In consistent with genes and proteins expressions in lignin biosynthesis, the contents of lignin monomers precursors were significantly different in IFP and IVP. These results enable us to understand lotus petioles rigidity formation better and provide valuable candidate genes information on further investigation.

Keywords: sacred lotus; petiole rigidity; proteomics; cell wall; lignin biosynthesis

---

1. Introduction

Plant architecture is the embodiment of space utilization. There are numerous factors contribute to plant architecture including stem height, leaf and branching mode [1]. In crops, several factors such as stem rigidity, branching pattern, and plant height which strongly influences crop yields and efficiency of harvesting have been extensively studied [2,3]. The sacred lotus is a relic aquatic plant with two leaf types, namely the floating leaf and the vertical leaf. Floating leaf petiole exhibits flexibility and floating tendency despite it being longer than the depth of the water, while the vertical leaf petiole is rigid and erect (Figure 1). Moreover, there is no reported interconvertibility between these two kinds of leaves during the entire life span of lotus, meaning that lotus is not a heterophyllous plant. Stem or petiole rigidity is species-specific and plastic [4] controlled by environmental factors including light, temperature, water and nutrients, among others.

Cell wall makes plant cell different from animal cell. Cell wall contributes to plant architecture, organ development, transport of water and nutriment and defense. In order to accomplish variety function mentioned above, cell in plant was specialized into collenchyma, sclerenchyma and xylem [5–7]. Generally, cell wall can be divided into middle lamella, primary wall and secondary wall if the cell wall has secondary thickening. The main components in secondary cell wall are polysaccharide and lignin [8]. Polysaccharide mainly includes cellulose, hemicellulose and pectin. Lignin is compounded of three primary monolignols including coniferyl alcohol, sinapyl alcohol and p-coumaryl alcohol in varieties of proportions [9]. The components of cell wall are diversified depending on the plant taxa, age, tissue and cell type [8]. Studies have verified that many genes are associated with cell wall biosynthesis and secondary cell wall thickening in *Arabidopsis* [10], Populus [6] and other species [11]. Typically, cellulose synthase [12,13], glycosyltransferases and glycosyl hydrolases [8,14] play a crucial role in polysaccharide biosynthesis, and enzymes involved in phenylalanine deamination, hydroxylation, methylation and redox reactions in phenylpropanoid pathway play a key role in lignin biosynthesis [15]. Transcription factor families are engaged more in the upstream of cell wall biosynthesis and modification, including NAC domain (VND6/7:vascular-related NAC domain 6/7, SND1: secondary wall-associated NAC domain 1 and NST1: NAC secondary wall thickening promoting factor) and HD-ZIP homeobox, which were reported that they play sufficient role in influencing secondary cell wall synthesis in xylem vessel [16–18]. Members in MYB family are also informed as regulators involved in secondary cell wall formation [19], while members in E2F family were affirmed as a fundamental upstream transcriptional regulator of VND6/7 and other secondary cell wall biosynthesis genes [10].

Despite the increasing understanding on cell wall biosynthesis in the past two decades, a lot still remain elusive regarding related gene regulation. The findings in heterophyllous plants recommend that light and temperature are involved in transformation of heterophyllous leaves [20], and internal factors including phytohormones, such as gibberellin (GA), abscisic acid (ABA) and ethylene play a dominant role in the induction of heterophylly in plants [21]. Besides, some studies focused on *Arabidopsis* petioles indicating light quality, phytohormones and crosstalk between light and phytohormones mediate petiole elongation [22], and it is ascertained that numerous genes are involved in cell wall formation and secondary cell wall thickening in *Arabidopsis* [23], rice [24] and Populus [25]. Since there is little research done, it is unclear whether lotus has the same mechanism of gene network regulation as typical heterophyllous plants. Furthermore, lotus is the only aquatic plant which has a combination of floating and vertical leaves, and it is a good example to probe the mechanism that controls differentiation of these two leaves. These findings will facilitate us to understand better the mechanism of cell wall formation.

High-throughput profiling techniques used in transcriptome and proteome are effective in exploring complex biologic processes at the overall level [26]. The expression levels of mRNAs and proteins can be accurately measured using RNA-sequencing, iTRAQ (isobaric tags for relative and absolute quantitation) and TMT (tandem mass tag) technologies. As we known, the regulation of genes at transcriptional, translational and post-translational levels are different, integrated analysis of these data are necessary [27]. In *Arabidopsis*, transcriptome and proteome technologies were used to identified genes involved in cell wall biosynthesis. It also indicates no clear correlation was found between the omics data demonstrating complementary in different methods [28,29].

In this study, the IFP (initial floating leaves' petiole) and IVP (initial vertical leaves' petiole) were used to as the research material for analysis through RNA sequencing and protein labeling quantification with tandem mass tags (TMT) technology to explore the mechanism that controls differentiation of these two types of leaves in lotus. By analyzing DEGs (genes differentially expressed between IFP and IVP), DEPs (proteins differentially expressed between IFP and IVP) and the abundance of metabolite in the IFP and IVP, we found that genes highly expressed in IVP in several central biologic processes, such as cell wall biosynthesis, organization, assembly and lignin biosynthesis were associated with lotus petiole rigidity. The integrated analysis of transcriptomic and proteomic data provides a clue to understand the formation of different petioles in lotus.

2. Results

2.1. Phenotypic Evaluation of Lotus Petioles at IFP and IVP Stages

Lotus, as an herbaceous perennial plant, sprouts as temperature rises toward spring. At the beginning, several initial leaves (three to five leaves) grow and float on, but not rise above the water surface in spite of long enough petioles. Then the succeeding leaves stand out of the water (Figure 1a,b, Figure S1). Generally speaking, initial leaf floats on the water surface all the time. Vertical leaf rises above the water surface once the petiole length is larger than water depth (Figure 1c, Figure S1). Moreover, both the folded and unfolded vertical leaves can rise above the water surface (Figure 1d, Figure S1). We have not observed any interconvertibility between IFP and IVP under green house and field conditions. Numbers of lateral branches grow continuously during its growing season, and each branch develops the floating leaf first, followed by the vertical leaf.

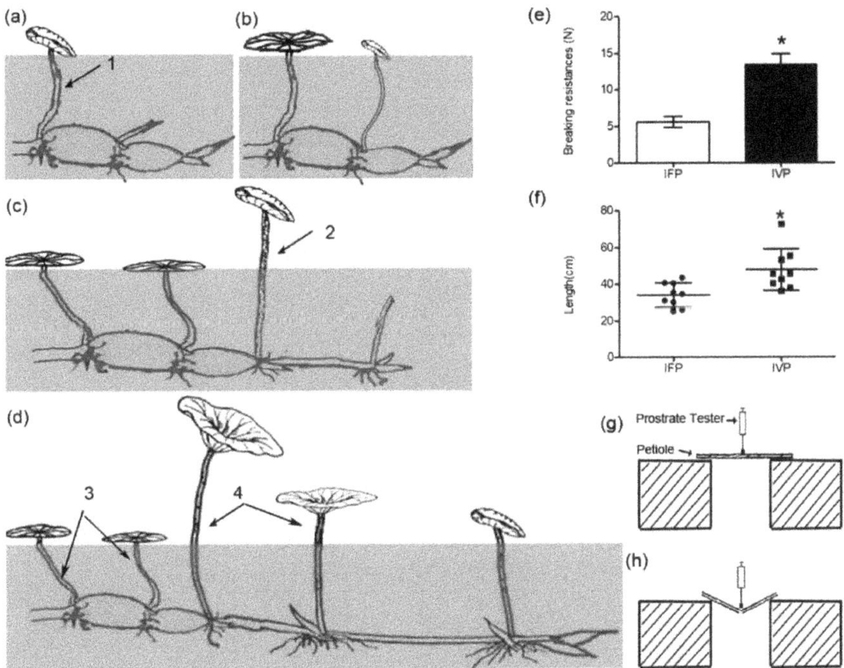

Figure 1. Schematic diagram of growing processes of floating leaf and vertical leaf. (**a–d**) stand different growth stages of leaves. Arrow 1 indicates initial floating leaves' petiole (IFP), arrow 2 indicates initial vertical leaves' petiole (IVP), arrow 3 indicates mature floating leaves' petiole (MFP) and arrow 4 indicates mature vertical leaves' petiole (MVP), respectively; (**e**) measurements of breaking resistances in two type petioles were performed using a prostrate tester; (**g,h**) show status of petioles before and after putting pressure on it; (**f**) shows the length of petioles which were cultured in pot with 20 cm depth water. Data are means ± SD from 9 independent petioles (IFP and IVP). Asterisks indicate significant changes according to Student's t-test (* $p < 0.05$).

For the purpose of quantifying the phenotype difference between the floating and vertical leaves, the breaking resistance of two type petioles was measured. The breaking resistance of IVP was significantly higher than that in IFP (Figure 1e). Furthermore, we observed two types of petioles in lotus with low water depth. The minimum length of IFP is over water depth at that condition, and IVP was greater than IFP. Even though the petiole length of floating leaf was greater than water depth as it matures, it still could not stand erect on water (Figure 1f). This result suggests the breaking resistance, but not the petiole length being the main factor affecting the leaf emergence from water surface.

2.2. Anatomic Structure Analysis of Petioles at IFP and IVP Stages

To explore the structure difference of petioles in floating and vertical leaf, we studied the sections of IFP and IVP. There were four primarily air cavities (the numbers of air cavities will increase as the petioles grow) in both transection of IFP and IVP. Numbers of small air cavities were scattered around the chief air cavities (Figure 2a,b), and some of them could develop into the chief air cavities along with petioles growth. The vascular bundles, which increase cell wall resistance and rigidity, were scattered around the big air cavities as well. These data showed the number of vascular bundles in IVP is more than those in IFP (Figure 2a,b,e). Furthermore, the epidermal cell had high level of cutinization in IVP than that in IFP. The outer cortex cells in IVP have specialized into collenchyma by uneven thickening of cell wall to strength the rigidity of petioles. In IFP, by contrast, the cortex cells had no cell wall thickening. (Figure 2c,d). These vascular bundles and collenchyma could contribute to cell wall resistance and rigidity thus supporting petioles emergence from water in IVP.

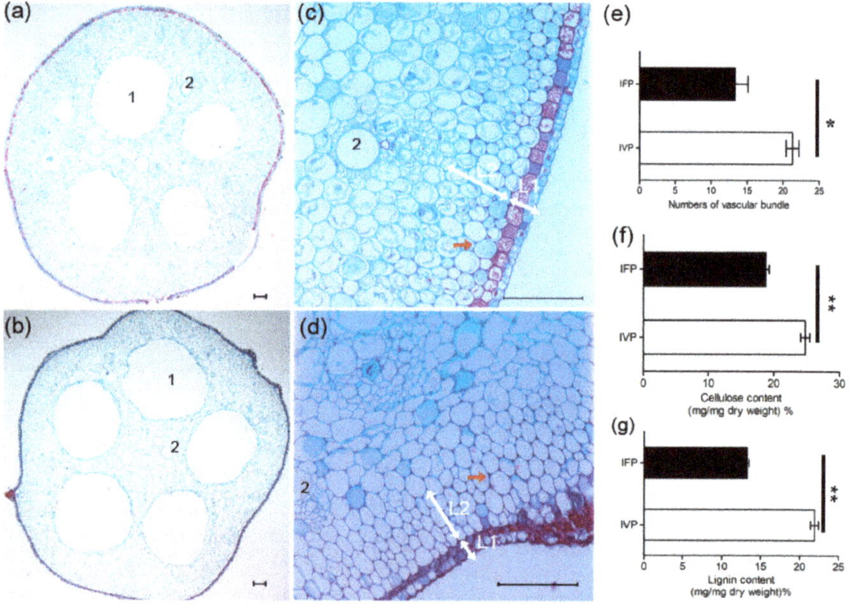

Figure 2. Transverse sections of IFP and IVP stained with fast green and the counterstain safranin. (**a,c**) and (**b,d**) show transverse sections of IFP and IVP, respectively. Arabic numerals 1 and 2 indicate big air cavities and xylem vessel. L1 and L2 indicate the epidermal and cortex. Red arrow shows collenchyma. Bars in figures indicate 100 μm; (**e**) statistics of vascular bundles number in IFP and IVP, data are means ± SD from 10 independent transverse sections (IFP and IVP); (**f**) determination of crude cellulose and (**g**) lignin. Data are means ± SD from 3 independent biologic repeats. Asterisks and double asterisks indicate significant changes compared to the control as assessed according to Student's *t*-test (* $p < 0.05$ and ** $p < 0.01$).

To further study the structure character of petiole, thickness of xylem vessels cell wall was measured (Figure 3a–d). It showed that xylem vessels cell wall in IFP was 0.4 μm compared with IVP (0.8 μm) (Figure 3g). This result suggests that xylem vessel in IVP may provide more strength to support IVP to remain upright. The main components of cell wall are polysaccharide and lignin, and the content of cellulose and lignin in cell wall were measured. This result noted that the cellulose contents were 18.7% and 24.8% in IFP and IVP, and the lignin contents were 13.3% and 21.9% in IFP and IVP, respectively (Figure 2f,g). Both cellulose and lignin contents were significantly higher in IVP than that in IFP. This result is consistent with morphologic results which displayed more vascular bundles, collenchyma and cell wall thickening in IVP compared to IFP.

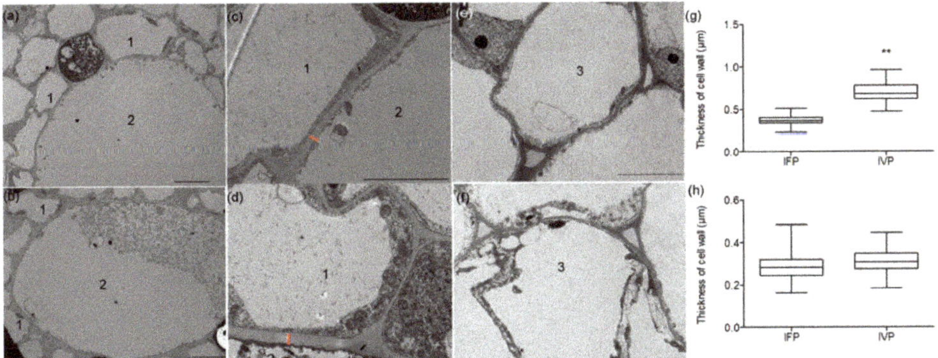

Figure 3. Thickness of cell wall in lotus petioles was observed using a transmission electron microscope. (**a,c**) shows cell wall thickness of xylem vessels in IFP; (**b,d**) shows cell wall thickness of xylem vessels in IVP; (**e,f**) show thickness of parenchymatous cell wall of IFP and IVP, respectively. The Arabic numerals 1 indicates sclerenchyma cell; Arabic numerals 2 indicates xylem cell, The Arabic numerals 3 indicates parenchymatous cell; (**g,h**) statistics of thickness of the cell wall of xylem vessels and parenchymatous cell in IFP and IVP, respectively. Data are means ± SD from 10 independent transverse sections (IFP and IVP). Double asterisks indicate significant changes compared to IFP as assessed according to Student's *t*-test (** $p < 0.01$). Bars in figures indicate 10 µm.

2.3. Overview of Transcriptomic Analysis and Transcriptomic Data Validation

To explore the different genes involved in cell wall formation in IFP and IVP, comparative transcriptomic experiment was done. RNA-seq was performed for IFP and IVP groups on the Illumina platform and each group included three biologic replicates. Generally, 87–90% of the clean reads were mapped to the genome of the sacred lotus (https://bioinformatics.psb.ugent.be/plaza/versions/plaza_v4_dicots/). The expression estimates were calculated as fragments per kilobase per million reads (FPKM). Pearson's correlation coefficient was used to evaluate repeatability of individual sample (Figure S2a). A total of 20,109 genes or transcripts (about 75% predicted genes from genome) were obtained from the six samples (Figure 4a). Differentially expressed genes (DEGs) were screened based on an absolute fold change value of |log2 ratio| ≥ 1 and false discovery rate (FDR) ≤ 0.01. A total of 1422 DEGs were found, including 401 upregulated genes and 1021 downregulated genes (IVP vs. IFP) (Figure 4b, Figure S2b, Table S1).

The expression levels of 14 genes involved in the cell wall biosynthesis and 8 genes with different expression patterns in IFP and IVP were determined by quantitative reverse transcription polymerase chain reaction (qRT-PCR) to validate the transcriptomic data. A total of 19 genes shared the same expression pattern between transcriptomic and qRT-PCR results (Figure 4c, Table S2). The transcriptomic data and qRT-PCR results of these genes were highly correlated (r = 0.8339; Figure 4d). These results further proved that the transcriptomic data were reliable.

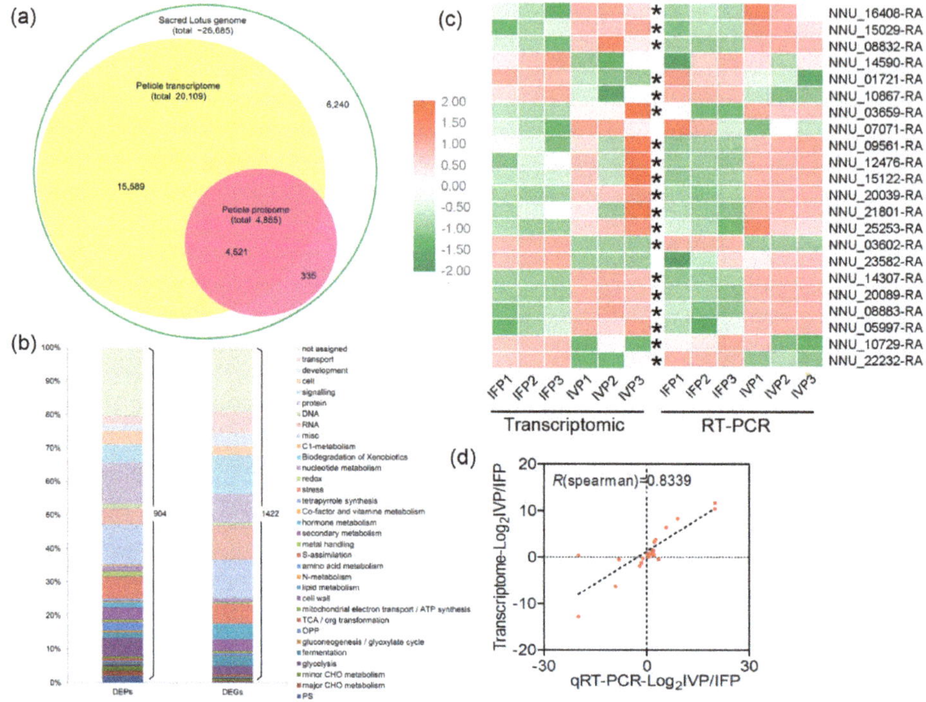

Figure 4. Comparison of protein and transcript abundance in lotus petioles. (**a**) Congruency between the detected transcripts and proteins of lotus petioles; (**b**) functional classification and distribution of differentially expressed genes (DEGs) and differentially expressed proteins (DEPs); (**c**) heatmap of 22 selected genes performed by transcriptomic and RT-PCR data. Asterisk indicates genes have a similar expression pattern in transcriptomic and RT-PCR experiments; (**d**) correlations of gene expression between transcriptomic and RT-PCR data. R represents the Pearson correlations coefficient.

2.4. Overview of Quantitative Proteomics Analysis and Data Validation Using PRM

To scrutinize the different proteins involved in cell wall formation of IFP and IVP, quantitative and comparative proteomics was done. A total of 25,529 peptides belonging to 5808 proteins (about 22% predicted genes from genome) were identified in the petiole samples. Among these identified proteins, 4855 (about 18% predicted genes from genome) of them were quantified (Figure 4a). Pearson's correlation coefficient was used to evaluate repeatability of individual sample (Figure S3a). After filtering with an expression fold change of >1.5 and *p*-value < 0.05, 904 out of the quantified proteins were defined as differentially expressed (DEPs) between IFP and IVP. Among these DEPs, the abundance of 421 proteins in IFP was 1.5 times higher than in IVP, and the abundance of 483 proteins in IVP was 1.5 times higher than in IFP (Table S3).

To validate the reliability of tandem mass tag-labeling (TMT-labeling) protein quantification in this study, parallel reaction monitoring (PRM) method was used. The results showed that 11 out of 15 selected proteins were successfully quantified using PRM methods, and the changes in all 11 proteins abundances between IFP and IVP were consistent with TMT-labeling protein quantification (Figure S4a, Table S4). The PRM data and proteomics results of these proteins were highly correlated (r = 0.6301, Figure S4b). This result confirms that our protein quantification result was reliable.

2.5. Integrated Analysis of Transcriptome and Proteome Data

To analyze the correlation of DEGs and DEPs identified in IFP and IVP, data from transcriptome and proteome were compared. Globally, a total of 20,109 transcripts (approximate 75.36% genome) were identified in lotus petioles, and transcripts were detected for 93.12% of the proteins (Figure 4a). To explore the relationship between protein abundance and their corresponding gene expression, we conducted a correlation analysis using Spearman's rank correlation coefficient method between the transcriptome and quantitative proteome data. After analyzing, the expression levels of all the transcripts and their corresponding quantified proteins from IVP vs. IFP showed weak correlation (r = 0.2765, Figure 5a). The weak correlation was also observed in various multi-omics researches in *Arabidopsis*, rice and eggplant [26,30,31]. At the same time, a stronger correlation was discerned between the DEGs and their corresponding DEPs (r = 0.6148, Figure 5d). The expression ratio of proteins and their corresponding mRNAs with the same or opposite trend (both upregulated or both downregulated) were also plotted, and higher positive or negative correlation was indicated (Figure 5b,c,e,f).

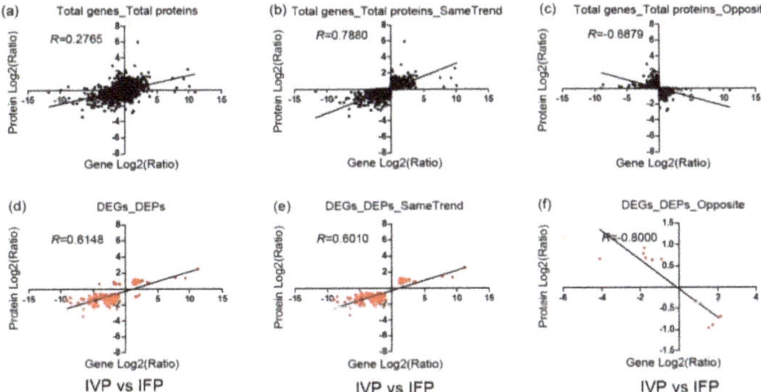

Figure 5. Concordance between changes in the abundance of mRNA and its encoded protein in IFP and IVP. R, Pearson correlations coefficient of the comparisons between fold changes of proteins and transcripts. (**a**) the expression levels of all the transcripts and their corresponding quantified proteins from IVP vs. IFP showed weak correlation (r = 0.2765). (**d**) a stronger correlation was discerned between the DEGs and their corresponding DEPs (r = 0.6148). The expression ratio of proteins and their corresponding mRNAs with the same or opposite trend (both upregulated or both downregulated) were also plotted, and higher positive or negative correlation was indicated (**b,c,e,f**).

In the present study, a total of 1422 DEGs and 904 DEPs were found. Both the DEGs and DEPs were classified into various functional groups (Figure 4b). Among these DEGs and DEPs, 135 of them had quantitative information both in transcript and protein level. For ease of description, the 135 genes both identified in transcriptional and translational levels were named as core genes. MapMan analysis showed the 135 core genes were classified in 22 functional groups. Except for unknown function genes, 72% of core genes were sorted into five larges groups including miscellaneous, stress, signaling, protein and cell wall (Figure S5a). Among the 135 core genes, 126 of them showed same trend in transcriptional and translational level, while only 9 genes showed the opposite trend at the two levels (Figure S5b, Tables S1 and S3). This suggests that these core genes play constant role at both RNA and protein levels in petioles development. Besides, a total of 67 transcription factors were identified in DEGs and 3 transcription factors were identified in DEPs. Only AP2/ERF and B3 domain-containing transcription factor RAV1-like both identified in DEGs and DEPs (Table S5). Combining MapMan and GO annotation results, 11 non-redundant genes out of 135 core genes were predicted as being involved in cell wall biosynthesis, degradation and assembly (Table 1). GO and KEGG (Kyoto Encyclopedia of Genes and Genomes) pathway enrichment analysis of the 135 core genes showed that majority genes were enriched in phenylpropanoid biosynthesis and chitin catabolic process (Figure S5c,d).

This suggested that these genes may play an important and sustained role in lotus petioles formation. However, the few core genes identified both in transcriptional and translational levels may be due to the fact that the genes involved in lotus petioles formation do not express synchronously. As reported by Casas-Vila et al. in *Drosophila melanogaster*, they found unusual behavior of RNA and protein during embryogenesis [32]. This indicates a strong post-translational regulation and the necessity of joint analysis of the transcriptome and proteome in biology.

Table 1. DEPs and DEGs involved in cell wall biosynthesis.

ID	Name	Log2(IVP/IFP)	Log2(IVP/IFP)
cell wall precursor synthesis			
NNU_03659-RA	UDP-glucose 6-dehydrogenase 4-like	0.9	Null
NNU_04520-RA	UDP-glucose 6-dehydrogenase 1	0.7	Null
NNU_07386-RA	UDP-glucose 6-dehydrogenase 1-like	0.9	Null
[a] NNU_08832-RA	probable rhamnose biosynthetic enzyme 1	1.2	1.3
NNU_10172-RA	probable rhamnose biosynthetic enzyme 1	0.8	Null
NNU_12302-RA	UDP-glucuronic acid decarboxylase 6	0.8	Null
NNU_15122-RA	UDP-glucuronic acid decarboxylase 6	1.2	Null
NNU_25253-RA	hypothetical protein OsI_05369	0.6	Null
NNU_26559-RA	probable rhamnose biosynthetic enzyme 1	1.0	Null
NNU_00650-RA	probable arabinose 5-phosphate isomerase	Null	1.4
NNU_21054-RA	bifunctional UDP-glucose 4-epimerase and UDP-xylose 4-epimerase 1	Null	1.9
cell wall proteins			
NNU_01080-RA	glucomannan 4-beta-mannosyltransferase 2-like	0.9	Null
NNU_25605-RA	fasciclin-like arabinogalactan protein 17	0.7	Null
[a] NNU_11213-RA	fasciclin-like arabinogalactan protein 13	−1.6	−3.2
NNU_12269-RA	fasciclin-like arabinogalactan protein 4	Null	1.4
NNU_15965-RA	fasciclin-like arabinogalactan protein 7	Null	1.8
NNU_06301-RA	leucine-rich repeat extensin-like protein 6	Null	−8.3
NNU_16861-RA	leucine-rich repeat extensin-like protein 4	Null	3.0
NNU_24457-RA	leucine-rich repeat extensin-like protein 4	Null	3.0
NNU_25277-RA	glucomannan 4-beta-mannosyltransferase 9	Null	1.7
cell wall degradation			
NNU_05055-RA	probable polygalacturonase	0.9	Null
NNU_11467-RA	hypothetical protein PHAVU_009G016100 g	2.2	Null
NNU_11761-RA	probable polygalacturonase	1.3	Null
NNU_23253-RA	probable pectate lyase 18	0.7	Null
NNU_23813-RA	GDSL esterase/lipase At5g14450 isoform X3	0.9	Null
NNU_00300-RA	probable polygalacturonase isoform X1	−0.7	Null
NNU_05224-RA	probable rhamnogalacturonate lyase B isoform X1	−0.8	Null
[a] NNU_10867-RA	alpha-L-arabinofuranosidase 1-like	−1.2	−2.1
NNU_11580-RA	polygalacturonase inhibitor-like	−0.9	Null
NNU_13918-RA	putative beta-D-xylosidase	−0.9	Null
NNU_22026-RA	probable pectate lyase 18	−2.9	Null
NNU_07943-RA	probable polygalacturonase	Null	−2.3
NNU_11529-RA	probable polygalacturonase isoform X2	Null	−1.9
NNU_11581-RA	polygalacturonase inhibitor-like	Null	−9.0
NNU_18600-RA	lysosomal beta glucosidase-like isoform X4	Null	−1.0
NNU_19090-RA	polygalacturonase inhibitor-like	Null	−2.8
NNU_13976-RA	probable polygalacturonase	Null	1.3
NNU_23205-RA	probable polygalacturonase isoform X1	Null	1.3
NNU_23792-RA	probable polygalacturonase non-catalytic subunit JP650	Null	1.8
NNU_26608-RA	alpha-L-fucosidase 1-like	Null	1.1
cell wall modification			
[a] NNU_15029-RA	xyloglucan endotransglucosylase/hydrolase protein 22-like	1.1	1.9
[a] NNU_16408-RA	probable xyloglucan endotransglucosylase/hydrolase protein 8	0.9	1.6
NNU_17562-RA	expansin-A13-like	0.7	Null
NNU_25629-RA	probable xyloglucan endotransglucosylase/hydrolase protein 6	1.0	Null
NNU_12232-RA	expansin-A8-like	−1.1	Null
NNU_12958-RA	expansin-A4	−0.8	Null
NNU_24404-RA	expansin-A8-like	−1.1	Null
NNU_24832-RA	probable xyloglucan endotransglucosylase/hydrolase protein 23	−0.8	Null
NNU_20658-RA	pectinesterase-like	0.9	Null
[a] NNU_01721-RA	probable pectinesterase/pectinesterase inhibitor 51	−1.1	−1.6
NNU_05006-RA	pectinesterase	−1.9	Null
NNU_05086-RA	pectinesterase	−1.9	Null
NNU_08272-RA	protein notum homolog	−1.1	Null
NNU_11705-RA	pectinesterase-like	−1.5	Null
NNU_15192-RA	pectinesterase-like	−0.8	Null
NNU_18238-RA	pectinesterase 2-like	−0.7	Null

Table 1. *Cont.*

ID	Name	Log2(IVP/IFP)	Log2(IVP/IFP)
NNU_05158-RA	putative expansin-A17 isoform X1	Null	−8.6
NNU_05160-RA	putative expansin-A17	Null	−4.6
NNU_23652-RA	expansin-A15-like	Null	−2.9
NNU_15361-RA	brassinosteroid-regulated protein BRU1-like	Null	1.6
NNU_16495-RA	xyloglucan endotransglucosylase/hydrolase protein 9	Null	2.3
NNU_24956-RA	probable xyloglucan endotransglucosylase/hydrolase protein 33	Null	4.4
NNU_12324-RA	probable pectinesterase/pectinesterase inhibitor 41	Null	−2.6
NNU_14002-RA	pectinesterase-like	Null	−1.7
NNU_14557-RA	pectinesterase 2-like	Null	−5.6
NNU_24388-RA	probable pectinesterase/pectinesterase inhibitor 41	Null	−1.4
NNU_05007-RA	pectinesterase/pectinesterase inhibitor PPE8B	Null	2.3
NNU_18245-RA	probable pectinesterase/pectinesterase inhibitor 41	Null	3.3
NNU_18519-RA	L-ascorbate oxidase homolog	Null	1.7
cellulose synthase			
NNU_09561-RA	probable cellulose synthase A catalytic subunit 5	2.4	Null
NNU_12044-RA	cellulose synthase A catalytic subunit 7	1.3	Null
NNU_21632-RA	probable cellulose synthase A catalytic subunit 1	1.2	Null
NNU_07451-RA	protein COBRA-like isoform X1	1.4	Null
NNU_07455-RA	COBRA-like protein 4	1.3	Null
NNU_13962-RA	COBRA-like protein 7	1.3	Null
NNU_21801-RA	endoglucanase 9-like	1.1	Null
NNU_18140-RA	cellulose synthase-like protein G3	Null	1.5
NNU_20039-RA	cellulose synthase A catalytic subunit 2	Null	1.8
NNU_10059-RA	endoglucanase 12-like	Null	−5.2
NNU_07071-RA	xyloglucan glycosyltransferase 4 isoform X1	Null	3.4
hemicellulose synthesis			
NNU_01719-RA	putative UDP-glucuronate:xylan alpha-glucuronosyltransferase 3	1.2	Null
[a] NNU_10542-RA	xyloglucan galactosyltransferase KATAMARI1-like	0.7	−1.7
NNU_12476-RA	probable beta-1,4-xylosyltransferase IRX9 isoform X1	2.0	Null
NNU_13626-RA	probable beta-1,4-xylosyltransferase IRX14H	1.5	Null
lignin biosynthesis			
NNU 03759-RA	laccase-4-like	0.7	Null
NNU_03827-RA	phenylalanine ammonia-lyase	1.5	Null
NNU_04966-RA	caffeic acid 3-*O*-methyltransferase 1	1.6	Null
NNU_06036-RA	isoflavone reductase homolog	0.6	Null
NNU_12048-RA	shikimate *O*-hydroxycinnamoyltransferase-like	0.9	Null
NNU_12868-RA	phenylalanine ammonia-lyase-like	1.1	Null
NNU_13598-RA	caffeoyl-CoA *O*-methyltransferase-like	1.0	Null
NNU_14758-RA	4-coumarate–CoA ligase 2-like	1.0	Null
NNU_17055-RA	caffeoyl-CoA *O*-methyltransferase	1.5	Null
NNU_18746-RA	laccase-17-like	0.6	Null
NNU_19318-RA	cinnamoyl-CoA reductase 1-like	1.5	Null
NNU_21321-RA	phenylalanine ammonia-lyase	1.4	Null
NNU_23025-RA	cytochrome P450 98A2	0.8	Null
NNU_23365-RA	cytochrome P450 84A1-like	1.6	Null
[a] NNU_23877-RA	probable cinnamyl alcohol dehydrogenase 6	0.8	3.6
NNU_24517-RA	cinnamoyl-CoA reductase 2	−0.8	Null
NNU_04106-RA	phenylalanine ammonia-lyase-like	Null	−2.1
NNU_05129-RA	phenylalanine ammonia-lyase-like	Null	−3.6
NNU_07568-RA	cinnamoyl-CoA reductase 2 isoform X2	Null	−1.4
NNU_11085-RA	laccase-17-like	Null	−5.6
NNU_12126-RA	cinnamoyl-CoA reductase 1-like	Null	−5.0
NNU_18647-RA	laccase-7-like	Null	−7.3
NNU_08076-RA	cytochrome P450 84A1-like	Null	1.8
NNU_22838-RA	cinnamoyl-CoA reductase 2-like	Null	2.1
major CHO metabolism			
NNU_02421-RA	maltose excess protein 1-like, chloroplastic	0.6	Null
NNU_04943-RA	alkaline/neutral invertase CINV2	1.0	Null
[a] NNU_17880-RA	probable fructokinase-1	1.2	1.6
[a] NNU_18248-RA	beta-fructofuranosidase, soluble isoenzyme I-like	0.9	2.3
NNU_19077-RA	sucrose synthase	1.0	Null
NNU_04529-RA	alpha-1,4 glucan phosphorylase L-2 isozyme, chloroplastic/amyloplastic	−0.8	Null
NNU_05767-RA	sucrose synthase 2-like	−1.1	Null
[a] NNU_11941-RA	beta-fructofuranosidase, insoluble isoenzyme CWINV3-like isoform X1	−0.9	−2.7
NNU_13572-RA	pentatricopeptide repeat-containing protein At2g04860	−1.7	Null
NNU_18912-RA	phosphoglucan phosphatase DSP4, amyloplastic-like	−0.7	Null
NNU_08846-RA	probable fructokinase-7	Null	1.8
NNU_09096-RA	alkaline/neutral invertase CINV2	Null	1.3

Superscript a indicates 11 core genes in DEPs (differentially expressed proteins) and DEGs (differentially expressed genes). Null means proteins or transcripts were not found.

2.6. Functional Enrichment of Quantified DEGs and DEPs

Excepting the 135 core genes, the function of other DEGs and DEPs was analyzed by MapMan software, GO and KEGG annotation tools. The DEPs were classified into 4 clusters according to the fold changes. An abundance of 164 proteins in IFP was 2-fold higher than IVP (Q1), while the abundance of 257 proteins in IFP was 1.5–2-fold higher than IVP (Q2). In addition, the abundance of 346 proteins in IVP was 1.5–2-fold higher than IFP (Q3), while the abundance of 137 proteins in IVP was 2-fold higher than IFP (Q4) (Figure S3b). Functional enrichment including GO enrichment, protein domain enrichment, and KEGG pathway enrichment of the DEPs in four clusters was analyzed. Among these 164 proteins in cluster Q1, 13 non-redundant proteins were enriched in cell wall formation and degradation. Among these 137 proteins in cluster Q4, 7 non-redundant proteins were enriched in cell wall biogenesis, cell wall organization and cell wall assembly. In cluster Q4, a total of 9 non-redundant proteins were enriched in polysaccharide including cellulose and hemicellulose biosynthetic process. Nonetheless, in cluster Q2 and Q3, no protein was enriched in cell wall formation and degradation related biologic processes (Figure 6a). This protein functional enrichment analysis coincides with the anatomic results and it indicates that cell wall biosynthesis in IFP and IVP may be different.

Generally, the GO enrichment analysis of these DEGs were shown in (Figure S6a,b). The DEGs were analyzed in upregulated and downregulated gene clusters. It showed that a total of 219, 141 and 87 upregulated genes in IVP were enriched in anatomic structure development, anatomic structure morphogenesis and cell growth. The principal point is that a total of 92 upregulated genes in IVP were enriched in cell wall organization or biogenesis. However, the upregulated genes in IFP were not enriched in pathways related to cell wall organization or biogenesis (Figure S2c). In MapMan analysis, a total of 46 non-redundant DEGs and 63 non-redundant DEPs were identified in cell wall biosynthesis and degradation pathway. Moreover, a total of 10 DEGs and 16 DEPs were identified in lignin biosynthesis. More DEPs than DEGs were identified in cell wall biosynthesis and lignin biosynthesis pathway may indicate proteins contribute more to the difference among samples.

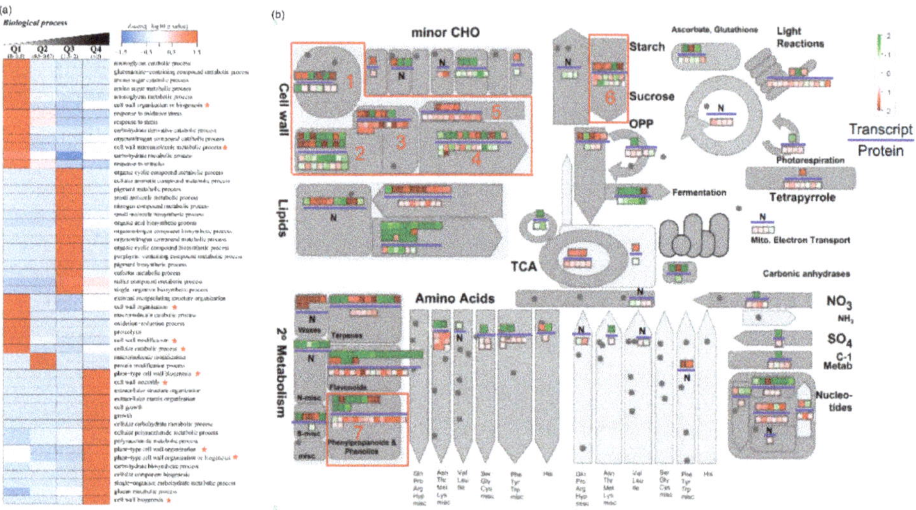

Figure 6. GO and MapMan analysis the function of DEGs and DEPs in lotus. (**a**) Function enrichment analysis of proteins with significant changes in abundance between IFP and IVP. Abundance of 164 proteins in IFP was 2-fold higher than IVP (Q1) and abundance of 257 proteins in IFP was 1.5–2-fold higher than IVP (Q2). Also, abundance of 346 proteins in IVP was 1.5–2-fold higher than IFP (Q3) and

abundance of 137 proteins in IVP was 2-fold higher than IFP (Q4). Red star indicates cell wall related process; (**b**) MapMan analysis shows that DEGs and DEPs in different metabolic pathways in lotus petioles. Red squares indicate enhancement and green squares were the opposite. Squares above the blue line represent the transcripts and squares below the blue line represent the proteins.N means not found. Arabic numerals 1 to 7 in red frame represent cell wall proteins, pectin esterases, cellulose synthesis, cell wall degradation, cell wall precursor synthesis, major CHO metabolism and lignin biosynthesis pathways.

2.7. Polysaccharide and Lignin Biosynthesis Pathway Were Highly Activated in IVP Compared to IFP

Plant cell walls are composed of polysaccharides, including cellulose, hemicellulose and pectin, lignin, proteins and minerals. Here numerous genes involved in cellulose, hemicellulose, and pectin biosynthesis pathways were identified with a changed expression both at transcriptional and protein level (Table 1). In the DEGs and DEPs identified in this study, 12 genes (3 transcripts and 9 proteins) involved in cell wall precursor synthesis were identified, and all the 12 genes showed expression increment in IVP. A total of 18 genes (6 transcripts and 12 proteins) involved in cellulose and hemicellulose synthesis were identified, and 16 of them showed expression increment in IVP except two transcripts. Moreover, a total of 26 genes (14 transcripts and 12 proteins) involved in cell wall biosynthesis and modifications were identified, and 15 of them showed expression increment in IVP. Nonetheless, a total of 24 genes (11 transcripts and 13 proteins) involved in cell wall degradation were identified and 13 of them showed expression decline in IVP (Table 1, Figure 6b). It indicated that more genes involved in cell wall precursor synthesis, cellulose and hemicellulose synthesis expressed highly in IVP may play foremost roles in its rigidity formation. Cell wall biosynthesis related genes like cellulose synthases (CESA), UDP-glucose 6-dehydrogenases (UGD), UDP-glucuronic acid decarboxylases (UXS), irregular xylem (IRX), and sucrose synthases (SUSY) showed significant higher expression level in IVP than that in IFP (Figure S6c).

In the KEGG pathway enrichment analysis, a total of 114 proteins were enriched in 6 different pathways such as starch and sucrose metabolism, amino sugar and nucleotide sugar metabolism, glycosphingolipid biosynthesis, phenylpropanoid biosynthesis, flavonoid biosynthesis and biosynthesis of amino acids (Table 2). Remarkably, 14 proteins highly expressed in IVP and 29 proteins highly expressed in IFP were enriched in phenylpropanoid biosynthesis. Furthermore, the abundance of 10 enzymes including phenylalanine ammonia-lyases (PAL, 3 PAL proteins appear in DEPs), cinnamate 4-hydroxylase (C4H, 1 C4H protein appears in DEPs), 4-(hydroxy)cinnamoyl CoA ligase (4CL, 1 4CL protein appears in DEPs), cinnamoyl CoA reductase (CCR, 1 CCR protein appears in DEPs), cinnamyl alcohol dehydrogenase/sinapyl alcohol dehydrogenase (CAD/SAD, 1 CAD protein appears in DEPs), caffeic acid/5-hydroxyferulic acid *O*-methyltransferase (COMT, 1 COMT protein appears in DEPs), ferulate 5-hydroxylase (F5H, 1 F5H protein appears in DEPs) and hydroxycinnamoyl CoA: shikimate hydroxycinnamoyltransferase/hydroxycinnamoyl CoA: quinate hydroxycinnamoyltransferase (CST/CQT, 1 CST protein appears in DEPs), which are enzymes activated on the upstream of lignin biosynthesis, were higher in IVP than that in IFP. Peroxidase is considered as the last enzyme to catalyze substrate in lignin biosynthesis. In this study, two of peroxidases were highly expressed in IVP compared to IFP and 25 of them were on the contrary. Moreover, the abundance of beta-glucosidase (BGLU, 3 BGLU proteins appear in DEPs) and aldehyde dehydrogenase (REF1, 1 REF protein appears in DEPs), which consume substrates without lignin output in lignin biosynthesis, were highly expressed in IFP (Figure 7). Among 43 identified proteins exhibiting significant changes in lignin biosynthesis pathway between IFP and IVP, 17 of them were selected for qRT-PCR to check their RNA expression. All the seventeen examined genes showed higher mRNA abundance in IVP than that in IFP except gene NNU_03827-RA. Tendencies of transcription level in genes were coincided well to their translation level excluding gene NNU_03827-RA, NNU_19318-RA and NNU_03385-RA (Figure 7). These results indicated that the genes regulating lignin biosynthesis pathway in petioles are regulated similarly at the transcription and translation level, and more genes in lignin biosynthesis pathway were more regulated in IVP compared to IFP.

To evaluate the consequence of genes in lignin biosynthesis pathway being more regulated at transcription and translation level in IVP, metabolites of lignin biosynthesis pathway were quantified in the two petioles types. As shown in (Figure 7, Table S6), a total of thirteen detected metabolites were quantified in IFP and IVP. Among these metabolites, nine of them exhibited significant changes in abundance between IFP and IVP. The changes abundance of metabolites between IFP and IVP was not exactly coinciding with genes expression, but the precursor abundance of syringyl lignin monomers and hydroxyphenyl lignin monomer were significantly higher in IVP compared to IFP. Considering the higher lignin content in IVP than IFP (Figure 2), the abundance of syringyl lignin and hydroxyphenyl lignin in IVP could be higher than in IFP, and the abundance of guaiacyl lignin may not change.

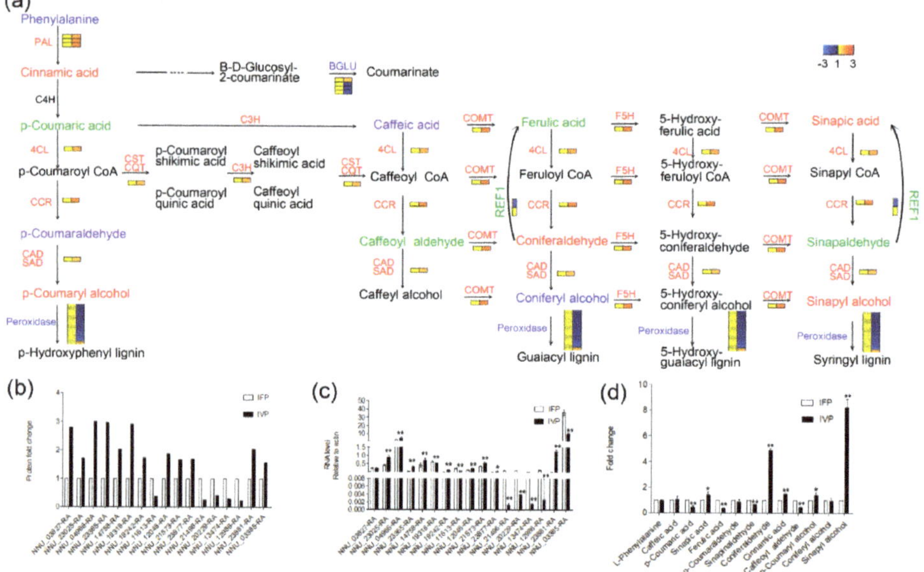

Figure 7. Overview of lignin biosynthesis pathway on transcription level, translation level and metabolite level. (**a**) Substrate, product and enzymes in whole pathway are written in red, green, blue and black color. Red color means abundance of substrate, product or enzymes were higher in IVP than IFP, green color means abundance of substrate, product or enzymes were higher in IFP than IVP and blue color means abundance of substrate and product were similar in two petioles or enzymes which catalyze same action show opposite abundance between IVP and IFP. Color bar shows changes in abundance of enzymes which were quantified in proteomics; (**b,c**) abundance of selected enzymes in lignin biosynthesis pathway and mRNA expression level of genes encoding selected enzymes in lignin biosynthesis pathway; (**d**) abundance of metabolites in lignin biosynthesis pathway. Data are means ± SD from 3 independent biologic repeats. Asterisks and double asterisks indicate significant changes compared to control as assessed according to Student's *t*-test (* *p* < 0.05 and ** *p* < 0.01).

Table 2. Proteins involved in six biologic processes were significantly enriched.

Protein Accession	Protein Name	VvsF Ratio	Regulated Type	*p*-Value
	Biosynthesis of amino acids			
NNU_09156-RA	S-adenosylmethionine synthase 5	2.5	Up	0
NNU_25903-RA	S-adenosylmethionine synthase 2	2.3	Up	0
NNU_07815-RA	5-methyltetrahydropteroyltriglutamate–homocysteine methyltransferase 2-like	2.3	Up	0.001
NNU_06116-RA	glutamine synthetase leaf isozyme, chloroplastic	2.2	Up	0
NNU_15496-RA	pyrroline-5-carboxylate reductase isoform X1	2.2	Up	0.002
NNU_16927-RA	S-adenosylmethionine synthase 1	1.9	Up	0
NNU_22690-RA	phospho-2-dehydro-3-deoxyheptonate aldolase 1, chloroplastic-like	1.9	Up	0
NNU_02525-RA	glutamate synthase 1	1.8	Up	0
NNU_16800-RA	serine acetyltransferase 5-like	1.8	Up	0.003
NNU_18211-RA	3-phosphoshikimate 1-carboxyvinyltransferase 2	1.7	Up	0.001
NNU_21805-RA	5-methyltetrahydropteroyltriglutamate–homocysteine methyltransferase	1.7	Up	0
NNU_18019-RA	serine hydroxymethyltransferase 4	1.7	Up	0
NNU_01496-RA	glutamate synthase 1	1.7	Up	0.031
NNU_13370-RA	phospho-2-dehydro-3-deoxyheptonate aldolase 2, chloroplastic-like	1.7	Up	0
NNU_04572-RA	chorismate mutase 3, chloroplastic-like	1.7	Up	0
NNU_16636-RA	2,3-bisphosphoglycerate-independent phosphoglycerate mutase	1.6	Up	0
NNU_17273-RA	probable fructose-bisphosphate aldolase 3, chloroplastic	1.6	Up	0
NNU_18207-RA	transketolase, chloroplastic	1.6	Up	0
NNU_20134-RA	shikimate kinase, chloroplastic isoform X1	1.6	Up	0
NNU_26622-RA	indole-3-glycerol phosphate synthase, chloroplastic-like isoform X1	1.5	Up	0.015
NNU_01941-RA	aspartokinase 2, chloroplastic-like isoform X1	1.5	Up	0
NNU_14707-RA	D-3-phosphoglycerate dehydrogenase 1, chloroplastic-like	1.5	Up	0
NNU_13158-RA	chorismate synthase, chloroplastic isoform X1	1.5	Up	0.015
NNU_20724-RA	glutamine synthetase cytosolic isozyme 1	1.5	Up	0
NNU_21817-RA	phosphoserine aminotransferase 1, chloroplastic-like	1.5	Up	0
	Flavonoid biosynthesis			
NNU_16100-RA	naringenin, 2-oxoglutarate 3-dioxygenase-like	4.4	Up	0
NNU_24753-RA	flavonol synthase/flavanone 3-hydroxylase-like	2.7	Up	0.002
NNU_19543-RA	flavonol synthase/flavanone 3-hydroxylase	2.0	Up	0
NNU_12048-RA	shikimate O-hydroxycinnamoyltransferase-like	1.9	Up	0
NNU_04498-RA	flavonoid 3′-monooxygenase-like	1.7	Up	0
NNU_23025-RA	cytochrome P450 98A2	1.7	Up	0
NNU_08856-RA	leucoanthocyanidin dioxygenase-like	1.6	Up	0

Table 2. *Cont.*

Protein Accession	Protein Name	VvsF Ratio	Regulated Type	*p*-Value
	Phenylpropanoid biosynthesis			
NNU_04966-RA	caffeic acid 3-*O*-methyltransferase 1	3.0	Up	0
NNU_23365-RA	cytochrome P450 84A1-like	3.0	Up	0
NNU_19318-RA	cinnamoyl-CoA reductase 1-like	2.9	Up	0.001
NNU_03827-RA	phenylalanine ammonia–lyase	2.8	Up	0
NNU_21321-RA	phenylalanine ammonia–lyase	2.6	Up	0.001
NNU_12868-RA	phenylalanine ammonia–lyase-like	2.2	Up	0.001
NNU_23881-RA	peroxidase 64	2.0	Up	0
NNU_14758-RA	4-coumarate–CoA ligase 2-like	2.0	Up	0
NNU_12048-RA	shikimate *O*-hydroxycinnamoyltransferase-like	1.9	Up	0
NNU_23025-RA	cytochrome P450 98A2	1.7	Up	0
NNU_23877-RA	probable cinnamyl alcohol dehydrogenase 6	1.7	Up	0.001
NNU_21573-RA	caffeoylshikimate esterase	1.7	Up	0
NNU_03385-RA	peroxidase 42-like	1.6	Up	0.006
NNU_13717-RA	peroxidase 73-like	0.6	Down	0.001
NNU_16422-RA	peroxidase 27-like	0.6	Down	0
NNU_04050-RA	peroxidase P7-like	0.6	Down	0
NNU_20058-RA	peroxidase 12-like	0.6	Down	0
NNU_18799-RA	peroxidase 4-like	0.6	Down	0
NNU_23337-RA	cationic peroxidase 1-like	0.6	Down	0
NNU_04265-RA	peroxidase 27-like	0.6	Down	0.012
NNU_04268-RA	peroxidase 3-like	0.5	Down	0
NNU_22685-RA	peroxidase 17-like	0.5	Down	0
NNU_13190-RA	peroxidase 21-like	0.5	Down	0
NNU_20096-RA	cationic peroxidase 1-like	0.5	Down	0
NNU_02934-RA	peroxidase N-like	0.5	Down	0
NNU_13360-RA	peroxidase 17-like	0.5	Down	0.003
NNU_24553-RA	peroxidase 47-like	0.5	Down	0.001
NNU_12989-RA	peroxidase 2-like	0.5	Down	0
NNU_06410-RA	peroxidase 12-like	0.5	Down	0
NNU_20132-RA	peroxidase 3-like	0.5	Down	0.004
NNU_01736-RA	peroxidase 43-like isoform X1	0.4	Down	0
NNU_20229-RA	peroxidase N1-like	0.4	Down	0

Table 2. *Cont.*

Protein Accession	Protein Name	VvsF Ratio	Regulated Type	*p*-Value
NNU_02064-RA	peroxidase 51	0.4	Down	0
NNU_11196-RA	peroxidase P7-like	0.4	Down	0
NNU_00369-RA	peroxidase 57-like	0.3	Down	0
NNU_13474-RA	peroxidase 10	0.3	Down	0
NNU_21496-RA	aldehyde dehydrogenase family 2 member C4-like	0.3	Down	0
NNU_04048-RA	peroxidase P7-like	0.2	Down	0.003
NNU_12986-RA	peroxidase P7-like	0.2	Down	0
Amino sugar and nucleotide sugar metabolism				
NNU_05331-RA	glucose-1-phosphate adenylyltransferase large subunit 1-like	0.7	Down	0.012
NNU_12353-RA	beta-hexosaminidase 1 isoform X1	0.6	Down	0.001
NNU_09609-RA	basic endochitinase-like	0.6	Down	0
NNU_10055-RA	beta-hexosaminidase 3-like	0.6	Down	0
NNU_06174-RA	glucose-1-phosphate adenylyltransferase large subunit 3, chloroplastic/amyloplastic	0.6	Down	0
NNU_10728-RA	acidic endochitinase-like	0.5	Down	0
NNU_12150-RA	G-type lectin S-receptor-like serine/threonine–protein kinase At1g11300	0.5	Down	0
NNU_17908-RA	endochitinase PR4-like	0.5	Down	0
NNU_17907-RA	endochitinase PR4-like	0.4	Down	0
NNU_10867-RA	alpha-ʟ-arabinofuranosidase 1-like	0.4	Down	0
NNU_20770-RA	acidic endochitinase-like	0.4	Down	0
NNU_14306-RA	basic endochitinase-like	0.4	Down	0
NNU_22938-RA	chitinase 5-like, partial	0.3	Down	0
NNU_11632-RA	acidic endochitinase-like	0.3	Down	0
NNU_09610-RA	endochitinase A-like	0.3	Down	0
NNU_17910-RA	endochitinase PR4-like	0.3	Down	0
NNU_12151-RA	acidic mammalian chitinase-like	0.3	Down	0
NNU_22939-RA	chitinase 5-like	0.2	Down	0
NNU_24291-RA	acidic endochitinase-like, partial	0.2	Down	0
Starch and sucrose metabolism				
NNU_24463-RA	glucan endo-1,3-beta-glucosidase 6-like	0.7	Down	0.001
NNU_05331-RA	glucose-1-phosphate adenylyltransferase large subunit 1-like	0.7	Down	0.012
NNU_26480-RA	probable sucrose-phosphate synthase 1 isoform X1	0.6	Down	0

Table 2. *Cont.*

Protein Accession	Protein Name	VvsF Ratio	Regulated Type	*p*-Value
NNU_04529-RA	alpha-1,4 glucan phosphorylase L-2 isozyme, chloroplastic/amyloplastic	0.6	Down	0
NNU_06174-RA	glucose-1-phosphate adenylyltransferase large subunit 3, chloroplastic/amyloplastic	0.6	Down	0
NNU_11941-RA	beta-fructofuranosidase, insoluble isoenzyme CWINV3-like isoform X1	0.5	Down	0.001
NNU_07396-RA	beta-glucosidase 40-like	0.5	Down	0
NNU_11617-RA	beta-glucosidase 12-like	0.5	Down	0
NNU_05767-RA	sucrose synthase 2-like	0.5	Down	0
NNU_11613-RA	beta-glucosidase 12-like	0.4	Down	0
NNU_13572-RA	pentatricopeptide repeat-containing protein At2g04860	0.3	Down	0
	Glycosphingolipid biosynthesis			
NNU_12353-RA	beta-hexosaminidase 1 isoform X1	0.6	Down	0.001
NNU_10055-RA	beta-hexosaminidase 3-like	0.6	Down	0
NNU_08403-RA	alpha-galactosidase-like	0.6	Down	0
NNU_24417-RA	alpha-galactosidase isoform X1	0.5	Down	0

3. Discussion

The molecular mechanism underlying petiole rigidity formation in lotus is important, but far from being fully clarified. Genome sequencing and analysis promote molecular biology research [33]. High throughput sequencing was commonly used in lotus to analyze rhizome formation, flowering time, SNPs and alternative splicing finding and evolution [34,35]. In the present study, gene expression at RNA and protein levels in two petioles types was studied using RNA sequencing and TMT techniques. Based on the integrative analysis of anatomic, transcriptomic and proteomic data, certain genes involved in polysaccharide and lignin biosynthesis pathway were deduced as potentially playing important role in petioles rigidity formation both in transcription and translation level. However, more work needs to be done to explore key genes involved in lotus petioles rigidity formation in future.

3.1. Genes Involved in Lotus Petioles Formation Exhibit No Synchrony at Transcriptional and Translation Level

In this study, approximately 75.36% and 18.19% annotated genes in lotus genome have found their transcripts and protein evidence in petioles (Figure 4a). A total of 93.12% proteins had their respective transcripts evidenced in transcriptome data and this ratio is similar to that in other model plant tissue such as *Arabidopsis* root and rice grain [26,30]. Undetected transcripts and proteins may be due to the fact that genes have different spatiotemporal expression characters or the limitation of RNA and protein extraction as well as the detection techniques. Moreover, a total of 335 proteins from proteomics data had no relevant transcripts in transcriptomics data. It was speculated that these genes may have a brief transcription time while their proteins have relatively longer life span in petioles. Similar to a previous study, our results also pointed out the un-synchronous in transcriptional and translation of genes in lotus [32]. The poor correlation of total identified transcripts and proteins suggest that a proportion of transcripts may not be translated into proteins and the higher correlation of DEGs and their corresponding DEPs suggest these genes have similar function in petioles development (Figure 5a,d). Inconsistency of genes in mRNA and protein level was similar with other previous studies [30–32]. Furthermore, it suggests that the transcriptomics and proteomics analysis are strategically complementary and have equal importance in lotus petioles development analysis [28,29].

3.2. Anatomic Evaluation of Lotus Petioles Indicated Cell Wall Thickness and Cell Differentiation Affect Petiole Rigidity

In spite of some anatomic evaluations being presented on Sacred lotus, no research has specifically explored the difference of the two types of petioles. In this present study, we found that IFP and IVP have similar anatomic structure. However, there are some differences between IFP and IVP. Usually, IVP had more collenchyma tissue, more vascular bundles and thicker xylem vessel cell wall (Figure 2), which enable the plant has more strength to keep its architecture [36]. Genes involved in cell wall polysaccharides biosynthesis may play key roles in its anatomic structure formation. In this study, 5 cellulose synthases, which plays important roles in cell wall formation [37,38], showed higher abundance in IVP than that in IFP. Specifically, there were more layers of collenchyma cell in IVP than that in IFP to support petioles stand upright. The vascular bundles and the thicker xylem vessel cell wall are the usual structure in stem or petioles of *Arabidopsis* [39], rice [40], Populus [41,42], as well as bamboo [43] to provide mechanical support. The more abundance of these structures in IVP indicate that more mechanical strength is generated. Additionally, the different amount of cellulose and lignin and the diverse structure of the cell wall of IVP and IFP may suggest that the cell wall structure could be the main reason for different petioles observed in lotus.

3.3. Abundance of Lignin Biosynthesis Related Proteins Significantly Higher in IVP Compared to IFP Supports Contribution of Lignin Biosynthesis in Petiole Rigidity

It has been established that a constant energy supply is a prerequisite for plant growth. Therefore, high carbohydrate metabolism in cell is necessary for growth of petioles. Despite the rapid growth, the floating leaf petiole was phenotypically weak, indicating less deposition of cell wall strengthening compounds and biologic process enrichment gives more clues for cell wall biosynthesis or organization related proteins, demonstrating that they are highly expressed in IVP.

Most importantly, after KEGG pathway enrichment analysis, numerous proteins related to cell wall polysaccharides and lignin biosynthesis were identified from 904 significantly differentially expressed proteins (Figure 6, Figure S3). Phenylpropanoid and other flavonoids were reported to respond to both abiotic and biotic stimuli [44]. Additionally, flavonoids biosynthesis and lignin biosynthesis were reported have some connections in plant [45,46]. Lignin gets deposited in the matrix of cellulose micro fibril, which strengthens the cell wall. This study showed the significantly enriched phenylpropanoid biosynthesis/lignin biosynthesis pathway and identified major enzymes involved in lignin biosynthesis (Figure 7, Table 2). Furthermore, 10 abundant enzymes that trigger lignin biosynthesis at the beginning until the formation of lignin monomer were higher in IVP than IFP, which indicates that IVP had more effective lignin deposition compared to IFP.

The transcription level of the selected enzymes genes in lignin biosynthesis pathway were checked in petioles. The results suggested that the genes involved in the regulation of lignin biosynthesis pathway appear at both transcription and translation levels exhibiting similar tendencies in both. Consequently, analyses of metabolites in lignin biosynthesis imply that more lignin was synthesized in IVP compared to IFP (Figures 2 and 7).

3.4. Abundance of Cell Wall Polysaccharides Biosynthesis Related Proteins Significantly Higher in IVP Than That in IFP

Cell wall polysaccharides biosynthesis is more complicated than lignin biosynthesis. Biologic processes relevant to cellulose and hemicellulose biosynthesis affect cell wall polysaccharides biosynthesis [47,48]. Pathways such as starch and sucrose metabolism, amino sugar and nucleotide sugar metabolism, and biosynthesis of amino acids were significantly enriched in this study (Figure 7, Table 2). Cellulose is the most abundant biopolymer deposited on the plant cell wall. Approximately 40–50% cellulose stored in secondary cell wall of wood tissue, which mainly contributes to the bodily structure [47]. Studies in *Arabidopsis*, Populus and other plants suggest that various genes involved in cellulose biosynthesis play fundamental roles in cell wall formation and thickening. A total of 10 cellulose synthases [49] mediate cellulose synthesis in both primary [50] and secondary walls in *Arabidopsis* [51,52]. Previous researches showed that some cellulose synthases in Populus participate in cell wall formation [25,53,54] and other genes including KORRIGAN1 (KOR1), sucrose synthase gene and transcription factors are also involved [42,55–57]. Specifically, the abundance of four CESAs (cellulose synthases) including NNU_09561-RA, NNU_12044-RA, NNU_21632-RA and NNU_01080-RA were significantly expressed higher in IVP compared to IFP, suggesting that more cellulose is deposited in IVP (Figures 2 and 3). Similarly, abundance of three UGD (UDP-glucose 6-dehydrogenase) and one SUSY (sucrose synthase), for instance, NNU_03659-RA, NNU_07386-RA, NNU_04520-RA and NNU_19077-RA were significantly enriched higher in IVP. These proteins have similar functions in cellulose biosynthesis as their homologous proteins in *Arabidopsis* and Populus [58,59]. Furthermore, hemicellulose biosynthesis related proteins such as HEX (hexokinase), UXS (UDP-glucuronic acid decarboxylase), IRXs (irregular xylem), GATL (galacturonosyltransferase) and UXT (UDP-xylose transporter) have been reported to be involved in cell wall hemicellulose biosynthesis [39,60–62]. In our study, several UXS (NNU_15122-RA, NNU_12302-RA, NNU_10172-RA and NNU_25253-RA), IRX (NNU_20069-RA, NNU_12476-RA, NNU_13626-RA, NNU_13619-RA, NNU_22319-RA, NNU_03759-RA, NNU_18746-RA and NNU_11544-RA) and UXT (NNU_12576-RA, NNU_24078-RA) were identified and most of them showed significantly higher abundance in IVP than in IFP, suggesting the huge contribution of hemicellulose biosynthesis in cell wall formation. The function of these proteins involved in polysaccharides biosynthesis need more experiments in vivo to confirm because the presence of glycoside hydrolases does not mean that the polysaccharides are degraded [63]. Their expression of these genes at mRNA and protein level may have more complicated relation with polysaccharides deposition in cell wall.

4. Materials and Methods

4.1. Plant Growth and Petiole Collection

The sacred lotus cultivar "Ancient Chinese lotus" was cultivated in Wuhan Botanical Garden, Chinese Academy of Sciences (N30°32′44.02″, E114°24′52.18″). There are two ecotypes of N. nucifera: temperate lotus and tropical lotus (N19°~N43° in China) [64]. "Ancient Chinese lotus" is temperate lotus. Its seed starts to germinate in April, flower blooms from June to August, leaf withers in September and October and dormant buds remains from November to March of the following year with an enlarged rhizome [65]. In middle May, the dormant buds sprout underground. In terms of leaf development, it could be divided into two stages. At floating leaf development stage, the newly developed leaf is the floating leaf, which is folded until emerging above the water surface. As the first floating leaf grows into maturity, the second leaf appears. After about several floating leaves have emergence, the succeeding leaves are vertical leaves. At the vertical leaf development stage, the newly developed vertical leaf is folded, which will grow bigger and unfold when the petiole is long enough to support leaf far above the water surface. Four kinds of petioles samples were collected at different stages. At the floating leaf development stage, initial floating leaves' petiole (IFP) was collected from floating leaf when it just floats on water surface. Mature floating leaves' petiole (MFP) was collected from floating leaf when its shape and size had stopped changing. At the vertical leaf development stage, initial vertical leaves' petiole (IVP) was collected when the folded leaf just emerges out of water surface. Mature vertical leaves' petiole (MVP) was collected when its shape and size stopped changing (Figure S1). In this study, rhizome of "Ancient Chinese lotus" was planted in concrete container (length, 2 m; width,1 m; depth, 80 cm) with loam soil is about 50 cm in depth on 5 April 2016. The containers were constructed outside under natural light conditions. Fifty grams of fertilizer (SAN AN; STANLEY, Shandong, China) was applied to container every 2 weeks during the growing season. In field, lotus usually was cultured in farmland or pond with different depth of water (20 cm to 100 cm). The two different types of petioles were always observed in different depth of water in filed. In order to exclude the effect of water depth on ability of leaf stand erect on water, lotus was cultivated in low water depth. this means the length of two types of petioles was longer than water depth.

4.2. Morphologic Observation

4.2.1. Petiole Microscopic Observation

Petiole segments (the middle section of the petiole) were cut and fixed in FAA (formaldehyde–acetic acid–alcohol solution) solution (5% glacial acetic acid, 5% formaldehyde and 70% ethanol) at room temperature for 24 h. The segments were then rinsed with tap water and dehydrated in ethanol baths ranges of 70% ethanol to 100% ethanol. After dehydration, ethanol was replaced by chloroform progressively. Then finely ground paraffin was added into chloroform and incubated it at 36 °C for 2 h. The small segments were transferred into xylene solution containing 50% and 75% paraffin and incubated at 42 °C and 50 °C for 2 h successively. After that, the small segments were transfer into pure paraffin solution at 58 °C for 1 h and this process was repeated three times. Finally, the sample were embedded using pure paraffin. Samples were cut with a rotary microtome Leica RM2265 (Leica, Benshein, Germany). Sections (10 μm) were stained using fast green and counterstained using safranin solution (Biosharp, Hefei, China) and observed under an optical microscope (Olympus BX53, Tokyo, Japan). Safranin appears red in lignified, suberized or cutinized cell walls. Fast Green (Biosharp, Hefei, China) presents green in cytoplasm and cellulosic cell walls. Petiole segments from three independent plants were collected, and at least 3 sections per plant were employed for microscopy observation.

4.2.2. Breaking Resistance Measurement

Petioles (petiole length was about 30 cm) with same diameter (5 mm) from IFP and IVP were cut into equally long stem segment (20 cm). The breaking resistance of the middle point of petiole was measured using a prostrate tester (DIK 7400, daiki rika kogyo co. ltd., Tokyo, Japan). The distance between tester fulcra was set at 10 cm (Figure 1g,h). The petiole was pressed using prostrate tester until petiole breaking point and the test readings were recorded. Breaking resistance was represented by the manual force added on petiole. Breaking resistance (F) = (test reading × 39.2 N) ÷ 40 [66,67]. Six biologic petioles of each type of petiole were tested in this experiment.

4.2.3. Transmission Electron Microscope

Petiole segment (the middle section of the petiole) was infiltrated in 4% (*v/v*) glutaraldehyde in phosphate buffered solution at 4 °C overnight (PBS, 33-mM Na_2HPO_4, 1.8-mM NaH_2PO_4 and 140-mM NaCl [pH 7.2]). Then the petiole segment was fixed in 1% (*w/v*) osmium tetroxide (Biosharp, Hefei, China) and dehydrated through a gradient of ethanol and eventually embedded in Spurr's resin (Biosharp, Hefei, China). Petiole sections (100 nm) were stained using uranyl acetate and lead citrate. Lastly, the petiole sections from three independent petioles from three biologic lotus were visualized using HT7700 transmission electron microscope (HITACHI, Tokyo, Japan) and the cell-wall thickness was measured from ten cells on each section using Hitachi TEM system microscopy software (Gatan, Pleasanton, CA, USA).

4.2.4. Determination of Crude Cellulose and Lignin

Crude cellulose was measured by acid digestion method using cellulose detective kit (Sinobestbio, Shanghai, China). The weighed samples (the middle section of the petiole) were grinded with 80% ethanol and washed with ethanol and acetone. After the collected residue was digested by concentrated sulfuric acid and the supernatant was used to detect cellulose content. Lignin content was measured according to Klason lignin methods [68]. Six biologic petioles of each type of petiole were tested in this experiment.

4.3. RNA Isolation and Illumina Sequencing

4.3.1. Total RNA Extraction

Total RNA was extracted from about 0.5 g of samples with TRIzol reagent (Invitrogen, Carlsbad, CA, USA) according to the standard protocol. RNA quality and concentration were evaluated using the 6000 Pico LabChip of the Agilent 2100 Bioanalyzer (Agilent, Santa Clara, CA, USA) and a NanoDrop spectrophotometer (Thermo Scientific, Waltham, MA, USA), respectively.

4.3.2. cDNA Synthesis and Illumina Sequencing

The cDNA was synthesized by cDNA synthesis kit (Bio-Rad, Hercules, CA, USA) according to the manufacturer's instructions. After purification, fragments were enriched by PCR amplification for cDNA library construction. Then, library products were sequenced with Illumina HiSeq™ 2000 platform (San Diego, CA, USA). Each sample had three biologic replicates from three individual plants. The raw data of this project was deposited in NCBI (www.ncbi.nlm.nih.gov/bioproject) and can be downloaded with identifier PRJNA642670.

4.3.3. Bioinformatics Analysis

RNA sequencing data processing, gene expression calculation, differentially expressed genes (DEGs) identification, gene ontology (GO) enrichment analysis and pathway enrichment were performed in BMK Cloud (Beijing, China, http://www.biocloud.net/). A false discovery rate (FDR) <1% and an absolute value of log2Ratio >1 was set as the threshold to judge the significance of gene expression difference between two petioles types. Generally, the raw RNA sequencing data were cleaned by removing adapter sequences, reads containing ploy-N and filtered with low-quality sequences (Q < 20) from raw data. All the downstream analyses were based on clean data with high quality. Clean reads were aligned to the reference genome sequence using the algorithm Tophat2 [69].

Tolerance parameters in Tophat2 were set as default with no more than two bases mismatches. Then gene function was annotated based on the following databases: Nr (NCBI non-redundant protein sequences); Nt (NCBI non-redundant nucleotide sequences); Pfam (Protein family); KOG/COG (Clusters of Orthologous Groups of proteins);Swiss-Prot (A manually annotated and reviewed protein sequence database); KO (KEGG Ortholog database); GO (Gene Ontology). Gene expression levels were estimated by FPKM (fragments per kilobase of transcript per million fragments mapped). Differential expression analysis of two groups was performed using the DESeq R package (1.10.1). DESeq provide statistical routines for determining differential expression in digital gene expression data using a model based on the negative binomial distribution. The resulting *p*-values were adjusted using the Benjamini and Hochberg's approach for controlling the false discovery rate.

4.4. Proteomics Analysis

4.4.1. Protein Extraction and Trypsin Digestion

Petioles were ground and transferred to a 5-mL centrifuge tube containing homogenizing buffer (8-M urea, 2-mM ethylenediaminetetraacetic acid (EDTA), 10-mM DL-dithiothreitol (DTT) and 1% protease inhibitor cocktail). The protein was precipitated using cold TCA-acetone solution for 2 h at −20 °C. Then, the remaining precipitate was washed with cold acetone. The protein was re-dissolved in lysis buffer (8-M urea, 100-mM triethylammonium bicarbonate (TEAB), pH 8.0) and the protein concentration was measured by BCA methods [70]. The protein solution was reduced by 10-mM DTT for 30 min at 37 °C and alkylated with 20-mM iodoacetamide (IAM) at 37 °C in darkness for 45 min. Trypsin was added with a 1:50 trypsin-to-protein mass ratio for overnight digestion. All regents used in this section were supplied by Sangon Biotech Company (Shanghai, China). Six biologic petioles of each type of petiole were used in this experiment.

4.4.2. TMT Labeling

After trypsin digestion, the peptide was desalted using ZipTip C18 column (Millipore, Bedford, MA, USA). Peptide was dissolved in 0.5-M TEAB and processed according to the manufacturer's protocol for 6-plex TMT kit (Thermo Fisher Scientific, Waltham, MA, USA). The solvable peptides were desalted using a MonoSpin C18 column (GL 217 Sciences, Tokyo, Japan) [70].

4.4.3. Quantitative Proteomic Analysis by LC-MS/MS

Peptides in each sample were separated into 18 fractions and dried by vacuum centrifugation (Labconco, Kansas, MO, USA). Then peptides were dissolved in 0.1% formic acid (FA) and loaded directly onto the reversed-phase pre-column (Acclaim PepMap 100, Thermo Scientific, Waltham, MA, USA). The peptides were separated by the reverse phase analysis column (Acclaim PepMap RSLC, Thermo Scientific, Waltham, MA, USA) according to the following conditions. The 6% to 22% solvent B (0.1% FA in 98% ACN, acetonitrile) for 22 min, 22% to 36% for 10 min, escalating to 85% in 5 min and held at 85% for the last 3 min. Flow rate was held at 300 nL/min on an EASY-nLC 1000 ultra-performance liquid chromatography system (UPLC)(Thermo Scientific, Waltham, MA, USA). Peptides were detected in the Orbitrap (Thermo Scientific, Waltham, MA, USA) at a resolution of 70,000 and instrument parameters were set as follows: Normalized collision energy (NCE) value was 30; ion fragments resolution was 17,500. The electrospray voltage was 2.0 kV. The *m/z* scan range is 350 to 1800. Fixed first mass is 100 *m/z* [70]. A data-dependent analysis method that alternated between primary MS scan followed by 20 MS/MS scans was applied for the top 20 precursor ions above a threshold ion count of 1E4 in the MS survey scan with 30.0 s dynamic exclusion.

4.4.4. Database Searching

The acquired MS/MS data were processed using Proteome Discoverer (version 2.1.0.81, Thermo Scientific, Waltham, MA, USA). Tandem mass spectra were searched against the sacred lotus genome database with mascot search engine (https://bioinformatics.psb.ugent.be/plaza/versions/plaza_v4_dicots/). The parameters were set as follows: Trypsin/P was chosen as enzyme and the maximum missing cleavages was 2. The number of modification and charge in each peptide were set up to 5. Mass error tolerance was set to 10 ppm and 0.02 Da for precursor and fragment ions, respectively. Fixed modification included carbamidomethylation on Cys and TMT-6-plex on Lys and N-term. Variable modification contained oxidation on Met and FDR (false discovery rate) threshold value 1% was applied in protein, peptide and modification site identification. Minimum peptide length was set at 7.

4.4.5. Protein Annotation, Functional Classification and Enrichment

Proteins were classified by GO annotation method. GO annotation proteome was derived from the UniProt-GOA database (http://www.ebi.ac.uk/GOA/) [71,72]. KEGG database was used in the identification of enriched pathways [71,72]. A two-tailed Fisher's exact test was employed to test the enrichment of DEPs against all identified proteins. The GO or pathway with a corrected p-value < 0.05 was considered significant.

4.5. Metabolism Analysis

4.5.1. Metabolite Extraction

Freeze-dried petiole was crushed using a mixer mill (MM 400, Retsch, Germany). One hundred milligrams of powder were weighed and stirred in 70% methanol (1.0 mL). After centrifugation, the supernatant containing extracts were absorbed (CNWBOND Carbon–GCB SPE Cartridge, 250 mg, 3 mL, Shanghai, China and filtrated before LC-MS analysis. Six biologic petioles of each type of petiole were tested in this experiment.

4.5.2. Metabolite Analysis

Metabolite profiling was carried out using a widely targeted metabolome method by Wuhan Metware Biotechnology Co., Ltd. (Wuhan, China, http://www.metware.cn/). The sample extracts were analyzed with LC-ESI-MS/MS system. The analytical conditions were as follows. Column (Waters, 1.8 μm, 2.1 mm × 100 mm) and buffer A (water containing 0.04% acetic acid) and buffer B (acetonitrile containing 0.04% acetic acid) were used. The gradient program in HPLC system (SCIEX, Redwood City, CA, USA) was set as follows: At the first 12 min, mobile phase containing buffer A 5% and buffer B 95%. In the next three min, mobile phase was changed into buffer A 95% and buffer B 5%. The flow rate was set as 0.40-mL/min and temperature was set at 40 °C. A total of 5-μL sample was injected in each run. The eluate was alternatively connected to MS (API 4500 Q TRAP LC/MS/MS System, SCIEX, Redwood City, CA, USA).

4.6. Validation of Transcriptomic and Proteomic Data Using qRT-PCR and PRM Analysis

4.6.1. qRT–PCR

Total RNA was extracted by TRIzol reagent (Invitrogen, Carlsbad, CA, USA) and 2 μg total RNA was used for reverse transcription using Rever Tra Ace-α-First Strand cDNA synthesis Kit (TOYOBO, Osaka, Japan). qRT-PCR was performed on CFX96 Real Time System (BIO-RAD, Hercules, CA, USA) with SYBR green fluorescence. The data were normalized based on relative transcript level of actin (NNU_24864). Three biologic petioles of each type of petiole were tested in this experiment.

4.6.2. Parallel Reaction Monitoring (PRM)

The digested peptides were dissolved in 0.1% formic acid and directly loaded onto a reversed-phase analytical column (15 cm length, 75 μm i.d.). The parameters of UPLC (Thermo Scientific, Waltham, MA, USA) were set as mentioned above. The parameters of MS were set as follows: The electrospray voltage was 2.0 kV. The *m/z* scan range was 350 to 1000 for full scan and the resolution was 35,000. Peptides were then selected for MS/MS using NCE setting at 27 and the resolution of fragments was 17,500. The maximum IT was set at 20 milliseconds for full MS and auto for MS/MS. The isolation window for MS/MS was set at 2.0 *m/z*. The acquired MS data were identified and quantified using Skyline software (v.3.6) [73]. MS proteomics data in this study were deposited in the ProteomeXchange Consortium (http://proteomecentral.proteomexchange.org) via the PRIDE partner repository [74]. Data are available via ProteomeXchange with identifier PXD009136.

4.7. Statistical Analysis

The data were analyzed using the Student's *t*-test to evaluate the statistical significance to compare the two groups, and one-way ANOVA test to compare multiple groups ($p < 0.05$).

5. Conclusions

In lotus growth, the floating leaves and erect rigid leaves develop subsequently. These two petioles types appear like lateral branches growing during its growing season. Due to the fact that floating leaf and vertical leaf experience similar ambient, the differentiation of petioles rigidity in lotus is not affected by environmental changes. To investigate the mechanism controlling the rigidity of lotus petiole, we performed integrated omics analysis in this study. Anatomically, more vascular bundles, collenchyma and thicker cell wall were observed in IVP than that in IFP. Many genes involved in cell wall biosynthesis and lignin biosynthesis exhibited differentially expression at both mRNA and protein levels. Gene function and pathway enrichment analysis displayed that DEGs and DEPs were significantly enriched in cell wall biosynthesis and lignin biosynthesis. Interestingly, the bulk of identified DEPs in lignin biosynthesis were up regulated in IVP, suggesting that the differences in lignin biosynthesis in the two types of lotus petioles may be responsible for the observed different rigidity. Despite our results are still far from underlying the rigidity of lotus construction, these findings afford novel clues for better understanding of the gene network involved in lotus petioles formation. In addition, roles of decisive candidate gene involved in cell wall and lignin biosynthesis in lotus should be accurately interpreted in future studies.

Supplementary Materials: The following are available online at http://www.mdpi.com/1422-0067/21/14/5087/s1, Figure S1: Phenotype of lotus live in a box and pond. Lotus was cultured in pot and field. 1, IFP; 2, MFP; 3, IVP; 4, MVP, Figure S2: Cluster analysis and gene function enrichment analysis of 1422 DEGs. (a) Pearson's correlation coefficient was used to evaluate repeatability of individual sample. (b) Cluster analysis of DEGs. F1–F3 means transcripts abundance in IFP, V1–V3 mean transcripts in IFP. (c) Gene function enrichment analysis of 1422 DEGs. Asterisks indicate genes enriched in anatomical structure and cell wall related process, Figure S3: Pathway enrichment analysis of 905 DEPs. (a) Pearson's correlation coefficient was used to evaluate repeatability of individual sample. (b)The 905 DEPs were classified into four groups. Abundance of 164 proteins in IFP was two-fold higher than IVP (Q1) and abundance of 257 proteins in IFP was 1.5–2-fold higher than IVP (Q2). Also, abundance of 346 proteins in IVP was 1.5–2-fold higher than IFP (Q3), and abundance of 137 proteins in IVP was two-fold higher than IFP (Q4); (c) KEGG pathway enrichment analysis. A total of 105 proteins were enriched in six different pathways such as starch and sucrose metabolism, amino sugar and nucleotide sugar metabolism, glycosphingolipid biosynthesis, phenylpropanoid biosynthesis, flavonoid biosynthesis and biosynthesis of amino acids, Figure S4: Validation of proteomics data using PRM. (a) Comparison of proteomics data and PRM validation; (b) Correlation of proteomics data and PRM validation; (c–f) show that the abundance of two peptides (NNU_21632-RA andNNU_03827-RA) belongs to cellulose synthase and phenylalanine ammonia–lyase, Figure S5: Function analysis of 135 core genes. (a) Gene function analysis of 135 core genes by MapMan bincodes; (b) Cluster analysis of 135 core genes. Asterisks indicate that DEGs and DEPs have the opposite expression of RNA and protein level. Dot indicates that core genes involved in cell wall biosynthesis. Red dot means DEGs and DEPs have the opposite expression of RNA and protein level, and black dots mean DEGs and DEPs have the same expression of RNA and protein level; (c,d) indicate the function and pathway enrichment analysis of 135 core genes, Figure S6: GO enrichment analysis of DEGs and DEPs. (a,b) GO enrichment analysis of DEPs and DEGs; (c) DEPs involved in pathways related to cellulose and hemicellulose biosynthesis. Red color characters mean abundance of proteins were higher in IVP than IFP, green color characters mean abundance of proteins were higher in IFP than IVP, Table S1: A total of 1422 DEGs in IFP and IVP were identified, Table S2: Expression of 22 selected genes in transcriptome and RT-PCR, Table S3: 904 out of the quantified DEPs in IFP and IVP were identified, Table S4: Protein quantified by TMT-labeling and PRM methods, Table S5: Transcription factors were identified in DEGs and DEPs, Table S6: All sample metabolomics data.

Author Contributions: Conceived and designed the experiments: P.Y. and M.L.; Performed the experiments: M.L. and I.H.; Analyzed the data: M.L., D.C. and D.H.; Wrote the study: M.L. All authors have read and agreed to the published version of the manuscript.

Funding: This research received no external funding.

Acknowledgments: The authors sincerely thank Rebecca Njeri Damaris at Hubei University and Tonny Maraga Nong'a at University of New Mexico for English writing optimization.

Conflicts of Interest: The authors declare no conflict of interest.

Abbreviations

ABA	abscisic acid
ACN	acetonitrile
CESA	cellulose synthase
DEGs	differentially expressed genes
DEPs	differentially expressed proteins
DTT	DL-dithiothreitol
EDTA	ethylenediaminetetraacetic acid
FAA	formaldehyde–acetic acid–alcohol solution
FA	formic acid
FDR	false discovery rate
GA	gibberellin
IAM	iodoacetamide
IFP	initial floating leaves' petiole
IRX	irregular xylem
IVP	initial vertical leaves' petiole
KEGG	Kyoto Gene Ontology and Encyclopedia of Genes and Genomes
MFP	mature floating leaves' petiole
MVP	mature vertical leaves' petiole
NCE	normalized collision energy
PBS	phosphate buffered solution
qRT-PCR	quantitative reverse transcription polymerase chain reaction
SUSY	sucrose synthase
TEAB	triethylammonium bicarbonate
TMT	tandem mass tag
UGD	UDP-glucose 6-dehydrogenase
UPLC	ultra-performance liquid chromatography system
UXS	UDP-glucuronic acid decarboxylase

References

1. Wang, B.; Smith, S.; Li, J. Genetic Regulation of Shoot Architecture. *Annu. Rev. Plant Biol.* **2018**, *69*, 437–468. [CrossRef] [PubMed]
2. Gao, J.; Yang, S.; Cheng, W.; Fu, Y.; Leng, J.; Yuan, X.; Jiang, N.; Ma, J.; Feng, X. GmILPA1, Encoding an APC8-like Protein, Controls Leaf Petiole Angle in Soybean. *Plant Physiol.* **2017**, *174*, 1167–1176. [CrossRef] [PubMed]
3. Peng, J.; Richards, D.E.; Hartley, N.M.; Murphy, G.P.; Devos, K.M.; Flintham, J.E.; Beales, J.; Fish, L.J.; Worland, A.J.; Pelica, F.; et al. 'Green revolution' genes encode mutant gibberellin response modulators. *Nature* **1999**, *400*, 256–261. [CrossRef] [PubMed]
4. Jo, I.S.; Han, D.U.; Cho, Y.J.; Lee, E.J. Effects of Light, Temperature, and Water Depth on Growth of a Rare Aquatic Plant, *Ranunculus kadzusensis*. *J. Plant Biol.* **2010**, *53*, 88–93. [CrossRef]
5. Cosgrove, D.J. Growth of the plant cell wall. *Nat. Rev. Mol. Cell Biol.* **2005**, *6*, 850–861. [CrossRef]
6. Zhang, R.; Li, L. Research progress of the plant cell wall signaling. *Plant Physiol. J.* **2018**, *54*, 1254–1262. [CrossRef]
7. Zhang, B.; Zhou, Y. Plant Cell Wall Formation and Regulation. *Sci. Sin. Vitae* **2015**, *45*, 544–556. [CrossRef]

8. Burton, R.A.; Gidley, M.J.; Fincher, G.B. Heterogeneity in the chemistry, structure and function of plant cell walls. *Nat. Chem. Biol.* **2010**, *6*, 724–732. [CrossRef]

9. Vanholme, R.; Cesarino, I.; Rataj, K.; Xiao, Y.; Sundin, L.; Goeminne, G.; Kim, H.; Cross, J.; Morreel, K.; Araujo, P.; et al. Caffeoyl shikimate esterase (CSE) is an enzyme in the lignin biosynthetic pathway in Arabidopsis. *Science* **2013**, *341*, 1103–1106. [CrossRef]

10. Taylor-Teeples, M.; Lin, L.; de Lucas, M.; Turco, G.; Toal, T.W.; Gaudinier, A.; Young, N.F.; Trabucco, G.M.; Veling, M.T.; Lamothe, R.; et al. An *Arabidopsis* gene regulatory network for secondary cell wall synthesis. *Nature* **2015**, *517*, 571–575. [CrossRef]

11. Wolf, S. Plant cell wall signalling and receptor-like kinases. *Biochem. J.* **2017**, *474*, 471–492. [CrossRef] [PubMed]

12. Gonneau, M.; Desprez, T.; Guillot, A.; Vernhettes, S.; Höfte, H. Catalytic Subunit Stoichiometry within the Cellulose Synthase Complex. *Plant Physiol.* **2014**, *166*, 1709–1712. [CrossRef] [PubMed]

13. Hill, J.; Hammudi, M.B.; Tien, M. The Arabidopsis Cellulose Synthase Complex: A Proposed Hexamer of CESA Trimers in an Equimolar Stoichiometry. *Plant Cell* **2014**, *26*, 4834–4842. [CrossRef] [PubMed]

14. Grantham, N.; Wurman Rodrich, J.; Terrett, O.; Lyczakowski, J.; Stott, K.; Iuga, D.; Simmons, T.; Durand-Tardif, M.; Brown, S.P.; Dupree, R.; et al. An even pattern of xylan substitution is critical for interaction with cellulose in plant cell walls. *Nat. Plants* **2017**, *3*, 859–865. [CrossRef]

15. Humphreys, J.M.; Chapple, C. Rewriting the lignin roadmap. *Curr. Opin. Plant Biol.* **2002**, *5*, 224–229. [CrossRef]

16. Yamaguchi, M.; Kubo, M.; Fukuda, H.; Demura, T. Vascular-related NAC-DOMAIN7 is involved in the differentiation of all types of xylem vessels in *Arabidopsis* roots and shoots. *Plant J.* **2008**, *55*, 652–664. [CrossRef]

17. Zhong, R.; Richardson, E.A.; Ye, Z.-H. Two NAC domain transcription factors, SND1 and NST1, function redundantly in regulation of secondary wall synthesis in fibers of *Arabidopsis*. *Planta* **2007**, *225*, 1603–1611. [CrossRef]

18. Zhong, R.; Demura, T.; Yu, L. SND1, a NAC Domain Transcription Factor, Is a Key Regulator of Secondary Wall Synthesis in Fibers of *Arabidopsis*. *Plant Cell* **2006**, *18*, 3158–3170. [CrossRef] [PubMed]

19. Kim, W.-C.; Ko, J.-H.; Han, K.-H. Identification of a cis-acting regulatory motif recognized by MYB46, a master transcriptional regulator of secondary wall biosynthesis. *Plant Mol. Biol.* **2012**, *78*, 489–501. [CrossRef]

20. Sato, M.; Tsutsumi, M.; Ohtsubo, A.; Nishii, K.; Kuwabara, A.; Nagata, T. Temperature-dependent changes of cell shape during heterophyllous leaf formation in *Ludwigia arcuata* (Onagraceae). *Planta* **2008**, *228*, 27–36. [CrossRef]

21. Jackson, M.B. Ethylene-promoted Elongation: An Adaptation to Submergence Stress. *Ann. Bot.* **2008**, *101*, 229–248. [CrossRef] [PubMed]

22. Sasidharan, R.; Keuskamp, D.H.; Kooke, R.; Voesenek, L.A.C.J.; Pierik, R. Interactions between Auxin, Microtubules and XTHsMediate Green Shade- Induced Petiole Elongation in *Arabidopsis*. *PLoS ONE* **2013**, *9*, e90587. [CrossRef]

23. Mitsuda, N.; Seki, M.; Shinozaki, K.; Ohme-Takagi, M. The NAC transcription factors NST1 and NST2 of *Arabidopsis* regulate secondary wall thickenings and are required for anther dehiscence. *Plant Cell* **2005**, *17*, 2993–3006. [CrossRef] [PubMed]

24. Lin, F.; Manisseri, C.; Fagerstrom, A.; Peck, M.L.; Vega-Sanchez, M.E.; Williams, B.; Chiniquy, D.M.; Saha, P.; Pattathil, S.; Conlin, B.; et al. Cell Wall Composition and Candidate Biosynthesis Gene Expression During Rice Development. *Plant Cell Physiol.* **2016**, *57*, 2058–2075. [CrossRef] [PubMed]

25. Xi, W.; Song, D.; Sun, J.; Shen, J.; Li, L. Formation of wood secondary cell wall may involve two type cellulose synthase complexes in *Populus*. *Plant Mol. Biol.* **2017**, *93*, 419–429. [CrossRef] [PubMed]

26. Lan, P.; Li, W.; Schmidt, W. Complementary Proteome and Transcriptome Profiling in Phosphate-Deficient *Arabidopsis* Roots Reveals Multiple Levels of Gene Regulation. *Mol. Cell. Proteom.* **2012**, *11*, 1156–1166. [CrossRef] [PubMed]

27. Moreno-Risueno, M.A.; Busch, W.; Benfey, P. Omics meet networks—Using systems approaches to infer regulatory networks in plants. *Curr. Opin. Plant Biol.* **2010**, *13*, 126–131. [CrossRef]

28. Jamet, E.; Roujol, D.; San-Clemente, H.; Irshad, M.; Soubigou-Taconnat, L.; Renou, J.P.; Pont-Lezica, R. Cell wall biogenesis of *Arabidopsis thaliana* elongating cells: Transcriptomics complements proteomics. *BMC Genom.* **2009**, *10*, 505. [CrossRef]

29. Minic, Z.; Jamet, E.; San-Clemente, H.; Pelletier, S.; Renou, J.P.; Rihouey, C.; Okinyo, D.P.; Proux, C.; Lerouge, P.; Jouanin, L. Transcriptomic analysis of Arabidopsis developing stems: A close-up on cell wall genes. *BMC Plant Biol.* **2009**, *9*, 6. [CrossRef]

30. Lin, Z.; Wang, Z.; Zhang, X.; Liu, Z.; Li, G.; Wang, S.; Ding, Y. Complementary Proteome and Transcriptome Profiling in Developing Grains of a Notched-Belly Rice Mutant Reveals Key Pathways Involved in Chalkiness Formation. *Plant Cell Physiol.* **2017**, *58*, 560–573. [CrossRef]

31. Li, J.; Ren, L.; Gao, Z.; Jiang, M.; Liu, Y.; Zhou, L.; He, Y.; Chen, H. Combined transcriptomic and proteomic analysis constructs a new model for light-induced anthocyanin biosynthesis in eggplant (*Solanum melongena* L.). *Plant Cell Environ.* **2017**, *40*, 3069–3087. [CrossRef] [PubMed]

32. Casas-Vila, N.; Bluhm, A.; Sayols, S.; Dinges, N.; Dejung, M.; Altenhein, T.; Kappei, D.; Altenhein, B.; Roignant, J.Y.; Butter, F. The developmental proteome of *Drosophila melanogaster*. *Genome Res.* **2017**, *27*, 1273–1285. [CrossRef] [PubMed]

33. Ming, R.; Vanburen, R.; Liu, Y.; Yang, M.; Han, Y.; Li, L.-T.; Zhang, Q.; Kim, M.-J.; Schatz, M.C.; Campbell, M.; et al. Genome of the long-living sacred lotus (*Nelumbo nucifera* Gaertn.). *Genome Biol.* **2013**, *14*, R41. [CrossRef] [PubMed]

34. Zhang, Y.; Nyong, A.T.; Shi, T.; Yang, P. The complexity of alternative splicing and landscape of tissue-specific expression in lotus (*Nelumbo nucifera*) unveiled by Illumina- and single-molecule real-time-based RNA-sequencing. *DNA Res.* **2019**, *26*, 301–311. [CrossRef]

35. Yang, M.; Xu, L.; Liu, Y.; Yang, P. RNA-Seq Uncovers SNPs and Alternative Splicing Events in Asian Lotus (*Nelumbo nucifera*). *PLoS ONE* **2015**, *10*, e0125702. [CrossRef]

36. Chen, D.; Melton, L.D.; Zujovic, Z.; Harris, P.J. Developmental changes in collenchyma cell-wall polysaccharides in celery (*Apium graveolens* L.) petioles. *BMC Plant Biol.* **2019**, *19*, 81. [CrossRef]

37. Khan, G.A.; Persson, S. Cell Wall Biology: Dual Control of Cellulose Synthase Guidance. *Curr. Biol.* **2020**, *30*, R232–R234. [CrossRef]

38. Endler, A.; Persson, S. Cellulose synthases and synthesis in *Arabidopsis*. *Mol. Plant* **2011**, *4*, 199–211. [CrossRef]

39. Kuang, B.; Zhao, X.; Zhou, C.; Zeng, W.; Ren, J.; Ebert, B.; Beahan, C.T.; Deng, X.; Zeng, Q.; Zhou, G.; et al. Role of UDP-Glucuronic Acid Decarboxylase in Xylan Biosynthesis in *Arabidopsis*. *Mol. Plant* **2016**, *9*, 1119–1131. [CrossRef]

40. Xu, P.; Kong, Y.; Li, X.; Li, L. Identification of molecular processes needed for vascular formation through transcriptome analysis of different vascular systems. *BMC Genom.* **2013**, *14*, 217. [CrossRef]

41. Zhu, Y.; Song, D.; Sun, J.; Wang, X.; Li, L. PtrHB7, a class III HD-Zip gene, plays a critical role in regulation of vascular cambium differentiation in *Populus*. *Mol. Plant* **2013**, *6*, 1331–1343. [CrossRef] [PubMed]

42. Zhao, Y.; Sun, J.; Xu, P.; Zhang, R.; Li, L. Intron-mediated alternative splicing of WOOD-ASSOCIATED NAC TRANSCRIPTION FACTOR1B regulates cell wall thickening during fiber development in Populus species. *Plant Physiol.* **2014**, *164*, 765–776. [CrossRef] [PubMed]

43. Wang, X.; Ren, H.; Zhang, B.; Fei, B.; Burgert, I. Cell wall structure and formation of maturing fibres of moso bamboo (*Phyllostachys pubescens*) increase buckling resistance. *J. R. Soc. Interface* **2012**, *9*, 988–996. [CrossRef] [PubMed]

44. La Camera, S.; Gouzerh, G.; Dhondt, S.; Hoffmann, L.; Fritig, B.; Legrand, M.; Heitz, T. Metabolic reprogramming in plant innate immunity: The contributions of phenylpropanoid and oxylipin pathways. *Immunol. Rev.* **2004**, *198*, 267–284. [CrossRef] [PubMed]

45. Besseau, S.; Hoffmann, L.; Geoffroy, P.; Lapierre, C.; Pollet, B.; Legrand, M. Flavonoid accumulation in *Arabidopsis* repressed in lignin synthesis affects auxin transport and plant growth. *Plant Cell* **2007**, *19*, 148–162. [CrossRef]

46. Lan, W.; Lu, F.; Regner, M.; Zhu, Y.; Rencoret, J.; Ralph, S.A.; Zakai, U.I.; Morreel, K.; Boerjan, W.; Ralph, J. Tricin, a flavonoid monomer in monocot lignification. *Plant Physiol.* **2015**, *167*, 1284–1295. [CrossRef]

47. McFarlane, H.E.; Doring, A.; Persson, S. The cell biology of cellulose synthesis. *Annu. Rev. Plant Biol.* **2014**, *65*, 69–94. [CrossRef]

48. Guerriero, G.; Fugelstad, J.; Bulone, V. What do we really know about cellulose biosynthesis in higher plants? *J. Integr. Plant Biol.* **2010**, *52*, 161–175. [CrossRef]

49. Richmond, T.A.; Somerville, C.R. The cellulose synthase superfamily. *Plant Physiol.* **2000**, *124*, 495–498. [CrossRef] [PubMed]

50. Wang, J.; Elliott, J.E.; Williamson, R.E. Features of the primary wall CESA complex in wild type and cellulose-deficient mutants of *Arabidopsis thaliana*. *J. Exp. Bot.* **2008**, *59*, 2627–2637. [CrossRef]
51. Chithrani, B.D.; Chan, W.C.W. Elucidating the mechanism of cellular uptake and removal of protein-coated gold nanoparticles of different sizes and shapes. *Nano Lett.* **2007**, *7*, 1542–1550. [CrossRef] [PubMed]
52. Desprez, T.; Juraniec, M.; Crowell, E.F.; Jouy, H.; Pochylova, Z.; Parcy, F.; Hofte, H.; Gonneau, M.; Vernhettes, S. Organization of cellulose synthase complexes involved in primary cell wall synthesis in *Arabidopsis thaliana*. *Proc. Natl. Acad. Sci. USA* **2007**, *104*, 15572–15577. [CrossRef] [PubMed]
53. Song, D.; Shen, J.; Li, L. Characterization of cellulose synthase complexes in *Populus* xylem differentiation. *New Phytol.* **2010**, *187*, 777–790. [CrossRef]
54. Mutwil, M.; Debolt, S.; Persson, S. Cellulose synthesis: A complex complex. *Curr. Opin. Plant Biol.* **2008**, *11*, 252–257. [CrossRef] [PubMed]
55. Yu, L.; Chen, H.; Sun, J.; Li, L. PtrKOR1 is required for secondary cell wall cellulose biosynthesis in *Populus*. *Tree Physiol.* **2014**, *34*, 1289–1300. [CrossRef] [PubMed]
56. Yu, L.; Sun, J.; Li, L. PtrCel9A6, an endo-1,4-beta-glucanase, is required for cell wall formation during xylem differentiation in *Populus*. *Mol. Plant* **2013**, *6*, 1904–1917. [CrossRef]
57. Kim, W.C.; Ko, J.H.; Kim, J.Y.; Kim, J.; Bae, H.J.; Han, K.H. MYB46 directly regulates the gene expression of secondary wall-associated cellulose synthases in *Arabidopsis*. *Plant J.* **2013**, *73*, 26–36. [CrossRef]
58. Coleman, H.D.; Yan, J.; Mansfield, S.D. Sucrose synthase affects carbon partitioning to increase cellulose production and altered cell wall ultrastructure. *Proc. Natl. Acad. Sci. USA* **2009**, *106*, 13118–13123. [CrossRef]
59. Klinghammer, M.; Tenhaken, R. Genome-wide analysis of the UDP-glucose dehydrogenase gene family in *Arabidopsis*, a key enzyme for matrix polysaccharides in cell walls. *J. Exp. Bot.* **2007**, *58*, 3609–3621. [CrossRef]
60. Ebert, B.; Rautengarten, C.; Guo, X.; Xiong, G.; Stonebloom, S.; Smith-Moritz, A.M.; Herter, T.; Chan, L.J.; Adams, P.D.; Petzold, C.J.; et al. Identification and Characterization of a Golgi-Localized UDP-Xylose Transporter Family from *Arabidopsis*. *Plant Cell* **2015**, *27*, 1218–1227. [CrossRef]
61. Jensen, J.K.; Johnson, N.R.; Wilkerson, C.G. *Arabidopsis thaliana* IRX10 and two related proteins from psyllium and *Physcomitrella patens* are xylan xylosyltransferases. *Plant J.* **2014**, *80*, 207–215. [CrossRef] [PubMed]
62. Zhang, Y.; Zhen, L.; Tan, X.; Li, L.; Wang, X. The involvement of hexokinase in the coordinated regulation of glucose and gibberellin on cell wall invertase and sucrose synthesis in grape berry. *Mol. Biol. Rep.* **2014**, *41*, 7899–7910. [CrossRef]
63. Frankova, L.; Fry, S.C. Biochemistry and physiological roles of enzymes that 'cut and paste' plant cell-wall polysaccharides. *J. Exp. Bot.* **2013**, *64*, 3519–3550. [CrossRef] [PubMed]
64. Zhang, X.; Wang, Q. Preliminary study of the ecotypes of genetic resources of tropical lotus. *Landsc. Plants* **2006**, *20*, 82–85.
65. Yang, M.; Zhu, L.; Xu, L.; Pan, C.; Liu, Y. Comparative transcriptomic analysis of the regulation of flowering in temperate and tropical lotus (*Nelumbo nucifera*) by RNA-Seq. *Ann. Appl. Biol.* **2014**, *165*, 73–95. [CrossRef]
66. Hai, L.; Guo, H.; Xiao, S.; Jiang, G.; Zhang, X.; Yan, C.; Xin, Z.; Jia, J. Quantitative trait loci (QTL) of stem strength and related traits in a doubled-haploid population of wheat (*Triticum aestivum* L.). *Euphytica* **2005**, *141*, 1–9. [CrossRef]
67. Yadav, S.; Singh, U.M.; Naik, S.M.; Venkateshwarlu, C.; Ramayya, P.J.; Raman, K.A.; Sandhu, N.; Kumar, A. Molecular Mapping of QTLs Associated with Lodging Resistance in Dry Direct-Seeded Rice (*Oryza sativa* L.). *Front. Plant Sci.* **2017**, *8*, 1431. [CrossRef]
68. Lin, S.Y.; Dence, C.W. *Methods in Lignin Chemistry*; Springer: Berlin, Germany, 1992.
69. Trapnell, C.; Roberts, A.; Goff, L.; Pertea, G.; Kim, D.; Kelley, D.R.; Pimentel, H.; Salzberg, S.L.; Rinn, J.L.; Pachter, L. Differential gene and transcript expression analysis of RNA-seq experiments with TopHat and Cufflinks. *Nat. Protoc.* **2012**, *7*, 562–578. [CrossRef]
70. Guerreiro, A.C.; Benevento, M.; Lehmann, R.; van Breukelen, B.; Post, H.; Giansanti, P.; Maarten Altelaar, A.F.; Axmann, I.M.; Heck, A.J. Daily rhythms in the cyanobacterium synechococcus elongatus probed by high-resolution mass spectrometry-based proteomics reveals a small defined set of cyclic proteins. *Mol. Cell. Proteom.* **2014**, *13*, 2042–2055. [CrossRef]
71. Moriya, Y.; Itoh, M.; Okuda, S.; Yoshizawa, A.C.; Kanehisa, M. KAAS: An automatic genome annotation and pathway reconstruction server. *Nucleic Acids Res.* **2007**, *35*, W182–W185. [CrossRef]
72. Aoki-Kinoshita, K.F.; Kanehisa, M. *Gene Annotation and Pathway Mapping in KEGG*; Humana Press Inc.: Totowa, NJ, USA, 2007; Volume 396.

73. MacLean, B.; Tomazela, D.M.; Shulman, N.; Chambers, M.; Finney, G.L.; Frewen, B.; Kern, R.; Tabb, D.L.; Liebler, D.C.; MacCoss, M.J. Skyline: An open source document editor for creating and analyzing targeted proteomics experiments. *Bioinformatics* **2010**, *26*, 966–968. [CrossRef] [PubMed]

74. Vizcaino, J.A.; Cote, R.G.; Csordas, A.; Dianes, J.A.; Fabregat, A.; Foster, J.M.; Griss, J.; Alpi, E.; Birim, M.; Contell, J.; et al. The PRoteomics IDEntifications (PRIDE) database and associated tools: Status in 2013. *Nucleic Acids Res.* **2013**, *41*, D1063–D1069. [CrossRef] [PubMed]

International Journal of
Molecular Sciences

Article

Label-Free Comparative Proteomic Analysis Combined with Laser-Capture Microdissection Suggests Important Roles of Stress Responses in the Black Layer of Maize Kernels

Quanquan Chen [1,†], Ran Huang [1,†], Zhenxiang Xu [1], Yaxin Zhang [1], Li Li [1], Junjie Fu [2], Guoying Wang [2], Jianhua Wang [1], Xuemei Du [1,*] and Riliang Gu [1,*]

[1] Center for Seed Science and Technology, Beijing Innovation Center for Seed Technology (MOA), Key Laboratory of Crop Heterosis Utilization (MOE), College of Agronomy and Biotechnology, China Agricultural University, Beijing 100193, China; znchenquanquan@163.com (Q.C.); huangran086@126.com (R.H.); dlxuzhenxiang@163.com (Z.X.); zhangyaxin1994@126.com (Y.Z.); lili2016@cau.edu.cn (L.L.); wangjh63@cau.edu.cn (J.W.)
[2] Institute of Crop Sciences, Chinese Academy of Agricultural Sciences, Beijing 100081, China; fujunjie@caas.cn (J.F.); wangguoying@caas.cn (G.W.)
*   Correspondence: duxuemei1986@yeah.net (X.D.); rilianggu@cau.edu.cn (R.G.)
†   These authors contribute equally to this work.

Received: 6 January 2020; Accepted: 16 February 2020; Published: 18 February 2020

Abstract: The black layer (BL) is traditionally used as an indicator for kernel harvesting in maize, as it turns visibly dark when the kernel reaches physiological maturity. However, the molecular roles of BL in kernel development have not been fully elucidated. In this work, microscopy images showed that BL began to appear at a growth stage earlier than 10 days after pollination (DAP), and its color gradually deepened to become dark as the development period progressed. Scanning electron microscopy observations revealed that BL is a tissue structure composed of several layers of cells that are gradually squeezed and compressed during kernel development. Laser-capture microdissection (LCM) was used to sample BL and its neighboring inner tissue, basal endosperm transfer layer (BETL), and outer tissue, inner epidermis (IEP), from 20 DAP of kernels. Matrix-assisted laser desorption/ionization time-of-flight mass spectrometry profiling (MALDI-TOF MS profiling) detected 41, 104, and 120 proteins from LCM-sampled BL, BETL, and IEP, respectively. Gene ontology (GO) analysis indicated that the 41 BL proteins were primarily involved in the response to stress and stimuli. Kyoto Encyclopedia of Genes and Genomes (KEGG) pathway analysis found that the BL proteins were enriched in several defense pathways, such as the ascorbate and aldarate metabolic pathways. Among the 41 BL proteins, six were BL-specific proteins that were only detected from BL. Annotations of five BL-specific proteins were related to stress responses. During kernel development, transcriptional expression of most BL proteins showed an increase, followed by a decrease, and reached a maximum zero to 20 DAP. These results suggest a role for BL in stress responses for protecting filial tissue against threats from maternal sides, which helps to elucidate the biological functions of BL.

Keywords: maize; black layer; seed development; proteomic analysis; stress response; MALDI-TOF MS

---

1. Introduction

Maize is one of the most important grain crops with a total production of more than one billion tons, accounting for ~30% of the world's food supply (https://www.statista.com). Harvesting in proper time is a key factor to obtain high-quality kernels and avoid yield loss in maize production [1–3].

Black layer (BL) development has been widely used as an indicator of kernel maturity in agronomy, as it turns visibly dark when the kernel reaches physiological maturity (maximum kernel weight) in maize. The physiological relationship between BL formation and kernel maturity has been intensively investigated in recent decades [4–9], but the developmental processes, as well as the molecular roles, of BL have not been fully elucidated.

BL comprises several layers of cells located between the filial endosperm and the maternally derived pericarp and belongs to a major bridge structure, known as the placento-chalazal (P-C) layer. BL is derived from the inner integument and the integumental P-C (iP-C), which is located immediately below another P-C layer, nucellar P-C (nP-C) which is immediately subtended from the endosperm and is derived from the nucellus epidermis [10]. At the early kernel development stage (before 10 days after pollination, DAP), programmed cell death process (PCD)-based cell death was observed in BL [11], which leads to cell degradation into a narrow band of crushed cells with accumulated brown pigment and subsequently leads to the emergence of visible BL [11,12]. These visible BLs are generally used as kernel maturity indicators by agronomists and farmers.

Cells above the P-C layer include several layers of endosperm cells called the basal endosperm transfer layer (BETL), and cells below the P-C layer are the inner epidermal cells of the pedicel (IEP). The P-C layer, together with the BETL and pedicel, function as the transfer of photoassimilates and nutrients from the mother plant to the filial caryopsis. In maize, phloem tissue is transported in the pedicel. However, this tissue does not extend beyond the pedicel. Thus, a different transport system is expected to exist in the P-C layer and BETL after the phloem termini in the pedicel [11–13]. Nevertheless, whether BL is involved in this critical role of post-phloem transport has not been determined.

Although described in various plant species, a P-C region is known to exhibit a high level of anatomical variability in its structural adaptations [14]. P-C layers are believed to possess the best developed structure in tropical crops, such as maize and sorghum, which show assimilated transport from only the base of the caryopsis [11,15]. Thus, investigating BL development and predicting its biological roles are particularly interesting in maize.

To date, the rapid development of proteomic technologies provides a notable opportunity for plant proteomic profiling [16]. Proteomics studies in maize whole kernel development have been intensively conducted at different scales of genotype, growth stage, and treatment [17–23]. Proteomics has also been performed in specific kernel tissue of BETL and cellular component of protein body to investigate their special function and developmental process [24,25]. However, no attempt has been made to attach to BL tissue. In addition, most of the abovementioned proteomics analyses were generally conducted by the two-dimensional (2D) gel-based mass spectrometry method. Recently, matrix-assisted laser desorption/ionization time-of-flight mass spectrometry profiling (MALDI-TOF MS profiling) is an emerging approach based on rapid and high-throughput screening of ions from molecules directly detected in biological samples [26], which has been applied in plant protein identification [27].

In this work, cells of BL together with its neighboring tissues, BETL and IEP, were obtained by laser-capture microdissection, and their accumulated proteins were identified by the MALDI-TOF MS method. The networks and putative functions of BL-expressed proteins were further discussed. These results could help to elucidate the mechanisms of BL formation and the biological roles of this special structure.

2. Results

2.1. Observation of Black Layer Development in Maize Kernels

The kernel development was investigated 10, 20, 30, 40, and 50 DAP (Figure 1). The embryo structure was unobservable 10 DAP, while its structure outline could be observed 20 DAP. This result is consistent with previous findings that embryo structure is differentiated from the endosperm 10 to 20 DAP [28,29].

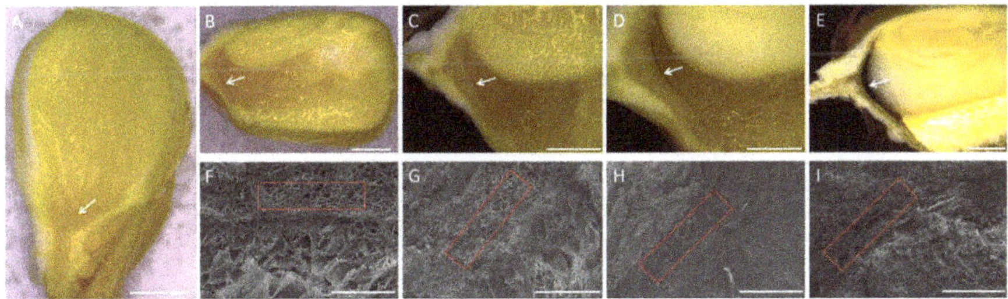

Figure 1. Observation of the black layer in maize kernels at different developmental stages. (**A–E**) Longitudinal section of maize kernels 10, 20, 30, 40, and 50 days after pollination (DAP). The arrowhead indicates the position of the black layer. Bar = 2 mm; (**F-I**) Scanning electron microscopy analysis of the black layer of maize kernels 20, 30, 40, and 50 DAP. Regions surrounded by red boxes indicate the structure of the black layer. Bar = 200 μm.

Ten DAP, a band with a different color from the surrounding area of the endosperm could be observed at the base of the kernel, which was likely to form BL during the latter development (Figure 1A). From 10 to 40 DAP, the color of this band gradually changed to black (Figure 1A–D), while from 40 to 50 DAP, the color changed dramatically, and a dark black band emerged 50 DAP (Figure 1D, E). These results suggested that BL development could start earlier than embryo differentiation at a stage before 10 DAP and progress to seed maturity 50 DAP.

From the electron-microscope analysis, it was observed that BL was composed of several layers of cells. This structure was gradually compressed and finally became a small band along with kernel development from 20 DAP to 50 DAP (Figure 1F–I). This result raised the possibility that the color change of BL from light to dark was partially due to a concentration effect of brown pigment cells.

2.2. Isolation of Black Layer Cells

Since BL begins to form as early as 10 DAP, 20 DAP kernels were selected for proteomic analysis because the proteins involved in BL development are expected to accumulate at this stage. Furthermore, the longitudinal section of the whole kernel indicated that BL, BETL, and IEP could be clearly distinguished at 20 DAP (Figure 2A–D). Thus, BL, as well as its neighboring tissues BETL and IEP, were successfully collected after subjecting kernel section and Laser-Capture Microdissection (LCE) under high microscope resolution conditions (Figure 2E–G).

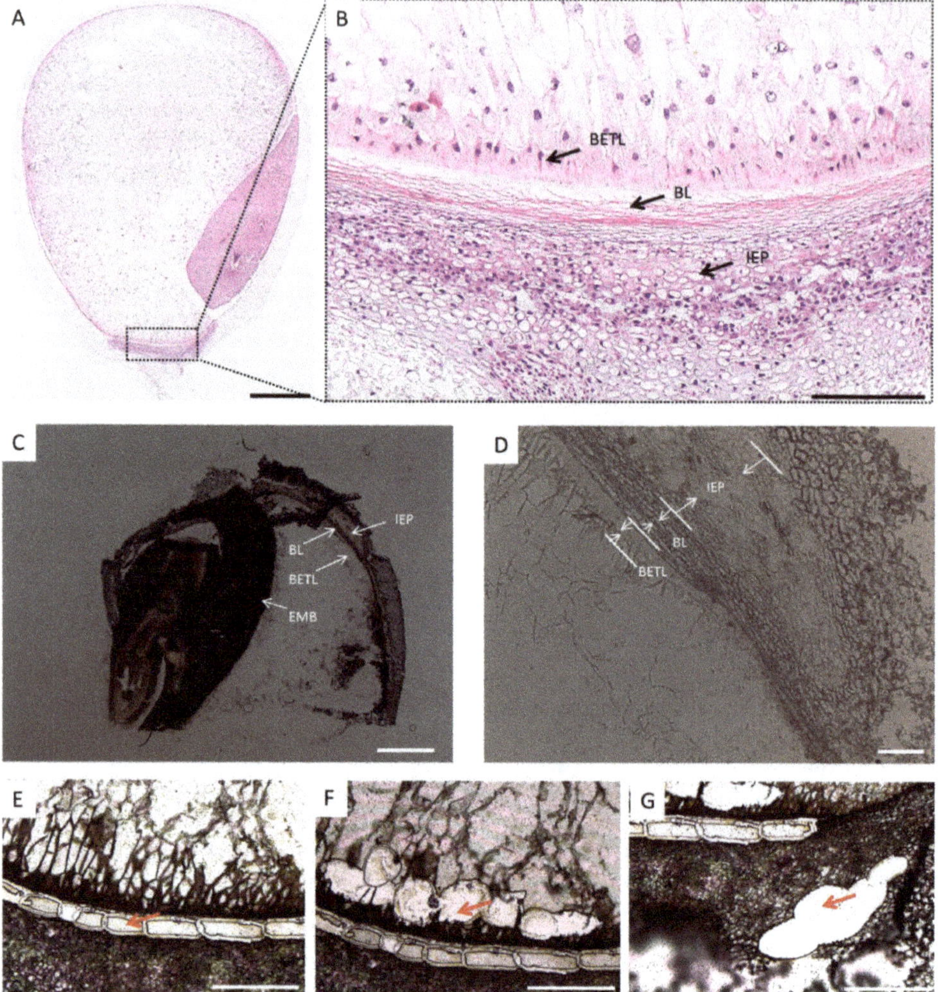

Figure 2. Isolation of the black layer (BL), inner epidermis of the pedicel (IEP), and basal endosperm transfer layer (BETL) from maize kernels 20 DAP by laser-capture microdissection. (**A**) Longitudinal paraffin sections of maize kernels 20 DAP. Bar = 1.25 mm; (**B**) High-magnification longitudinal sections of maize kernels 20 DAP. Arrows indicate the position of BL, IEP, and BETL. Bar = 200 μm; (**C**) and (**D**) Tissues from 8 μm thick cryosections before LCM cutting, showing the IEP, BL, BETL, and EMB. EMB, embryo. (**C**) Bar = 1000 μm. (**D**) Bar = 100 μm; (**E**–**G**) Tissues from 8 μm thick cryosections after LCM cutting. The red arrow indicates the successfully cut tissues of BL (**E**), BETL (**F**), and IEP (**G**). Bar = 200 μm.

2.3. Proteomic Analysis of the Black Layer Cells

Proteins were identified by quadrupole time-of-flight mass spectrometry. The MS data were searched against the UniProt Plant protein database containing 87,406 sequences (http://www.uniprot.org) and the NCBI maize database containing 280,238 sequences (http://www.ncbi.nlm.nih.gov), which resulted in 210, 800, and 653 peptides from BL, IEP, and BETL, respectively (Table S1). These peptides represented 1020 non-redundant peptides (Figure 3C). Using these peptides, PCA analysis showed that the three biological replicates for each tissue were closely related and grouped together (Figure 3A). PC1 accounted for 85.9% of the total component. From this component, the BETL and IEP groups showed a closed relationship, but they were located far from the BL group. Meanwhile, the correlation analysis showed notably high positive correlations among the replicates within a tissue (R^2 = 0.99 to 1.00) and high positive correlations across tissues (R^2 = 0.75 to 0.93). Furthermore, correlations between BL and either BETL (R^2 = 0.75 to 0.77) or IEP (R^2 = 0.81 to 0.86) were lower than those between BETL and IEP (R^2 = 0.92 to 0.93) (Figure 3B). The PCA and correlation analyses indicated that the protein accumulation in BL was significantly different from those in the other two tissues.

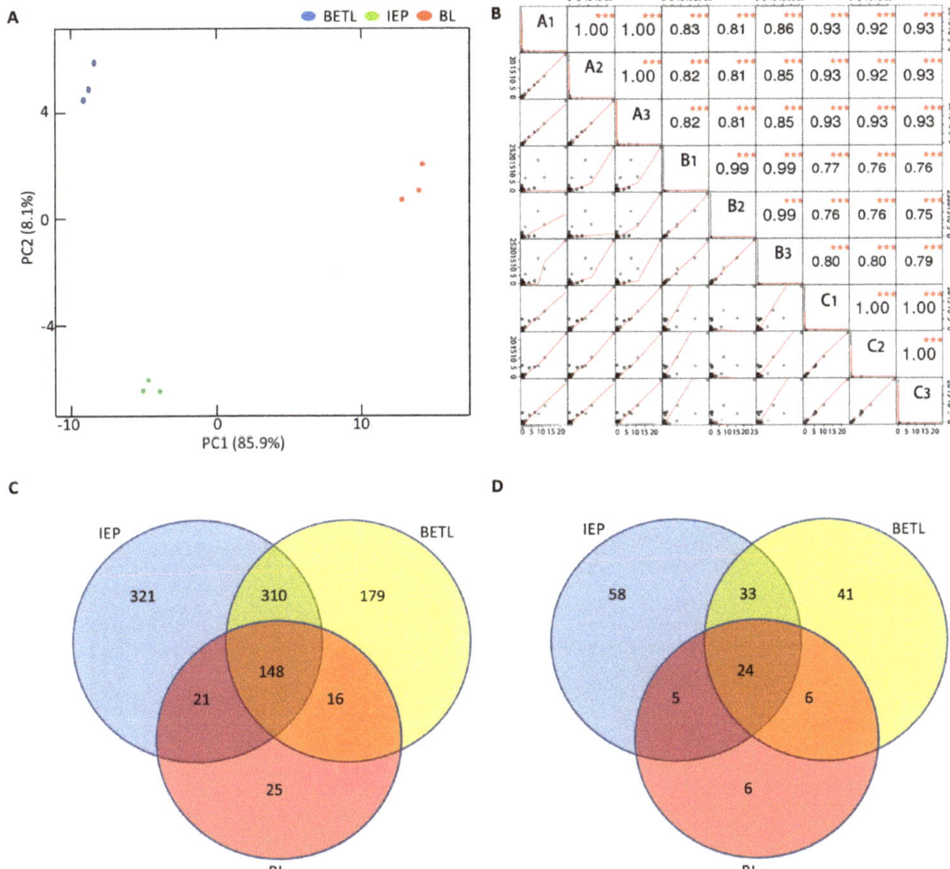

Figure 3. Proteomic analysis of peptides and protein identified from the black layer (BL), inner epidermis of pedicel (IEP), and basal endosperm transfer layer (BETL), of maize kernel 20 DAP. (**A**) Principle component analysis (PCA) of the identified peptides from the three replicated samples of IEP, BL, and BETL; (**B**) Coefficiency correlation analysis of the identified peptides from the three replicated samples of IEP, BL, and BETL. A (1, 2, 3), B (1, 2, 3), and C (1, 2, 3) represent the three replicates of IEP, BL, and BETL, respectively. (**C**) Venn diagram of the identified peptides from IEP, BL, and BETL. (**D**) Venn diagram of the identified proteins from IEP, BL, and BETL.

The 1020 peptides represented 173 nonredundant protein entrances (Table S2). Among these proteins, 41 were recognized in BL, which was considerably lower than the number detected in IEP (120) and BETL (104) (Figure 3D). Of the 41 BL proteins, six uniquely accumulated in BL, and 24 constitutively accumulated in all three tissues from a heatmap analysis of their protein accumulation (Figure 4A). The lowest total protein number and tissue-specific protein number detected from BL suggested a lower metabolic activity of this tissue as compared with its neighboring tissues.

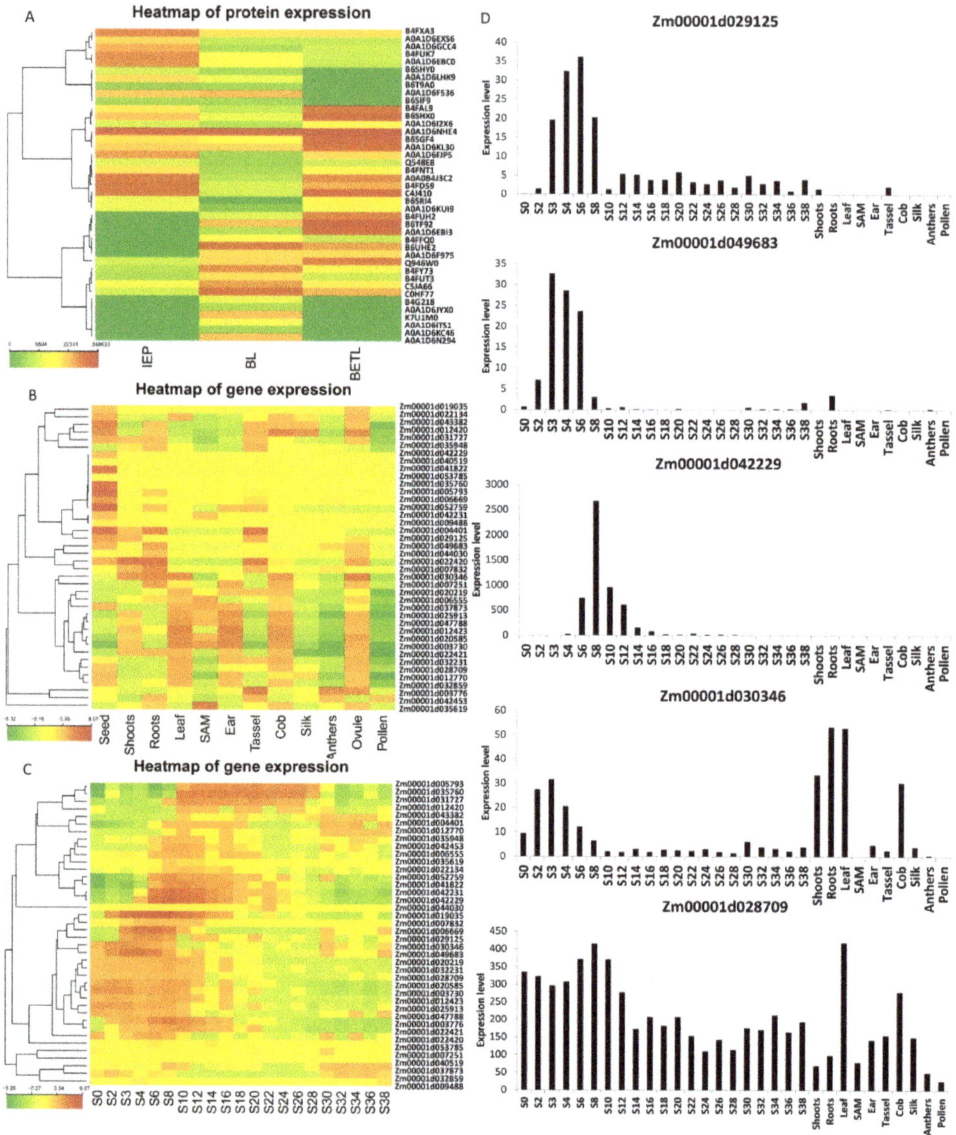

Figure 4. Expression patterns of proteins and genes identified from the black layer (BL). (**A**) Heatmap cluster of proteins identified from BL. Expression pattern analysis of BL genes in maize tissues (**B**) and during seed development (**C**). (**D**) Expression of the 5 BL-specific expression genes in maize tissues and during seed development. The normalized values after $\log_2$ transformation of reads per kilobase per million (RPKM) were used for (**B**) and (**C**). S represents seed and number represents days after pollination. Expression data were extracted from Chen et al. (2014) [30].

2.4. Gene ontology Annotation Function and Kyoto Encyclopedia of Genes and Genomes Analysis of Proteins Identified from BL

Gene ontology (GO) (http://systemsbiology.cau.edu.cn/agriGOv2/) analysis of the 41 BL accumulated proteins revealed six GO terms for cellular components and biological processes and four for molecular functions (Figure 5A and Table S3). In the biological process category, the largest subcategory was response to stimulus followed by response to stress. For the cellular component category, the subcategories were significantly enriched at the organelle part, plastid, nucleolus, plasma membrane, macromolecular complex, and cell wall. For molecular function, the top four subcategories with highly significant p-values were nucleic acid binding, protein binding, DNA binding, and translation factor activity.

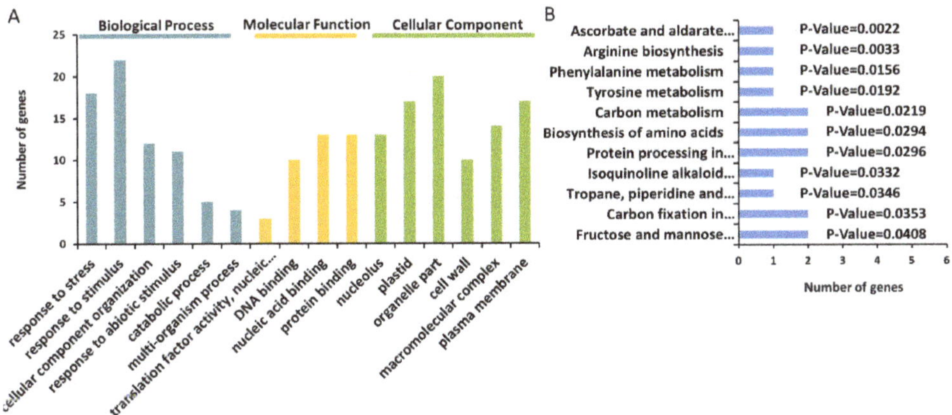

Figure 5. Gene ontology (GO) annotation function (**A**) and Kyoto Encyclopedia of Genes and Genomes (KEGG) (**B**) analyses of the identified proteins from the black layer.

Kyoto Encyclopedia of Genes and Genomes (KEGG) pathway analysis was further performed for the 41 BL proteins, which mapped to 11 KEGG pathways (Figure 5B). Five pathways were significantly enriched in different metabolic processes, including a pathway in the ascorbate and aldarate metabolic pathways that played protective roles in plants, particularly under stress conditions.

Except for one unknown protein, five out of the six BL-specific proteins were cytosolic ascorbate peroxidase protein, pathogenesis-related protein, salt stress-induced protein, desiccation-related protein, and small heat shock protein. All five of these proteins have been reported to be involved in stress responses.

2.5. Gene Expression of BL Accumulated Proteins

According to a heatmap analysis for the 41 BL proteins, the accumulation of 14 proteins showed higher levels in BL than in IEP and BETL. Of these 14 proteins, six were specifically expressed in BL (Figure 4A). From a published dataset [30], 38 out of the 41 BL genes had expression records. Interestingly, most of the BL genes showed high expression levels in seeds as compared with other plant organs (Figure 4B and Table S4). In addition, 33 genes showed an increasing and, then, a decreasing expression pattern along with the seed development process, with 17 genes showing the highest level at the early seed development stage (zero to 10 DAP) and 16 at the middle stage (six to 22 DAP) (Figure 4C). As gene transcription always started slightly earlier than protein accumulation, the high transcription levels of these genes from zero to 22 DAP were consistent with the results of protein profiling at 20 DAP.

Concerning approximately the five BL-specific genes, Zm00001d029125, Zm00001d049683, and Zm00001d042229 showed exclusive expression in seeds, particularly at seeds three to 12 DAP

(Figure 4D). Zm00001d028709 and Zm00001d030346 showed constitutive expression in many plant organs, including seeds. Within seeds, Zm00001d028709 was expressed during the whole seed developmental process, while Zm00001d030346 showed high levels at the early developmental stage.

3. Discussion

As proteins are the critical executors that directly participate in life activities, the study of proteomics can make significant contributions to elucidating the mechanism of important biological processes. The identification of proteins in previous works was mainly performed by two-dimensional electrophoresis, which has been considered a method with low specificity, efficiency, and accuracy [31]. In recent years, liquid chromatography-mass spectrometry (LC-MS) technology has become the preferred method for spatial proteomics (the localizations of proteins and their dynamics at the subcellular level), which could show the quantitative state of a proteome. On the basis of the MS method, several new technologies have been developed for quantitative proteomics [32], including label-free iTRAQ, SILAC, MRM (MRMHR), and SWATH [33–37]. In this paper, the label-free iTRAQ method was used for MALDI-TOF MS profiling.

Maize kernel is a single-seeded fruit that consists of a large embryo and a triploid endosperm encased in a maternally originated pericarp. BL is located between the filial endosperm and the maternally derived pericarp and belongs to a major bridge structure called the P-C layer. Neighboring inside the BL is an endosperm tissue called BETL, and outside is a pericarp IEP. The proteome of whole maize seeds, embryos, and endosperm has been intensively investigated in recent decades [23,25,38–44]. Due to the difficulty of separating tissues from kernels, the proteome for some targeted seed parts has seldom been the focus, except for BETL, whose functions have been implied to mediate embryo–endosperm interactions and play a role in plant defense by this method [45–48] To date, no attempt has been focused on BL, and very little is known about its developmental processes, as well as the physiological and molecular roles of this tissue in kernel development. In this study, we obtained BL tissue by the LCE method and analyzed its protein accumulation by MALDI-TOF MS. BL was located between BETL and IEP from the observation of microscope image (Figure 1), but its protein accumulation was significantly different from its neighboring tissues IEP and BETL (Figure 3A,B). In addition, detectable protein from BL was considerably lower than those from the neighboring tissues (Figure 3C,D), suggesting a distinct behavior of lower metabolic activity for BL.

BETL has the functions of transporting nutrients from the maternal to the endosperm and embryo of the seed. As BL is located outside the BETL, we speculated that BL plays a connecting role between BETL and IEP. Compared to several transport proteins, e.g., ion channel protein and aminotransferase protein, identified from BETL, no transport-related proteins could be found from BL (Table S2). However, three genes were found to be related to cell death. Zm00001d043382 was found by GO functional annotation with an encoded protein involved in the senescence pathway (GO:0010149) [49]. Zm00001d037873 and Zm00001d012770 encode proteins with GTPase activity (GO:0003924), which has been reported to be associated with cell death [50,51]. This finding was consistent with those of previous publications, where PCD-based cell death was previously observed at a stage as early as 10 DAP in BL [11]. As cell death would lead BL cells to lose their cellular activity during the later seed development stage, the death of BL cells would block nutrient flow through this tissue. Thus, these results suggested that BL could have a special role, other than nutrient transport, in seed development.

GO and KEEG analyses indicated that the BL-accumulated proteins showed functions related to the stress response (Figure 5A,B). Of the six BL-specific proteins, Zm00001d028709 encodes the cytosolic ascorbate peroxidase Apx1 and Zm00001d049683 encodes a homolog of rat L-gulono-1,4-lactone (L-GulL) oxidase that is involved in the biosynthesis of L-ascorbic acid. Ascorbic acid reacts with H_2O_2 under the action of ascorbic peroxidase, thereby, eliminating the toxicity of H_2O_2 [52–55]. This process plays a significant role in the stress resistance of plants. Zm00001d030346 encodes a small heat shock protein 21 (Hsp21), which plays a crucial role in protecting plants against stress by re-establishing cellular homeostasis in the abiotic stress response, plant disease resistance, and oxidative stress [56–59].

As Apx1 and Hsp can be involved in the same metabolic network and play a protective role in plants [52,60], these results suggest that Zm00001d028709, Zm00001d049683, and Zm00001d030346 function in stress responses under a connected pathway within BL.

Zm00001d042229 was predicted to encode a PR (pathogenesis-related) protein, which possesses antimicrobial properties involved in the biotic and abiotic stress responses of plants. A protective role for the embryo surrounding the region of the maize endosperm was previously demonstrated by the characterization of *ZmESR-6*, a defensin gene specifically expressed in this region [48]. A similar protective role for the filial surrounding region of the maternal tissue could be verified after further functional characterization of the BL-specific accumulation gene Zm00001d042229.

Zm00001d029125 encodes the desiccation-related protein PCC13-62 precursor. Cellular desiccation regulates the maturation of seeds [61–63]. Thus, the desiccation of BL cells along with PCD-based cell death blocks the transport of nutrients between BETL and IEP tissues, which plays a role in protecting the endosperm and embryo of seeds. This possibility was supported by evidence from previous works. For example, the basal layer-type antifungal protein 2 (BAP2) protein accumulated in BL and exhibited broad-range activity against a range of filamentous fungi [64]. Similar cellular autolytic events have also been reported in the nucellar projection (a tissue similar to BL) in barley [65].

4. Materials and Methods

4.1. Plant Materials and Sample Collection

PH6WC, the maternal line of an elite hybrid XY335 [3], was grown in 2017 at Zhuozhou, Hebei Province (39.486°N 115.974°E). Ears were manually self-pollinated and harvested 10, 20, 30, 40, and 50 days after pollination (DAP). For each stage, the harvests were repeated three times serving as the three biological replicates. Seeds from the middle part of each cob were carefully sampled for further analysis. Seeds were cut longitudinally with a scalpel and observed under a photomicroscope (Leica S9 i; Leica, Wetzlar, Germany). Seeds were cut into small cubes (approximately 10×10 mm) by a dissecting scalpel for laser-capture microdissection (LCM) following a method described by Zhu et al. [66]. These cubes were immediately submerged in a precooled fixative solution (75% (v/v) ethanol and 25% (v/v) acetic acid) at a 1:10 volume ratio of cube ice for 15 min and transferred into a new fixing solution overnight with gentle rotation at 4 °C. Cubes were transferred into 10% w/v sucrose in phosphate buffered saline (PBS) buffer (137 mM NaCl, 8 mM Na_2HPO_4, 2.7 mM KCl, and 1.5 mM KH_2PO_4, pH = 7.3) containing protease inhibitor (1:1000 volume ratio dilution) (Sigma, St Louis, MO, USA), followed by infiltration under vacuum on ice for 15 min. Cubes were transferred to 20% w/v sucrose in the same PBS and protease inhibitor buffer for another 15 min of infiltration. Then, the cubes were washed with optimum cutting temperature (OCT) medium (Tissue-Tek, Sakura, Japan), transferred into Eppendorf tubes supplied with OCT medium, and frozen with liquid nitrogen. The frozen tissue cubes were placed on ice for immediate microsection or stored at −80 °C.

4.2. Light and Scanning Electron Microscopy

Kernels were collected 20 DAP and cut along the longitudinal axis for imaging under light microscopy. Tissues containing the embryo part were fixed for 3 days at room temperature in FAA solution (38% formaldehyde 5 mL/glacial acetic acid 5 mL/70% ethanol 90 mL). The material was embedded in paraffin by dehydration in an ethanol gradient series (70%, 80%, 95%, and 100% ethanol) and subsequently cut into 8 μm sections. The sections were stained with toluidine blue and observed using a Nikon Ti microscope (Nikon, Melville, NY, USA).

Kernels were collected 20, 30, 40, and 50 DAP as samples for scanning electron microscopy. Kernels were critically dried, and sputter coated with gold. Gold-coated samples were observed with a scanning electron microscope S-3400N (Hitachi, Tokyo, Japan).

4.3. Laser Capture Microdissection

The tissues were sectioned at 8 μm in a cryostat (CM3050S; Leica, Wetzlar, Germany) and mounted on an adhesive-coated slide at −25 °C, as described by Nakazono et al. [67]. The sections were immediately incubated in 70% (*v/v*) ethanol at −20 °C for 1 min and washed with precooled ddH$_2$O for 30 s followed by dehydration steps of 1 min each in 95% ethanol and 100% ethanol and 2 min twice in xylene. The sections were air-dried and used intermediately for LCM. During the LCM process, BETL, BL, and IEP cells were isolated using the following parameters: 7.5 μm laser spot size, 50 mW laser power, and 550 to 650 μs laser pulse duration in the PixCell II LCM system (Arcturus Bioscience, Carlsbad, CA, USA) [67].

4.4. Protein Extraction

The isolated tissues were transferred to 0.2 mL tubes containing 30 μL extraction solution (6 M urea, 50 mM dithiothreitol, 0.5 M Tris-HCl, pH 8.0, 1:1000 diluted protease inhibitor). Proteins were dissolved by shaking the tube and, then, incubated on ice for 15 min. For efficient protein extraction, the tissue sample was cracked by ultrasonication on ice under an ultrasonic parameter of 3 seconds 30 Watt power supply interrupted by 6 seconds 50 times. After centrifugation at 4 °C and 14,000 g for 40 min, the supernatant was transferred to a clean 2 mL tube for protein digestion, which was performed using the FASP procedure described by Wisniewski et al. [68]. DTT was added into the tube to a final concentration of 10 mM for incubation at 37 °C for 1 h. IAA was added to a final concentration of 30 mM for reaction at 37 °C for 30 min in the dark. Then, 50 mM NH$_4$HCO$_3$ was added to a final volume of 1 mL. Five micrograms of trypsin were added to this solution for protein digestion at 37 °C overnight. After adding 10% (*v/v*) formic acid to a final concentration of 1% and reacting for 5 min, the supernatant containing peptides was collected by centrifugation at 4 °C and 14,000 g for 10 min, which served for desalt filtration by C18 cartridges (Empore™ SPE Cartridges C18 standard density, bed I.D. 7 mm, volume 3 mL, Sigma). Then, the peptides were concentrated by vacuum and redissolved in 20 μL 0.1% (*v/v*) formic acid. The concentration of peptides was determined by UV spectrometry at 280 nm [69].

4.5. Proteomics Analysis

Proteins were identified in a quadrupole time-of-flight mass spectrometer (Agilent model 6500, Wilmington, DE, USA) following the user manual of MALDI-TOF MS. Using MaxQuant software 1.3.0.5, the MS data were searched against the UniProt Plant protein database containing 87,406 sequences (http://www.uniprot.org) and the NCBI maize database containing 280,238 sequences (http://www.ncbi.nlm.nih.gov). The search parameters were set as: ± 20 ppm for peptide mass tolerance, 0.1 Da fragment mass tolerance, and 2 maximum missed cleavages. The cutoff for the global false discovery rate (FDR) in peptide identification was 0.01. To reduce the probability of false identification, only peptides with significance scores at the 95% confidence interval were counted as identified peptides, and peptides repeatedly identified from at least two out of the three replicates were considered proteins that were expressed in the corresponding tissue. Intensity-based absolute quantification (iBAQ) in MaxQuant was performed to quantify peptide abundance for each replicate, with the cutoff values p = <0.05 and FDR = <0.05. The reproducibility of the triplicates for all tissues was analyzed using the iBAQ data by coefficiency correlation analysis (Pearson's) at a significance level of p < 0.05 and principal component analysis (PCA). Then, the mean iBAQ data from the three replicates served as the abundant data for each peptide, which were used for identification of the corresponding protein under threshold of fold changes of >1.5 or < 0.67 at p value < 0.05.

Functional pathways were analyzed using gene ontology (GO) functional enrichment analysis (https://david.ncifcrf.gov/), where the GO categories included biological processes, molecular function, and subcellular locations. In addition, the significant biological process-associated proteins were selected to map to the pathways in KEGG (https://www.genome.jp/kegg/).

4.6. Expression Analysis of Candidate Genes

Raw datasets of RNA-Seq from different maize tissues were extracted from RNA-seq libraries and used for heatmap analysis of candidate genes [30]. The details about the data sources are described in Table S2. The raw reads were aligned to the B73 reference genome (RefGen_v2) using Tophat 2.0.6 (http://ccb.jhu.edu/software/tophat/index.shtml; Trapnell et al., 2009) with maximum intron length set to 30 kb, with default settings for other parameters. The number of uniquely mapped reads for each gene model in B73 was calculated by parsing the alignment output files from Tophat and, then, normalizing the resulting read counts by reads per kilobase per million (RPKM) to measure the gene expression level. Uniquely mapped reads were used to estimate the normalized transcription level, and the normalized value after $\log_2$ transformation of RPKM was used for Figure 4.

5. Conclusions

In this work, microscopy observation revealed that BL began to appear at a growth stage earlier than 10 DAP. Cells of BL, BETL, and IEP were successfully collected from 20 DAP kernels by LCM and used for protein profiling by the MALDI-TOF MS method. The BL-accumulated proteins were primarily enriched in functions of stress responses. By comparing BETL and IEP, six BL-specific proteins were identified, with five showing high gene expression at the early kernel development stage and homology to previously reported plant stress-responsive genes. Thus, the results of this study suggest a special role for BL in protective functions.

Supplementary Materials: The following are available online at http://www.mdpi.com/1422-0067/21/4/1369/s1.

Author Contributions: G.W., J.W. and R.G. designed this study; R.H., Z.X., Y.Z., and L.L. performed the experiments; Q.C. and J.F. analyzed the data; Q.C. and R.H. drafted the manuscript; X.D., and R.G. revised the manuscript. All authors have read and agreed to the published version of the manuscript.

Funding: This research was supported by The National Key R & D Program of China (2018YFD0100903), the Ministry of Agriculture of China (2016ZX08003-002), the National Natural Science Foundation of China (31771891), and the China Agriculture Research System (CARS-02-10).

Conflicts of Interest: The authors declare no conflicts of interest.

Abbreviations

BL	black layer
DAP	days after pollination
BETL	basal endosperm transfer layer
IEP	inner epidermis cells of the pedicel
GO	Gene ontology
KEGG	Kyoto Encyclopedia of Genes and Genomes
P-C	placento-chalazal
iP-C	integumental P-C
nP-C	nucellar P-C
PCD	program cell death process
LCM	laser-capture microdissection
PBS	phosphate buffered saline
OCT	optimum cutting temperature
FDR	false discovery rate
iBAQ	Intensity-based absolute quantification
PCA	principal component analysis
L-GulL	L-gulono-1,4-lactone
Hsp21	heat shock protein21
BAP2	basal layer-type antifungal protein
MALDI-TOF MS	Matrix-assisted laser desorption/ionization time-of-flight mass spectrometry profiling

Int. J. Mol. Sci. **2020**, *21*, 1369

References

1. Joazeiro, C.A.P.; Wing, S.S.; Huang, H.; Leverson, J.D.; Hunter, T.; Liu, Y.-C. The Tyrosine Kinase Negative Regulator c-Cbl as a RING-Type, E2-Dependent Ubiquitin-Protein Ligase. *Science* **1999**, *286*, 309–312. [CrossRef] [PubMed]
2. Panison, F.; Sangoi, L.; Kolling, D.F.; Coelho, C.M.M.; Durli, M.M. Harvest time and agronomic performance of maize hybrids with contrasting growth cycles. *Acta Sci. Agron.* **2016**, *38*, 219. [CrossRef]
3. Gu, R.; Li, L.; Liang, X.; Wang, Y.; Fan, T.; Wang, Y.; Wang, J. The ideal harvest time for seeds of hybrid maize (Zea mays L.) XY335 and ZD958 produced in multiple environments. *Sci. Rep.* **2017**, *7*, 17537. [CrossRef] [PubMed]
4. Daynard, T.B.; Duncan, W.G. The Black Layer and Grain Maturity in Corn. *Crop Sci.* **1969**, *9*, 473. [CrossRef]
5. Rench, W.E.; Shaw, R.H. Black Layer Development in Corn. *Agron. J.* **1971**, *63*, 303–305. [CrossRef]
6. Daynard, T.B. Relationships Among Black Layer Formation, Grain Moisture Percentage, and Heat Unit Accumulation in Corn1. *Agron. J.* **1972**, *64*, 716. [CrossRef]
7. Carter, M.W.; Poneleit, C.G. Black Layer Maturity and Filling Period Variation Among Inbred Lines of Corn (Zea mays L.). *Crop Sci.* **1973**, *13*, 436. [CrossRef]
8. Hunter, J.L.; TeKrony, D.M.; Miles, D.F.; Egli, D.B. Corn Seed Maturity Indicators and their Relationship to Uptake of Carbon-14 Assimilate. *Crop Sci.* **1991**, *31*, 1309. [CrossRef]
9. Gunn, R.B.; Christensen, R. Maturity Relationships Among Early to Late Hybrids of Corn (Zea mays L.). *Crop Sci.* **1965**, *4*.
10. Esau, K. *Anatomy of Seed Plants*, 2nd ed.; John Wiley & Sons: New York, NY, USA, 1977.
11. Kladnik, A.; Chamusco, K.; Dermastia, M.; Chourey, P. Evidence of Programmed Cell Death in Post-Phloem Transport Cells of the Maternal Pedicel Tissue in Developing Caryopsis of Maize. *Plant Physiol.* **2004**, *136*, 3572–3581. [CrossRef]
12. Felker, F.C.; Shannon, J.C. Movement of C-labeled Assimilates into Kernels of Zea mays L. *Plant Physiol.* **1980**, *65*, 7. [CrossRef] [PubMed]
13. Schel, J.H.N.; Kieft, H.; Lammeren, A.A.M.V. Interactions between embryo and endosperm during early developmental stages of maize caryopses (*Zea mays*). *Can. J. Bot.* **1984**, *62*, 2842–2853. [CrossRef]
14. Thorne, J.H. Phloem Unloading of C and N Assimilates in Developing Seeds. *Plant Physiol.* **1985**, *36*, 27. [CrossRef]
15. Lowe, J.; Nelson, O.E. Miniature seed-A study in the development of a defective caryopsis in maize. *Genetics* **1946**, *31*, 11.
16. Oeljeklaus, S.; Meyer, H.E.; Warscheid, B. Advancements in plant proteomics using quantitative mass spectrometry. *J. Proteom.* **2009**, *72*, 545–554. [CrossRef]
17. Méchin, V.; Thévenot, C.; Le Guilloux, M.; Prioul, J.-L.; Damerval, C. Developmental Analysis of Maize Endosperm Proteome Suggests a Pivotal Role for Pyruvate Orthophosphate Dikinase. *Plant Physiol.* **2007**, *143*, 1203–1219. [CrossRef]
18. Huang, H.; Møller, I.M.; Song, S.-Q. Proteomics of desiccation tolerance during development and germination of maize embryos. *J. Proteom.* **2012**, *75*, 1247–1262. [CrossRef]
19. Jin, X.; Fu, Z.; Ding, D.; Li, W.; Liu, Z.; Tang, J. Proteomic Identification of Genes Associated with Maize Grain-Filling Rate. *PLoS ONE* **2013**, *8*, e59353. [CrossRef]
20. Sabelli, P.A.; Liu, Y.; Dante, R.A.; Lizarraga, L.E.; Nguyen, H.N.; Brown, S.W.; Klingler, J.P.; Yu, J.; LaBrant, E.; Layton, T.M.; et al. Control of cell proliferation, endoreduplication, cell size, and cell death by the retinoblastoma-related pathway in maize endosperm. *Proc. Natl. Acad. Sci. USA* **2013**, *110*, E1827–E1836. [CrossRef]
21. Balsamo, G.M.; de Mello, C.S.; Arisi, A.C.M. Proteome Comparison of Grains from Two Maize Genotypes, with Colorless Kernel Pericarp (*P1-ww*) and Red Kernel Pericarp (*P1-rr*). *Food Biotechnol.* **2016**, *30*, 110–122. [CrossRef]
22. Yu, T.; Li, G.; Dong, S.; Liu, P.; Zhang, J.; Zhao, B. Proteomic analysis of maize grain development using iTRAQ reveals temporal programs of diverse metabolic processes. *BMC Plant Biol.* **2016**, *16*, 241. [CrossRef] [PubMed]

23. Zhang, L.; Dong, Y.; Wang, Q.; Du, C.; Xiong, W.; Li, X.; Zhu, S.; Li, Y. iTRAQ-Based Proteomics Analysis and Network Integration for Kernel Tissue Development in Maize. *Int. J. Mol. Sci.* **2017**, *18*, 1840. [CrossRef] [PubMed]

24. Silva-Sanchez, C.; Chen, S.; Zhu, N.; Li, Q.-B.; Chourey, P.S. Proteomic comparison of basal endosperm in maize miniature1 mutant and its wild-type Mn1. *Front. Plant Sci.* **2013**, *4*, 211. [CrossRef] [PubMed]

25. Wang, G.; Wang, G.; Wang, J.; Du, Y.; Yao, D.; Shuai, B.; Han, L.; Tang, Y.; Song, R. Comprehensive proteomic analysis of developing protein bodies in maize (*Zea mays*) endosperm provides novel insights into its biogenesis. *J. Exp. Biol.* **2016**, *67*, 6323–6335. [CrossRef]

26. Albrethsen, J. The first decade of MALDI protein profiling: A lesson in translational biomarker research. *J. Proteom.* **2011**, *74*, 765–773. [CrossRef]

27. Mehta, A.; Silva, L.P. MALDI-TOF MS profiling approach: How much can we get from it? *Front. Plant Sci.* **2015**, *6*, 184. [CrossRef]

28. Olsen, O.-A. ENDOSPERM DEVELOPMENT: Cellularization and Cell Fate Specification. *Annu. Rev. Plant. Physiol. Plant. Mol. Biol.* **2001**, *52*, 233–267. [CrossRef]

29. Sabelli, P.A.; Larkins, B.A. The Development of Endosperm in Grasses. *Plant Physiol.* **2009**, *149*, 14–26. [CrossRef]

30. Chen, J.; Zeng, B.; Zhang, M.; Xie, S.; Wang, G.; Hauck, A.; Lai, J. Dynamic Transcriptome Landscape of Maize Embryo and Endosperm Development. *Plant Physiol.* **2014**, *166*, 252–264. [CrossRef]

31. Wittmann-Liebold, B.; Graack, H.-R.; Pohl, T. Two-dimensional gel electrophoresis as tool for proteomics studies in combination with protein identification by mass spectrometry. *PROTEOMICS* **2006**, *6*, 4688–4703. [CrossRef]

32. Ruedi Aebersold; Matthias Mann Mass spectrometry-based proteomics. *Nature* **2003**, *422*, 198–207. [CrossRef] [PubMed]

33. Ong, S.E.; Blagoev, B.; Kratchmarova, I. Stable isotope labeling by amino acids in cell culture, SILAC, as a simple and accurate approach to expression proteomics. *Mol. Cell Proteomics* **2002**, *1*, 376–386. [CrossRef] [PubMed]

34. Mani, D.R.; Abbatiello, S.E.; Carr, S.A. Statistical characterization of multiple-reaction monitoring mass spectrometry (MRM-MS) assays for quantitative proteomics. *BMC Bioinformatics* **2012**, *13*, S9. [CrossRef] [PubMed]

35. Ludwig, C.; Gillet, L.; Rosenberger, G.; Amon, S.; Collins, B.C.; Aebersold, R. Data-independent acquisition-based SWATH-MS for quantitative proteomics: A tutorial. *Mol. Syst. Biol.* **2018**, *14*, e8126. [CrossRef]

36. Cox, J.; Hein, M.Y.; Luber, C.A. Accurate Proteome-wide Label-free Quantification by Delayed Normalization and Maximal Peptide Ratio Extraction, Termed MaxLFQ. *Mol. Cell Proteomics* **2014**, *13*, 2513–2526. [CrossRef]

37. Köcher, T.; Pichler, P.; Schutzbier, M.; Stingl, C.; Kaul, A.; Teucher, N.; Hasenfuss, G.; Penninger, J.M.; Mechtler, K. High Precision Quantitative Proteomics Using iTRAQ on an LTQ Orbitrap: A New Mass Spectrometric Method Combining the Benefits of All. *J. Proteome Res.* **2009**, *8*, 4743–4752. [CrossRef]

38. Ge, F.; Hu, H.; Huang, X. Metabolomic and Proteomic Analysis of Maize Embryonic Callus induced from immature embryo. *Sci. Rep.* **2017**, *7*, 1004. [CrossRef]

39. Liu, S.; Zenda, T.; Dong, A.; Yang, Y.; Liu, X.; Wang, Y.; Li, J.; Tao, Y.; Duan, H. Comparative Proteomic and Morpho-Physiological Analyses of Maize Wild-Type Vp16 and Mutant vp16 Germinating Seed Responses to PEG-Induced Drought Stress. *Int. J. Mol. Sci.* **2019**, *20*, 5586. [CrossRef]

40. Jiang, Z.; Jin, F.; Shan, X. iTRAQ-Based Proteomic Analysis Reveals Several Strategies to Cope with Drought Stress in MaizeSeedlings. *Int. J. Mol. Sci.* **2019**, *20*, 5956. [CrossRef]

41. Yu, G.; Lv, Y.; Shen, L.; Wang, Y.; Qing, Y.; Wu, N.; Li, Y.; Huang, H.; Zhang, N.; Liu, Y.; et al. The Proteomic Analysis of Maize Endosperm Protein Enriched by Phos-tagtm Reveals the Phosphorylation of Brittle-2 Subunit of ADP-Glc Pyrophosphorylase in Starch Biosynthesis Process. *Int. J. Mol. Sci.* **2019**, *20*, 986. [CrossRef]

42. Prioul, J.L.; Méchin, V.; Lessard, P.; Thévenot, C.; Grimmer, M.; Chateau-Joubert, S.; Coates, S.; Hartings, H.; Kloiber-Maitz, M.; Murigneux, A.; et al. A joint transcriptomic, proteomic and metabolic analysis of maize endosperm development and starch filling. *Plant Biotechnol. J.* **2008**, *6*, 855–869. [CrossRef] [PubMed]

43. Tan, Y.; Tong, Z.; Yang, Q.; Sun, Y.; Jin, X.; Peng, C.; Guo, A.; Wang, X. Proteomic analysis of phytase transgenic and non-transgenic maize seeds. *Sci. Rep.* **2017**, *7*, 9246. [CrossRef] [PubMed]

44. Niu, L.; Ding, H.; Zhang, J.; Wang, W. Proteomic Analysis of Starch Biosynthesis in Maize Seeds. *Starch - Stärke* **2019**, *71*. [CrossRef]
45. Bonello, J.-F.; Sevilla-Lecoq, S.; Berne, A.; Risueño, M.-C.; Dumas, C.; Rogowsky, P.M. Esr proteins are secreted by the cells of the embryo surrounding region. *J. Exp. Biol.* **2002**, *53*, 1559–1568. [CrossRef]
46. Sharma, V.K.; Ramirez, J.; Fletcher, J.C. The ArabidopsisCLV3-like (CLE) genes are expressed in diverse tissues and encode secreted proteins. *Plant Mol. Biol.* **2003**, *51*, 415–425. [CrossRef]
47. Bate, N.J.; Niu, X.; Wang, Y.; Reimann, K.S.; Helentjaris, T.G. An Invertase Inhibitor from Maize Localizes to the Embryo Surrounding Region during Early Kernel Development. *Plant Physiol.* **2004**, *134*, 246–254. [CrossRef]
48. Balandín, M.; Royo, J.; Gómez, E.; Muniz, L.M.; Molina, A.; Hueros, G. A protective role for the embryo surrounding region of the maize endosperm, as evidenced by the characterisation of ZmESR-6, a defensin gene specifically expressed in this region. *Plant Mol. Biol.* **2005**, *58*, 269–282. [CrossRef]
49. Tian, T.; Liu, Y.; Yan, H.; You, Q.; Yi, X.; Du, Z.; Xu, W.; Su, Z. agriGO v2.0: A GO analysis toolkit for the agricultural community, 2017 update. *Nucleic Acids Res.* **2017**, *45*, W122–W129. [CrossRef]
50. Liu, J.; Park, C.H.; He, F.; Nagano, M. The RhoGAP SPIN6 Associates with SPL11 and OsRac1 and Negatively Regulates Programmed Cell Death and Innate Immunity in Rice. *PLoS Pathog.* **2015**, *11*, e1004629. [CrossRef]
51. Thao, N.P.; Chen, L.; Nakashima, A.; Hara, S. RAR1 and HSP90 Form a Complex with Rac/Rop GTPase and Function in Innate-Immune Responses in Rice. *Plant Cell* **2007**, *19*, 4035–4045. [CrossRef]
52. Pnueli, L.; Liang, H.; Rozenberg, M.; Mittler, R. Growth suppression, altered stomatal responses, and augmented induction of heat shock proteins in cytosolic ascorbate peroxidase (Apx1)-deficient Arabidopsis plants. *Plant J.* **2003**, *34*, 187–203. [CrossRef] [PubMed]
53. Sharma, P.; Dubey, R.S. Ascorbate peroxidase from rice seedlings: Properties of enzyme isoforms, effects of stresses and protective roles of osmolytes. *Plant Sci.* **2004**, *167*, 541–550. [CrossRef]
54. Davletova, S.; Rizhsky, L.; Liang, H.; Shengqiang, Z.; Oliver, D.J.; Coutu, J.; Shulaev, V.; Schlauch, K.; Mittler, R. Cytosolic Ascorbate Peroxidase 1 Is a Central Component of the Reactive Oxygen Gene Network of Arabidopsis. *Plant Cell* **2005**, *17*, 268–281. [CrossRef] [PubMed]
55. Miller, G.; Suzuki, N.; Rizhsky, L.; Hegie, A.; Koussevitzky, S.; Mittler, R. Double Mutants Deficient in Cytosolic and Thylakoid Ascorbate Peroxidase Reveal a Complex Mode of Interaction between Reactive Oxygen Species, Plant Development, and Response to Abiotic Stresses. *Plant Physiol.* **2007**, *144*, 1777–1785. [CrossRef]
56. Lu, R. High throughput virus-induced gene silencing implicates heat shock protein 90 in plant disease resistance. *EMBO J.* **2003**, *22*, 5690–5699. [CrossRef]
57. Wang, W.; Vinocur, B.; Shoseyov, O.; Altman, A. Role of plant heat-shock proteins and molecular chaperones in the abiotic stress response. *Trends Plant Sci.* **2004**, *9*, 244–252. [CrossRef]
58. Neta-Sharir, I.; Isaacson, T.; Lurie, S.; Weiss, D. Dual Role for Tomato Heat Shock Protein 21: Protecting Photosystem II from Oxidative Stress and Promoting Color Changes during Fruit Maturation. *Plant Cell* **2005**, *17*, 1829–1838. [CrossRef]
59. Nishizawa, A.; Yabuta, Y.; Yoshida, E.; Maruta, T.; Yoshimura, K.; Shigeoka, S. Arabidopsis heat shock transcription factor A2 as a key regulator in response to several types of environmental stress. *Plant J.* **2006**, *48*, 535–547. [CrossRef]
60. Kotak, S.; Larkindale, J.; Lee, U.; von Koskull-Döring, P.; Vierling, E.; Scharf, K.-D. Complexity of the heat stress response in plants. *Curr. Opin. Plant Biol.* **2007**, *10*, 310–316. [CrossRef]
61. Leprince, O.; Hendry, G.A.F.; McKersie, B.D. The mechanisms of desiccation tolerance in developing seeds. *Seed Sci. Res.* **1993**, *3*, 231–246. [CrossRef]
62. Oliver, M.J. Desiccation tolerance in vegetative plant cells. *Physiol. Plant.* **1996**, *97*, 779–787. [CrossRef]
63. Angelovici, R.; Galili, G.; Fernie, A.R.; Fait, A. Seed desiccation: A bridge between maturation and germination. *Trends Plant Sci.* **2010**, *15*, 211–218. [CrossRef] [PubMed]
64. Serna, A.; Maitz, M.; O'Connell, T.; Santandrea, G.; Thevissen, K.; Tienens, K.; Hueros, G.; Faleri, C.; Cai, G.; Lottspeich, F.; et al. Maize endosperm secretes a novel antifungal protein into adjacent maternal tissue: Maize basal endosperm antifungal protein. *Plant J.* **2001**, *25*, 687–698. [CrossRef] [PubMed]
65. Linnestad, C.; Doan, D.N.P.; Brown, R.C.; Lemmon, B.E.; Meyer, D.J.; Jung, R.; Olsen, O.-A. Nucellain, a Barley Homolog of the Dicot Vacuolar-Processing Protease, Is Localized in Nucellar Cell Walls. *Plant Physiol.* **1998**, *118*, 1169–1180. [CrossRef] [PubMed]

66. Zhu, Y.; Li, H.; Bhatti, S.; Zhou, S.; Yang, Y.; Fish, T.; Thannhauser, T.W. Development of a laser capture microscope-based single-cell-type proteomics tool for studying proteomes of individual cell layers of plant roots. *Hortic. Res.* **2016**, *3*, 16026. [CrossRef]
67. Nakazono, M.; Qiu, F.; Borsuk, L.A.; Schnable, P.S. Laser-Capture Microdissection, a Tool for the Global Analysis of Gene Expression in Specific Plant Cell Types: Identification of Genes Expressed Differentially in Epidermal Cells or Vascular Tissues of Maize. *Plant Cell* **2003**, *15*, 583–596. [CrossRef]
68. Wiśniewski, J.R.; Zougman, A.; Mann, M. Combination of FASP and StageTip-Based Fractionation Allows In-Depth Analysis of the Hippocampal Membrane Proteome. *J. Proteome Res.* **2009**, *8*, 5674–5678. [CrossRef]
69. Whitaker, J.R.; Granum, P.E. An absolute method for protein determination based on difference in absorbance at 235 and 280 nm. *Anal. Biochem.* **1980**, *109*, 156–159. [CrossRef]

International Journal of
Molecular Sciences

Article

S-Nitroso-Proteome Revealed in Stomatal Guard Cell Response to Flg22

Sheldon R. Lawrence II [1,2]**, Meghan Gaitens** [2]**, Qijie Guan** [2]**, Craig Dufresne** [3] **and Sixue Chen** [1,2,4,]*****

[1] Plant Molecular and Cellular Biology Program, University of Florida, Gainesville, FL 32610, USA; s.lawrence@ufl.edu

[2] Department of Biology, University of Florida Genetics Institute, Gainesville, FL 32611, USA; mgaitens@ufl.edu (M.G.); qijie.guan@ufl.edu (Q.G.)

[3] Thermo Fisher Scientific, 1400 Northpoint Parkway, West Palm Beach, FL 33407, USA; Craig.dufresne@thermofisher.com

[4] Proteomics and Mass Spectrometry, Interdisciplinary Center for Biotechnology Research, University of Florida, Gainesville, FL 32610, USA

***** Correspondence: schen@ufl.edu

Received: 10 February 2020; Accepted: 28 February 2020; Published: 1 March 2020

Abstract: Nitric oxide (NO) plays an important role in stomata closure induced by environmental stimuli including pathogens. During pathogen challenge, nitric oxide (NO) acts as a second messenger in guard cell signaling networks to activate downstream responses leading to stomata closure. One means by which NO's action is achieved is through the posttranslational modification of cysteine residue(s) of target proteins. Although the roles of NO have been well studied in plant tissues and seedlings, far less is known about NO signaling and, more specifically, protein S-nitrosylation (SNO) in stomatal guard cells. In this study, using iodoTMTRAQ quantitative proteomics technology, we analyzed changes in protein SNO modification in guard cells of reference plant *Arabidopsis thaliana* in response to flg22, an elicitor-active peptide derived from bacterial flagellin. A total of 41 SNO-modified peptides corresponding to 35 proteins were identified. The proteins cover a wide range of functions, including energy metabolism, transport, stress response, photosynthesis, and cell–cell communication. This study creates the first inventory of previously unknown NO responsive proteins in guard cell immune responses and establishes a foundation for future research toward understanding the molecular mechanisms and regulatory roles of SNO in stomata immunity against bacterial pathogens.

Keywords: *Arabidopsis thaliana*; redox proteomics; flg22; iodoTMTRAQ; nitric oxide; stomatal immunity

---

1. Introduction

Crop stress due to bacterial pathogens results in sizeable losses in economic revenue annually and threatens global food security [1,2]. Bacteria are unable to penetrate the plant epidermis and enter plants primarily through stomatal pores. They establish themselves in the apoplast where they utilize plants' nutrient resources and suppress plant immune defense. Stomatal pores are controlled by pairs of specialized guard cells, which swell and shrink as a result of the influx or efflux of ions and water in order to adjust the stomatal aperture in response to environmental stimuli such as light, CO_2, pathogen and drought [3]. Because of the importance of stomata as a plant's first line of defense against bacterial pathogens, guard cell signal transduction is essential for bacterial invasion and plant–pathogen interactions.

To rapidly respond to environmental stimuli such as bacterial pathogens, guard cells employ posttranslational modifications (PTMs) of key proteins in signaling pathways [4]. The PTMs of these proteins function as molecular switches to modulate protein functions, thus adding a level of regulation to plant defense processes [5,6]. Increasing studies have been conducted to elucidate the function of PTMs on specific plant proteins [7–10]. Among the PTMs, S-nitrosylation (SNO) is one of the most common modifications found in plants and its specific involvement in plant stress signaling has been under much investigation over the years [11–13].

Nitric oxide (NO) is an essential signaling molecule that functions in myriad physiological processes in plants, including hormone signaling, flowering, growth and development, stomata closure and cell death [14–18]. In plants, NO can be produced via different routes that involve enzymatic or nonenzymatic reactions and a number of different cellular compartments, including chloroplast, mitochondria, peroxisomes, cytosol and plasma membrane [19–23]. The most well understood source of NO is the cytosolic localized nitrite-dependent nitrate reductase pathway. This enzymatic source of NO is produced through the reduction of nitrate to nitrite catalyzed by the enzyme nitrate reductase (NR), which uses NAD(P)H as an electron donor [24]. Nitrite can further be reduced to NO following a reaction similar to that of nitric oxide synthase found in animals. Under normal physiological conditions, NO is present in the cell at low and controlled levels, where it acts as a crucial regulator involved in every stage of normal plant growth and development. However, under abiotic and biotic stresses, NO levels can increase to harmful levels leading to cellular damage. In plants, NO is known to modulate the expression of genes involved in primary metabolism, hormonal signaling and stress responses [25,26].

Over the years, the role of NO as a component of plant immunity against pathogens has become increasingly evident [27–30]. Pathogen infection triggers NO production. At the early stage of infection, the burst of NO results in S-nitrosothiols that lead to enhanced cell death [31]. However, at later stages the increased levels of NO decrease cell death via nitrosylation of respiratory burst oxidase homolog (RBOH), leading to a reduction in its activity and oxidative stress. This highlights the versatility of NO in the fine-tuning of cell death in response to stress. Additionally, other studies have helped to support NO function in plant defense via SNO. For example, SNO of the transcriptional co-regulator nonexpresser of PR gene 1 (NPR1) at cysteine 156 promotes its localization in the cytoplasm as inactive oligomers [32]. As well, the sucrose non-fermenting-1-related protein kinase (SnRK 2.6)/open stomata 1 (OST1), a positive regulator of abscisic acid (ABA)-induced stomata closure, is regulated via SNO [33]. SNO of OST1 at the cysteine residue (Cys137) abolished the kinase activity, indicating a negative feedback mechanism by which NO helps fine-tune control of ABA signaling in guard cells. In addition, S-nitrosoglutathione reductase 1 (GSNOR1), an important enzyme involved maintaining NO and glutathione levels in the cell, was shown to be regulated by nitrosylation [34]. These studies highlight the importance of SNO in plant defense. However, little is known about NO regulation via protein nitrosylation in guard cell functions. In addition, many redox proteomic studies do not address the issue of protein turnover, which often leads to misleading results of redox regulation [35,36].

In an effort to gain a greater understanding of the role of nitrosylation in plant responses to bacterial stress and guard cell signaling, we have employed a quantitative proteomics approach in analyzing the S-nitroso-proteome of *Arabidopsis thaliana* guard cells. Using a double-labeling strategy (iTRAQ and iodoTMT) termed iodoTMTRAQ, in which redox changes such as SNO and protein level changes can be monitored in one experiment, we identified a total of 41 SNO-modified peptides, which corresponded to 35 SNO-responsive proteins, and they were significantly changed in response to flg22 treatment. The proteins function in energy metabolism, transport, stress response, photosynthesis, and cell–cell communication. These results reveal an inventory of SNO-responsive proteins and their associated pathways in guard cell flg22 responses and will contribute to our understanding of molecular mechanisms underlying guard cell pathogen signaling.

2. Results and Discussion

2.1. Enriched Arabidopsis Guard Cells Are Viable

To maintain the stomatal movement output that enables morphological, physiological and other biological studies, a method for isolating intact stomatal guard cells rather than guard cell protoplasts was developed. In this method, *A. thaliana* leaves were partitioned between two portions of transparent Scotch tape and peeled apart to separate the abaxial side. The abaxial side of the peels was transferred to an enzyme solution to digest away the epidermal cells. After 20 minutes of digestion, the stomatal guard cells were left to recover for 60 min under light. Peels were then assessed for guard cell viability and purity using neutral red and fluorescein diacetate (FDA), respectively. The guard cells were the dominant cell type (> 90%) on the peels and were intact and viable (Figure 1).

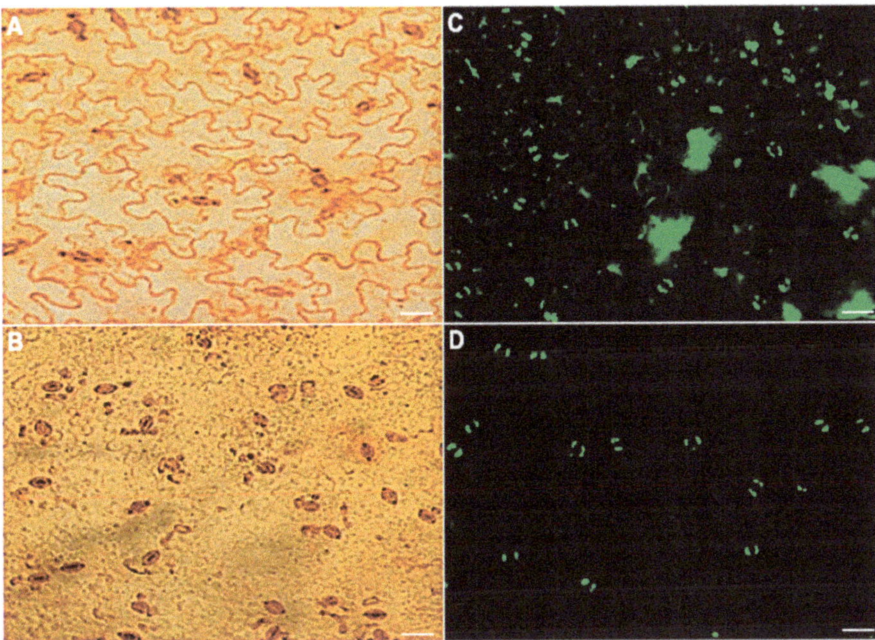

Figure 1. Stomatal guard cell purity and viability before and after the removal of mesophyll and epidermal cells. Neutral red viability staining of epidermal peels: (**A**) before digestion and (**B**) after digestion. FDA viability staining of the peels: (**C**) before digestion and (**D**) 60 min after digestion. Scale bar: 30 μm.

2.2. Flg22 Induction of Stomatal Closure and Production of Reactie Oxygen Species (ROS) and NO

The bacterial flagellin peptide flg22 is a well-known plant defense elicitor that induces stomatal closure [37]. To test whether this response was present in the enriched *A. thaliana* guard cells using the above method (Figure 1), 10 μM flg22 was added to the guard cell peels and the stomatal aperture was monitored. Upon flg22 treatment, stomata begin closing as early as 15 min after treatment, and after 60 min stomata aperture was decreased by more than half (Figure 2). Next, the ability of the enriched stomatal guard cells to produce ROS and NO was tested. The redox indicator compounds $H_2DCF\text{-}DA$ and DAF-2DA were used to measure ROS and NO levels, respectively. We show that 10 μM flg22 treatment induced ROS (Figure 3) and NO production (Figure 4) in the enriched stomatal guard cells. Guard cells were able to produce ROS and NO with the highest levels at 15 minutes and 30 minutes, respectively, after the treatments. These results suggest that ROS, NO, and/or overall guard cell redox state play significant roles in guard cell signaling that leads to stomatal closure. Furthermore, they show that guard cell molecular and biochemical responses remain intact in the enriched guard cell peels.

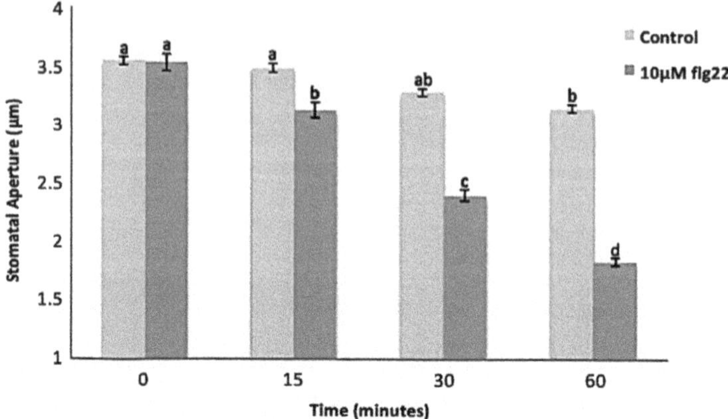

Figure 2. Stomatal movement in response to 10 µM flg22. Data were obtained from 180 stomata from three independent experiments and presented as means ± SE. Different letters indicate significantly different mean values at $p < 0.05$.

A

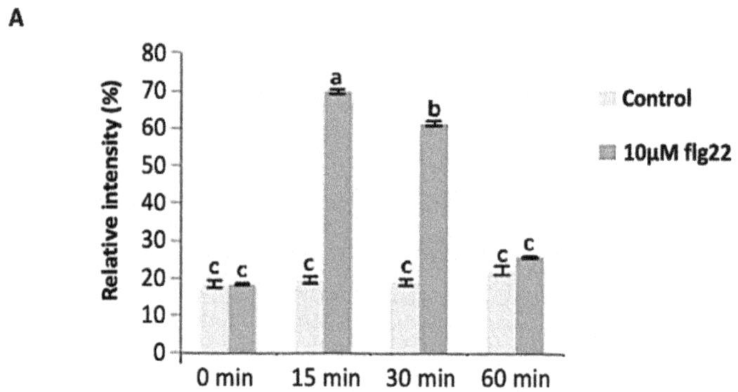

B

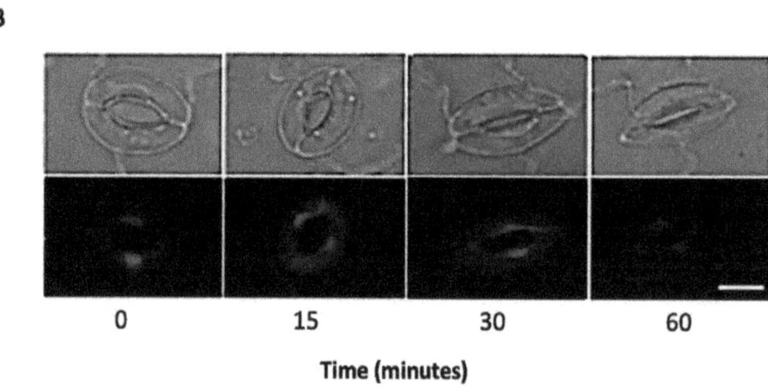

Figure 3. Reactive oxygen species (ROS) sproduction in guard cells in response to 10 µM flg22. (**A**) ROS levels measured from a total of 180 stomata from three independent experiments and presented as means ± SE. Different letters indicate significantly different mean values at $p < 0.05$. (**B**) Representative images of stomatal guard cells at each time point are shown. Scale bar: 10 µm.

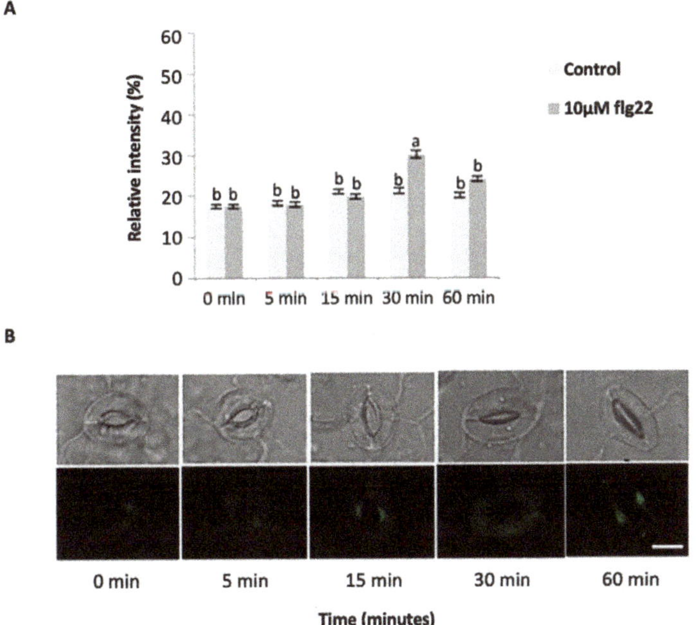

Figure 4. Guard cell nitric oxide (NO) levels in response to 10 μM flg22. (**A**) NO levels measured from a total of 180 stomata from three independent experiments and presented as means ± SE. Different letters indicate significantly different mean values at $p < 0.05$. (**B**) Representative images of stomatal guard cells at each time point are shown. Scale bar: 10 μm.

2.3. Identification of flg22-Regulated Proteins and Nitrosylated Proteins

To identify nitrosylated proteins and uncover their roles in flg22-triggered stomatal closure, we used a double-labeling strategy with iodoTMT and iTRAQ, which allows for simultaneous analysis of cysteine redox changes and total protein level change [38,39]. In this strategy, we conducted redox proteomic analysis of control and treatment samples at 15, 30, and 60 min time points. Proteins were extracted in the presence of N-ethylmaleimide (NEM) to irreversibly block free thiols and prevent artificial cysteine oxidation during protein extraction. Then, the nitrosylated cysteine residues were reduced with sodium ascorbate, and the protein thiols were labeled with iodoTMT tags for quantification. The iodoTMT proteins were digested with trypsin and the peptides were labeled with iTRAQ tags. This reverse-labeling strategy allows for the maintenance of the redox state of the proteins after treatment and prevents artificial oxidation during sample preparation. Therefore, the differences in iodoTMT signals from specific peptides derived from treated samples compared to control samples indicate the presence of redox-sensitive nitrosylated cysteine residues. Furthermore, the differences in iTRAQ tags from control compared to treatment samples indicate the change in total protein level between the two during the stomatal closure process. In this study, a total of 433, 578, and 463 unique cysteine-containing peptides corresponding to 229, 283, and 237 proteins were confidently identified (FDR 0.01) among three biological replicates of the samples treated with 10 μM flg22 for 15, 30, and 60 min, respectively. In order to determine the flg22-redox sensitive cysteines, we first compared the relative peak intensity in the control and treatment samples. A p-value smaller than 0.05 and a threshold of greater than 1.2 or less than 0.8 were set as criteria to determine significant differences between the control and treatment. A total of 41 SNO modified peptides corresponding to 6, 12, and 21 potential NO regulated redox proteins were identified at 15, 30, and 60 min, respectively. Among these proteins, 6, 2, and 17 underwent oxidation, while 0, 10, and 4 underwent reduction after flg22 treatment at 15, 30, and 60 min, respectively (Table 1).

Table 1. S-nitrosylated proteins identified in guard cells after flg22 treatment for 15 min, 30 min and 60 min. The tick under DiANNA indicates the predicted cysteine residue involved in disulfide bond formation.

Protein Accession	Peptide	Protein Description	Fold Change	*p*-value (FDR-adj.)	Biological Function	DiANNA
	15 min after flg22 treatment					
AT2G10940.1	NPPPGYTCSI	Lipid-transfer protein (LTP) II	4.72	0.016	Lipid binding	√
AT2G10940.1	ATCPIDTLK	LTP II	4.67	0.025	Lipid binding	√
AT3G23810.1	FDNLYGCR	S-adenosyl-L-homocysteine hydrolase 2 (SAHH2)	3.29	0.024	Carbon metabolic process	√
AT3G44310.2	IGAAICWENR	Nitrilase (NIT) 1	3.18	0.003	Nitrogen compound metabolic process	√
AT3G08560.1	IVCENTLDAR	Vacuolar H$^+$-ATPase (VHA) E2	2.94	0.001	Hydrogen ion transport	√
AT1G78830.1	CLGYFYK	Curculin-like lectin protein (CLP)	1.91	0.020	Response to cytokinin, Stress response	√
AT1G78850.1	GLLGWDETCK	Mannose binding lectin (MBL) 1	1.90	0.010	Mannose binding	√
AT1G78850.1	SPSLASCDPK	MBL 1	1.67	0.039	Mannose binding	√
AT5G54270.1	WAMLGAFGCITPEVLQK	Chlorophyll a-b binding protein (CBP) 3	1.48	0.032	Photosynthesis	–
	30 min after flg22 treatment					
AT2G10940.1	NPPPGYTCSI	LTP II	2.22	0.005	Lipid binding	√
AT5G17220.1	LYGQVTAACPQR	Glutathione S-transferase (GST) phi 12	1.53	0.049	Nitrogen compound metabolism, Stress response	√
AT1G26850.3	CLIPWGANDGMYLMEVDR	S-adenosyl-L-methionine-methyltransferase (SAMMT)	0.64	0.035	Methylation	√
AT5G54270.1	WAMLGAFGCITPEVLQK	CBP 3	0.57	0.047	Photosynthesis	–
AT1G78850.1	GLLGWDETCK	MBL1	0.55	0.023	Mannose binding	√
AT1G65590.1	VVPFEPGSCLAQ	Beta-hexosaminidase (HAD) 3	0.54	0.021	Lipid metabolic process	√
AT3G01500.1	VCPSHVLDFQPGDAFVVR	Beta-carbonic anhydrase (CA) 1	0.49	0.047	Carbon utilization, Stress response	√
AT4G39710.1	SGLGFCDLDVGFGDEAPR	FK506-binding protein (FKBP) 16-2	0.43	0.012	Photosynthesis	√
AT5G63800.1	SPDAPDPVINTCNGMK	Beta-galactosidase (GALD) 6	0.38	0.035	Carbohydrate metabolism	√
AT3G52500.1	YLCSGCDFSGLDPTLIPR	Aspartyl protease (APP)	0.30	0.039	Cellular process	√
AT4G28520.3	VVPGCAETFMDSQPMQGQQQGQPWQGR	Cruciferin (CRU) 3	0.15	0.045	Response to abscisic acid	√
AT5G44120.3	VIPGCAETFQDSSEFQPR	CRU 1	0.10	0.027	Response to abscisic acid	√

Table 1. *Cont.*

Protein Accession	Peptide	Protein Description	Fold Change	p-value (FDR-adj.)	Biological Function	DiANNA
60 min after flg22 treatment AT3G44310.2 AT3G44310.3	CIWGQGDGSTIPVYDTPIGK	NIT 1	6.31	0.049	Nitrogen compound metabolic process	√
AT3G23810.1	FDNLYGCR	SAHH2	3.29	0.024	Cysteine and methionine metabolism	√
AT4G11150.1	IVCENTLDAR	VHA E1	2.55	0.003	ATP synthesis	√
AT2G05520.1	QGGGGSGGSYCR	Glycine-rich protein (GRP) 3	2.12	0.028	Stress response	√
ATCG01060.1	CESACPTDFLSVR	PSI PsaC subunit	2.11	0.031	Photosynthesis	√
AT5G38410.3	QVQCISFIAYKPPSFTEA	Ribulose bisphosphate carboxylase (Rubisco) small chain (RBCS)	2.03	0.049	Photosynthesis	√
AT2G30970.1	IAAVQTLSGTGACR	Aspartate aminotransferase (APAT)1	1.99	0.023	Amino acid metabolic process	√
AT2G21060.1	ECSQGGGGYSGGGGGGR	GRP 2	1.96	0.030	Stress response	√
AT3G62940.3	LKPLGLTVSEIKPDGHCLYR	Cysteine proteinase (CP)	1.91	0.008	-	√
AT5G16390.1	QLDCELVIR	Acetylcoenzyme A carboxylase (AAC) 1	1.71	0.001	Fatty acid biosynthetic process	√
ATCG00490.1	VALEACVQAR	Rubisco Large chain (RBCL)	1.71	0.019	Photosynthesis	√
AT2G20360.1	YIQVSCLGASVSSPSR	NAD (P)-binding protein (NBP)	1.45	0.018	Electron transport, Stress response	–
AT5G08680.1	CALVYGQMNEPPGAR	ATP synthase subunit (ASS) beta-3	1.45	0.046	ATP synthesis	–
AT5G23890.1	VIETDTQPSDLCTR	Unknown protein (UNK)	1.45	0.007	–	√
AT3G09820.1	AGCYASNVVIQR	Adenosine kinase (ADK) 1	1.42	0.010	Purine metabolism, response to stimuli	√
AT5G03340.1	YTQGFSGADITEICQR	Cell division cycle (CDC) 48	1.30	0.041	Protein transport, cell division	√
AT4G13010.1	LANAHVTATCGAR	Oxidoreductase (OR)	1.28	0.046	Oxidation and reduction	√
AT1G42970.1	TNPADEECKVYD	Glyceraldehyde-3-phosphate dehydrogenase (GAPDH) B	0.79	0.031	Glucose metabolism, Oxidation and reduction	√
AT5G36700.1	ENPGCLFIATNR	2-phosphoglycolate phosphatase (PGP) 1	0.76	0.043	Photorespiration	√
AT3G01500.3	YMVFACSDSR	CA 1	0.69	0.021	Carbon utilization	–
AT5G14740.1	VLAESESSAFEDQCGR	CA 2	0.69	0.007	Carbon utilization	–

As previously stated, redox-regulated proteins can be challenging to identify due to protein level changes. With the double-labeling strategy, in addition to monitoring redox changes we were able to monitor changes in protein levels. In total, 2614, 3804, and 3327 peptides corresponding to 1429, 2012, and 1745 proteins were confidently identified. When the control and treatment samples were compared via iTRAQ labels, which quantify protein level changes using the same criteria as iodoTMT, a total of 9, 28, and 41 proteins were identified to be significantly changed in samples treated with flg22 at 15, 30, and 60 min, respectively (Supplemental Table S1–S3). Of these 7, 16, and 31 were shown to increase while 2, 12, and 10 decreased at 15, 30, and 60 min after treatment respectively. In this study, iTRAQ data was used to determine if the potential redox-responsive proteins identified by iodoTMT were in fact due to redox changes between the control and treated samples or a result of protein level changes. After considering the protein level changes, 6, 10, and 19 redox proteins were determined to be bona fide nitrosylated proteins that were responsive to the flg22 treatment at 15, 30, and 60 min, respectively.

2.3.1. Guard Cell S-nitroso-proteomic Changes in Response to flg22 at Early Stage of Stomatal Closure

In this study, early stomatal responses were defined at the 15- and 30-minute time-points. They reflect the highest levels of ROS and NO, respectively. At 15 minutes after flg22 treatment, there were seven nitrosylated proteins identified. They include curculin-like lectin family protein, mannose binding lectin protein, bifunctional inhibitor/lipid-transfer protein/seed storage 2S albumin, vacuolar H^+-ATP synthase subunit E1, nitrilase 1 (NIT1), S-adenosyl-l-homocysteine hydrolase 2 (SAHH2), and chlorophyll B-binding protein (Table 1). Analysis of gene ontology (GO) terms revealed that these proteins are involved in protein transport, carbohydrate binding, and nitrogen compound metabolic process. It should be noted that several other proteins were initially thought to be redox-regulated. However, after protein fold change was taken into account, those proteins were found to show significant protein level change. Therefore, the observed redox-fold change in those may have been a result of change in protein levels rather than the cysteine redox response, thus they were discarded as potential nitrosylated proteins. There were nine cysteine-containing peptides from the seven different proteins that showed changes in redox state among control and treated samples. Four of the nine peptides derived from two proteins (D-mannose binding lectin protein and bifunctional inhibitor/lipid-transfer protein/seed storage 2S albumin), while the other five derived from only a single peptide. The subcellular localization of the six proteins was predicted using both internet (YLoc and LocTree3) and literature information. In total, five of the six proteins were predicted to be localized to the cell wall or plasma membrane in addition to one or two different organelles. Individual analysis revealed that five of these proteins, D-mannose binding lectin protein, SAHH2, vacuolar H^+-ATP synthase subunit E1, NIT1, and curculin-like lectin family protein were previously reported to undergo S-nitrosylation following salt or cold stress [40–42]. The sequences of the identified nitrosylated proteins were analyzed for possible intra-molecular disulfide bond formation using DiANNA software (http://clavius.bc.edu/~{}clotelab/DiANNA/). All six of the redox-sensitive proteins were predicted to form intra-molecular disulfide bonds (Table 1). A few of the predicted cysteine(s) were the same as those identified as having the SNO modification. It is important to note that disulfide bond formation is another possible modification. However, in this study we detected SNO modification instead of disulfides.

At 30 min post treatment, 12 nitrosylated proteins were identified. Interestingly, only two of them (glutathione S-transferase phi 12 and bifunctional inhibitor/lipid-transfer protein/seed storage 2S albumin) were found to be in oxidized states, while the other ten proteins (fK506-binding protein, beta-hexosaminidase, d-mannose binding lectin protein, RmlC-like cupins, S-adenosyl-L-methionine-dependent methyltransferase, glycosyl hydrolase family 35 protein, eukaryotic aspartyl protease, cruciferin, chlorophyll B-binding protein, and Beta-carbonic anhydrase) were in a reduced state compared to the control samples (Table 1). Two proteins (d-mannose binding lectin protein and bifunctional inhibitor/lipid-transfer protein/seed storage 2S albumin) were redox-regulated at both 15- and 30-min time points. Analysis of protein level changes revealed that neither of the proteins showed significant protein level changes. Analysis of potential intra-molecular disulfide bond formation revealed that eleven of the twelve potentially nitrosylated proteins were predicted to form intra-molecular disulfide bonds (Table 1). Here, several of the predicted cysteine(s) were the same as those identified as having the SNO modification. Biological processes of these potential redox-regulated proteins include lipid metabolic process, photosynthesis, and nitrogen compound metabolic process (Figure 5). Further analysis revealed that the SNO-proteins identified were localized to the chloroplast, cytoplasm, cell wall, or vacuole with the majority showing localization to the chloroplast and having diverse functions in photosynthetic electron transfer and carbon utilization. In addition, several, including bifunctional inhibitor/lipid-transfer protein/seed storage 2S albumi, glutathione S-transferase, mannose-binding lectin protein, chlorophyll B-binding protein, and Beta-carbonic anhydrase, have been implicated in stomatal movement in response to stress [43–47]. This suggests that these proteins have important roles in stomatal function that may be regulated via S-nitrosylation. In addition, these findings are in agreement with the idea that photosynthesis is highly redox-regulated [48], and it is interesting to know that the regulation in guard cells is via protein nitrosylation.

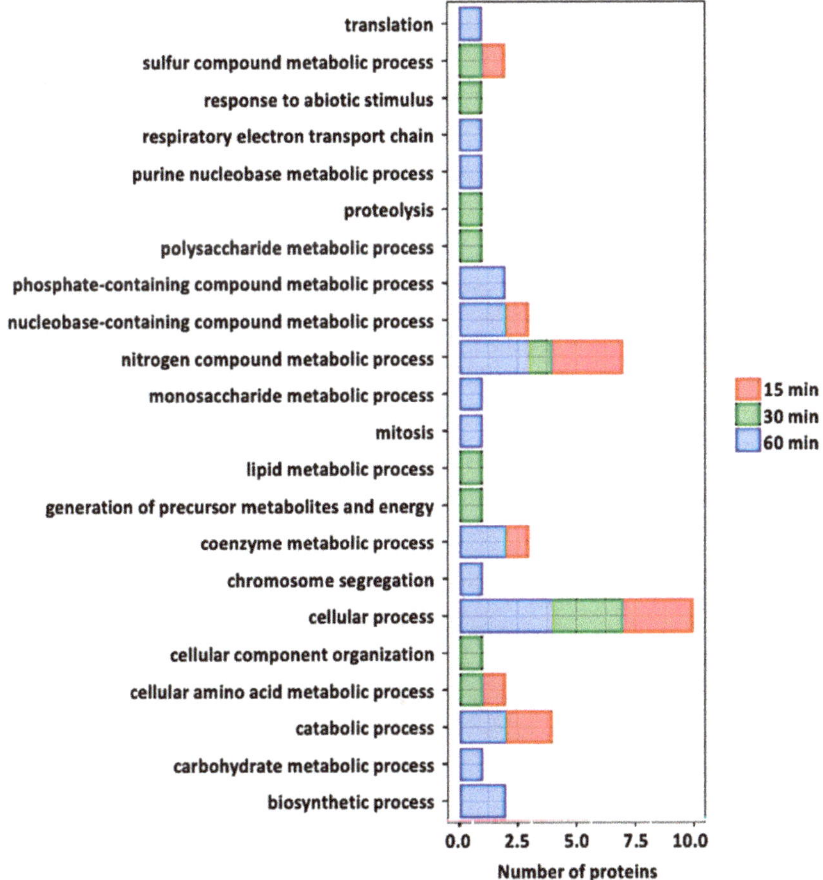

Figure 5. Gene ontology analysis of the redox proteins after flg22 treatment. Relevant biological processes are shown on the *y*-axis. Number of proteins significantly redox-regulated at each time point after treatment is shown on the *x*-axis.

2.3.2. Guard Cell S-nitroso-proteomic Changes in Response to flg22 at Late Stage of Stomatal Closure

At 60 min of flg22 treatment, 21 nitrosylated proteins were identified (Table 1). Seventeen proteins (oxidoreductase/zinc-binding dehydrogenase, cell division cycle 48, adenosine kinase 1, GPI-anchored adhesin-like protein, ATP synthase, NAD (P)-binding protein, ribulose-bisphosphate carboxylase (Rubisco) small subunit, Rubisco large subunit, acetylcoenzyme A carboxylase 1, cysteine proteinase, glycine-rich protein 2B, aspartate aminotransferase 1, PSI iron-sulfur center subunit PsaC, glycine-rich protein 3, vacuolar H^+-ATP synthase subunit E1, NIT1, and SAHH2) were in oxidized states, whereas the other four (carbonic anhydrase 1 and 2, 2-phosphoglycolate phosphatase 1, and glyceraldehyde-3-phosphate dehydrogenase (GAPDH)) were in reduced states compared to control samples (Table 1). It is worth mentioning that carbonic anhydrase was identified to be reduced in both 30 and 60 minutes after the treatment. In addition, NIT1 and SAHH2, previously reported to undergo protein nitrosylation in *Arabidopsis* seedlings [31], were identified in guard cells and were oxidized at both early and late stages of stomatal closure. None of these proteins showed significant protein level changes. Analysis of potential intra-molecular disulfide bonds revealed that 17 of the 21 potential nitrosylated proteins were predicted to form intra-molecular disulfide bonds. Again, several of the predicted cysteine(s) were the same as those identified as having the SNO modification. Analysis of GO terms revealed that the biological processes of the 21 potentially nitrosylated proteins include defense response, transcription, carbohydrate metabolic process, response to ABA, proteolysis, oxidation–reduction process, response to hypoxia, lipid metabolic process, and electron transport, amino acid biosynthetic process (Figure 5).

2.3.3. Proteomic Changes in the Course of flg22-induced Stomatal Closure

Stomatal closure was significant at as early as 15 minutes after flg22 treatment. At 15 min, seven proteins showed a significant increase in expression levels after treatment (Supplemental Table S1). These proteins are involved in cellular processes of cell wall biogenesis/degradation, oxidoreductase process, lipid catabolic process, and stomata regulation. Only two of the total significantly changing proteins showed decreased levels after the treatment. They were light harvesting complex photosystem II (PSII) subunit 6 and chlorophyll a-b binding protein 3, which function in the energy capture process of photosynthesis. The small number of proteins that showed significant changes at 15 minutes after treatment correlates with the number identified to be potentially redox-regulated. At 30 min, 16 proteins showed increased expression levels (Supplemental Table S2). The proteins fell into biological processes that include oxidative–reductive process (malate dehydrogenase 1, bifunctional dTDP-4-dehydrorhamnose 3,5-epimerase/dTDP-4-dehydrorhamnose reductase, pyruvate dehydrogenase E1 subunit alpha-3, formate dehydrogenase, GAPDH, isocitrate dehydrogenase, UDP-glucose 6-dehydrogenase 3, 3-ketoacyl-CoA thiolase 2, and citrate synthase 4), response to cytokinin (60S acidic ribosomal protein P0-2 and ABC transporter B4), response to salt stress (malate dehydrogenase 1, 40S ribosomal protein Sa-1, 60S acidic ribosomal P0-2, isocitrate dehydrogenase, and ras-related protein). It is clear that most of the proteins with increased levels belong to the oxidative—reductive process. These proteins may be critical to maintaining guard cell redox homeostasis in response to stresses. The 12 proteins showing decreased levels after flg22 treatment are involved in processes that include photosynthesis (light harvesting complex PSII subunit 6, PSII CP47 reaction center protein, and cytochrome b6), metabolic energy generation (fructose-bisphosphate aldolase 5, isocitrate dehydrogenase catalytic subunit 6, PSII CP47 reaction center protein, and cytochrome b6), oxidation–reduction process (isocitrate dehydrogenase subunit 6, glutamate dehydrogenase-1, peroxidase, and thioredoxin M4) (Supplementary Table S2).

At 60 min after flg22 treatment, 31 proteins with increased levels were identified (Supplementary Table S3). Analysis of the biological processes revealed that many of the proteins are involved in plant response to stress. For example, leucine-rich repeat (LRR) protein, dehydrin ERD10, 3-ketoacyl-CoA thiolase 2, calcium-dependent protein kinase 9, ascorbate peroxidase 1, and abscisic acid receptor PYL2 are responsive to biotic stress and endogenous stimuli. In addition, many proteins belong to the oxidation-reduction process, and they include GAPDH, malate dehydrogenase 1, succinate dehydrogenase [ubiquinone] iron-sulfur subunit 1, glutamate dehydrogenase 2, and cytochrome P450. Other proteins that showed increases after flg22 treatment are involved in biological processes such as carbohydrate metabolic process (biotin carboxyl carrier protein of acetyl-CoA carboxylase 1, beta-D-xylosidase 1, cellulose synthase A catalytic subunit 1, malate dehydrogenase 1, and GAPDH), as well as cellular process (clathrin light chain 1, calmodulin-like protein 12, cytochrome b6-f complex iron-sulfur subunit). Only a small portion of the identified proteins showed decreased levels at 60 min after flg22 treatment, and they are categorized mostly into metabolic process (beta-D-glucopyranosyl abscisate beta-glucosidase, calreticulin-3, bifunctional 3-dehydroquinate dehydratase/shikimate dehydrogenase, chlorophyll a-b binding protein, transportin-1, fructose-bisphosphate aldolase 1, and cytosolic enolase 3) (Supplementary Table S3).

2.4. Functional Classification of Nitrosylated Proteins

To explore the regulatory roles of protein nitrosylation in specific biological processes, we performed GO analysis to functionally classify the S-nitrosylated proteins identified in this study (Figure 5). We identified 35 nitrosylated proteins that covered a wide range of biological processes, including metabolism, transport, stress response, photosynthetic and other cellular processes. Among the 35 proteins, the most enriched category was related to several metabolic processes, suggesting that metabolism is actively regulated by SNO. As previously stated, the role(s) of a number of the proteins identified in this study have been described in stomatal guard cell function during stress, thus it would be interesting to know whether SNO is central to their role(s).

In addition, a significant number of proteins involved in plant response to stresses were identified to be nitrosylated. These proteins include oxidoreductase, beta-carbonic anhydrase, cruciferin 1 and 3, glycine-rich protein 2, adenosine kinase 1, glutathione S-transferase phi 12, and a curculin-like lectin. Interestingly, a lipid transfer protein LTP-II was identified by two unique peptides as a potential NO-regulated protein (Figure 6). However, its role in plant defense is unclear and future studies to elucidate its role are necessary. We also identified SAHH2 and NIT1, two NO-regulated proteins known to be involved in nitric oxide metabolism and regulating the cellular redox state [41]. This suggests the involvement of redox signaling in the stomatal immune responses. To gain insight to the dynamics of nitrosylation in the time-course experiments, a heat-map of the 35 nitrosylated proteins was generated (Figure 7). As previously stated, the majority of the proteins belonged to metabolic processes. Here a large number were involved in primary metabolism such as carbohydrate, amino acid, and lipid processes. Although no distinct pattern was observed between the time points, it was clear that SNO is represented in many guard cell processes and reveals increases in protein regulation as early as 15 minutes after flg22 treatment. This further provides support that SNO is an important PTM in stomatal responses to bacterial pathogens. In addition, several proteins revealed by this study were previously functionally characterized as NO targets, suggesting the utility and reliability of this iodoTMTRAQ method in the identification of NO target proteins. Despite the identification of many previously unknown NO responsive proteins in guard cells, additional studies will be needed to build a functional network of NO signaling in stomatal immunity against bacterial pathogens. Based on the results presented, it can be reasoned that many of the identified proteins may be important players in guard cell perception of NO during pathogen infection.

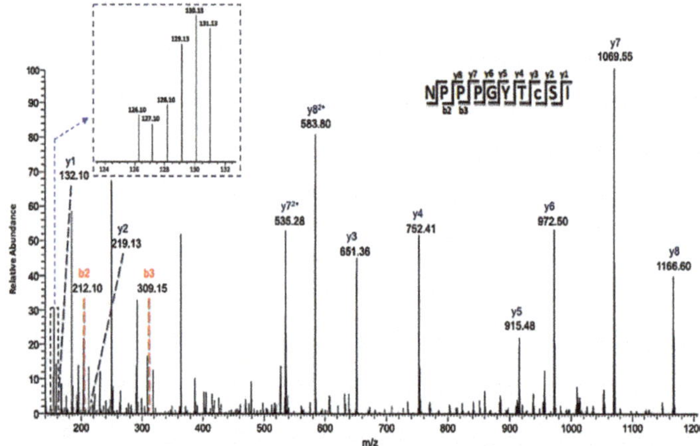

Figure 6. Annotated mass spectrum of a representative cysteine-containing peptide (C8-iodoTMT) showing differential redox status from control and treated samples. The MS/MS ions used to identify the peptide were labeled, and the intensity of the individual iodoTMT peaks for quantification was inserted in the upper left corner. This peptide is one of two that was used to identify lipid-transfer protein isoform II as a potential NO regulated protein.

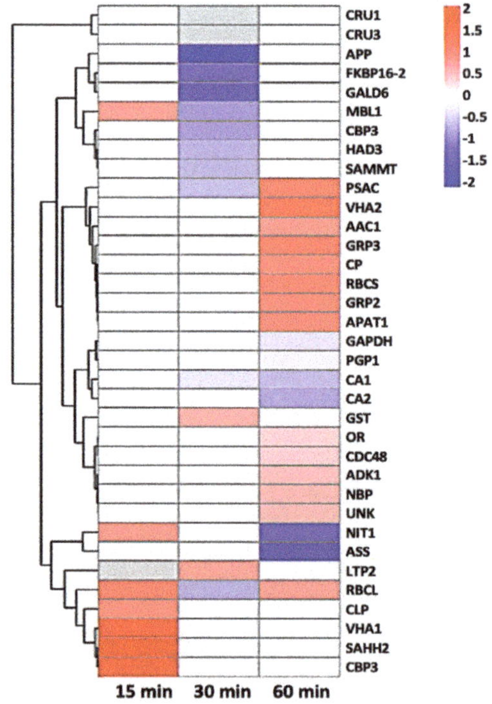

Figure 7. Heat-map of 35 nitrosylated proteins obtained by hierarchical clustering. The columns represent different time point ratios of treatment/control. The rows represent individual proteins. Protein AGI numbers are listed to the right. The increased and decreased proteins are represented in red or green, respectively. The color intensity increases with increasing differences, as shown in the scale bar. Please refer to Table 1 for detailed information.

3. Materials and Methods

3.1. Plant Material

Arabidopsis thaliana ecotype Columbia (Col-0) seeds were obtained from the Arabidopsis Biological Resource Center (ABRC, Columbus, OH, USA). Seeds were germinated in a Metro-Mix 500 potting mixture (The Scotts Co., Marysville, OH, USA). The seedlings were grown in a growth chamber at 140 µmol photons m^{-2}s^{-1} with a photoperiod of 8-hour light at 22 °C and 16-hour dark at 18 °C for 5 weeks. Fully expanded leaves were used for stomatal movement assay and guard cell enrichment.

3.2. Preparation of Epidermal Peels for Stomatal Movement Assay

Epidermal peels were prepared as previously described with minor modifications [49]. Small squares (0.5 × 0.5 cm^2) of leaf sections from the 5-week-old *A. thaliana* plants were fixed abaxial side down onto coverslips coated with a medical adhesive (Hollister, Libertyville, IL, USA). Adaxial epidermis and mesophyll layers were removed with a scalpel. After washing twice with distilled water to remove cellular debris, the coverslips were incubated with a cell wall digesting enzyme mixture at 26 °C, 140 excursions per min on a reciprocal shaker for 15 min. The enzyme mixture was constituted as follows: 0.7% cellulase R-10 (Yakult Honsha Co., Ltd, Tokyo, Japan), 0.025% macerozyme R-10 (Yakult Honsha Co., Ltd, Tokyo, Japan), 0.1% (*w/v*) polyvinylpyrrolidone-40 (Calbiochem, Billerica, MA, USA), and 0.25% (*w/v*) bovine serum albumin (Research Products International Corp., Mt Prospect, Illinois, USA) in a basic solution (0.55 M sorbitol, 0.5 mM CaCl$_2$, 0.5 mM MgCl$_2$, 0.5 mM ascorbic acid, 10 µM KH$_2$PO$_4$, 5 mM 4-morpholineethanesulfonic acid (MES), pH 5.5 adjusted with 1 M KOH). The coverslips were then incubated in an opening buffer (10 mM KCl, 50 µM CaCl$_2$, 10 mM MES-KOH, adjusted to pH 6.15 with 1 M KOH) under light (110 µmol m^{-2}s^{-1}) for an hour. For flg22 experiments, the stomata were treated with either 10 µM flg22 or water (mock). A total of 60 stomatal apertures for each individual experiment were measured with imaging software Image J version 1.50 (NIH, USA). Three independent experiments were conducted. Standard error and significance at a *p*-value < 0.05 were calculated using Microsoft Excel.

3.3. Stomatal ROS and NO Measurement

For ROS measurement, 50 µM of the redox-sensitive 2′-7′-dihydro-dichlorofluorescein diacetate (H$_2$DCF-DA) was added to the opening buffer and incubated the epidermal peels for 30 min. After washing with opening buffer to remove excess dye, the epidermal peels were transferred to a Petri dish containing water (control) or water with 10 µM flg22 (treatment). Images of stomata were captured using the Leica DM 6000 B fluorescence microscope (Leica, Buffalo Grove, IL, USA) with excitation 450–490 nm and emission 500–550 nm. For NO measurement, the same procedure was followed using a NO-sensitive fluorescent dye 4,5-diamino fluorescein diacetate (DAF-2DA) with excitation 495 nm and emission 515–560 nm. The fluorescence emission levels of 150 stomata from three independent experiments were analyzed using ImageJ software (National Institutes of Health, Bethesda, MD, USA).

3.4. Large-Scale Preparation of Stomatal Guard Cells for Proteomics

Stomatal guard cells from *A. thaliana* plants were prepared using a Scotch tape method as previously described [50]. Briefly, 5-week-old leaves were attached to clear Scotch tapes (3M, St Paul, MN, USA) with the abaxial side facing down. Another piece of tape was used to separate the adaxial layer containing the mesophyll cells from the abaxial side containing pavement and guard cells. A total of 100 epidermal peels constituted a biological replicate, and three replicates were collected. The peels were digested to remove the pavement cells as aforementioned. The samples were treated with 10 µM flg22 or with water and collected at 0, 15, 30, and 60 min. The samples were quickly frozen in liquid nitrogen and stored at −80 °C before protein extraction.

3.5. Protein Extraction

The enriched guard cells were ground in liquid nitrogen. For every 100 peels, 6 mL of Tris saturated phenol (pH 8.8) and 6 mL protein extraction buffer (0.9 M sucrose, 0.1 M Tris-HCl, 0.01 M EDTA, 1 mM PMSF, and 20 mM NEM were added and the peels were ground for 5 mins in a fume hood. The homogenate was transferred to Oakridge centrifuge tubes (Thermo Fisher Scientific, Waltham, MA, USA) and agitated on a shaker at room temperature (RT) for 1 h, followed by centrifugation at 15,000 g at 4 °C for 15 min. The supernatants were added to 5 volume of 100 mM ammonium acetate/methanol and kept at −20 °C overnight. After centrifugation at 15,000 g for 15 min at 4 °C, the pellets were washed with 100 mM ammonium acetate/methanol twice, cold 80% acetone twice, and 100% acetone once. The pellets were dissolved in a protein dissolution buffer (6 M urea, 1 mM EDTA, 50 mM Tris-HCL (pH 8.5) and 1% SDS). Protein samples were prepared from three independent biological replicates, each containing 100 peels.

3.6. iodoTMT Labeling and Digestion

The iodoTMT labeling of thiols was performed as described previously [51]. Reduced thiols for reverse labeling were generated by incubation with 10 mM sodium ascorbate (to specifically reduce SNO-modified cysteine residues) at RT for 1 h. For this study we labeled the 15-, 30-, and 60-min control samples with 126, 127, and 128 mass tags, and the flg22-treated samples with 129, 130, and 131 mass tags, respectively. Labeling was performed with protection from light for 2 h at 37 °C. The reaction was quenched with 0.5 M DTT for 15 min at 37 °C. The samples were then separated on a 12% SDS-PAGE gel at 100 V for 5 min and then 80 V for 10 min. This step is to clean up the protein samples and prepare them for in-gel digestion with trypsin. Trypsin digestion was performed at 37 °C overnight as recommended by the iodoTMT manual (Thermo Scientific, San Jose, CA, USA). Excess trypsin was removed, and peptides were cleaned up using C18 desalting columns (Thermo Scientific, San Jose, CA, USA) and lyophilized to dryness.

3.7. iTRAQ Labeling, Strong Cation Exchange Fractionation and Reverse Phase LC-MS/MS

For iTRAQ labeling, the lyophilized samples were resuspended in 0.5 M triethylammonium bicarbonate. The peptides were labeled with the iTRAQ reagents according to the manufacturer's manual (AB Sciex Inc., Framingham, MA, USA). A final concentration of 70% isopropanol was used for all the iTRAQ tags. We labeled the 15-, 30-, and 60-min control samples with 113, 114, and 115 mass tags, and the flg22-treated samples with 116, 117, and 118 mass tags, respectively. The labeling was carried out at 37 °C overnight. Labeled peptides were desalted with C18-solid phase extraction and dissolved in strong cation exchange (SCX) solvent A (25% *v/v* acetonitrile, 10 mM ammonium formate, and 0.1% *v/v* formic acid, pH 2.8). The peptides were fractionated using an Agilent high performance liquid chromatography (HPLC) 1260 with a SCX column (polysulfoethyl A 2.1 mm, 100 mm, 5 μm, 300 Å). Peptides were eluted with a linear gradient of 0% to 20% solvent B (25% *v/v* acetonitrile and 500 mM ammonium formate, pH 6.8) over 50 min followed by ramping up to 100% solvent B in 5 min. Peptide absorbance at 280 nm was monitored, and 20 fractions were collected, followed by desalting and lyophilization.

The fractionated peptides were resuspended in loading solvent A (3% *v/v* acetonitrile, 0.1% *v/v* acetic acid) and separated on an EASY-nLC1000 system coupled to a Q-Exactive Orbitrap Plus™ mass spectrometer (Thermo Scientific, Bremen, Germany). The peptides were loaded onto a C18 PepMap nanoflow column (20 mm × 75 μm; 3 μm-C18), and then separated on a PepMap RSLC EASY-column (250 mm × 75 μm; 2 μm-C18). The elution gradient started at 97% solvent A (0.1% *v/v* acetic acid, 3% *v/v* acetonitrile)/3% solvent B (0.1% *v/v* acetic acid, 96.9% *v/v* acetonitrile) and finished at 40% solvent A/60% solvent B within 90 min. After each run, the column was washed with 90% solvent B and re-equilibrated with solvent A. MS/MS analysis was carried out in positive mode, applying data-dependent MS scanning and MS/MS acquisition (Thermo Scientific, Bremen, Germany).

The chromatographic peak width was 4 s and the default charge state was 3. Survey full-scan MS spectra scan range was 400–2000 m/z, with a resolution of 70,000 at 200 m/z. The MS/MS resolution was 17,500. The first mass was fixed at 105 m/z to accommodate the lower m/z iTRAQ reporter ions and 445.12003 m/z (polysiloxane ion mass) was used for real-time mass calibration.

3.8. Database Searching and Data Analysis

Proteome Discoverer (PD) 2.2 (Thermo Scientific, Bremen, Germany) was used for protein identification based on searching the raw data against the uniprot *A. thaliana* database. PD nodes for spectrum grouper and spectrum selector were set to default with the spectrum properties filter set to a minimum and maximum precursor mass of 300 DA and 5 kDa, respectively. The SEQUEST HT algorithm was used for protein identification. Parameters were set to two maxima missed cleavage sites of trypsin digestion, absolute XCorr threshold of 0.4, and fragment ion cutoff percentage at 0.1. Tolerances were set to a 10 ppm precursor mass tolerance and a 0.02 Da fragment mass tolerance. Dynamic modifications included phosphorylation (+79,966 (S, T, Y)), oxidation (+15.995 Da (M)), N-ethylmaleimide (+124.048 Da (C)), iTRAQ8plex (+304.205 Da (N-terminus and K)), and iodoTMT6plex (+329.227 Da (C)). Percolator was used for protein identification with parameters of a strict target false discovery rate (FDR) of 0.01 and a relaxed target FDR of 0.05. The reporter ion quantifier node was set with a peak integration tolerance of 20 ppm, and the event detector used mass precision set to 2 ppm with a signal-to-noise ratio threshold of 3. Quantification was preformed using the iodoTMT and iTRAQ reporter ion peak intensities. Peptides were filtered to include only those with high confidence of identification (1% FDR).

Unique peptides were used for the relative protein quantification. Peak intensities for each iodoTMT label (126–131) were exported, and ratios were calculated accordingly from the median-normalized peak intensity values. Student's t-test (two-tailed) on the log2-transformed treated/control ratios was performed. To control false discovery rates, Benjamini–Hochberg correction of p-values [52] was performed by using p.adjust function in the R-package. A peptide with a p-value 0.05 was considered to be statistically significant. The iTRAQ data were searched using the same database to generate information on protein level changes. For the iTRAQ data, peptides were normalized to the total summation of their intensities based on the 113-reporter channel. Protein grouping was performed by taking the sum of repeated peptides followed by median calculation for the summed peptides belonging to the same protein. The ratios between the control and treated samples for each protein were obtained, a second normalization step for all the ratios to the median value was carried out. Log transformation of the normalized ratios and t-test (two-tailed) for significant differences between the control and treatment samples were conducted. The significant peptides labeled with iodoTMT were compared with the significant proteins quantified via iTRAQ. Student's t-test was conducted between the fold change of the iodoTMT-labeled peptides and the fold change of the corresponding proteins quantified via iTRAQ. A correction factor was applied to the fold change of iodoTMT-labeled peptides, taking into account the fold change of the protein quantified via iTRAQ. The proteomics data were submitted to MASSIVE (MSV000084409) and can be accessed via username s_lawrence and password 797960sS!.

3.9. Functional Annotation and Hierarchical Clustering

We used Araport (https://apps.araport.org/thalemine) annotations, Panther GO (http://pantherdb.org/) tools, and Fisher's Exact with false-discovery-rate to perform functional analysis of significant differentially redox-regulated proteins. For hierarchical clustering analysis the log (base2) transformed data, treatment and control ratios were used (http://bonsai.hgc.jp/$\sim$mdehoon/software/cluster/software.htm). Using a tree algorithm, proteins were organized based on similarities in their expression profile. Short branches join proteins that are very similar to each other while longer branches join proteins that are less similar to each other.

4. Conclusion

We present the utility of iodoTMTRAQ in the identification and quantification of cysteine SNO modifications in an important plant single-cell type, guard cell. We identified 35 potential SNO-regulated guard cell proteins in response to the bacterial peptide elicitor flg22 after taking into consideration protein level changes. The nitrosylated proteins, most of which have not been identified in previous redox proteomic experiments, were largely involved in metabolism, stress and defense and may provide further insight to the importance of this oxidative PTM in guard cell immunity against pathogens. This study provides the first inventory of potential SNO-regulated proteins in stomatal guard cells. The action of NO in stomatal immunity against bacterial pathogens is not fully understood. For example, many protein targets of NO signaling in guard cells are still unknown. The identification of these targets will lead to a deeper understanding of how guard cells perceive and respond to NO, thus the results of this work are useful. Future studies using biochemical and genetic tools are necessary to functionally characterize the nitrosylated proteins revealed in this study. The result of future studies will provide insight to the biological relevance of these findings and improve the overall knowledge of NO signaling in stomatal innate immunity against bacterial pathogens.

Supplementary Materials: Supplementary Materials can be found at http://www.mdpi.com/1422-0067/21/5/1688/s1. Supplemental Table S1. List of proteins significantly changed in guard cells at 15 min after exposure to flg22. Supplemental Table S2. List of proteins significantly changed in guard cells at 30 min after exposure to flg22. Supplemental Table S3. List of proteins significantly changed in guard cells at 60 min after exposure to flg22.

Author Contributions: S.R.L.II and S.C. designed the experiment. S.R.L.II carried out the redox proteomics experiments with technical assistance from M.G. Q.G. and S.R.L.II analyzed all the data. C.D. helped with mass spectrometry data acquisition and analysis. S.R.L.II wrote the manuscript draft and S.C. revised the manuscript. All authors contributed to the final version of manuscript.

Funding: This research was funded the U.S. National Science Foundation grants [0818051] and [1412547] to S.C., and a University of Florida Graduate School Fellowship to S.R.L. II.

Conflicts of Interest: The authors declare no conflict of interest.

References

1. Fletcher, J.; Bender, C.; Budowle, B.; Cobb, W.T.; Gold, S.E.; Ishimaru, C.A.; Luster, D.; Melcher, U.; Murch, R.; Scherm, H.; et al. Plant pathogen forensics: Capabilities, needs, and recommendations. *Microbiol. Mol. Biol. Rev.* **2006**, *70*, 450–471. [CrossRef] [PubMed]
2. Velásquez, A.C.; Castroverde, C.D.M.; He, S.Y. Plant-pathogen warfare under changing climate conditions. *Curr. Biol.* **2018**, *28*, 619–634. [CrossRef] [PubMed]
3. Assmann, S.M. Signal transduction in guard cells. *Annu. Rev. Cell Biol.* **1993**, *9*, 345–375. [CrossRef] [PubMed]
4. Agurla, S.; Gayatri, G.; Raghavendra, A.S. Nitric oxide as a secondary messenger during stomatal closure as a part of plant immunity response against pathogens. *Nitric Oxide* **2014**, *43*, 89–96. [CrossRef] [PubMed]
5. Lothrop, A.P.; Torres, M.P.; Fuchs, S.M. Deciphering post-translational modification codes. *FEBS Lett.* **2013**, *587*, 1247–1257. [CrossRef] [PubMed]
6. Waszczak, C.; Akter, S.; Jacques, S.; Huang, J.; Messens, J.; Van Breusegem, F. Oxidative post-translational modifications of cysteine residues in plant signal transduction. *J. Exp. Bot.* **2015**, *66*, 2923–2934. [CrossRef] [PubMed]
7. Zhu, M.M.; Zhu, N.; Song, W.Y.; Harmon, A.C.; Assmann, S.M.; Chen, S. Thiol-based redox proteins in abscisic acid and methyl jasmonate signaling in *Brassica napus* guard cells. *Plant J.* **2014**, *78*, 491–515. [CrossRef]
8. Yang, H.; Mu, J.; Chen, L.; Feng, J.; Hu, J.; Li, L.; Zhou, J.M.; Zuo, J. S-nitrosylation positively regulates ascorbate peroxidase activity during plant stress responses. *Plant Physiol.* **2015**, *167*, 1604–1615. [CrossRef]
9. Arnaud, D.; Hwang, I. A sophisticated network of signaling pathways regulates stomatal defenses to bacterial pathogens. *Mol. Plant* **2015**, *8*, 566–581. [CrossRef]
10. Balmant, K.M.; Parker, J.; Yoo, M.J.; Zhu, N.; Dufresne, C.; Chen, S. Redox proteomics of tomato in response to *Pseudomonas syringae* infection. *Hortic. Res.* **2015**, *2*, 15043. [CrossRef]

11. Romero-Puertas, M.C.; Campostrini, N.; Mattè, A.; Righetti, P.G.; Perazzolli, M.; Zolla, L.; Roepstorff, P.; Delledonne, M. Proteomic analysis of S-nitrosylated proteins in *Arabidopsis thaliana* undergoing hypersensitive response. *Proteomics* **2008**, *8*, 1459–1469. [CrossRef] [PubMed]

12. Gupta, K.J.; Fernie, A.R.; Kaiser, W.M.; van Dongen, J.T. On the origins of nitric oxide. *Trends Plant Sci.* **2011**, *16*, 160–168. [CrossRef] [PubMed]

13. Gould, N.; Doulias, P.T.; Tenopoulou, M.; Raju, K.; Ischiropoulos, H. Regulation of protein function and signaling by reversible cysteine S-nitrosylation. *J. Biol. Chem.* **2013**, *288*, 26473–26479. [CrossRef] [PubMed]

14. Besson-Bard, A.; Pugin, A.; Wendehenne, D. New insights into nitric oxide signaling in plants. *Annu. Rev. Plant Biol.* **2008**, *59*, 21–39. [CrossRef] [PubMed]

15. Camejo, D.; Ortiz-Espín, A.; Lázaro, J.J.; Romero-Puertas, M.C.; Lázaro-Payo, A.; Sevilla, F.; Jiménez, A. Functional and structural changes in plant mitochondrial PrxII F caused by NO. *J. Proteomics* **2015**, *119*, 112–125. [CrossRef] [PubMed]

16. Desikan, R.; Griffiths, R.; Hancock, J.; Neill, S. A new role for an old enzyme: Nitrate reductase-mediated nitric oxide generation is required for abscisic acid-induced stomatal closure in *Arabidopsis thaliana*. *Proc. Natl. Acad. Sci. USA* **2002**, *99*, 16314–16318. [CrossRef]

17. Khokon, A.R.; Okuma, E.; Hossain, M.A.; Munemasa, S.; Uraji, M.; Nakamura, Y.; Mori, I.C.; Murata, Y. Involvement of extracellular oxidative burst in salicylic acid-induced stomatal closure in Arabidopsis. *Plant Cell Environ.* **2011**, *34*, 434–443. [CrossRef]

18. Yu, M.; Lamattina, L.; Spoel, S.H.; Loake, G.J. Nitric oxide function in plant biology: A redox cue in deconvolution. *New Phytol.* **2014**, *202*, 1142–1156. [CrossRef]

19. Gould, K.S.; Lamotte, O.; Klinguer, A.; Pugin, A.; Wendehenne, D. Nitric oxide production in tobacco leaf cells: A generalized stress response? *Plant Cell Environ.* **2003**, *26*, 1851–1862. [CrossRef]

20. Gupta, K.J.; Stoimenova, M.; Kaiser, W.M. In higher plants, only root mitochondria, but not leaf mitochondria reduce nitrite to NO, *in vitro* and *in situ*. *J. Exp. Bot.* **2005**, *56*, 2601–2609. [CrossRef]

21. Corpas, F.J.; Hayashi, M.; Mano, S.; Nishimura, M.; Barroso, J.B. Peroxisomes are required for *in vivo* nitric oxide accumulation in the cytosol following salinity stress of Arabidopsis plants. *Plant Physiol.* **2009**, *151*, 2083–2094. [CrossRef] [PubMed]

22. Rockel, P.; Strube, F.; Rockel, A.; Wildt, J.; Kaiser, W.M. Regulation of nitric oxide (NO) production by plant nitrate reductase *in vivo* and *in vitro*. *J. Exp. Bot.* **2002**, *53*, 103–110. [CrossRef] [PubMed]

23. Stöhr, C.; Strube, F.; Marx, G.; Ullrich, W.R.; Rockel, P. A plasma membrane-bound enzyme of tobacco roots catalyses the formation of nitric oxide from nitrite. *Planta* **2001**, *212*, 835–841. [CrossRef] [PubMed]

24. Yamasaki, H.; Sakihama, Y. Simultaneous production of nitric oxide and peroxynitrite by plant nitrate reductase: *In vitro* evidence for the NR-dependent formation of active nitrogen species. *FEBS Lett.* **2000**, *468*, 89–92. [CrossRef]

25. Besson-Bard, A.; Astier, J.; Rasul, S.; Wawer, I.; Dubreuil-Maurizi, C.; Jeandroz, S.; Wendehenne, D. Current view of nitric oxide-responsive genes in plants. *Plant Sci.* **2009**, *177*, 302–309. [CrossRef]

26. Grün, S.; Lindermayr, C.; Sell, S.; Durner, J. Nitric oxide and gene regulation in plants. *J. Exp. Bot.* **2006**, *57*, 507–516. [CrossRef]

27. Delledonne, M.; Xia, Y.; Dixon, R.A.; Lamb, C. Nitric oxide functions as a signal in plant disease resistance. *Nature* **1998**, *394*, 585–588. [CrossRef]

28. Wang, Y.Q.; Feechan, A.; Yun, B.W.; Shafiei, R.; Hofmann, A.; Taylor, P.; Xue, P.; Yang, F.Q.; Xie, Z.S.; Pallas, J.A.; et al. S-nitrosylation of AtSABP3 antagonizes the expression of plant immunity. *J. Biol. Chem.* **2009**, *284*, 2131–2137. [CrossRef]

29. Slaymaker, D.H.; Navarre, D.A.; Clark, D.; del Pozo, O.; Martin, G.B.; Klessig, D.F. The tobacco salicylic acid-binding protein 3 (SABP3) is the chloroplast carbonic anhydrase, which exhibits antioxidant activity and plays a role in the hypersensitive defense response. *Proc. Natl. Acad. Sci. USA* **2002**, *99*, 11640–11645. [CrossRef]

30. Leitner, M.; Vandelle, E.; Gaupels, F.; Bellin, D.; Delledonne, M. NO signals in the haze: Nitric oxide signalling in plant defence. *Curr. Opin. Plant Biol.* **2009**, *12*, 451–458. [CrossRef]

31. Yun, B.W.; Feechan, A.; Yin, M.; Saidi, N.B.; Le Bihan, T.; Yu, M.; Moore, J.W.; Kang, J.G.; Kwon, E.; Spoel, S.H.; et al. S-nitrosylation of NADPH oxidase regulates cell death in plant immunity. *Nature* **2011**, *478*, 264–268. [CrossRef] [PubMed]

32. Tada, Y.; Spoel, S.H.; Pajerowska-Mukhtar, K.; Mou, Z.; Song, J.; Wang, C.; Zuo, J.; Dong, X. Plant immunity requires conformational charges of NPR1 via S-nitrosylation and thioredoxins. *Science* **2008**, *321*, 952–956. [CrossRef] [PubMed]

33. Wang, P.; Du, Y.; Hou, Y.-J.; Zhao, Y.; Hsu, C.-C.; Yuan, F.; Zhu, X.; Tao, W.A.; Song, C.-P.; Zhu, J.-K. Nitric oxide negatively regulates abscisic acid signaling in guard cells by S-nitrosylation of OST1. *Proc. Natl. Acad. Sci. USA* **2015**, *112*, 613–618. [CrossRef] [PubMed]

34. Frungillo, L.; Skelly, M.J.; Loake, G.J.; Spoel, S.H.; Salgado, I. S-nitrosothiols regulate nitric oxide production and storage in plants through the nitrogen assimilation pathway. *Nat. Commun* **2014**, *5*, 5401. [CrossRef]

35. Muthuramalingam, M.; Matros, A.; Scheibe, R.; Mock, H.P.; Dietz, K.J. The hydrogen peroxide-sensitive proteome of the chloroplast *in vitro* and *in vivo*. *Front. Plant Sci.* **2013**, *4*, 54. [CrossRef]

36. Go, Y.M.; Roede, J.R.; Orr, M.; Liang, Y.; Jones, D.P. Integrated redox proteomics and metabolomics of mitochondria to identify mechanisms of cd toxicity. *Toxicol. Sci.* **2014**, *139*, 59–73. [CrossRef]

37. Melotto, M.; Underwood, W.; Koczan, J.; Nomura, K.; He, S.Y. Plant stomata function in innate immunity against bacterial invasion. *Cell* **2006**, *126*, 969–980. [CrossRef]

38. Parker, J.; Balmant, K.; Zhu, F.; Zhu, N.; Chen, S. cysTMTRAQ-An integrative method for unbiased thiol-based redox proteomics. *Mol. Cell Proteomics* **2015**, *14*, 237–242. [CrossRef]

39. Yin, Z.; Balmant, K.; Geng, S.; Zhu, N.; Zhang, T.; Dufresne, C.; Dai, S.; Chen, S. Bicarbonate induced redox proteome changes in Arabidopsis suspension cells. *Front. Plant Sci.* **2017**, *8*, 58. [CrossRef]

40. Fares, A.; Rossignol, M.; Peltier, J.B. Proteomics investigation of endogenous S-nitrosylation in Arabidopsis. *Biochem. Biophys. Res. Commun.* **2011**, *416*, 331–336. [CrossRef]

41. Hu, J.; Huang, X.; Chen, L.; Sun, X.; Lu, C.; Zhang, L.; Wang, Y.; Zuo, J. Site-specific nitrosoproteomic identification of endogenously S-nitrosylated proteins in Arabidopsis. *Plant Physiol.* **2015**, *167*, 1731–1746. [CrossRef] [PubMed]

42. Puyaubert, J.; Fares, A.; Rézé, N.; Peltier, J.B.; Baudouin, E. Identification of endogenously S-nitrosylated proteins in Arabidopsis plantlets: Effect of cold stress on cysteine nitrosylation level. *Plant Sci.* **2014**, *215*, 151–156. [CrossRef] [PubMed]

43. Xu, Y.; Zheng, X.; Song, Y.; Zhu, L.; Yu, Z.; Gan, L.; Zhou, S.; Liu, H.; Wen, F.; Zhu, C. NtLTP4, a lipid transfer protein that enhances salt and drought stresses tolerance in *Nicotiana tabacum*. *Sci. Rep.* **2018**, *8*, 8873. [CrossRef] [PubMed]

44. Chen, J.H.; Jiang, H.W.; Hsieh, E.J.; Chen, H.Y.; Chien, C.T.; Hsieh, H.L.; Lin, T.P. Drought and salt stress tolerance of an arabidopsis glutathione S-transferase U17 knockout mutant are attributed to the combined effect of glutathione and abscisic acid. *Plant Physiol.* **2012**, *158*, 340–351. [CrossRef] [PubMed]

45. Van Hove, J.; De Jaeger, G.; De Winne, N.; Guisez, Y.; Van Damme, E.J.M. The Arabidopsis lectin EULS3 is involved in stomatal closure. *Plant Sci.* **2015**, *238*, 312–322. [CrossRef]

46. Xu, Y.H.; Liu, R.; Yan, L.; Liu, Z.Q.; Jiang, S.C.; Shen, Y.Y.; Wang, X.F.; Zhang, D.P. Light-harvesting chlorophyll a/b-binding proteins are required for stomatal response to abscisic acid in Arabidopsis. *J. Exp. Bot.* **2012**, *63*, 1095–1106. [CrossRef]

47. Hu, H.; Rappel, W.J.; Occhipinti, R.; Ries, A.; Böhmer, M.; You, L.; Xiao, C.; Engineer, C.B.; Boron, W.F.; Schroeder, J.I. Distinct cellular locations of carbonic anhydrases mediate carbon dioxide control of stomatal movements. *Plant Physiol.* **2015**, *169*, 1168–1178. [CrossRef]

48. Lawson, T. Guard cell photosynthesis and stomatal function. *New Phytol.* **2009**, *181*, 13–34. [CrossRef]

49. Geng, S.; Misra, B.B.; de Armas, E.; Huhman, D.V.; Alborn, H.T.; Sumner, L.W.; Chen, S. Jasmonate-mediated stomatal closure under elevated CO. *Plant J.* **2016**, *88*, 947–962. [CrossRef]

50. Lawrence, S.; Pang, Q.; Kong, W.; Chen, S. Stomata Tape-Peel: An Improved Method for Guard Cell Sample Preparation. *J. Vis Exp* **2018**, 137. [CrossRef]

51.	Qu, Z.; Meng, F.; Bomgarden, R.D.; Viner, R.I.; Li, J.; Rogers, J.C.; Cheng, J.; Greenlief, C.M.; Cui, J.; Lubahn, D.B.; et al. Proteomic quantification and site-mapping of S-nitrosylated proteins using isobaric iodoTMT reagents. *J. Proteome Res* **2014**, *13*, 3200–3211. [CrossRef] [PubMed]
52.	Benjamini, Y.; Hochberg, Y. Controlling the false discovery rate: A practical and powerful approach to multiple testing. *J. R. Stat. Soc. Ser. B* **1995**, *57*, 289–300. [CrossRef]

International Journal of
Molecular Sciences

Article

Effects of Excess Manganese on the Xylem Sap Protein Profile of Tomato (*Solanum lycopersicum*) as Revealed by Shotgun Proteomic Analysis

Laura Ceballos-Laita [1,†], Elain Gutierrez-Carbonell [1,‡], Daisuke Takahashi [2,§], Andrew Lonsdale [3], Anunciación Abadía [1], Monika S. Doblin [4], Antony Bacic [4], Matsuo Uemura [2,5], Javier Abadía [1,*] and Ana Flor López-Millán [1]

[1] Plant Stress Physiology Group, Plant Nutrition Department, Aula Dei Experimental Station, CSIC, P.O. Box 13034, 50080 Zaragoza, Spain; ceballos.laita@gmail.com (L.C.-L.); Elain.Gutierrez@sciex.com (E.G.-C.); mabadia@eead.csic.es (A.A.); anaflorlopez@gmail.com (A.F.L.-M.)

[2] United Graduate School of Agricultural Sciences, Iwate University, Morioka 020-8550, Japan; dtakahashi@mail.saitama-u.ac.jp (D.T.); uemura@iwate-u.ac.jp (M.U.)

[3] School of Biosciences, The University of Melbourne, Parkville, VIC 3052, Australia; a.lonsdale@student.unimelb.edu.au

[4] La Trobe Institute for Agriculture & Food, Department of Animal, Plant & Soil Sciences, AgriBio Building, La Trobe University, Bundoora, VIC 3086, Australia; M.Doblin@latrobe.edu.au (M.S.D.); t.bacic@latrobe.edu.au (A.B.)

[5] Department of Plant-bioscience, Faculty of Agriculture, Iwate University, Morioka 020-8550, Japan

[*] Correspondence: jabadia@eead.csic.es

[†] Current address: BIFI, University of Zaragoza, BIFI-IQFR-CSIC Joint Unit, Mariano Esquillor s/n, Campus Rio Ebro, Edificio I+D, 50018 Zaragoza, Spain.

[‡] Current address: SCIEX S.L. c/Valgrande 8, Alcobendas, 28108 Madrid, Spain.

[§] Current address: Graduate School of Science & Engineering, Saitama University, 255 Shimo-Okubo, Sakura-ku, Saitama City, Saitama 338-8570, Japan.

Received: 15 October 2020; Accepted: 19 November 2020; Published: 23 November 2020

Abstract: Metal toxicity is a common problem in crop species worldwide. Some metals are naturally toxic, whereas others such as manganese (Mn) are essential micro-nutrients for plant growth but can become toxic when in excess. Changes in the composition of the xylem sap, which is the main pathway for ion transport within the plant, is therefore vital to understanding the plant's response(s) to metal toxicity. In this study we have assessed the effects of exposure of tomato roots to excess Mn on the protein profile of the xylem sap, using a shotgun proteomics approach. Plants were grown in nutrient solution using 4.6 and 300 µM $MnCl_2$ as control and excess Mn treatments, respectively. This approach yielded 668 proteins reliably identified and quantified. Excess Mn caused statistically significant (at $p \leq 0.05$) and biologically relevant changes in relative abundance ($\geq$2-fold increases or $\geq$50% decreases) in 322 proteins, with 82% of them predicted to be secretory using three different prediction tools, with more decreasing than increasing (181 and 82, respectively), suggesting that this metal stress causes an overall deactivation of metabolic pathways. Processes most affected by excess Mn were in the oxido-reductase, polysaccharide and protein metabolism classes. Excess Mn induced changes in hydrolases and peroxidases involved in cell wall degradation and lignin formation, respectively, consistent with the existence of alterations in the cell wall. Protein turnover was also affected, as indicated by the decrease in proteolytic enzymes and protein synthesis-related proteins. Excess Mn modified the redox environment of the xylem sap, with changes in the abundance of oxido-reductase and defense protein classes indicating a stress scenario. Finally, results indicate that excess Mn decreased the amounts of proteins associated with several signaling pathways, including fasciclin-like arabinogalactan-proteins and lipids, as well as proteases, which may be involved in the release of signaling peptides and protein maturation. The comparison of the proteins changing in abundance in xylem sap and roots indicate the existence of tissue-specific and systemic responses to excess Mn. Data are available via ProteomeXchange with identifier PXD021973.

Int. J. Mol. Sci. **2020**, *21*, 8863

Keywords: xylem sap; manganese toxicity; proteome; tomato; shotgun proteomics

---

1. Introduction

Manganese (Mn) is an essential micronutrient that plays important roles in many plant physiological processes [1]. Manganese is required for photosynthesis as a constituent of the O_2 evolving complex in photosystem II, and acts as a cofactor of multiple enzymes in metabolic pathways such as the Krebs cycle and in redox homeostasis [2–4]. However, when in excess, Mn can be toxic to plants [3]. Manganese occurs in soils largely in oxide forms, and is taken up by plants as Mn^{2+}, with its availability being influenced by soil parameters such as soil structure and composition, water content, microbial activity and especially pH. Plant-available Mn can increase considerably in acidic soils, and this may lead to Mn toxicity [1,5]. Increases in soil acidity are driven by the use of intensive horticulture and acidifying fertilizers [3,6,7], as well as by climate events such as elevated ozone levels, increased ambient temperatures and emission of acidic gases [8–10].

Excess Mn concentrations in roots and shoots of plants leads to leaf chlorosis and stunted growth, increases in the concentration of reactive oxygen species (ROS), and consequently oxidative stress, and perturbations in carbohydrate, amino acid, protein and nucleic acid metabolism [11–16]. Excess Mn also alters photosynthesis and photosynthetic pigments and hormone levels [17,18]. To cope with excess Mn, plants elicit several stress-response processes, including decreases in the translocation of this metal from roots to shoots [19–21] and regulation of the mechanisms involved in its acquisition and distribution inside the plant, including sequestration in the vacuole [22–24]. The plant vasculature plays an essential role in these homeostatic mechanisms, since it constitutes the main conduit for long distance transport, distribution and delivery of Mn.

The vascular system is composed by the phloem and xylem conduits, which transport a myriad of molecules, with the apoplast compartment acting as an intercellular space between them and also between cells [25,26]. The xylem is an essential component of this system, whose primary role is the transport from roots to shoots of water and nutrients, including Mn. The xylem sap flows through the treachery elements that compose the xylem vessels, being driven by the negative pressure created by transpiration and/or by the positive root pressure created by the differences in water potential between the soil and the root system [27]. Although the most abundant components of the xylem sap are water and minerals, it also contains many other compounds in low quantities, including carbohydrates, hormones, secondary metabolites, peptides and also proteins [28–33]. The protein profile of the xylem sap is unique, and includes many proteins containing N-terminal signal peptides. These proteins are synthesized in the root and subsequently loaded to the xylem sap via the so-called conventional secretory pathway [33–36]. However, not all proteins present in the xylem sap are secreted by this pathway. Other proteins without a signal peptide are thought to be secreted via different mechanisms; the pathway for proteins that are non-classically secreted is called unconventional protein secretion (UPS) [37]. Different software tools have been developed for predicting which proteins are likely to undergo UPS, first based on bacterial and mammalian protein data [38], whose applicability to plant systems is under debate [39] and, more recently, on plant protein data [40]. The presence of proteins originating from neighboring tissues has also been commonly reported in the xylem sap proteome, suggesting some degree of cytoplasmic contamination [33].

Proteomic approaches, either gel (2-DE) or non-gel-based, are powerful tools to understand the responses that plants elicit when facing metal stresses [41,42]. The xylem sap proteome is expected to reflect changes in plant metabolism upon excess Mn and could also be an essential key for understanding Mn homeostasis. Limited information is still available on the effects of excess Mn on the xylem sap proteome, with current findings being mainly restricted to whole root and leaf tissues. In roots, proteomic analyses on Mn toxicity have been carried out by 2-DE using two *Citrus* species [20] and *Glycine max* [21], and by shotgun proteomics and 2-DE using *Solanum lycopersicum* (tomato) [42].

These studies have suggested that excess Mn causes alterations in the structure and lignin composition of root cell walls and an impairment of metabolic pathways involved in energy production, which are probably responsible for the growth inhibition observed in the roots of plants grown in the presence of excess Mn [21,42]. Other alterations in roots include changes in protein turnover and an activation of plant defense and ROS protection mechanisms [21,43–45]. In leaves, proteomic studies have suggested that Mn toxicity causes decreases in CO_2 fixation and large increases in ROS [20,44,46]. On the other hand, studies on xylem sap obtained from plants subjected to other biotic and abiotic stresses, including Mn deficiency, indicate that the proteins in xylem sap might play important roles in nutrient stress signaling [32,41,47–49].

In this study we have tested the hypothesis that excess Mn causes changes in the xylem sap protein profile of tomato, using a shotgun proteomics approach. Tomato was chosen as the model plant because its genome has been sequenced and the combination of root pressure and turgid stems allows for sufficient xylem sap collection by de-topping. Additionally, we have compared the changes in the protein profiles of the xylem sap and roots (the latter published in [42]), with the aim of finding changes that may constitute a systemic response in contrast with those specific to the roots or the xylem sap. Results support that excess Mn elicited an overall deactivation of metabolic pathways, with alterations in proteins associated with the cell wall, protein turnover the redox environment and signaling pathways.

2. Results

2.1. Xylem Sap Collection and Mineral Composition

Hydroponically grown tomato plants (26 days old) were treated with either 4.6 µM (control) or excess Mn (300 µM $MnCl_2$). Plants grown in excess Mn showed toxicity symptoms 6 days after the treatment onset. Eight days after imposing the treatments, plants were de-topped, and the xylem sap was collected. Visual toxicity symptoms at sampling time included brown spots in stems and leaf veins, 50% chlorophyll decreases in young leaves, and increases in Mn concentrations of 17- and 21-fold in roots and leaves, respectively, whereas the concentrations of Fe, Zn and Cu were not affected [42]. Xylem bleeding rates were similar in plants treated with excess Mn and in the controls (0.52 and 0.56 mL g^{-1} DW h^{-1}, respectively) whereas the protein concentration in the xylem was on average 3-fold higher in plants grown in the presence of high Mn compared to the controls (27 and 9 ng protein µL^{-1}, respectively). In addition, leaf transpiration rates were reduced by 52% in the high Mn-grown plants when compared to the controls (10.9 and 5.2 mmol H_2O m^{-2} s^{-1}, respectively) (Table 1).

Table 1. Exudation rates, protein yields, leaf transpiration rates and Mn, Fe, Zn and Cu concentrations in the xylem sap of *Solanum lycopersicum* plants grown in control and excess Mn conditions for 8 days. Data are means ± SE (n = 6 for collection parameters and n = 4 for micronutrient concentrations). Statistically significant differences were calculated with the Student's *t*-test: * $p \leq 0.05$; ** $p \leq 0.01$; *** $p \leq 0.001$).

Collection Parameters	Control	Excess Mn
Exudation rate (mL g^{-1} DW h^{-1})	0.56 ± 0.11	0.52 ± 0.21
Protein yield (ng protein µL^{-1} xylem)	9.0 ± 3.4	27.0 ± 12.5 *
Leaf transpiration rate (mmol H_2O m^{-2} s^{-1})	10.9 ± 0.5	5.2 ± 0.1 ***
Micronutrient concentrations (µM)		
Mn	6.5 ± 0.8	385.7 ± 16.1 ***
Fe	47.8 ± 12.5	15.7 ± 2.5 **
Zn	9.5 ± 1.3	8.0 ± 0.9 *
Cu	3.6 ± 0.8	3.1 ± 0.5

Manganese concentration in the xylem sap was 59-fold higher in the plants grown in excess Mn than in the controls (386 and 7 µM, respectively; Table 1), slightly higher than the concentration within the hydroponic solution in which they were grown. On the other hand, the concentrations of Fe and Zn were 67% and 16% lower, respectively, in the high Mn grown plants when compared to the controls, and no significant differences were found for Cu.

2.2. Proteins Detected, Class Annotation and PCA Analysis

The shotgun proteomics (LC-MS/MS) analysis detected 1783 proteins in tomato xylem sap, and 1230 were initially quantified by the Progenesis software (Table S1). Of these, 668 proteins were present in all six biological replicates of at least one treatment and reliably identified and quantified with at least 2 peptides. Only these protein species were considered in this study. The full list of proteins detected is shown in Table S1, and the list of the corresponding peptides is available in the ProteomeXchange Consortium via the Pride partner repository within the dataset identifier PXD021973. All identified proteins were annotated using GO and Uniprot (see Section 4.5) into one of ten different functional classes: polysaccharide related, oxido-reductases, protein metabolism, carbohydrate metabolism, lipid metabolism, signaling/regulation, defense, nutrient reservoir, unknown and a miscellaneous group containing categories not belonging to the previous groups.

All proteins showing statistically significant changes (ANOVA; $p \leq 0.05$) were used for a PCA analysis. This analysis showed a good separation between treatments, with the first and second components (PC1 and PC2, respectively) explaining approximately 94.9 and 1.9% of the variation, respectively (Figure 1; Table S2). Proteins with large positive PC1 scores (13 proteins) were from the protein metabolism, oxido-reductases, defense, lipid, carbohydrate and polysaccharide classes (3, 2, 2, 2, 1 and 1 proteins, respectively) and the miscellaneous group (2 proteins), whereas those with large negative contributions (16 proteins) were from the polysaccharide, defense, oxido-reductases and protein classes (8, 3, 3 and 2 proteins, respectively) (Table 2).

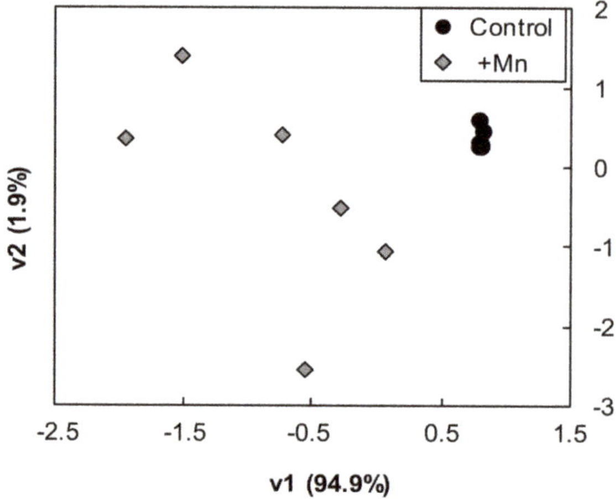

Figure 1. Score scatter PCA (Principal Component Analysis) plot of the Mn-excess treated samples when compared to the controls. PCA was carried out using SPSS Statistical software (v. 24.0) and included proteins showing statistically significant changes (ANOVA, $p \leq 0.05$) when compared to the controls. Black dots and grey diamonds depict control and Mn-treated samples, respectively.

Table 2. List of proteins contributing to component 1 in the PCA. Standardized component scores were obtained in the PCA analysis of the differential proteins (ANOVA, $p \leq 0.05$). Accession and description correspond to identifier and name in the ITAG3.2 database, respectively. Fold-changes (Fold +Mn/Control) were calculated by dividing the mean of normalized abundances (obtained with Progenesis QI v.2.5.1) in plants affected by excess Mn by that of control plants. The single protein with an asterisk after the ITAG database identifier was not included in Table S3 because it was out of the biological threshold range.

Accession	Comp. 1 Score	Comp. 2 Score	Description	Fold +Mn/ Control
			Polysaccharide metabolism (9 proteins)	
Solyc01g104950.3.1	0.001	0.013	LEXYL2 (GH3)	0.5
Solyc12g056960.2.1	−0.002	−0.061	glucan endo-1,3-β-D-glucosidase (GH17)	4.0
Solyc10g079860.2.1	−0.008	−0.034	glucan endo-1,3-β-D-glucosidase (GH17)	51.6
Solyc01g008620.3.1	−0.024	0.169	glucan endo-1,3-β-glucosidase (GH17)	233.6
Solyc01g059965.1.1	−0.492	−2.371	glucan endo-1,3- β-glucosidase (GH17)	188.7
Solyc02g086700.3.1	−0.001	−0.018	glucan endo-1,3-β-glucosidase-like (GH17)	48.8
Solyc05g050130.3.1	−0.001	0.002	acidic endochitinase (GH18)	18.3
Solyc10g055800.2.1	−0.004	−0.041	endochitinase (GH19)	51.1
Solyc02g082920.3.1	−0.001	0.005	chitinase (GH19)	91.3
			Protein metabolism (5 proteins)	
Solyc04g078110.1.1	0.002	0.014	serine protease SBT1 (MEROPS peptidase family S8)	0.5
Solyc06g074350.3.1	0.001	0.006	carboxypeptidase (MEROPS peptidase family S10)	0.3
Solyc08g067100.2.1	0.024	0.047	eukaryotic aspartyl protease family protein (MEROPS peptidase family A1)	0.5
Solyc08g079370.3.1	−0.011	0.072	Subtilisin (MEROPS peptidase family S8)	74.3
Solyc08g079900.3.1	−0.390	3.028	subtilisin-like protease (MEROPS peptidase family S8)	57.1
			Oxido-reductases (5 proteins)	
Solyc06g076630.3.1	0.007	0.093	peroxidase	0.3
Solyc12g094620.2.1	0.015	0.247	catalase	0.2
Solyc01g105070.3.1	−0.012	−0.205	LECEVI16G peroxidase precursor	14.1
Solyc06g005940.3.1	−0.001	−0.001	protein disulfide-isomerase-like	11.5
Solyc09g009390.3.1	−0.001	−0.004	monodehydroascorbate reductase	2.2

Table 2. *Cont.*

Accession	Comp. 1 Score	Comp. 2 Score	Description	Fold +Mn/ Control
			Defense (5 proteins)	
Solyc01g097240.3.1	−0.003	0.011	pathogenesis-related protein PR-4	153.6
Solyc00g174340.2.1	−0.001	0.003	pathogenesis-related protein PR-1	27.2
Solyc09g090980.3.1	0.003	−0.024	major allergen Mal d 1	25.7
Solyc12g005720.1.1	−0.001	−0.012	cysteine-rich receptor-like kinase protein	9.8
Solyc09g091000.3.1	0.003	−0.016	major allergen d 1	54.3
			Lipid metabolism (2 proteins)	
Solyc03g079880.3.1	0.003	0.043	protease inhibitor/seed storage/lipid transfer family protein	0.4
Solyc08g067500.1.1	0.002	0.030	non-specific lipid-transfer protein	0.2
			Carbohydrate metabolism (1 protein)	
Solyc06g073190.3.1 *	0.003	0.66	fructokinase-2	0.7
			Miscellaneous (2 proteins)	
Solyc01g111170.3.1	0.002	0.032	diageotropica/Peptidyl-prolyl cis-trans isomerase	0.2
Solyc08g074682.1.1	0.001	−0.015	polyphenol oxidase precursor	22.3

* Chloroplastic protein.

2.3. Effect of Excess Mn on the Xylem Sap Protein Profile

Excess Mn caused statistically significant (ANOVA, $p \leq 0.05$) and biologically relevant (defined in this study as having $\geq$2-fold increases or $\geq$50% decreases) changes in 322 protein species in the xylem sap proteome. From these, 224 (70%) decreased in abundance (Table S3), whereas 98 (30%) increased in abundance (Table S4). Changes found are shown in a volcano plot (Figure 2A), which depicts the relationship between statistical $[-\log_{10}(p\text{-value})]$ and biological $[\log_2(\text{fold-change})]$ significances.

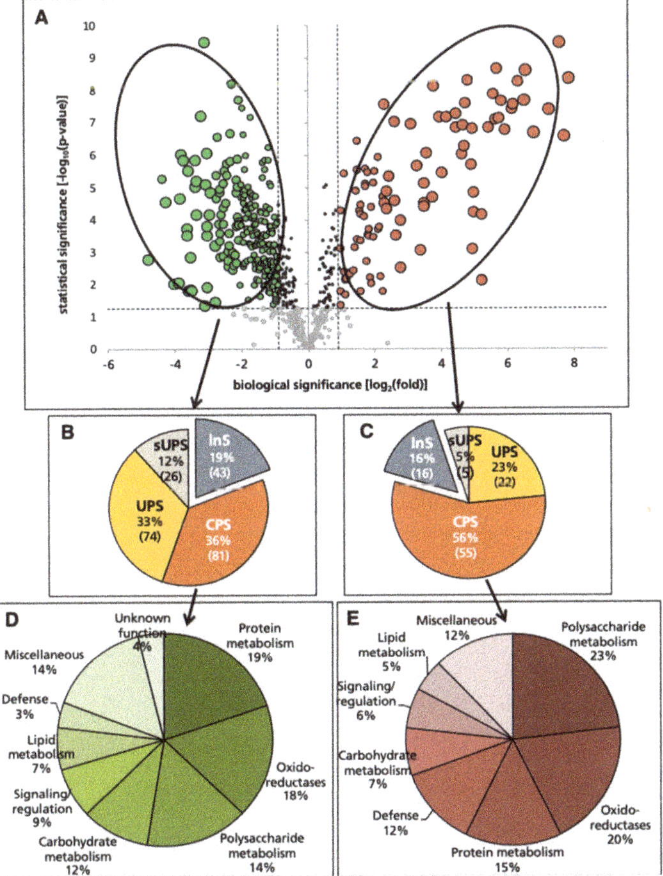

Figure 2. Effect of excess Mn on the xylem sap protein profile using label-free shotgun proteomic analyses. (**A**) Volcano scatter plot with the 668 identified and quantified proteins (peptides assigned to a protein and used for quantification $\geq$2). Proteins decreasing and increasing in relative abundance (ANOVA, $p \leq 0.05$) are in green and red, respectively, whereas those whose relative abundance was unaffected are in grey. Light and dark colors are used for proteins meeting only the statistical threshold (ANOVA, $p \leq 0.05$) and both the statistical and biological (>2-fold increase [98 proteins] or $\geq$50% decrease [224 proteins]) thresholds, respectively. The dot size is proportional to the fold-change. (**B,C**) Pie charts depicting the classification using different software tools (see Materials and Methods) for proteins showing decreases (**B**) and increases (**C**) in relative abundance. CPS: conventional secretory pathway; UPS: unconventional secretory pathway; sUPS: suggested unconventional secretory pathway; lnS: likely non-secretory. (**D,E**) Functional classification based on GO biological process and domain annotations of secretory (CPS + UPS + sUPS) proteins showing decreases (**D**) and increases (**E**) in relative abundance.

All these proteins were grouped into one of four categories using different software tools to distinguish secretory proteins from intracellular (cytoplasmic) contaminants. Those with a N-terminal leader signal sequence detected by SignalP, TargetP and/or SecretomeP were identified first and labelled conventional protein secretory (CPS). Proteins without a classical signal peptide (SP) that can be secreted through unconventional pathways were predicted in multiple ways. As pointed out previously, the accuracy of SecretomeP in predicting such proteins in plants is not optimal, being significantly less than observed for mammalian proteins [39], but this tool is still commonly used in plant studies. Thus, to complement this strategy, we used a new plant-based-tool, LSPpred [40], considered to be an improvement in UPS prediction in the absence of a well-defined plant unconventional secretory protein dataset. Those proteins lacking a SP but predicted to be secretory by at least one of the LSPpred or SPLpred modules of LSPpred were labelled as unconventional protein secretory (UPS). Proteins only predicted by the mammalian-based SecretomeP tool were assigned to the category of suggested unconventional secretory proteins (sUPS). A fourth group with proteins not predicted to be secretory by any prediction tool were labelled likely non-secretory (lnS) (Figure 2B,C; protein species are listed in Tables S3 and S4). A second volcano plot, with the distribution of the different categories, is shown in Figure S1.

As expected, the different tools used to predict unconventional secretory proteins gave contrasting results, with a total of 142 proteins being predicted to be so by at least one tool: 36 by LSPpred only, 5 by SPLpred only, 37 by LSPpred and Secretome P, 11 by SPLpred and Secretome P, 9 by LSPpred and SPLpred, 7 by all three algorithms and 37 only by the non-plant-based algorithm SecretomeP (Figure S2). Forty-four more proteins were considered as likely to be non-secretory because they were not selected by any of the algorithms (Tables S3 and S4). Among the proteins considered by the prediction tools to be secretory, 15 (5 by LSPpred only, 1 by SPLpred only, 1 by SPLpred and Secretome P, 2 by LSPpred and SPLpred and 6 only by SecretomeP) were judged to be contaminants from their clear intracellular localization (e.g., chloroplastic) and added to the lnS group (Tables S3 and S4), adjusting the final number of possible UPS proteins to 127 (96 UPS and 31 sUPS).

2.4. Protein Species Decreasing in Abundance

Among the 224 proteins decreasing in relative abundance (dark green dots in Figure 2A; Table S3), 181 (81%) were classified as secretory (including 81 CPS, 74 UPS and 26 sUPS; Figure 2B). The remaining 43 proteins were in the lnS group and were excluded from the biological interpretation, since they could originate from cytoplasmic contamination from neighboring cells. The most highly represented classes based on GO biological process in these secretory proteins (CPS + UPS + sUPS) were protein metabolism (18%), oxido-reductases (18%) and polysaccharide metabolism (14%), followed by carbohydrate (12%), signaling/regulation (9%), lipid metabolism (7%) and defense (3%) (Figure 2D). Decreases in abundance were generally moderate (83% of them ranging between 50 and 80%), with only 23 proteins, scattered among the different metabolic classes, decreasing more than 90% as a result of excess Mn (Table 3).

Table 3. Xylem sap proteins showing large changes in abundance (>25-fold increases or 90% decreases) among the total 263 secretory proteins affected by excess Mn (ANOVA, $p \leq 0.05$ and fold-change ≥ 2 or ≤ 0.5). Accession ITAG3.2 and UniProt indicates the database entry. Abundance changes (change +Mn/Control) were calculated by dividing the mean of normalized abundances (obtained with Progenesis QI v.2.5.1) in plants affected by excess Mn by that in control plants. Description includes the protein name according to the ITAG3.2 and UniProt databases. The column Category indicates the classification (CPS, UPS or sUPS), and the column Prediction tool indicates the algorithms providing true values (SignalP, TargetP, SecretomeP, LSP and SPL). Detailed information about functional classification, identification and quantification is given in Tables S3 and S4.

#	Accession ITAG3.2	Accession UniProt	Change +Mn/Control	Description	Category	Prediction Tool
				Polysaccharide metabolism (13 proteins)		
1	Solyc09g092170.2.1	A0A3Q7IA20	0.1	β-galactosidase STBG2 (GH35)	CPS	TarP, SecP
2	Solyc07g062210.3.1	A0A3Q7HE74	0.1	protein trichome birefringence-like 41	CPS	SigP, TarP, SecP
3	Solyc04g016470.3.1	A0A3Q7GST1	111.6	glucan endo-1,3-β-D-glucosidase (GH17)	CPS	TarP, SecP
4	Solyc10g079860.2.1	A0A3Q7IKF2	51.6	glucan endo-1,3-β-D-glucosidase (GH17)	CPS	SigP, TarP, SecP
5	Solyc01g008620.3.1	A0A3Q7E938	233.6	glucan endo-1,3-β-glucosidase (GH17)	CPS	SigP, TarP, SecP
6	Solyc01g059965.1.1	Q01413	188.7	glucan endo-1,3-β-glucosidase (GH17)	CPS	SigP, TarP, SecP
7	Solyc02g086700.3.1	A0A3Q7FVX4	48.8	glucan endo-1,3-β-glucosidase-like (GH17)	CPS	TarP, SecP
8	Solyc07g005100.3.1	A0A3Q7H377	30.4	Chitinase/lysozyme (GH18)	CPS	SigP, TarP, SecP
9	Solyc10g055800.2.1	A0A3Q7IHS3	51.1	endochitinase (GH19)	CPS	SigP, TarP, SecP
10	Solyc10g055810.2.1	Q05538	80.9	basic 30 kDa endochitinase (GH19)	CPS	SigP, TarP, SecP
11	Solyc02g082920.3.1	Q05539	91.3	acidic extracellular 26 kD chitinase (GH19)	CPS	SigP, TarP, SecP
12	Solyc10g055820.2.1	A0A3Q7IIQ3	44.4	endochitinase (GH19)	CPS	SigP, TarP, SecP
13	Solyc11g005480.2.1	A0A3Q7IP30	26.1	citrate-binding protein-like	CPS	SigP, TarP, SecP
				Protein metabolism (12 proteins)		
14	Solyc00g005000.3.1	A0A494G8A2	0.1	eukaryotic aspartyl protease family protein	CPS	SigP, TarP, SecP
15	Solyc07g064590.3.1	A0A3Q7HFD0	0.1	inducible plastid-lipid associated protein	UPS	LSP, SecP
16	Solyc06g008170.3.1	K4CUW3	0.1	50S ribosomal protein L14	sUPS	SecP
17	Solyc09g066430.3.1	A0A3Q7J0J5	0.1	60S ribosomal protein L14	sUPS	SecP
18	Solyc08g079870.3.1	-	74.3	subtilisin (MEROPS S8, clan SB)	CPS	SigP, TarP, SecP
19	Solyc08g079900.3.1	-	57.1	subtilisin-like protease (MEROPS S8, clan SB)	CPS	SigP, TarP, SecP
20	Solyc06g008620.1.1	A0A3Q7GN27	31.7	tolB protein-like protein (MEROPS S9)	UPS	SPL, SecP
21	Solyc01g087850.2.1	O82777	31.0	serine protease SBT3 (MEROPS I13, clan IG)	CPS	SigP, TarP, SecP
22	Solyc03g020010.1.1	O48625	93.3	Miraculin (MEROPS I13, clan IG)	CPS	SigP, TarP, SecP
23	Solyc03g019690.1.1	A0A3Q7FGU5	37.8	kunitz-type protease inhibitor (MEROPS I13, clan IG)	CPS	SigP, TarP, SecP
24	Solyc03g098740.1.1	A0A3Q7FNG4	208.5	kunitz trypsin inhibitor (MEROPS I3, clan IC)	CPS	SigP, TarP, SecP
25	Solyc08g080630.3.1	A0A3Q7IQ00	31.8	ethylene-responsive proteinase inhibitor 1 (MEROPS I3, clan IC)	CPS	SigP, TarP, SecP

Table 3. *Cont.*

#	Accession ITAG3.2	Accession UniProt	Change +Mn/Control	Description	Category	Prediction Tool
				Polysaccharide metabolism (13 proteins)		
				Oxido-reductases (6 proteins)		
26	Solyc06g005160.3.1	Q3I5C4	0.1	Ascorbate peroxidase	UPS	LSP
27	Solyc04g074740.3.1	A0A3Q7G526	0.0	blue copper protein-like	CPS	SigP, TarP, SecP
28	Solyc08g066740.3.1	A0A3Q7HR33	0.1	early nodulin-like protein 1-like	CPS	SigP, TarP, SecP
29	Solyc08g028690.3.1	A0A3Q7HMQ9	0.0	NAD(P)-binding Rossmann-fold superfamily protein	UPS	LSP, SPL, SecP
30	Solyc03g006700.3.1	A0A3Q7FFR5	28.7	peroxidase (AtPrx52)	CPS	SigP, TarP, SecP
31	Solyc04g071890.3.1	-	71.6	peroxidase	CPS	TarP, SecP
				Defense (5 proteins)		
32	Solyc01g097240.3.1	P32045	153.6	pathogenesis-related protein PR-4	CPS	SigP, TarP, SecP
33	Solyc00g174340.2.1	A0A494GA45	27.2	pathogenesis-related protein 1	CPS	SigP, TarP, SecP
34	Solyc01g106620.2.1	B2LW68	61.8	pathogenesis-related protein 1	CPS	SigP, TarP, SecP
35	Solyc07g005380.3.1	A0A3Q7H2K6	37.7	pathogenesis-related PR-10-related/norcoclaurine synthase-like protein	UPS	SPL, SecP
36	Solyc09g090980.3.1	-	25.7	major allergen Mal d 1	UPS	SPL
				Carbohydrate metabolism (4 proteins)		
37	Solyc01g094200.3.1	A0A3Q7EKQ6	0.1	malic enzyme	UPS	LSP, SecP
38	Solyc01g101040.3.1	A0A3Q7EPC2	0.1	ATP-citrate synthase	UPS	LSP
39	Solyc01g058390.3.1	A0A3Q7EEN4	0.1	galactokinase	sUPS	SecP
40	Solyc02g088690.3.1	A0A3Q7F9B8	0.1	UDP-glucose 6-dehydrogenase family protein	UPS	LSP, SecP
				Signaling/regulation (5 proteins)		
41	Solyc02g063090.3.1	A0A3Q7FJZ3	0.1	T-complex protein 1 subunit zeta 1	UPS	LSP
42	Solyc05g056310.3.1	A0A3Q7GLM4	0.1	T-complex protein 1 subunit gamma	UPS	LSP
43	Solyc01g086920.3.1	A0A3Q7F2I5	0.0	leucine-rich repeat receptor-like protein kinase family	CPS	SigP, TarP, SecP
44	Solyc10g050110.1.1	A0A3Q7IFT3	0.1	leucine-rich repeat receptor-like protein kinase family	CPS	SigP, TarP, SecP
45	Solyc08g082820.3.1	A0A3Q7HX02	31.3	tomato BiP (binding protein)/grp78 (HSP70)	CPS	SigP, TarP, SecP
				Lipid metabolism (1 protein)		
46	Solyc01g107990.3.1	A0A3Q7FCU3	0.1	PI-PLC X domain-containing protein	CPS	SigP, TarP, SecP
				Miscellaneous (4 proteins)		
47	Solyc03g115630.3.1	A0A3Q7FR13	0.1	carbamoyl-phosphate synthase	UPS	LSP, SPL, SecP
48	Solyc06g005360.3.1	A0A3Q7GNQ9	0.1	actin-depolymerizing factor family protein	sUPS	SecP
49	Solyc07g064160.3.1	A0A3Q7HFB9	0.1	thiamine thiazole synthase	sUP	SecP
50	Solyc10g049970.2.1	A0A3Q7IFZ6	24.9	kynurenine formamidase	CPS	SigP, TarP, SecP
				Unknown function (1 protein)		
51	Solyc06g035920.3.1	A0A3Q7GU06	0.1	remorin	sUPS	SecP

In the protein metabolism class, decreases in relative abundance were measured in 27 proteolysis-related proteins (18 CPS, 6 UPS and 3 sUPS) and five ribosome structural components (one of them UPS and four sUPS) (Table S3). Proteolytic proteins included four subtilisins and two serine endopeptidases (MEROPS family S8), three serine carboxy-peptidases (MEROPS family S10), two cysteine proteases and one glycine decarboxylase (MEROPS family C1), one aspartic peptidase and three aspartyl proteases (MEROPS family A1), one processing peptidase (MEROPS family M16), one lipid associated peptidase (MEROPS family C85), a neproxin (MEROPS family U74) and seven proteins related to proteasome [50]. The last protein in this group was a CPS (12g014270.2.1) involved in glycopeptide hydrolysis (all accession numbers mentioned in the text are Solyc from the tomato ITAG3.2 database).

In the oxidoreductase class (33 proteins: 17 CPS, 11 UPS and 5 sUPS; Table S3), the most represented families were peroxidases, Cu oxidases and aldehyde/histidinol dehydrogenases (seven proteins each). The twelve remaining proteins in this class included six miscellaneous reductases (10g081440.2.1, 02g087230.3.1, 01g108630.3.1, 06g062280.3.1, 06g083690.3.1 and 08g028690.3.1), a thioredoxin peroxidase (06g049080.3.1), a germin family member (01g102390.3.1), a carbonic anhydrase (04g080570.3.1) and three miscellaneous oxidases (10g005110.3.1, 08g068420.3.1 and 12g008640.2.1).

In the polysaccharide metabolism class (26 proteins: 21 CPS and 4 UPS and 1 sUPS; Table S3), most proteins (20) belong to several glycoside hydrolase (GH) subfamilies (GH 3, 5, 10, 16, 17, 19, 20, 27, 28, 32, 35, 79 and 127), with GH17 and GH 28 being the most highly represented (four and three proteins, respectively). The remaining six proteins in this group included two pectin acetyl esterases (10g038130.2.1 and 07g062210.3.1), a pectin methyl esterase inhibitor (PMEI; 11g005820.1.1), a rhamnogalacturonate lyase (04g076660.3.1) and two family 75 glycosyl transferases (GTs) (04g005340.2.1 and 05g012070).

In the carbohydrate class (22 proteins: 19 UPS and 3 sUPS; Table S3), there were six proteins involved in glycolysis (12g095760.2.1, 01g110360.3.1, 09g009260.3.1, 10g083570.2.1, 04g082630.3.1 and 04g009030.3.1), eight related to TCA (12g009400.2.1, 07g064800.3.1, 01g100360.3.1, 01g094200.3.1, 08g077920.3.1, 10g074500.2.1, 01g101040.3.1 and 12g005080.2.1), two participating in the pentose phosphate shunt (05g010260.3.1 and 03g121720.2.1) and six involved in monosaccharide metabolism: two related to galactose (01g058390.3.1 and 05g024415.1.1), two related to glucose (02g088690.3.1 and 11g011960.2.1) and one each to rhamnose (08g080140.3.1), and xylose (11g066720.2.1).

In the signaling/regulation class (16 proteins: 9 CPS, 6 UPS and 1 sUPS; Table S3), five proteins were fasciclin-like arabinogalactan-proteins (FLAs), five more were involved in chaperone-mediated protein folding (four of them with a Cpn60 domain (02g063090.3.1, 05g056310.3.1, 06g065520.3.1 and 11g069000.2.1) and four were kinases (three of them with a leucine-rich repeat receptor: 01g086920.3.1, 10g050110.1.1 and 03g111670.3.1). The remaining two proteins were an abscisic acid receptor (08g076960.1.1) and an adenylyl cyclase-associated protein (10g051340.2.1).

In the lipid metabolism class (13 proteins: 7 CPS, 5 UPS and 1 sUPS; Table S3), eight were esterases and lipases, three were lipid-transfer proteins (03g079880.3.1, 03g093360.3.1 and 08g067500.1.1) and the remaining two were a fatty acid dehydratase (01g105060.3.1) and an acyl-transferase (01g006980.3.1).

Six more proteins were in the defense class (5 CPS and 1 UPS; Table S3), including three osmotins belonging to the pathogenesis-related (PR) family 5 (PR-5; 02g087520.3.1, 04g081550.3.1 and 05g053020.3.1), two stress related proteins (07g007760.3.1 and 03g096460.3.1) and a lectin family protein with a Ricin B-like domain (01g010750.3.1).

A miscellaneous class included 24 proteins (19 UPS and 5 sUPS; Table S3), with eleven involved in amino acid metabolism, and two each involved in purine metabolism, ATP hydrolysis coupled proton transport and protein transport, among others.

Finally, eight proteins (4 CPS, 3 UPS and 1 sUPS; Table S3) were classified as having an unknown function.

Marked decreases in abundance (>90%) were found in 23 proteins (Table 3): a β-galactosidase (GH35; 09g092170.2.1) and the trichome birefringence-like protein classified as a pectin acetyl esterase (07g062210.3.1) in the polysaccharide class; an aspartyl protease (from MEROPS family A1), a inducible plastid-lipid associated protein (07g064590.3.1 from MEROPS family C85) and two ribosome structural components in the protein metabolism class; an ascorbate peroxidase (06g005160.3.1), two cupredoxins (04g074740.3.1 and 08g066740.3.1) and a NAD(P)-binding Rossmann-fold protein with SDR domain (08g028690.3.1) in the oxido-reductase class; proteins related to TCA (01g094200.3.1 and 01g101040.3.1), galactose and glucose metabolism (01g058390.3.1 and 02g088690.3.1, respectively), in the carbohydrate metabolism class; two proteins with chaperone (Cpn60) activity (02g063090.3.1 and 05g056310.3.1) and two kinases with a leucine-rich repeat receptor (01g086920.3.1 and 10g050110.1.1) in the signaling/regulation class; a phospholipase in the lipid metabolism class (01g107990.3.1), three more proteins in the miscellaneous class and a remorin having an unknown function (06g035920.3.1). The most remarkable decreases were observed in the blue copper oxidase (04g074740.3.1), the NAD (P)-binding Rossmann-fold protein (08g028690.3.1) and the leucine-rich repeat receptor (01g086920.3.1), which were barely detectable under excess Mn (Table 3).

2.5. Protein Species Increasing in Abundance

Among the 98 proteins increasing in relative abundance (dark red dots in Figure 2A; Table S4), 82 (84%) were classified as secretory (including 55 CPS, 22 UPS and 5 sUPS; Figure 2C). The remaining 16 were in the lnS group and therefore were excluded from this analysis. The most represented classes in these secretory proteins (CPS + UPS + sUPS) were polysaccharide metabolism (23%) and oxido-reductases (20%), followed by protein metabolism (16%), defense (11%), signaling/regulation (9%) and carbohydrate and lipid processes (6% each) (Figure 2E). The three most highly represented classes were similar to those found for proteins decreasing in abundance; however, within a given class, proteins increasing and decreasing belonged to different subfamilies.

In the polysaccharide-related metabolic class (19 proteins, all CPS; Table S4), GHs were predominant (16 of 19 proteins), with the most abundant subfamilies being GH17, GH18 and GH19 (six, three and four proteins, respectively), and with the subfamilies GH5, GH16 and GH20 accounting for one protein each. The subfamily GH18 was not found among those decreasing in abundance upon excess Mn. The remaining three proteins in this group were a pectin acetyl esterase (08g005800.3.1), a polysaccharide lyase family 7 protein (11g005480.2.1) and an invertase inhibitor (12g099200.2.1).

In the oxido-reductase class (16 proteins: 10 CPS, 5 UPS and 1 sUPS; Table S4), most (88%) presented moderate increases between 2- to 14-fold. This class included eight peroxidases, two aldehyde/histidinol dehydrogenases and two thioredoxins, whereas Cu oxidases were absent. The four remaining oxido-reductases (3 UPS and 1 sUPS) were a monodehydroascorbate reductase (MDAR; 09g009390.3.1), a glutathione reductase (09g091840.3.1), a quinone reductase (02g079750.3.1) and a phenylacetaldehyde reductase (01g008550.4.1).

As for proteins with reduced abundance, most protein species in the protein metabolism class (13 proteins: 10 CPS, 2 UPS and 1 sUPS; Table S4) with increased abundance belong to proteolytic processes. These included proteins of the MEROPS subfamilies S8 (two), S9 (one) and A1 (three) and five protease inhibitors from the MEROPS families I3 and I13 (these two subfamilies and subfamily S9 were not among those decreasing in abundance). Finally, this group also included the UPS elongation factor (11g072190.2.1) and an alanine-tRNA ligase (03g097290.3.1) involved in protein synthesis.

In the defense class (9 proteins: 7 CPS and 2 UPS; Table S4), there were seven cysteine-rich secreted proteins (CRSPs), with either Barwin (01g097240.3.1 and 01g097270.3.1), allergen V5 (00g174340.2.1 and 01g106620.2.1), MLP (07g005380.3.1 and 09g090980.3.1; both UPS) or Gnk2 (12g005720.1.1) domains (Table S4). The remaining two proteins were an osmotin of the PR-5 family (08g080585.1.1), and a stress up-regulated Nod 19 protein 12g042380.2.1).

Other metabolic classes accounted for one third of the total proteins increasing in abundance: signaling/regulation (four protein folding related proteins, two kinases and a nuclease; 5 CPS and 2 UPS); carbohydrate metabolism (two proteins from the glycolytic pathway, two from the pentose phosphate shunt and one involved in xylose metabolism; 1 CPS and 4 UPS); lipid metabolism (two carboxylesterases, an esterase/lipase, an enoyl reductase and a thiolase; 1 CPS, 3 UPS and 1 sUPS); amino acid metabolism (four amino transferases, a cysteine synthase and a kynurenine formidase; 1 CPS, 4 UPS and 1 sUPS), one CPS acid phosphatase and one sUPS adenylate kinase (Table S4).

Marked increases in abundance (≥25-fold) were found in 28 proteins (Table 3): ten glycosyl hydrolases (from GH families 17, 18 and 19) and a citrate binding protein with an alginate lyase domain (11g005480.2.1) in the polysaccharide class; two subtilisin-like proteases (MEROPS S8) (08g079870.3.1 and 08g079900.3.1), the tolB protein (06g008620.1.1, MEROPS S9) and five protease inhibitors from the MEROPS families I3 and I13 in the protein metabolism class; the peroxidases 03g006700.3.1 and 04g071890.3.1 in the oxido-reductase class, and five PR proteins in the defense class. The most remarkable increases were observed in the CPS β-1,3-glucanases 01g008620.3.1 and 01g059965.1.1 (234- and 189-fold, respectively), the MEROPS family I3 protease inhibitor 03g098740.1.1 (209-fold) and the PR protein 01g097240.3.1 (154-fold) (Table 3).

Thirteen more secretory proteins (11 CPS and 2 UPS) had changes in relative abundance between 8.3- and 25-fold, and therefore will have increased ≥25-fold if the basis used for comparison was total protein concentration in the xylem sap (that increases 3-fold with excess Mn) instead of relative abundance (Table S5). These proteins belong to the oxido-reductases, defense, signaling-regulation, polysaccharide metabolism and lipid metabolism classes (3, 3, 2, 1 and 1 proteins, respectively), with 3 more being included in the amino acid metabolism class.

2.6. Comparison of Changes Observed in the Xylem and Root Protein Profiles in Response to Excess Mn

The changes in the tomato xylem sap protein profiles as a result of excess Mn were compared with those observed in the root protein profiles of tomato grown in the same conditions, which were assessed using a combined 2-DE and shotgun proteomics approach [42] (Figure 3). This comparison yielded 30 proteins (13 CPS, 10 UPS, 2 sUPS and 5 lnS) in common between xylem sap and roots, with most (28) following the same trends in both tissues as a result of excess Mn (16 and 12 proteins decreasing and increasing in abundance in both tissues, respectively). The only proteins changing trends were a CPS GDSL esterase/lipase (12g017460.1.1), which increased 14-fold in the xylem sap and decreased by 50% in roots, and the UPS heat shock protein 70 (01g106210.3.1), which increased 3.4-fold in xylem sap but decreased 60% in roots (Table 4).

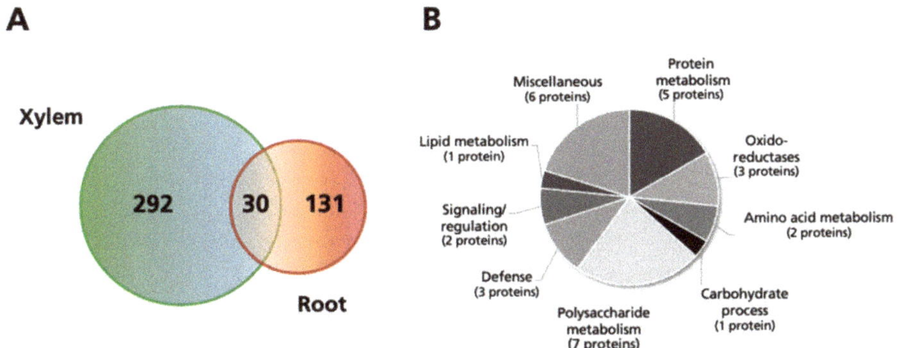

Figure 3. (**A**) Venn diagram comparing the number of proteins changing as a result of excess Mn in roots and xylem sap of tomato plants (ANOVA, $p \leq 0.05$), identified and quantified with at least two peptides, and above the threshold level (fold change ≥2 or ≤0.5). (**B**) Functional classification of the common proteins (30) showing changes in xylem and root samples of plants grown with excess Mn.

Table 4. Comparison of the changes in the xylem and root protein profiles. List of proteins showing statistically significant (ANOVA, $p \leq 0.05$) and biologically relevant (fold ≥ 2 or ≤ 0.5) changes in abundance in both root and xylem sap proteomes when tomato plants grown in excess Mn were compared to controls. The column Category indicates the classification (CPS, UPS or sUPS), and the column Prediction tool indicates the algorithms providing true values (SignalP, TargetP, SecretomeP, LSP and SPL). Proteins changing in abundance in roots as a result of excess Mn were described in [42].

			Category	Prediction Tool	Xylem sap Fold +Mn/Control	Root Fold +Mn/Control
		Proteins decreasing in both proteomes (16 proteins)				
Polysaccharide metabolism	Solyc01g104950.3.1	β -D-xylosidase 2 precursor (LEXYL2)	CPS	SigP, TarP, SecP	0.5	0.4
	Solyc05g012070.3.1	UDP-glucose:protein transglucosylase-like protein	UPS	SPL	0.2	0.5
Protein metabolism	Solyc02g068740.3.1	glycine cleavage system H family protein (MEROPS peptidase family C1)	CPS	SecP	0.3	0.5
	Solyc12g008630.2.1	mitochondrial-processing peptidase subunit α-like (MEROPS peptidase family M16)	UPS	LSP, SecP	0.4	0.3
	Solyc02g081700.1.1	proteasome subunit α type (MEROPS peptidase family T1A)	UPS	LSP, SecP	0.5	0.4
	Solyc06g073310.3.1	60S ribosomal protein l9	sUPS	SecP	0.5	0.5
Oxido-reductases	Solyc06g005160.3.1	ascorbate peroxidase	UPS	LSP	0.1	0.2
	Solyc01g107590.3.1	cinnamyl alcohol dehydrogenase	UPS	LSP	0.2	0.3
Amino acid metabolism	Solyc02g080810.3.1	aminomethyltransferase	UPS	LSP, SecP	0.4	0.5
	Solyc01g080280.3.1 *	glutamine synthetase	lnS	LSP	0.4	0.5
Carbohydrate metabolism	Solyc08g080140.3.1	3,5-epimerase/4-reductase	sUPS	SecP	0.3	0.2
Signaling/regulation	Solyc06g065520.3.1	T-complex protein 1 subunit eta	UPS	LSP	0.3	0.2
Miscellaneous	Solyc02g087300.1.1	transducin/WD40 repeat-like superfamily protein	UPS	SPL, SecP	0.4	0.5
	Solyc01g006280.3.1	formate-tetrahydrofolate ligase	UPS	LSP, SPL	0.4	0.3
	Solyc01g109660.2.1	meloidogyne-induced giant cell protein DB275 (glycine-rich RNA-binding protein)	lnS	None	0.3	0.5
	Solyc12g098150.2.1	Aldo/keto reductase	lnS	None	0.2	0.5

Table 4. *Cont.*

			Category	Prediction Tool	Xylem sap Fold +Mn/Control	Root Fold +Mn/Control
Proteins increasing in both proteomes (12 proteins)						
Polysaccharide metabolism	Solyc01g059965.1.1	β-1,3-glucanase (GH17)	CPS	SigP, TarP, SecP	188.7	8.3
	Solyc07g005100.3.1	chitinase/lysozyme (GH18)	CPS	SigP, TarP, SecP	30.4	3.4
	Solyc10g055800.2.1	chitinase (GH19)	CPS	SigP, TarP, SecP	51.1	3.6
	Solyc10g055810.2.1	chitinase Z15140 (GH19)	CPS	SigP, TarP, SecP	80.9	4.9
	Solyc10g055820.2.1	chitinase (GH19)	CPS	SigP, TarP, SecP	44.4	9.5
Protein metabolism	Solyc03g019690.1.1	Kunitz-type protease inhibitor (MEROPS I3, clan IC)	CPS	SigP, TarP, SecP	37.8	105.3
Oxido-reductases	Solyc03g006700.3.1	peroxidase	CPS	SigP, TarP, SecP	28.7	8.4
Defense	Solyc01g097240.3.1	pathogenesis-related protein PR-4	CPS	SigP, TarP, SecP	153.6	12.2
	Solyc01g106620.2.1	pathogenesis-related protein 1	CPS	SigP, TarP, SecP	61.8	4.3
	Solyc09g091000.3.1	Major allergen d 1	lnS	None	54.3	10.5
Miscellaneous	Solyc10g049970.2.1	kynurenine formamidase	CPS	SigP, TarP, SecP	24.9	32.6
	Solyc09g090430.3.1	cyanate hydratase	lnS	None	3.4	2.4
Proteins with opposite responses in root and xylem sap proteomes (2 proteins)						
Lipid metabolism	Solyc12g017460.1.1	GDSL esterase/lipase At1g28590-like	CPS	SigP, TarP, SecP	13.5	0.5
Signaling/regulation	Solyc01g106210.3.1	heat shock protein 70	UPS	LSP	3.4	0.4

* Chloroplastic protein.

Many of the proteins decreasing in abundance in both tissues were in the UPS and sUPS categories (69%; 9 and 2 proteins, respectively), whereas most of those increasing (83%; 10 proteins) were in the CPS category (Table 4). The magnitude of changes was similar for protein species decreasing in abundance in the roots and xylem sap, whereas increases in abundance were, generally speaking, more marked in the xylem than in roots. Secretory proteins increasing in abundance in both tissues were classified mainly in four metabolic classes (polysaccharide metabolism, proteolysis, oxido-reductases and defense) and included five GHs, a peroxidase, a Kunitz-type protease inhibitor and three PR proteins, whereas those decreasing in abundance were scattered among different metabolic classes, with protein metabolism being the most represented, with four proteins.

3. Discussion

Excess Mn caused large Mn concentration increases in the xylem sap when compared to those reported in the roots and leaves of plants grown in the same system [42], indicating the presence of an active translocation of this metal from roots to shoots. Such a system has been proposed to be related to the relative tolerance to this stress in some plant species [19,51].

The proteomic approach used in this study has allowed for the reliable identification and quantification of a large set of protein species (668) in xylem sap, with 82% (263 out of 322) of the proteins changing significantly being classified as secretory using several prediction tools (CPS + UPS + sUPS). A total of 181 and 82 secretory proteins showed decreases and increases in relative abundance, respectively, suggesting a relative deactivation of metabolic pathways upon this nutritional stress.

Our data suggest that 18% of the proteins identified and quantified (59) in this study may have an intracellular origin and thus reflect cytoplasmic and organelle contamination. The percentage of secretory proteins found in the present study (82%) falls within the range of those reported (70–90%) using a similar shotgun approach in xylem sap of several plant species [36,41,49]. Furthermore, the separation between samples was largely unaffected by contamination, since only two proteins contributing significantly to component 1 in the PCA analysis were InS. However, total cytoplasmic/organelle contamination may be up to 28% if the 31 proteins (12% of the total secretory; classified as sUPS) predicted to be secreted via alternative secretion pathways by SecretomeP alone, a tool not designed for and displaying suboptimal performance with plant proteins [39,40], are included. Alternatively, some proteins not predicted as secretory by the algorithms can still be secreted to the xylem by still unknown mechanisms.

On the other hand, the use of plant-specific algorithms to identify secretory proteins used in this study detected 49 new proteins not found by SecretomeP, therefore highlighting the limitations of the use of a non-plant-based software in plant research.

3.1. Excess Mn Affects the Metabolism of the Cell Wall

When considering the changes in the secretory (CPS + UPS + sUPS) xylem sap protein profile upon excess Mn, polysaccharide metabolism was among the most represented categories, accounting for 17% of the total protein abundance changes (45 proteins; 26 decreasing and 19 increasing). This indicates that excess Mn caused significant alterations in the xylem sap cell wall, since most of these enzymes are extracellular and their function is to hydrolyze and/or rearrange glycoside bonds [52]. Most of these proteins were GHs (36), similar to previous xylem sap proteomic studies [41]. Protein decreases measured in GHs were relatively moderate, whereas GH increases were among the highest measured in this study (up to 234-fold). This suggests that even though the number of GH proteins increasing or decreasing in abundance was fairly similar, those increasing may play a more relevant role in the excess Mn response, leading to an overall degradation of the cell wall. Proteins in the polysaccharide category contributed to a large extent to the separation of treatments (nine proteins; Table 2), with five of them being glucan-endo-glucosidases (including 01g008620.3.1, one of the largest increases found), supporting the importance of the endo-hydrolyzing enzymes in the response to excess Mn. Similar increases in glucan-endo-glucosidases have been described in roots of tomato plants exposed to excess Mn [42]. Two GH subfamilies (GH 5 and 17 [53]) containing endo-glucanases, mainly acting in primary cell wall degradation, had different members increasing or decreasing as a result of excess Mn. However, other GH families had a distinct behavior. On one hand, subfamilies presenting mostly increases in abundance were GH 18 (three proteins) and GH 19 (four out of five), which are N-acetyl-hexosaminidases mainly involved in N-glycan degradation [54]. Therefore, in addition to cell wall degradation, the large increases in those proteins indicate a decrease in N-glycosylation, which is assumed to be a major post-translational modification of secreted proteins.

On the other hand, GH subfamilies with members having only decreases in abundance were glycosylases with very diverse catalytic and substrate specificities, acting in the degradation of xylans (xylosidases, GH3; endoxylanases GH10), mannans (GH5), galactosides (GH 27, GH28 and GH 35), β glucuronidases (GH 79) and arabinofuranosides (GH127) [54]. Overall, these decreases, although quantitatively less important than the increases in other GHs, suggest that certain processes involving cell wall degradation may be reduced in the xylem sap of plants exposed to excess Mn, consequently leading to modifications to the cell wall. A similar situation has been described in the xylem sap of Mn-deficient plants [41]. These complex changes in GHs, in conjunction with the abundance changes measured in pectin acyl esterases (one increasing and two decreasing) and pectin methyl esterase inhibitors (one increasing and another decreasing), support that excess Mn likely causes an increased degradation of the primary cell wall, which occurs along with a re-arrangement of cell wall glycosides. A disorganization of the xylem vessels was observed using optical microscopy in *Glycine max* affected by Mn toxicity [55]. The fact that the protein concentration in the xylem sap of plants affected by excess Mn was 3-fold higher than in the controls, whereas the xylem flow rates were similar, may be in line with an increased cell wall permeability in these plants. It should be noted that since GHs can have effects in primary, secondary and/or both cell walls, functional studies are needed to confirm these hypotheses.

A significant proportion (19%; 49) of proteins changing in relative abundance in xylem sap with excess Mn were found in the oxido-reductase category, and some of these changes could be associated to Mn-induced modifications in the secondary cell wall. Among them, 13 proteins were secretory peroxidases (seven decreasing and eight increasing). Those increasing were orthologues of Prx 12, 25, 52, 53 and 73 from *Arabidopsis*, which are known to be involved in lignin formation and play a key role in controlling its deposition in the vascular tissue [56–61]. However, among the seven peroxidases with decreased abundance, five have been described as stress-response proteins [62–64]. These results indicate that excess Mn also affects the secondary cell wall via peroxidases and reveals the complexity of the observed protein changes, since some lignin-related isoforms increased and some decreased. Peroxidases increasing in abundance may also have a ROS-protecting role by depleting H_2O_2 and therefore diminishing lipid peroxidation and protein oxidation [61,65]. Increased peroxidase activity has also been described to modulate Mn oxidation and compartmentalization, and thus affect Mn tolerance, in the leaf apoplast of *Vigna unguiculata* [66].

An alteration of lignin deposition and composition in the secondary cell wall is also supported by the decreases measured in seven Cu-oxidases, including six blue-Cu proteins with cupredoxin and phytocyanin domains, and the multi-Cu oxidoreductase laccase 3-like 06g082260.3.1. Increases in blue-Cu and multi-Cu proteins have been related to increases in lignification [67–69] and therefore, decreases in these proteins may lead to reduced lignin deposition, therefore affecting the cell wall structure.

3.2. Excess Mn Alters Protein Turnover

Approximately 17% of the secretory proteins affected by excess Mn were related to protein metabolism (46 proteins), and their changes suggest a decrease in proteolysis in the xylem sap. Nineteen of those decreasing in abundance were MEROPS proteolytic enzymes [50,70] and seven more were related to the proteasome 26S protein degradation pathway, indicating that both proteolytic pathways are affected by excess Mn. In agreement with these results, five proteins from the MEROPS

I3/I3 families involved in proteolysis inhibition presented substantial increases (between 31- and 210-fold). The deactivation of proteolytic processes is in line with the high protein concentrations measured in the xylem sap of plants grown with high Mn. In addition, excess Mn caused decreases in six ribosomal structural components (1 UPS and 5 sUPS), suggesting that protein synthesis might be compromised. These decreases would also be consistent with the large number of proteins decreasing in abundance in xylem sap as a result of excess Mn. Similar results have been described in the root proteome of tomato plants subjected to excess Mn, and it was hypothesized that decreased protein synthesis might be balanced with a less proteolytic environment, with this altered protein turnover leading to an accumulation of damaged proteins that could be responsible for some of the deleterious effects of Mn toxicity [42].

However, three proteases (two MEROPS S8B and one MEROPS S9) presented large increases (31- to 74-fold) in relative abundance with excess Mn, with the two MEROPS S8B contributing to a large extent to the separation of the treatments in the PCA analysis (Table 2). Moreover, three aspartic peptidases A1 also presented moderated increases (2- to 5-fold) in relative abundance, indicating that specific proteolytic events are induced upon excess Mn. These increases might play roles in either the degradation of ROS-damaged proteins as is the case under other nutritional stresses [71,72], in the maturation of cell wall proteins or in signaling systems mediated by peptide elicitors [73,74]. An alteration in protein turnover can also be inferred from the increases in three HSP70 chaperones and the decreases in four Cpn60 chaperones involved in protein folding.

3.3. Excess Mn Alters Defense Mechanisms

As discussed above, changes observed in the oxido-reductase category (16 proteins increasing and 33 decreasing) can alter the redox environment of the xylem sap and could potentially lead to oxidative stress in combination with the high Mn concentrations. Proteins increasing in abundance included several with antioxidant activity, such as two thioredoxins, a MDAR, as well as glutathione, quinone and phenylacetaldehyde reductases. Thioredoxins act as antioxidants by reducing proteins with disulfide bridges and participate in redox signaling [75], whereas MDAR is active in the ascorbate-glutathione cycle that detoxifies H_2O_2. In addition, peroxidases showing increases in abundance can also have a role protecting from excess Mn by removing H_2O_2, although in this process they would generate ROS. Indeed, ROS produced by extracellular peroxidases have been proposed to play a role in the IAA signaling pathway [76]. Increases in MDAR have also been previously described in roots of *C. sinensis* [20] upon Mn toxicity, although a decrease in the ascorbate-glutathione cycle was described in roots of *S. lycopersicum*, suggesting that the response may depend on the plant species [42]. It is worth mentioning that a decreased abundance was measured in a Mn-binding protein identified as CPS germin protein containing a cupin domain (01g102390.3.1) and in a thioredoxin peroxidase 1 (06g049080.3.1) with MnSOD activity.

In addition to oxidative stress-related proteins, 15 defense-related proteins (nine increasing, mostly cysteine-rich proteins with as yet unknown function, and six decreasing) were affected by excess Mn, with five of them having >25-fold increases in abundance. These changes indicate that excess Mn elicits a high level of stress that affects the xylem sap proteome. Increases in proteins of this type have previously been observed in *Nicotiana tabacum* [77] and *V. unguiculata* [44]. Two of the proteins with increased abundance (01g097240.3.1 and 01g097270.3.1) contain RlpA-like domains involved in lytic transglycosylase activity (InterPro entry), suggesting a role in cell wall modification, whereas two others (00g174340.2.1 and 01g106620.2.1) contain Tpx1 and CAP domains, described to regulate ion channel activity [78,79]. The latter may suggest an involvement in Mn homeostasis, perhaps to sequester Mn in the xylem sap, which was very high (386 µM) in plants grown in excess Mn. A protein in the defense category (09g090980.3.1; 26-fold increase) and an oxido-reductase (06g076630.3.1; 70% decrease) had large positive contributions to the separation of the treatments in the PCA analysis, again supporting the existence of a high stress environment affecting the xylem sap proteome of plants grown in excess Mn. On the other hand, large increases in the PR-10 (07g005380.3.1) and the major allergen (09g090980.3.1) proteins in the defense category, both containing a lipid-binding domain (START) that is found in signaling proteins, suggests the existence of a lipid-based signaling system (see discussion below).

3.4. Excess Mn Effects on Signaling and Regulation Mechanisms

Decreases in relative abundance of proteins within the signaling/regulation category suggest that several regulatory and signaling mechanisms may be repressed under high Mn. Among the 16 proteins decreasing in this category, one third (five proteins) are FLAs that belong to a subclass of arabinogalactan-proteins (AGPs), which participate in cell-adhesion and are involved in the regulation of stem development and response to abiotic stress [80,81]. These proteins affect stem biomechanics (strength and elasticity) by altering cellulose deposition in the stem secondary cell wall [81–84], and therefore their decreases are in line with the changes in peroxidases and Cu-oxidases affecting the secondary cell wall. Similar decreases in FLAs and Cu-oxidases were observed in the xylem sap of Mn-deficient plants [41], suggesting that alterations in Mn homeostasis may have a marked impact in stem biomechanics via FLAs.

Relative protein abundance changes were also observed in seven receptor proteins, including six leucine rich repeat receptor-like kinases (LRR; four decreasing 01g086920.3.1, 10g050110.1.1, 12g055720.2.1 and 03g111670.3.1 and two increasing 01g107670.2.1 and 10g052880.1.1) and a decrease in the PYL1 abscisic acid receptor 08g076960.1.1. The *Arabidopsis* orthologue (SRF6; At1g53730) of the LRR receptor 12g055720.2.1 participates in stress-related processes and probably in the positive regulation of leaf size [85], and therefore its decrease would be in line with the small leaf size of plants grown with Mn toxicity. The PYL1 receptor is required for ABA mediated responses such as stomatal closure [86–88], and therefore its decrease may imply an inhibition of the ABA signaling pathway. This category also contained changes in eight chaperones (three increasing and five decreasing) involved in protein folding and an increase in one protein folding activator (09g057670.3.1), whose changes are in line with the proposed alteration in the protein turnover.

Results also suggest the possible existence of a lipid-based signaling system as has already been proposed to exist in the xylem sap of Mn-deficient *S. lycopersicum* [41], as well as in the phloem sap of other species [89,90]. Changes in the abundance of lipases and lipid transfer proteins have been found in the xylem sap of plants growing with other nutritional deficiencies [32,41]. Excess Mn caused decreases in eight proteins with either lipase, esterase or phospholipase activities, including the PI-PLC X domain-containing lipase 01g107990.3.1, which shows one of the largest relative protein abundance decreases measured (Table 3). However, two more lipases increased, the UPS pepper-esterase-like 02g069800.1.1 and the GDSL esterase/lipase 12g017460.1.1. The large number of lipases decreasing in abundance would suggest a reduction of lipid catabolism, but since lipases and lipid-transfer proteins play important roles in signal transduction [91,92], these changes may also indicate the existence of a lipid-based signal transduction pathway which may be repressed or induced depending on the specific lipase acting in the signaling cascade. Furthermore, decreases in three lipid transfer proteins (03g079880.3.1, 08g067500.1.1 and 01g006980.3.1) and the increases in proteins containing the START lipid-binding domain (07g005380.3.1 and 09g090980.3.1) mentioned above also support that lipid metabolism plays a role in the responses to excess Mn.

Finally, the existence of a redox-based signaling mechanism can be supported by the large percentage of differential proteins belonging to the oxido-reductase category (18%), as it has been proposed to occur in other stress situations [93].

3.5. Excess Mn Affects General Metabolism

Excess Mn affected a significant number of carbohydrate metabolism-related proteins (27 in total), with most of them predicted to be UPS (23) or sUPS (2). Nevertheless, overall changes indicate a decrease in the TCA cycle (eight TCA proteins decreasing) which might imply a decrease in the production of reducing power and energy which has also been observed in roots of *S. lycopersicum* plants exposed to excess Mn [42]. The decreases in malate and isocitrate dehydrogenases could also potentially lead to an accumulation of malate, which has been reported to chelate Mn(II) in Mn-hyperaccumulating species [94,95]. Other energy-related pathways, including glycolysis (six proteins decreasing and two increasing) and the pentose phosphate shunt (two decreasing and two increasing), were also affected by excess Mn, indicating the complexity of the metabolic regulation in this stress.

Two proteins in the carbohydrate-related category, a UDP-glucose dehydrogenase (02g088690.3.1) and a galactokinase (01g058390.3.1) that participates in galactose catabolism, displayed remarkable decreases in relative abundance (>90%). UDP-glucose dehydrogenases are glucosyl donors that play an important role in C partitioning between sucrose synthesis and cell wall formation [96,97], and therefore this decrease also suggests the occurrence of excess Mn-induced changes in cell wall biosynthesis. Decreases in UDP-glucose dehydrogenases have also been observed in roots of *S. lycopersicum* plants grown with excess Mn [42].

Finally, as commented above, excess Mn also affected a large number of lipid-related proteins (18), with most of the changes (nine) being decreases in lipases, suggesting a decrease in lipid catabolism.

3.6. Comparison of Changes Induced by Excess Mn in the Root and Xylem Sap Proteomes

The list of xylem sap proteins identified in this study was compared to the root sap proteome generated in a previous study with *S. lycopersicum* plants grown in excess Mn conditions [42] to identify any commonalities. Indeed, 30 proteins were observed to be shared between the two proteomes, accounting for 19 and 9% of the proteins changing in relative abundance in the roots and in the xylem sap, respectively. This indicates that the majority of changes (81 and 91%, respectively) are specific to the roots or the xylem sap. However, a significant number of proteins change in abundance in both samples, with most (28 out of 30) following the same response upon excess Mn, suggesting a systemic response. As expected, most of the common proteins (25) were secretory, suggesting first that the contamination of the xylem sap is relatively low, and second that excess Mn induces changes in the extracellular space/cell wall in both tissues, with changes mostly following the same trend as commented above. Of the 16 common proteins decreasing in abundance in both proteomes 75% were UPS + sUPS (11; Table 4). These proteins have a diverse array of metabolic functions, suggesting that Mn toxicity causes a deactivation of a wide range of metabolic responses in both tissues via common players.

Most of the common proteins increasing in abundance in the root and xylem sap proteomes as a result of excess Mn were CPS secretory (10 out of 12), and distributed in four main metabolic classes, including polysaccharide metabolism, protein metabolism, defense and oxido-reductases (Table 4). Results indicate that these proteins most likely participate in extracellular processes elicited by excess Mn that may be part of overall specific responses. Proteins increasing in the polysaccharide and oxido-reductase classes were a 1,3-glucanase, four chitinases and a suberization-associated peroxidase that may be involved in cell wall modifications occurring upon excess Mn in both xylem sap and roots, whereas common proteins in the defense category included three pathogenesis related proteins. One of these, the lnS major allergen d 1 (09g091000.3.1), showed increases amongst the highest measured in both proteomic studies. This is a PR10 protein containing a hydrophobic pocket able to bind hormones and siderophores [98], which could be involved in hormone signaling or even Mn sequestration, and therefore constitutes a good candidate for future studies.

Finally, it is worth mentioning that upon treatment with excess Mn, proteins decreasing in relative abundance in both tissues were mainly non-classical secretory (UPS + sUPS), whereas those increasing were CPS. The implications of these findings would deserve future studies.

4. Materials and Methods

4.1. Plant Material and Sampling

Tomato (*Solanum lycopersicum*, cv. Tres Cantos) plants were grown in hydroponics in a controlled environment chamber (Fitoclima 10.000 EHHF, Aralab, Albarraque, Portugal) with a photosynthetic photon flux density (PPFD) of 400 μmol m^{-2} s^{-1} photosynthetically active radiation at leaf level, 80% relative humidity and a photoperiod of 16 h, 23 °C/8 h, 18 °C day/night regime. Seeds were germinated in vermiculite for 13 days in half-strength Hoagland nutrient solution containing 4.6 μM $MnCl_2$. Seedlings were then transplanted to 10 L plastic buckets (16–18 plants per bucket) containing half-strength Hoagland nutrient solution and grown for an additional 13-day period. After this time, solutions were renewed, and control (4.6 μM $MnCl_2$) and high Mn (300 μM $MnCl_2$) treatments were imposed. The same treatments were used before to obtain root proteome profiles [42]. Xylem sap was collected eight days after treatment onset. Plants were de-topped approximately five mm above the mesocotyl using a carbon steel disposable scalpel (Nahita, Beriain, Spain) and the exuded fluid was collected from the cut surface. The sap collected during the first five min was discarded to minimize contamination with other plant fluids and broken cells, and the xylem sap sample was collected for 30 min using a micropipette tip. Samples were kept on ice during the entire collection period and proteins immediately precipitated. Samples were then stored at −80 °C until proteomic analysis.

4.2. Experimental Design

The experiment was repeated six times with independent batches of plants. Each batch of plants consisted of one 10 L bucket per treatment with 16–18 plants per bucket. In each batch of plants, xylem sap fluid from all plants in a given treatment was pooled together and considered as a biological replicate. Therefore, six biological replicates (n = 6) were used for LC-MS/MS, with all of them being used for protein identification and quantification.

4.3. Mineral Analysis

Micronutrients (Fe, Mn, Cu and Zn) in xylem sap were measured in four of the six biological replicates obtained as mentioned above (n = 4). The concentrations of micronutrients in the collected fluid were determined by ICP-MS (Inductively Coupled Plasma Mass Spectrometry; model Agilent 7500ce; Agilent Technologies, Tokyo, Japan) after digestion with 1% HNO_3 (TraceSELECT Ultra, Sigma-Aldrich, Madrid, Spain), using mono-elemental standard solutions for ICP-MS (Inorganic Ventures, Christiansburg, VA, USA). Recovery and limit of detection were 95.8% and 2 μg Mn L^{-1}, respectively.

4.4. Protein Extraction

Xylem sap proteins were solubilized as described in detail in [99,100]. Protein was quantified in diluted samples (Bradford kit, Sigma-Aldrich, St. Louis, MO, USA) using a microtiter plate spectrophotometer (Asys UVM 340, Biochrom Ltd., Cambridge, UK) and bovine serum albumin (Sigma) as standard.

4.5. Label Free Liquid Chromatography-Tandem Mass Spectrometry (LC-MS/MS)

Sample preparation for label free LC-MS/MS shotgun analysis was carried out as described previously [99,100]. Briefly, 5 μg of total proteins were subjected to 1-DE (for 10–15 min, in a precast Laemmli gel PAGEL NPU-10L, ATTO Corporation, Tokyo, Japan) to remove non-protein compounds, and the resulting gel band was cut into six pieces and pooled in a microtube. Proteins were in gel digested with trypsin and peptides were extracted subsequently.

Peptide separation was performed in an ADVANCE UHPLC system (Michrom Bioresources, Auburn, CA, USA) as described in [41]. Mass spectrometry analysis was carried out on an LTQ Orbitrap

XL device (Thermo Fisher Scientific, Waltham, MA, USA), carrying out peptide ionization with a spray voltage of 1.8 kV and an ADVANCE spray source (Michrom Bioresources). Data acquisition parameters were set as in [101], and Xcalibur v. 2.0.7 (Thermo Fisher Scientific) was used as instrument control software.

Mass data analysis was performed as described previously [41,42,100–102]. Protein identification was carried out using the full peptide list with the Mascot search engine (version 2.4.1, Matrix Science, London, UK) and ITAG3.2 database (35,768 sequences; 11,956,401 residues). Search parameters were peptide mass tolerance ± 5 ppm, MS/MS tolerance ± 0.6 Da, one allowed missed cleavage, allowed fixed modification carbamidomethylation (Cys) and variable modification oxidation (Met), with peptide charges being set to +1, +2 and +3. Positive protein identification was assigned with at least two unique top-ranking peptides with scores above the threshold level ($p < 0.05$). Protein data were exported from Mascot .xml format and imported to Progenesis QI proteomics software (v. 2.0, Nonlinear Dynamics, Newcastle upon Tyne, UK), which then associates peptide and protein information. The mass spectrometry proteomics data have been deposited to the ProteomeXchange Consortium via the PRIDE [103] partner repository with the dataset identifier PXD021973 (in this dataset, samples 1–6 are for controls and 19–24 for plants grown with excess Mn).

Relative quantification of proteins was carried out using non-conflicting peptides, with protein abundances being calculated in every run from the sum of all unique normalized peptide ion abundances corresponding to that protein. A protein was considered in the study when present in all six biological replicates in at least one treatment. To assess the effect of excess Mn in the protein profile of tomato xylem sap, we calculated the ratio of normalized protein abundance between treatment and control samples (n = 6). Only changes with a $p \leq 0.05$ (ANOVA) and a ratio (fold change) ≥ 2 or ≤ 0.5 were considered as statistically significant and biologically relevant, respectively. Multivariate statistical analyses (Principal Component Analysis; PCA) were carried out using SPSS Statistical software (v. 24.0), including only proteins showing statistically significant changes (ANOVA; $p \leq 0.05$) as a result of the excess Mn treatment.

The GO biological process annotation [104] and domain annotations described in the UniProt database were used for classification of each protein identified into one of ten different functional classes: polysaccharide related, oxido-reductases, protein metabolism, carbohydrate metabolism, lipid metabolism, signaling/regulation, defense, nutrient reservoir, unknown and a miscellaneous group containing categories not belonging to the previous groups. The presence of a signal peptide (conventional secretory pathway) was assessed using SignalP (v.5.0) [105], TargetP (v.2.0) [106] and SecretomeP (v.2.0) [107,108]. To assign proteins to unconventional protein secretion (UPS) pathways, three tools were used: SecretomeP2.0, based on mammalian protein data, and the new plant-based tools LSPpred (based on a curated list of likely unconventionally secreted proteins in Arabidopsis) and SLPpred (a SecretomeP-like tool, based on classically secreted proteins, from Arabidopsis with their signal peptide removed) [40,109].

4.6. Root Protein Profiling and Comparison with the Xylem Sap

The analysis of root proteins was carried out by shotgun and 2-DE techniques as described in detail in [42]. This combined analysis gave a list of 161 root proteins being reliably identified and quantified with at least two peptides and showing significant changes (ANOVA $p \leq 0.05$ and fold >2). These proteins were classified in the same categories used for the xylem sap, and both protein databases (xylem sap and roots) were compared to find commonalities.

5. Conclusions

The concentration of Mn in the xylem sap increased when compared to that present in the nutrient solution, indicating the presence of an active translocation of this metal from roots to shoots. Excess Mn caused statistically significant and biologically relevant changes in the relative abundance of 322 proteins, with more decreasing than increasing, suggesting that this metal stress causes an overall deactivation of metabolic pathways. Processes most affected by excess Mn were in the oxido-reductase, polysaccharide and protein metabolism classes. Excess Mn induced changes in relative abundance in proteins involved in cell wall degradation and lignin formation, consistent with the existence of alterations in the cell wall. Protein turnover was also affected, as indicated by the decrease in proteolytic enzymes and protein synthesis-related proteins. Excess Mn modified the redox environment of the xylem sap, with changes in the abundance of oxido-reductase and defense protein classes indicating a stress scenario. Results also indicate that excess Mn decreased the amounts of proteins associated with several signaling pathways, including fasciclin-like arabinogalactan-proteins and lipids, as well as those of proteases that may be involved in the release of signaling peptides and protein maturation.

The prediction tools used to identify unconventional secretory proteins gave contrasting results. From a total of 142 proteins, 37 (26%) were predicted as secretory by the bacterial/mammalian protein data-based SecretomeP, but not by the plant data-based tools. This highlights the complexity of the issue and the need to conduct further research towards the development of plant-based tools for predicting protein secretion.

The comparison of the proteins changing in abundance in the roots and xylem sap indicate the response to Mn excess includes both tissue-specific and systemic changes.

Supplementary Materials: The following are available online at http://www.mdpi.com/1422-0067/21/22/8863/s1, Figure S1: Effect of excess Mn on the xylem sap protein profile using label-free shotgun proteomic analyses, showing the different protein categories. Figure S2: UpSet plot showing the number of proteins lacking a signal peptide, but predicted to be secretory by the three different prediction tools used in the study. Table S1. List of proteins detected in tomato xylem sap by shotgun proteomics and Progenesis LC-MS analyses. Table S2. List of proteins contributing to component 1 and component 2 in the PCA. Standardized component scores were obtained in the PCA analysis of the differential proteins (ANOVA, $p \leq 0.05$). Table S3. List of proteins that presented significant decreases (ANOVA, $p \leq 0.05$; quantification with at least 2 peptides; fold ≤ 0.5) when xylem sap samples of plants grown in control conditions were compared to those grown in Mn toxicity by shotgun proteomics. Table S4. List of proteins that presented significant increases (ANOVA, $p \leq 0.05$; quantification with at least 2 peptides; fold ≥ 2) when xylem sap samples of plants grown in control conditions were compared to those grown in Mn excess by shotgun proteomics. Table S5. List of proteins that presented increases between 8,33-25 (ANOVA, $p \leq 0.05$; quantification with at least 2 peptides; fold ≥ 2) when xylem sap samples of plants grown in control conditions were compared to those grown in Mn-excess by shotgun proteomics.

Author Contributions: Design of the experiment and methodology, L.C.-L., A.F.L.-M., D.T. and E.G.-C.; data analysis, L.C.-L., A.L., M.S.D., A.B., A.F.L.-M. and D.T.; writing—original draft preparation, L.C.-L. and A.F.L.M.; writing—review and editing, L.C.-L., A.F.L.M., A.A. and J.A.; supervision and project administration, J.A. and M.U.; and funding acquisition, J.A. and M.U. All authors have read and agreed to the published version of the manuscript.

Funding: Supported by the Spanish State Research Agency (AEI) co-financed by the European Regional Development Fund (FEDER) (project AGL2016-75226-R; AEI/FEDER, UE) and the Aragón Government (group A09-20R). Research conducted in Iwate University was in part supported by JSPS KAKENHI Grant Numbers 24-7373, 22120003, and 24370018. E.G.-C. was supported by a JAE Pre-CSIC contract and L.C.-L was supported by a FPI-MINECO contract.

Conflicts of Interest: The authors declare no conflict of interest.

References

1. Marschner, P. *Marschner's Mineral Nutrition of Higher Plants*; Academic Press: Boston, MA, USA, 2012.
2. Pittman, J. Managing the manganese: Molecular mechanisms of manganese transport and homeostasis. *New Phytol.* **2008**, *167*, 733–742. [CrossRef]
3. Millaleo, R.; Reyes-Díaz, M.; Ivanov, A.G.; Mora, M.L.; Alberdi, M. Manganese as essential and toxic element for plants: Transport, accumulation and resistance mechanisms. *J. Soil Sci. Plant Nutr.* **2010**, *10*, 470–481. [CrossRef]
4. Socha, A.L.; Guerinot, M.L. Mn-euvering manganese: The role of transporter gene family members in manganese uptake and mobilization in plants. *Front. Plant Sci.* **2014**, *5*, 106. [CrossRef]
5. Mukhopadhyay, M.; Sharma, A. Manganese in cell metabolism of higher plants. *Bot. Rev.* **1991**, *57*, 117–149. [CrossRef]

6. Graham, M.H.; Haynes, R.J.; Meyer, J.H. Changes in soil chemistry and aggregate stability induced by fertilizer applications, burning and trash retention on a long-term sugarcane experiment in South Africa. *Eur. J. Soil Sci.* **2002**, *53*, 589–598. [CrossRef]

7. Driscoll, C.T.; Lawrence, G.B.; Bulger, A.J.; Butler, T.J.; Cronan, C.S.; Eagar, C.; Lambert, K.F.; Likens, G.E.; Stoddard, J.L.; Weathers, K.C. Acidic deposition in the Northeastern United States: Sources and inputs, ecosystem effects, and management strategies. *BioScience* **2001**, *51*, 180–198. [CrossRef]

8. Foy, C.D. Physiological effects of hydrogen, aluminum and manganese toxicities in acid soils. In *Soil Acidity and Liming*, 2nd ed.; Adams, J., Ed.; American Society of Agronomy: Madison, WI, USA, 1984; pp. 57–97.

9. Guo, J.H.; Liu, X.J.; Zhang, Y.; Shen, J.L.; Han, W.X.; Zhang, W.F.; Christie, K.W.; Goulding, K.W.; Vitousek, P.M.; Zhang, F.S. Significant acidification in major Chinese croplands. *Science* **2010**, *327*, 1008–1010. [CrossRef] [PubMed]

10. Fernando, D.R.; Lynch, J.P. Manganese phytotoxicity: New light on an old problem. *Ann. Bot.* **2010**, *116*, 313–319. [CrossRef] [PubMed]

11. Shi, Q.; Zhu, Z.; Xu, M.; Qian, Q.; Yu, J. Effect of excess manganese on the antioxidant system in *Cucumis sativus* L. under two light intensities. *Environ. Exp. Bot.* **2006**, *58*, 197–205. [CrossRef]

12. Venkatesan, S.; Hemalatha, K.V.; Jayaganesh, S. Characterization of manganese toxicity and its influence on nutrient uptake, antioxidant enzymes and biochemical parameters in tea. *Res. J. Phytochem.* **2007**, *1*, 52–60.

13. Führs, H.; Behrens, C.; Gallien, S.; Heintz, D.; Van Dorsselaer, A.; Braun, H.P.; Horst, W.J. Physiological and proteomic characterization of manganese sensitivity and tolerance in rice (*Oryza sativa*) in comparison with barley (*Hordeum vulgare*). *Ann. Bot.* **2010**, *105*, 1129–1140.

14. Gangwar, S.; Singh, V.P.; Prasad, S.M.; Maurya, J.N. Modulation of manganese toxicity in *Pisum sativum* L. seedlings by kinetin. *Sci. Hortic.* **2010**, *126*, 467–474. [CrossRef]

15. Führs, H.; Specht, A.; Erban, A.; Kopka, J.; Horst, W.J. Functional associations between the metabolome and manganese tolerance in *Vigna unguiculata*. *J. Exp. Bot.* **2012**, *63*, 329–340. [CrossRef] [PubMed]

16. Yao, Y.; Xu, G.; Mou, D.; Wang, J.; Ma, J. Subcellular Mn compartation, anatomic and biochemical changes of two grape varieties in response to excess manganese. *Chemosphere* **2012**, *89*, 150–157. [CrossRef] [PubMed]

17. Srivastava, A.K.; Singh, S. Biochemical markers and nutrient constraints diagnosis in Citrus: A perspective. *J. Plant Nutr.* **2006**, *29*, 827–855. [CrossRef]

18. Millaleo, R.; Reyes-Díaz, M.; Alberdi, M.; Ivanov, A.G.; Krol, M.; Huner, N.P. Excess manganese differentially inhibits photosystem I versus II in *Arabidopsis thaliana*. *J. Exp. Bot.* **2013**, *64*, 343–354. [CrossRef]

19. Zhou, C.P.; Qi, Y.P.; You, X.; Yang, L.T.; Guo, P.; Ye, X.; Zhou, X.X.; Ke, F.J.; Chen, L.S. Leaf cDNA-AFLP analysis of two citrus species differing in manganese tolerance in response to long-term manganese-toxicity. *BMC Genom.* **2013**, *14*, 621. [CrossRef]

20. You, X.; Yang, L.T.; Lu, Y.B.; Li, H.; Zhang, S.Q.; Chen, L.S. Proteomic changes of citrus roots in response to long-term manganese toxicity. *Trees* **2014**, *28*, 1383–1399. [CrossRef]

21. Chen, Z.; Yan, W.; Sun, L.; Tian, J.; Liao, H. Proteomic analysis reveals growth inhibition of soybean roots by manganese toxicity is associated with alteration of cell wall structure and lignification. *J. Proteom.* **2016**, *143*, 151–160. [CrossRef]

22. Wu, Z.; Liang, F.; Hong, B.; Young, J.C.; Sussman, M.R.; Harper, J.F.; Sze, H. An endoplasmic reticulum-bound Ca^{2+}/Mn^{2+} pump, ECA1, supports plant growth and confers tolerance to Mn^{2+} stress. *Plant Physiol.* **2002**, *130*, 128–137. [CrossRef]

23. Delhaize, E.; Kataoka, T.; Hebb, D.M.; White, R.G.; Ryan, P.R. Genes encoding proteins of the cation diffusion facilitator family that confer manganese tolerance. *Plant Cell* **2003**, *15*, 1131–1142. [CrossRef] [PubMed]

24. Dou, C.; Fu, X.; Chen, X.; Shi, J.; Chen, Y. Accumulation and interaction of calcium and manganese in *Phytolacca*. *Am. Plant Sci.* **2009**, *177*, 601–606. [CrossRef]

25. Oparka, K.J.; Cruz, S.S. The great escape: Phloem transport and unloading of macromolecules. *Annu. Rev. Plant Biol.* **2000**, *51*, 323–347. [CrossRef] [PubMed]

26. Lucas, W.J.; Groover, A.; Lichtenberger, R.; Furuta, K.; Yadav, S.R.; Helariutta, Y.; He, X.Q.; Fukuda, H.; Kang, J.; Brady, S.M.; et al. The plant vascular system: Evolution, development and functions. *J. Integr. Plant Biol.* **2013**, *55*, 294–388. [CrossRef] [PubMed]

27. Fisher, D.B. Long-distance transport. In *Biochemistry and Molecular Biology of Plants*; Buchanan, B., Gruissem, W., Jones, R., Eds.; Wiley Blackwell: Rockville, MD, USA, 2000; pp. 729–784.

28. Dickson, R.E. Xylem translocation of amino-acids from roots to shoots in cottonwood plants. *Can. J. For. Res.* **1979**, *9*, 374–378. [CrossRef]

29. Friedman, R.; Levin, N.; Altman, A. Presence and identification of polyamines in xylem and phloem exudates of plants. *Plant Physiol.* **1986**, *82*, 1154–1157. [CrossRef] [PubMed]

30. López-Millán, A.F.; Morales, F.; Abadía, A.; Abadía, J. Effects of iron deficiency on the composition of the leaf apoplastic fluid and xylem sap in sugar beet. Implications for iron and carbon transport. *Plant Physiol.* **2000**, *124*, 873–884. [CrossRef]

31. Escher, P.; Eiblmeier, M.; Hetzger, I.; Rennenberg, H. Seasonal and spatial variation of carbohydrates in mistletoes (*Viscum album*) and the xylem sap of its hosts (*Populus x euamericana* and *Abies alba*). *Physiol. Plant.* **2004**, *120*, 212–219. [CrossRef]

32. Carella, P.; Wilson, D.C.; Kempthorne, C.J.; Cameron, R.K. Vascular sap proteomics: Providing insight into long-distance signaling during stress. *Front. Plant Sci.* **2016**, *7*, 651. [CrossRef]

33. Rodríguez-Celma, J.; Ceballos-Laita, L.; Grusak, M.A.; Abadía, J.; López-Millán, A.F. Plant fluid proteomics: Delving into the xylem sap, phloem sap and apoplastic fluid proteomes. *BBA-Proteins Proteom.* **2016**, *1864*, 991–1002. [CrossRef]

34. Fukuda, H. Xylogenesis: Initiation, progression, and cell death. *Annu. Rev. Plant Biol.* **1996**, *47*, 199–325. [CrossRef] [PubMed]

35. Alvarez, S.; Goodger, J.Q.D.; Marsh, E.L.; Chen, S.; Asirvatham, V.S.; Schachtman, D.P. Characterization of the maize xylem sap proteome. *J. Proteome Res.* **2006**, *5*, 963–972. [CrossRef] [PubMed]

36. Ligat, L.; Lauber, E.; Albenne, C.; Clemente, H.S.; Valot, B.; Zivy, M.; Pont-Lezica, R.; Arlat, M.; Jamet, E. Analysis of the xylem sap proteome of *Brassica oleracea* reveals a high content in secreted proteins. *Proteomics* **2011**, *11*, 1798–1813. [CrossRef] [PubMed]

37. Wang, X.; Chung, K.; Lin, W.; Jiang, L. Protein secretion in plants: Conventional and unconventional pathways and new techniques. *J. Exp. Bot.* **2018**, *69*, 21–37. [CrossRef]

38. Bendtsen, J.D.; Kiemer, L.; Fausbøll, A.; Brunak, S. Non-classical protein secretion in bacteria. *BMC Microbiol.* **2005**, *5*, 58. [CrossRef]

39. Lonsdale, A.; Davis, M.J.; Doblin, M.S.; Bacic, A. Better than nothing? Limitations of the prediction tool SecretomeP in the search for leaderless secretory proteins (LSPs) in plants. *Front. Plant Sci.* **2016**, *7*, 1451. [CrossRef]

40. Lonsdale, A. Computational biology methods for identifying leaderless secretory proteins in Arabidopsis thaliana and other plant species. Ph.D. Thesis, University of Melbourne, Parkville, VIC, Australia, 2019.

41. Ceballos-Laita, L.; Gutierrez-Carbonell, E.; Takahashi, D.; Abadía, A.; Uemura, M.; Abadía, J.; López-Millán, A.F. Effects of Fe and Mn deficiencies on the protein profiles of tomato (*Solanum lycopersicum*) xylem sap as revealed by shotgun analyses. *J. Proteom.* **2018**, *170*, 117–129. [CrossRef]

42. Ceballos-Laita, L.; Gutierrez-Carbonell, E.; Imai, H.; Abadía, A.; Uemura, M.; Abadía, J.; López-Millán, A.F. Effects of manganese toxicity on the protein profile of tomato (*Solanum lycopersicum*) roots as revealed by two complementary proteomic approaches, two-dimensional electrophoresis and shotgun analysis. *J. Proteom.* **2018**, *185*, 51–63. [CrossRef]

43. González, A.; Steffen, K.L.; Lynch, J.P. Light and excess manganese. Implications for oxidative stress in common bean. *Plant Physiol.* **1998**, *118*, 493–504.

44. Fecht-Christoffers, M.M.; Braun, H.P.; Lemaitre-Guillier, C.; Van Dorsselaer, A.; Horst, W.J. Effect of manganese toxicity on the proteome of the leaf apoplast in cowpea. *Plant Physiol.* **2003**, *133*, 1935–1946. [CrossRef]

45. Chen, Z.; Sun, L.; Liu, P.; Liu, G.; Tian, J.; Liao, H. Malate synthesis and secretion mediated by a manganese-enhanced malate dehydrogenase confers superior manganese tolerance in *Stylosanthes guianensis*. *Plant Physiol.* **2014**, *167*, 176–188. [CrossRef] [PubMed]

46. Führs, H.; Hartwig, M.; Molina, L.E.; Heintz, D.; Van Dorsselaer, A.; Braun, H.; Horst, W.J. Early manganese-toxicity response in *Vigna unguiculata* L.—A proteomic and transcriptomic study. *Proteomics* **2008**, *8*, 149–159. [CrossRef] [PubMed]

47. Kaida, R.; Serada, S.; Norioka, N.; Norioka, S.; Neumetzler, L.; Pauly, M.; Sampedro, J.; Zarra, I.; Hayashi, T.; Kaneko, T.S. Potential role for purple acid phosphatase in the dephosphorylation of wall proteins in tobacco cells. *Plant Physiol.* **2010**, *153*, 603–610. [CrossRef] [PubMed]

48. Del Vecchio, H.A.; Ying, S.; Park, J.; Knowles, V.L.; Kanno, S.; Tanoi, K.; She, Y.-M.; Plaxton, W.C. The cell wall-targeted purple acid phosphatase AtPAP25 is critical for acclimation of *Arabidopsis thaliana* to nutritional phosphorus deprivation. *Plant J.* **2014**, *80*, 569–581. [CrossRef]

49. Zhang, Z.; Xin, W.; Wang, S.; Zhang, X.; Dai, H.; Sun, R.; Frazier, T.; Zhang, B.; Wang, Q. Xylem sap in cotton contains proteins that contribute to environmental stress response and cell wall development. *Funct. Integr. Genom.* **2015**, *15*, 17–26. [CrossRef]

50. MEROPS the Peptidase Database. Available online: https://www.ebi.ac.uk/merops/ (accessed on 7 October 2020).

51. Vose, P.B.; Randall, P.J. Resistance to aluminum and manganese toxicities in plants related to variety and cation-exchange capacity. *Nature* **1962**, *196*, 85–86. [CrossRef]

52. Minic, Z. Physiological roles of plant glycoside hydrolases. *Planta* **2008**, *227*, 723–740. [CrossRef]

53. CAZy. Carbohydrate-Active enZYmes. Available online: http://www.cazy.org/ (accessed on 7 October 2020).

54. Cantarel, B.L.; Coutinho, P.M.; Rancurel, C.; Bernard, T.; Lombard, V.; Henrissat, B. The Carbohydrate-Active EnZymes database (CAZy): An expert resource for Glycogenomics. *Nucleic Acids Res.* **2009**, *37*, 233–238. [CrossRef]

55. Lavres, J.; Malavolta, E.; Nogueira, N.L.; Moraes, M.F.; Reis, A.R.; Rossi, M.L.; Cabral, C.P. Changes in anatomy and root cell ultrastructure of soybean genotypes under manganese stress. *Rev. Bras. Cienc. Solo* **2009**, *33*, 395–403. [CrossRef]

56. Freudenberg, K. Biosynthesis and constitution of lignin. *Nature* **1959**, *183*, 1152–1155. [CrossRef]

57. Castillo, F.J.; Greppin, H. Balance between anionic and cationic extracellular peroxidase activities in *Sedum album* leaves after ozone exposure. Analysis by high-performance liquid chromatography. *Physiol. Plant.* **1986**, *68*, 201–208. [CrossRef]

58. Fernández-Pérez, F.; Pomar, F.; Pedreño, M.A.; Novo-Uzal, E. The suppression of AtPrx52 affects fibers but not xylem lignification in *Arabidopsis* by altering the proportion of syringyl units. *Physiol. Plant.* **2015**, *154*, 395–406. [CrossRef] [PubMed]

59. Fernández-Pérez, F.; Pomar, F.; Pedreño, M.A.; Novo-Uzal, E. Suppression of *Arabidopsis* peroxidase 72 alters cell wall and phenylpropanoid metabolism. *Plant Sci.* **2015**, *239*, 192–199. [CrossRef] [PubMed]

60. Fernández-Pérez, F.; Vivar, T.; Pomar, F.; Pedreño, M.A.; Novo-Uzal, E. Peroxidase 4 is involved in syringyl lignin formation in *Arabidopsis thaliana*. *J. Plant Physiol.* **2015**, *175*, 86–94. [CrossRef] [PubMed]

61. Shigeto, J.; Tsutsumi, Y. Diverse functions and reactions of class III peroxidases. *New Phytol.* **2016**, *209*, 1395–1402. [CrossRef] [PubMed]

62. Kreps, J.A.; Wu, Y.; Chang, H.S.; Zhu, T.; Wang, X.; Harper, J.F. Transcriptome changes for *Arabidopsis* in response to salt, osmotic, and cold stress. *Plant Physiol.* **2002**, *130*, 2129–2141. [CrossRef] [PubMed]

63. Tokunaga, N.; Kaneta, T.; Sato, S.; Sato, Y. Analysis of expression profiles of three peroxidase genes associated with lignification in *Arabidopsis thaliana*. *Physiol. Plant.* **2009**, *136*, 237–249. [CrossRef]

64. Kim, S.Y.; Kim, B.H.; Lim, C.J.; Lim, C.O.; Nam, K.H. Constitutive activation of stress-inducible genes in a *brassinosteroid-insensitive 1 (bri1)* mutant results in higher tolerance to cold. *Physiol. Plant.* **2010**, *138*, 191–204. [CrossRef]

65. Santandrea, G.; Pandolfini, T.; Bennici, A. A physiological characterization of Mn-tolerant tobacco plants selected by in vitro culture. *Plant Sci.* **2000**, *150*, 163–170. [CrossRef]

66. Fecht-Christoffers, M.M.; Führs, H.; Braun, H.P.; Horst, W.J. The role of hydrogen peroxide-producing and hydrogen peroxide-consuming peroxidases in the leaf apoplast of cowpea in manganese tolerance. *Plant Physiol.* **2006**, *140*, 1451–1463. [CrossRef]

67. Dean, J.F.; Eriksson, K.E.L. Laccase and the deposition of lignin in vascular plants. *Holzforschung* **1994**, *48*, 21–33. [CrossRef]

68. Ezaki, B.; Sasaki, K.; Matsumoto, H.; Nakashima, S. Functions of two genes in aluminium (Al) stress resistance: Repression of oxidative damage by the AtBCB gene and promotion of efflux of Al ions by the NtGDI1gene. *J. Exp. Bot.* **2005**, *56*, 2661–2671. [CrossRef] [PubMed]

69. Ma, J.; Zhang, Z.; Yang, G.; Mao, J.; Xu, F. Ultrastructural topochemistry of cell wall polymers in *Populus nigra* by transmission electron microscopy and Raman imaging. *BioResources* **2011**, *6*, 3944–3959.

70. Rawlings, N.D.; Barrett, A.J.; Thomas, P.D.; Huang, X.; Bateman, A.; Finn, R.D. The MEROPS database of proteolytic enzymes, their substrates and inhibitors in 2017 and a comparison with peptidases in the PANTHER database. *Nucleic Acids Res.* **2018**, *46*, D624–D632. [CrossRef]

71. Buhner-Zaharieva, T.; Abadía, J. Iron deficiency enhances the levels of ascorbate, glutathione, and related enzymes in sugar beet roots. *Protoplasma* **2003**, *221*, 269–275. [CrossRef]

72. Lan, P.; Li, W.; Wen, T.N.; Shiau, J.Y.; Wu, Y.C.; Lin, W.; Schmidt, W. iTRAQ protein profile analysis of *Arabidopsis* roots reveals new aspects critical for iron homeostasis. *Plant Physiol.* **2011**, *155*, 821–834. [CrossRef]

73. Schaller, A. A cut above the rest: The regulatory function of plant proteases. *Planta* **2004**, *220*, 183–197. [CrossRef]

74. Xia, Y.; Suzuki, H.; Borevitz, J.; Blount, J.; Guo, Z.; Patel, K.; Dixon, R.A.; Lamb, C. An extracellular aspartic protease functions in *Arabidopsis* disease resistance signaling. *EMBO J.* **2004**, *23*, 980–988. [CrossRef]

75. Meng, L.; Wong, J.H.; Feldman, L.J.; Lemaux, P.G.; Buchanan, B.B. A membrane-associated thioredoxin required for plant growth moves from cell to cell, suggestive of a role in intercellular communication. *Proc. Natl. Acad. Sci. USA* **2010**, *107*, 3900–3905. [CrossRef]

76. Kawano, T. Roles of the reactive oxygen species-generating peroxidase reactions in plant defense and growth induction. *Plant Cell Rep.* **2003**, *21*, 829–837. [CrossRef]

77. Edreva, A.M. Induction of 'pathogenesis-related' proteins in tobacco leaves by physiological (non-pathogenic) disorders. *J. Exp. Bot.* **1990**, *41*, 701–703. [CrossRef]

78. Milne, T.J.; Abbenante, G.; Tyndall, J.D.A.; Halliday, J.; Lewis, R.J. Isolation and characterization of a cone snail protease with homology to CRISP proteins of the pathogenesis-related protein superfamily. *J. Biol. Chem.* **2003**, *278*, 31105–31110. [CrossRef] [PubMed]

79. Gibbs, G.M.; Scanlon, M.J.; Swarbrick, J.; Curtis, S.; Gallant, E.; Dulhunty, A.F.; O'Bryan, M.K. The cysteine-rich secretory protein domain of Tpx-1 is related to ion channel toxins and regulates ryanodine receptor Ca^{2+} signaling. *J. Biol. Chem.* **2006**, *281*, 4156–4163. [CrossRef] [PubMed]

80. MacMillan, C.P.; Mansfield, S.D.; Stachurski, Z.H.; Evans, R.; Southerton, S.G. Fasciclin-like arabinogalactan proteins: Specialization for stem biomechanics and cell wall architecture in Arabidopsis and Eucalyptus. *Plant J.* **2010**, *62*, 689–703. [CrossRef]

81. Seifert, G.J.; Xue, H.; Acet, T. The Arabidopsis thaliana FASCICLIN LIKE ARABINOGALACTAN PROTEIN 4 gene acts synergistically with abscisic acid signalling to control root growth. *Ann. Bot.* **2014**, *114*, 1125–1133. [CrossRef]

82. Lafarguette, F.; Leple, J.C.; Dejardin, A.; Laurans, F.; Costa, G.; Lesage-Descauses, M.C.; Pilate, G. Poplar genes encoding fasciclin-like arabinogalactan proteins are highly expressed in tension wood. *New Phytol.* **2004**, *164*, 107–121. [CrossRef]

83. Brown, D.M.; Zeef, L.A.H.; Ellis, J.; Goodacre, R.; Turner, S.R. Identification of novel genes in Arabidopsis involved in secondary cell wall formation using expression profiling and reverse genetics. *Plant Cell* **2005**, *17*, 2281–2295. [CrossRef]

84. Persson, S.; Wei, H.; Milne, J.; Page, G.P.; Somerville, C.R. Identification of genes required for cellulose synthesis by regression analysis of public microarray data sets. *Proc. Natl. Acad. Sci. USA* **2008**, *102*, 8633–8638. [CrossRef]

85. Eyüboglu, B.; Pfister, K.; Haberer, G.; Chevalier, D.; Fuchs, A.; Mayer, K.F.X.; Schneitz, K. Molecular characterisation of the STRUBBELIG-RECEPTOR FAMILY of genes encoding putative leucine-rich repeat receptor-like kinases in *Arabidopsis thaliana*. *BMC Plant Biol.* **2007**, *7*, 16. [CrossRef]

86. Miyazono, K.-I.; Miyakawa, T.; Sawano, Y.; Kubota, K.; Kang, H.-J.; Asano, A.; Miyauchi, Y.; Takahashi, M.; Zhi, Y.; Fujita, Y.; et al. Structural basis of abscisic acid signalling. *Nat. Cell Biol.* **2009**, *462*, 609. [CrossRef]

87. Hao, Q.; Yin, P.; Li, W.; Wang, L.; Yan, C.; Lin, Z.; Wu, J.Z.; Wang, J.; Yan, S.F.; Yan, N. The molecular basis of ABA-independent inhibition of PP2Cs by a subclass of PYL proteins. *Mol. Cell.* **2011**, *42*, 662–672. [CrossRef] [PubMed]

88. Zhang, X.; Jiang, L.; Wang, G.; Yu, L.; Zhang, Q.; Xin, Q.; Wu, W.; Gong, Z.; Chen, Z. Structural insights into the abscisic acid stereospecificity by the ABA receptors PYR/PYL/RCAR. *PLoS ONE* **2013**, *8*, e67477. [CrossRef] [PubMed]

89. Benning, U.F.; Tamot, B.; Guelette, B.S.; Hoffmann-Benning, S. New aspects of phloem-mediated long-distance lipid signaling in plants. *Front. Plant Sci.* **2012**, *3*, 53. [CrossRef] [PubMed]

90. Barbaglia, A.M.; Tamot, B.; Greve, V.; Hoffmann-Benning, S. Phloem proteomics reveals new lipid-binding proteins with a putative role in lipid-mediated signaling. *Front. Plant Sci.* **2016**, *7*, 563. [CrossRef]

91. Marshall, S.D.G.; Putterill, J.J.; Plummer, K.M.; Newcomb, R.D. The carboxylesterase gene family from *Arabidopsis thaliana*. *J. Mol. Evol.* **2003**, *57*, 487–500.

92. Ko, M.K.; Jeon, W.B.; Kim, K.S.; Lee, H.H.; Seo, H.H.; Kim, Y.S.; Oh, B.J. A *Colletotrichum gloeosporioides*-induced esterase gene of nonclimacteric pepper (*Capsicum annuum*) fruit during ripening plays a role in resistance against fungal infection. *Plant Mol. Biol.* **2008**, *58*, 529–541. [CrossRef]

93. Suzuki, N.; Koussevitzky, S.; Mittler, R.O.N.; Miller, G.A.D. ROS and redox signalling in the response of plants to abiotic stress. *Plant Cell Environ.* **2012**, *35*, 259–270. [CrossRef]

94. Bidwell, S.D.; Woodrow, I.E.; Batianoff, G.N.; Sommer-Knudsen, J. Hyperaccumulation of manganese in the rainforest tree *Austromyrtus bidwillii* (Myrtaceae) from Queensland, Australia. *Funct. Plant Biol.* **2002**, *29*, 899–905. [CrossRef]

95. Fernando, D.R.; Mizuno, T.; Woodrow, I.E.; Baker, A.J.M.; Collins, R.N. Characterization of foliar manganese (Mn) in Mn (hyper)accumulators using X-ray absorption spectroscopy. *New Phytol.* **2010**, *188*, 1014–1027. [CrossRef]

96. Tenhaken, R.; Thulke, O. Cloning of an enzyme that synthesizes a key nucleotide-sugar precursor of hemicellulose biosynthesis from soybean. UDP-glucose dehydrogenase. *Plant Physiol.* **1996**, *112*, 1127–1134. [CrossRef]

97. Klinghammer, M.; Tenhaken, R. Genome-wide analysis of the UDP-glucose dehydrogenase gene family in Arabidopsis, a key enzyme for matrix polysaccharides in cell walls. *J. Exp. Bot.* **2007**, *58*, 3609–3621. [CrossRef] [PubMed]

98. Mogensen, J.E.; Wimmer, R.; Larsen, J.N.; Spangfort, M.D.; Otzen, D.E. The major birch allergen, Bet v 1, shows affinity for a broad spectrum of physiological ligands. *J. Biol. Chem.* **2002**, *277*, 23684–23692. [CrossRef] [PubMed]

99. Li, B.; Takahashi, D.; Kawamura, Y.; Uemura, M. Comparison of plasma membrane proteomic changes of *Arabidopsis* suspension-cultured cells (T87 Line) after cold and ABA treatment in association with freezing tolerance development. *Plant Cell Physiol.* **2012**, *53*, 543–554. [CrossRef] [PubMed]

100. Takahashi, D.; Li, B.; Nakayama, T.; Kawamura, Y.; Uemura, M. Shotgun proteomics of plant plasma membrane and microdomain proteins using nano-LC-MS/MS. In *Plant Proteomics: Methods and Protocols*; Jorrin-Novo, J.V., Komatsu, S., Weckwerth, W., Wienkoop, S., Eds.; Humana Press: Totowa, NJ, USA, 2014; pp. 481–498.

101. Gutierrez-Carbonell, E.; Takahashi, D.; Lüthje, S.; González-Reyes, J.A.; Mongrand, S.; Contreras-Moreira, B.; Abadía, A.; Uemura, M.; Abadía, J.; López-Millán, A.F. A shotgun proteomic approach reveals that Fe deficiency causes marked changes in the protein profiles of plasma membrane and detergent-resistant microdomain preparations from *Beta vulgaris* roots. *J. Proteome Res.* **2014**, *15*, 2510–2524. [CrossRef] [PubMed]

102. Ceballos-Laita, L.; Gutierrez-Carbonell, E.; Takahashi, D.; Abadía, A.; Uemura, M.; Abadía, J.; López-Millán, A.F. Data on xylem sap from Mn- and Fe-deficient tomato plants. *Data Brief* **2018**, *17*, 512–516. [CrossRef] [PubMed]

103. Perez-Riverol, Y.; Csordas, A.; Bai, J.; Bernal-Llinares, M.; Hewapathirana, S.; Kundu, D.J.; Inuganti, A.; Griss, J.; Mayer, G.; Eisenacher, M.; et al. The PRIDE database and related tools and resources in 2019: Improving support for quantification data. *Nucleic Acids Res.* **2019**, *47*, D442–D450. [CrossRef] [PubMed]

104. Gene Ontology Resource. Available online: http://geneontology.org/ (accessed on 24 September 2020).

105. SignalP-5.0 Server. Available online: http://www.cbs.dtu.dk/services/SignalP/ (accessed on 15 September 2020).

106. TargetP-2.0 Server. Available online: http://www.cbs.dtu.dk/services/TargetP/ (accessed on 15 September 2020).

107. SecretomeP 2.0 Server. Available online: http://www.cbs.dtu.dk/services/SecretomeP/ (accessed on 15 September 2020).

108. Emanuelsson, O.; Brunak, S.; von Heijne, G.; Nielsen, H. Locating proteins in the cell using TargetP, SignalP and related tools. *Nat. Protoc.* **2007**, *2*, 953–971. [CrossRef]

109. LSPpred. Leaderless Secretory Protein Predictor for Plants. Available online: http://lsppred.lspdb.org/ (accessed on 7 October 2020).

Publisher's Note: MDPI stays neutral with regard to jurisdictional claims in published maps and institutional affiliations.

International Journal of
Molecular Sciences

Article

Reduction of Allergenic Potential in Bread Wheat RNAi Transgenic Lines Silenced for *CM3*, *CM16* and *0.28 ATI* Genes

Raviraj M. Kalunke [1,2,†], Silvio Tundo [1,3,†], Francesco Sestili [1], Francesco Camerlengo [1], Domenico Lafiandra [1], Roberta Lupi [4], Colette Larré [4], Sandra Denery-Papini [4], Shahidul Islam [5], Wujun Ma [5], Stefano D'Amico [6] and Stefania Masci [1,*]

[1]   Department of Agriculture and Forest Science (DAFNE), University of Tuscia, 01100 Viterbo, Italy; rkalunke@gmail.com (R.M.K.); silvio.tundo@unipd.it (S.T.); francescosestili@unitus.it (F.S.); f.camerlengo@unitus.it (F.C.); lafiandr@unitus.it (D.L.)

[2]   Institute of Plant and Microbial Biology, Academia Sinica, Taipei 11529, Taiwan

[3]   Department of Land, Environment, Agriculture and Forestry (TESAF), University of Padova, 35020 Legnaro, Italy

[4]   INRAE UR1268 BIA, 44000 Nantes, France; robertinalupi@hotmail.it (R.L.); colette.larre@inrae.fr (C.L.); sandra.denery@inrae.fr (S.D.-P.)

[5]   Australia China Centre for Wheat Improvement, College of Science Health Engineering and Education, Murdoch University, Murdoch, WA 6150, Australia; S.Islam@murdoch.edu.au (S.I.); w.ma@murdoch.edu.au (W.M.)

[6]   Institute for Animal Nutrition and Feed, AGES-Austrian Agency for Health and Food Safety, 1220 Vienna, Austria; stefano.d-amico@ages.at

[*]   Correspondence: masci@unitus.it; Tel.: +39-0761-357-255

[†]   These authors contribute equally to this work.

Received: 23 July 2020; Accepted: 11 August 2020; Published: 13 August 2020

Abstract: Although wheat is used worldwide as a staple food, it can give rise to adverse reactions, for which the triggering factors have not been identified yet. These reactions can be caused mainly by kernel proteins, both gluten and non-gluten proteins. Among these latter proteins, α-amylase/trypsin inhibitors (ATI) are involved in baker's asthma and realistically in Non Celiac Wheat Sensitivity (NCWS). In this paper, we report characterization of three transgenic lines obtained from the bread wheat cultivar Bobwhite silenced by RNAi in the three ATI genes *CM3*, *CM16* and *0.28*. We have obtained transgenic lines showing an effective decrease in the activity of target genes that, although showing a higher trypsin inhibition as a pleiotropic effect, generate a lower reaction when tested with sera of patients allergic to wheat, accounting for the important role of the three target proteins in wheat allergies. Finally, these lines show unintended differences in high molecular weight glutenin subunits (HMW-GS) accumulation, involved in technological performances, but do not show differences in terms of yield. The development of new genotypes accumulating a lower amount of proteins potentially or effectively involved in allergies to wheat and NCWS, not only offers the possibility to use them as a basis for the production of varieties with a lower impact on adverse reaction, but also to test if these proteins are actually implicated in those pathologies for which the triggering factor has not been established yet.

Keywords: wheat; RNAi silencing; α-amylase/trypsin inhibitor (ATI); allergy; non celiac wheat sensitivity (NCWS)

---

1. Introduction

Wheat (*Triticum* spp.) is one the "big three" cereal crops, together with maize and rice, cultivated worldwide thanks to its adaptability, nutritional value and versatility. This is mostly due to the unique viscoelastic properties of its dough, capable to give rise to a wide range of products, such as bread, pasta, noodles, biscuits, some of which are characteristic of specific geographical areas.

Such properties derive from wheat grain proteins, classified into four major groups based on solvent solubility [1]: albumins (water), globulins (dilute salt solution), prolamins including gliadins (alcohol/water mixture), and finally glutelins, including glutenins (diluted acid or alkaline solutions). At present, gliadins and glutenins are both considered prolamins, because they are soluble in alcohol/water mixtures once glutenins are present in the reduced form.

Gliadins and glutenins make up the gluten, defined as the viscoelastic mass obtained after full flour hydration and washing out of water-soluble components, composed mostly by starch and non-prolamin proteins, namely albumins and globulins (A/G). Among gluten proteins, glutenins play the major role and, in particular, their size and amount are major determinants of dough technological quality [2].

Compared to the gliadins and glutenins, few studies have been carried out on non-prolamins so far. This is probably because their role in flour quality is not as well defined as that of gluten proteins. Nevertheless, A/G constitute around 15–20% of total flour protein [3]. They are a mixture of structural, metabolic and storage proteins [4]. A/G are mostly located in the seed coat, the aleurone cells and the germ; they are relatively scarce in the starchy endosperm [5]. Their amino acid compositions are relatively well balanced because of higher lysine content as compared to the prolamin fraction. Predominant A/G components such as α-amylase/trypsin inhibitors (ATI), serpins and purothionins have multiple functions; indeed, they serve as nutrient reserves for the germinating embryo and as inhibitors of insects and fungal pathogens before germination [6].

Wheat proteins can cause different adverse reactions, some of which are better characterized, such as in Celiac Disease (CD), Wheat Allergies including Food Allergy to Wheat (FAW), Wheat-Dependent Exercise-Induced Anaphylaxis (WDEIA), or in Baker's Asthma (BA) [7]. Differently, the role of wheat components in Irritable Bowel Syndrome (IBS) or Non Celiac Wheat/Gluten Sensitivity (NCWS or NCGS, respectively) is still not clear. In particular, this can be deduced by the use of the two names, NCWS or NCGS to describe a pathology that includes both gastrointestinal and non-gastrointestinal symptoms caused by wheat ingestion, but that excludes CD and FAW. Because specific serological markers are not present so far, this is actually a self-reported condition, whose diagnosis is based on double-blind placebo-controlled wheat challenge [8]. This situation makes it even more difficult to establish the triggering factor, that initially was identified in gluten, mostly for analogies with CD, but that at present indicates rather ATI or fermentable oligosaccharides disaccharides monosaccharides and polyols (FODMAPs), reason why it is currently preferred to use the name NCWS, rather than NCGS. Since the prevalence worldwide is in the range 0.6–13%, it is important to identify the real culprit of such pathology.

Wheat ATI are among the putative triggering factors of NCWS and are unquestionably involved in BA, the most common occupational respiratory disease in Western countries, affecting about 10% of flour workers [9]. Moreover, this class of wheat proteins seems involved in some wheat-related food-allergies, and, to a minor extent, with WDEIA [10]. In this regard, recently, Tundo et al. [11] tested three heterologously expressed ATI proteins, named *CM3*, *CM16* and 0.28 in basophils degranulation assay against human sera of patients with FAW. Although all the three proteins induced degranulation, the most effective one was *CM3*, accounting for the important role played by this protein in triggering allergic reactions. Moreover, *CM3* has an important role in innate immune response, at least in monocytes, macrophages, and dendritic cells [12,13].

Most ATI proteins belong to the so-called CM protein fraction of wheat, because they are soluble in chloroform and methanol solutions [14]. Three classes of ATI are typically described, that correspond to monomeric, dimeric and tetrameric forms, with different specificities against various heterologous α-amylases. In particular, the 12 kDa monomeric inhibitors, also known as 0.28 proteins, are encoded by genes on the short arms of the group 6 chromosomes; the 24 kDa homodimeric inhibitors, also known as the 0.19 and 0.53 proteins, are encoded by genes on the short arms of the group 3 chromosomes; the third group is constituted of the 60 kDa heterotetrameric inhibitors composed of one copy of either CM1 or CM2, encoded by genes on chromosomes 7D or 7B, plus one copy of either CM16 or CM17, encoded by genes on chromosomes 4B or 4D, plus two copies of *CM3*, also encoded on chromosomes 4B or 4D [15].

The availability of genotypes with reduced amounts of ATI proteins might allow either the development of novel wheat cultivars with a lower triggering potential, or a better understanding of the role of these components in such wheat-related pathologies. Recently, our group [16] has reported the application of CRISPR-Cas9 silencing technology to silence *CM3* and *CM16* ATI genes in durum wheat, whereas in this paper we report the development and the extensive characterization, including the allergenic potential and qualitative evaluations, of wheat lines derived from the bread wheat cultivar Bobwhite, in which *CM3*, *CM16* and *0.28* ATI genes have been silenced by RNAi.

2. Results

2.1. RNAi Transformed Wheat Lines Show Silencing of ATI Genes During the Kernel Development Stages

In total, 1669 immature embryos of *Triticum aestivum* L. cv. Bobwhite were co-transformed using three RNAi constructs (*pRDPT-CM3*, *pRDPT-CM16* and *pRDPT-0.28*) and the plasmid pUBI::BAR. Eight T_0 independent transgenic lines resistant to Bialaphos were obtained which represents 0.44% transformation efficiency. The presence of the transgenes was verified by PCR analysis on genomic DNA from regenerated plants (data not shown). No significant differences in morphology or growth were observed between the "null" genotype that had lost the transgene by segregation (null-segregant), and the untransformed plants (data not shown), thus we have used the untransformed plant (indicated as either BW or WT) for the experiments.

In order to get homozygous transgenic line, T_1 progenies from T_0 plants containing all three transgenes as well as the *bar* marker gene were used in segregation analysis. PCR analyses were performed on DNA extracted from half-seeds using primers specific for the *CM3*, *CM16*, *0.28* and *bar* transgenes. Lines containing all four transgenes were used for further seeds production to obtain homozygous transgenic lines showing silencing of the three target genes. In T_4 generation, homozygous transgenic line plants were obtained and used for the analyses, in comparison with WT.

Relative gene expression analyses were performed on the lines 10-10a, 22-2 and on the Bobwhite control plants to investigate the reduction of transcripts of the genes encoding *CM3*, *CM16* and *0.28* subunits during kernel development stages. In this case, we could not use line 24-1 because we did not get enough developing grains to extract mRNA.

Data obtained from RNA extracted at 10 Days Post Anthesis (DPA) showed an immediate reduction in the first development stages of all transcripts investigated. The decrease in the gene expression of the *CM3*, *CM16* and *0.28* subunits was more evident in line 22-2, but it was also significant for in line 10-10a. The reduction of the transcripts, in fact, ranged from 80% to almost 100% in line 22-2 for all the genes analyzed, whereas in line 10-10a, the reduction ranged from 60% to 90%; a similar result was obtained from the analysis carried out with RNA extracted at 20 DPA with a slight increase in the transcripts compared to 10 DPA. The almost total silencing of ATI genes was quite clear at 30 DPA for both RNAi lines (Figure 1).

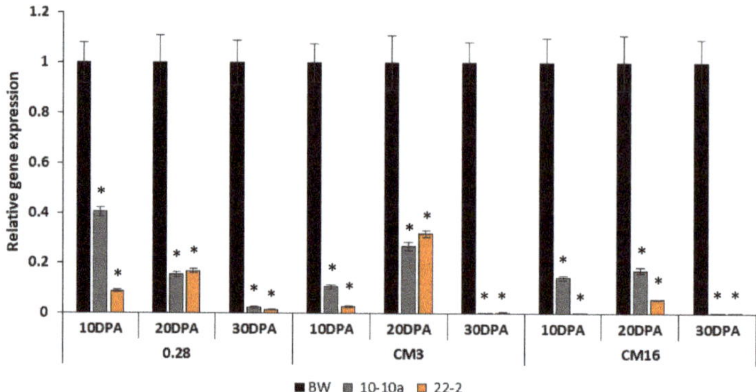

Figure 1. Expression analysis of *CM3, CM16, 0.28* α-amylase/trypsin inhibitors (ATI) genes by qRT-PCR in different caryopses growth stages (10, 20 and 30 days post anthesis (DPA)) in the bread wheat cv Bobwhite (BW) and RNAi-silenced lines 10-10a and 22-2. The relative gene expression is reported as the fold increase in the transcripts compared to Bobwhite control plants. Standard error is shown above each bar, along with asterisk to indicate where the value differed significantly ($p < 0.05$).

2.2. Targeted Proteins CM16, 0.28 and CM3 in the Transgenic Lines Are Accumulated to a Lower Extent with Respect to the Control

Differential expression of the grain proteins across the silenced lines and in comparison with the parental cultivar Bobwhite, were analyzed using a 4-plex iTRAQ experiment. A total of 3915 distinct peptides have been detected at >95% confidence level. The total number of detected proteins was 794 with ≥1 peptide matching, while the threshold of ≥2 peptides matching downsized the protein numbers to 550. Compared to the parental cultivar, 80, 80 and 74 proteins showed differential expression in the silenced lines 24-1, 22-2 and 10-10a, respectively. Overall, the protein expression changes across the silenced lines compared to the parental cultivar were consistent.

The results demonstrated that the expression of the targeted three ATI, namely *CM16* (TraesCS4B01G328000), *0.28* (TraesCS6D01G000200) and *CM3* (TraesCS4B01G328100) were reduced remarkably in the silenced lines with proper statistical significance (Table S1). In the silenced line 24-1, all the three target proteins expression reduced more than the other two silenced lines. Proteins *CM16, 0.28* and *CM3* showed 91-, 79- and 72-fold reductions in expression, respectively, in the silenced line 24-1, compared to the parental cultivar (Table S1). On the other hand, in lines 22-2, and 10-10a, those protein's expressions were reduced by 25, 24 and 25 folds; and 17, 23 and 24 folds, respectively. In addition, *CM17* which has 85% sequence similarity with CM16, showed 59, 28 and 29 folds reduced expression in silenced lines 24-1, 22-2 and 10-10a, respectively (Table S1). Furthermore, some other closely related proteins, namely dimeric α-amylase inhibitor, trypsin inhibitor CMc, Bowman–Birk type trypsin inhibitor and thaumatin-like xylanase inhibitor also showed reduced expression in the silenced lines which ranged from a 5- to a 14-fold reduction (Table S1). Several other trypsin inhibitor proteins also demonstrated reduced expression but were excluded from the interpretation due to the higher *p* value than the threshold of 0.05.

2.3. Some of Non-Target Proteins Show Differential Accumulation in Silenced Lines Compared to the Control

The expression patterns of the all identified glutenin and gliadin proteins were analyzed across the silenced lines and in comparison with the parental cultivar. A total of 21 proteins of these groups showed differential expression across the silenced lines (Table S2), although only four of them were statistically significant, considering the cut off *p* value of 0.05. The high molecular weight glutenin subunits (HMW-GS) identified as 1Dx1.6t, likely corresponding to 1Dx5, showed a 64-, 22- and 17-fold reduction in expression in silenced lines 24-1, 22-2 and 10-10a, respectively (Table S2). However, another HMW-GS corresponding to 1Dy10 subunit (Uniprot accession No. A9YSK3) showed 19.23-fold reduced expression in the line 10-10a only. On the other hand, an s-type LMW-GS (Uniprot accession No. F8SGN6) showed 3.63-, 0.14- and 4.29-fold reductions in expression in silenced lines 24-1, 22-2 and 10-10a, respectively. Furthermore, another LMW-GS (Uniprot accession No. R4JBK0), but of m-type, showed 6.73- and 7.31-fold reductions in expression in silenced lines 22-2 and 10-10a, respectively, although it did not show significant reduction in line 24-1. Differential expression of all the gliadin proteins and other LMW and HMW-GSs were statistically non-significant (Table S2).

However, several non-target proteins demonstrated differential expression (by both over expression and reduced expression) in the silenced lines compared to the parental cultivar (Table S3). An uncharacterized protein (Uniprot accession No. A0A3B6RB62) whose molecular function is nutrient reservoir activity, showed reduced expression by 44-, 17- and 15-fold in the silenced lines 24-1, 22-2 and 10-10a, respectively. The other major down regulated proteins in the silenced lines included Avenin-like b2, Puroindoline b, AAI domain-containing protein, grain softness proteins and farinin protein (Table S3). On the other hand, an uncharacterized protein (Uniprot accession No. W5GLX4) whose molecular function is ATPase activity, showed a 99-fold increased expression in all the three lines compared to the parental cultivar (Table S3). The other major upregulated proteins in all the silenced lines included DUF89 domain-containing protein, 60S acidic ribosomal protein P2B, WHy domain-containing protein and Eukaryotic translation initiation factor 5A.

2.4. ATI Reacting with Anti-ATI Antibody Was Lower in All the Three Transgenic Lines Compared to Bobwhite Control

In order to assess whether the transformation with silencing constructs of bread wheat could lead to a reduction of ATI content in kernels, we performed indirect enzyme-linked immunosorbent assay (ELISA) experiments using a polyclonal antibody against α-amylase inhibitors on A/G fraction.

The ELISA analysis revealed that the amount of ATI reacting with anti-ATI antibody was lower in all the three transgenic lines compared to Bobwhite control plants (Figure 2). The levels of antigen were significantly different between transgenic lines and Bobwhite control plants in all primary antibody dilutions, except for the lower ratio (1:64,000), for which no difference was observed.

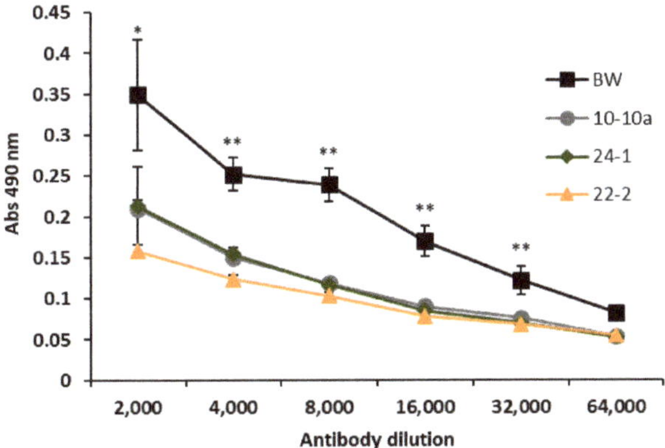

Figure 2. Enzyme-linked immunosorbent assay (ELISA) performed on albumins and globulins (A/G) fraction of Bobwhite control plants (BW) and transgenic lines with polyclonal anti ATI antibody. Values represent the average of the absorbance at 490 nm ± standard error of three biological and three technical replicates. All data were subjected to ANOVA analysis and Tukey test. ** indicates difference between RNAi transgenic lines and Bobwhite control plants at $p < 0.01$ level of significance, * indicates $p < 0.05$ level of significance.

2.5. Sera of Wheat Allergic Patients Show a Lower Reactivity to the Transgenic Lines

All the twenty-two sera selected for their content in IgE specific to the A/G fraction of the bread wheat cultivar Recital reacted with all the genotypes. A broad variability (from 10 to 500 ng/mL) in the IgE binding capacity between genotypes was measured, in all cases the IgE binding of Bobwhite was higher than that of RNAi silenced lines. In general, more than half of the sera showed lower reactions to the silenced lines with respect to Bobwhite (Figure 3). Two-way ANOVA was performed to investigate the effect of the three ATI silenced lines with respect to Bobwhite and the effect of sera on IgE binding capacity (In Supplementary Files). The two factors (genotype and sera) and their interaction were significant ($p < 0.05$). The genotypes account for 7.27% of the total variance, whereas sera account for 86.30% of the total variance. This latter result is related to the well-known variation existing between individual reactions against the same antigen.

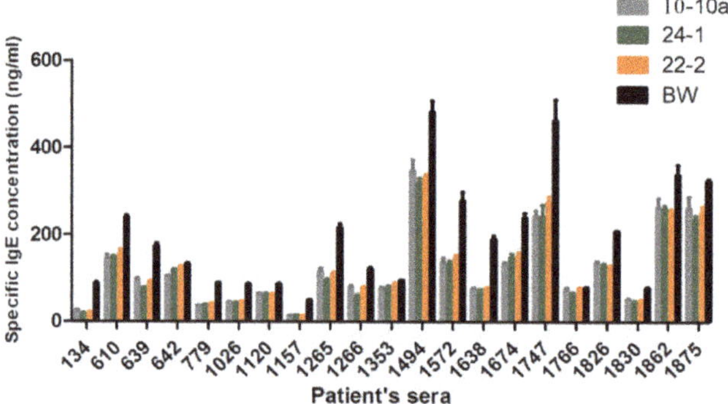

Figure 3. Human IgE binding to salt soluble proteins from the three RNAi transgenic lines and from the Bobwhite control plants (BW), detected in the sera from 22 allergic patients by fluorescent (F)-ELISA assay. Results are expressed as the means of specific IgE concentration. For major clarity, ANOVA analysis is reported in the Supplementary File.

2.6. Transgenic Wheat Lines Do Not Affect Trypsin Inhibition Activity, but Mostly α-Amylase Activity

The presence of trypsin inhibitors is health-related because they are involved in several gastro-intestinal and non-intestinal disorders. Thus, we have performed a measure of TIA (trypsin inhibitor activity) in the wheat lines in which specific α-amylase and trypsin inhibitors were silenced.

In Table 1, TIA is reported both as absolute and normalized values according to the amount of soluble protein tested. Noteworthy, the total protein amount in the silenced lines almost doubled that of Bobwhite, and this is reflected in the amount of soluble proteins as well, including ATI. This is likely a compensation effect, already known for other wheat genotypes in which specific gene expression is suppressed either by transgenesis [17] or by gene deletions [18]. TIA results were coherent with protein amounts, being increased in silenced lines, but, when normalized, significant lower values were observed in line 22-2 and 10-10a.

Table 1. Concentration of total protein, salt-water soluble proteins and trypsin inhibition activity (TIA).

Sample	Total Protein (g/100 g)	Soluble Protein (g/100 g)	TIA (mg/Kg)	TIA (mg/g Soluble Protein)
Bobwhite	10.08 ± 0.03 a	1.28 ± 0.13 a	107.89 ± 1.92 a	8.43 ± 0.15 c
24-1	21.10 ± 0.07 b	2.40 ± 0.32 b	207.54 ± 3.14 d	8.65 ± 0.13 c
22-2	22.18 ± 0.05 c	2.78 ± 0.05 b	170.75 ± 3.50 b	6.13 ± 0.12 a
10-10a	20.83 ± 0.06 b	2.52 ± 0.14 b	197.82 ± 2.71 c	7.86 ± 0.10 b

Small letters indicate homogeneous subgroups based on ANOVA ($p \leq 0.05$) and post hoc test according to Scheffe.

This result was also confirmed in the progress of trypsin inhibition by using increasing sample extract volumes. In Figure S1 it is shown that, in the three silenced lines, the maximum of trypsin inhibition is reached with lower sample extract volumes with respect to the reference genotype Bobwhite, indicating that a higher amount of trypsin inhibitors might be present in the transgenic lines. MALDI TOF spectra (Figure S2) clearly indicated lower signal intensities in the molecular weight range of ATI and the peak at about 15.5 kDa of *CM3* disappeared almost completely. Previously [19], it has been shown that a broad variety of proteins are bound to trypsin and thus play a role for trypsin inhibition. Furthermore, α-amylase/subtilisin inhibitor was slightly increased in modified samples as seen in Table S1. Nevertheless, mentioned facts cannot explain completely variations in TIA. Other proteins, probably uncharacterized ones like W5GLX4 and A0A453HBR5 (Table S3) might play a key role for the measured trypsin inhibition levels because they increased strongly (up to 99%). An uncharacterized protein, A0A077RSX3_WHEAT, was found to inhibit trypsin as well [19]. However, since the other results we have obtained clearly indicate that the three target genes have been silenced and absence of most ATI, as shown also by LC-MS/MS data, the results got about TIA indicate that the three target genes do not have a large trypsin inhibition activity, but they likely inhibit α-amylase activity. Furthermore, aggregation state of ATI might be important for their biological functionality, but this characterization was not performed.

2.7. Influence of Silencing of the Three Target ATI Genes on Predictive Quality Parameters

Thousand kernel weight (TKW) and coleoptile length are two parameters correlated with wheat yield and have been measured to test if possible pleiotropic effects of the silencing may influence such important aspect. No significant differences were detected in TKW between RNAi transgenic lines and Bobwhite control plants (Figure S3). The coleoptile lengths of the WT and transgenic lines were compared at two different timepoints: 5 and 7 days post germination (DPG). No significant difference was observed in any of the two time points (Figure S4). The absence of significant differences in both the tested parameters showed that yield was not affected by the silencing of the three ATI genes.

Since higher amounts of flour are needed to perform rheological tests, we used percentage of unextractable polymeric proteins (%UPP) micro-test to predict gluten quality of RNAi-silenced plants. Figure 4 reports the histograms representing the %UPP values found for Bobwhite control plants and the three transgenic lines. ANOVA analysis showed that there is a dramatic decrease in the %UPP value in all silenced lines, as compared to Bobwhite control plants.

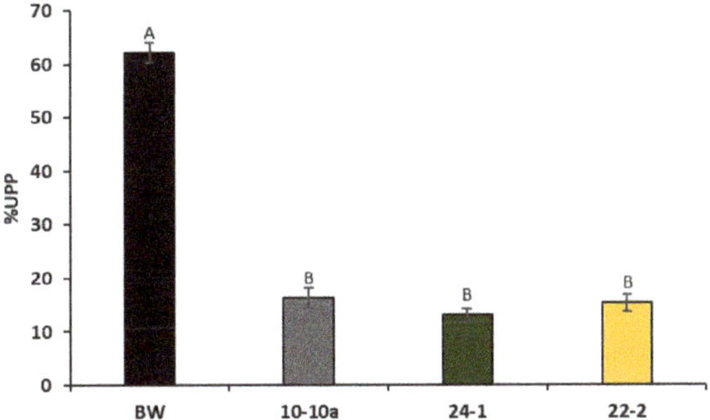

Figure 4. Determination of percentage of unextractable polymeric proteins (%UPP) in Bobwhite control plants (BW) and the three RNAi transgenic lines. Values represent the average of three biological and three technical replicates. All data were subjected to ANOVA analysis. Letters above the histograms correspond to ranking of Tukey test at 0.95 confidence and $p < 0.01$ level of significance.

Afterwards, we analyzed the electrophoretic pattern of glutenin subunits, whose structure influences the formation of glutenin polymers, the major parameter determining %UPP.

The SDS-PAGE pattern of total kernel proteins is reported in Figure 5. Although already indicated by gene expression analyses and LC-MS/MS results, this gel clearly shows that HMW-GS have been an off-target of RNAi silencing, in particular the *Glu-D1* coded pair 5+10, mostly responsible of qualitative characteristics of bread wheat. This observation could explain the decrease in %UPP.

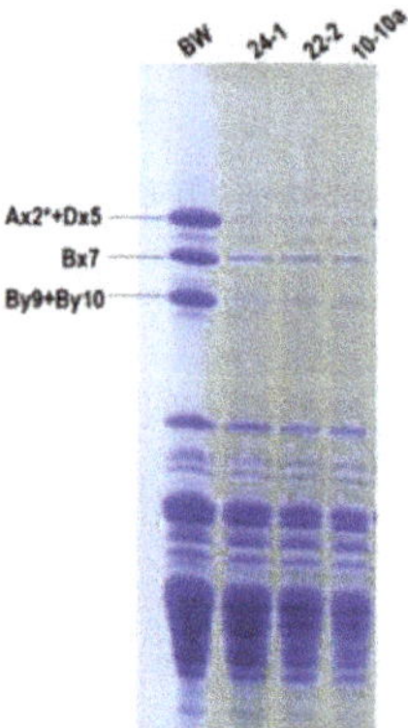

Figure 5. SDS-PAGE of total wheat kernel protein subunits extracted from the three RNAi transgenic lines and Bobwhite control plants. High molecular weight glutenin subunits (HMW-GS) identification is reported.

The expression analysis of *HMW-GS Dx5* gene was tested and revealed a significant reduction of Dx5 transcripts already at 10 DPA, especially for the line 22-2. At 30 DPA, the amount of the *Dx5* transcripts was drastically reduced in both transgenic lines, approximately 90% less (Figure 6). In this case also, only two transgenic lines have been analyzed because of the lack of enough developing grains of line 24-1 (Figure 6).

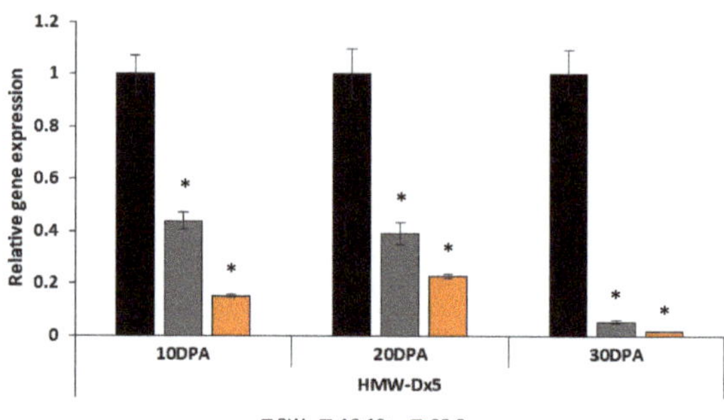

Figure 6. Expression analysis of *HMW-GS Dx5* gene by qRT-PCR in different caryopses growth stages (10, 20 and 30 DPA) in the control line BW and RNAi-silenced lines 10-10a and 22-2. The relative gene expression is reported as the fold increase in the transcripts compared to Bobwhite control plants. Standard error is shown above each bar, along with asterisk to indicate where the value differed significantly ($p < 0.05$).

To determine if non-targeted silencing of *HMW-GS* was due to possible homologies among *ATI* and *HMW-GS* genes, we made a bioinformatics comparison among the three target *ATI* genes and the genes coding for the *HMW-GS* genes present in cultivar Bobwhite, that are *Ax2**, *Dx5*, *Bx7*, *By9* and *Dy10*.

The analysis allowed us to identify the presence of a conserved region in *0.28* and *HMW-GS By9* and *Dy10* (TGCTGCCAGCAGCT), that might explain the off-target silencing for at least of these two subunits. Moreover, the alignment between *0.28* and *HMW-GS Dy10* showed a further homology of some nucleotides separated from the conserved region by a guanine pair (Figure S5). The conserved sequence has been submitted to a further analysis with PlantGRN (http://plantgrn.noble.org), containing information on plant transcriptional regulation. By using the psRNATarget and ppsRNAi tools, the presence of siRNA and miRNA is suggested. psRNATarget highlights 16 miRNA sequences with sequence homologies with 0.28 and HMW-GS Dy10. By using the ppsRNAit tool, 18 possible siRNA have been identified (Figure S6).

3. Discussion

Although RNAi is a transgenic procedure, it can be used as a proof of concept before developing new genotypes with non-transgenic techniques. Regardless of the procedure used, in fact, any new genetic combination can potentially give rise to unintended effects. Silencing of specific genes can be pursued by exploiting natural variation present in genetic resources, or by generating pseudogenes by classical or advanced mutagenesis, but all the classical methods require time consuming procedures that, in case undesirable unintended effects are produced, make the work performed useless.

The possibility to develop new wheat genotypes accumulating a lower amount of proteins potentially or effectively involved in such pathologies, not only offers the possibility to use them as a basis for the creation of wheat varieties with a lower impact on adverse reaction, but also to test if these proteins are actually implicated in those pathologies for which the triggering factor has not been established yet.

Recently we have obtained durum wheat lines in which the ATI genes *CM3* and *CM16* have been silenced by CRISPR-Cas9 multiple editing [16], and that are now under multiplication and characterization. Differently, in the present work, we report the characterization, in terms of protein and gene expression, of allergenic potential, and of parameters related to technological performances of bread wheat lines in which three genes coding for ATI proteins, namely *CM3*, *CM16* and *0.28*, involved in BA and likely in NCWS, have been silenced by RNAi.

We have obtained different transgenic lines showing an effective decrease in the target genes, and in the corresponding protein products. Very interesting was the observation that, differently from what was expected, the assay of trypsin inhibition (TIA) showed higher levels for the transgenic lines. These findings are confirmed by the study of Call et al. [20], which found no significant correlation between ATI concentration and TIA. As already mentioned, besides ATI, other non-gluten proteins probably play a key role in trypsin inhibition.

Whatever the explanation, such a result clearly highlights the role of the three target genes in allergic reactions, because the silenced lines result in effectively less reactivity with respect to the untransformed line, when tested with sera of patients allergic to wheat. Especially *CM3* might be the most important factor in triggering allergies, as suggested also by Tundo et al. [11], that, after testing heterologously produced *CM3*, *CM16* and *0.28* in basophils degranulation assay against human sera of patients with FAW, found that *CM3* was the most effective, giving results comparable to those of a whole soluble protein extract.

This plant material is available for testing the role of the silenced proteins in triggering either allergic reactions or NCWS, because it does not show significant differences in terms of yield, although, from the technological point of view, it is rather poor, due to the non-target effect on HMW-GS, mostly influencing rheological properties of wheat doughs. Another possible explanation might reside in epigenetic silencing mechanisms due to the elevated number of *Dx5* promoters in ATI transgenic lines. A silencing mechanism, known as homology-dependent gene silencing (HDGS), can arise when multiple copies of the same or homologous sequences are introduced in a genome [21,22]. Jauvion et al. [23] distinguished two types of HDGS, based on the stage at which it occurs: the first is the transcriptional gene silencing (TGS) that is associated with transcription and promoter modification;

the second one is the post-transcriptional gene silencing (PTGS), which occurs after the formation of mRNA. In our transgenic plants, the RNAi constructs and the endogenous *HMW-GS* gene have a high homology in the promoter regions, so the activation of a TGS mechanism is plausible, that can lead to the methylation of the promoter or coding regions of endogenous *HMW-GS*. Similar results were previously observed in transgenic plants, in which a non-symmetrical methylation caused the silencing of endogenous genes [22]. If the explanation of such an untargeted effect resides in epigenetic silencing mechanisms due to the high number of Dx5 promoters in ATI transgenic lines, this could be circumvented by back-crossing the transgenic lines with the WT genotype, in order to decrease the number of Dx5 promoters and possibly restore the proper HMW-GS composition.

4. Materials and Methods

4.1. Plant Material and Transformation

RNAi constructs preparation, wheat transformation by the biolistic procedure and the selection of the transgenic lines were carried out as previously described by Sestili et al. [24]. Briefly, three vectors *pRDPT-CM3*, *pRDPT-CM16* and *pRDPT-0.28* were prepared for RNAi cassettes of *ATI* genes *CM3*, *CM16* and *0.28*, respectively. These genes were amplified using cDNA template generated from RNA extracted from immature embryos (28 Days Post Anthesis, DPA) of *Triticum aestivum*. L. cv. Bobwhite. The cDNA synthesis was performed by using the Quanti-Tect Reverse Transcription Kit (Qiagen, Hilden, Germany) following the manufacturer instructions. The vector pRDPT contains promoter and terminator of the *Dx5* high molecular weight glutenin subunit gene and intron of starch branching enzymes of class II (*SBEIIa*). Sense and antisense sequences of the target region of *ATI* gene were inserted in the plasmid pRDPT separated by the *SBEIIa* intron with help of restriction sites introduced in the oligonucleotides (denoted in primer name). Restriction sites *SalI* and *ClaI* were used for the insertion of sense sequences, whereas *XhoI* and *XbaI* were used for antisense sequences. The following primers were used (sequences are listed in Table S4) to amplify the three target genes: for *CM3* sense sequence iRNA_Sal-CM3-F and iRNA ClaI-CM3-R; *CM3* anti-sense sequence: iRNA XhoI-CM3-F, and iRNA XbaI-CM3-R. For *CM16* sense sequence: iRNA SalI-CM16-F, 5′- and iRNA ClaI-CM16-R; *CM16* Anti-sense sequence: iRNA XhoI-CM16-F and iRNA XbaI-CM16-R. For 0.28 sense sequence: iRNA Sal-AAI0.28-F and iRNA ClaI-0.28 R; *CM0.28* Anti-sense sequence: iRNA XhoI-0.28_F, and iRNA XbaI-0.28-R.

The immature embryo derived calli of the bread wheat cultivar Bobwhite were co-bombarded with the three RNAi plasmids (*pRDPT-CM3, pRDPT-CM16* and *pRDPT-0.28*) and the plasmid *pUBI::BAR* (carrying the *bar* gene as selection marker) mixed in a 3:1 molar ratio following the biolistic procedure.

The presence of the construct in Bialaphos-resistant plants and their progeny was verified by PCR on genomic DNA obtained from young leaves of the regenerated plants [25] using specific primer pairs (sequences are listed in Table S4) for *CM3* (iRNA_Sal-CM3-F, and Intron R), *CM16* (Dx5_Promo 2F and iRNA ClaI-CM16-R), 0.28 (Dx5_Promo 2F and iRNA ClaI-0.28 R) and *bar* gene (UBI-49F and BAR 2). The *actin* gene amplification was used as positive control to confirm the presence of genomic DNA template in PCR reaction, using actin-specific primer pair: Actin 77F and Actin 312R. PCR amplifications were carried out according to the procedure for GoTaq PCR Master Mix (Promega, Madison, WI, USA) and the amplification conditions were as follows: 1 cycle at 95 °C for 2 min; 35 cycles at 95 °C for 1 min, 60 °C for 1 min, and 72 °C for 1 min; and a final 3-min step at 72 °C. The PCR products were loaded into 1.2% (*w/v*) agarose gels and subjected to electrophoresis followed by visualization under UV light (data not shown).

Three homozygous transgenic lines (named as 24-1, 22-2 and 10-10a) showing silencing of the three genes have been advanced to the T_4 generation, thus T_4 plants were used for the reported analyses, in comparison with the control untransformed line (WT).

4.2. Quantitative Analyses of ATI Transcripts by qRT-PCR

All the lines were cultivated in growth chambers in the same conditions. The qRT-PCR was performed on total RNA extracted from the caryopses of the three RNAi lines along with the wild type line at 10 DPA, 24 DPA and 30 DPA, using the Spectrum Plant Total RNA kit (Sigma-Aldrich, St. Louis, MO, USA) and following manufacturer's instructions. First strand complementary DNA was synthesized using the Quanti-Tect Reverse Transcription Kit (Qiagen, Hilden, Germany) according to the manufacturer's instructions from one µg of total RNA of each sample. The experiment was carried out with the same primer used for amplification of target genes reported above. For *Dx5* transcripts, the following primers were used: Dx5F2: 5′-GACCAACAGCTCCGAGACA-3′ and Dx5R2: 5′-GTATGAAACCTGCTGCGGAC-3′. The *actin* gene was used as reference gene. The analyses were performed in the CFX 96 Real-Time PCR Detection System device (Bio-Rad, Hercules, CA, USA). The reactions were prepared in a final volume of 15 µL, consisting in 7.5 µL SsoAdvUniver SYBR GRN SMX (Bio-Rad, Hercules, CA, USA), 0.5 µM of each primer and 1 µL of cDNA, following a three-step PCR protocol: 95 °C for 30 sand 39 cycles at 95 °C for 10 s, 60 °C for 25 s and 72 °C for 15 s. Three biological replicates with three technical replicas were carried out for each reaction and the maximum difference accepted between Ct values of technical replicates was 0.5. The relative quantification normalized to the reference gene was performed using the $2^{-\Delta\Delta Ct}$ method.

4.3. Protein Analysis by iTRAQ

Total proteins were obtained by crashed kernels by extraction with a buffer containing 0.5 M Tris-HCl buffer pH 8.8, 1% dithiothreitol (DTT) in 50% propanol for 1 h at room temperature, and precipitating with 4 volumes of cold acetone, by repeating this latter step three times, in order to obtain pure wheat proteins free from salts and other reagents. Eventually proteins present in the pellet were freeze dried. Before iTRAQ analysis, total protein samples were further acetone precipitated, reduced, alkylated, trypsin digested [26] and labelled according to the manufacturer's iTRAQ protocol (SCIEX, Framingham, MA, USA). The samples were then labelled using the iTRAQ reagents (Table S5).

All labelled samples were combined to make a pooled sample. Peptides were desalted on a Strata-X 33 µm polymeric reversed phase column (Phenomenex, Torrance, CA, USA) and dissolved in a buffer containing 2% acetonitrile 0.1% formic acid, before separation by High pH on an Agilent 1100 HPLC system using a Zorbax C18 column (2.1 × 150 mm) (Agilent Technologies, Palo Alto, CA, USA). Peptides were eluted with a linear gradient of 20 mM ammonium formate, 2% ACN to 20 mM ammonium formate, 90% ACN at 0.2 mL/min. The 95 fractions were concatenated into 12 fractions and dried down. Each fraction was analyzed by electrospray ionization mass spectrometry using a Thermo UltiMate 3000 nanoflow UHPLC system (Thermo Scientific, Waltham, MA, USA) coupled to a Q Exactive HF mass spectrometer (Thermo Scientific, Waltham, MA, USA). Peptides were loaded onto an Acclaim™ PepMap™ 100 C18 LC Column, 2 µm particle size x 150mm (Thermo Scientific, Waltham, MA, USA) and separated with a linear gradient of water/acetonitrile/0.1% formic acid (*v/v*). Amount of pooled sample loaded on the mass spectrometer: 9 µg.

4.3.1. Data Analysis

All the raw MS data were analyzed using ProteinPilotTM software versions 5.0 (AB Sciex) for protein identification and quantification. ProteinPilot applies a hybrid approach to integrate de novo sequencing with a traditional database searching approach in identifying proteins. Firstly, high-confidence small peptide sequence segments are identified, called taglets; taglets are then used to improve the downstream database searching procedure [27]. ProteinPilot is a well-established protein identification algorithm in the proteomics community. Spectral data were analyzed against the database "UniProt Taxonomy: *Viridiplantae* (Green plants), Version: August 2019". The number of the sequences in the database was 8,344,090. iTRAQ impurity was automatically corrected following the protocol provided by ABI Sciex for ProteinPilot.

4.3.2. Quality Control and Filtering

The statistical analysis was performed at protein levels. The protein identifications were filtered following several criteria:

1. All the artificially introduced false positive identifications by ProteinPilot's FDR analysis were filtered;
2. Proteins identified below 95% confidence by ProteinPilot standard (identification score < 1.3) were discarded from downstream analysis;
3. Proteins unused for quantification were discarded from the analysis, including those having discordant peptide type, quantification signal too weak, and with low confidence;
4. The proteins that have at least one unique peptide assigned were accepted.

4.4. SDS-PAGE

The electrophoretic control of lines was performed on the albumin and globulin fraction, containing ATI, including the CM proteins, in all lines obtained, as well as on total gluten proteins.

A/G extraction was performed on pools of mature grains, that were crushed and mixed with extraction buffer containing 0.05M phosphate buffer/0.1M NaCl, pH 7.8 (3 g: 80 mL). After centrifugation at 8000 g for 15 min at 4 °C, the supernatant was precipitated with acetone; the pellet, corresponding to the A/G proteins was then rinsed two times. Protein concentrations of the A/G proteins were determined with the 'Bio-Rad Protein assay' kit [28]. The A/G fraction was solubilized in Tris-HCl buffer, pH 6.8, containing 2% SDS, 10% glycerol, 0.1% 2-mercaptoethanol and bromophenol blue. SDS-PAGE was performed using stacking gels (T = 3.75%; C = 2.67%) to concentrate the proteins and running gels (T = 15%; C = 0.9%) to separate the proteins based on their molecular weight.

SDS-PAGE was performed using the Mini-PROTEAN Tetra Vertical Electrophoresis Cell (Bio-Rad) and the gels were 1.0 mm thick. For each sample, 15 µg of protein extract was loaded into the gel that was run at a constant voltage of 200V until the bromophenol blue dye front disappeared.

Total protein subunits were extracted from 10 mg of flour of RNAi transgenic lines and Bobwhite control plants in a 1:10 ratio w/v in 70 mM Tris-HCl, pH 6.8, 2% SDS, 10% glycerol, 0.02% γ-pyronine and 1% DTT, stirring for 1 h at room temperature. After centrifugation at 13,000 g for 10 min, the extracts were quantified with the 'Bio-Rad Protein assay' kit and 10 µg of protein was used for SDS-PAGE analysis through SE 600 Hoefer (Fisher Scientific, Waltham, Massachusetts, USA) device. The main gel was T = 11 and C = 1.28, while the stacking gel was T = 3.75 and C = 2.67. The electrophoresis was performed at 40 mA per gel at 10 °C and stopped as the tracking dye reached the bottom of the gel.

In both cases, gels were stained according to Neuhoff et al. [29].

4.5. Immunological Tests

4.5.1. Sera from Wheat Allergic Patients and Controls

Sera from patients with clinically documented FAW were obtained with the informed consent of the patients and with Ethic Committee approbation. Eight control sera were obtained from healthy volunteers. Twenty-two sera were selected based on their concentrations in specific IgE for albumins/globulins measured by F-ELISA (Table S6). Sera were obtained from the Biological Resource Center (BB-0033-00038) of Clinical Immunology and Allergy Service of Angers University Hospital (France) with the informed consent of the patients.

4.5.2. Indirect ELISA with Anti-ATI Antibodies

The enzyme-linked immunosorbent assay (ELISA) was performed for quantifying ATI in the A/G fraction of the WT and transgenic lines. For each genotype, three replicates of each sample (A/G fraction) were tested, except for the two transgenic lines 24-1 and 22-2, whose samples were tested in duplicate. The wells on microtiter plate (ELISA plate 82.1581.100, Sarstedt, Nümbrecht,

Germany) were coated with the A/G fractions at 5 µg/mL in 100 mM carbonate (pH 9.6) for 1h at room temperature. After three washes with PBS-0.05% Tween 20, the plate was blocked with PBS-4% milk for 1h at 37 °C. The microplate was washed three times with PBS-0.05% Tween 20 and then incubated for 1h at 37 °C with a serial dilution of primary antibody from 1:2000 to 1:64,000 in PBS-2% milk; a monoclonal antibody against ATI, properly developed in mouse at the INRA, Nantes (France). After three washes with PBS-0.05% Tween 20, the plate was incubated for 1h at 37 °C with the secondary antibody (anti-rabbit IgG labelled with the enzyme HRP). After three additional washes with PBS-0.05% Tween 20, the colorimetric substrate for HRP OPD (o-Phenylenediamine) in 0.05M citrate buffer plus H_2O_2 were introduced to have a yellow-orange product detectable at 490 nm. The OPD reaction was stopped after 30 min with H_2SO_4 4N. Three biological and three technical replicates were used. All data were subjected to ANOVA analysis by using SYSTAT12 software (Systat Software Incorporated, San Jose, CA, USA). When significant F values were observed, a pairwise analysis was carried out by the Tukey Honestly Significant Difference test (Tukey test).

4.5.3. ELISA Test with Sera of Allergic Patients

Sera obtained from 22 patients with allergy to wheat were characterized by ELISA using salt soluble protein extracts (Table S6), as described by Battais et al. [30]. Specific IgE concentrations against the soluble fraction from Bobwhite and RNAi lines were determined by fluorescent (F)-ELISA on white 384-well plates (NUNC 460372, Fisher Scientific, Waltham, MA, USA), using alkaline-phosphatase-conjugated goat anti-human IgE and 4-MUP as a substrate (Sigma A3525 and M3168, respectively, Saint-Quentin Fallavier, France) as described by and Lupi et al. [31]. Patient sera were diluted from 1:10 to 1/50 according to the serum reactivity.

The concentration of specific IgE binding to the antigen was calculated using a standard curve, as described by Bodinier et al. [32]. Three replicates were performed for each serum. Data are presented as mean values with their standard errors.

Statistical analyses were performed using GraphPad Prism 5.02 software (GraphPad Software, La Jolla, CA, USA), and p values below 0.05 were considered significant. Analyses of variance (Two-way ANOVA) with subsequent Bonferroni's multiple-comparisons tests between means of the Bobwhite vs. RNAi lines were performed.

4.6. Trypsin Inhibition Assay

Trypsin inhibition was measured according to Call et al. [19]. An amount of 2 g of flour was extracted with 10 mL 150 mM NaCl solution containing 0.02 M phosphate buffer (pH 7.0) by magnetic stirring for 5 min at room temperature followed by centrifugation (4000 rpm for 10 min.). Equilibration was performed for 20 min at 37 °C and enzyme/substrate ratio of 1:100 was used. For determination of the linear region four to seven different sample volumes of salt-water extracts were tested, depending on the strength of inhibition. Each sample was extracted three times, and each extract was measured in duplicate with an optimized sample volume. TIA values were expressed in mg/kg and mg/g protein based on protein concentration determined by the Bradford assay. The total protein content was analyzed by the Dumas combustion method with a Dumaster equipment from Büchi (Uster, Switzerland) and a conversion factor of 5.7 was applied.

MALDI TOF MS of salt-water extracts was performed according Call et al. [20] after purification with C18 ZipTips (Millipore, Merck, Germany).

4.7. Parameters Related to Yield and Calculation of the Percentage of Unextractable Polymeric Proteins (%UPP)

Thousand kernel weight (TKW) was used as a yield indicator. It was calculated by weighing and counting the kernels. Besides, coleoptile length was used as a parameter to measure seedling vigor and growth. It was measured at 5 and 7 days after imbibition (DPI) by using a scale. For both parameters, two independent experiments were performed. Data were subjected to ANOVA analysis.

The percentage of unextractable polymeric proteins (%UPP) was determined according to Gupta et al. [2]. Soluble glutenin polymers were extracted from 10 mg of flour by using 50% ACN 0.05% TFA in a 1:10 (*w/v*) ratio. The samples were left under stirring for 30 min at room temperature and subsequently centrifuged for 15 min at 10,000 *g*. The supernatant was filtered through 0.45-μm filters (Ultrafree-MC, centrifugal filters, PVDF, Millipore Corp., Billerica, MA) by centrifugation at 16,900 *g* for 10 min. The insoluble polymers were extracted by adding to the pellet from the previous extraction 50% ACN 0.05% TFA, in a 1:10 ratio. The extraction of insoluble polymers was carried out by sonication (Vibra cell, Sonics Materials, INC), with a 5-mm probe with the following settings: tune 50, output control 25 for 20 sec. The supernatant was filtered through PVDF 0.45-μm filters by centrifugation at 16,900 *g* for 10 min. For the Size Exclusion Chromatography (SEC) analysis of soluble and insoluble polymers, HPLC System Gold 126 NM (Beckman Coulter) was used with management and acquisition software 32 Karat 7.0, equipped with UV detector Spectra Series UV150 (Thermo Separation Products) set at 214 nm and chromatographic 30 cm column molecular exclusion BioSep-SEC-S 4000 (Phenomenex) with an internal diameter of 4.6 mm. The samples were injected using 100-μL loop connected to a six-way valve, elution occurred under isocratic conditions using 50% ACN 0.05% TFA eluent, at a flow rate of 1 mL/min for a duration of the stroke chromatography of 10 min. The column was kept at the temperature of 30 °C. Three biological and three technical replicates were used. All data were subjected to ANOVA analysis by using SYSTAT12 software (Systat Software Incorporated, San Jose, CA, USA). When significant F values were observed, a pairwise analysis was carried out by the Tukey Honestly Significant Difference test (Tukey test).

4.8. Bioinformatic Comparison of HMW-GS Gene Sequences

The NCBI (National Centre for Biotechnology Information https://www.ncbi.nlm.nih.gov.) data bank was used to compare *0.28* (accession number AJ223492.1), *CM3* and *CM16* (accession numbers AY436554.1 and X17573.1, respectively), with the following *HMW-GS* genes: *Dx5* (accession number DQ907161.1), *Dy10* (accession number X12929.2), *Ax2** (accession number M22208.2), *Bx7* (accession number BK006773.1) e *By9* (accession number X61026.1).

Alignment was performed with the software ClustalOmega (https://www.ebi.ac.uk/Tools/msa/clustalo/r).

Supplementary Materials: Supplementary Materials can be found at http://www.mdpi.com/1422-0067/21/16/5817/s1.

Author Contributions: Conceptualization, S.M.; Data curation, R.M.K., S.T., F.S., F.C., R.L., C.L., S.D.-P., S.I., S.D. and S.M.; Formal analysis, R.M.K., S.T., F.S., F.C., R.L., C.L., S.D.-P., S.I., W.M. and S.D.; Funding acquisition, D.L. and S.M.; Investigation, S.M.; Methodology, R.M.K., S.T., F.S., F.C., R.L., C.L., S.D.-P., W.M. and S.D.; Project administration, S.M.; Supervision, S.M.; Validation, R.M.K., S.T., F.S., F.C., R.L., C.L., S.D.-P., S.I. and S.D.; Visualization, D.L.; Writing—original draft, R.M.K., S.T., F.S., F.C., R.L., C.L., S.D.-P., S.I., S.D. and S.M.; Writing—review and editing, R.M.K., S.T., F.S., F.C., D.L., R.L., C.L., S.D.-P., S.I., S.D. and S.M. All authors have read and agreed to the published version of the manuscript.

Funding: This research was partially funded by the Italian Ministry of Education, University, and Research (MIUR) in the frame of the MIUR initiative "Departments of excellence", Law 232/2016.

Acknowledgments: The authors wish to thank Francesco Solimei who performed the bioinformatic comparison among High Molecular Weight Glutenin Subunits genes as a part of his undergraduate degree dissertation. This paper is dedicated to the dear memory of Renato D'Ovidio who started this project with Stefania Masci.

Conflicts of Interest: The authors declare no conflict of interest. The funders had no role in the design of the study; in the collection, analyses, or interpretation of data; in the writing of the manuscript; or in the decision to publish the results.

Abbreviations

A/G	Albumins and Globulins
ATI	α-Amylase/Trypsin Inhibitors
BA	Baker's Asthma
BW	Bobwhite
CD	Celiac Disease
DPA	Days Post Anthesis
DPG	Days Post Germination
DPI	Days Post Imbibition
FAW	Food Allergy to Wheat
FODMAPs	Fermentable Oligosaccharides Disaccharides Monosaccharides And Polyols
HDGS	Homology Dependent Gene Silencing
HMW-GS	High Molecular Weight Glutenin Subunits
IBS	Irritable Bowel Syndrome
NCGS	Non Celiac Gluten Sensitivity
NCWS	Non Celiac Wheat Sensitivity
PTGS	Post-Transcriptional Gene Silencing
qRT-PCR	Quantitative Real-Time PCR
TGS	Transcriptional Gene Silencing
TIA	Trypsin Inhibitor Activity
TKW	Thousand Kernel Weight
%UPP	Percentage of Unextractable Polymeric Proteins
WDEIA	Wheat-Dependent Exercise-Induced Anaphylaxis
WT	Wild Type (the untransformed control genotype corresponding to BW)

References

1. Osborne, T.B. The vegetable proteins. In *Monographs on Biochemistry*; Longmans & Green: London, UK, 1924.
2. Gupta, R.B.; Khan, K.; MacRitchie, F. Biochemical basis of flour properties in bread wheats. I. Effects of variation in the quantity and size distribution of polymeric protein. *J. Cereal Sci.* **1993**, *18*, 3–41. [CrossRef]
3. Pence, J.W.; Weinstein, N.E.; Mecham, D. The albumin and globulin contents of wheat flour and their relationship to protein quality. *Cereal Chem.* **1954**, *31*, 303–311.
4. Goesaert, H.; Gebruers, K.; Courtin, C.M.; Brijs, K.; Delcour, J.A. Enzymes in breadmaking. In *Bakery Products: Science and Technology*; Hui, Y.H., Ed.; Blackwell: Ames, IA, USA, 2006; pp. 337–364.
5. Šramková, Z.; Gregová, E.; Šturdík, E. Chemical composition and nutritional quality of wheat grain. *Acta Chim. Slov.* **2009**, *2*, 115–138.
6. Dupont, F.M.; Altenbach, S.B. Molecular and biochemical impacts of environmental factors on wheat grain development and protein synthesis. *J. Cereal Sci.* **2003**, *38*, 133–146. [CrossRef]
7. Elli, L.; Branchi, F.; Tomba, C.; Villalta, D.; Norsa, L.; Ferretti, F.; Roncoroni, L.; Bardella, M.T. Diagnosis of gluten related disorders: Celiac disease, wheat allergy and non-celiac gluten sensitivity. *World J. Gastroenterol.* **2015**, *21*, 7110–7119. [CrossRef]
8. Mansueto, P.; Soresi, M.; Iacobucci, R.; La Blasca, F.; Romano, G.; D'Alcamo, A.; Carroccio, A. Non-celiac wheat sensitivity: A search for the pathogenesis of a self-reported condition. *Ital. J. Med.* **2019**, *13*, 15–23. [CrossRef]
9. Salcedo, G.; Quirce, S.; Diaz-Perales, A. Wheat allergens associated with Baker's asthma. *J. Investig. Allergol. Clin. Immunol.* **2011**, *21*, 81–92.
10. Zapatero, L.; Martínez, M.I.; Alonso, E.; Salcedo, G.; Sánchez-Monge, R.; Barber, D.; Lombardero, M. Oral wheat flour anaphylaxis related to wheat α-amylase inhibitor subunits CM3 and CM16. *Allergy* **2003**, *58*, 956. [CrossRef]
11. Tundo, S.; Lupi, R.; Lafond, M.; Giardina, T.; Larré, C.; Denery-Papini, S.; Morisset, M.; Kalunke, R.; Sestili, F.; Masci, S. Wheat alpha-amylase and trypsin inhibitors (ATI) CM3, CM16 and 0.28 allergens produced in *Pichia pastoris* display a different eliciting potential in food allergy to wheat. *Plants* **2018**, *7*, 101. [CrossRef]

12. Junker, Y.; Zeissig, S.; Kim, S.J.; Barisani, D.; Wieser, H.; Leffler, D.A.; Zevallos, V.; Libermann, T.A.; Dillon, S.; Freitag, T.L.; et al. Wheat amylase trypsin inhibitors drive intestinal inflammation via activation of toll-like receptor 4. *J. Exp. Med.* **2012**, *209*, 2395–2408. [CrossRef]

13. Cuccioloni, M.; Mozzicafreddo, M.; Bonfili, L.; Cecarini, V.; Giangross, M.; Falconi, M.; Saitoh, S.I.; Eleuteri, A.M.; Angeletti, M. Interfering with the high-affinity interaction between wheat amylase trypsin inhibitor CM3 and toll-like receptor 4: In silico and biosensor based studies. *Sci. Rep.* **2017**, *7*, 13169. [CrossRef] [PubMed]

14. Sanchez-Monge, R.; Gomez, L.; Garcia-Olmedo, F.; Salcedo, G. New dimeric inhibitor of heterologous α-amylases encoded by a duplicated gene in the short arm of chromosome 3B of wheat (*Triticum aestivum* L.). *Eur. J. Biochem.* **1989**, *183*, 37–40. [CrossRef] [PubMed]

15. Garcia-Olmedo, F.; Salcedo, G.; Sanchez-Monge, R.; Gómez, L.; Royo, J.; Carbonero, P. Plant proteinaceous, inhibitors of proteinases and 3-amylases. In *Oxford Surveys of Plant Molecular and Cell Biology*; Miflin, B., Ed.; Oxford University Press: Oxford, UK, 1987; Volume 4, pp. 275–334.

16. Camerlengo, F.; Frittelli, F.; Sparks, C.; Doherty, A.; Martignago, D.; Larré, C.; Lupi, R.; Sestili, F.; Masci, S. CRISPR-Cas9 multiplex editing of the α-amylase/trypsin inhibitor genes to reduce allergen proteins in durum wheat. *Front. Sustain. Food Syst.* **2020**, *4*, 104. [CrossRef]

17. García-Molina, M.D.; Muccilli, V.; Saletti, R.; Foti, S.; Masci, S.; Barro, F. Comparative proteomic analysis of two transgenic low-gliadin wheat lines and non-transgenic wheat control. *J. Proteom.* **2017**, *165*, 102–112. [CrossRef]

18. Camerlengo, F.; Sestili, F.; Silvestri, M.; Colaprico, G.; Margiotta, B.; Ruggeri, R.; Lupi, R.; Masci, S.; Lafiandra, D. Production and molecular characterization of bread wheat lines with reduced amount of α-type gliadins. *BMC Plant. Biol.* **2017**, *17*, 248. [CrossRef]

19. Call, L.; Reiter, E.V.; Wenger-Oehn, G.; Strnad, I.; Grausgruber, H.; Schoenlechner, R.; D'Amico, S. Development of an enzymatic assay for the quantitative determination of trypsin inhibitory activity in wheat. *Food Chem.* **2019**, *299*, 125038. [CrossRef]

20. Call, L.; Kapeller, M.; Grausgruber, H.; Reiter, E.; Schoenlechner, R.; D'Amico, S. Effects of species and breeding on wheat protein composition. *J. Cereal Sci.* **2020**, *93*, 102794. [CrossRef]

21. Meyer, P.; Saedler, H. Homology-Dependent gene silencing in plants. *Annu. Rev. Plant. Physiol. Plant. Mol. Biol.* **1996**, *47*, 23–48. [CrossRef]

22. Rajeevkumar, S.; Anunanthini, P.; Sathishkumar, R. Epigenetic silencing in transgenic plants. *Front. Plant. Sci.* **2015**, *6*, 693. [CrossRef]

23. Jauvion, V.; Rivard, M.; Bouteiller, N.; Elmayan, T.; Vaucheret, H. RDR2 partially antagonizes the production of RDR6-dependent siRNA in sense transgene-mediated PTGS. *PLoS ONE* **2012**, *7*, e29785. [CrossRef]

24. Sestili, F.; Janni, M.; Doherty, A.; Botticella, E.; D'Ovidio, R.; Masci, S.; Jones, H.D.; Lafiandra, D. Increasing the amylose content of durum wheat through silencing of the SBEIIa genes. *BMC Plant. Biol.* **2010**, *10*, 144. [CrossRef] [PubMed]

25. Tai, T.H.; Tanksley, S.D. A rapid and inexpensive method for isolation of total DNA from dehydrated plant tissue. *Plant. Mol. Biol. Rep.* **1990**, *8*, 297–303. [CrossRef]

26. Casey, T.M.; Khan, J.M.; Bringans, S.D.; Koudelka, T.; Takle, P.S.; Downs, R.A.; Livk, A.; Syme, R.A.; Tan, K.C.; Lipscombe, R.J. Analysis of reproducibility of proteome coverage and quantitation using Isobaric Mass Tags (iTRAQ and TMT). *J. Proteome Res.* **2017**, *16*, 384–392. [CrossRef] [PubMed]

27. Shilov, I.V.; Seymour, S.L.; Patel, A.A.; Loboda, A.; Tang, W.H.; Keating, S.P.; Hunter, C.L.; Nuwaysir, L.M.; Schaeffer, D.A. The Paragon algorithm, a Next Generation search engine that uses sequence temperature values and feature probabilities to identify peptides from Tandem Mass Spectra. *Mol. Cell. Proteom.* **2007**, *6*, 1638. [CrossRef]

28. Bradford, M.M. A rapid and sensitive method for the quantitation of microgram quantities of protein utilizing the principle of protein-dye binding. *Anal. Biochem.* **1976**, *72*, 248–254. [CrossRef]

29. Neuhoff, V.; Arold, N.; Taube, D.; Ehrhardt, W. Improved staining of proteins in polyacrylamide gels including isoelectric focusing gels with clear background at nanogram sensitivity using Coomassie Brilliant Blue G-250 and R-250. *Electrophoresis* **1988**, *9*, 255–262. [CrossRef]

30. Battais, F.; Courcoux, P.; Popineau, Y.; Kann, Y.G.; Moneret-Vautrin, D.A.; Denery-Papini, S. Food allergy to wheat: Differences in immunoglobulin E-binding proteins as a function of age or symptoms. *J. Cereal Sci.* **2005**, *42*, 109–117. [CrossRef]

31. Lupi, R.; Masci, S.; Rogniaux, H.; Tranquet, O.; Brossard, C.; Lafiandra, D.; Moneret-Vautrin, D.; Denery-Papini, S.; Larré, C. Assessment of the allergenicity of soluble fractions from GM and commercial genotypes of wheats. *J. Cereal Sci.* **2014**, *60*, 179–186. [CrossRef]

32. Bodinier, M.; Brossard, C.; Triballeau, S.; Morisset, M.; Guerin-Marchand, C.; Pineau, F.; de Coppet, P.; Moneret-Vautrin, D.A.; Blank, U.; Denery-Papini, S. Evaluation of an in vitro mast cell degranulation test in the context of food allergy to wheat. *Int. Arch. Allergy Immunol.* **2008**, *146*, 307–320. [CrossRef]

International Journal of
Molecular Sciences

Article

CRISPR/Cas9 Directed Mutagenesis of *OsGA20ox2* in High Yielding Basmati Rice (*Oryza sativa* L.) Line and Comparative Proteome Profiling of Unveiled Changes Triggered by Mutations

Gul Nawaz [1], Babar Usman [1], Neng Zhao [1], Yue Han [1], Zhihua Li [1], Xin Wang [1], Yaoguang Liu [2,*] and Ronghai Li [1,*]

[1] College of Agriculture, State Key Laboratory for Conservation and Utilization of Subtropical Agro-Bioresources, Guangxi University, Nanning 530004, China; gulnawazmalik@yahoo.com (G.N.); babarusman119@gmail.com (B.U.); nengzhao_gxu@163.com (N.Z.); hanyue0624@126.com (Y.H.); lizhihua-88@163.com (Z.L.); xinwang0112@126.com (X.W.)

[2] State Key Laboratory for Conservation and Utilization of Subtropical Agricultural Bioresources, South China Agricultural University, Guangzhou 510642, China

* Correspondence: ygliu@scau.edu.cn (Y.L.); lirongbai@126.com (R.L.);
   Tel.: +86-20-8528-1908 (Y.L.); +86-136-0009-4135 (R.L.)

Received: 16 July 2020; Accepted: 23 August 2020; Published: 26 August 2020

Abstract: In rice, semi-dwarfism is among the most required characteristics, as it facilitates better yields and offers lodging resistance. Here, semi-dwarf rice lines lacking any residual transgene-DNA and off-target effects were generated through CRISPR/Cas9-guided mutagenesis of the *OsGA20ox2* gene in a high yielding Basmati rice line, and the isobaric tags for relative and absolute quantification (iTRAQ) strategy was utilized to elucidate the proteomic changes in mutants. The results indicated the reduced gibberellins (GA_1 and GA_4) levels, plant height (28.72%), and flag leaf length, while all the other traits remained unchanged. The *OsGA20ox2* expression was highly suppressed, and the mutants exhibited decreased cell length, width, and restored their plant height by exogenous GA_3 treatment. Comparative proteomics of the wild-type and homozygous mutant line (GXU43_9) showed an altered level of 588 proteins, 273 upregulated and 315 downregulated, respectively. The identified differentially expressed proteins (DEPs) were mainly enriched in the carbon metabolism and fixation, glycolysis/gluconeogenesis, photosynthesis, and oxidative phosphorylation pathways. The proteins (Q6AWY7, Q6AWY2, Q9FRG8, Q6EPP9, Q6AWX8) associated with growth-regulating factors (*GRF2, GRF7, GRF9, GRF10,* and *GRF11*) and GA (Q8RZ73, Q9AS97, Q69VG1, Q8LNJ6, Q0JH50, and Q5MQ85) were downregulated, while the abscisic stress-ripening protein 5 (*ASR5*) and abscisic acid receptor (*PYL5*) were upregulated in mutant lines. We integrated CRISPR/Cas9 with proteomic screening as the most reliable strategy for rapid assessment of the CRISPR experiments outcomes.

Keywords: rice; gibberellins; plant height; CRISPR/Cas9; *OsGA20ox2*; proteomic analysis

1. Introduction

Rice is a very important crop for the developing world and is considered a staple food for half of the world's population [1–4]. Basmati rice has a unique specialty with good palatability, longer shelf life, easy digestibility, and good aroma [5]. The tall stature of traditional Basmati varieties is a big drawback as such varieties are easily lodged, which negatively affects the rice yield [6]. Plant height (PH) is an important character owing to critical roles in the plant architecture and adaptability to the environmental conditions that are directly linked with yield [7,8]. The number and length of the internodes determine the PH [9], which results in improved lodging resistance capacity and better biomass production due to the short stature [10]. Gibberellin (GA) is a plant growth hormone and many genes are vital for GA biosynthesis and signaling, which plays decisive roles in the metabolism of brassinosteroid and influences the stem length regulation in plants [11]. Mutations in components participating in GA signaling and metabolism pathways exhibit notable dwarf or semi-dwarf plant phenotype [12]. The tall phenotype is controlled by the *SD1* (semi dwarf1) allele while its recessive allele (*sd1*) controls the semi-dwarf phenotype [13,14]. The *OsGA20ox2* has 3 exons and 2 introns, a total of 389 amino acids are encoded by this gene and GA20 oxidase2 is a major enzyme for GA synthesis [13]. Mostly, GAs, consist of large groups of diterpenoid carboxylic acids such as GA1 and GA4, and display a vital role in the development and growth of plants [15,16]. *OsGA20ox2* is mainly expressed in the stem, and mutations in *OsGA20ox2* resulted in the production of a moderate amount of GA, and its mutants with loss-of-function exhibit a semi-dwarf phenotype [10,17,18]. The miracle rice cultivar IR8 has a short plant height because of the *SD1* gene mutation and enabled extraordinary yield increases and facilitated the prevention of food shortages [13,18,19]. The *OsGA20ox2* mutants have no negative effects on the morphological and yield-related traits and extensive efforts have been made by the research community for achieving the semi-dwarf phenotype in rice through molecular breeding and PH can be restored in mutants similar to that of WT (wild type) plants by exogenous GA$_3$ treatment [20–22].

The targeted and effective gene mutations have been successfully achieved by the CRISPR/Cas9 (clustered regularly interspaced short palindromic repeats/CRISPR-associated protein 9) system in rice for improving the existing varieties and to develop new mutant lines by targeting one or multiple genes [23–25]. CRISPR/Cas9 is a tremendous technique for precise and targeted editing of the genome of plants and animals. ZFNs (zinc finger nucleases) and TALEN (transcriptional activator-like effector nuclease) genome editing techniques were established before the CRISPR/Cas9, but because of the simplicity and flexibility, rapidness, multiplexing capacity, high efficiency, and mutation frequency, this system gained worldwide popularity and is widely accepted by researchers [26,27]. The CRISPR/Cas9 technique has been successfully applied in Arabidopsis [28,29], rice [28,30–32], maize [33], wheat [31,34], sorghum [28], soybean [35–37], and tomato [38], and the resulted mutations are transmitted to the next generations according to the classical inheritance principles [39].

The development of the proteomics over the last few decades has greatly contributed in omics and is now accepted widely to study various species [40–44]. To comprehend the effects of mutations on the genome of plants, whole genome-wide profiling of proteins or transcripts is an efficient way to investigate the distinct variations in diverse biological and molecular processes [45]. The dramatic improvement in the methods used in molecular biology, extensive profiling of transcripts, and iTRAQ (isobaric tags for relative and absolute quantification)-based proteomic analysis can provide insights into a universal view of genes and protein expression patterns and can be helpful to understand the potential molecular mechanism behind the mutagenesis [46,47]. The iTRAQ has been applied successfully in plant species to recognize the diverse biological processes, including Arabidopsis [48], rice [49,50], wheat [51], and maize [52]. It is an effective and high-throughput approach with high sensitivity and accuracy, multiplexing capacity, and repeatability. However, this tremendous method has not yet been utilized to investigate which mechanisms are underlying the semi-dwarf phenotype. In our study, we have exploited the CRISPR/Cas9 technology to knock out the *OsGA20ox2* gene in Basmati rice, and earmark and delicate mutations in the *OsGA20ox2* gene and homozygous T$_1$ semi-dwarf mutant lines were achieved effectively, with significantly reduced PH and GA content without disturbing other agronomic characters. To further understand the functional roles of the *OsGA20ox2* gene, comparative iTRAQ analysis was performed to elucidate the effects of mutations on the protein level. Our results provide new hints in understanding *OsGA20ox2* functions and suggest that the proteomic screening is a reliable tool for assessment of CRISPR experiments. By this tactic, our aim was to infer the mechanisms that are more closely associated with rice dwarfism.

2. Results

2.1. Editing of OsGA20ox2 Gene and Identification of Transgene-Free Plants

We used 165 calli for the *A. tumefaciens* transformation and a total of 30 rice seedlings were achieved. Mutant lines were screened by using hygromycin phosphotransferase (HPT-F/R) primers, and the final product was amplified and confirmed in mutants (Supplementary file 1, Table S1, Figure S1A). A high rate of mutation was observed in both targets of the *OsGA20ox2* gene, with a total mutation efficiency of 70% (73.33% for T_1 and 66.33% for T_2). Among 30 plants, there were 8 (26.67%) homozygous, 9 (30%) biallelic (heterozygous), 4 (13.33%) heterozygous (mono-allelic), and 1 (3.33%) chimeric. A total of 30 mutant lines were obtained for T_2, including 7 (23.33%) homozygous, 10 (33.33%) biallelic (heterozygous), 1 (3.33%) heterozygous (mono-allelic), and 2 (6.67%) were chimeric (Figure 1A,B). Mutated alleles were amplified from the genomic DNA of T_0 mutant lines and the sequencing chromatograms with overlapping traces were decoded (Figure 1C).

We selected four transgene-DNA-free (T-DNA-free) homozygous lines (GXU43_2, GXU43_4, GXU43_9, and GXU43_19) for agronomic traits evaluation. GXU43_2 showed homozygous mutations with 1 bp insertion and 3 bp deletion, the mutant GXU43_4 showed 27 bp and 1 bp deletions, the homozygous mutant GXU43_9 revealed 172 bp and 12 bp deletions, and finally, the homozygous mutant line GXU43_19 resulted in 4 bp deletion and 1 bp insertion on the first and second target sites, respectively. The sequencing analysis of these homozygous mutants showed that the mutations were stable and inheritable in T_1, T_2, and subsequent T_3 generations (Figure 1D). The plant numbers and the corresponding mutations in both target sites are mentioned in Supplementary file 1, Table S2. The thirty plants were selected, and the DNA was amplified for 5 loci of each target site with the highly ranked off-target potential, and no secondary off-target mutations were detected in sequencing results (Supplementary file 1, Table S3).

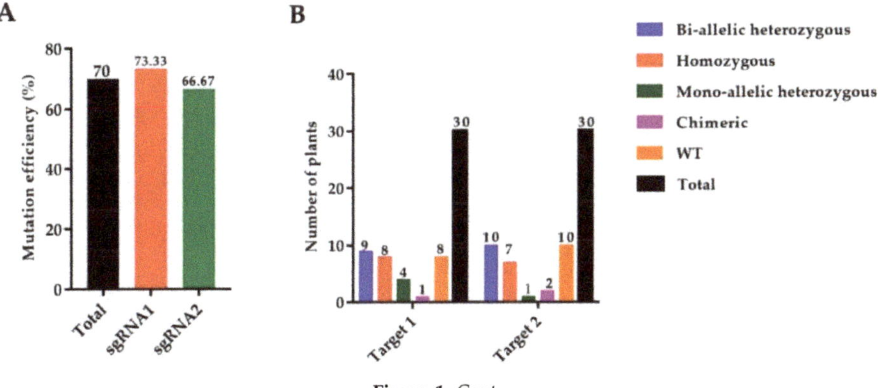

Figure 1. *Cont.*

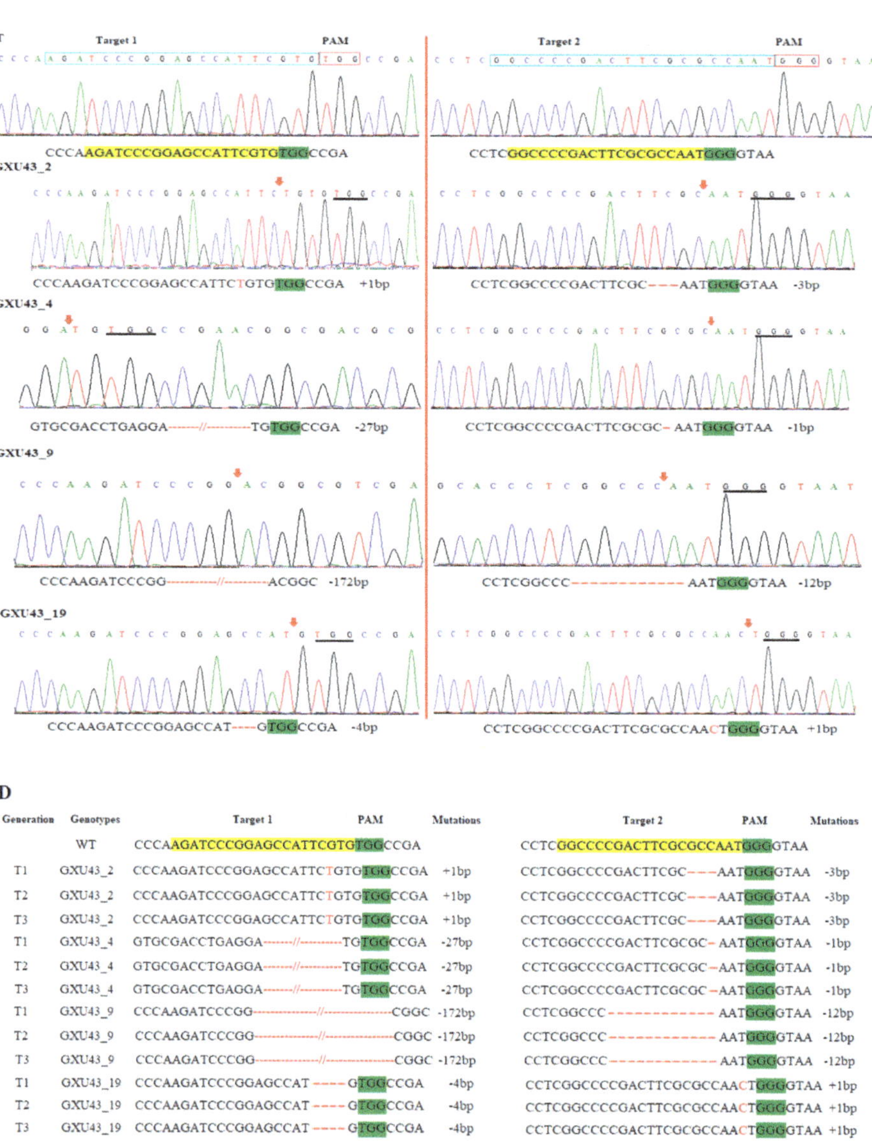

Figure 1. Identification of mutations generated by the clustered regularly interspaced short palindromic repeats/CRISPR-associated protein 9 (CRISPR/Cas9) system. (**A**) The mutation efficiency of sgRNAs (single guided RNAs), (**B**) The mutation rate in T_0 (Transgenic) generation, (**C**) sequencing chromatograms of WT (wild type) and homozygous mutant lines for Target 1 and Target 2, and (**D**) the alignment of sequences for T_1, T_2, and T_3 generations, respectively. The target sequence is painted in yellow, while the PAM (protospacer adjacent motif) is in the green background, and insertions/deletions are represented by red hyphens and letters. The analysis was carried out in three replications for each line.

2.2. T-DNA-Free Mutants and Segregation Ratio in the T_1 Generation

Homozygous mutants of T_0 and T_1 were grown and 50 plants were evaluated to analyze the transformation patterns and thus to detect the exogenous DNA in the mutant lines. The plants were believed to be T-DNA-free if they failed to amplify against HPT and Cas9-specific primers (Supplementary file 1, Table S1). The results showed that 30 plants were amplified to the Cas9 vector sequence and 20 lines showed no amplification, and therefore, were considered as T-DNA-free. The frequency of such plants was recorded at 40% (Supplementary File 1, Figure S1B). The heterozygous (mono and biallelic) T_1 plants of GXU43_8 were segregated according to Mendelian inheritance (1:2:1). All the T_1 plants obtained from T_0 homozygous plants (GXU43_2, GXU43_4, GXU43_9, and GXU43_19) also showed homozygosity for the same mutations, which indicated the stable transmission of mutations to the subsequent generations (Supplementary File 1, Table S4).

2.3. Endogenous GA Content and PH in T_1, T_2, and T_3 Generations

The mutant lines exhibited decreased PH (plant height) and GA content (GA_1, GA_4) as compared with WT plants. The mutant line GXU43_9 showed a minimum PH of 114.77, 115.55, and 113.98 cm in T_1, T_2, and T_3 generations respectively, and the GA_1 and GA_4 content (0.95 and 0.84 in T_1, 0.91 and 0.81 in T_2, and 0.88 and 0.87 µg/kg fresh weight (FW) in T_3 generation, respectively), while WT showed maximum plant height (161.31 cm) and gibberellins (1.66 and 1.54 µg/kg FW) content. The results revealed a significant and positive correlation among endogenous GA content and PH of WT and transgenic plants (Table 1).

Table 1. Plant height (PH) and gibberellin (GA) content (µg/kg FW) in T_1, T_2, and T_3 mutant lines.

Gen	Lines	GA_1	GA_4	PH
	WT	1.66 ± 0.14	1.54 ± 0.16	161.31 ± 3.9
	GXU43_2	1.19 ± 0.09 **	1.12 ± 0.12 **	128.25 ± 4.2 **
T_1	GXU43_4	1.10 ± 0.10 **	1.05 ± 0.11 **	118.25 ± 2.9 **
	GXU43_9	0.95 ± 0.08 **	0.84 ± 0.10 **	114.77 ± 3.1 **
	GXU43_19	1.20 ± 0.12 **	1.24 ± 0.17 **	122.24 ± 3.1 **
	WT	1.59 ± 0.18	1.65 ± 0.15	164.45 ± 3.7
	GXU43_2	1.21 ± 0.11 **	1.24 ± 0.10 **	125.90 ± 4.3 **
T_2	GXU43_4	1.20 ± 0.09 **	1.15 ± 0.12 **	117.53 ± 2.8 **
	GXU43_9	0.91 ± 0.11 **	0.81 ± 0.12 **	115.55 ± 3.5 **
	GXU43_19	1.14 ± 0.13 **	1.30 ± 0.13 **	123.35 ± 3.3 **
	WT	1.60 ± 0.15	1.63 ± 0.17	163.31 ± 3.9
	GXU43_2	1.14 ± 0.10 **	1.29 ± 0.11 **	126.25 ± 4.2 **
T_3	GXU43_4	1.20 ± 0.12 **	1.35 ± 0.13 **	116.25 ± 2.9 **
	GXU43_9	0.88 ± 0.13 **	0.87 ± 0.12 **	113.98 ± 3.1 **
	GXU43_19	1.21 ± 0.15 **	1.22 ± 0.14 **	121.24 ± 3.1 **

Gen: generations, FW: fresh weight, PH: plant height, GA: gibberellins. Mean data for three independent replicates. ** represent a significant difference at $p \leq 0.01$.

2.4. Performance of Agronomic and Quality Traits

The data for the major agronomic traits of mutant and WT plants were recorded at 120 days after growing. The mean results for morphological traits expressed a significant difference amongst WT and mutant plants in PH (Figure 2A) and FLL (flag leaf length), while semi-dwarf lines showed slightly shorter PL (panicle length), with no difference in PN (panicle numbers), FLW (flag leaf width), GNPP (grain number per panicle), SSR (seed setting rate), and GW(1000-grain weight), YPP (yield per plant), GL (grain length), and GWD (grain width). We selected the four homozygous mutant lines GXU43_2, GXU43_4, GXU43_9, and GXU43_19, and the results of data for three consecutive generations revealed that the PH and FLL of mutants were significantly decreased (Table 2). However, grain appearance and shape were not altered (Figure 2B). Mutant lines showed shortened internodal length than WT plants (Figure 2C).

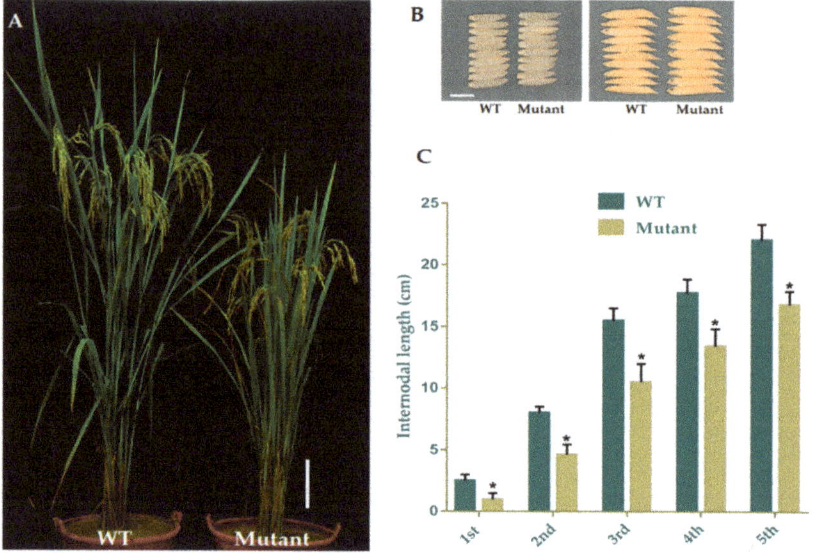

Figure 2. Phenotypic appearance of *OsGA20ox2* in GXU43_9 and WT. (**A**) Plant height of mutant lines and WT after the heading stage. Bar = 15 cm. (**B**) Grain phenotype of mutant line and WT; Bar = 5 mm. (**C**) The lengths of internodes in mutant line and WT. The "*" denotes the significant difference at $p < 0.01$. Values are means ± standard deviation (SD) ($n = 10$ plants).

Table 2. Performance of major agronomic traits in WT and mutant lines.

Gen	Line	PN	PL (cm)	FLL (cm)	FLW (cm)	GNPP	SSR (%)	GW (g)	YPP (g)	GL mm	GWD mm
T_1	WT	9.91 ± 0.8	27.16 ± 1.4	49.97 ± 3.39	1.75 ± 0.1	145 ± 12	87.12 ± 5.9	29.34 ± 1.3	31.53 ± 2.1	8.52 ± 0.4	2.41 ± 0.2
	GXU43_2	9.17 ± 0.3 [ns]	26.41 ± 1.3 [ns]	38.63 ± 2.40 **	1.61 ± 0.2 [ns]	148 ± 10 [ns]	92.10 ± 4.1 [ns]	29.63 ± 1.8 [ns]	31.85 ± 1.3 [ns]	8.19 ± 0.3 [ns]	2.49 ± 0.3 [ns]
	GXU43_4	10.28 ± 0.5 [ns]	26.15 ± 2.5 [ns]	39.13 ± 2.13 **	1.71 ± 0.3 [ns]	147 ± 11 [ns]	91.33 ± 5.1 [ns]	30.27 ± 1.1 [ns]	30.41 ± 1.2 [ns]	8.61 ± 0.4 [ns]	2.35 ± 0.4 [ns]
	GXU43_9	10.29 ± 0.6 [ns]	24.98 ± 1.9 [ns]	37.43 ± 2.15 **	1.64 ± 0.1 [ns]	149 ± 09 [ns]	90.44 ± 5.3 [ns]	30.29 ± 2.2 [ns]	32.53 ± 1.4 [ns]	9.10 ± 0.4 [ns]	2.40 ± 0.4 [ns]
	GXU43_19	10.75 ± 0.4 [ns]	25.30 ± 1.3 [ns]	40.33 ± 1.85 **	1.55 ± 0.3 [ns]	146 ± 12 [ns]	93.03 ± 5.7 [ns]	30.43 ± 1.6 [ns]	30.95 ± 1.5 [ns]	8.98 ± 0.5 [ns]	2.45 ± 0.3 [ns]
T_2	WT	10.81 ± 0.9	28.12 ± 1.8	48.59 ± 3.88	1.90 ± 0.2	147 ± 13	88.62 ± 6.5	29.72 ± 1.5	31.21 ± 1.8	8.95 ± 0.5	2.39 ± 0.3
	GXU43_2	10.63 ± 0.5 [ns]	26.63 ± 1.9 [ns]	39.84 ± 2.69 **	1.75 ± 0.3 [ns]	146 ± 11 [ns]	93.30 ± 5.4 [ns]	29.96 ± 1.7 [ns]	32.44 ± 1.9 [ns]	8.27 ± 0.3 [ns]	2.33 ± 0.2 [ns]
	GXU43_4	9.71 ± 0.7 [ns]	25.55 ± 2.1 [ns]	38.23 ± 2.52 **	1.65 ± 0.5 [ns]	149 ± 12 [ns]	90.95 ± 6.2 [ns]	30.42 ± 1.3 [ns]	31.77 ± 1.5 [ns]	8.06 ± 0.5 [ns]	2.29 ± 0.3 [ns]
	GXU43_9	10.29 ± 0.8 [ns]	26.88 ± 1.3 [ns]	36.70 ± 2.58 **	1.55 ± 0.2 [ns]	150 ± 12 [ns]	92.51 ± 6.9 [ns]	30.88 ± 2.1 [ns]	32.56 ± 1.7 [ns]	8.87 ± 0.7 [ns]	2.48 ± 0.3 [ns]
	GXU43_19	10.85 ± 0.6 [ns]	24.99 ± 1.7 [ns]	39.73 ± 2.85 **	1.71 ± 0.3 [ns]	148 ± 10 [ns]	91.43 ± 6.6 [ns]	29.54 ± 1.9 [ns]	31.73 ± 1.6 [ns]	8.61 ± 0.6 [ns]	2.38 ± 0.4 [ns]
T_3	WT	10.50 ± 0.8	27.96 ± 1.6	50.97 ± 3.39	1.85 ± 0.1	146 ± 12	86.85 ± 5.9	30.14 ± 1.3	31.96 ± 2.6	8.34 ± 0.7	2.27 ± 0.2
	GXU43_2	10.90 ± 0.3 [ns]	27.41 ± 1.3 [ns]	40.63 ± 2.40 **	1.67 ± 0.2 [ns]	149 ± 10 [ns]	90.10 ± 6.1 [ns]	30.33 ± 1.8 [ns]	31.15 ± 1.3 [ns]	8.45 ± 0.5 [ns]	2.31 ± 0.3 [ns]
	GXU43_4	10.77 ± 0.5 [ns]	24.86 ± 1.5 [ns]	41.33 ± 2.13 **	1.76 ± 0.3 [ns]	147 ± 11 [ns]	92.33 ± 5.8 [ns]	29.92 ± 1.7 [ns]	30.49 ± 1.2 [ns]	8.62 ± 0.4 [ns]	2.42 ± 0.4 [ns]
	GXU43_9	10.45 ± 0.6 [ns]	24.18 ± 1.9 [ns]	38.40 ± 2.15 **	1.62 ± 0.1 [ns]	150 ± 09 [ns]	93.44 ± 5.3 [ns]	30.50 ± 2.2 [ns]	31.85 ± 1.4 [ns]	8.50 ± 0.6 [ns]	2.41 ± 0.4 [ns]
	GXU43_19	10.99 ± 0.4 [ns]	25.30 ± 1.7 [ns]	39.33 ± 2.50 **	1.59 ± 0.3 [ns]	149 ± 11 [ns]	89.03 ± 5.7 [ns]	29.93 ± 1.6 [ns]	31.29 ± 1.5 [ns]	8.68 ± 0.5 [ns]	2.36 ± 0.3 [ns]

Gen: Generation; WT: wild-type; PL: panicle length, PN: panicle numbers; FLW: flag leaf width, FLL: flag leaf length, GNPP: grain number per panicle; SSR: seed setting rate; GW: 1000-grain weight; YPP: yield per plant; GWD: grain width; GL: grain length. Mean data of three replicates. ** represent the significant difference and [ns] represents the non-significant difference at $p < 0.01$.

2.5. Effect of Exogenous GA₃

We applied 10 µM GA₃ to analyze the response of mutant and WT plants at the seedling stage. The data for plant height was recorded after 25 days. The mutant plants responded to exogenous GA₃ significantly and restored the PH identically to that of the WT (Figure 3A,B). The homozygous mutant line GXU43_9 exhibited the lowest PH (15.52 cm) in controlled conditions as compared to the PH of wild type plants (25.85 cm), which was significantly higher than the mutant plants. Under the GA₃ application, the T_1 mutant line GXU43_9 restored the PH (28.95 cm), nearly equal to that of WT plants (29.24 cm). These results clearly show that the GA₃ application promoted the growth of WT and mutant plants (Figure 3C).

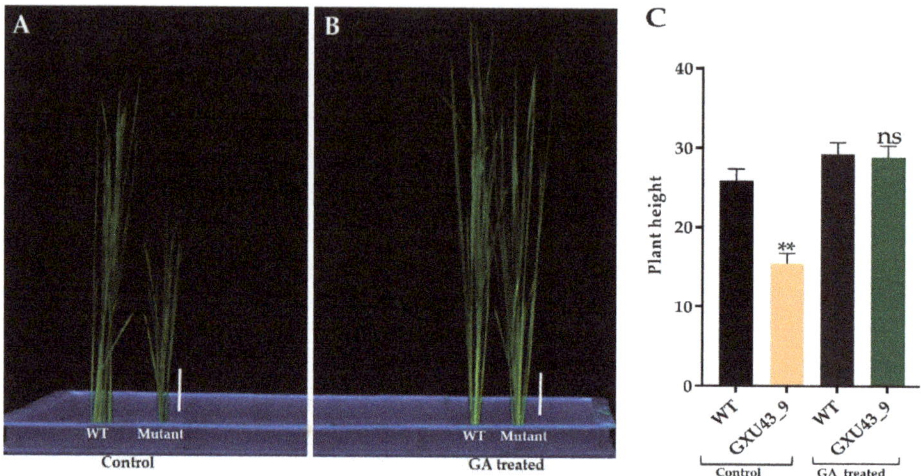

Figure 3. The seedling phenotype of the homozygous mutant GXU43_9 and WT. (**A**) Seedlings phenotype without GA, (**B**) Seedlings treated with GA, and (**C**) PH (plant height) of GA3-treated and control (*n* = 15), Bars = 3 cm. Data are mean ± SD. "**" and "ns" represent a significant and non-significant difference respectively at $p \leq 0.01$.

2.6. Section Analysis of Culm Cells

The microscopic analysis of the culm cell of mutant line GXU43_9 showed irregularly shaped cells with thin walls, significantly smaller in size, and an increase in cell layers was observed as compared to WT (Figure 4A,B). From these results, we can conclude that the development of the stalk was affected in GXU43_9. The length and width of the cells in mutants was significantly smaller as compared to WT (Figure 4C,D).

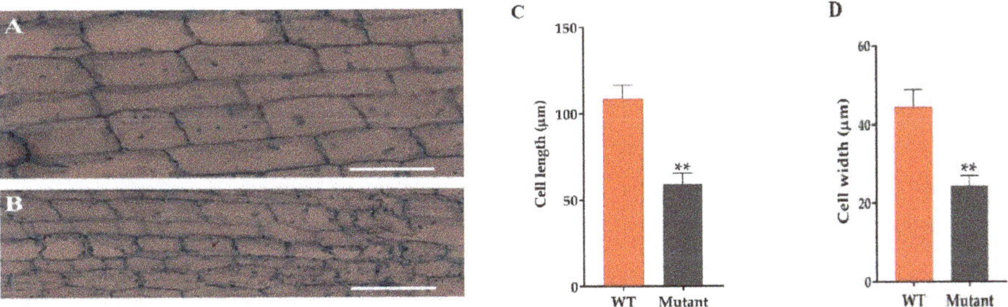

Figure 4. The microscopic analysis of culm cells of WT and mutant line GXU43_9. (**A**) WT second intermodal longitudinal length of the cell and (**B**) mutant plant GXU43_9, respectively. (**C,D**) Quantitative measurement of the length and width of the cells of mutant line (GXU43_9) and wild-type plants (n = 15), Bars = 100 μm. Data are the mean ± SD, "**" shows the significant difference at $p \leq 0.01$.

2.7. Proteomics Analysis

A total of 267,114 spectra was generated, and after the analysis of these spectra, we identified 68,489 known spectra with 24,230 peptides and 4003 proteins, respectively (Supplementary Figure S2A). The protein mass distribution is represented in Supplementary Figure S2B. The proteins having 20–40 kDa represented more protein numbers, and proteins with 141–150 kDa showed less protein numbers. The number of peptides identified in the proteins is shown in Supplementary Figure S2C. The information about the distribution of peptide length and protein sequence coverage is shown in Supplementary Figure S2D,E. After implementing the analysis with a minimum fold change (FC) of ≥ 1.2 and p-value adjusted to ≤ 0.05, a total of 588 DEPs (273 upregulated and 315 downregulated) were obtained (Figure 5A). The detailed information of all the proteins identified in this analysis is given in Supplementary file 2.

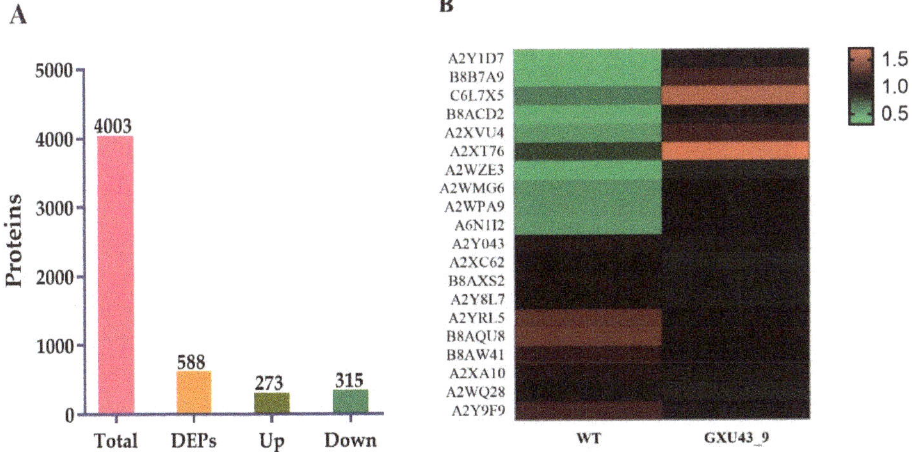

Figure 5. Proteomic analysis information of the CRISPR mutant GXU43_9 and its WT. (**A**) Upregulated and downregulated differentially expressed proteins (DEPs), and (**B**) Heatmap of the top twenty DEPs. Red color denotes the higher while the green color represents a lower level of expression.

The top 30 up- and down-regulated proteins were selected based on the highest log2 FC value, of which the top 10 upregulated DEPs were Metallothionein-like protein 3B (A2Y1D7), Uncharacterized protein (B8B7A9), Ethylene response factor (C6L7X5), Uncharacterized protein (B8ACD2),

Uncharacterized protein (A2XVU4), Abscisic stress-ripening protein 5 (Q53JF7), Uncharacterized protein (A2WZE3), Salt stress root protein (A2WMG6), Peroxidase (A2WPA9), and ATP-dependent clp protease proteolytic subunit (A6N1I2). While the Peroxidase (A2Y043), Uncharacterized protein (A2XC62), Uncharacterized protein (B8AXS2), Peptidase A1 domain-containing protein (A2Y8L7), Growth-regulating factor 9 (Q9FRG8), MFS domain-containing protein (B8AQU8), NAD(P)-bd_dom domain-containing protein (B8AW41), Uncharacterized protein (A2XA10), Ent-copalyl diphosphate synthase 2 (Q5MQ85), and ATP-dependent 6-phosphofructokinase (A2Y9F9) were downregulated in the mutant line (Figure 5B).

We further searched the DEPs related to GA and plant growth. We found that five DEPs (Q6AWY7, Q6AWY2, Q9FRG8, Q6EPP9, and Q6AWX8) related to growth-regulating factors (*GRF2*, *GRF7*, *GRF9*, *GRF10*, and *GRF11*) were downregulated in this report. The DEPs related to gibberellin response modulator-like proteins (Q8RZ73 and Q9AS97), Chitin-inducible gibberellin-responsive proteins (Q69VG1 and Q339D4), Putative gibberellin oxidase (Q8LNJ6), Ent-copalyl diphosphate synthase 2 (Q5MQ85), gibberellin 20 oxidase 2 (Q0JH50), ATP synthase subunit beta chloroplastic (J7EYN3), and fructose-1,6-bisphosphatase, chloroplastic (A2XEX2) were also downregulated in mutant lines. However, the DEPs (Abscisic stress-ripening protein 5) and Q6I5C3 (Abscisic acid receptor) were upregulated in mutant plants (Table 3).

Table 3. Differentially expressed proteins related to GA and plant growth.

Protein ID	Locus/Gene Name	Annotation	Regulate
Q6AWY7	*Os06g0204800/GRF2*	Growth-regulating factor 2	Down
Q6AWY2	*Os12g0484900/GRF7*	Growth-regulating factor 7	Down
Q9FRG8	*Os03g0674700/GRF9*	Growth-regulating factor 9	Down
Q6EPP9	*Os02g0678800/GRF10*	Growth-regulating factor 10	Down
Q6AWX8	*Os07g0467500/GRF11*	Growth-regulating factor 11	Down
Q8RZ73	*B1065G12.22*	Gibberellin response modulator-like proteins	Down
Q9AS97	*Os01g0646300*	Gibberellin response modulator-like	Down
Q69VG1	*Os07g0545800/CIGR1*	Chitin-inducible gibberellin-responsive protein 1	Down
Q339D4	*LOC_Os10g22430*	Chitin-inducible gibberellin-responsive protein 2	Down
Q8LNJ6	*OSJNBb0028C01.33*	Putative gibberellin oxidase	Down
Q0JH50	*Os01g0883800/GA20ox2*	Gibberellin 20 oxidase 2	Down
Q5MQ85	*CPS2*	Ent-copalyl diphosphate synthase 2	Down
P0C511	*RBCS*	Ribulose bisphosphate carboxylase large chain	Down
A2YQT7	*GAPC*	Glyceraldehyde-3-phosphate dehydrogenase, cytosolic	Down
Q6ZG90	*Os02g0131300*	ATP synthase	Down
J7EYN3	*atpB*	ATP synthase subunit beta, chloroplastic	Down
A2XEX2	*OsI_10887*	Fructose-1,6-bisphosphatase, chloroplastic	Down
C6L7X5	*Snorkel2*	Ethylene response factor	Down
Q53JF7	*Os11g0167800/ASR5*	Abscisic stress-ripening protein	Up
Q6I5C3	*Os05g0213500/PYL5*	Abscisic acid receptor	Up

2.8. DEPs Functional Networks and Hub-Protein Analysis

The STRING database was used for retrieving the protein interaction networks; for this purpose, the confidence (score) cutoff was adjusted to 50 with 30 additional interactors. The nodes represent the proteins, and protein-protein interaction (PPI) modes are shown by the lines among the nodes. The analysis of network revealed a higher co-expression between catalase isozyme B (Q0D9C4), 2-Cys peroxiredoxin BAS1 (Q6ER94), protein coleoptile photomorphogenesis 2 (Q75KD7), adenine phosphoribosyltransferase 1, putative, expressed (Q2QMV8), adenine phosphoribosyltransferase 1, putative (Q2QMT1), linoleate 9S-lipoxygenase 2 (P29250), lipoxygenase 7, chloroplastic (P38419), lipoxygenase (Q0IS17), linoleate 9S-lipoxygenase 1 (Q76I22), probable indole-3-acetic acid-amido synthetase (Q5NAZ7), isocitrate dehydrogenase (NADP) (Q7F280), and phloem sap 13 kDa protein 1 (Q0D840), which showed a higher interaction score of ≥ 5 (Figure 6A).

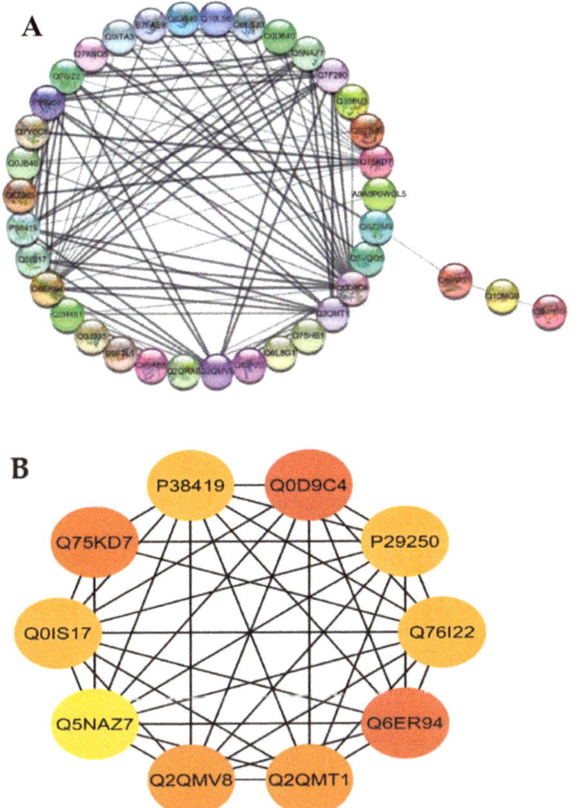

Figure 6. Functional networks of DEPs and hub-protein analysis of WT and mutant line GXU43_9. (**A**) STRING software predicted proteins' associations. The nodes represent differentially expressed proteins, while the edge denotes the interaction relationship between the nodes. (**B**) The top hub-proteins of WT and GXU43_9 comparison. The higher co-expression is denoted by red color.

Highly connected to 10 hub proteins with a higher degree of connectivity by the STRING database, the following were selected as candidate hub proteins: the adenine phosphoribosyltransferase 1, putative, expressed (Q2QMV8), adenine phosphoribosyltransferase 1 (Q2QMT1), linoleate 9S-lipoxygenase 2 (P29250), lipoxygenase 7, chloroplastic (P38419), lipoxygenase (Q0IS17), linoleate 9S-lipoxygenase 1 (Q76I22), probable indole-3-acetic acid-amido synthetase (Q5NAZ7), and 2-Cys peroxiredoxin BAS1, chloroplastic (Q6ER94) (Figure 6B).

2.9. Gene Ontology (GO) and Pathway Enrichment Analysis

GO annotations for DEPs related to BP (biological processes) were associated with the cellular process, metabolic process, cellular metabolic process, organic substance metabolic process, primary metabolic process, single-organism process, nitrogen compound metabolic process, biosynthetic process, single-organism cellular process, and cellular biosynthetic process. Proteins conferring CC (cellular components) were mainly involved in cell, cell part, intracellular (intracellular part, organelle, and membrane-bounded), organelle (primary and membrane-bounded), cytoplasm, and cytoplasmic part. Finally, for the MF (molecular functions) perspective, the DEPs took part in catalytic activity, oxidoreductase activity, RNA binding, tetrapyrrole binding, hydrolase activity (primary and cofactor binding), acting on acid anhydrides in phosphorus-containing anhydrides, acting on acid anhydrides structural molecule activity, pyrophosphatase activity, and structural constituent of ribosome (Figure 7A).

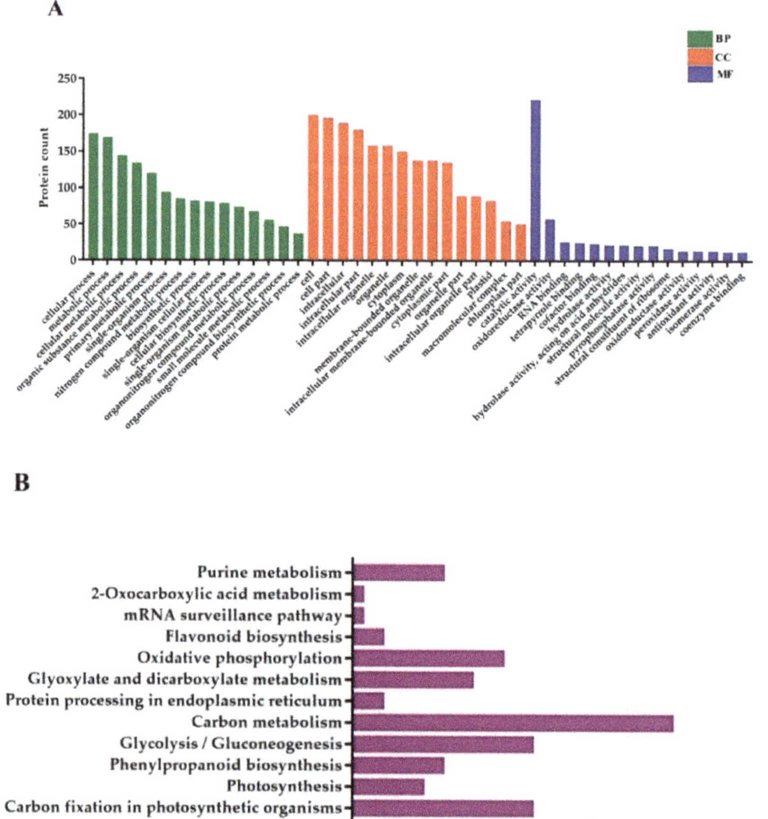

Figure 7. Gene ontology (GO) and KEGG (Kyoto Encyclopedia of Genes and Genomes) pathway enrichment analysis of DEPs. (**A**) GO annotations of the DEPs, and (**B**) the histogram of KEGG pathway enrichment, with the bar showing the number of proteins. BP; biological processes, CC; cellular components, and MF; molecular functions.

KEGG (Kyoto Encyclopedia of Genes and Genomes) pathway analysis uncovered that the DEPs were mostly enriched in carbon fixation in photosynthetic organisms, photosynthesis, phenylpropanoid biosynthesis, glycolysis/gluconeogenesis, protein processing in the endoplasmic reticulum, carbon metabolism, glyoxylate and dicarboxylate metabolism, oxidative phosphorylation, flavonoid biosynthesis, mRNA surveillance pathway, and 2-Oxocarboxylic acid metabolism (Figure 7B).

2.10. RT-qPCR Analysis of Target Gene and Proteomic Data Validation

The RT-qPCR was used to assess the relative expression of the *OsGA20ox2* gene in mutants and WT plants. To normalize the expression, the Rice *Actin* gene was used as a reference between the samples, and the expression of *OsGA20ox2* was significantly suppressed in all mutant lines ($p < 0.01$, Figure 8A). Ten genes associated with DEPs were selected for the validation of the proteomic data, and the qRT-PCR assay was performed for independent samples of WT and mutant lines.

In total, eight key genes encoding downregulated DEPs, including *GRF2*, *GRF7*, ethylene response factor gene (*Snorkel2*), ribulose bisphosphate carboxylase small-chain gene (*RBCS*), ATP synthase subunit beta (*atpB*), Ent-copalyl diphosphate synthase 2 gene (*CPS2*), *GRF10*, and Chitin-inducible gibberellin-responsive protein1 (*CIGR1*), and two genes encoding upregulated DEPs, including abscisic stress-ripening protein 5 gene (*ASR5*) and Abscisic acid receptor (*PYL5*), were chosen. The level of expression of selected genes was consistent with the proteomic analysis (Figure 8B). Primers used for RT-qPCR are given in Supplementary file 1, Table S1.

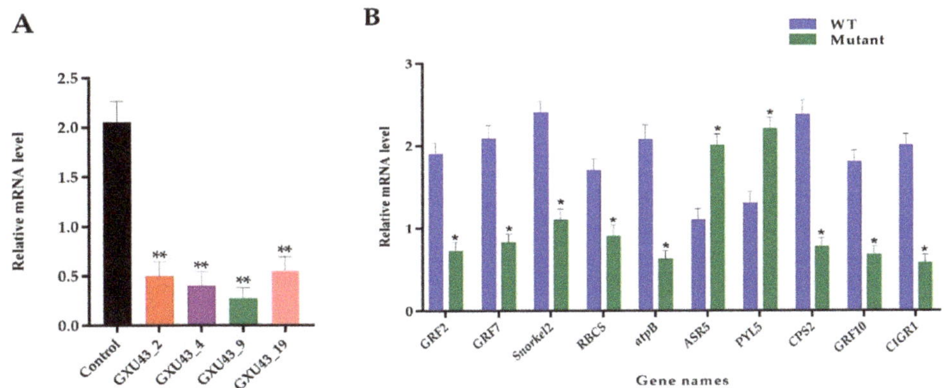

Figure 8. Real-time quantitative PCR assessment and validation of proteomics data. (**A**) Expression analysis of *OsGA20ox2*, in WT and T_1 mutant lines, (**B**) RT qPCR validation of the ten DEPs responsive genes. The SD is shown with error bars, "*" represents the significant differences at $p \leq 0.05$ and "**" represents the significant differences at $p \leq 0.01$ respectively; $n = 3$.

3. Discussion

The rise of modern molecular breeding technologies has provided fast and efficient means for plant breeding, which can directly modify desirable traits without changing and affecting other traits. CRISPR/Cas9 is determined by the easily and cheaply modified sgRNA (Single guided RNA) followed by a short PAM (protospacer adjacent motif) sequence to induce mutations with high accuracy and reduced probability of off-targets [53–55]. Crop improvement by using gene editing suggests decent prospects to generate mutants with preferred traits. However, there are very few examples of the utilization of SSNs (sequence-specific nucleases) for the generation of novel genotypes with desired plant type. The CRISPR/Cas9 technique has been extensively used in many species for the targeted mutagenesis because of its high efficiency. In this study, we have successfully used this system to edit the *OsGA20ox2* gene with a higher mutation efficiency. Mutants are very important for genetic research and crop breeding. However, the assimilation of desired genes into elite breeding varieties is still a challenging task.

Breeding productivity is restricted by the continuous selection process, which is the main drawback of conventional breeding [56,57]. The natural selection events occur at random and result in unguided mutagenesis that brings a little frequency of, mutational events at target loci [57]. On the other hand, CRISPR technology has a great potential to produce mutants by the targeted mutagenesis in pre-decided locus [49]. We employed the iTRAQ strategy to confirm the effects of CRISPR-based gene editing at the whole proteome level.

The CRISPR/Cas9 vector has two key functional parts, including a gRNA and Cas9 expression cassettes, driven by RNA polymerase III and 35S/ubiquitin promoters, respectively. Because the Pol III has a limited transcriptional capacity, previously, researchers used U3/U6 promoters from rice, which have specific transcriptional sites with nucleotides A and G, respectively [58,59]. We used U6 (OsU6a, OsU6) promoters to construct the expression cassette, the target sequences were selected with

5′-GN(19)NGG and, 5′-AN(19)NGG, respectively. Two sgRNAs were ligated successfully in the U6 promoters-driven expression cassette for targeting *OsGA20ox2*. In the present work, the CRISPR/Cas9 tool was used for editing the *OsGA20ox2* gene in the Basmati rice line. The heterozygous and homozygous mutation events were found frequently, however, the chimeric mutations were rare. The homozygous and compound heterozygous mutations are also found frequently in previous studies [60,61]. The homozygous mutant plants achieved in T_0 generation showed stable and inheritable mutations to the subsequent generations. In previously reported studies the homozygous mutations have also been achieved in T_0 generation [61]. In this study, the mean mutation efficiency obtained was 70%, and the mutant lines were obtained without any off-target mutations. The previously reported work also suggests that the Cas9 rarely tempts off-target mutations in rice [62,63], and we screened the T-DNA-free lines at a frequency of 40%. The segregation and predicted inheritance have great importance in molecular breeding. In our study, we found inheritable and highly stable mutations induced by Cas9. In segregation analysis, the homozygous mutations were transmitted stably, showing no inversions or new mutations in T_1 according to the Mendelian principle. Bi-allelic and heterozygous mutations were segregated at a 1:2:1 ratio, while the chimeric mutations were found to be unpredicted. We can conclude from the results that the new mutations are not induced by Cas9, while further experiments are needed to uncover the unpredictable segregations of chimeric mutations [61,64].

The GA biosynthesis and deactivation genes confer endogenous GA levels and response to GA metabolism [12,65–69]. Semi-dwarfism is a very important trait and rice varieties with this feature brought about significant improvements in grain production. GA is considered as the main plant growth-promoting hormone for triggering stem elongation, which experiences dynamic cell division for controlling plant development and growth [70]. The expression level of *OsGA20ox2* was significantly reduced and mutant lines exhibited semi-dwarf PH at seedlings and mature stages, with significantly lowered GA content. Studies have revealed that GA_3 promotes stem and sheath elongation and synthesis of many proteins related to plant growth [71–73]. The mutant plants restored their PH with exogenous GA_3 application. The previous studies also showed that actual height similar to WT can be restored through GA_3 treatment [74]. *OsGA20ox2* CRISPR mutants also showed increased number and layers of cells, but decreased cell length and width, which may be the main cause of shortened PH. Dwarf plants usually show compact cell size and this results in affected cell expansion [75].

A total of 588 DEPs were identified in GXU43_9 versus the WT comparison, and among them, 273 were upregulated and 315 were downregulated, respectively. Some highly expressed proteins controlling PH were also identified in mutant lines. The DEPs including Q6AWY7, Q6AWY2, Q9FRG8, Q6EPP9, and Q6AWX8 (*GRF2, 7, 9, 10, 11*) were downregulated in mutant lines. In rice, the GRF family of TFs (transcription factors) has been previously identified [76], which are known as plant-specific TFs that were firstly known for their developmental role in leaf and stem growth [77]. *OsGRF1* is a GA-induced gene in intercalary meristem internodes, and was the first GRF to be recognized in rice [78]. Some GRFs' expression level in Arabidopsis and rice is usually high in the tissues growing actively; besides, plants showed increased expression levels for several GRFs (*OsGRFs* 1, 2, … 12) after exogenous GA_3 treatment, whereas *OsGRF9* showed decreased expression [79–81]. *OsGRF6* shows higher expression in developing inflorescences, which showed that GRFs are also involved in organ growth and development related to floral parts [82]. The gibberellin response modulator-like proteins (Q8RZ73 and Q9AS97), Chitin-inducible gibberellin-responsive proteins (Q69VG1 and Q339D4), Putative gibberellin oxidase (Q8LNJ6 and), Ent-copalyl diphosphate synthase 2 (Q5MQ85), gibberellin 20 oxidase 2 (Q0JH50), ATP synthase subunit beta chloroplastic (J7EYN3), fructose-1,6-bisphosphatase, chloroplastic (A2XEX2), and Ethylene response factor (C6L7X5) were also downregulated in mutant lines. Chitin-inducible GA responsive proteins (CIGR1 and CIGR2), inducible by the potent elicitor *N*-acetylchitooligosaccharide (GN), are fast induced by exogenous GA. The expression of proteins is reliant on the quantity and biological activity of GA, showing that the expression of these genes is mediated by GA [83]. Ent-copalyl-diphosphate act in the biosynthesis of defensive phytoalexin and GA phytohormone, which generally functions in GA biosynthesis, as mutations in the gene controlling this protein result in weakened growth [84]. Gibberellin 20 oxidase 2 is a major oxidase enzyme for the GA biosynthesis that is responsible for the catalyzation process which converts GA53 to GA20 through an oxidation reaction at C-20 of the GA skeleton and takes part in the internodal elongation [19].

Ribulose bisphosphate carboxylase (large and small chain) was downregulated in the mutant line and it was reported that RuBisco plays a significant role in the accumulation of chlorophyll and photosynthesis, and its overexpression leads to increased photosynthetic activity to attain growth [85]. The TPSs (terpene synthases) family is responsible for terpene molecules in plants, which play a key role in primary metabolism. In bryophyte, only a single TPS gene, *CPS* (*copalyl synthase*), is a precursor of GA, which encodes a bi-functional enzyme-producing ent-kaurene [86]. Loss-of-function mutants of SOTs (sulfotransferases) and TPST (tyrosylprotein SOTs) showed various abnormal characteristics due to peptides or proteins' sulfurization related to growth and development [87,88]. The TPST activity was previously observed in rice during microsomal membrane preparations [89]. The ethylene response factor (product of *SNORKEL2*) triggers notable internode elongation via gibberellin [90]. However, the DEPs, A2WPN7 (Salt stress-induced protein), A2WMG6 (Salt stress root protein), Q53JF7 (Abscisic stress-ripening protein 5), and Q6I5C3 (Abscisic acid receptor), were upregulated in mutant line GXU43_9 as compared to WT. Abscisic stress-ripening protein 5 was engaged in the GA signaling pathway and plant growth regulation in the region extending to basal leaf sheaths. The expression regulation of various genes is also carried out by ASR5 that contributes to cell protection against aluminum stress in rice plants [91]. In rice, the overexpression of *PYL5* enhances abiotic stress tolerance and inhibits growth through gene expression modulation [92].

GO annotations of DEPs revealed that most of the proteins related to BP were associated with cellular and organic substance metabolic processes, the single-organism process, nitrogen compound metabolic process, and cellular biosynthetic process. Proteins conferring CC were associated with cell, cell part, intracellular (intracellular part, organelle, and membrane-bounded), organelle (primary and membrane-bounded), cytoplasm, and cytoplasmic part. Finally, for the MF perspective, the DEPs were mainly involved in catalytic activity, oxidoreductase activity, RNA binding, tetrapyrrole binding, cofactor binding hydrolase activity, hydrolase activity, pyrophosphatase activity, and structural constituent of ribosome. KEGG pathway analysis showed that the DEPs were enriched in carbon fixation in photosynthetic organisms, photosynthesis, and glycolysis/gluconeogenesis. In network analysis of hub-proteins, we found that most of the DEPs, including linoleate 9S-lipoxygenase 2, lipoxygenase 7, chloroplastic, lipoxygenase, and linoleate 9S-lipoxygenase 1, were related to metabolic processes that may be engaged in various diverse aspects associated with plant physiology including growth and development [93,94]. In the present study, the proteins related to plant growth and GA were downregulated, implying that they may be responsible for a series of biochemical and physiological changes related to plant growth and ultimately affecting plant height. In summary, the plant height reduction through the modification of *OsGA20ox2* expression levels in rice by affecting the biosynthesis of GA is potentially of great agronomic interest. This study showed that plant characteristics can be improved through genetic mutations. In this study, the successfully developed semi-dwarf mutant rice lines can be exploited for future breeding programs. Further studies regarding the cell signaling mechanisms owing to genome manipulations are warranted.

4. Material and Methods

4.1. Material Used and Field Conditions

Seeds of Basmati rice variety (VP-1643) were provided by the Wild Rice Group of Guangxi University (GXU) and plants were propagated in the experimental area of GXU, China, and at the Farm of Divisional Headquarters, Sanya Hainan, China, in the normal growing season and maintained consistently. The pYL CRISPR/Cas9Pubi-*H* vector (Supplementary file 1, Figure S3) and the promoters (OsU6a and OsU6b) (Supplementary file 1, Figure S4A–C) were used to construct plasmid. This expression vector possesses the HPT selectable marker with a sequence bordering *BsaI* sites for sgRNA expression cassette insertion (Supplementary file 1, Figure S5A) [95].

4.2. SgRNAs Selection, Vector Construction, and Transformation

The target sequences were selected after the confirmation of $(N)_{20}$ GG or $G(N)_{20}$ GG template in coding regions of the *OsGA20ox2* gene by employing the online website CRISPR-GE (http://skl.scau.edu.cn/) [96], with higher targeting specificity (Supplementary file 1, Table S5; Figure S5B). The two targets for *OsGA20ox2* were selected in the exon region from 128 to 147 bp and 541 to 560 bp, respectively. The structures of all sgRNA's were developed by using an online tool, CRISPR-P (http://crispr.hzau.edu.cn/cgi-bin/CRISPR2/CRISPR) (Supplementary file 1, Figure S6). The overlapping PCR reaction was performed to construct the expression cassette [95] and is represented in Supplementary Figure S7, and the primers used are mentioned in Supplementary Table S1. The amplified product was purified by using TaKaRa MiniBEST Purification Kit Ver.4.0. The transformation of expression cassette to competent cells of *E. coli* DH5-alpha was performed according to a previously established method [97]. Primers SP-L1 and SP-R were used to assess the correct size of amplified products and sequenced directly (Supplementary file 1, Table S1). The sequences of target sites were confirmed in a constructed expression cassette. The order of U6 promoter-driven sgRNA cassettes was as follows: LacZ–OsU6a–T_1–OsU6b–T_2 (Supplementary file 1, Figure S8A). The sizes of sgRNA cassettes after amplification were as follows: OsU6a–sgRNA1: 629 bp, OsU6b–sgRNA: 515 bp (Supplementary file 1, Figure S8B). After transforming the expression vector into competent cells of DH5α the PCR amplification was performed to detect positive colonies (Supplementary file 1, Figure S8C). The sequence containing the target region of *OsGA20ox2* in WT was amplified (Supplementary file 1, Figure S8D). Sequencing results successfully confirmed the targets assembly in the vector (Supplementary file 1, Figure S8E). The CRISPR/Cas9 binary vector was successfully constructed, which was considered suitable for rice transformation and target gene editing.

The transformation of embryonic calli was accomplished by the *Agrobacterium tumefaciens*-mediated co-cultivation method, as previously established by Hiei et al. [98].

4.3. Genotyping, Off-Target Analysis, and Identification of Transgene-Free Plants

The DNA extracted by the CTAB (cetyl trimethylammonium bromide) method [99] and target-specific primers (*SD1T1* F/R and *SD1T2* F/R) were designed for the amplification of both target sites of *OsGA20ox2* gene (Supplementary file 1, Table S1). Sequences were decoded with DSDecode (http://skl.scau.edu.cn/dsdecode/) [100]. The CRISPR-GE online tool (http://skl.scau.edu.cn/) was accessed for the identification of the off-target sites (Supplementary file 1, Table S3) for the target regions, and five putative off-targets having ≥2 nucleotide mismatches for each target site were tested. The primers were designed, and PCR products were sequenced directly (Supplementary file 1, Table S6). The genomic DNA in T_1 and T_2 (T-DNA-free and T-DNA) was extracted for genotyping and to study the inheritance patterns. The screening of T-DNA-free plants was performed by Cas9-F/Cas9-R and HPT-F/HPT-R primers in the T_1 generation. Those plants regarded as T-DNA-free which lack both HPT and Cas9 simultaneously. The T-DNA-free homozygous mutant plants were further analyzed to study different agronomic and biochemical parameters. The mutations transmission patterns were studied for three consecutive generations by following the strict self-pollination of mutant lines. Segregation analysis was performed in T_1 generation for T-DNA-free mutant lines. We conducted the chi-square test for the confirmation of the Mendelian inheritance.

4.4. Phenotyping and Quantification of GA_3

The data of major agronomic traits included: PH (plant height), PN (number of panicles), PL (length of panicle), SSR (rate of seed setting rate), GNPP (number of grain per panicle), FLW (width of flag leaf), FLL (length of flag leaf), GL (grains length), GW (1000-grain weight), GWD (grain width), and YPP (yield per plant). Endogenous GA levels were measured as previously described [75]. The measurement results were represented in µg/kg FW according to standard methods [101], with three replicates.

4.5. Microscopic Analysis and Application of GA₃

Longitudinal sections of 0.5 cm taken from second internodes of mutant and WT mature plants and stained by using calcofluor and crystal violet and anatomical observations were done as described previously [102]. Slices were placed on slides in longitudinal sections and the results were analyzed by a Zeiss Axio Scope A1 Microscope. After 7 days of germination, the aqueous solution of GA₃ with 10 µM concentration was applied by spraying on seedlings, with an equal quantity of pure water used as a control, and the plant height was recorded after 25 days.

4.6. Extraction, Digestion, and iTRAQ Labeling of Proteins

Proteins were extracted from 100 mg leaf samples of WT (VP-1643) and its CRISPR mutant GXU43_9, with three replicates, by grinding into liquid nitrogen, and immediately transferred to pre-cooled acetone ($-20\,°C$), having 65 mM dithiothreitol (DTT), and 10% (*v/v*) TCA (trichloroacetic acid) was added and mixed thoroughly, precipitated (2 h at $-20\,°C$), and centrifuged (16,000× *g* for 30 min at 4 °C). Supernatant was carefully removed, and the pellet was washed thrice with 20 mL pre-cooled acetone. It was centrifuged (20,000× *g* at 4 °C) and kept for half an hour at $-20\,°C$. The precipitate was vacuum-dried soon after collection and the pellets obtained were fused with SDT containing 4% SDS, 100 mM Tris-HCl, 100 mM DTT, pH 8.0, boiled for 5 min, and then sonicated. The resultant product was centrifuged and filtered via a 0.22 µm Millipore filter. The concentration of proteins in the lysate was measured by using a bicinchoninic acid (BCA) assay kit (Beyo time Institute of Biotechnology, Shanghai, China). The extracted proteins were digested by following the FASP (filter-aided sample preparation) procedure [103]. iTRAQ labeling was conducted by using iTRAQ Reagents 8PLEXKit (Applied Biosystems, Foster City, CA, USA), according to the directions of the manufacturers. Peptides were labeled with iTRAQ tags as (GXU43_9)-114 and (WT1643)-115 and were combined and vacuum-dried at room temperature.

4.7. High-pH Reversed-phased Chromatography Separation

The fractionation of peptides was accomplished in a 1100 Series HPLC (high-performance liquid chromatography) System with a well-equipped Gemini-NX (Phenomenex, Torrance, CA, USA 00F-4453-E0) column (3 µm, 110 Å, 4.6 × 150 mm). The flow rate of 0.8 mL/min was maintained for the peptide's elution. The composition of Buffer A was 10 mM mmonium acetate, and buffer B was 10 mM Ammonium acetate with a 90% *v/v* concentration of CAN and pH of 10, respectively. The gradient separation was used as follows: for 40 min, 100% buffer A, 3 min with 0–5% buffer B, 30 min with 5–35% buffer B, 10 min with 35–70% buffer B, 10 min with 70–75% buffer B, 7 min with 75–100% buffer B, 15 min with 100% buffer B, and finally, 15 min with 100% buffer A, and the absorbance was assessed at 214 nm. For every sample, 20 fractions were taken and merged to get ten fractions. After the vacuum centrifugation, fractions were reconstituted by trifluoroacetic 40 µL at a concentration of 0.1% *v/v*. The samples were stored at $-80\,°C$ until LC-MS/MS analysis.

4.8. LC-MS/MS Analysis

To analyze the peptides, an easy-nLC 1000 HPLC system attached to an Orbitrap Elite mass spectrometer (Thermo Fisher Scientific, San Jose, CA, USA) was used. The samples were loaded on the Thermo Scientific EASY column with an autosampler at 150 nL/min. Peptides separation was carried out by using a C18 trap column (inner diameter 100 µm × 2 cm) and a C18 analytical column (inner diameter 75 µm × 25 cm). After 120 min, segmented gradient ran with Buffer A (0.1% formic acid in water) to 35% Buffer B (0.1% formic acid in 100% ACN) for 100 min, followed by 35–90% Buffer B for 4 min, and finally, 90% Buffer B for 6 min. The MS was conducted at a high-resolution mode (60K), with MS scans ranging between 300 and 2000 m/z, and 20 signals were acquired from the MS spectra according to the abundance for MS/MS analysis. DDA (data-dependent acquisition) and HCD (higher-energy collisional dissociation) were exploited with a medium resolution (15K) in MS/MS. 50 ms (10×10^{-6}) was the uppermost ion injection time, which was utilized for the survey scanned at 150 ms (5×10^4) for the MS/MS scans respectively, while the duration of the dynamic exclusion was 30 s.

4.9. Data Analysis

We used Statistical Software Program SPSS 16.0 to analyze the data related to agronomic traits. Proteome Discoverer 2.1 was utilized for proteomics analysis against the Rice database (Oryza sativa subsp. Indica) on 17 September 2018, with 40,869 entries, by using the default parameters. Peptides with a global false discovery rate (FDR) <1% were used for further protein annotation. The DEPs were functionally annotated by the GO database (http://www.geneontology.org/) and Blast2go software (http://www.blast2go.com/b2ghome) and proteins were grouped according to their participation in the BP, CC, and MF. The DAPs (differentially accumulated proteins) were further assigned to the KEGG (http://www.genome.jp/kegg/pathway) database. Fisher's Exact Test was used to identify the enriched GO terms and cluster analysis was performed by using Cluster 3.0 software. The pathways with FDR-corrected p-values ≤ 0.05 were regarded as significant. The PPI was evaluated by the STRING (http://string-db.org/) database and hub-proteins were assessed by Cytoscape (version 3.7.2).

4.10. Target Gene Expression Analysis and Proteomic Data Validation

We took 30-day-old rice seedlings and the panicle tissues for RNA extraction. RT-qPCR, performed on Real-Time LC480 (Roche Applied Science, Penzberg, Germany), polymerase chain reaction (PCR), was carried out by 10 µL volume with 0.08 µM primers, a 0.3 µL of reversed-transcribed product, and 5 µL of ChamQTM Universal SYBR qPCR Master Mix (Vazyme Biotech Co.,Ltd, Nanjing, China). The Rice *Actin* gene used as an internal control to normalize the reaction and the primers used are listed in Supplementary file 1, Table S1, and the comparative expression was evaluated by the $2^{-\Delta\Delta CT}$ method, as established earlier [104].

5. Conclusions

In conclusion, we have explored that the guided mutations of *OsGA20ox2* through the CRISPR/Cas9 system is an effective strategy to develop semi-dwarf rice plants for sustainable production. Combining CRISPR/Cas9 and comprehensive proteomic analysis revealed new clues that will facilitate the understanding of the complex cellular and molecular events for important traits. In this study, fundamental resources are provided for identifying phytohormones, and some novel candidate proteins and metabolic pathways were found that could be involved in rice semi-dwarfism. The proteins crucial for various steps of GA and chlorophyll synthesis pathways were significantly repressed in semi-dwarf mutant lines. The results of this work revealed that the CRISPR/Cas9 technology is a very effective tool for the targeted gene editing, and these findings will contribute to an increased understanding of rice semi-dwarfism, and the generated mutant lines can be useful source material for future rice breeding programs.

Supplementary Materials: The following are available online at http://www.mdpi.com/1422-0067/21/17/6170/s1: Table S1: All the primers used in this study; Table S2: Type of mutations obtained in T_0 generation by two constructs of CRISPR/Cas9; Table S3: Mutations detection on the potential off-targets; Table S4: Segregation of mutations induced by CRISPR/Cas9 in target genes; Table S5: Positions and efficiency score of both the targets; Table S6: List of primers utilized for analyzing off-targets. Figure S1: PCR amplification of CRISPR/Cas9 T-DNA integration; Figure S2: Analysis of the proteome of wild type (WT) and CRISPR/Cas9 mutants of rice.; Figure S3: Structure of pYLCRISPR/Cas9Pubi-*H* binary vector with fragment containing a modified ccdB flanked by two BsaI sites; Figure S4: Embedded view of plasmids; Figure S5: Schematic diagram of Vector map and sgRNA target sites in *OsGA20ox2*; Figure S6: Schematic representation of secondary structures; Figure S7: Illustration of overlapping PCR for generation of expression cassette; Figure S8: Detection and amplification of CRISPR/Cas9 T-DNA integration and the *OsGA20ox2* target sequence assembly in vector.

Author Contributions: Conceptualization, G.N.; Data curation, G.N. and B.U.; Formal analysis, G.N. and B.U.; Funding acquisition, R.L.; Investigation, G.N., B.U., N.Z., and Z.L.; Methodology, G.N., N.Z., Y.H., Z.L., and X.W.; Project administration, R.L.; Resources, Y.L. and R.L.; Software, G.N. and B.U.; Supervision, R.L.; Validation, G.N.; Visualization, Y.L. and R.L.; Writing—original draft, G.N. and B.U.; Writing—review and editing, G.N. All authors have read and agreed to the published version of the manuscript.

Funding: This study was supported by the State Key Laboratory for Conservation and Utilization of Subtropical Agro-Bioresources (SKLCUSA-a201914).

Acknowledgments: We would like to thank Muhammad Haneef Kashif for the helpful discussion and invaluable comments to make this research meaningful.

Conflicts of Interest: The authors declare no conflict of interest.

References

1. Chandler, V.L.; Wessler, S. *Grasses. A Collective Model Genetic System*; American Society of Plant Biologists: Rockville, MD, USA, 2001.

2. Demont, M.; Stein, A. Global value of GM rice: A review of expected agronomic and consumer benefits. *New Biotechnol.* **2013**, *30*, 426–436. [CrossRef] [PubMed]

3. Khush, G.S. What it will take to Feed 5.0 Billion Rice consumers in 2030. *Plant Mol. Biol.* **2005**, *59*, 1–6. [CrossRef] [PubMed]

4. Sasaki, T.; Burr, B. International Rice Genome Sequencing Project: The effort to completely sequence the rice genome. *Curr. Opin. Plant Biol.* **2000**, *3*, 138–142. [CrossRef]

5. Singh, V.; Siddiq, E.; Zaman, F.; Sadananda, A. *Improved Basmati Donors [Rice Varieties; India]*; International Rice Research Newsletter: Los Baños, Philippines, 1988.

6. Singh, A.K.; Gopalakrishnan, S.; Singh, V.P.; Prabhu, K.V.; Mohapatra, T.; Singh, N.K.; Sharma, T.R.; Nagarajan, M.; Vinod, K.K.; Singh, D.; et al. Marker assisted selection: A paradigm shift in Basmati breeding. *Indian J. Genet. Plant Breed.* **2011**, *71*, 120–128.

7. Fischer, R.; Stapper, M. Lodging effects on high-yielding crops of irrigated semidwarf wheat. *Field Crop. Res.* **1987**, *17*, 245–258. [CrossRef]

8. Zhou, L.; Liu, S.; Wu, W.; Chen, D.; Zhan, X.; Zhu, A.; Zhang, Y.; Cheng, S.; Cao, L.; Lou, X.; et al. Dissection of genetic architecture of rice plant height and heading date by multiple-strategy-based association studies. *Sci. Rep.* **2016**, *6*, 29718. [CrossRef]

9. Yang, X.-C.; Hwa, C.-M. Genetic modification of plant architecture and variety improvement in rice. *Heredity* **2008**, *101*, 396–404. [CrossRef]

10. Sakamoto, T.; Matsuoka, M. Generating high-yielding varieties by genetic manipulation of plant architecture. *Curr. Opin. Biotechnol.* **2004**, *15*, 144–147. [CrossRef]

11. Zhang, J.; Liu, X.; Li, S.; Cheng, Z.; Li, C. The Rice Semi-Dwarf Mutant sd37, Caused by a Mutation in CYP96B4, Plays an Important Role in the Fine-Tuning of Plant Growth. *PLoS ONE* **2014**, *9*, e88068. [CrossRef]

12. Thomas, S.G.; Rieu, I.; Steber, C.M. Gibberellin Metabolism and Signaling. *Vitam. Horm.* **2005**, *72*, 289–338. [CrossRef]

13. Monna, L.; Kitazawa, N.; Yoshino, R.; Suzuki, J.; Masuda, H.; Maehara, Y.; Tanji, M.; Sato, M.; Nasu, S.; Minobe, Y. Positional Cloning of Rice Semidwarfing Gene, sd-1: Rice "Green Revolution Gene" Encodes a Mutant Enzyme Involved in Gibberellin Synthesis. *DNA Res.* **2002**, *9*, 11–17. [CrossRef] [PubMed]

14. Magome, H.; Nomura, T.; Hanada, A.; Takeda-Kamiya, N.; Ohnishi, T.; Shinma, Y.; Katsumata, T.; Kawaide, H.; Kamiya, Y.; Yamaguchi, S. CYP714B1 and CYP714B2 encode gibberellin 13-oxidases that reduce gibberellin activity in rice. *Proc. Natl. Acad. Sci. USA* **2013**, *110*, 1947–1952. [CrossRef] [PubMed]

15. Marciniak, K.; Kućko, A.; Wilmowicz, E.; Świdziński, M.; Kęsy, J.; Kopcewicz, J. Photoperiodic flower induction in Ipomoea nil is accompanied by decreasing content of gibberellins. *Plant Growth Regul.* **2017**, *84*, 395–400. [CrossRef]

16. Thomas, S.G.; Hedden, P.; Roberts, J.A.; Evan, D.; McManus, M.T.; Rose, J.K.C. Gibberellin Metabolism and Signal Transduction. *Annu. Plant Rev. Online* **2018**, 147–184. [CrossRef]

17. Itoh, H.; Tatsumi, T.; Sakamoto, T.; Otomo, K.; Toyomasu, T.; Kitano, H.; Ashikari, M.; Ichihara, S.; Matsuoka, M. A Rice Semi-Dwarf Gene, Tan-Ginbozu (D35), Encodes the Gibberellin Biosynthesis Enzyme, ent-Kaurene Oxidase. *Plant Mol. Biol.* **2004**, *54*, 533–547. [CrossRef] [PubMed]

18. Sasaki, A.; Ashikari, M.; Ueguchi-Tanaka, M.; Itoh, H.; Nishimura, A.; Swapan, D.; Ishiyama, K.; Saito, T.; Kobayashi, M.; Khush, G.S.; et al. A mutant gibberellin-synthesis gene in rice. *Nature* **2002**, *416*, 701–702. [CrossRef]

19. Spielmeyer, W.; Ellis, M.H.; Chandler, P.M. Semidwarf (sd-1), "green revolution" rice, contains a defective gibberellin 20-oxidase gene. *Proc. Natl. Acad. Sci. USA* **2002**, *99*, 9043–9048. [CrossRef]

20. Qiao, F.; Zhao, K.-J. The Influence of RNAi Targeting of OsGA20ox2 Gene on Plant Height in Rice. *Plant Mol. Biol. Rep.* **2011**, *29*, 952–960. [CrossRef]

21. Wu, B.; Hu, W.; Ayaad, M.; Liu, H.; Xing, Y. Intragenic recombination between two non-functional semi-dwarf 1 alleles produced a functional SD1 allele in a tall recombinant inbred line in rice. *PLoS ONE* **2017**, *12*, e0190116. [CrossRef]

22. Han, Y.; Teng, K.; Nawaz, G.; Feng, X.; Usman, B.; Wang, X.; Luo, L.; Zhao, N.; Liu, Y.; Li, R. Generation of semi-dwarf rice (*Oryza sativa* L.) lines by CRISPR/Cas9-directed mutagenesis of OsGA20ox2 and proteomic analysis of unveiled changes caused by mutations. *3 Biotech* **2019**, *9*, 387. [CrossRef]

23. Qi, W.; Zhu, T.; Tian, Z.; Li, C.; Zhang, W.; Song, R. High-efficiency CRISPR/Cas9 multiplex gene editing using the glycine tRNA-processing system-based strategy in maize. *BMC Biotechnol.* **2016**, *16*, 58. [CrossRef] [PubMed]

24. Zheng, X.; Yang, S.; Zhang, Y.; Zhong, Z.; Tang, X.; Deng, K.; Zhou, J.; Qi, Y.; Zhang, Y. Effective screen of CRISPR/Cas9-induced mutants in rice by single-strand conformation polymorphism. *Plant Cell Rep.* **2016**, *35*, 1545–1554. [CrossRef] [PubMed]

25. Nawaz, G.; Usman, B.; Peng, H.; Zhao, N.; Yuan, R.; Liu, Y.-G.; Li, R. Knockout of *Pi21* by CRISPR/Cas9 and iTRAQ-Based Proteomic Analysis of Mutants Revealed New Insights into *M. oryzae* Resistance in Elite Rice Line. *Genes* **2020**, *11*, 735. [CrossRef] [PubMed]

26. Jung, Y.-J.; Nogoy, F.M.; Lee, S.-K.; Cho, Y.-G.; Kang, K.-K. Application of ZFN for Site Directed Mutagenesis of Rice SSIVa Gene. *Biotechnol. Bioprocess Eng.* **2018**, *23*, 108–115. [CrossRef]

27. Gaj, T.; Gersbach, C.A.; Barbas, I.C.F. ZFN, TALEN and CRISPR/Cas-based methods for genome engineering. *Trends Biotechnol.* **2013**, *31*, 397–405. [CrossRef]

28. Jiang, W.; Zhou, H.; Bi, H.; Fromm, M.; Yang, B.; Weeks, D.P. Demonstration of CRISPR/Cas9/sgRNA-mediated targeted gene modification in Arabidopsis, tobacco, sorghum and rice. *Nucleic Acids Res.* **2013**, *41*, e188. [CrossRef]

29. Li, J.-F.; Norville, J.E.; Aach, J.; McCormack, M.; Zhang, D.; Bush, J.; Church, G.M.; Sheen, J. Multiplex and homologous recombination–mediated genome editing in Arabidopsis and Nicotiana benthamiana using guide RNA and Cas9. *Nat. Biotechnol.* **2013**, *31*, 688–691. [CrossRef]

30. Mao, Y.; Zhang, H.; Xu, N.; Zhang, B.; Gou, F.; Zhu, J.-K. Application of the CRISPR–Cas System for Efficient Genome Engineering in Plants. *Mol. Plant* **2013**, *6*, 2008–2011. [CrossRef]

31. Shan, Q.; Wang, Y.; Li, J.; Zhang, Y.; Chen, K.; Liang, Z.; Zhang, K.; Liu, J.; Xi, J.J.; Qiu, J.; et al. Targeted genome modification of crop plants using a CRISPR-Cas system. *Nat. Biotechnol.* **2013**, *31*, 686–688. [CrossRef]

32. Nawaz, G.; Han, Y.; Usman, B.; Liu, F.; Qin, B.; Li, R. Knockout of OsPRP1, a gene encoding proline-rich protein, confers enhanced cold sensitivity in rice (*Oryza sativa* L.) at the seedling stage. *3 Biotech* **2019**, *9*, 1–18. [CrossRef]

33. Liang, Z.; Zhang, K.; Chen, K.; Gao, C. Targeted Mutagenesis in Zea mays Using TALENs and the CRISPR/Cas System. *J. Genet. Genom.* **2014**, *41*, 63–68. [CrossRef] [PubMed]

34. Wang, Y.; Cheng, X.; Shan, Q.; Zhang, Y.; Liu, J.; Gao, C.; Qiu, J. Simultaneous editing of three homoeoalleles in hexaploid bread wheat confers heritable resistance to powdery mildew. *Nat. Biotechnol.* **2014**, *32*, 947–951. [CrossRef] [PubMed]

35. Jacobs, T.B.; Lafayette, P.R.; Schmitz, R.J.; Parrott, W.A. Targeted genome modifications in soybean with CRISPR/Cas9. *BMC Biotechnol.* **2015**, *15*, 16. [CrossRef] [PubMed]

36. Li, Z.; Liu, Z.-B.; Xing, A.; Moon, B.P.; Koellhoffer, J.P.; Huang, L.; Ward, R.T.; Clifton, E.; Falco, S.C.; Cigan, A.M. Cas9-Guide RNA Directed Genome Editing in Soybean. *Plant Physiol.* **2015**, *169*, 960–970. [CrossRef]

37. Sun, X.; Hu, Z.; Chen, R.; Jiang, Q.; Song, G.; Zhang, H.; Xi, Y. Targeted mutagenesis in soybean using the CRISPR-Cas9 system. *Sci. Rep.* **2015**, *5*, 10342. [CrossRef]

38. Brooks, C.; Nekrasov, V.; Lippman, Z.; Van Eck, J. Efficient Gene Editing in Tomato in the First Generation Using the Clustered Regularly Interspaced Short Palindromic Repeats/CRISPR-Associated9 System1. *Plant Physiol.* **2014**, *166*, 1292–1297. [CrossRef]

39. Xu, R.-F.; Li, H.; Qin, R.-Y.; Li, J.; Qiu, C.-H.; Yang, Y.-C.; Ma, H.; Li, L.; Wei, P.; Yang, J.-B. Generation of inheritable and "transgene clean" targeted genome-modified rice in later generations using the CRISPR/Cas9 system. *Sci. Rep.* **2015**, *5*, 11491. [CrossRef]

40. Kok, E.; Kuiper, H.A. Comparative safety assessment for biotech crops. *Trends Biotechnol.* **2003**, *21*, 439–444. [CrossRef]

41. Chassy, B.M. Food safety evaluation of crops produced through biotechnology. *J. Am. Coll. Nutr.* **2002**, *21*, 166–173. [CrossRef]

42. Brandão, A.; Barbosa, H.; Arruda, M. Image analysis of two-dimensional gel electrophoresis for comparative proteomics of transgenic and non-transgenic soybean seeds. *J. Proteom.* **2010**, *73*, 1433–1440. [CrossRef]

43. Wang, L.; Wang, X.; Jin, X.; Jia, R.; Huang, Q.; Tan, Y.; Guo, A. Comparative proteomics of Bt-transgenic and non-transgenic cotton leaves. *Proteome Sci.* **2015**, *13*, 1–15. [CrossRef]

44. Wang, Y.; Xu, W.; Zhao, W.; Hao, J.; Luo, Y.; Tang, X.; Zhang, Y.; Huang, K. Comparative analysis of the proteomic and nutritional composition of transgenic rice seeds with Cry1ab/ac genes and their non-transgenic counterparts. *J. Cereal Sci.* **2012**, *55*, 226–233. [CrossRef]

45. Yang, Y.; Chen, X.; Xu, B.; Li, Y.; Ma, Y.; Wang, G. Phenotype and transcriptome analysis reveals chloroplast development and pigment biosynthesis together influenced the leaf color formation in mutants of Anthurium andraeanum 'Sonate'. *Front. Plant Sci.* **2015**, *6*, 139. [CrossRef]

46. Hellinger, R.; Koehbach, J.; Soltis, U.E.; Carpenter, E.J.; Wong, G.K.-S.; Gruber, C.W. Peptidomics of Circular Cysteine-Rich Plant Peptides: Analysis of the Diversity of Cyclotides from Viola tricolor by Transcriptome and Proteome Mining. *J. Proteome Res.* **2015**, *14*, 4851–4862. [CrossRef]

47. Kamal, A.H.M.; Cho, K.; Choi, J.-S.; Bae, K.-H.; Komatsu, S.; Uozumi, N.; Woo, S.-H. The wheat chloroplastic proteome. *J. Proteom.* **2013**, *93*, 326–342. [CrossRef]

48. Wang, F.-X.; Luo, Y.-M.; Ye, Z.-Q.; Cao, X.; Liang, J.-N.; Wang, Q.; Wu, Y.; Wu, J.-H.; Wang, H.-Y.; Zhang, M.; et al. iTRAQ-based proteomics analysis of autophagy-mediated immune responses against the vascular fungal pathogen Verticillium dahliae in Arabidopsis. *Autophagy* **2018**, *14*, 598–618. [CrossRef]

49. Wang, J.; Islam, F.; Li, L.; Long, M.; Yang, C.; Jin, X.; Ali, B.; Mao, B.; Zhou, W. Complementary RNA-Sequencing Based Transcriptomics and iTRAQ Proteomics Reveal the Mechanism of the Alleviation of Quinclorac Stress by Salicylic Acid in *Oryza sativa* ssp. japonica. *Int. J. Mol. Sci.* **2017**, *18*, 1975. [CrossRef]

50. Liao, S.; Qin, X.; Luo, L.; Han, Y.; Wang, X.; Usman, B.; Nawaz, G.; Zhao, N.; Liu, Y.-G.; Li, R. CRISPR/Cas9-Induced Mutagenesis of Semi-Rolled Leaf1,2 Confers Curled Leaf Phenotype and Drought Tolerance by Influencing Protein Expression Patterns and ROS Scavenging in Rice (*Oryza sativa* L.). *Agronomy* **2019**, *9*, 728. [CrossRef]

51. Chen, S.; Chen, J.; Hou, F.; Feng, Y.; Zhang, R. iTRAQ-based quantitative proteomic analysis reveals the lateral meristem developmental mechanism for branched spike development in tetraploid wheat (*Triticum turgidum* L.). *BMC Genom.* **2018**, *19*, 228. [CrossRef]

52. Bu, T.-T.; Shen, J.; Chao, Q.; Shen, Z.; Yan, Z.; Zheng, H.-Y.; Wang, B.-C. Dynamic N-glycoproteome analysis of maize seedling leaves during de-etiolation using Concanavalin A lectin affinity chromatography and a nano-LC–MS/MS-based iTRAQ approach. *Plant Cell Rep.* **2017**, *36*, 1943–1958. [CrossRef]

53. Chen, L.; Wang, S.; Zhang, Y.-H.; Li, J.; Xing, Z.; Yang, J.; Huang, T.; Cai, Y.-D. Identify Key Sequence Features to Improve CRISPR sgRNA Efficacy. *IEEE Access* **2017**, *5*, 26582–26590. [CrossRef]

54. Sauer, N.J.; Narváez-Vásquez, J.; Mozoruk, J.; Miller, R.B.; Warburg, Z.; Woodward, M.J.; Mihiret, Y.A.; Lincoln, T.A.; Segami, R.E.; Sanders, S.L.; et al. Oligonucleotide-Mediated Genome Editing Provides Precision and Function to Engineered Nucleases and Antibiotics in Plants. *Plant Physiol.* **2016**, *170*, 1917–1928. [CrossRef]

55. Li, J.; Meng, X.; Zong, Y.; Chen, K.; Zhang, H.; Liu, J.; Li, J.; Gao, C. Gene replacements and insertions in rice by intron targeting using CRISPR–Cas9. *Nat. Plants* **2016**, *2*, 16139. [CrossRef]

56. Soyk, S.; Lemmon, Z.H.; Oved, M.; Fisher, J.; Liberatore, K.L.; Park, S.J.; Goren, A.; Jiang, K.; Ramos, A.; Van Der Knaap, E.; et al. Bypassing Negative Epistasis on Yield in Tomato Imposed by a Domestication Gene. *Cell* **2017**, *169*, 1142–1155.e12. [CrossRef]

57. Jacob, P.; Avni, A.; Bendahmane, A. Translational Research: Exploring and Creating Genetic Diversity. *Trends Plant Sci.* **2018**, *23*, 42–52. [CrossRef]

58. Cong, L.; Ran, F.A.; Cox, D.; Lin, S.; Barretto, R.; Habib, N.; Hsu, P.D.; Wu, X.; Jiang, W.; Marraffini, L.A.; et al. Multiplex Genome Engineering Using CRISPR/Cas Systems. *Science* **2013**, *339*, 819–823. [CrossRef]

59. Shan, Q.; Wang, Y.; Chen, K.; Liang, Z.; Li, J.; Zhang, Y.; Zhang, K.; Liu, J.; Voytas, D.F.; Zheng, X.; et al. Rapid and Efficient Gene Modification in Rice and Brachypodium Using TALENs. *Mol. Plant* **2013**, *6*, 1365–1368. [CrossRef]

60. Nelson, D.R.; Schuler, M.A.; Paquette, S.M.; Werck-Reichhart, D.; Bak, S. Comparative Genomics of Rice and Arabidopsis. Analysis of 727 Cytochrome P450 Genes and Pseudogenes from a Monocot and a Dicot1[w]. *Plant Physiol.* **2004**, *135*, 756–772. [CrossRef]

61. Zhou, H.; Liu, B.; Weeks, D.P.; Spalding, M.H.; Yang, B. Large chromosomal deletions and heritable small genetic changes induced by CRISPR/Cas9 in rice. *Nucleic Acids Res.* **2014**, *42*, 10903–10914. [CrossRef]

62. Han, Y.; Luo, D.; Usman, B.; Nawaz, G.; Zhao, N.; Liu, F.; Li, R. Development of High Yielding Glutinous Cytoplasmic Male Sterile Rice (*Oryza sativa* L.) Lines through CRISPR/Cas9 Based Mutagenesis of Wx and TGW6 and Proteomic Analysis of Anther. *Agronomy* **2018**, *8*, 290. [CrossRef]

63. Usman, B.; Nawaz, G.; Zhao, N.; Liu, Y.-G.; Li, R. Generation of High Yielding and Fragrant Rice (*Oryza sativa* L.) Lines by CRISPR/Cas9 Targeted Mutagenesis of Three Homoeologs of Cytochrome P450 Gene Family and *OsBADH2* and Transcriptome and Proteome Profiling of Revealed Changes Triggered by Mutations. *Plants* **2020**, *9*, 788. [CrossRef]

64. Zhang, H.; Zhang, J.; Wei, P.; Zhang, B.; Gou, F.; Feng, Z.; Mao, Y.; Yang, L.; Zhang, H.; Xu, N.; et al. The CRISPR/Cas9 system produces specific and homozygous targeted gene editing in rice in one generation. *Plant Biotechnol. J.* **2014**, *12*, 797–807. [CrossRef]

65. Yamaguchi, S. Gibberellin Metabolism and its Regulation. *Annu. Rev. Plant Biol.* **2008**, *59*, 225–251. [CrossRef]

66. Hedden, P.; Thomas, S.G. Gibberellin biosynthesis and its regulation. *Biochem. J.* **2012**, *444*, 11–25. [CrossRef]

67. Fukazawa, J.; Mori, M.; Watanabe, S.; Miyamoto, C.; Ito, T.; Takahashi, Y. DELLA-GAF1 Complex Is a Main Component in Gibberellin Feedback Regulation of GA20 Oxidase 2. *Plant Physiol.* **2017**, *175*, 1395–1406. [CrossRef]

68. Boden, S.A.; Weiss, D.; Ross, J.J.; Davies, N.W.; Trevaskis, B.; Chandler, P.M.; Swain, S.M.; Krajinski, F.; Courty, P.-E.; Sieh, D.; et al. *EARLY FLOWERING3* Regulates Flowering in Spring Barley by Mediating Gibberellin Production and *FLOWERING LOCUS T* Expression. *Plant Cell* **2014**, *26*, 1557–1569. [CrossRef]

69. Fukazawa, J.; Nakata, M.; Ito, T.; Matsushita, A.; Yamaguchi, S.; Takahashi, Y. bZIP transcription factor RSG controls the feedback regulation ofNtGA20ox1via intracellular localization and epigenetic mechanism. *Plant Signal. Behav.* **2011**, *6*, 26–28. [CrossRef]

70. Yang, G.; Inoue, A.; Takasaki, H.; Kaku, H.; Akao, S.; Komatsu, S. A Proteomic Approach to Analyze Auxin- and Zinc-Responsive Protein in Rice. *J. Proteome Res.* **2005**, *4*, 456–463. [CrossRef]

71. Komatsu, S.; Zang, X.; Tanaka, N. Comparison of Two Proteomics Techniques Used to Identify Proteins Regulated by Gibberellin in Rice. *J. Proteome Res.* **2006**, *5*, 270–276. [CrossRef]

72. Shen, S.; Sharma, A.; Komatsu, S. Characterization of proteins responsive to gibberellin in the leaf-sheath of rice (*Oryza sativa* L.) seedling using proteome analysis. *Biol. Pharm. Bull.* **2003**, *26*, 129–136. [CrossRef]

73. Tanaka, N.; Konishi, H.; Khan, M.M.K.; Komatsu, S. Proteome analysis of rice tissues by two-dimensional electrophoresis: An approach to the investigation of gibberellin regulated proteins. *Mol. Genet. Genom.* **2003**, *270*, 485–496. [CrossRef]

74. Hedden, P. The genes of the Green Revolution. *Trends Genet.* **2003**, *19*, 5–9. [CrossRef]

75. Li, J.; Jiang, J.; Qian, Q.; Xu, Y.; Zhang, C.; Xiao, J.; Du, C.; Luo, W.; Zou, G.; Chen, M.; et al. Mutation of Rice BC12/GDD1, Which Encodes a Kinesin-Like Protein That Binds to a GA Biosynthesis Gene Promoter, Leads to Dwarfism with Impaired Cell Elongation. *Plant Cell* **2011**, *23*, 628–640. [CrossRef]

76. Kuijt, S.; Greco, R.; Agalou, A.; Shao, J.; Hoen, C.C.J.; Overnäs, E.; Osnato, M.; Curiale, S.; Meynard, D.; van Gulik, R.; et al. Interaction between the GROWTH-REGULATING FACTOR and KNOTTED1-LIKE HOMEOBOX Families of Transcription Factors1[W]. *Plant Physiol.* **2014**, *164*, 1952–1966. [CrossRef]

77. Omidbakhshfard, M.A.; Proost, S.; Fujikura, U.; Mueller-Roeber, B. Growth-Regulating Factors (GRFs): A Small Transcription Factor Family with Important Functions in Plant Biology. *Mol. Plant* **2015**, *8*, 998–1010. [CrossRef]

78. van der Knaap, E.; Kim, J.H.; Kende, H. A novel gibberellin-induced gene from rice and its potential regulatory role in stem growth. *Plant Physiol.* **2000**, *122*, 695–704. [CrossRef]

79. Kim, J.H.; Choi, D.; Kende, H. The AtGRF family of putative transcription factors is involved in leaf and cotyledon growth in Arabidopsis. *Plant J.* **2003**, *36*, 94–104. [CrossRef]

80. Bao, M.; Bian, H.; Zha, Y.; Li, F.; Sun, Y.; Bai, B.; Chen, Z.; Wang, J.; Zhu, M.; Han, N. miR396a-Mediated Basic Helix–Loop–Helix Transcription Factor bHLH74 Repression Acts as a Regulator for Root Growth in Arabidopsis Seedlings. *Plant Cell Physiol.* **2014**, *55*, 1343–1353. [CrossRef]

81. Choi, D.; Kim, J.H.; Kende, H. Whole genome analysis of the OsGRF gene family encoding plant-specific putative transcription activators in rice (*Oryza sativa* L.). *Plant Cell Physiol.* **2004**, *45*, 897–904. [CrossRef] [PubMed]

82. Liu, H.; Guo, S.; Xu, Y.; Li, C.; Zhang, Z.; Zhang, D.; Xu, S.; Zhang, C.; Chong, K. OsmiR396d-regulated OsGRFs function in floral organogenesis in rice through binding to their targets OsJMJ706 and OsCR4. *Plant Physiol.* **2014**, *165*, 160–174. [CrossRef]

83. Day, B.; Tanabe, S.; Koshioka, M.; Mitsui, T.; Itoh, H.; Ueguchi-Tanaka, M.; Matsuoka, M.; Kaku, H.; Shibuya, N.; Minami, E. Two Rice GRAS Family Genes Responsive to N-Acetylchitooligosaccharide Elicitor are Induced by Phytoactive Gibberellins: Evidence for Cross-Talk between Elicitor and Gibberellin Signaling in Rice Cells. *Plant Mol. Biol.* **2004**, *54*, 261–272. [CrossRef] [PubMed]

84. Prisic, S.; Xu, M.; Wilderman, P.R.; Peters, R.J. Rice Contains Two Disparate ent-Copalyl Diphosphate Synthases with Distinct Metabolic Functions1. *Plant Physiol.* **2004**, *136*, 4228–4236. [CrossRef]

85. Salvucci, M.E.; Ogren, W.L. The mechanism of Rubisco activase: Insights from studies of the properties and structure of the enzyme. *Photosynth. Res.* **1996**, *47*, 1–11. [CrossRef] [PubMed]

86. Chen, F.; Tholl, D.; Bohlmann, J.; Pichersky, E. The family of terpene synthases in plants: A mid-size family of genes for specialized metabolism that is highly diversified throughout the kingdom. *Plant J.* **2011**, *66*, 212–229. [CrossRef] [PubMed]

87. Hanai, H.; Nakayama, D.; Yang, H.; Matsubayashi, Y.; Hirota, Y.; Sakagami, Y. Existence of a plant tyrosylprotein sulfotransferase: Novel plant enzyme catalyzing tyrosine O-sulfation of preprophytosulfokine variants in vitro. *FEBS Lett.* **2000**, *470*, 97–101. [CrossRef]

88. Komori, R.; Amano, Y.; Ogawa-Ohnishi, M.; Matsubayashi, Y. Identification of tyrosylprotein sulfotransferase in Arabidopsis. *Proc. Natl. Acad. Sci. USA* **2009**, *106*, 15067–15072. [CrossRef]

89. Jin, Y.; Zhang, C.; Liu, W.; Tang, Y.; Qi, H.; Chen, H.; Cao, S. The Alcohol Dehydrogenase Gene Family in Melon (Cucumis melo L.): Bioinformatic Analysis and Expression Patterns. *Front. Plant Sci.* **2016**, *7*, 670. [CrossRef]

90. Hattori, Y.; Nagai, K.; Furukawa, S.; Song, X.-J.; Kawano, R.; Sakakibara, H.; Wu, J.; Matsumoto, T.; Yoshimura, A.; Kitano, H.; et al. The ethylene response factors SNORKEL1 and SNORKEL2 allow rice to adapt to deep water. *Nature* **2009**, *460*, 1026–1030. [CrossRef]

91. Takasaki, H.; Mahmood, T.; Matsuoka, M.; Matsumoto, H.; Komatsu, S. Identification and characterization of a gibberellin-regulated protein, which is ASR5, in the basal region of rice leaf sheaths. *Mol. Genet. Genom.* **2008**, *279*, 359–370. [CrossRef] [PubMed]

92. Kim, H.; Lee, K.; Hwang, H.; Bhatnagar, N.; Kim, D.-Y.; Yoon, I.S.; Byun, M.-O.; Kim, S.T.; Jung, K.-H.; Kim, B.-G. Overexpression of PYL5 in rice enhances drought tolerance, inhibits growth, and modulates gene expression. *J. Exp. Bot.* **2014**, *65*, 453–464. [CrossRef] [PubMed]

93. Welti, R. Plant lipidomics: Discerning biological function by profiling plant complex lipids using mass spectrometry. *Front. Biosci. J. Virtual Libr.* **2007**, *12*, 2494–2506. [CrossRef] [PubMed]

94. Rojas, C.M.; Senthil-Kumar, M.; Tzin, V.; Mysore, K.S. Regulation of primary plant metabolism during plant-pathogen interactions and its contribution to plant defense. *Front. Plant Sci.* **2014**, *5*. [CrossRef] [PubMed]

95. Ma, X.; Zhang, Q.; Zhu, Q.; Liu, W.; Chen, Y.; Qiu, R.; Wang, B.; Yang, Z.; Li, H.; Lin, Y.; et al. A Robust CRISPR/Cas9 System for Convenient, High-Efficiency Multiplex Genome Editing in Monocot and Dicot Plants. *Mol. Plant* **2015**, *8*, 1274–1284. [CrossRef]

96. Xie, X.; Ma, X.; Zhu, Q.; Zeng, D.; Li, G.; Liu, Y.-G. CRISPR-GE: A Convenient Software Toolkit for CRISPR-Based Genome Editing. *Mol. Plant* **2017**, *10*, 1246–1249. [CrossRef]

97. Ma, X.; Liu, Y. CRISPR/Cas9-Based Multiplex Genome Editing in Monocot and Dicot Plants. *Curr. Protoc. Mol. Biol.* **2016**, *115*, 31.6.1–31.6.21. [CrossRef] [PubMed]

98. Hiei, Y.; Ohta, S.; Komari, T.; Kumashiro, T. Efficient transformation of rice (*Oryza sativa* L.) mediated by Agrobacterium and sequence analysis of the boundaries of the T-DNA. *Plant J.* **1994**, *6*, 271–282. [CrossRef] [PubMed]
99. Xu, X.; Kawasaki, S.; Fujimura, T.; Wang, C. A protocol for high-throughput extraction of DNA from rice leaves. *Plant Mol. Biol. Rep.* **2005**, *23*, 291–295. [CrossRef]
100. Ma, X.; Chen, L.; Zhu, Q.; Chen, Y.; Liu, Y.-G. Rapid Decoding of Sequence-Specific Nuclease-Induced Heterozygous and Biallelic Mutations by Direct Sequencing of PCR Products. *Mol. Plant* **2015**, *8*, 1285–1287. [CrossRef]
101. Suge, H. Ethylene and Gibberellin: Regulation of Internodal Elongation and Nodal Root Development in Floating Rice. *Plant Cell Physiol.* **1985**, *26*, 607–614. [CrossRef]
102. Peng, P.; Liu, L.; Fang, J.; Zhao, J.; Yuan, S.; Li, X. The rice TRIANGULAR HULL1 protein acts as a transcriptional repressor in regulating lateral development of spikelet. *Sci. Rep.* **2017**, *7*, 13712. [CrossRef]
103. Chen, L.; Huang, Y.; Xu, M.; Cheng, Z.; Zhang, D.; Zheng, J. iTRAQ-Based Quantitative Proteomics Analysis of Black Rice Grain Development Reveals Metabolic Pathways Associated with Anthocyanin Biosynthesis. *PLoS ONE* **2016**, *11*, e0159238. [CrossRef]
104. Pfaffl, M.W. A new mathematical model for relative quantification in real-time RT-PCR. *Nucleic Acids Res.* **2001**, *29*, e45. [CrossRef]

International Journal of
Molecular Sciences

Article

Ultrastructural and Photosynthetic Responses of Pod Walls in Alfalfa to Drought Stress

Hui Wang [1,2], Qingping Zhou [2] and Peisheng Mao [1,*]

[1] Forage Seed Laboratory, Key Laboratory of Pratacultural Science, Beijing Municipality, China Agricultural University, Beijing 100193, China; huiwang@swun.edu.cn

[2] College of Qinghai-Tibetan Plateau, Southwest Minzu University, Chengdu 610041, China; qpingzh@aliyun.com

* Correspondence: maops@cau.edu.cn; Tel.: +86-010-6273-3311

Received: 3 June 2020; Accepted: 22 June 2020; Published: 23 June 2020

Abstract: Increasing photosynthetic ability as a whole is essential for acquiring higher crop yields. Nonleaf green organs (NLGOs) make important contributions to photosynthate formation, especially under stress conditions. However, there is little information on the pod wall in legume forage related to seed development and yield. This experiment is designed for alfalfa (*Medicago sativa*) under drought stress to explore the photosynthetic responses of pod walls after 5, 10, 15, and 20 days of pollination (DAP5, DAP10, DAP15, and DAP20) based on ultrastructural, physiological and proteomic analyses. Stomata were evidently observed on the outer epidermis of the pod wall. Chloroplasts had intact structures arranged alongside the cell wall, which on DAP5 were already capable of producing photosynthate. The pod wall at the late stage (DAP20) still had photosynthetic ability under well-watered (WW) treatments, while under water-stress (WS), the structure of the chloroplast membrane was damaged and the grana lamella of thylakoids were blurry. The chlorophyll a and chlorophyll b concentrations both decreased with the development of pod walls, and drought stress impeded the synthesis of photosynthetic pigments. Although the activity of ribulose-1,5-bisphosphate carboxylase (RuBisCo) decreased in the pod wall under drought stress, the activity of phosphoenolpyruvate carboxylase (PEPC) increased higher than that of RuBisCo. The proteomic analysis showed that the absorption of light is limited due to the suppression of the synthesis of chlorophyll a/b binding proteins by drought stress. Moreover, proteins involved in photosystem I and photosystem II were downregulated under WW compared with WS. Although the expression of some proteins participating in the regeneration period of RuBisCo was suppressed in the pod wall subjected to drought stress, the synthesis of PEPC was induced. In addition, some proteins, which were involved in the reduction period of RuBisCo, carbohydrate metabolism, and energy metabolism, and related to resistance, including chitinase, heat shock protein 81-2 (Hsp81-2), and lipoxygenases (LOXs), were highly expressed for the protective response to drought stress. It could be suggested that the pod wall in alfalfa is capable of operating photosynthesis and reducing the photosynthetic loss from drought stress through the promotion of the C4 pathway, ATP synthesis, and resistance ability.

Keywords: pod wall; nonleaf green organs; ultrastructure; proteomic; alfalfa

---

1. Introduction

Photosynthesis is considered as the most important chemical reaction and provides over 90% of dry matter for crop yield formation [1,2]. Increasing crop yield by promoting photosynthesis has been the research hotspot until now [3]. Green leaves are commonly focused as the main source for producing photosynthate. However, nonleaf green organs (NLGOs) have been proven to be practically or potentially capable of assimilating CO_2. Many scientists have previously reported that the silique shell of oil rape (*Brassica napus*) [4]; the boll shell of castor (*Ricinus communis*) [5]; the pod wall of legume crops, including chickpea (*Cicer arietinum*) [6], soybean (*Glycine max*) [7], and alfalfa (*Medicago sativa*) [8]; ears of cereal including rice (*Oryza sativa*) [9], barley (*Hordeum vulgare*) [10], and wheat (*Triticum turgidum*) [11]; flowers [12], stems [13], and roots [14] in some plants could photosynthesize and make an important contribution to yield formation. In addition, under drought conditions, the photosynthetic contribution of NLGOs turn greater, and NLGOs even become the primary photosynthetic organs for grain-filling [9,15].

Photosynthesis in leaves is sensitive to water deficit. Under water-deficit conditions, the flag leaves of cereal crops will wilt and senesce, while NLGOs are able to maintain relatively high water content due to some special features, including xeromorphic anatomy [9], lower stomatal conductance and transpiration rate [16], and a higher ability for osmotic adjustment [17]. In addition, NLGOs show better photosynthetic performance in the change of stomatal densities [18], chlorophyll concentration [5,19], and photosynthetic enzyme abilities [20] than leaves in response to water deficit. NLGOs have been proved to play an important role in regulating carbon partitioning during grain-filling to compensate for the reduction due to the decrease of photosynthetic ability in leaves [7]. Shreds of evidence based on anatomical, physiological, and molecular research have shown that the high photosynthetic efficiency pathway, C4-like or C3-C4 intermediate photosynthesis, might exist in the NLGOs of C3 crops [11]. Kranz anatomy is considered a crucial characteristic in C4 crops. The ear organs, including glume, lemma, and awn in C3 cereals, have two types of chloroplasts existing, respectively, in two types of cells, mesophyll cells and the cells arranged around the vascular bundles, similar to maize (*Zea mays*) leaves [11]. One of the key photosynthetic enzymes, phosphoenolpyruvate carboxylase (PEPC), has been detected with activity in NLGOs, and PEPC has a higher ability than ribulose-1,5-bisphosphate carboxylase (RuBisCo) under drought stress [21–23]. Besides PEPC, other enzymes, including NAD-dependent malic enzyme (NAD-ME), NADP-dependent malic enzyme (NADP-ME), and NADP-dependent malate dehydrogenase (NADP-MDH) involved in the C4 photosynthetic cycle, have been induced in NLGOs under drought stress [21]. Some genes, including *ppc, aat, mdh, me2, gpt,* and *ppdk,* specific to NAD-ME type-C4 photosynthesis, have been identified in wheat caryopsis [24]. However, other scientists have proposed the negative hypothesis that C4 photosynthesis is lacking in C3 crops. Singal et al. reported that CO_2 was assimilated by C4 photosynthesis in the pericarp of wheat, but not in awn and glume [25]. A C4 photosynthesis metabolism occurring in C3 crops depends on ontogeny differences, cultivars, and environments like high or low CO_2 and heat or drought stress [24,26]. Above all, the response mechanism of carbon fixation in NLGOs to drought stress is still unclear.

Alfalfa is widely cultured around the world to produce high-quality hay for feeding livestock, especially dairy cows. Seed producers have long focused on alfalfa seed yield increase. Moderate drought contributes to achieving higher seed yield during the flowering and seed maturation period. Nevertheless, little is known on the physiological response and the photosynthetic contribution of the pod wall in alfalfa under drought stress. Investigating and increasing the photosynthetic ability of NLGOs, especially under stress conditions, is a novel way to increase the photosynthetic ability of the whole plant and finally increase the grain yield. In this study, physiological, ultrastructural, and proteomic analyses were carried out to (1) investigate the photosynthetic characteristics of the pod wall in alfalfa, and (2) research the response mechanism of photosynthesis in the pod wall to drought stress.

2. Results

2.1. Changes of the Surface Characteristics and Ultrastructure of Pod Wall under Drought Stress

Stomata and epidermal hair were distinctly observed in the outer surface of the pod wall (Figure 1A,B). Stomata were composed of two semilunar guard cells encircled by several subsidiary cells. Stomata were open under both WW and WS, and the thick inner wall of stomata, the bright color part, could be clearly observed. In addition, a hump occurred around the base of epidermal hair, and lots of dots existed on the epidermal hair. Cells of the inner surface of the pod wall were tightly arranged together (Figure 1C).

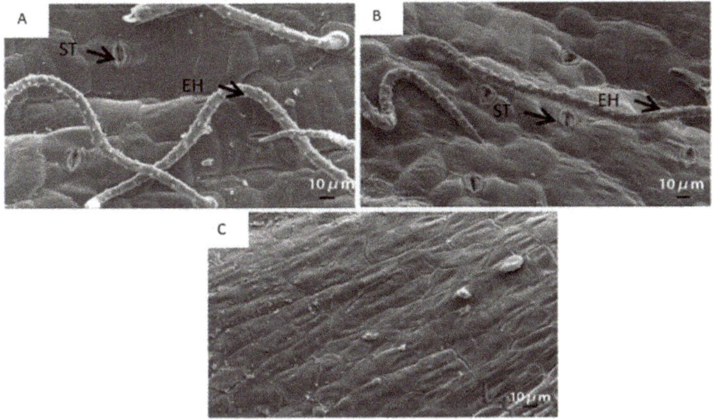

Figure 1. The scanning electron micrograph of the outer (**A,B**) and the inner surface (**C**) of the pod wall on the 10th day after pollination (DAP10) under well-watered (**A**) and water-stressed treatments (**B**). ST, stoma; EH, epidermal hair; D, dots; H, hump.

Under WW, chloroplasts in the pod wall had the ability to photosynthesize from DAP5 to DAP20. Chloroplasts on DAP5 existed with the intact structure and were arranged close to the cell wall (Figure 2A). Chloroplast membrane structure was intact, and grana lamella was arrayed along the long axis of the chloroplast, some of which had already produced starch grains. More and bigger starch grains were produced in the chloroplasts on DAP10 (Figure 2B) and DAP15 (Figure 2C). The pod wall on DAP20 still had photosynthetic activity, while the cells had started to age and the nuclei were degrading. Few osmiophilic granules were found in cells (Figure 2D).

Under WS, chloroplasts were able to produce photosynthate on DAP5 and DAP10, while the structure of chloroplasts was gradually damaged from DAP15 to DAP20. The chloroplasts had intact membrane structures and had already started to produce starch grains on DAP5 (Figure 2E), and they produced more and bigger starch grains on DAP10 (Figure 2F). Lots of starch grains could still be observed on DAP15, while the evident changes occurred in the structure of chloroplasts, i.e., the membrane was partly broken, and the grana lamellae of thylakoids became blurry (Figure 2G). Few starch grains existed on DAP20, while lots of osmiophilic granules were presented. The membrane of chloroplasts was seriously broken, and the structure of thylakoids was blurring (Figure 2H).

Except for chloroplasts, the structure of other organelle or tissues changed under drought stress as well. The central vacuole was bigger in the cell under WW (Figure 2D), while the gap between the central vacuole and the cell wall become wider under WS (Figure 2H). The membrane structure of the mitochondrion was intact and clear on DAP5 under WS (Figure 2E), while it was broken and blurred on DAP20 (Figure 2H).

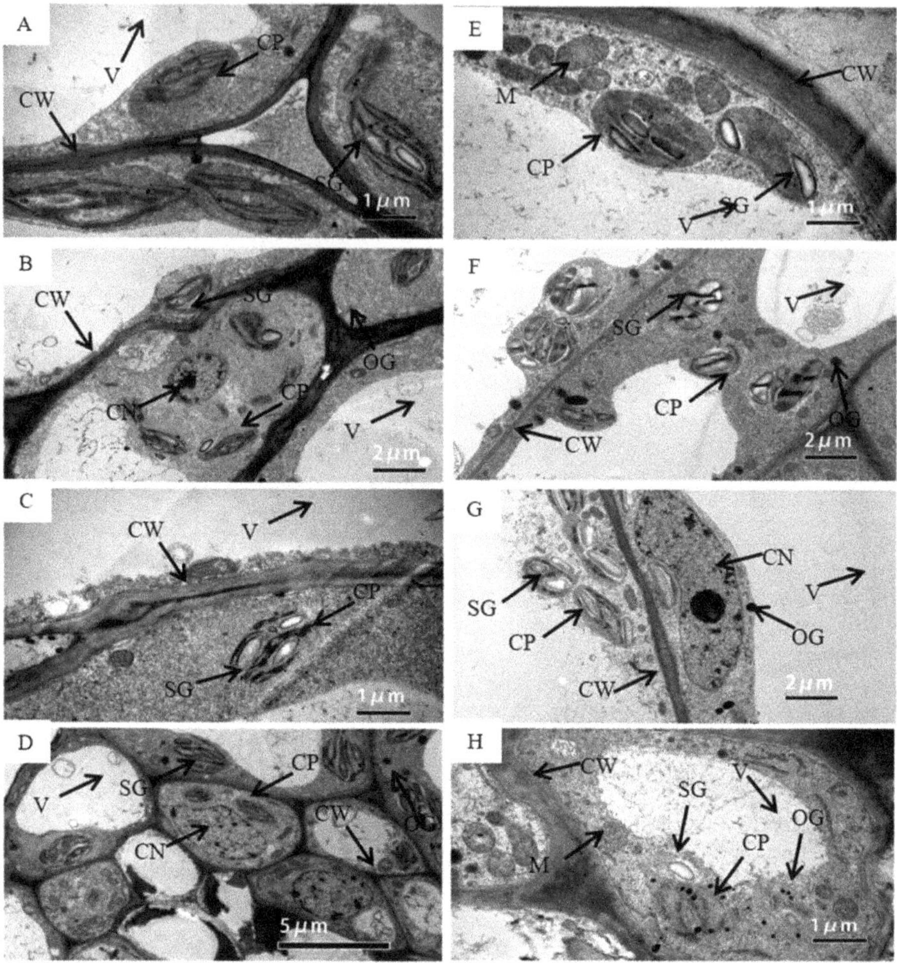

Figure 2. The transmission electron micrograph of cells in the pod wall on DAP5 (**A,E**), DAP10 (**B,F**), DAP15 (**C,G**), and DAP20 (**D,H**) under well-watered (**A–D**) and water-stressed treatment (**E–H**). CW, cell wall; CP, chloroplast; SG, starch grain; OG, osmiophilic granules; V, central vacuole; CN, cell nucleus; M, mitochondrion; T, thylakoid.

2.2. Changes of Chlorophyll Concentration in Pod Wall under Drought Stress

With the development of the pod wall, the concentration of chlorophyll a, chlorophyll b, and total chlorophyll decreased under both WW and WS treatments, and the concentration of chlorophyll a was higher than that of chlorophyll b, respectively. Furthermore, the concentration of chlorophyll a, chlorophyll b, and total chlorophyll in the treatment of WS decreased significantly ($p < 0.05$) compared with WW (Table 1).

Table 1. Effect of drought stress on the concentration of chlorophyll in the pod wall.

Days after Pollination	Chlorophyll a (mg g^{-1})		Chlorophyll b (mg g^{-1})		Total Chlorophyll (mg g^{-1})	
	WW	WS	WW	WS	WW	WS
DAP5	0.314 [aA]	0.141 [aB]	0.137 [aA]	0.055 [aB]	0.451 [aA]	0.196 [aB]
DAP10	0.138 [bA]	0.085 [abB]	0.064 [bA]	0.044 [abA]	0.202 [bA]	0.128 [bB]
DAP15	0.094 [bA]	0.041 [bcB]	0.055 [bA]	0.025 [bcB]	0.149 [bA]	0.066 [cB]
DAP20	0.024 [cA]	0.012 [cB]	0.019 [cA]	0.010 [cB]	0.043 [cA]	0.022 [dB]

WW, well-watered; WS, water-stressed. Different small letters in the same column and different capitals in the same row meant a significant difference at the 0.05 probability level.

2.3. Changes of Photosynthetic Enzyme Activities in Pod Wall under Drought Stress

PEPC and RuBisCo activities in pod walls both present a declining trend with the pod development (Figure 3). However, there were different responses for PEPC and RuBisCo in the treatment of WS. As a comparison with WW, PEPC activities could be increased significantly ($p < 0.05$) in the treatment of WS, while RuBisCo activities were decreased.

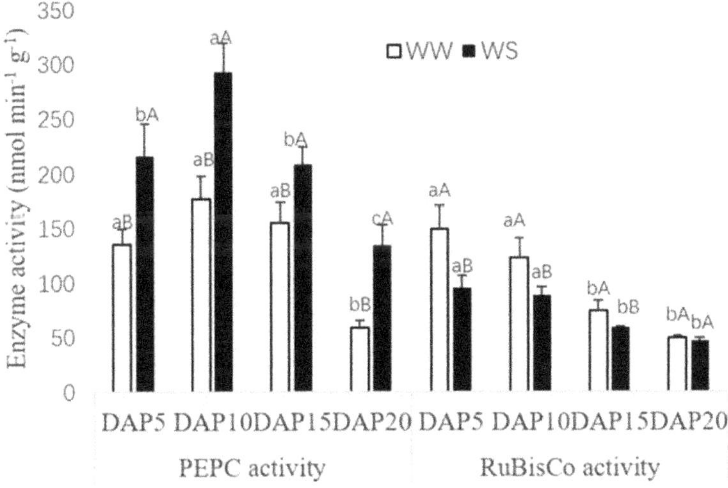

Figure 3. Effect of drought stress on photosynthetic enzyme activity (nmol min^{-1} g^{-1}) in the pod wall. Different small letters up the white bar and different capitals up the black bar within one photosynthetic enzyme mean a significant difference at the 0.05 probability level. WW, well-watered; WS, water-stressed.

2.4. Proteomic Analysis on the Response of Pod Wall to Drought Stress

According to the Medicago database, a total of 4215 proteins were identified in the samples of pod walls (Supplementary File 1). Under the WW, 373 proteins were significantly upregulated (fold change > 2.0, $p < 0.05$) for WW15 vs. WW10, and 101 proteins were significantly downregulated for WW20 vs. WW10 (fold change < 0.5, $p < 0.05$, Figure 4). For the WS, there were 184 upregulated proteins obtained in WS20 vs. WS10, and the number of downregulated proteins was lowest in WS20 vs. WS15. In addition, 144 proteins were significantly downregulated in WS10 vs. WW10, while 48 proteins were significantly downregulated in WS15 vs. WW15.

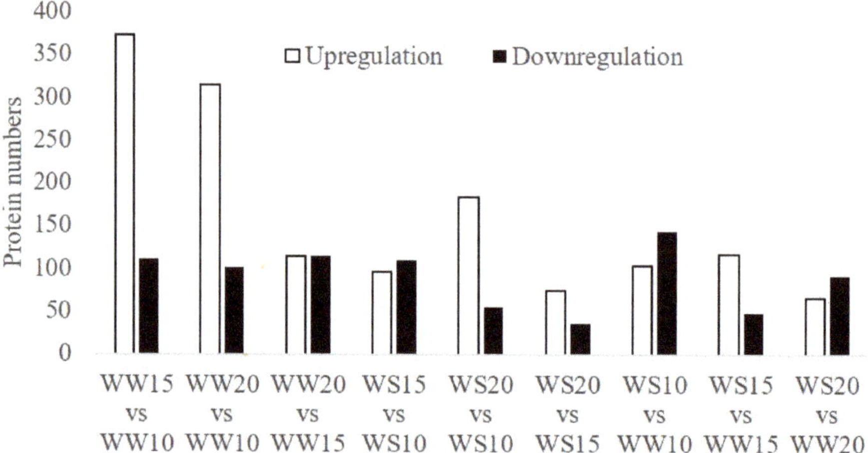

Figure 4. Number of proteins differently expressed in pod walls under drought stress. The white and black squares represent, respectively, significant upregulation and downregulation at the 0.05 probability level.

According to the Kyoto Encyclopedia of Genes and Genomes (KEGG) enrichment analysis, during pod development, most of the identified chlorophyll a–b binding proteins involved in the photosynthesis–antenna proteins were significantly downregulated in pod walls under WW or WS (Figure 5, Supplementary File 2). The synthesis of chlorophyll a–b binding proteins in the pod wall was restricted by the drought stress on DAP10 (Figure 5, Supplementary File 3).

Under the treatment of WS, some proteins, including ribulose-phosphate 3-epimerase, ctosolic fructose-1 6-bisphosphatase, fuctose-1, 6-bisphosphatase, fructose-bisphosphate aldolase, glyoxysomal malate dehydrogenase, malate dehydrogenase, malic enzyme, and sedoheptulose-1,7-bisphosphatase, which are involved in carbon fixation of photosynthetic organisms, were significantly upregulated among different growth stages after pollination (Figure 5, Supplementary File 2). In contrast, some proteins, including 26S proteasome on-ATPase regulatory subunit 6, aspartate aminotransferase, and glutamate-glyoxylate aminotransferase, were significantly downregulated at different durations after pollination. Furthermore, under the treatment of WW, proteins such as fructose-1, 7-bisphosphatase, fructose-bisphosphate aldolase, malic enzyme, and PEPC were significantly upregulated at different durations after pollination, while proteins of cytosolic triosephosphate isomerase, glyceraldehyde-3-phosphate dehydrogenase, and phosphoglycerate kinase were significantly downregulated.

The synthesis of cytosolic fructose-1 6-bisphosphatase, fructose-1, 6-bisphosphatase, fructose-bisphosphate aldolase, and malic enzyme was inhibited by the drought stress on DAP10, and the synthesis of glyoxysomal malate dehydrogenase and ribose-5-phosphate isomerase A were both inhibited on DAP15 and DAP20, respectively (Figure 5, Supplementary File 3). Nevertheless, aspartate aminotransferase on DAP15 and cytosolic triosephosphate isomerase and phosphoglycerate kinase on DAP20 were significantly induced by drought stress.

Proteins involved in photosynthesis, including cytochrome b6-f complex iron-sulfur subunit, F0F1 ATP synthase subunit gamma, oxygen-evolving complex/thylakoid lumenal 25.6 kDa protein, and photosystem II oxygen-evolving enhancer protein, were significantly upregulated in the pod wall at different durations after pollination under WS, while light-harvesting complex I chlorophyll a/b binding protein, oxygen-evolving enhancer protein, photosystem I P700 chlorophyll a apoprotein a2, and photosystem II D2 protein were significantly downregulated (Figure 5, Supplementary File 2). Under WW, there were some significantly downregulated proteins, including cytochrome b559 subunit alpha, cytochrome b6-f complex iron-sulfur subunit, light-harvesting complex I chlorophyll a/b binding protein, oxygen-evolving enhancer protein, and photosystem II D2 protein at different durations after pollination.

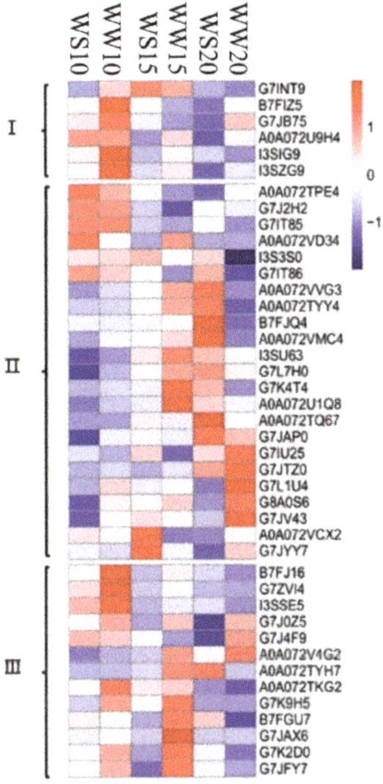

Figure 5. Cluster analysis of proteins involved in the significant pathway related to photosynthesis in pod walls at different durations after pollination under WW and WS. I, photosynthesis—antenna proteins; II, carbon fixation in photosynthetic organisms; III, photosynthesis.

For the treatment of WS, some proteins, including oxygen-evolving enhancer protein, photosystem II oxygen-evolving enhancer protein, photosystem I reaction center subunit II, and photosystem I reaction center subunit N in the pod wall on DAP10, were significantly upregulated (Figure 5, Supplementary File 3). However, there were some proteins presenting downregulation in the pod wall on DAP15, such as cytochrome b559 subunit alpha, cytochrome b6-f complex iron-sulfur subunit, ATP synthase subunit gamma, oxygen-evolving complex/thylakoid lumenal 25.6 kDa protein, photosystem I P700 chlorophyll a apoprotein a2, and photosystem I reaction center subunit II.

Meanwhile, some proteins were expressed differently and significantly in the amino sugar and nucleotide sugar metabolism, the ascorbate and aldarate metabolism, the beta-alanine metabolism, the carbon metabolism, the starch and sucrose metabolism, the citrate cycle, the glycine, serine, and threonine metabolism, the linoleic acid metabolism, oxidative phosphorylation, phagosome, plant–pathogen interaction, proteasome, and the alpha-linolenic acid metabolism in pod walls under drought stress (Table 2).

Table 2. Different expression of proteins involved in some pathways in the pod wall under drought stress.

KEGG	Accession	Proteins	Fold		
			WS10 vs. WW10	WS15 vs. WW15	WS20 vs. WW20
Amino sugar and nucleotide sugar metabolism	A0A072UKS2	PfkB family carbohydrate kinase	NS	NS	0.4
	A0A072VQZ5	UDP-D-apiose/UDP-D-xylose synthase	NS	NS	0.2
	G7JUS9	UDP-glucuronic acid decarboxylase	NS	3.3	0.4
	G7ID31	Chitinase	NS	NS	5.0
	G7LA76	Chitinase (Class Ib)/Hevein	NS	NS	3.3
Ascorbate and aldarate metabolism	A0A072TLF4	Myo-inositol oxygenase	0.4	NS	NS
	A0A072U2G7	NAD-dependent aldehyde dehydrogenase family protein	0.2	3.3	NS
	A0A072UQP6	UDP-glucose 6-dehydrogenase	2.5	NS	NS
	G7L571	UDP-glucose 6-dehydrogenase	NS	2.5	NS
	A0A072V120	UTP-glucose-1-phosphate uridylyltransferase	NS	2.5	NS
	A0A072V151	L-ascorbate oxidase	5.0	NS	NS
	A0A072VNM9	GME GDP-D-mannose-3, 5-epimerase	NS	3.3	0.1
	G7L1 × 0	GME GDP-D-mannose-3, 5-epimerase	2.5	NS	0.3
	G7JTZ5	Aldo/keto reductase family oxidoreductase	0.5	NS	NS
	G7KAG7	Thylakoid lumenal 29 kDa protein	0.3	NS	NS
beta-Alanine metabolism	A0A072UCM6	Glutamate decarboxylase	NS	NS	0.2
Carbon metabolism	G7IT85	Phosphoglycerate kinase	NS	NS	3.3
	G7IT86	Phosphoglycerate kinase	NS	NS	3.3
	G7KJZ8	Glucose-6-phosphate isomerase	NS	NS	0.4
	G7L1U4	Ribose-5-phosphate isomerase A	NS	NS	0.4
	I3S3S0	Cytosolic triosephosphate isomerase	NS	NS	2.0
	A0A072VS77	Methylenetetrahydrofolate reductase	NS	3.3	0.3
Starch and sucrose metabolism	A0A072UCM8	Phosphotransferase	0.3	NS	NS
	A0A072UKS2	PfkB family carbohydrate kinase	NS	NS	0.4
	A0A072UU47	Glycoside hydrolase family 1 protein	3.3	NS	NS
	A0A072VLQ9	Starch synthase	2.5	NS	NS
	G7IJV7	Glycoside hydrolase family 3 protein	0.3	NS	NS
	G7KJZ8	Glucose-6-phosphate isomerase	NS	NS	0.4

Table 2. *Cont.*

KEGG	Accession	Proteins	Fold		
			WS10 vs. WW10	WS15 vs. WW15	WS20 vs. WW20
Citrate cycle (TCA cycle)	G7KVS0	E1 subunit-like 2-oxoglutarate dehydrogenase	NS	NS	0.3
	G7JYQ8	Aconitate hydratase	NS	NS	0.3
	B7FJJ4	Pyruvate dehydrogenase E1 beta subunit	NS	2.0	2.5
	G7KHI5	Isocitrate dehydrogenase [NADP]	NS	NS	0.2
	A2Q2V1	ATP-citrate lyase/succinyl-CoA ligase	NS	2.5	0.1
Glycine, serine and threonine metabolism	A0A072URB1	Amine oxidase	NS	NS	5.0
	A0A072V290	Amine oxidase	NS	NS	3.3
	G7J7B0	Amine oxidase	NS	NS	5.0
	A9YWS0	Serine hydroxymethyltransferase	NS	NS	0.2
	G7I9Z0	Glycine dehydrogenase [decarboxylating] protein	NS	NS	0.4
	G7JJ96	Aminomethyltransferase	NS	NS	0.3
	G7JNS2	NAD-dependent aldehyde dehydrogenase family protein	NS	NS	0.4
	G7L9H1	Phosphoserine aminotransferase	NS	NS	0.3
Linoleic acid metabolism	A0A072UMH4	Lipoxygenase	0.5	NS	NS
	G7J629	Lipoxygenase	0.4	NS	NS
	G7LIX7	Lipoxygenase	5.0	NS	NS
	G7LIY0	Lipoxygenase	5.0	NS	0.4
	G7LIY2	Lipoxygenase	10.0	NS	NS
	G7J632	Lipoxygenase	2.5	2.5	NS
Oxidative phosphorylation	A0A072URM9	Archaeal/vacuolar-type H+-ATPase subunit A	NS	2.0	NS
	A0A072V4G2	F0F1 ATP synthase subunit gamma	NS	0.5	NS
	A0A072W1H5	ATP synthase subunit beta	NS	2.5	NS
	A0A126TGR5	ATP synthase subunit alpha	NS	2.0	NS
	B7FN64	NADH dehydrogenase	NS	0.4	NS
	G7JIL4	V-type proton ATPase subunit a	NS	2.5	0.3
	G7I9M9	ATP synthase D chain	NS	5.0	NS

Table 2. *Cont.*

KEGG	Accession	Proteins	Fold		
			WS10 vs. WW10	WS15 vs. WW15	WS20 vs. WW20
Phagosome	A0A072VSL4	Archaeal/vacuolar-type H+-ATPase subunit B	NS	NS	0.4
	B7FMK2	Archaeal/vacuolar-type H+-ATPase subunit E	NS	NS	3.3
	G7KSI7	Archaeal/vacuolar-type H+-ATPase subunit B	NS	NS	0.3
	G7LIN7	Tubulin beta-1 chain	NS	NS	5.0
Plant–pathogen interaction	B7FNA2	EF hand calcium-binding family protein	NS	NS	2.5
	G7I7Q4	Heat shock protein 81-2	NS	2.5	0.2
	G7IDZ4	Heat shock protein 81-2	NS	NS	0.1
	A0A072U9J1	Heat shock protein 81-2	2.5	5.0	0.3
Proteasome	A0A072TQB8	Glyceraldehyde-3-phosphate dehydrogenase	NS	NS	10
	B7FGZ8	Proteasome subunit beta type	NS	NS	2.5
	G7JTX3	6S proteasome regulatory subunit S2 1B	NS	5.0	0.3
	I3RZQ6	Proteasome subunit alpha type	NS	NS	3.3
	I3SSX1	Proteasome subunit alpha type	NS	NS	2.5
alpha-linolenic acid metabolism	G7J5N1	Uncharacterized protein	3.3	NS	NS
	Q711Q9	Allene oxide cyclase	2.5	0.1	NS

Fold change over 2.0 means significant ($p < 0.05$) upregulation and below 0.5 means significant ($p < 0.05$) downregulation. NS, nonsignificant.

3. Discussion

3.1. Observation of Surface and Ultrastructure in the Pod Wall

Stomata were distinctly observed on the outer surface of the pod wall (Figure 1). Stoma acted in respiration and transpiration and allowed CO_2 to enter for operating photosynthesis as well. Previous research has shown that stoma was also found on other NLGOs, such as the exposed peduncles of wheat [22], the silique shell of oilseed rape [4], and the capsule wall of castor [5]. In addition, chloroplast, the important site for doing the light reaction of photosynthesis, was found in cells of the pod wall (Figure 2). At the early stage of pod development, the structure of chloroplasts in the pod wall is well-organized and intact, and the photosynthate, starch grains, are already observed in the chloroplasts (Figure 2A,E). Similar results were reported when observing the ultrastructure of the pod wall of pea [27] and chickpea [6]. Under WW, the pod wall at the late stage even could produce starch grains (Figure 2D). NLOG could maintain functional activity at the late stage when the photosynthetic activity of leaves declined [8,22,28]. However, drought stress could damage the structure of chloroplast membranes and thylakoids, and few starch grains were produced (Figure 2H). In addition, the epidermal hair, existing on the outer surface of the pod wall, likely acts to prevent damage from direct sunlight and protect against water loss.

3.2. Response of Chlorophyll Concentration and Photosynthetic Enzyme Activities

In the pod wall of alfalfa, chlorophyll a and chlorophyll b could be detected, and the concentration of chlorophyll a, chlorophyll b and total chlorophyll decreased with the development of pods (Table 1). Similarly, the concentration of chlorophyll a, chlorophyll b, and total chlorophyll in cotton (*Gossypium hirsutum*) leaves and NLGOs, including bracts, stems and boll shells, decreased with bolls developing, and the decreasing rate of the concentration in stems and boll shells was lower than that in leaves at the late stage of boll [28]. The chlorophyll biosynthesis was inhibited in the pod wall under drought stress (Table 1), while the content change was lower in NLOG than in leaves to maintain relatively high photosynthetic capacity [15,20,23]. Except for chlorophyll, photosynthetic enzymes are crucial for operating photosynthesis as well. The activities of key enzymes in the C3 and C4 cycles in the pod wall were determined in the present study and the activities of PEPC and RuBisCo both decreased with the development of pods (Figure 3). In addition, the activity of PEPC was higher than that of RuBisCo, and drought stress could induce the activity of PEPC (Figure 3). Previous studies reported that PEPC could make more contributions to photosynthesis than RuBisCo in NLGOs [22]. Although the activity of RuBisCo decreased under drought stress in NLGOs, the increasing activity of PEPC could, in part, compensate to ensure dry matter production [21,23].

3.3. The Differential Expression of Proteins under Drought Stress

3.3.1. Photosynthesis-Antenna Proteins

Six proteins identified in the pod wall were chlorophyll a–b binding proteins, which are the apoproteins of the light-harvesting complex of photosystem II that existed on the membrane of chloroplasts [29]. I3SZG9 (Lhca3) is the PSI inner antenna protein (LCHI); Lhcb 1, Lhcb 4, Lhcb 5, and Lhcb 6 belong to the PSII inner antenna proteins (LCHII). In the plants, 50% of chlorophyll associated with LCHII play important roles in the regulation of light energy distribution and photoprotective reaction. In this study, two types of antenna proteins identified in the pod wall were downregulated with the development of pod, and their synthesis was limited by the drought stress (Figure 5; Table 1). Similarly, the antenna proteins were downregulated in sugarcane (*Saccharum officinarum*) [30] and cucumber (*Cucumis sativus*) [31] under drought stress, which implied that drought stress impeded the synthesis of antenna proteins, suppressing the absorption of light. In addition, the downregulation of these proteins could decrease energy and substance consumption to promote the operation of other physiological activities for resisting drought stress [32].

3.3.2. Photosynthesis

According to KEGG enrichment analysis for photosynthesis, the identified protein complex participates in the reactions taking place on the thylakoid membrane. Proteins, including PetC, PsbS, PsbQ, PsbO, and PsbC, were downregulated under WW (Figure 5, Supplementary File 2), which means that the photosynthetic ability of the pod wall decreases as the pod developes. In addition, the PSII components, including PsbE, PsbP, and PsbO, were downregulated under drought stress (Figure 5, Supplementary File 2). PsbE is cytochrome b559 subunit α, and the set of these three proteins, PsbP, PsbO, and PsbQ, was bound to the luminal surface of PSII, oxidizing water molecules to release O_2 [33]. PsaD, PsaN, and PsaB are subunits of the PSI complex, and their synthesis is suppressed by the drought stress as well. F0F1 ATP synthase subunit γ (F0F1-ATPases), largely existing in chloroplasts, mitochondria, and cell nuclei, was downregulated in this study, which resulted in the reduction of ATP synthesis [34] and the decline of photosynthetic ability [35]. The downregulation of F0F1-ATPases by drought stress was also found in poplar (*Populus yunnanensis*) [36] and soybean [37]. Drought stress damages the electron transfer system on the thylakoid to reduce the photosynthetic ability of the pod wall through suppressing the synthesis of F0F1-ATPases, cytochrome b6-f complex, and the proteins involved in the photosystem [37].

3.3.3. Carbon Fixation in Photosynthetic Organisms

Under WW, PEPC, the key enzyme in the C4 cycle, was significantly upregulated in the pod wall on DAP15, compared with that on DAP10 (Figure 5, Supplementary File 2). Drought stress inhibited the synthesis of RuBisCo in leaves in alfalfa [38]. The induced proteins in the C4 cycle could compensate for the decrease of photosynthetic ability resulting from the inhibition of proteins in the C3 cycle under drought stress [39]. NLGOs of C3 plants might operate C3–C4 intermediate photosynthesis or C4-similar photosynthesis [11]. Two cell types (mesophyll and Kranz cells) were localized in the ears of wheat [11], in which the activity of PEPC was higher than in flag leaves, and the activity of PEPC was higher than that of RuBisCo [40]. In addition, compared with WW, another C4-pathway enzyme, aspartate aminotransferase, was significantly upregulated in the pod wall on DAP15 under WS (Figure 5, Supplementary File 3). The upregulation of aspartate aminotransferase contributed to increasing the stress resistance of plants and maintaining high photosynthetic abilities under stress [41]. Some enzymes participating in the reduction period of the C3 cycle were upregulated under drought stress, while others involved in the regeneration period of RuBisCo were downregulated (Figure 5, Supplementary File 3). Fructose-1, 6-bisphosphatase and fructose-bisphosphate aldolase were downregulated in pod walls on DAP10 under WS in comparison with WW, while triosephosphate isomerase and phosphoglycerate kinase were upregulated in pod walls on DAP20. Phosphoglycerate kinase, belonging to an upstream acting enzyme in the C3 cycle, had interaction with PEPC and aspartate aminotransferase in the C4 cycle under drought stress, which was found in maize [42]. Drought stress could induce the synthesis of triosephosphate isomerase in rice [43] and maize [44] to ensure the operation of photosynthesis.

3.3.4. Carbohydrate Metabolism

Starch grains were observed to be filling the chloroplasts in the pod wall on DAP10 under WS (Figure 2), which meant that starch synthase was induced in pod walls on DAP10 by drought stress (Table 2). The recent research reported that total carbohydrate content decreased in the pod wall of soybean under drought stress, but not the starch content [7]. Enhance expression of starch synthase presented the protective response to drought stress in the pod wall.

Identified proteins, including pyruvate dehydrogenase (PDH) E1 beta subunit and ATP-citrate lyase/succinyl (ACL)-CoA ligase involved in the citrate cycle, were upregulated in pod walls on DAP15 under WS (Table 2). PDH produces chemical energy, and drought stress could promote the expression of relative encoding genes in rice [45]. PDH is one component of the pyruvate dehydrogenase complex

(PDC) that can oxidize pyruvate into acetyl-CoA and NADH. In addition, the overexpression of the *ACL* gene could enhance drought resistance in tobacco (*Nicotiana tabacum*) [46]. Some other proteins involved in the citrate cycle, including methylenetetrahydrofolate reductase, E1 subunit-like 2-oxoglutarate dehydrogenase, aconitate hydratase, isocitrate dehydrogenase [NADP] and ATP-citrate lyase/succinyl-CoA ligase, were downregulated in pod walls on DAP20 under drought stress (Table 2). The downregulation of these proteins impedes the carbohydrate metabolism in the pod wall at the late growth stage. Previous research has shown that the stagnate of carbohydrate metabolism in the plants under drought stress could cause the accumulation of sugar [47], which contributes to improving the osmotic potential to enhance drought tolerance [38].

3.3.5. Energy Metabolism

Identified proteins, including archaeal/vacuolar-type H+-ATPase subunit A, ATP synthase subunit beta, ATP synthase subunit alpha, and ATP synthase D chain, were upregulated in pod walls on DAP15 (Table 2). Budak et al. reported that ATP synthase subunit CF1 and ATP synthase subunit alpha had a higher expression level in wild wheat, with stronger drought resistance [48]. Under drought stress, the upregulation of these proteins involved in ATP synthesis could ensure energy metabolism in the plants for maintaining the operation of main physiological activities [49]. Photosynthesis is sensitive to drought stress, while respiration is not. The normal operation of respiration could provide the necessary energy for decreasing the damage from drought stress [7].

3.3.6. Other Metabolism

Chitinase plays an important role in resistance to stress [50,51]. The activity of chitinase is low in the plant, but drought stress can induce the expression of the chitinase gene in wheat [50] and faba beans (*Vicia faba*) [51]. Similar results were found in the present study, where chitinase was upregulated in the pod wall on DAP15 under drought stress (Table 2). Heat shock protein 90 (Hsp90) complex regulated proteins fold and degrade and maintain stable plant cells [52]. Heat shock protein 81-2 (Hsp81-2) is a member of Hsp90 family and can be induced by drought stress in arabidopsis (*Arabidopsis thaliana*) [53]. In the present study, Hsp81-2 involved in plant–pathogen interaction was identified and upregulated in the pod wall on DAP10 and DAP15 under drought stress. Most of the lipoxygenases (LOXs) identified were upregulated in the pod wall on DAP10 (Table 2). LOX participates in many activities in the plants, and drought stress caused the rapid accumulation of LOX mRNA in barley [54]. The overexpression of *CaLOX1* enhanced the resistance to drought stress in arabidopsis [55].

In summary, the structure of chloroplast in the pod wall is damaged at the late stage of development under drought stress, but not at the early stage. The synthesis of some proteins involved in photosystem I, photosystem II, and the regeneration period of RuBisCo in the pod wall at the early stage and TCA cycle at the late stage are impeded under drought stress (Figure 6). Nevertheless, drought stress can induce the activity of PEPC and promote the synthesis of some proteins participating in the pathway of the C4 cycle and energy metabolism at the early stage and the reduction period of RuBisCo at the late stage.

This study provides the ultrastructural, physiological, and proteomic changes in alfalfa pod walls under drought stress. The results suggest that the pod wall shows the capability of conducting photosynthesis and regulating the C4 photosynthetic pathway, ATP synthesis, and resistance metabolism to ensure the operation of physiological reactions under drought stress.

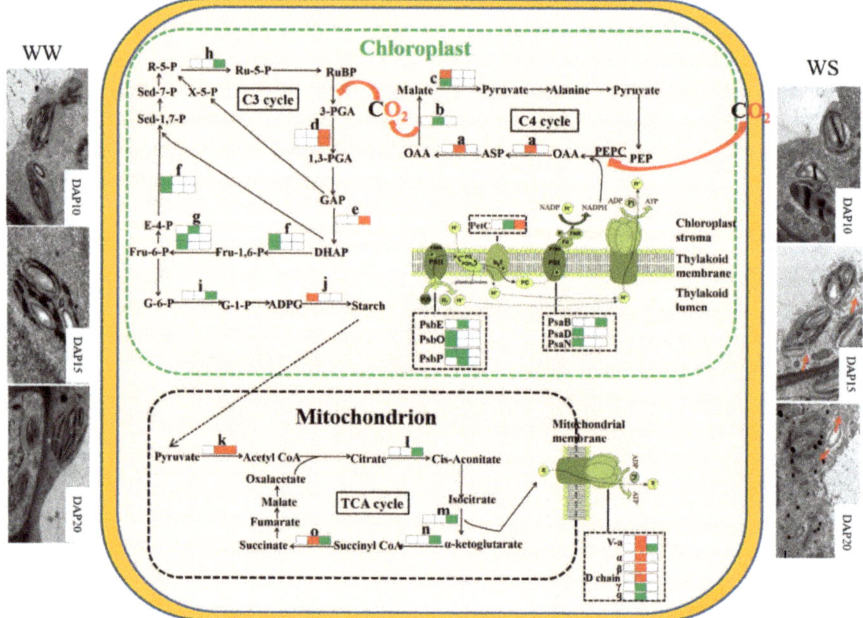

Figure 6. The pathways of proteomic mechanisms in the pod wall under drought stress. a, Aspartate aminotransferase; b, malate dehydrogenase; c, malic enzyme; d, phosphoglycerate kinase; e, triosephosphate isomerase; f, fructose-1,6-bisphosphate aldolase; g, fructose-1,6-bisphosphatase; h, ribose-5-phosphate isomerase; i, glucose-6-phosphate isomerase; j, starch synthase; k, pyruvate dehydrogenase E1 beta subunit; l, aconitate hydratase; m, isocitrate dehydrogenase (NADP); n, E1 subunit-like 2-oxoglutarate dehydrogenase, o, ATP-citrate lyase/succinyl-CoA ligase. Three squares from left to right represent WS10 vs. WW10, WS15 vs. WW15, and WS20 vs. WW20, respectively. The red squares represent significant upregulation at the 0.05 probability level. The green squares represent significant low-regulation at the 0.05 probability level. The white squares represent no significance. The red arrows show the damaged part of the chloroplast membrane in the pod wall under WS. Define the Dotted black arrow if possible

4. Material and Method

4.1. Material

M. sativa cv. Zhongmu No. 2 seeds were sown in the plastic pots (weight 0.5 kg, height 30 cm, base diameter 19 cm, and top diameter 25 cm) in a greenhouse in October 2015. Each pot was filled with a mixture soil of vermiculite, peat, and black soil by 2:1:1 (total soil weight 3.5 kg, water content 23%). Four seedlings with a similar growth status were kept in each pot when the seedlings' height was around 10 cm. They were equivalently and adequately watered every two days. All plants were cut till 10 cm, and then pots were moved out of and nearby the greenhouse, without any shelter, in April 2016. Every day, each pot was weighed with an electronic scale and watered to 7 kg. From 18 April (before visible bud stage), a drought stress treatment, denoted by WS, was started by watering each pot to 4.5 kg, while the control, denoted by WW, was still watered to 7 kg. Plants in WW and WS treatments were pollinated artificially every 5 days from 6 June till 21 June. After pollination, the flowers pollinated were marked with hang tags. On 26 June, the pod walls of marked pods with 3 replicates were collected on day 5 (DAP5), day 10 (DAP10), day 15 (DAP15), and day 20 (DAP20) after pollination under WW and WS treatments.

4.2. Surface and Ultrastructure Characteristics Observation for Pod Wall

The pod walls of DAP10 under WW and WS treatments were used as samples to observe the surface characteristics by using a scanning electron microscope (manufacture Hitachi S-570, city Japan). The pod walls of DAP5, DAP10, DAP15, and DAP20 under WW and WS treatments were used as samples for taking images with a transmission electron microscope (manufacture Hitachi H-7500, city Japan) to observe ultrastructure characteristics. Pretreatments of the pod walls for observation by using the scanning and transmission electron microscopes were conducted according to [56].

4.3. Chlorophyll Concentration of Pod Walls Measurement

The pod walls on DAP5, DAP10, DAP15, and DAP20 under WW and WS treatments were sampled and cut into filaments, respectively. Chlorophyll a and chlorophyll b concentrations in each sample with three repetitions were determined by soaking in extracting solution, filling in the 10 mL centrifuge tube. The extracting solution was the mixture of acetone and absolute ethyl alcohol by the volume rate of 2:1. Absorbancy of extracting solution at 663 and 645 nm was determined by using a ultraviolet spectrophotometer (manufacture UH5300, city Japan) to calculate the concentration of chlorophyll a and chlorophyll b. The gross chlorophyll concentration was the sum of chlorophyll a and chlorophyll b.

4.4. Photosynthetic Enzyme Activities Assays

The pod walls on DAP5, DAP10, DAP15, and DAP20 under WW and WS treatments were sampled with three repetitions. Each sample was homogenized in extracting solutions using a pestle and mortar in ice. The extracting solution contained 1 M Tris-H_2SO_4 (pH 7.8), 5% glycerol, 7 mM DTT, and 1 mM EDTA. The homogenate was filtered, and then the filtrate was centrifuged at 8000× g for 10 min at 4 °C. The supernatant was saved in an ice bath for subsequent enzyme assay. Photosynthetic enzyme activities were determined using an ultraviolet spectrophotometer (UH5300, Hitachi, Japan) at 340 nm. The reaction solution for phosphoenolpyruvate carboxylase (PEPC) contained 0.1 mol L^{-1} Tris-H_2SO_4 (pH 9.2), 0.1 M $MgCl_2$, 100 mM $NaHCO_3$, 40 mM PEP, 1 mg mL^{-1} NADH and MDH [57]. The reaction mixture for ribulose-1,5-bisphosphate carboxylase (RuBisCo) contained 100 mM Tris-HCl (pH 7.8), 160 U/mL CPK, 160 U mL^{-1} GAPDH, 50 mM ATP, 50 mM phosphocreatine, and 160 U mL^{-1} phosphoglyceric kinase [58].

4.5. Proteomic Analysis

The pod walls on DAP10, DAP15, and DAP20 under WW (denoted by WW10, WW15, and WW20, respectively) and WS (denoted by WS10, WS15, and WS20, respectively) treatments were collected and stored at −80 °C. Then, 0.05 g tissues of each sample with 3 repetitions were ground in liquid nitrogen before the addition of 200 μL plant total protein lysis buffer containing 20 mM Tris-HCl (pH7.5), 250 mM sucrose, 10 mM EGTA, 1% Triton X-100, protease inhibitor, and 1 M DTT. The mixture was incubated on ice for 20 min. Plant cell debris was removed via centrifugation at a speed of relative centrifugal force of 15,000× g for 15 min at 4 °C. The supernatant was collected, and the rest was centrifugated again, as above. The protein concentration was determined with a Bio-Rad Protein Assay kit based on the Bradford method, using BSA as a standard at a wavelength of 595 nm. All independent protein extractions were performed.

Briefly, 60 μg protein samples were reduced with 5 μL 1 M DTT for 1 h at 37 °C, alkylated with 20 μL 1 M iodoacetamide (IAA) for 1 h in the dark, and then digested with sequencing-grade modified trypsin (Promega) for 20 h at 37 °C. Digested peptides were separated with chromatography using an Easy-nLC1000 system (Thermo Scientific) autosampler. The peptide mixture was loaded on a self-made C18 trap column (C18 3 μm, 0.10 × 20 mm) in solution A (0.1% formic acid), then separated with a self-made Capillary C18 column (1.9 μm, 0.15 × 120 mm), with a gradient solution B (100% acetonitrile and 0.1% formic acid) at a flow rate of 600 nL/min. The gradient consisted of the following steps: 0–10% solution B for 16 min, 10–22% for 35 min, 22–30% for 20 min, then increasing to 95% solution B in 1 min and holding for 6 min. Separated peptides were examined in the Orbitrap Fusion mass spectrometer (Thermo Scientific, Waltham, MA, USA), with a Michrom captive spray nanoelectrospray ionization (NSI) source. Spectra were scanned over the m/z range 300–4000 Da at 120,000 resolution. An 18-s exclusion time and 32% normalization collision energy were set at the dynamic exclusion window.

4.6. Statistical Analysis

The significance of differences between mean values of physiological parameters, including the chlorophyll concentration and the enzyme activities under WW and WS, were analyzed using an LSD test by software SAS version 8.0.

RAW files of mass spectrometry were extracted using the MASCOT version 2.3.02 (Matrix Science, London, UK). Mass spectrometry data were searched, identified, and quantitatively analyzed using the software of Sequest HT and Proteome Discover 2.0 (Thermo Scientific). The database used in this study was uniprot-Medicago.fasta. Protein species with at least two unique peptides were selected for protein species quantitation, and the relative quantitative protein ratios between the two samples were calculated by comparing the average abundance values (three biological replicates). Protein species detected in only one material (A-line or B-line), with at least two replicates considered to be presence/absence protein species. Additionally, Student's *t*-tests were performed to determine the significance of changes between samples. A fold-change of >2 and *p*-value < 0.05 in at least two replicates were used as the thresholds to define differently accumulated protein species.

Supplementary Materials: Supplementary materials can be found at http://www.mdpi.com/1422-0067/21/12/4457/s1.

Author Contributions: H.W. and P.M. designed the experiments; H.W. did the experiments wrote the manuscript. Q.Z. and P.M. reviewed and edited the manuscript. All authors have read and agreed to the published version of the manuscript.

Funding: This research was financially supported by the China Agriculture Research System (CARS-34) and the earmarked fund for the Beijing Common Construction Project.

Conflicts of Interest: The authors declare no conflict of interest.

References

1. Loomis, R.S.; Williams, W.A. Maximum crop productivity: An extimate. *Crop Sci.* **1963**, *3*, 67–72. [CrossRef]
2. Shen, Y. *The Most Improtant Chemistry Reaction on the Earth-Photosynthesis*; Tsinghua University Press: Beijing, China, 2000; pp. 2–13.
3. Long, S.; Marshall-Colon, A.; Zhu, X. Meeting the global food demand of the future by engineering crop photosynthesis and yield potential. *Cell* **2015**, *161*, 56–66. [CrossRef] [PubMed]
4. Wang, C.; Yang, J.; Hai, J.; Chen, W.; Zhao, X. Photosynthetic features of leaf and silique of 'Qinyou 7' oilseed rape (*Brassica napus* L.) at reproductive growth stage. *Oil Crop. Sci.* **2018**, *3*, 176.
5. Zhang, Y.; Mulpuri, S.; Liu, A. Photosynthetic capacity of the capsule wall and its contribution to carbon fixation and seed yield in castor (*Ricinus communis* L.). *Acta Physiol. Plant.* **2016**, *38*, 245. [CrossRef]
6. Furbank, R.T.; White, R.; Palta, J.A.; Turner, N.C. Internal recycling of respiratory CO_2 in pods of chickpea (*Cicer arietinum* L.): The role of pod wall, seed coat, and embryo. *J. Exp. Bot.* **2004**, *55*, 1687–1696. [CrossRef]
7. Sengupta, D.; Kariyat, D.; Marriboina, S.; Reddy, A.R. Pod-wall proteomics provide novel insights into soybean seed-filling process under chemical-induced terminal drought stress. *J. Sci. Food Agric.* **2019**, *99*, 2481–2493. [CrossRef] [PubMed]
8. Wang, H.; Hou, L.; Wang, M.; Mao, P. Contribution of the pod wall to seed grain filling in alfalfa. *Sci. Rep.* **2016**, *6*, 26586. [CrossRef]
9. Tambussi, E.A.; Bort, J.; Guiamet, J.J.; Nogués, S.; Araus, J.L. The photosynthetic role of ears in C3 cereals: Metabolism, water use efficiency and contribution to grain yield. *Crit. Rev. Plant Sci.* **2007**, *26*, 1–16. [CrossRef]
10. Jiang, Q.; Roche, D.; Durham, S.; Hole, D. Awn contribution to gas exchanges of barley ears. *Photosynthetica* **2006**, *44*, 536–541. [CrossRef]

11. Baluar, N.; Badicean, D.; Peterhaensel, C.; Mereniuc, L.; Vorontsov, V.; Terteac, D. The peculiarities of carbon metabolism in the ears of C3 cereals CO2 exchange kinetics, chloroplasts structure and ultra-structure in the cells from photosynthetic active components of the ear. *J. Tissue Cult. Bioeng.* **2018**, *1*, 1–14.

12. AuBuchon-Elder, T.; Coneva, V.; Goad, D.M.; Allen, D.K.; Kellogg, E.A. Sterile spikelets assimilate carbon in sorghum and related grasses. *BioRxiv* **2018**. [CrossRef]

13. Ávila-Lovera, E.; Zerpa, A.J.; Santiago, L.S. Stem photosynthesis and hydraulics are coordinated in desert plant species. *New Phytol.* **2017**, *216*, 1119–1129. [CrossRef] [PubMed]

14. Kitaya, Y.; Yabuki, K.; Kiyota, M.; Tani, A.; Hirano, T.; Aiga, I. Gas exchange and oxygen concentration in pneumatophores and prop roots of four mangrove species. *Trees-Struct. Funct.* **2002**, *16*, 155–158. [CrossRef]

15. Zhang, C.; Zhan, D.; Luo, H.; Zhang, Y.; Zhang, W. Photorespiration and photoinhibition in the bracts of cotton under water stress. *Photosynthetica* **2016**, *54*, 12–18. [CrossRef]

16. Li, Y.; Li, H.; Li, Y.; Zhang, S. Improving water-use efficiency by decreasing stomatal conductance and transpiration rate to maintain higher ear photosynthetic rate in drought-resistant wheat. *Crop J.* **2017**, *5*, 231–239. [CrossRef]

17. Abebe, T.; Melmaiee, K.; Berg, V.; Wise, R.P. Drought response in the spikes of barley: Gene expression in the lemma, palea, awn, and seed. *Funct. Interg. Genomic.* **2010**, *10*, 191–205. [CrossRef] [PubMed]

18. Zhang, Y.; Wang, Z.; Wu, Y.; Zhang, X. Stomatal characteristics of different green organs in wheat under different irrigation regimes. *Acta Agron. Sin.* **2006**, *32*, 70–75.

19. Hu, Y.; Zhang, Y.; Yi, X.; Zhan, D.; Luo, H.; Soon, C.W.; Zhang, W. The relative contribution of non-foliar organs of cotton to yield and related physiological characteristics under water deficit. *J. Integr. Agric.* **2014**, *13*, 975–989. [CrossRef]

20. Lou, L.; Li, X.; Chen, J.; Li, Y.; Tang, Y.; Lv, J. Photosynthetic and ascorbate-glutathione metabolism in the flag leaves as compared to spikes under drought stress of winter wheat (*Triticum aestivum* L.). *PLoS ONE* **2018**, *13*, e0194625. [CrossRef]

21. Wei, A.; Wang, Z.; Zhai, Z.; Cong, Y. Effect of soil drought on C4 photosynthetic enzyme activities of flag leaf and ear in wheat. *Agric. Sci. China* **2003**, *36*, 508–512.

22. Kong, L.; Wang, F.; Feng, B.; Li, S.; Si, J.; Zhang, B. The structural and photosynthetic characteristics of the exposed peduncle of wheat (*Triticum aestivum* L.): An important photosynthate source for grain-filling. *BMC Plant Biol.* **2010**, *10*, 141. [CrossRef]

23. Jia, S.; Lv, J.; Jiang, S.; Liang, T.; Liu, C.; Jing, Z. Response of wheat ear photosynthesis and photosynthate carbon distribution to water deficit. *Photosynthetica* **2015**, *53*, 95–109. [CrossRef]

24. Rangan, P.; Furtado, A.; Henry, R.J. New evidence for grain specific C4 photosynthesis in wheat. *Sci. Rep.* **2016**, *6*, 31721. [CrossRef] [PubMed]

25. Singal, H.R.; Sheoran, I.S.; Singh, R. In vitro enzyme activities and products of $^{14}CO_2$ assimilation in flag leaf and ear parts of wheat (*Triticum aestivum* L.). *Photosynth. Res.* **1986**, *8*, 113–122. [CrossRef]

26. Gu, S.; Yin, L.; Wang, Q. Phosphoenolpyruvate carboxylase in the stem of the submersed species Egeria densa may be involved in an inducible C 4-like mechanism. *Aquat. Bot.* **2015**, *125*, 1–8. [CrossRef]

27. Atkins, C.A.; Kuo, J.; Pate, J.S.; Flinn, A.M.; Steele, T.W. Photosynthetic pod wall of pea (*Pisum sativum* L.) distribution of carbon dioxide-fixing enzymes in relation to pod structure. *Plant Physiol.* **1977**, *60*, 779–786. [CrossRef]

28. Hu, Y.; Zhang, Y.; Luo, H.; Li, W.; Oguchi, R.; Fan, D.; Soon, C.W.; Zhang, W. Important photosynthetic contribution from the non-foliar green organs in cotton at the late growth stage. *Planta* **2012**, *235*, 325–336. [CrossRef]

29. Liu, R.; Xu, Y.; Jiang, S.; Lu, K.; Lu, Y.; Feng, X.; Zhen, W.; Shan, L.; Yu, Y.; Wang, X.; et al. Light-harvesting chlorophyll a/b-binding proteins, positively involved in abscisic acid signalling, require a transcription repressor, WRKY40, to balance their function. *J. Exp. Bot.* **2013**, *64*, 5443–5456. [CrossRef] [PubMed]

30. Liang, P. Physiological and Molecular Basis of Drought Resistance Enhanced by Si Application in Sugarcane. Ph.D. Thesis, Guangxi University, Nanning, China, 2012.

31. Cui, Q. Analysis of the Chloroplast Proteome of Cucumber Leaves under Elevated CO_2 Concentration and Drought Stress. Masters' Thesis, Shandong Agricultural University, Taian, China, 2017.

32. Hossain, M.A.; Wani, S.H.; Bhattacharjee, S.; Burritt, D.J.; Tran, L.S.P. *Drought Stress Tolerance in Plants, Physiology and Biochemistry*; Springer: Berlin/Heidelberg, Germany, 2016; Volume 1, pp. 155–157.

33. Ifuku, K. Localization and functional characterization of the extrinsic subunits of photosystem II: An update. *Biosci. Biotechnol. Biochem.* **2015**, *79*, 1223–1231. [CrossRef] [PubMed]

34. Caruso, G.; Cavaliere, C.; Foglia, P.; Gubbiotti, R.; Samperi, R.; Laganà, A. Analysis of drought responsive proteins in wheat (*Triticum durum*) by 2D-PAGE and MALDI-TOF mass spectrometry. *Plant Sci.* **2009**, *177*, 570–576. [CrossRef]

35. Xiao, X.; Yang, F.; Zhang, S.; Korpelainen, H.; Li, C. Physiological and proteomic responses of two contrasting *Populus cathayana* populations to drought stress. *Physiol. Plant.* **2009**, *136*, 150–168. [CrossRef]

36. Li, X.; Yang, Y.; Sun, X.; Lin, H.; Chen, J.; Ren, J.; Hu, X.; Yang, Y. Comparative physiological and proteomic analyses of poplar (*Populus yunnanensis*) plantlets exposed to high temperature and drought. *PLoS ONE* **2014**, *9*, e107605. [CrossRef]

37. Das, A.; Eldakak, M.; Paudel, B.; Kim, D.W.; Hemmati, H.; Basu, C.; Rohila, J.S. Leaf proteome analysis reveals prospective drought and heat stress response mechanisms in soybean. *BioMed Res. Int.* **2016**, 6021047. [CrossRef]

38. Aranjuelo, I.; Molero, G.; Erice, G.; Avice, J.C.; Nogués, S. Plant physiology and proteomics reveals the leaf response to drought in alfalfa (*Medicago sativa* L.). *J. Exp. Bot.* **2010**, *62*, 111–123. [CrossRef]

39. Zhang, Y.; Zhang, Y.; Wang, Z. Photosynthetic diurnal variation characteristics of leaf and non-leaf organs in winter wheat under different irrigation regimes. *Acta Ecol. Sin.* **2011**, *31*, 1312–1322.

40. Feng, B. Physiological Basis and Response Mechanisms of Different Green Organs under High Temperature after Anthesis in Winter Wheat. Ph.D. Thesis, Shandong Agricultural University, Taian, China, 2014.

41. Pei, C.; Zhang, Z.; Ma, J. Differentially expressed proteins analysis of seedling leaf of southern type alfalfa (*Medicago sativa* 'Millenium') under salt stress. *J. Agric. Biotechnol.* **2016**, *24*, 1629–1642.

42. Li, H. Phosphoproteomic Differentiation of Maize Seedlings in Response to Drought Stress and Abscisic and Induction. Masters' Thesis, Sichuan Agricultural University, Chengdu, China, 2016.

43. Umeda, M.; Hara, C.; Matsubayashi, Y.; Li, H.; Liu, Q.; Tadokoro, F.; Aotsuka, S.; Uchimiya, H. Expressed sequence tags from cultured cells of rice (*Oryza sativa* L.) under stressed conditions: Analysis of transcripts of genes engaged in ATP-generating pathways. *Plant Mol. Biol.* **1994**, *25*, 469–478. [CrossRef]

44. Riccardi, F.; Gazeau, P.; de Vienne, D.; Zivy, M. Protein changes in response to progressive water deficit in maize: Quantitative variation and polypeptide identification. *Plant Physiol.* **1998**, *117*, 1253–1263. [CrossRef]

45. Silveira, R.D.D.; Abreu, F.R.M.; Mamidi, S.; McClean, P.E.; Vianello, R.P.; Lanna, A.C.; Carneiro, N.P.; Brondani, C. Expression of drought tolerance genes in tropical upland rice cultivars (*Oryza sativa*). *Genet. Mol. Res.* **2015**, *14*, 8181–8200. [CrossRef]

46. Phan, T.T.; Li, J.; Sun, B.; Jia, Y.; Wen, H.; Chan, H.; Li, T.; Li, Y. ATP-citrate lyase gene (SOACLA-1), a novel ACLA gene in sugarcane, and its overexpression enhance drought tolerance of transgenic tobacco. *Sugar Technol.* **2017**, *19*, 258–269. [CrossRef]

47. Stitt, M.; Gibon, Y.; Lunn, J.E.; Piques, M. Multilevel genomics analysis of carbon signalling during low carbon availability: Coordinating the supply and utilisation of carbon in a fluctuating environment. *Funct. Plant Biol.* **2007**, *34*, 526–549. [CrossRef]

48. Budak, H.; Akpinar, B.A.; Unver, T.; Turktas, M. Proteome changes in wild and modern wheat leaves upon drought stress by two-dimensional electrophoresis and nanoLC-ESI–MS/MS. *Plant Mol. Biol.* **2013**, *83*, 89–103. [CrossRef] [PubMed]

49. Wang, X.; Cai, X.; Xu, C.; Wang, Q.; Dai, S. Drought-responsive mechanisms in plant leaves revealed by proteomics. *Int. J. Mol. Sci.* **2016**, *17*, 1706. [CrossRef] [PubMed]

50. Gregorova, Z.; Kovacik, J.; Klejdus, B.; Maglovski, M.; Kuna, R.; Hauptvogel, P.; Matušíková, I. Drought-induced responses of physiology, metabolites, and PR proteins in Triticum aestivum. *J. Agric. Food Chem.* **2015**, *63*, 8125–8133. [CrossRef]

51. Li, P.; Zhang, Y.; Wu, X.; Liu, Y. Drought stress impact on leaf proteome variations of faba bean (*Vicia faba* L.) in the Qinghai–Tibet Plateau of China. *3 Biotech* **2018**, *8*, 110. [CrossRef]

52. Xu, Z.; Li, Z.; Chen, Y.; Chen, M.; Li, L.; Ma, Y. Heat shock protein 90 in plants: Molecular mechanisms and roles in stress responses. *Int. J. Mol. Sci.* **2012**, *13*, 15706–15723. [CrossRef]

53. Hasegawa, Y.; Seki, M.; Mochizuki, Y.; Heida, N.; Hirosawa, K.; Okamoto, N.; Sakurai, T.; Satou, M.; Akiyama, K.; Iida, K.; et al. A flexible representation of omic knowledge for thorough analysis of microarray data. *Plant Methods* **2006**, *2*, 5. [CrossRef]

54. Ashoub, A.; Beckhaus, T.; Berberich, T.; Karas, M.; Brüggemann, W. Comparative analysis of barley leaf proteome as affected by drought stress. *Planta* **2013**, *237*, 771–781. [CrossRef]
55. Lim, C.W.; Han, S.W.; Hwang, I.S.; Kim, D.S.; Hwang, B.K.; Lee, S.C. The pepper lipoxygenase CaLOX1 plays a role in osmotic, drought and high salinity stress response. *Plant Cell Physiol.* **2015**, *56*, 930–942. [CrossRef]
56. Kang, L. *Biological Electron Microscopy Techniques*; China Science and Technology University Press: Hefei, China, 2004.
57. Blanke, M.M.; Ebert, G. Phosphoenolpyruvate carboxylase and carbon economy of apple seedlings. *J. Exp. Bot.* **1992**, *43*, 965–968. [CrossRef]
58. Lilley, R.M.C.; Walker, D.A. An improved spectrophotometric assay for ribulosebisphosphate carboxylase. *Biochim. Biophys. Acta* **1974**, *358*, 226–229. [CrossRef]

International Journal of
Molecular Sciences

Article

Comparative Analysis of the Effect of Inorganic and Organic Chemicals with Silver Nanoparticles on Soybean under Flooding Stress

Takuya Hashimoto [1,†], Ghazala Mustafa [1,2,†], Takumi Nishiuchi [3] and Setsuko Komatsu [1,*]

[1] Faculty of Environment and Information Sciences, Fukui University of Technology, Fukui 910-8505, Japan;
     regios2400@yahoo.co.jp (T.H.); mghazala@qau.edu.pk (G.M.)
[2] Department of Plant Sciences, Quaid-i-Azam University, Islamabad 45320, Pakistan
[3] Institute for Gene Research, Kanazawa University, Kanazawa 920-8640, Japan; tnish9@staff.kanazawa-u.ac.jp
[*] Correspondence: skomatsu@fukui-ut.ac.jp; Tel.: +81-766-29-2466
[†] These authors contributed equally to this work.

Received: 15 January 2020; Accepted: 11 February 2020; Published: 14 February 2020

Abstract: Extensive utilization of silver nanoparticles (NPs) in agricultural products results in their interaction with other chemicals in the environment. To study the combined effects of silver NPs with nicotinic acid and potassium nitrate (KNO_3), a gel-free/label-free proteomic technique was used. Root length/weight and hypocotyl length/weight of soybean were enhanced by silver NPs mixed with nicotinic acid and KNO_3. Out of a total 6340 identified proteins, 351 proteins were significantly changed, out of which 247 and 104 proteins increased and decreased, respectively. Differentially changed proteins were predominantly associated with protein degradation and synthesis according to the functional categorization. Protein-degradation-related proteins mainly consisted of the proteasome degradation pathway. The cell death was significantly higher in the root tips of soybean under the combined treatment compared to flooding stress. Accumulation of calnexin/calreticulin and glycoproteins was significantly increased under flooding with silver NPs, nicotinic acid, and KNO_3. Growth of soybean seedlings with silver NPs, nicotinic acid, and KNO_3 was improved under flooding stress. These results suggest that the combined mixture of silver NPs, nicotinic acid, and KNO_3 causes positive effects on soybean seedling by regulating the protein quality control for the mis-folded proteins in the endoplasmic reticulum. Therefore, it might improve the growth of soybean under flooding stress.

Keywords: proteomics; soybean; flooding; silver nanoparticles; chemicals

---

1. Introduction

Industrial revolution drastically increased atmospheric concentration of carbon dioxide that leads to global warming and changed the precipitation pattern [1]. Changing climatic conditions caused tragic losses in crop productivity [2]. Under these changing climatic conditions, plants are at the forefront of different kinds of abiotic stresses including drought [3], cold [4,5], salinity [6], and heat stress [2]. Flooding is one of the most widely spread abiotic stresses that affects all the terrestrial plants by limiting the carbon dioxide, oxygen, ethylene, and nitric oxide those play important roles in signal transduction cascades [7]. Flooding also affects the soil chemical characteristics including soil pH and redox potential [8]; due to which the availability of soil nutrients is hindered, which results in accumulation of phytotoxins [9]. Flooding itself acts as a complex stress that hampers the plant growth [10]. Based on these reports, flooding severely reduced crop growth and productivity.

Soybean is one of the most important legume crops due to its high protein level and oil contents. Soybean is important for biodiesel production because it emits zero nitrogen, which is highly beneficial [11]. Its yield was decreased by 17–43% at vegetative stage and 50–56% at reproductive stage due to flooding [12]. Yield reduction was due to various changes caused by flooding at seedling stage in soybeans [13]. At the early stage, flooding induced the alcohol fermentation and ethylene biosynthesis-related proteins in soybean [14]. At the seedling stage of soybean, flooding stress caused sucrose accumulation, cell wall loosening, mitochondrial impairment, and proteasome-mediated proteolysis [15]. Soybean is highly susceptible to flooding stress especially at the early stage.

One of the nanoparticles (NPs), silver NPs, enhanced shoot and root length, leaf area, and biochemical attributes such as chlorophyll, carbohydrate, protein contents, and antioxidant enzymes in *Brassica juncea* [16,17]. Silver NPs enhanced seed germination and seedling growth in *Boswellia ovalifoliolata* and reduced the root length in *Arabidopsis thaliana* [18,19]. In *Eruca sativa*, silver NPs stimulated plant growth [20]. Transcription of antioxidant and aquaporin genes was altered by silver NPs in *A. thaliana* [21]. In soybean, silver NPs helped to mitigate the flooding stress condition by regulating the proteins related to fermentation, glycolysis, amino acid synthesis, and wax formation [22,23]. Silver NPs caused variable effects on plants; therefore, the molecular mechanisms underlying these effects demand investigation.

Within the environment, NPs come in contact with other materials like chemical compounds [24,25]. These interactions could be of additive, synergistic, and antagonistic type [26]. Nicotinic acid is involved in the primary and secondary metabolism of plants. It acts as a building block for pyridine compounds like trigonelline, nicotine, anabasin, and ricinine [27]. Nicotinamide disturbs the cytosolic adenosine triphosphate concentration, and regulated period length adjustment, meristem activation, and root growth [28]. Application of nicotinic acid protected spruce seedlings from weevil attack through epigenetic regulation [29]. Nicotinic acid reduced the oxidative stress in plant cells by regulating the aconitase, fumarase, and glutathione metabolism [30]. On the other hand, potassium nitrate (KNO_3) is an inorganic chemical used for seed priming and dormancy breaking in tomato and maize [31,32]. Potassium nitrate reduced the germination time and enhanced the germination rate in tomato by regulating the nitrate reductase, catalase, and superoxide dismutase activities [31]. Individual application of organic/inorganic chemicals and NPs accelerated the growth of different plants; however, molecular mechanisms altered by their combined applications are still not clear. In the present study, the alterations induced by the silver NPs with nicotinic acid and KNO_3 were evaluated using gel-free/label-free proteomic technique. In addition, molecular and biochemical analyses were performed to confirm the proteomics results.

2. Results

2.1. Growth Response of Soybean to Silver NPs Mixed with Organic and Inorganic Chemicals

In order to investigate the effects of silver NPs with and without nicotinic acid and KNO_3, two-day-old soybeans were treated without or with 5 ppm silver NPs, 8 µM nicotinic acid, 0.1 mM KNO_3, and 5 ppm silver NPs/8 µM nicotinic acid/0.1mM KNO_3. After treatments, root length/weight and hypocotyl length/weight were measured at two days of stress. Root length and weight were increased under the silver NPs/nicotinic acid/KNO_3 treatment compared to control (Figure 1). On the other hand, the hypocotyl length was significantly increased under the silver NPs/nicotinic acid/KNO_3 treatment compared to other treatments and control. Hypocotyl length was increased with silver NPs, nicotinic acid, or KNO_3; however, change in hypocotyl length was insignificant among silver NPs, nicotinic acid, and KNO_3 treatments (Figure 1). Hypocotyl weight was significantly increased under the silver NPs/nicotinic acid/KNO_3 treatment compared to control and all other treatments; however, change in hypocotyl weight was insignificant among silver NPs, KNO_3 treatments, and nicotinic acid (Figure 1).

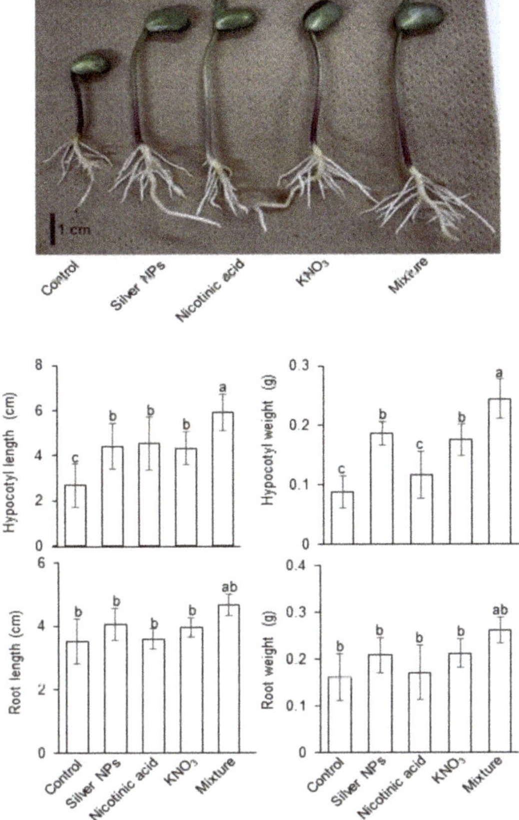

Figure 1. Effects of silver NPs, nicotinic acid, and KNO_3 on the morphology of soybean seedlings. Two-day-old soybeans were treated with 5 ppm silver NPs, 8 µM nicotinic acid, 0.1 mM KNO_3, and 5 ppm silver NPs/8 µM nicotinic acid/0.1mM KNO_3. After treatments, root length/weight and hypocotyl length/weight were measured at two days of stress. Data are presented as the mean ± S.D. from three independent biological replicates. Mean values of each data point with different letters are significantly different according to one-way ANOVA Duncan's Multiple Range test ($p < 0.05$). Scale bar indicates 1 cm.

2.2. Protein Responses in Soybean to Silver NPs Mixed with Organic and Inorganic Chemicals

To get a deep insight into the effects caused by silver NPs mixed with nicotinic acid and KNO_3 on soybean root under flooding stress, a gel-free/label-free proteomic technique was used. In total, 6340 proteins were identified in the proteomics analysis. In the differential analysis, 351 proteins were significantly changed. Out of these 351 proteins, 247 and 104 proteins increased and decreased, respectively (Supplemental Table S1, Figure 2).

To determine the functional role of these proteins, functional categorization was performed using MapMan bin codes. The majority of the significantly changed proteins in the soybean root, treated with silver NPs mixed with nicotinic acid and KNO_3 compared to control, were related to protein (60 proteins), stress (25 proteins), transport (14 proteins), RNA (19 proteins), and cell wall (12 proteins) (Figure 2). The differentially changed proteins related to the protein category were further divided into sub-categories. This sub-categorization revealed that more proteins related to proteins degradation were significantly changed; out of which, the abundance of 14 and 6 proteins was increased and decreased, respectively. Among the identified protein degradation-related proteins,

RRM domain-containing protein was accumulated at a higher degree than the proteasome subunit alpha and ubiquitin conjugating two domain-containing proteins (Supplemental Table S1). The second most changed proteins related to protein category were related to protein synthesis. In total, 18 protein synthesis-related proteins were identified; out of which, the abundance of 12 and 6 proteins was decreased and increased, respectively (Figure 2).

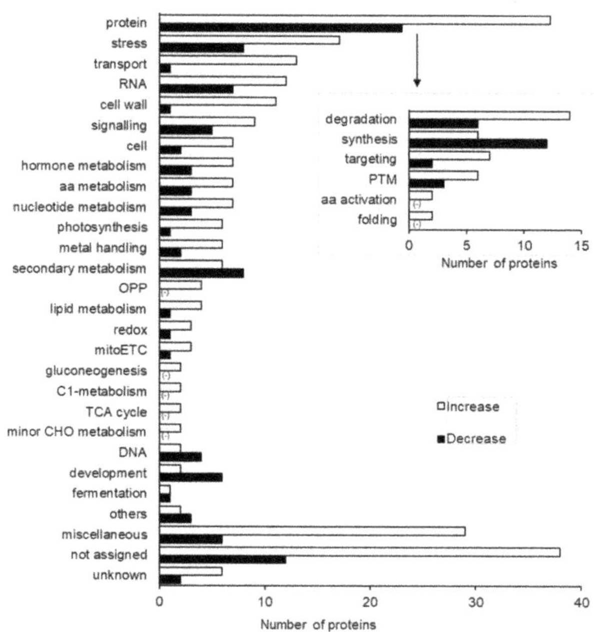

Figure 2. Functional categorization of proteins identified in flooding-stressed soybean treated with 5 ppm silver NPs/8 μM nicotinic acid/0.1 mM KNO_3. Proteins extracted from root were analyzed using a gel-free/label-free proteomic technique and significantly changed proteins were identified ($p < 0.05$). The identified proteins were functionally categorized using MapMan bin codes. The x-axis indicates the number of identified proteins. Abbreviations: aa, amino acid; protein, protein synthesis/degradation/post-translational modification/targeting/folding; cell, cell division/ organization/ vesicle transport; PTM, post-translational modification; RNA, RNA processing/ transcription/binding; OPP, oxidative pentose pathway; mito ETC, mitochondrial electron transport chain; C1 metabolism, carbon 1 metabolism; TCA, tricarboxylic acid cycle; CHO, carbohydrate. The negative sign shows zero proteins identified in the respective functional category.

The second most significantly changed protein category was stress. From the stress-related proteins, a total of 25 proteins were significantly changed, out of which 17 proteins increased in abundance while 8 proteins decreased. Among the identified stress-related proteins, the abundance of methyltransferases increased; however, the abundance of MLO like protein decreased. The third most significantly changed protein category was transport. From the transport category, a total of 14 proteins were significantly changed, out of which 13 proteins increased in abundance while one protein decreased. Among the identified transport-related proteins, aldo-ket-red-domain-containing protein increased; however, uncharacterized protein decreased (Supplemental Table S1). Principal component analysis (PCA) data of total proteins from six samples under control and mixture treatment depicted the closeness of the three independent biological replicates (Supplemental Figure S1).

2.3. Evaluation of Root-Tip Cell Death in Flooded Soybean Seedlings

To investigate the role of protein degradation-related proteins in the flooding stressed soybean seedlings treated with or without silver NPs/nicotinic acid/KNO$_3$, cell death analysis was performed. From the functional categorization of the significantly changed proteins, 60 proteins related to the protein category were identified, 20 of which were involved in protein degradation as part of the ubiquitin proteasome degradation pathway (Figure 2). Two-day-old soybeans were flooded with or without silver NPs/ nicotinic acid/ KNO$_3$ and stained with Evans-blue dye to assess the cell death (Figure 3). The degree of staining was higher in the root tip of flooded soybean compared to control; however, it was much higher in root tip treated with silver NPs/ nicotinic acid/ KNO$_3$ under flooding stress compared to silver NPs/ nicotinic acid/ KNO$_3$, flooded, and untreated soybean (Figure 3). There was no difference in the stain uptake between flooded and silver NPs/ nicotinic acid/ KNO$_3$-treated soybean.

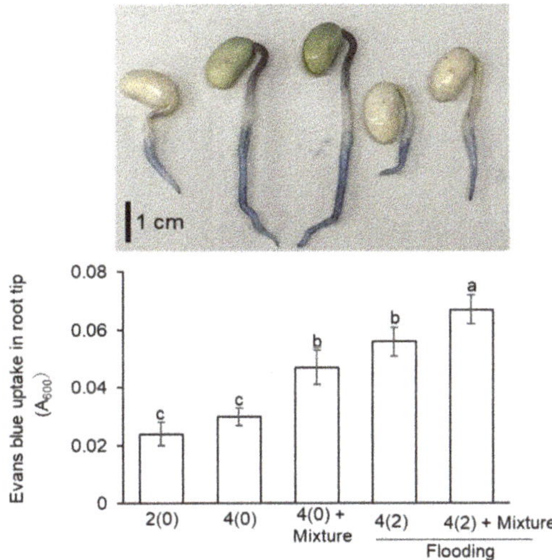

Figure 3. Evaluation of cell death in soybean roots treated with silver NPs/nicotinic acid/ KNO$_3$ under flooding stress. Two-day-old soybeans were flooded with or without 5 ppm silver NPs/8 µM nicotinic acid /0.1 mM KNO$_3$. After the treatments, soybean plants were stained with 0.25% Evans blue dye. The Evans blue dye was then extracted from the root tip and absorbance was measured at 600 nm. Data are presented as the mean ± S.D. from three independent biological replicates. Mean values of each data point with different letters are significantly different according to one-way ANOVA Duncan's Multiple Range test ($p < 0.05$). Scale bar indicates 1 cm.

2.4. Calreticulin/Calnexin Cycle and Glycolysis in Flooded Soybean Seedlings

To analyze the changes in the accumulation pattern of calreticulin and calnexin under flooding stress with or without silver NPs/ nicotinic acid/ KNO$_3$, immuno-blot analysis was performed. Two-day-old soybeans were flooded with or without silver NPs/nicotinic acid/ KNO$_3$, and proteins were extracted from root at two days of stress. Extracted proteins were separated on SDS-PAGE, transferred onto membranes, and cross reacted with calnexin, calreticulin, and ConcanavalinA antibodies (Figures 4 and 5). The relative band intensities were calculated.

Immuno-blot results identified two bands for calnexin. One was of 63 kDa and other was 45 kDa. In the first identified band, the accumulation of the calnexin was not significantly changed under untreated and treated samples. On the other hand, the accumulation of the second band significantly increased in soybean root under flooding stress with or without silver NPs/ nicotinic acid/KNO$_3$ treatment compared to untreated and silver NPs/nicotinic acid/KNO$_3$ treatment (Figure 4). In the immuno-blot analysis, the band for calreticulin appeared at 55 kDa. The accumulation of calreticulin significantly increased under flooding with silver NPs/nicotinic acid/KNO_3 treatment compared to untreated and silver NPs/nicotinic acid/KNO_3 treatment. It remained unchanged under flooding and silver NPs/ nicotinic acid/ KNO_3 treatment compared to control (Figure 4).

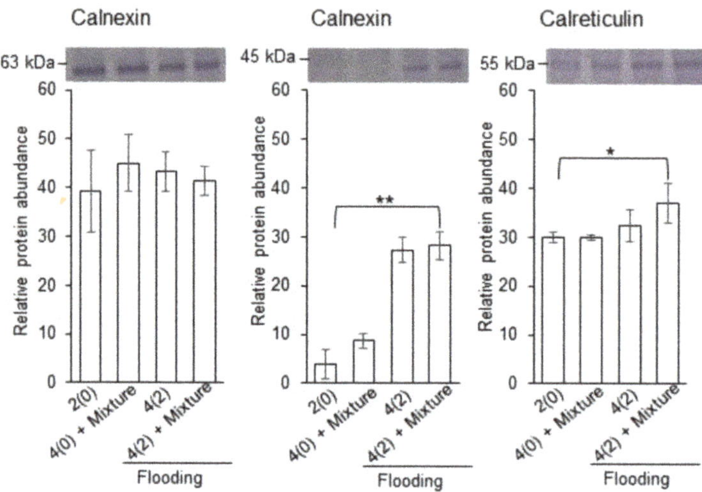

Figure 4. Accumulation of calnexin and calreticulin in soybean root under flooding with silver NPs/nicotinic acid/KNO_3. Two-day-old soybeans were flooded with or without 5 ppm silver NPs/8 µM nicotinic acid/ 0.1 mM KNO_3 for two days. Proteins (10 µg) extracted from roots were applied to gel electrophoresis and immuno-blot was performed with anti-calnexin and anti-calreticulin antibodies. The integrated densities of bands were calculated using Image J software. Data are presented as the mean ± S.D. from three independent biological replicates. Significance was calculated using Students *t*-test (* $p < 0.05$, ** $p < 0.01$).

From the functional categorization of the identified proteins, protein synthesis-related proteins were significantly changed under silver NPs/ nicotinic acid/ KNO_3 treatment (Supplemental Table S1, Figure 2). Based on these results, in order to check the protein synthesis, immuno-blot analysis was performed to investigate the level of glycosylation. The accumulation of glycoproteins was increased under flooding with silver NPs/nicotinic acid/KNO_3 treatment compared to control; however, it remained unchanged under flooding, silver NPs/nicotinic acid/ KNO_3, and control (Figure 5).

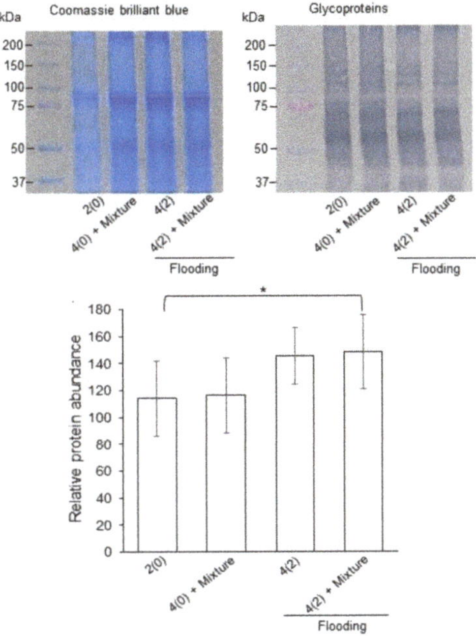

Figure 5. Accumulation of ConcanavalinA in soybean root under flooding stress with silver NPs/ nicotinic acid/KNO$_3$. Two-day-old soybeans were flooded with or without 5 ppm silver NPs/8 µM nicotinic acid /0.1 mM KNO$_3$ for two days. Proteins (10 µg) extracted from roots were applied to gel electrophoresis and immuno-blotting was performed with peroxidase-ConcanavalinA antibody (left side). The integrated densities of bands were calculated using Image J software. Coomassie brilliant blue is used as a loading control (right side). Data are presented as the mean ± S.D. from three independent biological replicates. Significance was calculated using Student's *t*-test (* $p < 0.05$).

2.5. Growth Response of Soybean to Silver NPs Mixed with Organic and Inorganic Chemicals under Flooding Stress

From the immuno-blot results, the accumulation of glycoproteins was increased under flooding with silver NPs/nicotinic acid/ KNO$_3$ treatment compared to control (Figure 5); therefore, the growth of soybean seedlings under flooding stress with or without silver NPs/nicotinic acid/KNO$_3$ treatment was analyzed. Two-day-old soybeans were flooded with or without silver NPs/ nicotinic acid/KNO$_3$ for two days. After treatments, length and weight of root including hypocotyl were measured (Figure 6). The length and weight of root including hypocotyl significantly increased under flooding with silver NPs/nicotinic acid/ KNO$_3$ treatment compared to flooding stress alone (Figure 6).

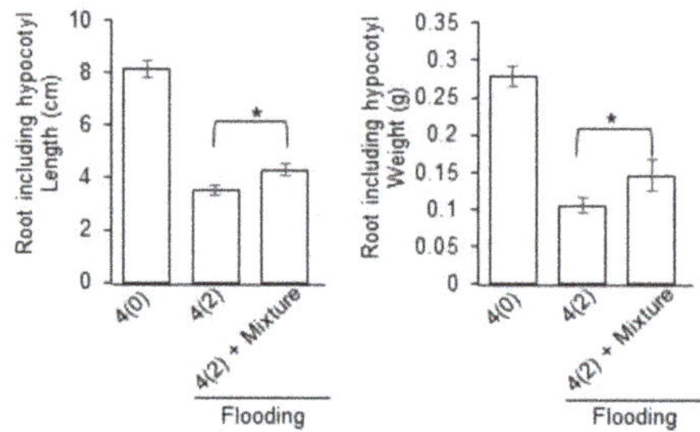

Figure 6. Effects of flooding stress with silver NPs/nicotinic acid/KNO$_3$ on the morphology of soybean seedlings. Two-day-old soybeans were flooded with or without 5 ppm silver NPs/8 µM nicotinic acid /0.1mM KNO$_3$. After treatments, length and weight of root including hypocotyl were measured at two days of stress. Data are presented as the mean ± S.D. from three independent biological replicates. Significance was calculated using Student's t test (* $p < 0.05$). Scale bar indicates 1 cm.

3. Discussion

3.1. Soybean Morphology Alters under Silver NPs, Nicotinic Acid, and KNO$_3$

To investigate the effects caused by silver NPs mixed with nicotinic acid and KNO$_3$ on soybean growth (Figure 1), soybean was treated with or without silver NPs mixed with nicotinic acid and KNO$_3$. Nanoparticles combined with different compounds caused variable effects on plants, which could be antagonistic, synergistic, or additive. Promotion of seed germination and inhibition of root growth of *Lepidium sativum*, flax, cucumber, and wheat were more under individual NPs compared to the combined mixture of copper and zinc NPs [33]. Mixture of copper and zinc NPs inhibited the growth of common bean compared to the individual application of NPs [34]. Binary mixture of copper and zinc NPs significantly reduced plant fresh weight and root length in spinach compared to individual NPs due to increased internal uptake of copper and zinc inside the cells [35]. Binary mixture of copper and zinc oxide NPs caused deleterious effects on radish seeds and inhibited germination. Moreover, the effect of binary mixture of copper and zinc NPs in radish root/shoot length was less toxic than individual copper NPs and more toxic than individual zinc NPs [36]. Binary mixtures of NPs cause opposite or combined effects to the individual NPs.

Zinc NPs combined with tetrabromobisphenol caused severe toxicity to the freshwater microalgae by causing accumulation of reactive oxygen species that leads to oxidative stress [37]. Gold NPs alone caused less toxicity in microalgae; however, when combined with microplastics, the toxicity increased [38]. Silver NPs combined with nicotinic acid and KNO$_3$ improved the growth and development of wheat seedlings [39]. In the present study, growth of soybean seedling significantly increased under the combined application of silver NPs, nicotinic acid, and KNO$_3$ (Figure 1). Nicotinic acid increased the plant growth by increasing the photosynthesis [40]; while KNO$_3$ is important for its role in photosynthesis, enzyme activation, and stress resistance [41]. In wheat, the silver NPs combined with nicotinic acid and KNO_3 maintained the redox homeostasis through regulation of glycolysis and increased activities of antioxidant enzymes that regulated energy metabolism [39]. These results suggest that silver NPs synergistically enhance soybean-plant growth with nicotinic acid and KNO_3 similar to the wheat. Although growth of soybean seedling significantly increased under the combined application of silver NPs, nicotinic acid, and KNO_3, the detailed mechanism of soybean under the combined treatment of silver NPs, nicotinic acid, and KNO_3 is still not clear. In this research, proteomic technique as well as molecular and biochemical analyses were performed to confirm the mechanism.

3.2. Role of Proteasome Degradation Proteins under Silver NPs, Nicotinic Acid, and KNO_3

To get an understanding of the proteins response under the silver NPs mixed with nicotinic acid and KNO_3 on soybean growth under flooding stress, the identified proteins were subjected to functional categorization (Figure 2). Within cells, protein synthesis and degradation are well balanced, because a small decrease in synthesis or increase in degradation results in cell death [42]. Similarly, in the present study, the abundance of protein degradation related proteins increased (Figure 2). Under stress conditions, cells maintain the cellular integrity by repairing the stress-induced damaged proteins by degradation. This degradation is important to restrict the accumulation of misfolded or heavily damaged proteins [43]. The ubiquitin proteasome system acts as a regulatory mechanism for the protein regulation in plants. Various studies support the evidence that ubiquitin proteasome system is primarily regulated under the stressful conditions [44]. The ubiquitin proteasome pathway facilitates the plants to alter their proteins to effectively combat the stress conditions. This regulation is an integral part of plant adaptation to the environmental stress. The ubiquitin proteasome system responds to the changing environmental conditions including stress by regulating the protein degradation [45]. In soybean, flooding stress induces ubiquitin/proteasome mediated proteolysis, which leads to the degradation and loss of the root tip [46]. The ubiquitin proteasome system responds to the stress conditions and mediates the protein regulation.

The ubiquitin proteasome system helps the plant to increase the oxidative stress tolerance by altering its proteins so that it efficiently perceives and responds to the stress [47]. In *Arabidopsis*, 26S proteasome regulatory particle mutants increased oxidative stress tolerance. In the present study, several proteins related to protein degradation, particularly ubiquitin-proteasome-related proteins increased under silver NPs mixed with nicotinic acid and KNO_3 treatment (Figure 2, Supplemental Table S1). Moreover, cell death increased in the soybean root-tip under silver NPs mixed with nicotinic acid and KNO_3 (Figure 3). In soybean, ascorbate peroxidase was suppressed under flooding stress [48]; thereby, it induced the oxidative stress [49]. These results suggest that ubiquitin proteasome system is regulated under silver NPs mixed with nicotinic acid and KNO_3 by increasing the oxidative stress tolerance in order to maintain the cellular integrity; suggesting that growth might be improved under flooding stress.

3.3. Calnexin/Calreticulin Cycle and Glycolysis Have Important Role in Soybean Seedling under Flooding with Silver NPs, Nicotinic Acid, and KNO_3

In order to investigate the level of protein synthesis in soybean root under the silver NPs mixed with nicotinic acid and KNO_3, the accumulation of calnexin/calreticulin and glycoproteins were accessed (Figures 4 and 5). Endoplasmic reticulum mediates the protein folding and assembly through a well-coordinated system of chaperones including calnexin, protein disulfide isomerase, and heat shock proteins [50]. Calnexin is involved in the protein folding and quality control [51]. In soybean, calnexin was increased during the first day while it decreased at the two-day flooding stress [15,52]. In soybean, calnexin and calreticulin decreased and it led to the reduced accumulation of glycoproteins and disruption of endoplasmic reticulum homeostasis under flooding and drought stresses [53]. On the other hand, heterologous expression of rice calnexin confers drought tolerance in tobacco [54]. Calreticulin, a major calcium binding chaperone in the endoplasmic reticulum, is a key component of the calreticulin/calnexin cycle, which is responsible for the folding of newly synthesized proteins, especially glycoproteins, for quality control and stability in the endoplasmic reticulum [55]. In the present study, accumulation of calnexin and calreticulin was higher in the soybean under the silver NPs mixed with nicotinic acid and KNO_3 compared to the control (Figure 4). These results suggest that calnexin/calreticulin cycle is enhanced with the silver NPs mixed with nicotinic acid and KNO_3 in order to regulate the misfolded proteins under flooding stress.

Protein glycosylation is an important post-translational modification with roles in biological activities of proteins, induction of correct folding, signal recognition, protein structure stability, protein interactions, and ultimately the growth of plant organs [56–58]. In order to maintain the protein quality control, endoplasmic reticulum is responsible for the protein degradation of misfolded proteins under stressful conditions [59]. The activity of ubiquitin conjugating enzyme enhances the endoplasmic reticulum for the degradation of misfolded proteins [60]. Similarly, in the present study, the ubiquitin conjugating protein increased under silver NPs mixed with nicotinic acid and KNO_3 (Figure 2 and Supplemental Table S1). The N-glycan degradation of glycoproteins in endoplasmic reticulum caused by extensive chilling stress leads to partial misfolding of proteins and its associated degradation, which eventually caused cell death of plants [61]. In soybean, flooding negatively affected the process of glycosylation of proteins related to stress glycoproteins and protein degradation while the glycolysis related proteins increased [62]. Overall accumulation of glycoproteins decreased under flooding stress. On the other hand, cold stress significantly decreased the glycosylation level of proteins especially calreticulin [14]. Under drought, the glycoproteins related to protein proteolysis increased [63]. In the present study, accumulation of glycoproteins significantly increased with silver NPs, nicotinic acid, and KNO_3 treatment under flooding stress, suggesting the regulation of the misfolded proteins in the endoplasmic reticulum, which were damaged by the flooding stress.

4. Materials and Methods

4.1. Plant Material and Treatments

Soybean (*Glycine max* L.) cultivar Enrei was used as the plant material for the present experiment. Seeds were first surface sterilized in 2% sodium hypochlorite solution and allowed to germinate on silica sand. Seedlings were maintained at 25 °C in a growth chamber illuminated with white fluorescent light (600 μmol m$^{-2}\cdot$s^{-1}, 16 h light period/day) and 70% relative humidity.

To study the effects of silver NPs combined with organic and inorganic chemicals under flooding stress, 15 nm silver NPs (US Research Nanomaterials, Houston, TX, USA), nicotinic acid (Sigma Aldrich, Darmstadt, Germany), and KNO_3 (Sigma Aldrich, Darmstadt, Germany) were used. Two-day-old soybeans were treated with flooding, 5 ppm silver NPs, 8 μM nicotinic acid, 0.1 mM KNO_3, 5 ppm silver NPs/8 μM nicotinic acid/0.1 mM KNO_3. Concentrations of organic and inorganic chemicals were selected based on preliminary experiment data and results of previous publication [39]. Soybean were grown in seedling cases that were kept in the stainless-steel tray. After 2 days of sowing, soybean was flooded by adding water up to 4cm in a stainless-steel tray (Supplemental Figure S2). After treatments, root length/weight and hypocotyl length/weight were measured at 2 days of stress. Four-day-old roots were collected for the proteomics, enzymatic, and biochemical assays. Three independent experiments were performed as biological replicates for all the experiments. Independent biological replicates were sown with one-day difference.

4.2. Protein Extraction

A portion (300 mg) of roots was ground with a mortar and pestle in 500 μL of lysis buffer, which contained 7 M urea, 2 M thiourea, 5% CHAPS, and 2 mM tributylphosphine. The suspension was centrifuged twice at 15,000$\times$ *g* for 10 min at 4 °C. The detergents from the supernatant were removed using the Pierce Detergent Removal Spin Column (Pierce Biotechnology, Rockford, IL, USA). The method of Bradford [64] was used to determine the protein concentration with bovine serum albumin used as the standard.

4.3. Protein Enrichment, Reduction, Alkylation, and Digestion

Extracted proteins (100 μg) were adjusted to a final volume of 100 μL. Methanol (400 μL) was added to each sample and mixed before addition of 100 μL of chloroform and 300 μL of water. After mixing and centrifugation at 20,000× g for 10 min to achieve phase separation, the upper phase was discarded and 300 μL of methanol was added to the lower phase, and then centrifuged at 20,000× g for 10 min. The pellet was collected as the soluble fraction [65]. Proteins were re-suspended in 50 mM ammonium bicarbonate, reduced with 50 mM dithiothreitol for 30 min at 56 °C in the dark, and alkylated with 50 mM iodoacetamide for 30 min at 37 °C in the dark. Alkylated proteins were digested with trypsin and lysyl endopeptidase (Wako, Osaka, Japan) at a 1:100 enzyme/ protein ratio for 16 h at 37 °C. Peptides were desalted with MonoSpin C18 Column (GL Sciences, Tokyo, Japan), acidified with 1% trifluoroacetic acid and analyzed by nano-liquid chromatography (LC) mass spectrometry (MS).

4.4. Protein Identification Using Nano-liquid Chromatography Mass Spectrometry

The LC conditions as well as the MS acquisition conditions were described in the previous study [66]. The peptides were loaded onto the LC system (EASY-nLC 1000; Thermo Fisher Scientific, San Jose, CA, USA) equipped with a trap column (Acclaim PepMap 100 C18 LC column, 3 μm, 75 μm ID × 20 mm; Thermo Fisher Scientific, San Jose, CA, USA), equilibrated with 0.1% formic acid, and eluted with a linear acetonitrile gradient (0–35%) in 0.1% formic acid at a flow rate of 300 nL min^{-1}. The eluted peptides were loaded and separated on the column (NANO-HPLC capillary column C18, 3 μm, 75 μm × 150 mm; Nikkyo Technos, Tokyo, Japan) with a spray voltage of 2 kV (Ion Transfer Tube temperature: 275 °C). The peptide ions were detected using MS (Orbitrap QE plus MS; Thermo Fisher Scientific, San Jose, CA, USA) in the data-dependent acquisition mode with the installed Xcalibur software (version 4.0; Thermo Fisher Scientific, San Jose, CA, USA). Full-scan mass spectra were acquired in the MS over 375–1500 m/z with resolution of 70,000. The most intense precursor ions were selected for collision-induced fragmentation in the linear ion trap at normalized collision energy of 35%. Dynamic exclusion was employed within 15 s to prevent repetitive selection of peptides.

4.5. Mass Spectrometry Data Analysis

The MS/MS searches were carried out using SEQUEST HT search algorithms against the UniprotKB Glycine max (Soybean) protein database (2017-10-25) using Proteome Discoverer (PD) 2.2 (Version 2.2.0.388; Thermo Fisher Scientific, San Jose, CA, USA). Label-free quantification was also performed with PD 2.2 using precursor ions detector nodes. The processing workflow included spectrum files RC, spectrum selector, SEQUEST HT search nodes, percolator, ptmRS, and minor feature detector nodes. Oxidation of methionine was set as a variable modification and carbamidomethylation of cysteine was set as a fixed modification. Mass tolerances in MS and MS/MS were set at 10 ppm and 0.6 Da, respectively. Trypsin was specified as protease and a maximum of two missed cleavages was allowed. Target-decoy database searches used for calculation of false discovery rate (FDR) and for peptide identification FDR was set at 1%. For MS data, RAW data, peak lists and result files have been deposited in the ProteomeXchange Consortium [67] via the jPOST [68] partner repository under data-set identifiers PXD016449.

4.6. Differential Analysis of Proteins using Mass Spectrometry Data

The consensus workflow included MSF files, Feature Mapper, precursor ion quantifier, PSM groper, peptide validator, peptide and protein filter, protein scorer, protein marker, protein FDR validator, protein grouping, and peptide in protein. Normalization of the abundances was performed using total peptide amount mode. Significance was assessed using Abundance Ratio Adjusted P-Value. PCA was performed with PD 2.2.

4.7. Functional Predictions

Protein functions were categorized using MapMan bin codes [69].

4.8. Evaluation of Cell Death Using Evans Blue

Cell death was evaluated by Evans blue staining as described by Baker and Mock [70]. Soybean seedlings were washed twice in water and stained in 0.25% (*w/v*) aqueous Evans blue (Wako, Osaka, Japan) for 15 min at room temperature. The stained seedlings were washed with water and immediately photographed. For quantitative assessment of staining, the terminal 1 cm of root tip was excised and immersed in 1 mL of N,N-dimethylformamide for 24 h at 4 °C. After the incubation, the absorbance of Evans blue released into the solvent was measured at 600 nm.

4.9. Immuno-Blot Analysis

SDS-sample buffer consisting of 60 mM Tris-HCl (pH 6.8), 2% SDS, 10% glycerol, and 5% dithiothreitol was added to protein samples [71]. Quantified proteins (10 µg) were separated by electrophoresis on a 10% SDS-polyacrylamide gel and transferred onto a polyvinylidene difluoride membrane using a semidry transfer blotter (Nippon Eido, Tokyo, Japan). The blotted membrane was blocked for 5 min in Bullet Blocking One reagent (Nacalai Tesque, Kyoto, Japan). After blocking, the membrane was cross reacted with a 1:1000 dilution of the primary antibodies for 1 h at room temperature. As primary antibodies, the followings were used: Anti-calnexin antibody [72], anti-calreticulin antibody [73], and peroxidase-ConcanavalinA antibody (Seikagaku, Tokyo, Japan). Anti-rabbit IgG conjugated with horseradish peroxidase (Bio-Rad, Hercules, CA, USA) was used as the secondary antibody. After 1 h incubation, signals were detected using TMB Membrane Peroxidase Substrate kit (Seracare, Milford, MA, USA) following the manufacturer's protocol. Coomassie brilliant blue (CBB) staining was used as loading control. The integrated densities of bands were calculated using Image J software (version 1.8, National Institutes of Health, Bethesda, MD, USA).

4.10. Statistical Analysis

The statistical significance of two groups was evaluated by the Student's *t*-test. The statistical significance of multiple groups was evaluated by one-way ANOVA test. SPSS 20.0 (IBM, Chicago, IL, USA) statistical software was used for the evaluation of the results. A *p*-value of less than 0.05 was considered as statistically significant.

5. Conclusions

Nanoparticles are engineered and extensively applied on the plants to improve their growth by modulating the metabolic pathways [74]; therefore, exposing the environment to interact with these NPs [25]. Based on this information, it is essential to understand the interaction of NPs with the chemicals in the environment. The present study investigated the combined effects of silver NPs with nicotinic acid and KNO_3 on soybean seedlings. The main findings of this study are as follows: (i) Root length/weight and hypocotyl length/weight of soybean were enhanced with silver NPs mixed with nicotinic acid and KNO_3; (ii) protein degradation- and synthesis-related proteins increased and decreased, respectively, with silver NPs mixed with nicotinic acid and KNO_3; (iii) the cell death was higher in the root tip under flooding stress with silver NPs mixed with nicotinic acid and KNO_3; (iv) accumulation of calnexin/calreticulin and glycoproteins increased; and (v) growth of soybean was improved under flooding stress with silver NPs mixed with nicotinic acid and KNO_3. In wheat, the silver NPs combined with nicotinic acid and KNO_3 maintained the redox homeostasis through regulation of glycolysis and enhanced activities of antioxidant enzymes, which regulated energy metabolism [39]. In gladiolus, the zinc oxide NPs, which induced oxidative stress, were reduced by improving the antioxidant potential and redox homeostasis [75]. These results with previous reports suggest that the combined mixture of silver NPs, nicotinic acid, and KNO_3 causes positive effects on soybean seedling by improving the redox homeostasis. Additionally, the regulation of protein quality control for the misfolded proteins in the endoplasmic reticulum might improve the growth of soybean under flooding stress by the combined mixture of silver NPs, nicotinic acid, and KNO_3.

Supplementary Materials: Supplementary materials can be found at http://www.mdpi.com/1422-0067/21/4/1300/s1. Table S1. List of identified proteins in flooding-stressed soybean treated with silver NPs, nicotinic acid, and KNO3 compared to control. Figure S1. Overview of data of total proteins from 6 samples based on PCA. Figure S2. Flooding treatment of soybean.

Author Contributions: S.K. conceived and designed the research; T.H. and S.K. performed the experiments; T.N. did the MS analysis; T.H., S.K., and G.M. did the data analysis; G.M. and S.K. wrote, reviewed, and edited; All authors have read and agreed to the published version of the manuscript.

Funding: This research was supported by grant (No. 2) from Fukui University of Technology, Japan.

Conflicts of Interest: The authors declare no conflict of interest.

Abbreviations

CBB	Coomassie brilliant blue
FDR	false discovery rate
LC	liquid chromatography
MS	mass spectrometry
NPs	Nanoparticles
PD	protein discoverer
PCA	principle component analysis.

References

1. Hao, X.Y.; Han, X.; Ju, H.; Lin, E.D. Impact of climatic change on soybean production: A review. *Ying Yong Sheng Tai Xue Bao* **2010**, *21*, 2697–2706. [PubMed]
2. Bita, C.E.; Gerats, T. Plant tolerance to high temperature in a changing environment: Scientific fundamentals and production of heat stress-tolerant crops. *Front. Plant Sci.* **2013**, *4*, 273. [CrossRef]
3. Khan, A.; Pan, X.; Najeeb, U.; Tan, D.K.Y.; Fahad, S.; Zahoor, R.; Luo, H. Coping with drought: Stress and adaptive mechanisms, and management through cultural and molecular alternatives in cotton as vital constituents for plant stress resilience and fitness. *Biol. Res.* **2018**, *51*, 47. [CrossRef] [PubMed]
4. Beck, E.H.; Heim, R.; Hansen, J. Plant resistance to cold stress: Mechanisms and environmental signals triggering frost hardening and dehardening. *J. Biosci.* **2004**, *29*, 449–459. [CrossRef]
5. Beck, E.H.; Fettig, S.; Knake, C.; Hartig, K.; Bhattarai, T. Specific and unspecific responses of plants to cold and drought stress. *J. Biosci.* **2007**, *32*, 501–510. [CrossRef] [PubMed]
6. Parvaiz, A.; Satyawati, S. Salt stress and phyto-biochemical responses of plants. *Plant Soil J.* **2008**, *54*, 89–99. [CrossRef]
7. Sasidharan, R.; Hartman, S.; Liu, Z.; Martopawiro, S.; Sajeev, N.; van Veen, H.; Yeung, E.; Voesenek, L.A.C.J. Signal Dynamics and Interactions during Flooding Stress. *Plant Physiol.* **2018**, *176*, 1106–1117. [CrossRef]
8. Dat, J.F.; Capelli, N.; Folzer, H.; Bourgeade, P.; Badot, P.M. Sensing and signaling during plant flooding. *Plant Physiol. Biochem.* **2004**, *42*, 273–282. [CrossRef]
9. Pezeshki, S.R.; DeLaune, R.D. Soil oxidation-reduction in wetlands and its impact onplant functioning. *Biology* **2012**, *1*, 196–221. [CrossRef]
10. Fukao, T.; Barrera-Figueroa, B.E.; Juntawong, P.; Peña-Castro, J.M. Submergence and waterlogging stress in plants: A review highlighting research opportunities and understudied aspects. *Front. Plant Sci.* **2019**, *10*, 340. [CrossRef]
11. Pimentel, D.; Patzek, T.W. Ethanol production using corn, switchgrass, and wood; biodiesel production using soybean and sunflower. *Nat. Resour. Res.* **2005**, *14*, 65–76. [CrossRef]
12. Oosterhuis, D.M.; Scott, H.D.; Hampton, R.E.; Wullschleger, S.D. Physiological responses of two soybean [*Glycine max* (L.) Merr] cultivars to short-term flooding. *Environ. Exp. Bot.* **1990**, *30*, 85–92. [CrossRef]
13. Wang, X.; Komatsu, S. Proteomic approaches to uncover the flooding and drought stress response mechanisms in soybean. *J. Proteom.* **2018**, *172*, 201–215. [CrossRef] [PubMed]

14. Komatsu, S.; Yamamoto, R.; Nanjo, Y.; Mikami, Y.; Yunokawa, H.; Sakata, K. A comprehensive analysis of the soybean genes and proteins expressed under flooding stress using transcriptome and proteome techniques. *J. Proteome Res.* **2009**, *8*, 4766–4778. [CrossRef] [PubMed]

15. Komatsu, S.; Hiraga, S.; Yanagawa, Y. Proteomics techniques for the development of flood tolerant crops. *J. Proteome Res.* **2012**, *11*, 68–78. [CrossRef] [PubMed]

16. Sharma, P.; Bhatt, D.; Zaidi, M.G.H.; Saradhi, P.P.; Khanna, P.K.; Arora, S. Ag nanoparticle-mediated enhancement in growth and antioxidant status of *Brassica juncea*. *Appl. Biochem. Biotechnol.* **2012**, *167*, 2225–2233. [CrossRef] [PubMed]

17. Zea, L.; Salama, H.M.H. Effects ofsilver nanoparticles in some crop plants, common bean (*Phaseolus vulgaris* L.) and corn. *Int. Res. J. Biotechnol.* **2012**, *3*, 190–197.

18. Savithramma, N.; Ankanna, S.; Bhumi, G. Effect of nanoparticles on seed germination and seedling growth of *Boswelliaovalifoliolata*—An endemic and endangered medicinal tree taxon. *Nano Vis.* **2012**, *2*, 61–68.

19. Geisler-Lee, J.; Wang, Q.; Yao, Y.; Zhang, W.; Geisler, M.; Li, K.; Huang, Y.; Chen, Y.; Kolmakov, A.; Ma, X. Phytotoxicity, accumulation and transport of silver nanoparticles by *A. thaliana*. *Nanotoxicology* **2013**, *7*, 323e337. [CrossRef]

20. Zaka, M.; Shah, A.; Zia, M.; Abbasi, B.H.; Rahman, L. Synthesis and characterization of metal NPs and their effects on seed germination and seedling growth in commercially important *Eruca sativa*. *IET Nanobiotechnol.* **2016**, *10*, 134e140. [CrossRef]

21. Qian, H.; Peng, X.; Han, X.; Ren, J.; Sun, L.; Fu, Z. Comparison of the toxicity ofsilver NPs and silver ions on the growth of terrestrial plant model *Arabidopsis thaliana*. *J. Environ. Sci.* **2013**, *25*, 1947–1956. [CrossRef]

22. Mustafa, G.; Sakata, K.; Hossain, Z.; Komatsu, S. Proteomic study on the effects of silver nanoparticles on soybean under flooding stress. *J. Proteom.* **2015**, *122*, 100–118. [CrossRef]

23. Mustafa, G.; Sakata, K.; Komatsu, S. Proteomic analysis of soybean root exposed to varying sizes of silver nanoparticles under flooding stress. *J. Proteom.* **2016**, *148*, 113–125. [CrossRef] [PubMed]

24. Impellitteri, C.A.; Harmon, S.; Silva, R.G.; Miller, B.W.; Scheckel, K.G.; Luxton, T.P.; Schupp, D.; Panguluri, S. Transformation of silver nanoparticles in fresh, aged, andincinerated biosolids. *Water Res.* **2013**, *47*, 3878–3886. [CrossRef] [PubMed]

25. Ma, R.; Levard, C.; Judy, J.D.; Unrine, J.M.; Durenkamp, M.; Martin, B.; Jefferson, B.; Lowry, G.V. Fate of zinc oxide and silver nps in a pilot wastewater treatment.plant and in processed biosolids. *Environ. Sci. Technol.* **2014**, *48*, 104–112. [CrossRef] [PubMed]

26. Ince, N.H.; Dirilgen, N.; Apikyan, I.G.; Tezcanli, G.; Üstün, B. Assessment of toxicinteractions of heavy metals in binary mixtures: A statistical approach. *Arch. Environ. Contam. Toxicol.* **1999**, *36*, 365–372. [CrossRef] [PubMed]

27. Zrenner, R.; Ashihara, H. Nucleotide metabolism. In *Plant Metabolism and Biotechnology*; Ashihara, H., Crozier, A., Komamine, A., Eds.; John Wiley & Sons, Ltd.: Chichester, UK, 2011; pp. 135–162.

28. Zhang, N.; Meng, Y.; Li, X.; Zhou, Y.; Ma, L.; Fu, L.; Schwarzländer, M.; Liu, H.; Xiong, Y. Metabolite-mediated TOR signaling regulates the circadian clock in *Arabidopsis*. *Proc. Natl. Acad. Sci. USA* **2019**, *116*, 25395–25397. [CrossRef]

29. Berglund, T.; Lindstrom, A.; Aghelpasand, H.; Stattin, E.; Ohlsson, A.B. Protection of spruce seedlings against pine weevil attacks by treatment of seeds or seedlings with nicotinamide, nicotinic acid and jasmonic acid. *Forestry* **2016**, *89*, 127–135. [CrossRef]

30. Berglund, T.; Wallström, A.; Nguyen, T.V.; Laurell, C.; Ohlsson, A.B. Nicotinamide; antioxidative and DNA hypomethylation effects in plant cells. *Plant Physiol. Biochem.* **2017**, *118*, 551–560. [CrossRef]

31. Lara, T.S.; Lira, J.M.S.; Rodrigues, A.C.; Rakocevic, M.; Alvarenga, A.A. Potassium nitrate priming affects the activity of nitrate reductase and antioxidant enzymes in tomato germination. *J. Agric. Sci.* **2014**, *6*, 72. [CrossRef]

32. Anosheh, H.P.; Sadeghi, H.; Emam, Y. Chemical priming with urea and KNO$_3$ enhances maize hybrids (*Zeamays* L.) seed viability under abiotic stress. *J. Crop Sci. Biotechnol.* **2011**, *14*, 289–295. [CrossRef]

33. Jośko, I.; Oleszczuk, P.; Skwarek, E. Toxicity of combined mixtures of nanoparticles to plants. *J. Hazard. Mater.* **2017**, *331*, 200–209. [CrossRef] [PubMed]

34. Dimkpa, C.O.; McLean, J.E.; Britt, D.W.; Anderson, A.J. Nano-CuO and interaction with nano-ZnO or soil bacterium provide evidence for the interference of nanoparticles in metal nutrition of plants. *Ecotoxicology* **2015**, *24*, 119–129. [CrossRef] [PubMed]

35. Singh, D.; Kumar, A. Quantification of metal uptake in *Spinaciaoleracea* irrigated with water containing a mixture of CuO and ZnO nanoparticles. *Chemosphere* **2019**, *243*, 125239. [CrossRef] [PubMed]

36. Singh, D.; Kumar, A. Assessment of toxic interaction of nano zinc oxide and nano copper oxide on germination of *Raphanussativus* seeds. *Environ. Monit. Assess.* **2019**, *191*, 703. [CrossRef] [PubMed]

37. Meng, Y.; Wang, S.; Wang, Z.; Ye, N.; Fang, H. Algal toxicity of binary mixtures of zinc oxide nanoparticles and tetrabromobisphenol A: Roles of dissolved organic matters. *Environ. Toxicol. Pharmacol.* **2018**, *64*, 78–85. [CrossRef] [PubMed]

38. Davarpanah, E.; Guilhermino, L. Are gold nanoparticles and microplastics mixtures more toxic to the marine microalgae *Tetraselmischuii* than the substances individually? *Ecotoxicol. Environ. Saf.* **2019**, *181*, 60–68. [CrossRef]

39. Jhanzab, H.M.; Razzaq, A.; Bibi, Y.; Yasmeen, F.; Yamaguchi, H.; Hitachi, K.; Tsuchida, K.; Komatsu, S. Proteomic analysis of the effect of inorganic and organic chemicals on silver nanoparticles in wheat. *Int. J. Mol. Sci.* **2019**, *20*, 4. [CrossRef]

40. Sajjad, Y.; Jaskani, M.; Ashraf, M.Y.; Ahmad, R. Response of morphological and physiological growth attributes to foliar application of plant growth regulators in gladiolus 'white prosperity'. *Pak. J. Agric. Sci.* **2014**, *51*, 123–129.

41. Wang, M.; Zheng, Q.; Shen, Q.; Guo, S. The critical role of potassium in plant stress response. *Int. J. Mol. Sci.* **2013**, *14*, 7370–7390. [CrossRef]

42. Mitch, W.E.; Goldberg, A.L. Mechanisms of muscle wasting. The role of the ubiquitin-proteasome pathway. *N. Engl. J. Med.* **1996**, *335*, 1897–1905. [CrossRef] [PubMed]

43. Flick, K.; Kaiser, P. Protein degradation and the stress response. *Semin. Cell Dev. Biol.* **2012**, *23*, 515–522. [CrossRef] [PubMed]

44. Xu, F.Q.; Xue, H.W. The ubiquitin-proteasome system in plant responses to environments. *Plant Cell Environ.* **2019**, *42*, 2931–2944. [CrossRef] [PubMed]

45. Dielen, A.S.; Badaoui, S.; Candresse, T.; German-Retana, S. The ubiquitin/26S proteasome system in plant-pathogen interactions: A never-ending hide-and-seek game. *Mol. Plant Pathol.* **2010**, *11*, 293–308. [CrossRef]

46. Yanagawa, Y.; Komatsu, S. Ubiquitin/proteasome-mediated proteolysis is involved in the response to flooding stress in soybean roots, independent of oxygen limitation. *Plant Sci.* **2012**, *185*, 250–258. [CrossRef]

47. Kurepa, J.; Toh-E, A.; Smalle, J.A. 26S proteasome regulatory particle mutants have increased oxidative stress tolerance. *Plant J.* **2008**, *53*, 102–114. [CrossRef]

48. Nishizawa, K.; Hiraga, S.; Yasue, H.; Chiba, M.; Tougou, M.; Nanjo, Y.; Komatsu, S. The synthesis of cytosolic ascorbate peroxidases in germinating seeds and seedlings of soybean and their behavior under flooding stress. *Biosci. Biotechnol. Biochem.* **2013**, *77*, 2205–2209. [CrossRef]

49. Caverzan, A.; Passaia, G.; Rosa, S.B.; Ribeiro, C.W.; Lazzarotto, F.; Margis-Pinheiro, M. Plant responses to stresses: Role of ascorbate peroxidase in the antioxidant protection. *Genet. Mol. Biol.* **2012**, *35*, 1011–1019. [CrossRef]

50. Healy, S.J.; Verfaillie, T.; Jäger, R.; Agostinis, P.; Samali, A. Biology of the endoplasmic reticulum. In *Endoplasmic Reticulum Stress in Health and Disease*; Agostinis, P., Samali, A., Eds.; Springer: Dordrecht, The Netherlands, 2012; pp. 3–22.

51. Bergeron, J.J.; Brenner, M.B.; Thomas, D.Y.; Williams, D.B. Calnexin: A membrane-bound chaperone of the endoplasmic reticulum. *Trends Biochem. Sci.* **1994**, *19*, 124–128. [CrossRef]

52. Nanjo, Y.; Skultety, L.; Ashraf, Y.; Komatsu, S. Comparative proteomic analysis of early-stage soybean seedlings responses to flooding by using gel and gel-free techniques. *J. Proteome Res.* **2010**, *9*, 3989–4002. [CrossRef]

53. Wang, X.; Komatsu, S. Gel-free/label-free proteomic analysis of endoplasmic reticulum proteins in soybean root tips under flooding and drought stresses. *J. Proteome Res.* **2016**, *15*, 2211–2227. [CrossRef] [PubMed]

54. Sarwat, M.; Naqvi, A.R. Heterologous expression of rice calnexin (OsCNX) confers drought tolerance in *Nicotianatabacum*. *Mol. Biol. Rep.* **2013**, *40*, 5451–5464. [CrossRef] [PubMed]

55. Michalak, M.; Robert Parker, J.M.; Opas, M. Ca2+ signaling and calcium binding chaperones of the endoplasmic reticulum. *Cell Calcium* **2002**, *32*, 269–278. [CrossRef] [PubMed]

56. Pan, S.; Chen, R.; Aebersold, R.; Brentnall, T.A. Mass spectrometry based glycoproteomics—From a proteomics perspective. *Mol. Cell Proteom.* **2011**, *10*, R110.003251. [CrossRef] [PubMed]

57. Bond, A.E.; Row, P.E.; Dudley, E. Post-translation modification of proteins; methodologies and applications in plant sciences. *Phytochemistry* **2011**, *72*, 975–996. [CrossRef] [PubMed]

58. Lindner, H.; Kessler, S.A.; Müller, L.M.; Shimosato-Asano, H.; Boisson-Dernier, A.; Grossniklaus, U. TURAN and EVAN mediate pollen tube reception in Arabidopsis Synergids through protein glycosylation. *PLoS Biol.* **2015**, *13*, e1002139. [CrossRef] [PubMed]

59. Strasser, R. Protein quality control in the endoplasmic reticulum of plants. *Annu. Rev. Plant Biol.* **2018**, *69*, 147–172. [CrossRef]

60. Hagiwara, M.; Ling, J.; Koenig, P.A.; Ploegh, H.L. Posttranscriptional regulation of glycoprotein quality control in the endoplasmic reticulum is controlled by the E2 UB-conjugating enzyme UBC6e. *Mol. Cell* **2016**, *63*, 753–767. [CrossRef]

61. Ma, J.; Wang, D.; She, J.; Li, J.; Zhu, J.K.; She, Y.M. Endoplasmic reticulum-associated N-glycan degradation of cold-upregulated glycoproteins in response to chilling stress in Arabidopsis. *New Phytol.* **2016**, *212*, 282–296. [CrossRef]

62. Mustafa, G.; Komatsu, S. Quantitative proteomics reveals the effect of protein glycosylation in soybean root under flooding stress. *Front. Plant Sci.* **2014**, *5*, 627. [CrossRef]

63. Zadražnik, T.; Moen, A.; Egge-Jacobsen, W.; Meglič, V.; Šuštar-Vozlič, J. Towards a better understanding of protein changes in common bean under drought: A case study of N-glycoproteins. *Plant Physiol. Biochem.* **2017**, *118*, 400–412. [CrossRef] [PubMed]

64. Bradford, M.M. A rapid and sensitive method for the quantitation of microgram quantities of protein utilizing the principle of protein-dye binding. *Anal. Biochem.* **1976**, *72*, 248–254. [CrossRef]

65. Komatsu, S.; Nanjo, Y.; Nishimura, M. Proteomic analysis of the flooding tolerance mechanism in mutant soybean. *J. Proteom.* **2013**, *79*, 231–250. [CrossRef] [PubMed]

66. Li, X.; Rehman, S.U.; Yamaguchi, H.; Hitachi, K.; Tsuchida, K.; Yamaguchi, T.; Sunohara, Y.; Matsumoto, H.; Komatsu, S. Proteomic analysis of the effect of plant-derived smoke on soybean during recovery from flooding stress. *J. Proteom.* **2018**, *181*, 238–248. [CrossRef]

67. Vizcaíno, J.A.; Côté, R.G.; Csordas, A.; Dianes, J.A.; Fabregat, A.; Foster, J.M.; Griss, J.; Alpi, E.; Birim, M.; Contell, J.; et al. The PRoteomicsIDEntifications (PRIDE) database and associated tools: Status in 2013. *Nucleic Acids Res.* **2013**, *41*, 1063–1069. [CrossRef]

68. Okuda, S.; Watanabe, Y.; Moriya, Y.; Kawano, S.; Yamamoto, T.; Matsumoto, M.; Takami, T.; Kobayashi, D.; Araki, N.; Yoshizawa, A.C.; et al. jPOSTrepo: An international standard data repository for proteomes. *Nucleic Acids Res.* **2017**, *45*, 1107–1111. [CrossRef]

69. Usadel, B.; Nagel, A.; Thimm, O.; Redestig, H.; Blaesing, O.E.; Rofas, N.P.; Selbig, J.; Hannemann, J.; Piques, M.C.; Steinhauser, D.; et al. Extension of the visualization tool MapMan to allow statistical analysis of arrays, display of corresponding genes and comparison with known responses. *Plant Physiol.* **2005**, *138*, 1195–1204. [CrossRef]

70. Baker, C.J.; Mock, N.M. An improved method for monitoringcell death in cell suspension and leaf disc assays using Evans blue. *Plant Cell Tissue Organ Cult.* **1994**, *39*, 7–12. [CrossRef]

71. Laemmli, U.K. Cleavage of structural proteins during the assembly of the head of bacteriophage T4. *Nature* **1970**, *227*, 680–685. [CrossRef]

72. Nouri, M.Z.; Hiraga, S.; Yanagawa, Y.; Sunohara, Y.; Matsumoto, H.; Komatsu, S. Characterization of calnexin in soybean roots and hypocotyls under osmotic stress. *Phytochemistry* **2012**, *74*, 20–29. [CrossRef]

73. Sharma, A.; Isogai, M.; Yamamoto, T.; Sakaguchi, K.; Hashimoto, J.; Komatsu, S. A novel interaction between calreticulin and ubiquitin-like nuclear protein in rice. *Plant. Cell Physiol.* **2004**, *45*, 684–692. [CrossRef] [PubMed]

74. Pradhan, S.; Mailapalli, D.R. Interaction of engineered nanoparticles with the agri-environment. *J. Agric. Food Chem.* **2017**, *65*, 8279–8294. [CrossRef] [PubMed]

75. Li, M.; Ahammed, G.J.; Li, C.; Bao, X.; Yu, J.; Huang, C.; Yin, H.; Zhou, J. Brassinosteroid ameliorates zinc oxide nanoparticles-induced oxidative stress by improving antioxidant potential and redox homeostasis in tomato seedling. *Front. Plant Sci.* **2016**, *7*, 615. [CrossRef] [PubMed]

International Journal of
Molecular Sciences

Article

Comparative Proteomic Analysis of Wild-Type *Physcomitrella Patens* and an OPDA-Deficient *Physcomitrella Patens* Mutant with Disrupted *PpAOS1* and *PpAOS2* Genes after Wounding

Weifeng Luo [1,†], Setsuko Komatsu [2], Tatsuya Abe [1,‡], Hideyuki Matsuura [1] and Kosaku Takahashi [1,3,*]

1   Division of Fundamental Agroscience Research, Research Faculty of Agriculture, Hokkaido University, Kita 9, Nishi 9, Kita-ku, Sapporo 060-8589, Japan; luoweifeng1989@outlook.com (W.L.); abet@nitten.co.jp (T.A.); matsuura@hokudai.ac.jp (H.M.)

2   Department of Environmental and Food Sciences, Faculty of Environmental and Information Sciences, Fukui University of Technology, 3-6-1 Gakuen, Fukui 910-8505, Japan; skomatsu@fukui-ut.ac.jp

3   Department of Nutritional Science, Faculty of Applied Bioscience, Tokyo University of Agriculture, 1-1-1 Sakuragaoka, Setagaya-ku, Tokyo 165-8502, Japan

*   Correspondence: kt207119@nodai.ac.jp

†   Present Address: Institute of Agrobiological Sciences, National Agriculture and Food Research Organization (NARO), Kannondai, Tsukuba 305-8602, Japan.

‡   Present Address: Research Center, Nippon Beet Sugar MFG. Co., Ltd. Inada-cho, Obihiro, Hokkaido 080-8031, Japan.

Received: 17 January 2020; Accepted: 17 February 2020; Published: 19 February 2020

Abstract: Wounding is a serious environmental stress in plants. Oxylipins such as jasmonic acid play an important role in defense against wounding. Mechanisms to adapt to wounding have been investigated in vascular plants; however, those mechanisms in nonvascular plants remain elusive. To examine the response to wounding in *Physcomitrella patens*, a model moss, a proteomic analysis of wounded *P. patens* was conducted. Proteomic analysis showed that wounding increased the abundance of proteins related to protein synthesis, amino acid metabolism, protein folding, photosystem, glycolysis, and energy synthesis. 12-Oxo-phytodienoic acid (OPDA) was induced by wounding and inhibited growth. Therefore, OPDA is considered a signaling molecule in this plant. Proteomic analysis of a *P. patens* mutant in which the *PpAOS1* and *PpAOS2* genes, which are involved in OPDA biosynthesis, are disrupted showed accumulation of proteins involved in protein synthesis in response to wounding in a similar way to the wild-type plant. In contrast, the fold-changes of the proteins in the wild-type plant were significantly different from those in the *aos* mutant. This study suggests that *PpAOS* gene expression enhances photosynthesis and effective energy utilization in response to wounding in *P. patens*.

Keywords: Allene oxide synthase; 12-oxo-phytodienoic acid; *Physcomitrella patens*; proteomic analysis; wounding

---

1. Introduction

The influence of abiotic and biotic stresses on plants has been particularly well studied [1]. Wounding is an abiotic stress that can cause severe damage to plants. Wounded tissues are more likely to be infected by pathogenic microorganisms, which cause serious damage to plants. However, to protect cells from irreversible damage caused by wounding, plants induce changes in the abundance of many proteins, indicating that plants have developed resistance mechanisms against stresses, including wounding.

Plant adaptations against wounding have been extensively investigated. Various responses to wounding have been elucidated, including activation of metabolic pathways, cell-wall modifications, and the production of pathogenesis-related proteins and proteinase inhibitors [2,3]. Moreover, plant stress hormones play a critical role in the defense against wounding. Jasmonic acid (JA) is among the most important signaling hormones in response to wounding in vascular plants [4]. A recent study indicated that JA also functions as a signaling compound in the model lycophyte *Selaginella moellendorffii* [5].

JA, which is synthesized from α-linolenic acid through the octadecanoid pathway, modulates the expression of various genes in response to abiotic and biotic stresses [4,6,7]. The JA signaling pathway is known in detail. The isoleucine conjugate of JA (JA-Ile) is a versatile compound in the JA signaling pathway [8–11]. The binding of JA-Ile with its receptor coronatine insensitive 1 (COI1) triggers various physiological responses in plants. 12-Oxo-phytodienoic acid (OPDA), an intermediate of JA biosynthesis, shows JA-dependent and JA-independent biological activities in plants [12–16]. OPDA induces expression of a set of genes in response to wounding in *Arabidopsis thaliana* and plays important roles in embryo development and seed germination in tomato [17–20]. However, a detailed OPDA signaling mechanism remains elusive in plants.

Bryophytes, including *Marchantiophyta* (liverworts), *Bryophyta* (mosses), and *Anthocerotophyta* (hornworts), are taxonomically positioned between algae and vascular plants and comprise an early-diverging lineage of land plants [21,22]. Therefore, bryophytes occupy a key evolutionary position and aid in our understanding of the molecular basis of the key innovations that allowed green plants to evolve from aquatic ancestors and adapt to the terrestrial environment [22]. *Physcomitrella patens*, a model moss, has been utilized to study basic plant physiology and development, as well as the molecular mechanisms of plant evolution from the hydrosphere to land [23]. The life cycle of *P. patens* is characterized by two generations: a haploid gametophyte and a diploid sporophyte. A spore develops into a filamentous structure called the protonema, which can differentiate into structurally complex gametophores with leaf-like structures, rhizoids, and the sexual organ [24]. As the protonema is distinctly different from the gametophore, the expression profiles of the genes and proteins in the protonema are different from those of the gametophore.

Transcriptomic and proteomic analyses have been conducted to understand the molecular basis of environmental stress tolerance in *P. patens* [25–28]. Specifically, drought tolerance studies in *P. patens* could help to elucidate how plants acquired drought resistance, thereby permitting their invasion of the terrestrial environment. However, the influence of wounding, a severe environmental stress in plants, has been overlooked in *P. patens* until recently. Previous research found that OPDA and methyl jasmonate (MeJA), but not JA, retards the growth of *P. patens* [29]. Unlike vascular plants, *P. patens* produces OPDA, but not JA, in response to wounding and pathogenic infection [29,30]. These findings strongly suggest that the first half of the octadecanoid pathway in chloroplasts is conserved in *P. patens* and that OPDA, not JA, functions as a signaling molecule in *P. patens*. The differences in the responses to JA and OPDA between vascular plants and bryophytes are a significant observation in the study of plant evolution. Moreover, understanding the functions of OPDA in *P. patens* would help to elucidate the OPDA signal transduction pathway in plants. Allene oxide synthase (AOS) is an OPDA biosynthetic enzyme. AOS converts 13(S)-hydroperoxyoctadecatrienoic acid (13-HPOT) to 12,13(S)-epoxyoctadecatrienoic acid (12,13-EOT), and this product is then cyclized by allene oxide cyclase (AOC) into OPDA. Two AOS genes, *PpAOS1* and *PpAOS2*, are present in the genome of *P. patens* [31]. To investigate the mechanism underlying the adaptation to wounding and the role of AOS gene expression leading to OPDA synthesis in the response to wounding, we used wild-type *P. patens* and an OPDA-deficient mutant with disrupted *PpAOS1* and *PpAOS2* in this research. For analysis of the mechanism underlying the adaptation to wounding of *P patens*, a gel-free/label-free proteomic technique was used. Furthermore, gene expression analysis was performed to confirm the proteins identified by proteomic analysis.

2. Results and Discussion

2.1. Generation of 12-Oxo-phytodienoic acid (OPDA)-deficient P. Patens Mutants with Disrupted PpAOS1 and PpAOS2 Genes

To study the influence of *AOS* gene expression in physiology of *P. patens*, we generated three *P. patens* OPDA-deficient mutants in which both *PpAOS1* and *PpAOS2* were disrupted (A5, A19, and A22). An ultra-performance liquid chromatography-tandem mass spectrometry (UPLC-MS/MS) analysis of OPDA in the three mutants revealed that wounding did not induce OPDA accumulation (Figure 1). Gametophores of these three double-knockout mutants were grown for three weeks under standard conditions, which led to the formation of colonies (Figure 2, Figure S1). Compared to wild type, we did not observe any differences in growth in the double-knockout mutants, which was similar to a previous report of a single knockout mutant [31].

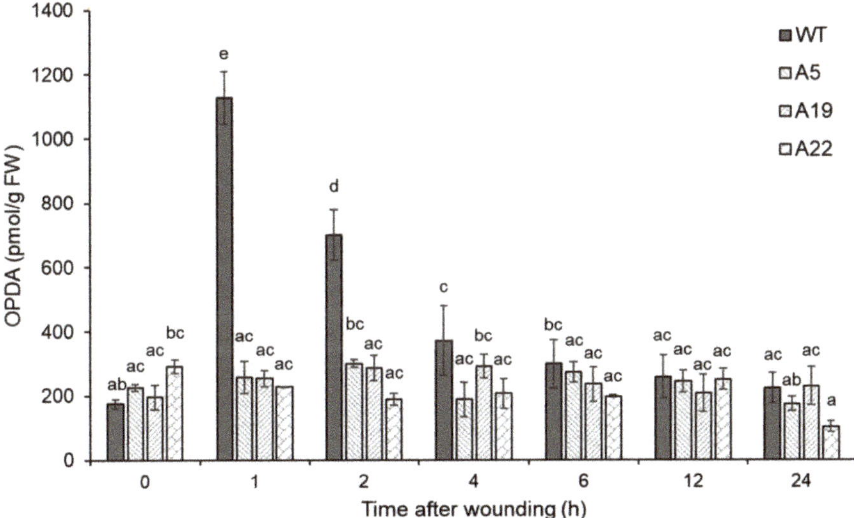

Figure 1. Accumulation of 12-oxo-phytodienoic acid (OPDA) in *P. patens* after wounding. The wild-type and *aos* mutants of *P. patens* were grown on BCDAT agar for 3 weeks and then treated with wounding. The samples were prepared at 1, 2, 4, 6, 12, and 24 h after wounding. Three independent experiments were conducted as biological replicates. The concentrations of OPDA in the wild-type and *aos* mutants (A5, A19, and A22) after wounding were analyzed using ultra-performance liquid chromatography-tandem mass spectrometry (UPLC-MS/MS). The values are the mean ± SD ($n = 3$). Different letters indicate that the change is significant as determined by one-way ANOVA according to Tukey's multiple comparison test ($p < 0.05$).

The OPDA concentration in these mutants was lower than that in the wild-type in the first 2 h after wounding; however, the disruption of these two *PpAOS* genes did not completely prevent OPDA synthesis (Figure 1). Because hydroperoxide lyases (PpHPLs) show weak AOS activities in *P. patens* [31], it is possible that the OPDA detected in the mutants was produced by PpHPLs. In a previous study, the disruption of *PpAOC1* and *PpAOC2* genes caused reduced fertility, aberrant sporophyte morphology and interrupted sporogenesis in *P. patens* [13]. However, the phenotype of the *PpAOS1* and *PpAOS2* mutants was not significantly different from that of wild-type under conventional growth conditions (Figure 2), indicating that the trace amount of OPDA detected in the mutants was apparently sufficient for the growth of *P. patens*.

As the phenotypes of the *P. patens* mutants in which *PpAOS1* and *PpAOS2* were disrupted (A5, A19, and A22) were almost the same, the A5 strain (referred to as the *aos* mutant hereafter) was utilized to investigate the influence of *PpAOS* gene disruption at the protein level in response to wounding.

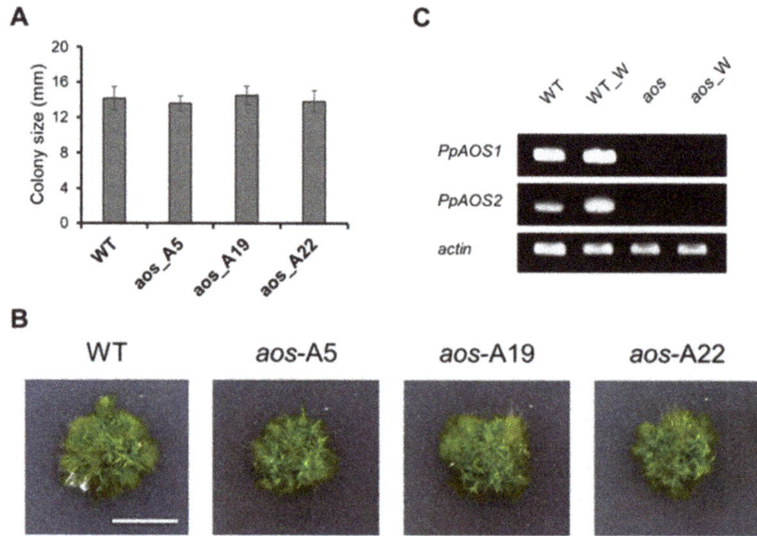

Figure 2. Phenotypic analysis of the wild-type and *aos* mutant of *P. patens*. The wild-type and the *aos* mutants of *P. patens* were grown on BCDAT agar for 3 weeks, and their phenotypes were compared. (**A**) Colony size of the wild-type and *aos* mutants of *P. patens* (the values are the mean ± SD, $n = 12$). A significant difference was not found between wild-type and mutants (Student's t-test). (**B**) Images of the wild-type and *aos* mutants of *P. patens*. White size bar represents 1 cm length. (**C**) Semi-quantitative RT-PCR of *PpAOS* gene expression after wounding in the wild-type and *aos* mutant (A5 strain). WT and aos as control, without wounding treatment; WT_W and aos_W as wounded group: gametophores of *P. patens* (WT and *aos* mutant) were wounded using tweezers and harvested 1 h later for RNA isolation. The gels were stained using ethidium bromide.

2.2. Identification of Proteins that are Differentially Accumulated in Response to Wounding

To identify *P. patens* proteins that are altered by wounding, gel-free/label-free proteomic analysis was performed. The outline of the procedure used for proteomic analysis in this study is illustrated in Figure S2. In reports of investigation of plant stress adaptation, samples were collected at each time point after stress (30 min to several days) [32]. A certain period is required for signal transduction and protein synthesis after stress. Twenty-four hours is one of the typical time points to analyze stress responses for proteome analyses. Thus, proteins were extracted from 3-week-old wild-type *P. patens* and the *aos* mutant at 24 h after wounding. The extracted proteins were digested, and the resulting peptides were analyzed using nano LC-MS/MS [33]. Three biological replicates were utilized in this study. The levels of 136 and 88 proteins with more than two matched peptides were significantly altered by more than 1.5-fold in response to wounding in the wild-type and the *aos* mutant, respectively ($p < 0.05$) (Tables S1 and S2). Wounding increased the abundance of 114 proteins in the wild-type and 88 proteins in the *aos* mutant compared with untreated plants, while only 22 proteins in the wild-type decreased due to wounding. Remarkably, none of the proteins in the *aos* mutant showed a significant decrease in abundance in response to wounding. It appears that *PpAOS* gene expression is related to the decrease in protein abundance observed in response to wounding.

2.3. Functional Categories of Identified Wounding-Responsive Proteins

To determine the functions of the identified proteins, the proteins were annotated using Phytozome (https://phytozome.jgi.doe.gov/pz/portal.html) and functionally categorized (Figure 3). To investigate the functional relationship of the proteins identified in response to wounding in *P. patens* (wild-type and *aos* mutant), significantly enriched Gene Ontology (GO) terms and Kyoto Encyclopedia of Genes and Genomes (KEGG) pathways were identified according to the *P* value and enrichment factor [34]. Proteins were significantly altered by wounding in wild-type and/or the *aos* mutant (Figure 3). In the wild-type plant, wounding predominantly increased the abundance of proteins related to photosystems, protein synthesis, amino acid metabolism, redox, and energy synthesis. The *aos* mutant largely accumulated proteins related to protein synthesis, stress, and protein degradation in response to wounding.

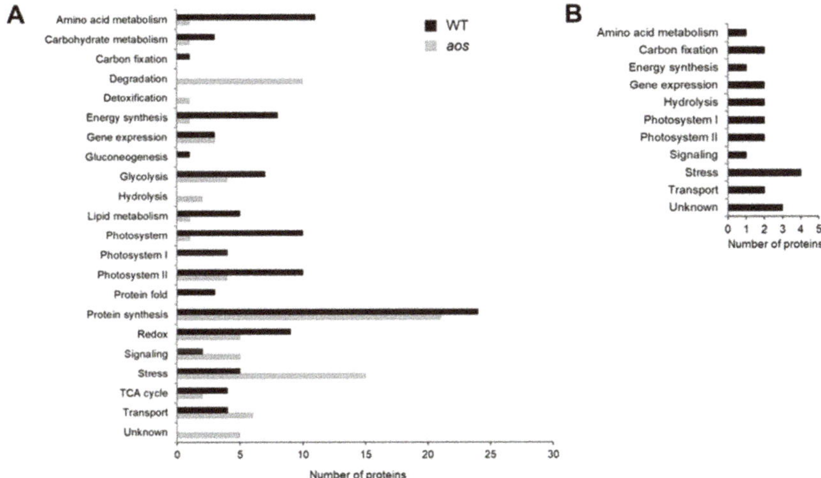

Figure 3. Functional categorization of proteins identified in response to wounding stress in *P. patens* wild-type and the *aos* mutant. (**A**) Functional categorization of proteins that accumulated in response to wounding in *P. patens* wild-type and the *aos* mutant. (**B**) Functional categorization of proteins that decreased in response to wounding in the wild-type *P. patens*. Three-week-old *P. patens* tissues were wounded using tweezers. After 24 h, proteins were extracted from the tissue and analyzed using a gel-free/label-free proteomic technique, and significantly changed proteins ($p < 0.05$, matched peptides >2) were identified using a student's *t*-test. The identified proteins were annotated using Phytozome ver. 11.0.9 (https://phytozome.jgi.doe.gov/pz/portal.html) and functionally categorized.

To further analyze the correlations of the differentially expressed proteins, protein–protein interactions were determined, and pathway prediction was performed. The Search Tool or the Retrieval of Interacting Genes (STRING) database was used to calculate all direct interactions between the 136 and 88 proteins identified in wild-type and the *aos* mutant. As shown in Figure 4, a complicated network of protein–protein interactions were found in the wild-type. The proteins categorized in each group were closely related each other. In contrast to the wild-type plant, the interactions between each protein group, which were clustered based on functions, were remotely related in the *aos* mutant. Comparison of *aos* mutant and WT suggested that fewer proteins interacted in the *aos* mutant due to wounding.

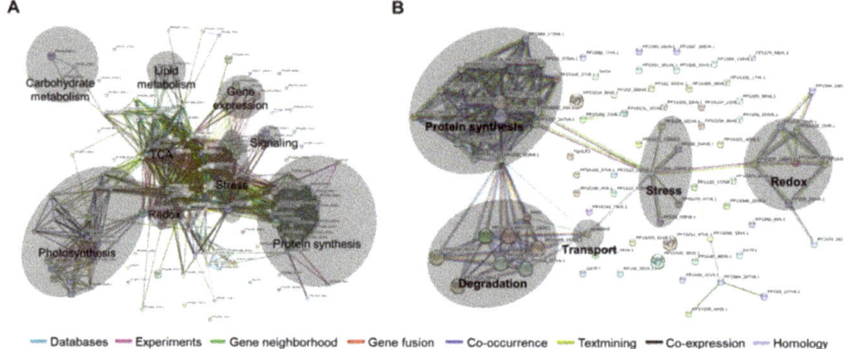

Figure 4. The Search Tool or the Retrieval of Interacting Genes (STRING) bioinformatic analysis of significantly accumulated proteins in *P. patens* after wounding. (**A**) Visualization of the protein interaction network of significantly differentially accumulated proteins in wild-type *P. patens* after wounding. The shaded analysis networks are for photosynthesis, carbohydrate metabolism, lipid metabolism, TCA, redox, stress, gene expression, signaling, and protein synthesis. Databases and text-mining were chosen as the active prediction methods. (**B**) Visualization of the protein interaction network of significantly differentially accumulated proteins in the *aos* mutant after wounding. The shaded analysis networks are for protein synthesis, degradation, transport, stress, and redox. Databases and text-mining were chosen as the active prediction methods.

2.4. Subcellular Localization of the Identified Proteins in Response to Wounding

To predict the subcellular localization of the proteins whose abundance was altered in the wild-type and the *aos* mutant under wounding stress, we conducted bioinformatic analysis [34]. Subcellular localization data revealed that more than 30% of the proteins with changes in abundance in response to wounding, including 52 proteins in the wild-type and 27 proteins in the *aos* mutant, were predicted to be localized to the chloroplast (Figure 5). Moreover, 15 proteins with increased abundance were predicted to be localized to the nucleus in the *aos* mutant, and 8 proteins were inferred to be localized to the nucleus in the wild-type. The bioinformatic findings suggested that proteins in chloroplasts play an important role in the physiological response to wounding in *P. patens*.

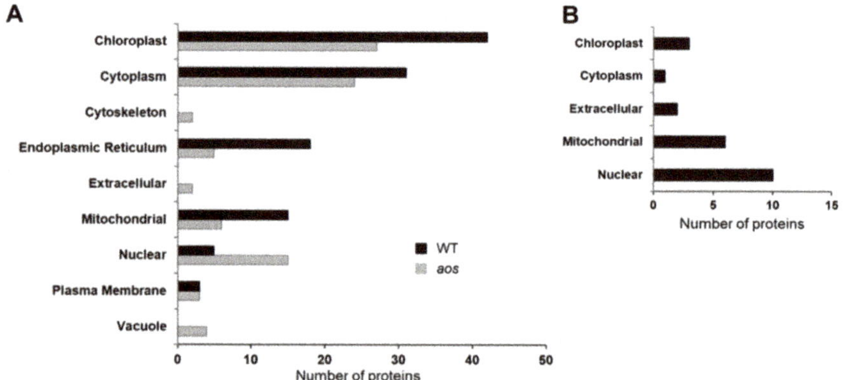

Figure 5. Subcellular localization of proteins that changed in abundance in response to wounding in *P. patens* wild-type and the *aos* mutant. (**A**) Increased proteins. (**B**) Decreased proteins. The subcellular localization of the identified proteins was predicted using TargetP, Bacello, and WoLF PSORT.

2.5. Protein Catalogs of the Identified Proteins

2.5.1. Proteins Involved in Protein Synthesis

A considerable number of proteins involved in protein synthesis increased in response to wounding in the wild-type plant (Figure 3, Table S1). Ribosomal proteins were mainly accumulated by wounding. The number of ribosomal proteins that accumulated in the *aos* mutant was comparable to that in the wild-type. These data indicated that *AOS* gene disruption was not directly related to the accumulation of proteins involved in protein synthesis.

The representative physiological reactions to wounding in plants include increases in cell-wall integrity, secondary metabolite synthesis, and stress-related plant hormone synthesis [35]. Furthermore, many metabolic pathways are activated in response to wounding. Before the induction of adaptive responses to wounding stress, the de novo production of proteins associated with those physiological events is required [35]. Therefore, increasing the abundance of proteins involved in protein synthesis is a significant physiological response in the adaptation to wounding in *P. patens*.

2.5.2. Proteins Involved in Protein Degradation

Interestingly, the levels of 10 identified proteins related to protein degradation (26S proteasome-related proteins) were increased in the *aos* mutant (Figure 3, Table S2). In contrast, increased abundance of proteins involved in protein degradation was not observed in the wild-type plant. Proteasomes are protein complexes that are related to protein degradation, which plays a major role in regulating cell physiology [36,37]. Because the increase in the 26S proteasome-related proteins was observed only in the *aos* mutant, *PpAOS* gene expression is suggested to suppress the accumulation of 26S proteasome-related proteins in the wild-type plant.

2.5.3. Proteins Involved in Amino Acid Metabolism

The abundance of more than 10 proteins related to amino acid metabolism was increased after wounding in the wild-type plant, including ketol-acid reductoisomerase, acetohydroxy acid isomeroreductase, dihydropyrimidine dehydrogenase, pyridoxal phosphate-dependent enzyme synthase, vitamin-B12 independent methionine 5-methyltetrahydropteroyltriglutamate-homocysteine, glutamate dehydrogenase, and amino acid binding protein (Figure 3, Table S1). In contrast, only one protein, asparagine synthetase, was accumulated in the *aos* mutant in response to wounding (Figure 3, Table S2). The number of proteins involved in amino acid synthesis whose abundance was increased was substantially greater in the wild-type than in the *aos* mutant. *PpAOS* gene expression is suggested to enhance the synthesis of amino acids in response to wounding. Amino acids are the basic units of proteins and are thus required for protein synthesis. It is considered that simultaneous increases in the abundance of proteins related to both amino acid production and protein synthesis is advantageous in efficient and adequate protein synthesis for stress adaptation.

2.5.4. Proteins Involved in Protein Folding

Proteins involved in the correct folding of other proteins, such as heat shock proteins (HSPs), which are classified as stress proteins in this report, accumulated in both the wild-type and the *aos* mutant of *P. patens* after wounding (Figure 3, Tables S1 and S2). These proteins are involved in spatial structural changes in a protein, which can trigger signaling that activates and/or deactivates a function [38]. Chaperonins, which are important for correct protein folding, were shown to be the highest-abundance proteins in only wild-type and were the most abundant among the identified wound-induced proteins (Table S1) [39]. Wound-induced accumulation of chaperonins in *P. patens* is inferred to depend on *PpAOS* expression. In contrast to the chaperonins detected in the wild-type plant, late embryogenesis abundant (LEA) proteins accumulated in the *aos* mutant (Table S2). LEA proteins were reported to suppress protein aggregation, which is promoted by abiotic stresses [40]. LEA proteins may be involved in protein folding in the *aos* mutant subjected to wounding.

As protein synthesis is activated in response to wounding stress, correct protein folding in wounded plants is more important than in untreated plants. This study revealed wound-stimulated increases in proteins involved in protein folding, accompanied by enhanced protein synthesis. Increased protein synthesis could require more proteins related to protein folding in wounded *P. patens*.

2.5.5. Proteins Involved in Photosystems

Proteins involved in photosystems were a major protein group induced by wounding in the wild-type plant (Figures 3 and 6, Table S1), including the photosystem I reaction center protein PsaF subunit III, the photosystem II manganese-stabilizing protein PsbO, oxygen-evolving enhancer protein I, light-harvesting complex II protein lhcb5 (chlorophyll a/b-binding protein), type III chlorophyll a/b-binding protein, and the small subunit of Rubisco.

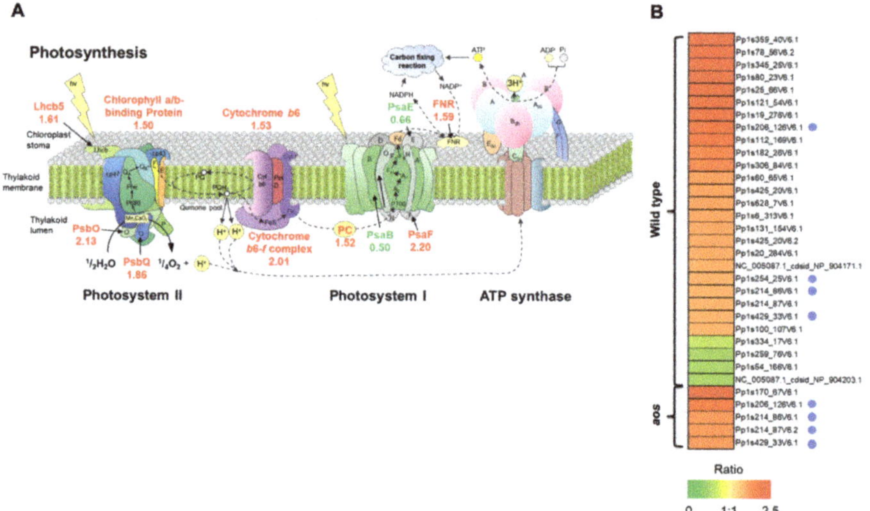

Figure 6. Differentially accumulated proteins involved in photosynthesis. (**A**) Proteins involved in photosynthesis were differentially accumulated in response to wounding stress in wild-type *P. patens*. Red and green numbers represent the relative abundance of proteins related to photosynthesis in response to wounding in the wild-type. Lhcb5 (Pp1s628_7V6.1, Pp1s6_313V6.1), light-harvesting complex II protein Lhcb5; chlorophyll a/b-binding protein (Pp1s214_86V6.1, Pp1s214_87V6.1, Pp1s429_33V6.1), type III chlorophyll a/b-binding protein; cytochrome b6 (NC_005087.1_cdsid_NP_904171.1), cytochrome b6; PsbO (Pp1s25_66V6.1, Pp1s306_84V6.1, Pp1s60_65V6.1), photosystem II manganese-stabilizing protein PsbO; PsbQ (Pp1s182_26V6.1), photosystem II oxygen-evolving complex protein PsbQ; cytochrome b6f complex (Pp1s112_169V6.1), cytochrome b6-f complex iron-sulfur subunit; PC (Pp1s254_25V6.1), plastocyanin; PsaB (NC_005087.1_cdsid_NP_904203.1), photosystem I P700 chlorophyll a apoprotein A2 (PsaB); PsaF (Pp1s345_25V6.1, Pp1s80_23V6.1, Pp1s121_54V6.1, Pp1s19_276V6.1), photosystem I reaction center protein PsaF; PsaE (Pp1s334_17V6.1), photosystem I reaction center subunit IV (PsaE); FNR (Pp1s131_154V6.1), ferredoxin-NADP+ reductase. (**B**) Heat map represents the profile of differentially accumulated photosystem-related proteins induced by wounding in *P. patens*. Red indicates high accumulation, whereas green indicates low accumulation. Blue circles indicate proteins that accumulated in both wild-type and the *aos* mutant.

Photosynthesis, which is among the most important plant physiological processes, converts light energy to chemical energy and produces carbohydrates, such as sucrose and starch, from water and carbon dioxide using light energy. The activation of the defense system, including protein synthesis, requires additional energy in plants [35]. Wounding is suggested to result in the accumulation of proteins involved in photosystems to provide energy for various stress responses. This study indicated that wounding induced the accumulation of fewer photosystem proteins in the *aos* mutant than in the wild-type. These results are accordance with the reported accumulation of proteins related to photosystems in OPDA-treated *P. patens* [30,41]. The genes encoding the photosystem proteins showing the greatest increases in response to wounding are present in the chloroplast genome. As OPDA is biosynthesized through AOS reaction in chloroplasts [4], OPDA may be important to regulate gene expression in chloroplasts under wounding stress. Thus, enhancement of the photosynthetic capacity in response to wounding is suggested to be crucial for stress adaptation in *P. patens*.

Additionally, the increased abundance of proteins related to photosystems leads to the production of reactive oxygen species (ROS), which can oxidize polyunsaturated lipids in plastid membranes. Hydroperoxidation of *a*-linolenic acid is required to increase the accumulation of OPDA [4]. Wound-induced proteins that are associated with photosystems may be involved in supplying oxygen for OPDA biosynthesis [4].

2.5.6. Proteins Involved in Glycolysis, the TCA Cycle, and Energy Synthesis

Wounding caused the accumulation of proteins involved in glycolysis, the TCA cycle, and energy synthesis in the wild-type plant (Figures 3 and 7, Table S1); these proteins included pyruvate kinase, UDP-glucose pyrophosphorylase, malate dehydrogenase, phosphoenolpyruvate carboxykinase, pyruvate dehydrogenase E1 component subunit β, glyceraldehyde-3-phosphate dehydrogenase (GAPDH), ATPase, ATP synthase β chain, and vacuolar ATPase β subunit. The number of proteins involved in glycolysis, the TCA cycle, and energy synthesis was greater in the wild-type plant than in the *aos* mutant in response to wounding (Figure 3, Tables S1 and S2). Glycolysis and the TCA cycle are required for the efficient conversion of photosynthetic products to ATP. Proteins involved in energy synthesis, such as ATPase and ATP synthase, are related to the release of energy. Wounding appears to induce proteins involved in glycolysis and the TCA cycle to readily provide usable chemical energy.

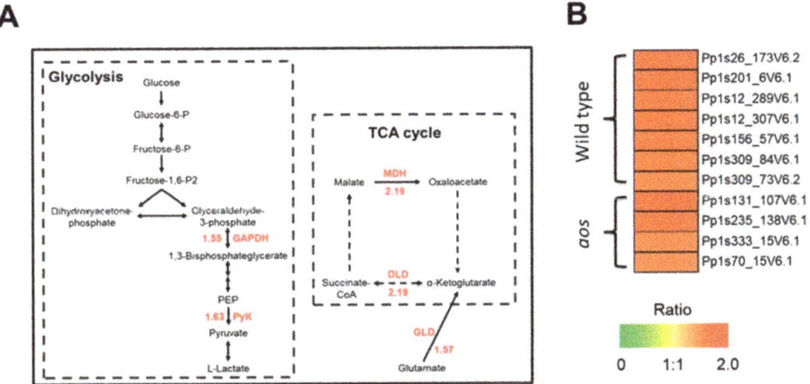

Figure 7. Differentially accumulated proteins involved in glycolysis and the TCA cycle. (**A**) Proteins involved in glycolysis and the TCA cycle were all up-regulated in response to wounding stress in wild-type *P. patens*. Red numbers represent the relative abundance of proteins related to glycolysis and the TCA cycle in response to wounding in wild-type. GAPDH (Pp1s309_84V6.1, Pp1s309_73V6.2), glyceraldehyde-3-phosphate dehydrogenase; PyK (Pp1s12_289V6.1, Pp1s12_307V6.1), pyruvate kinase; MDH (Pp1s38_300V6.1, Pp1s39_428V6.1, Pp1s79_110V6.1), malate dehydrogenase; DLD (Pp1s98_132V6.1), dihydrolipoamide dehydrogenase; GLD (Pp1s62_236V6.4), glutamate dehydrogenase. (**B**) Heat map representing the profile of differentially accumulated glycolysis and TCA cycle-related proteins induced by wounding in *P. patens*. Red indicates high accumulation, whereas green indicates low accumulation.

The accumulation in photosystem-related proteins causes increased energy production, and effective energy release is needed for various physiological responses to stress [35]. The orchestrated accumulation of proteins involved in glycolysis, the TCA cycle, and energy synthesis, accompanied by the accumulation of photosystem-related proteins, is suggested to promote the effective utilization of light energy in *P. patens.*

2.5.7. Proteins for Reactive Oxygen Scavenging

Wounding increased the abundance of superoxide dismutase in the wild-type plant but not in the *aos* mutant (Figure 3, Table S1). Superoxide dismutase catalyzes the dismutation of the superoxide radical (O_2^-) into oxygen (O_2) or hydrogen peroxide (H_2O_2) [42]. Photosynthesis is enhanced by wounding but also produces ROS, which is likely harmful to cells. The accumulation of superoxide dismutase, which acts as an antioxidant, protects the cellular components from oxidation by ROS [42]. Therefore, wounding likely induces superoxide dismutase to reduce the accumulated ROS produced by photosynthesis to decrease oxidative stress.

2.6. Quantitative RT-PCR Analysis of Genes Encoding Proteins Accumulated by Wounding

To investigate the correlation between protein accumulation and gene expression, we selected eight genes encoding proteins, which were accumulated in the wild-type plant subjected to wounding and performed qRT-PCR analyses of the genes (Figure 8). In eukaryotes, transcription occurs in the nucleus, and translation or protein synthesis take place in the cytoplasm. These processes are separated sequentially and spatially. Moreover, proteins are modified by each organelle such as phosphorylation and glycosylation, and then many of proteins become mature. In most cases, genes are expressed before accumulation of proteins. These processes are required for minutes to hours. Therefore, samples were collected at 3 h after wounding in this study. The results showed that wounding induced the transcriptional levels of four genes encoding proteins, chaperonin, PsbQ, malate dehydrogenase, and ATP synthase β chain, in the wild-type plant. In the case of these proteins, wound-induced gene expression was in accordance with wound-induced accumulation of proteins. In contrast, the expression of four other genes encoding proteins, histone H4, ketol-acid reductoisomerase, PsbO, and pyruvate dehydrogenase e1 component subunit β, was not provoked by wounding in the wild-type plant. These data indicated that protein accumulation did not always coincide with gene expression in wounded *P. patens.* When *P. patens* was subjected to wounding, the degradation of histone H4, ketol-acid reductoisomerase, PsbO, and pyruvate dehydrogenase e1 component subunit β was possibly suppressed.

2.7. Comparison of Proteomic Data in This Study with Those in P. Patens Treated with OPDA

OPDA treatment and wounding increased the levels of many proteins related to photosystems in the wild-type plant (Figure 3, Table S1) [30]; however, the abundance of photosystem-related proteins was not increased by wounding in the *aos* mutant (Figure 3, Table S2). OPDA treatment has previously been shown to increase the abundance of proteins related to photosynthesis in wild-type *P. patens* [30,41]. These data suggested that wounding induces proteins involved in photosynthesis through OPDA signaling in *P. patens.*

In contrast to wounded *P. patens*, proteins involved in protein synthesis and protein folding were decreased in OPDA-treated *P. patens* [30]. Wounding stimulates a wide variety of signaling systems. OPDA-mediated signaling is one of wound-induced physiological responses in *P. patens*. Various types of signaling triggered by wounding are probably involved in the crosstalk with OPDA signaling in *P. patens* (Figure 9). Therefore, the proteins accumulated in wounded *P. patens* do not completely correspond to the proteins whose abundance is altered in *P. patens* treated with OPDA.

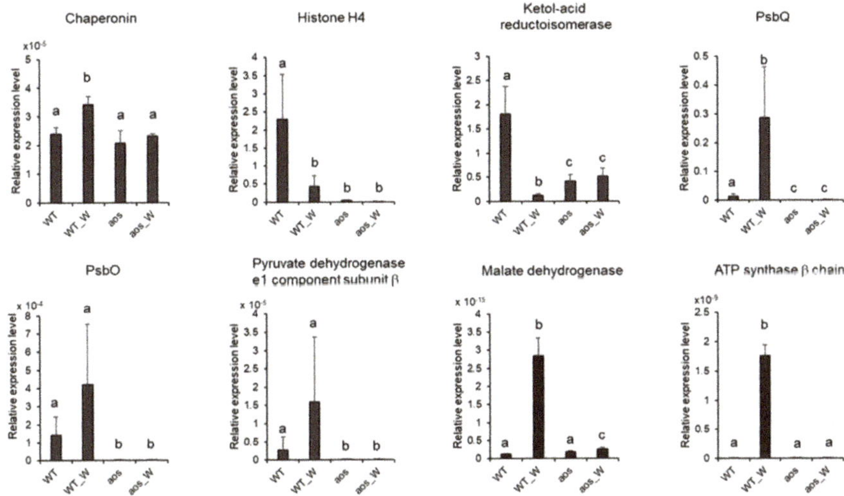

Figure 8. qRT-PCR analyses of genes encoding wound-accumulative proteins. The transcription levels of selected genes belonging to different groups were analyzed. The selected genes included the following: Chaperonin (Pp1s141_125V6.1); Histone H4 (Pp1s342_32V6.1); Ketol-acid reductoisomerase (Pp1s60_179V6.1); Photosystem II oxygen evolving complex Protein PsbQ (Pp1s182_26V6.1); Photosystem II manganese-stabilizing protein PsbO (Pp1s306_84V6.1); Pyruvate dehydrogenase e1 component subunit β (Pp1s156_57V6.1); Malate dehydrogenase (Pp1s79_110V6.1) and ATP synthase β chain (Pp1s310_30V6.1). WT, wild-type without treatment; WT_W, wild-type treated with wounding for 3 h; aos, *aos* mutant without treatment; aos_W, *aos* mutant treated with wounding for 3 h. The values are the mean ± SD ($n = 3$). Relative gene expression levels were normalized to actin. The data are shown as the mean ± SD from three independent biological replicates. Different letters indicate that the change is significant as determined by one-way ANOVA according to Tukey's multiple comparison test ($p < 0.05$).

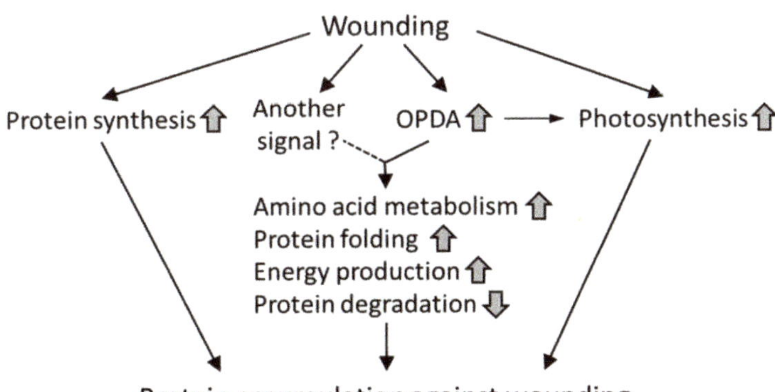

Figure 9. Proposed model for protein accumulation in response to wounding in *P. patens*. Upward or downward arrows indicate an increase or decrease, respectively.

In a previous study, *P. patens* was subjected to 10 μM OPDA for 24 h [30]. However, the OPDA concentration was transiently increased by wounding until 2 h after wounding and decreased to basal levels after 4 h. in *P. patens* (Figure 1). It appears that *P. patens* subjected to wounding was not significantly affected by OPDA at 4 h after wounding. The difference of the period under the influence of OPDA may cause differential accumulation of proteins between wounded *P. patens* and OPDA-treated *P. patens*.

3. Materials and Methods

3.1. Plant Growth Conditions and Treatment

The moss *Physcomitrella patens* Gransden 2004 strain was used as wild-type *P. patens* and was grown on 20 mL of BCDAT agar medium [43] for 3 weeks in a 9-cm Petri dish under continuous white fluorescent light (40 μmol photons $m^{-2}s^{-1}$) at 25°C. To generate wounding stress, whole gametophores were wounded with tweezers.

3.2. Analysis of OPDA Concentration in P. Patens

Gametophores of *P. patens* (approximately 200 mg), which were grown on BCDAT agar medium for 3 weeks, were frozen in liquid nitrogen and extracted with 10 mL of ethanol. The concentration of OPDA was analyzed by UPLC-MS/MS according to the method of Sato et al [44]. Four independent experiments were conducted as biological replicates. One-way analysis of variance (ANOVA) followed by Tukey's multiple comparison was used for comparison among multiple groups and conducted using R (version 3.6.0). A *p*-value of < 0.05 was considered as statistically significant.

3.3. Generation of a P. Patens Mutant with Disrupted PpAOS1 and PpAOS2

To construct a vector for *PpAOS1* gene disruption, a 1.0-kb genomic DNA fragment beginning 5′ to *PpAOS1* was amplified using KOD FX DNA polymerase (Toyobo, Osaka, Japan), and the primers PpAOS1KO5′-F and PpAOS1KO5′-R. A 1.0-kb genomic DNA fragment ending 3′ to *PpAOS1* was also amplified using primers PpAOS1KO3′-F and PpAOS1KO3′-R. Each fragment was cloned into the vector pBluescript SKII (+) (Merck, Darmstadt, Germany). The 5′ *PpAOS1* genomic fragment was digested with *Xba*I (Takara Bio Inc., Shiga, Japan) and *Eco*RI (Takara Bio Inc., Shiga, Japan) and inserted into pTN182, which carries a G418-resistant cassette, digested with *Xba*I (Takara Bio Inc., Shiga, Japan) and *Eco*RI (Takara Bio Inc., Shiga, Japan) to obtain pTN182-PpAOS1KO5′. Similarly, the 3′ *PpAOS1* genomic fragment was inserted into pTN182-PpAOS1KO5′, which had been digested with *Sph*I (Takara Bio Inc., Shiga, Japan) and *Nde*I, (Takara Bio Inc., Shiga, Japan) to yield pTN182-PpAOS1KO.

A vector for *PpAOS2* gene disruption was constructed using pTN186-PpAOS2KO. A 1.0-kb genomic DNA fragment beginning 5′ to *PpAOS2* was amplified using KOD FX DNA polymerase (Toyobo, Osaka, Japan) and primers PpAOS2KO5′-F and PpAOS2KO5′-R. A 1.0-kb genomic DNA fragment ending 3′ to *PpAOS2* was also amplified using primers PpAOS2KO3′-F and PpAOS2KO3′-R. Each fragment was cloned into pBluescript SKII (+) (Merck, Darmstadt, Germany). The 5′ *PpAOS2* genomic fragment was digested with *Kpn*I (Takara Bio Inc., Shiga, Japan) and *Hind*III (Takara Bio Inc., Shiga, Japan) and inserted into pTN186, which contained a hygromycin-resistance cassette and had been digested with *Kpn*I (Takara Bio Inc., Shiga, Japan) and *Hind*III (Takara Bio Inc., Shiga, Japan) to obtain pTN186-PpAOS2KO5′. The 3′ *PpAOS2* genomic fragment was inserted into pTN186-PpAOS2KO5′, which had been digested with *Sph*I (Takara Bio Inc., Shiga, Japan) and *Sac*I (Takara Bio Inc., Shiga, Japan), to yield pTN186-PpAOS2KO.

The plasmids pTN182-PpAOS1KO and pTN186-PpAOS2KO contained *Bam*HI and *Kpn*I restriction sites, respectively. Polyethylene glycol-mediated transformation was conducted as reported previously by Nishiyama et al [43]. The selected plants were incubated for an additional week without antibiotics and then transferred again onto selection medium. Stable transformants were chosen by PCR using appropriate primer sets (5′ end of *PpAOS1*: PpAOS2KO5′-F2 and Pcmv-R; 3′ end of *PpAOS1*: 35SPS-F and PpAOS2KO3′-R2; 5′ end of *PpAOS2*: PpAOS1KO5′-F2 and Pcmv-R; 3′ end of *PpAOS2*: 35SPS-F and PpAOS1KO3′-R2) to confirm the integration of the selectable marker into the targeted genes *PpAOS1* and *PpAOS2*. The primers used in this experiment are listed in Supplementary Table S3.

3.4. Protein Extraction

P. patens gametophores were grown for 3 weeks and then wounded with tweezers. At 24 h after wounding, approximately 500 mg of *P. patens* fresh tissue was ground into powder in liquid nitrogen using a mortar and pestle. The powder was transferred into a solution of 10% trichloroacetic acid and 0.07% 2-mercaptoethanol in acetone and mixed. The suspension was sonicated for 5 min and then incubated for 45 min at −20 °C. After this incubation, the suspension was centrifuged at $9,000 \times g$ for 20 min at 4 °C. The resulting supernatant was discarded, and the pellet was washed three times with 3 mL of acetone containing 0.07% 2-mercaptoethanol. The final pellet was dried using a vacuum pump. The pellet was resuspended by vortexing for 1 h at 25 °C in 5 mL of lysis buffer consisting of 100 mM Tris-HCl (pH 8.5), 2% SDS, and 50 mM dithiothreitol (DTT). The suspension was then centrifuged at $20,000 \times g$ for 20 min at 25 °C. The resulting supernatant was collected as the total protein solution. Three independent experiments were conducted as biological replicates. The concentration of the protein solution was measured using the Lowry method [45].

3.5. Digestion of Proteins

For in-solution digestion, 100 μg of protein was subjected to chloroform/methanol extraction [46]. The pellet was resuspended with 50 mM NH_4HCO_3. The solution was reduced with 50 mM DTT and then alkylated with 50 mM iodoacetamide. Proteins were digested using trypsin and lysyl endopeptidase at a 1:100 enzyme/protein ratio at 37 °C for 16 h [33].

3.6. Nanoliquid Chromatography-Tandem MS Analysis

Peptide separation and detection were performed using an Ultimate 3000 nano LC (Thermo Fisher Scientific, San Jose, CA, USA) and an LTQ Orbitrap mass spectrometer (Thermo Fisher Scientific, San Jose, CA, USA). The system was operated in data-dependent acquisition mode with XCalibur software (ver. 2.0.7, Thermo Fisher Scientific, San Jose, CA, USA). The peptides were loaded onto a C18 PepMap trap column (300 μm ID × 5 mm, Dionex, Thermo Fisher Scientific, San Jose, CA, USA). The peptides were eluted with a linear acetonitrile gradient (8–30% over 150 min) in 0.1% formic acid in acetonitrile at a flow rate of 200 nL/min and were loaded and separated on a C18 capillary tip column (75 μm ID × 120 mm, nano LC capillary column, NTTC-360/75-3, Nikkyo Technos, Tokyo, Japan) with a spray voltage of 1.5 kV. Elution was performed with a linear acetonitrile gradient (5-25% in 120 min) in 0.1% formic acid. Full-scan mass spectra were acquired in the Orbitrap over 400-1,500 *m/z* with a resolution of 30,000. A lock mass function was used to obtain high mass accuracy [47]. The top ten most intense precursor ions were selected for collision-induced fragmentation in the linear ion trap at a normalized collision energy of 35%. Dynamic exclusion was employed within 90 s to prevent the repetitive selection of the peptides [48].

3.7. Protein Identification Using Mascot

Proteins were identified from the acquired MS/MS spectra using Mascot software (ver. 2.5.1, Matrix Science, London, UK) and from the *P. patens* database (38,480 protein sequences) and a contaminant database (262 protein sequences) with Proteome Discoverer (ver. 1.4.0.288, Thermo Fisher Scientific, San Jose, CA, USA). The *P. patens* database was obtained from the Phytozome database (ver. 11. 0. 9, http://www.phytozome.net/). The parameters used in the Mascot searches were as follows: the carbamidomethylation of cysteine was set as a fixed modification; the oxidation of methionine was set as a variable modification; trypsin was specified as the proteolytic enzyme; and one missed cleavage was allowed. The peptide mass tolerance was set at 10 ppm. The fragment mass tolerance was set at 0.8 Da, and the peptide charge was set at +2, +3, and +4. An automatic decoy database search was performed within the search. The Mascot results were filtered using the percolator function in Proteome Discoverer to improve the accuracy and sensitivity of peptide identification [49]. False discovery rates for the identification of all searches were less than 1.0%. Peptides with a percolator ion score of more than 13 (student's *t*-test, $p < 0.05$) were used for protein identification.

3.8. Analysis of Differentially Accumulated Proteins using the Acquired MS Data

For differential analyses, the commercial label-free quantification package SIEVE (ver. 2.1, Thermo Fisher Scientific, San Jose, CA, USA) was used to compare the relative abundance of peptides and proteins between the control and experimental groups. The chromatographic peaks detected by MS were aligned, and the peptide peaks were detected as frames using the following settings: the frame time width was 5.0 min; the frame *m/z* width was 10 ppm, and frames were produced on all parent ions subjected to MS/MS scanning. The frames with MS/MS scans were matched to the imported Mascot results. In the analysis of differential protein abundance, the total ion current was used for normalization. The minimum requirement for the identification of a protein was two matched peptides and a *p* value of < 0.05.

3.9. Classification of Proteins and Bioinformatic Analysis

The functions of the identified proteins were categorized according to the annotations of the cosmoss.org *P. patens* database (http://www.cosmoss.org/) and the EU *Arabidopsis thaliana* genome project [50]. The functional interactions of the identified proteins were examined using STRING (version 10.5, http://string-db.org/) [51]. Briefly, the protein list was subjected to Blast searches against the *Physcomitrella patens* STRING database, which includes the physical and functional relationships of protein molecules, supported by associations derived from eight lines of evidence: the neighborhood in the genome; gene fusions; cooccurrence across the genome; coexpression; experimental/biochemical data; information in databases (associations in curated databases); text-mining (comentioned in PubMed abstracts); and homology [52]. The biological pathways of the identified proteins were deduced from KEGG analysis (http://www.genome.jp/kegg/). The results of GO functional and pathway enrichment analysis and the pathway enrichment of the identified proteins were analyzed using an international standardized gene functional classification system (http://www.geneontology.org/), employing settings in reference to previous reports [33,53]. Protein subcellular localization was predicted using TargetP (http://www.cbs.dtu.dk/services/TargetP/), Bacello (http://gpcr2.biocomp.unibo.it/bacello/index.htm) and WoLF PSORT (http://wolfspsort.org/) [54]. Multiple sequence alignments were conducted by using Clustal Omega (http://www.ebi.ac.uk/Tools/msa/clustalo/).

3.10. Quantitative RT-PCR

The coding sequences (CDSs) for the selected genes were used to design specific primers for quantitative RT-PCR (qRT-PCR). *P. patens* total RNA was extracted using the Isospin Plant RNA kit (Nippon gene, Tokyo, Japan) according to the manufacturer's instructions. First-strand cDNA prepared by M-MLV reverse transcriptase (Invitrogen, Carlsbad, CA, USA) was used as the template. To analyze gene expression levels of a chaperonin gene (Pp1s141_125V6.1), the KOD SYBR qPCR Mix (Toyobo, Osaka, Japan) was used according to the manufacturer's protocol. Each reaction mixture contained 12.5 μL of KOD SYBR qPCR Mix, 1 μL of each primer (10 mM), 1 μL of cDNA, and 9.5 μL of MilliQ water. qRT-PCR was performed on a Thermal Cycler Dice Real-Time system (TP800, Takara Bio Inc., Shiga, Japan). To analyze expression levels of other genes, the TB Green™ Premix Ex Taq™ II (Takara Bio Inc., Shiga, Japan) was used according to the manufacturer's protocol. Each reaction mixture contained 10 μL of TB Green *Premix Ex Taq* II (Takara Bio Inc., Shiga, Japan), 0.4 μL of each primer (10 mM), 2 μL of cDNA, 0.4 μL ROX Reference Dye, and 6.8 μL of MilliQ water. qRT-PCR was performed on a Thermal Cycler Dice Real-Time system (Applied Biosystem 7000, Thermo Fisher Scientific, United Kingdom). The qRT-PCR conditions were as follows: preincubation at 95°C for 30 s followed by 40 cycles of 95°C for 5 s and 60°C for 30 s. The specificity of each PCR amplicon was assessed with a dissociation curve (95°C for 15 s, 60°C for 30 s, and 95°C at 15 s). Actin (accession no: AW698983) was used as an internal standard for the normalization of gene expression, and the actin level was set to 1.0. The primers used in this experiment are listed in Supplementary Table S4.

4. Conclusions

This study revealed that wounding mainly promoted the accumulation of proteins involved in protein synthesis, amino acid synthesis, photosynthesis, protein folding, and glycolysis. Because these wounding-responsive proteins are also found in flowering plants, the accumulation of these proteins in response to wounding may be conserved in land plants. The comparison of proteomic data from the wild-type and the *aos* mutant suggests that *PpAOS* gene expression, which leads to an increase in OPDA, enhances photosynthesis and effective energy utilization in response to wounding in *P. patens*. The present data will help our understanding of adaptive signaling in response to wounding in land plants, as well as the effects of *AOS* gene expression for stress adaptation in *P. patens*.

Supplementary Materials: Supplementary materials can be found at http://www.mdpi.com/1422-0067/21/4/1417/s1.

Author Contributions: W.L. and K.T. conceived and designed the research. W.L., T.A., S.K., and K.T. conducted the experiments. W.L., S.K., H.M. and K.T. analyzed the data. W.L. and K.T. wrote the manuscript. All authors read and approved the manuscript.

Funding: This work was supported by a Grant-in-Aid for Scientific research (No. 21580123) from the Japan Society for the Promotion of Science (JSPS) to K.T.

Acknowledgments: We thank Y. Konishi, G. Tanaka and N. Kuwata of Hokkaido University for technical support. We also thank R. Imai of Institute of Agrobiological Sciences, National Agriculture and Food Research Organization (NARO), for his kind support. We are grateful to the Chinese Scholarship Council for scholarship to W.L. (CSC 201206880001).

Conflicts of Interest: The authors declare no conflict of interest.

Abbreviations

AOS	Allene oxide synthase
COI1	Coronatine insensitive 1
12,13-EOT	12,13(*S*)-epoxyoctadecatrienoic acid
GAPDH	Glyceraldehyde-3-phosphate dehydrogenase
GLD	Glutamate dehydrogenase
GO	Gene Ontology
HPL	Hydroperoxide lyase
13-HPOT	13(*S*)-hydroperoxyoctadecatrienoic acid
HSP	Heat shock protein
JA	Jasmonic acid
JA-Ile	Isoleucine conjugate of jasmonic acid
KEGG	Kyoto Encyclopedia of Genes and Genomes
LC-MS/MS	Liquid chromatography-tandem mass spectrometry
LEA	Late embryogenesis abundant
MDH	Malate dehydrogenase;
MeJA	Methyl jasmonate
OPDA	12-Oxo-phytodienoic acid
qRT-PCR	Quantitative real-time polymerase chain reaction
ROS	Reactive oxygen species
Rubisco	Ribulose-1,5-bisphosphate carboxylase/oxygenase
STRING	Search Tool or the Retrieval of Interacting Genes
UPLC-MS/MS	Ultra-performance liquid chromatography-tandem mass spectrometry

References

1. Ku, Y.S.; Sintaha, M.; Cheung, M.Y.; Lam, H.M. Plant hormone signaling crosstalks between biotic and abiotic stress responses. *Int. J. Mol. Sci.* **2018**, *19*, 3206. [CrossRef]
2. León, J.; Rojo, E.; Sánchez-Serrano, J.J. Wound signalling in plants. *J. Exp. Bot.* **2001**, *52*, 1–9. [CrossRef]
3. Nascimento, N.C.; Fett-Neto, A.G. Plant secondary metabolism and challenges in modifying its operation: an overview. *Methods Mol. Biol.* **2010**, *643*, 1–13. [PubMed]
4. Wasternack, C.; Hause, B. Jasmonates: Biosynthesis, perception, signal transduction and action in plant stress response, growth and development. An update to the 2007 review in Annals of Botany. *Ann. Bot.* **2013**, *111*, 1021–1058. [CrossRef]
5. Pratiwi, P.; Tanaka, G.; Takahashi, T.; Xie, X.; Yoneyama, K.; Matsuura, H.; Takahashi, K. Identification of jasmonic acid and jasmonoyl-isoleucine, and characterization of AOS, AOC, OPR and JAR1 in the model lycophyte *Selaginella moellendorffii*. *Plant Cell Physiol.* **2017**, *58*, 789–801. [CrossRef] [PubMed]
6. Browse, J.; Howe, G.A. New weapons and a rapid response against insect attack. *Plant Physiol.* **2008**, *146*, 832–838. [CrossRef] [PubMed]
7. Browse, J. Jasmonate passes muster: A receptor and targets for the defense hormone. *Ann. Rev. Plant Biol.* **2009**, *60*, 183–205. [CrossRef] [PubMed]
8. Sheard, L.B.; Tan, X.; Mao, H.; Withers, J.; Ben-Nissan, G.; Hinds, T.R.; Kobayashi, Y.; Hsu, F.; Sharon, M.; Browse, J.; et al. Jasmonate perception by inositol-phosphate-potentiated COI1-JAZ co-receptor. *Nature* **2010**, *468*, 400–405. [CrossRef]
9. Santino, A.; Taurino, M.; De Domenico, S.; Bonsegna, S.; Poltronieri, P.; Pastor, V.; Flors, V. Jasmonate signaling in plant development and defense response to multiple (a)biotic stresses. *Plant Cell Rep.* **2013**, *32*, 1085–1098. [CrossRef]
10. Fonseca, S.; Chico, J.M.; Solano, R. The jasmonate pathway: The ligand, the receptor and the core signalling module. *Curr. Opin. Plant Biol.* **2009**, *12*, 539–547. [CrossRef]
11. Fonseca, S.; Chini, A.; Hamberg, M.; Adie, B.; Porzel, A.; Kramell, R.; Miersch, O.; Wasternack, C.; Solano, R. (+)-7-*iso*-Jasmonoyl-L-isoleucine is the endogenous bioactive jasmonate. *Nat. Chem. Biol.* **2009**, *5*, 344–350. [CrossRef] [PubMed]
12. Bottcher, C.; Pollmann, S. Plant oxylipins: Plant responses to 12-oxo-phytodienoic acid are governed by its specific structural and functional properties. *FEBS J.* **2009**, *276*, 4693–4704. [CrossRef] [PubMed]
13. Stumpe, M.; Göbel, C.; Faltin, B.; Beike, A.K.; Hause, B.; Himmelsbach, K.; Bode, J.; Kramell, R.; Wasternack, C.; Frank, W.; et al. The moss *Physcomitrella patens* contains cyclopentenones but no jasmonates: mutations in allene oxide cyclase lead to reduced fertility and altered sporophyte morphology. *New Phytol.* **2010**, *188*, 740–749. [CrossRef]
14. Anterola, A.; Göbel, C.; Hornung, E.; Sellhorn, G.; Feussner, I.; Grimes, H. *Physcomitrella patens* has lipoxygenases for both eicosanoid and octadecanoid pathways. *Phytochemisty* **2009**, *70*, 40–52. [CrossRef]
15. Costa, C.L.; Arruda, P.; Benedetti, C.E. An Arabidopsis gene induced by wounding functionally homologous to flavoprotein oxidoreductases. *Plant Mol. Biol.* **2000**, *44*, 61–71. [CrossRef]
16. Buseman, C.M.; Tamura, P.; Sparks, A.A.; Baughman, E.J.; Maatta, S.; Zhao, J.; Roth, M.R.; Esch, S.W.; Shah, J.; Williams, T.D.; et al. Wounding stimulates the accumulation of glycerolipids containing oxophytodienoic acid and dinor-oxophytodienoic acid in Arabidopsis leaves. *Plant Physiol.* **2006**, *142*, 28–39. [CrossRef] [PubMed]
17. Taki, N.; Sasaki-Sekimoto, Y.; Obayashi, T.; Kikuta, A.; Kobayashi, K.; Ainai, T.; Yagi, K.; Sakurai, N.; Suzuki, H.; Masuda, T.; et al. 12-oxo-phytodienoic acid triggers expression of a distinct set of genes and plays a role in wound-induced gene expression in Arabidopsis. *Plant Physiol.* **2005**, *139*, 1268–1283. [CrossRef]
18. Dave, A.; Hernandez, M.L.; He, Z.; Andriotis, V.M.E.; Vaistij, F.E.; Larson, T.R.; Graham, I.A. 12-oxo-phytodienoic acid accumulation during seed development represses seed germination in *Arabidopsis*. *Plant Cell* **2011**, *23*, 583–599. [CrossRef]
19. Dave, A.; Vaistij, F.E.; Gilday, A.D.; Penfield, S.D.; Graham, I.A. Regulation of *Arabidopsis thaliana* seed dormancy and germination by 12-oxo-phytodienoic acid. *J. Exp. Bot.* **2016**, *67*, 2277–2284. [CrossRef]
20. Goetz, S.; Hellwege, A.; Stenzel, I.; Kutter, C.; Hauptmann, V.; Forner, S.; McCaig, B.; Hause, G.; Miersch, O.; Wasternack, C.; et al. Role of cis-12-oxo-phytodienoic acid in tomato embryo development. *Plant Physiol.* **2012**, *158*, 1715–1727. [CrossRef]

21. Qiu, Y.L.; Li, L.; Wang, B.; Chen, Z.; Knoop, V.; Groth-Malonek, M.; Dombrovska, O.; Lee, J.; Kent, L.; Rest, J.; et al. The deepest divergences in land plants inferred from phylogenomic evidence. *Proc. Natl. Acad. Sci. USA* **2006**, *103*, 15511–15516. [CrossRef] [PubMed]

22. Bowman, J.L.; Floyd, S.K.; Sakakibara, K. Green genes-comparative genomics of the green branch of life. *Cell* **2007**, *129*, 229–234. [CrossRef] [PubMed]

23. Rensing, S.A.; Lang, D.; Zimmer, A.D.; Terry, A.; Salamov, A.; Shapiro, H.; Nishiyama, T.; Perroud, P.F.; Lindquist, E.A.; Kamisugi, Y.; et al. The *Physcomitrella* genome reveals evolutionary insights into the conquest of land by plants. *Science* **2008**, *319*, 64–69. [CrossRef] [PubMed]

24. Roberts, A.W.; Roberts, E.M.; Haigler, C.H. Moss cell walls: Structure and biosynthesis. *Front. Plant Sci.* **2012**, *3*, 166. [CrossRef] [PubMed]

25. Wang, X.; Yang, P.; Gao, Q.; Liu, X.; Kuang, T.; Shen, S.; He, Y. Proteomic analysis of the response to high-salinity stress in *Physcomitrella patens*. *Planta* **2008**, *228*, 167–177. [CrossRef]

26. Wang, X.; Yang, P.; Zhang, X.; Xu, Y.; Kuang, T.; Shen, S.; He, Y. Proteomic analysis of the cold stress response in the moss, *Physcomitrella patens*. *Proteomics* **2009**, *9*, 4529–4538. [CrossRef]

27. Wang, X.; Kuang, T.; He, Y. Conservation between higher plants and the moss *Physcomitrella patens* in response to the phytohormone abscisic acid: A proteomics analysis. *BMC Plant Biol.* **2010**, *10*, 192. [CrossRef]

28. Wang, X.; Liu, Y.; Yang, P. Proteomic studies of the abiotic stresses response in model moss—*Physcomitrella patens*. *Front. Plant Sci.* **2012**, *3*, 258. [CrossRef]

29. Ponce de Leon, I.; Schmelz, E.A.; Gaggero, C.; Castro, A.; Álvarez, A.; Montesano, M. *Physcomitrella patens* activates reinforcement of the cell wall, programmed cell death and accumulation of evolutionary conserved defence signals, such as salicylic acid and 12-oxo-phytodienoic acid, but not jasmonic acid, upon *Botrytis cinerea* infection. *Mol. Plant Pathol.* **2012**, *13*, 960–974. [CrossRef]

30. Toshima, E.; Nanjo, Y.; Komatsu, S.; Abe, T.; Matsuura, H.; Takahashi, K. Proteomic analysis of *Physcomitrella patens* treated with 12-oxo-phytodienoic acid, an important oxylipin in plants. *Biosci. Biotechnol. Biochem.* **2014**, *78*, 946–953. [CrossRef]

31. Scholz, J.; Brodhun, F.; Hornung, E.; Herrfurth, C.; Stumpe, M.; Beike, A.K.; Faltin, B.; Frank, W.; Reski, R.; Feussner, I. Biosynthesis of allene oxides in *Physcomitrella patens*. *BMC Plant Biol.* **2012**, *12*, 228. [CrossRef] [PubMed]

32. Kosova, K.; Vitamvas, P.; Prasil, I.T.; Renaut, J. Plant proteome changes under abiotic stress–contribution of proteomics studies to understanding plant stress response. *J. Proteomics* **2011**, *74*, 1301–1322. [CrossRef]

33. Wang, X.; Komatsu, S. Proteomic analysis of calcium effects on soybean root tip under flooding and drought stresses. *Plant Cell Physiol.* **2017**, *58*, 1405–1420. [CrossRef]

34. Ashburner, M.; Ball, C.A.; Blake, J.A.; Botstein, D.; Butler, H.; Cherry, J.M.; Davis, A.P.; Dolinski, K.; Dwight, S.S.; Eppig, J.T.; et al. Gene ontology: Tool for the unification of biology. The Gene Ontology Consortium. *Nat. Genet.* **2000**, *25*, 25–29. [CrossRef] [PubMed]

35. Savatin, D.V.; Gramegna, G.; Modesti, V.; Cervone, F. Wounding in the plant tissues: The defense of a dangerous passage. *Front. Plant Sci.* **2014**, *5*, 470. [CrossRef] [PubMed]

36. Vierstra, R.D. The ubiquitin-26S proteasome system at the nexus of plant biology. *Nat. Rev. Mol. Cell Biol.* **2009**, *10*, 385–397. [CrossRef]

37. Sadanandom, A.; Bailey, M.; Ewan, R.; Lee, J.; Nelis, S. The ubiquitin-proteasome system: Central modifier of plant signalling. *New Phytol.* **2014**, *196*, 13–28. [CrossRef] [PubMed]

38. Jacob, P.; Hirt, H.; Bendahmane, A. The heat-shock protein/chaperone network and multiple stress resistance. *Plant Biotechnol. J.* **2017**, *15*, 405–414. [CrossRef]

39. Gruber, A.V.; Nisemblat, S.; Azem, A.; Weiss, C. The complexity of chloroplast chaperonins. *Trends Plant Sci.* **2013**, *18*, 688–694. [CrossRef]

40. Wise, M.J.; Tunnacliffe, A. POPP the question: What do LEA proteins do? *Trends Plant Sci.* **2014**, *9*, 13–17. [CrossRef]

41. Luo, W.; Nanjo, Y.; Komatsu, S.; Matsuura, H.; Takahashi, K. Proteomics of *Physcomitrella patens* protonemata subjected to treatment with 12-oxo-phytodienoic acid. *Biosci. Biotechnol. Biochem.* **2016**, *80*, 2357–2364. [CrossRef] [PubMed]

42. Alscher, R.G.; Erturk, N.; Heath, L.S. Role of superoxide dismutases (SODs) in controlling oxidative stress in plants. *J. Exp. Bot.* **2002**, *53*, 1331–1341. [CrossRef] [PubMed]

43. Nishiyama, T.; Hiwatashi, Y.; Sakakibara, K.; Kato, M.; Hasebe, M. Tagged mutagenesis and gene-trap in the moss, *Physcomitrella patens* by shuttle mutagenesis. *DNA Res.* **2000**, *7*, 9–17. [CrossRef] [PubMed]

44. Sato, C.; Seto, Y.; Nabeta, K.; Matsuura, H. Kinetics of the accumulation of jasmonic acid and its derivatives in systemic leaves of tobacco (*Nicotiana tabacum* cv. Xanthi nc) and translocation of deuterium-labeled jasmonic acid from the wounding site to the systemic site. *Biosci. Biotechol. Biochem.* **2009**, *73*, 1962–1970. [CrossRef]

45. Komatsu, S.; Han, C.; Nanjo, Y.; Altaf-Un-Nahar, M.; Wang, K.; He, D.; Yang, P. Label-free quantitative proteomic analysis of abscisic acid effect in early-stage soybean under flooding. *J. Proteome Res.* **2013**, *12*, 4769–4784. [CrossRef]

46. Nanjo, Y.; Skultety, L.; Uváčková, L.; Klubicová, K.; Hajduch, M.; Komatsu, S. Mass spectrometry-based analysis of proteomic changes in the root tips of flooded soybean seedlings. *J. Proteome Res.* **2012**, *11*, 372–385. [CrossRef]

47. Olsen, J.V.; de Godoy, L.M.F.; Li, G.; Macek, B.; Mortensen, P.; Pesch, R.; Makarov, A.; Lange, O.; Horning, S.; Mann, M. Parts per million mass accuracy on an orbitrap mass spectrometer via lock mass injection into a C-trap. *Mol. Cell Proteomics* **2005**, *4*, 2010–2021. [CrossRef]

48. Zhang, Y.; Wen, Z.; Washburn, M.P.; Florens, L. Effect of dynamic exclusion duration on spectral count based quantitative proteomics. *Anal. Chem.* **2009**, *81*, 6317–6326. [CrossRef]

49. Brosch, M.; Yu, L.; Hubbard, T.; Choudhary, J. Accurate and sensitive peptide identification with Mascot Percolator. *J. Proteome Res.* **2009**, *8*, 3176–3181. [CrossRef]

50. Bevan, M.; Bancroft, I.; Bent, E. Analysis of 1.9 Mb of contiguous sequence from chromosome 4 of *Arabidopsis thaliana*. *Nature* **1998**, *391*, 485–488.

51. Szklarczyk, D.; Franceschini, A.; Kuhn, M.; Simonovic, M.; Roth, A.; Minguez, P.; Doerks, T.; Stark, M.; Muller, J.; Bork, P.; et al. The STRING database in 2011: functional interaction networks of proteins, globally integrated and scored. *Nucl. Acids Res.* **2011**, *39* . [CrossRef] [PubMed]

52. Szklarczyk, D.; Morris, J.H.; Cook, H.; Kuhn, M.; Wyder, S.; Simonovic, M.; Santos, A.; Doncheva, N.T.; Roth, A.; Bork, P.; et al. The STRING database in 2017: quality-controlled protein-protein association networks, made broadly accessible. *Nucl. Acids Res.* **2017**, *45*, D362–D368. [CrossRef] [PubMed]

53. Xiao, L.; Zhang, L.; Yang, G.; Zhu, H.; He, Y. Transcriptome of protoplasts reprogrammed into stem cells in *Physcomitrella patens*. *PLoS ONE* **2012**, *7*, e35961. [CrossRef]

54. Martinez-Cortes, T.; Pomar, F.; Merino, F.; Novo-Uzai, E. A proteomic approach to *Physcomitrella patens* rhizoid exudates. *J. Plant Physiol.* **2014**, *171*, 1671–1678. [CrossRef]

International Journal of
Molecular Sciences

Article

Phenylpropanoids Are Connected to Cell Wall Fortification and Stress Tolerance in Avocado Somatic Embryogenesis

Carol A. Olivares-García [1,2], Martín Mata-Rosas [1], Carolina Peña-Montes [2,*],
Francisco Quiroz-Figueroa [3], Aldo Segura-Cabrera [4], Laura M. Shannon [5],
Victor M. Loyola-Vargas [6], Juan L. Monribot-Villanueva [7], Jose M. Elizalde-Contreras [7],
Enrique Ibarra-Laclette [7], Mónica Ramirez-Vázquez [7], José A. Guerrero-Analco [7]
and Eliel Ruiz-May [7,*]

[1] Red de Manejo Biotecnológico de Recursos, Instituto de Ecología A. C., Cluster BioMimic®,
   Carretera Antigua a Coatepec 351, Congregación el Haya, Xalapa, Veracruz CP 91073, Mexico;
   caro_aol@hotmail.com (C.A.O.-G.); martin.mata@inecol.mx (M.M.-R.)
[2] Tecnológico Nacional de México, Instituto Tecnológico de Veracruz, Unidad de Investigación y Desarrollo en
   Alimentos, Veracruz CP 91897, Mexico
[3] Instituto Politécnico Nacional, Centro Interdisciplinario de Investigación para el Desarrollo Integral
   Regional-Unidad Sinaloa, Boulevard Juan de Dios Bátiz Paredes # 250, Col. San Joachin, Guasave,
   Sinaloa 81101, Mexico; fquirozf@hotmail.com
[4] European Molecular Biology Laboratory, European Bioinformatics Institute, Wellcome Genome Campus,
   Hinxton, Cambridgeshire CB10 1SD, UK; asegura@ebi.ac.uk
[5] Department of Horticultural Science, University of Minnesota, Saint Paul, MN 55108, USA;
   lmshannon@umn.edu
[6] Unidad de Bioquímica y Biología Molecular de Plantas, Centro de Investigación Científica de Yucatán,
   Mérida, Yucatán CP 97205, Mexico; vmloyola@cicy.mx
[7] Red de Estudios Moleculares Avanzados, Instituto de Ecología A. C., Cluster BioMimic®,
   Carretera Antigua a Coatepec 351, Congregación el Haya, Xalapa, Veracruz CP 91073, Mexico;
   juan.monribot@inecol.mx (J.L.M.-V.); jose.elizalde@inecol.mx (J.M.E.-C.); enrique.ibarra@inecol.mx (E.I.-L.);
   monica.ramirez@inecol.mx (M.R.-V.); joseantonio.guerrero@inecol.mx (J.A.G.-A.)
* Correspondence: carolina.pm@veracruz.tecnm.mx (C.P.-M.); eliel.ruiz@inecol.mx (E.R.-M.)

Received: 28 June 2020; Accepted: 3 August 2020; Published: 8 August 2020

Abstract: Somatic embryogenesis (SE) is a valuable model for understanding the mechanism of plant embryogenesis and a tool for the mass production of plants. However, establishing SE in avocado has been complicated due to the very low efficiency of embryo induction and plant regeneration. To understand the molecular foundation of the SE induction and development in avocado, we compared embryogenic (EC) and non-embryogenic (NEC) cultures of two avocado varieties using proteomic and metabolomic approaches. Although Criollo and Hass EC exhibited similarities in the proteome and metabolome profile, in general, we observed a more active phenylpropanoid pathway in EC than NEC. This pathway is associated with the tolerance of stress responses, probably through the reinforcement of the cell wall and flavonoid production. We could corroborate that particular polyphenolics compounds, including *p*-coumaric acid and *t*-ferulic acid, stimulated the production of somatic embryos in avocado. Exogen phenolic compounds were associated with the modification of the content of endogenous polyphenolic and the induction of the production of the putative auxin-a, adenosine, cellulose and 1,26-hexacosanediol-diferulate. We suggest that in EC of avocado, there is an enhanced phenylpropanoid metabolism for the production of the building blocks of lignin and flavonoid compounds having a role in cell wall reinforcement for tolerating stress response. Data are available at ProteomeXchange with the identifier PXD019705.

Keywords: plant cell wall; embryogenic cultures; metabolomics; phenolic compounds; proteomics

---

1. Introduction

Avocado (*Persea americana* Mill.) is a crop of great economic importance. However, avocado production is threatened by different agents, mainly fungal phytopathogens, including *Colletotrichum* species, *Phytophthora cinnamomi*, and *Euwallacea kuroshio* [1]. The losses caused by these diseases run into millions of dollars and severely affect the health of the trees, production, and international trade of avocado [2]. There have been large-scale efforts to breed disease-resistant avocados [3,4]. However, these efforts have been stymied by long breeding cycles and the challenges of controlling pollination under dichogamy. Implementation of biotechnological approaches, such as in vitro cultures, offers the possibility of the mass production of avocado. So far, the most promising tool for the in vitro generation of avocado is somatic embryogenesis [5].

Although somatic embryogenesis (SE) has been studied for several decades and considerable effort has been invested in establishing an efficient in vitro propagation system for avocado, the efficiency of somatic embryo conversion into plants remains low to non-existent [2,6–8]. It has been assumed that this is primarily due to a failure of somatic embryos to mature [6,9–11]. Attempts to improve the maturation of avocado somatic embryos using culture media components have failed [9,12,13]. These studies were limited by the paucity of molecular and biochemical information available about SE in avocado, not only at the maturation stage, but also at each prior stage of SE development, including the induction of embryogenic potency (EP), transdifferentiation, and germination [14]. Although the failure to mature of the somatic embryos is the most visible indication of unsuccessful SE, the problems leading to this deficit may occur earlier in development. Avocado somatic embryos that can proceed through the maturation process displayed higher proportions of proteins associated with stress response [14]. Studies suggested that reactive oxygen species (ROS) homeostasis and growth regulator signal pathways are pivotal orchestrators of molecular signaling during SE induction [15].

There is limited information about the molecular and biochemical processes governing the induction and development of SE in avocado. In different species, epigenetic modifications have been suggested as modulators of morphogenetic properties of in vitro cultures [16–25]. Furthermore, the phenylpropanoid pathway (PP) has been implicated in SE [26]. Embryogenic cultures in sugar beet and sandalwood have a fortified cell wall with a higher proportion of phenolic metabolites and lignin deposition as compared to non-embryogenic cultures [16,27]. However, overaccumulation of other phenolic compounds has displayed a negative effect on the establishment of SE in some species [22,28–31]. Thus, highly coordinated biochemical regulation of the PP may dictate the proper metabolic flow to orchestrate the induction or repression of SE. The recent publication of the avocado genome provides an extraordinary tool to unravel the secrets of SE in this plant species [32]. However, detailed proteomic and metabolomic studies in avocado are also needed to determine the key molecular players of the induction and progression of SE.

Despite the great importance of establishing a successful in vitro avocado propagation protocol, so far, there has been no success. Therefore, we took a step back and focused on the comparative proteomics between embryogenic (EC) and non-embryogenic callus (NEC) from the Hass and Criollo avocado varieties. Previous studies suggested this is a crucial comparison that could lead to the identification of the molecular players and increasing the current understanding of the establishment of EP in avocado cultures. In our experiments, the EC cultures exhibited a higher accumulation of proteins associated with phenylpropanoid metabolism, flavonoids, cell wall, and stress-related processes compared to NEC. In addition, the EC treated with polyphenolic compounds produced an increased number of embryos as compared to the control treatments. The increase in the efficiency of the SE was associated with the tight regulation of stress/growth regulator signal pathways, where PP plays an intricate role.

2. Results

2.1. Establishment of Non-Embryogenic and Embryogenic Cultures of Hass and Criollo Avocado

Avocado embryogenic cultures were induced from immature zygotic embryo explants. In our study, we could induce direct SE in only 7% explants cultured in Murashige and Skoog medium supplemented with 0.1 mg L^{-1} picloram (MSP). This response was the same for both varieties. These successful cultures were then repeatedly subcultured in the same medium to develop embryogenic callus (EC). Only ~6% of the resulting callus lost embryogenic competence and became non-embryogenic callus (NEC, Figure 1), which is a well-known physiological feature of avocado cultures [2]. Light microscopy and SEM observations revealed that NEC is a white translucent friable mass (Figure 1A). On the other hand, pale yellow colored EC showed well-organized structures, including somatic embryos at an early globular stage (Figure 1B). The ECs also exhibited a layer resembling the extracellular matrix surface network (EMSN) observed in the EC of other plant species (Figure 1D,J, arrowhead). Close observation with TEM exhibited sharp differences in the cell walls between EC (Figure 1F,L) and NEC (Figure 1E,K). EC has a well-defined cell wall adjacent to the plasma membrane, while NEC displayed several detachments of the cell wall from the plasma membrane (Figure 1E,K, red dash lines).

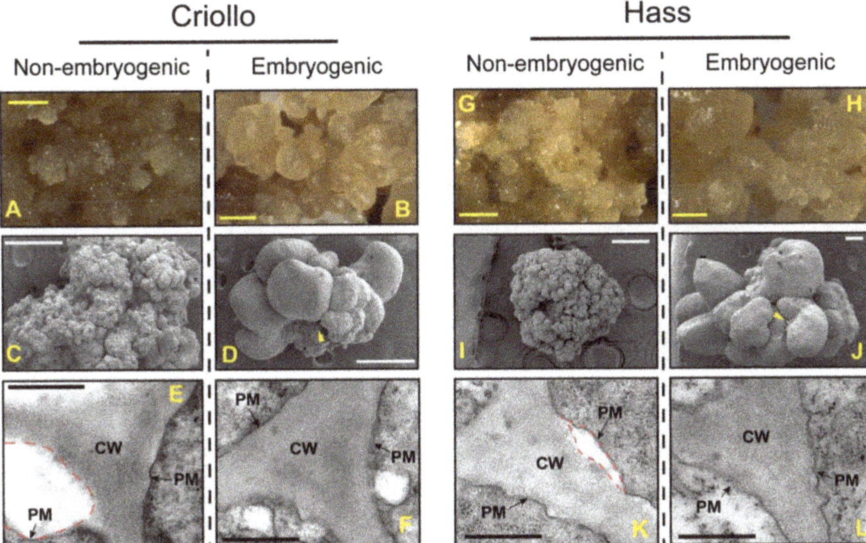

Figure 1. Phenotypes of non-embryogenic (NEC) and embryogenic (EC) cultures of avocado. Light microscopy (**A,B,G,H**), scan electron microscopy (SEM, **C,D,I,J**) and transmission electron microscopy micrographs (TEM, **E,F,K,L**) display contrasting differences between EC (**B,D,F,H,J,L**) and NEC (**A,C,E,G,I,K**) both in Criollo (**A–F**) and Hass (**G–L**) cultures. Yellow scale bars in (**A**), white scale bars in (**B**), and black scale bars in (**C**) denote 1 mm, 250 μm, and 500 nm, respectively. Yellow arrowhead indicates the extracellular matrix surface (EMSN) in (**D,J**). PM and CW indicate in (**C**) plasma membrane and cell wall, respectively. Dash red lines in (**C**) specified detachments of the PM from the cell wall.

2.2. Comparative Proteomics: Embryogenic vs. Non-Embryogenic Cultures

Our proteomic pipeline consisted of a peptide labeled with TMT6plex and synchronous precursor selection (SPS) MS3 (Figure S1). We could identify 1999 proteins in Hass samples while 1842 Criollo, among which 414 and 472 differentially expressed proteins by comparing EC and NEC in Hass and Criollo cultures, respectively (Figure 2A, Data S1, Data S2). Of those, 126 differentially accumulated proteins were identified in both varieties and indicated as core proteome in the following sections.

Gene ontology enrichment and clustering based on biological processes highlighted in the core proteome two main groups of proteins, including those associated with the response to stress stimuli and phenylpropanoid metabolism (Figure 2B). We were able to identify well-known stress-related proteins, which were overaccumulated in EC compared to NEC (Table S1, Figure S2). Furthermore, the analysis of differentially accumulated proteins in both avocado varieties exhibited more proteins associated with stress stimuli in Criollo compared to Hass in vitro cultures (Table S1, Figure S3).

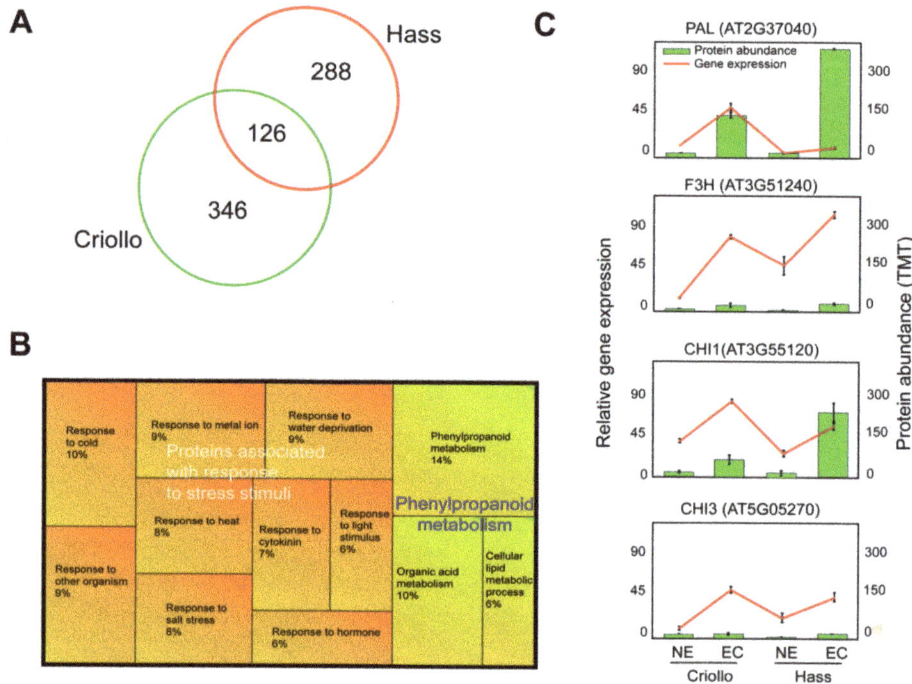

Figure 2. Differentially accumulated proteins identified by comparing embryogenic (EC) vs. non-embryogenic (NEC) cultures from Hass and Criollo avocado varieties. (**A**) The Venn diagram displays the number of differentially expressed proteins identified in Hass and Criollo cultures by comparing EC vs. NEC. (**B**) Core proteome representation-based gene ontology enrichment and clustering of biological process annotation (Data S2, **C**). Correlation between relative gene expression and protein abundance. The gene expression was normalized using Rubisco as an internal control. Points denote the mean fold expression as compared to the control ± three replicates of EC and NEC cultures of avocado cultivars "Hass" and "Criollo". Phenylalanine ammonia-lyase 1 (*PAL*, AT2G37040), flavanone 3-hydroxylase (*F3H*, naringenin 2_oxoglutarate 3-dioxygenase, AT3G51240), Chalcone flavonone isomerase 1 (*CHI1*, AT3G55120), probable chalcone-flavonone isomerase 3 (*CHI3*, AT5G05270).

The core proteome exhibited well-annotated PP enzymes (Figure 2B). For instance, the phenylalanine ammonia-lyase 1 (*PAL1*) was accumulated in both Criollo and Hass EC compared to NEC, detecting significant values in Hass (Figure 2C). Gene expression of *PAL1* corroborates the significant activation of *PAL* in Criollo, while in Hass EC, *PAL1* was slightly upregulated compared to NEC (Figure 2C). In addition, a cytochrome P450, family 71, subfamily B, polypeptide 35 (CYP7135), peroxidase 52 (POX52), 4-coumarate: CoA ligase 1 (4CL1), caffeoyl-CoA *O*-methyltransferase 1 (CCoAOMT1), flavone 3'-*O*-methyltransferase 1 (OMT1) and leucoanthocyanidin dioxygenase (LDOX) were overaccumulated in Hass and Criollo EC compared to NEC (Table S1, Figure S2). The probable caffeoyl-CoA *O*-methyltransferase (CCoAOMT) was accumulated in Hass EC, while in Criollo EC, it displayed contrasting values. Aldehyde dehydrogenase family 2-member C4 (ALH2C4) exhibited a similar pattern of accumulation as that of CCoAOMT. In addition, the probable cinnamyl alcohol dehydrogenase 9 (CAD9) was accumulated in NEC compared to EC (Table S1, Figure S2). In addition to PP metabolism, proteins associated with the flavonoid biosynthetic process were identified in a higher proportion in EC than NEC. These proteins include the chalcone-flavanone isomerases (CHI1 and CHI3) and flavanone 3-hydroxylase (F3H, Figure 2C).

The upregulation of gene expression of *CHI1*, *CHI3*, and *F3H* in EC compared to NEC underpinned our proteomics data (Figure 2C). In addition, the UDP-glycosyltransferase superfamily protein with an annotation of quercetin 3-*O*-glucosyltransferase (GO: 0080043) and quercetin 7-*O*-glucosyltransferase transferase activity (GO: 0080044) was overaccumulated in EC (Figure S2). Untargeted metabolomics, followed by an OPLS-DA, exhibited the overaccumulation of putative polyphenolics such as naringenin, flavanone, and epicatechin 3′O- glucuronide in EC (Figure S4).

Proteins related to PP metabolism were also mainly accumulated in either Hass or Criollo varieties (Table S1, Figures S2 and S3). In Hass EC, we were able to determine the overaccumulation of two cytochrome P450 family proteins (C3′H, CYP98A3, and CYP71A22), annotated with the phenylpropanoid biosynthetic process (GO: 0009699), coumarin biosynthetic process (GO: 0009805), flavonoid biosynthetic process (GO: 0009813) and lignin biosynthetic process (GO: 0009809). Two additional cytochrome P450 family proteins (CYP93D1 and CYP94D2) were identified in a higher proportion in EC compared to NEC, which were annotated with the secondary metabolite biosynthetic process (GO: 0044550). In addition, we observed the overaccumulation of cinnamate-4-hydroxylase (C4H, CYP73A5) and UDP-glucosyl transferase 73B5 annotated with the molecular function of GO: 0080043, GO: 0080044 and flavonol 3-*O*-glucosyltransferase activity (GO: 0047893). In Criollo EC, we could identify the PAL2 in a higher proportion than in NEC. A similar pattern of overaccumulation was observed with the peroxidase 72 (POX72) associated with the lignin biosynthetic process, UDP-glucosyl transferase 85A2, with the molecular function of quercetin 3-*O*-glucosyltransferase and quercetin 7-*O*-glucosyltransferase transferase activity as well as UDP-glycosyltransferase superfamily protein associated with the molecular function of lignin biosynthetic process. In contrast, proteins related to the L-phenylalanine catabolic process (Figure S3), such as *p*-hydroxyphenylpyruvate dioxygenase and tyrosine decarboxylase 1, were overaccumulated in Criollo NEC compared to EC.

Manual analysis of proteomic data also provided additional information on proteins associated with the cell wall biogenesis of avocado Hass and Criollo cultures. For example, the following proteins were identified in higher proportion in EC than NEC: probable pectinesterase/pectinesterase inhibitor 17, cellulose synthase-like E1, pectin lyase-like superfamily protein, probable glucan endo-1,3-β-glucosidases, and peroxidases were identified in higher proportion in EC than NEC (Table S1).

2.3. Phenolic Metabolites Improve the Production of Somatic Embryos in Hass and Criollo Avocado Embryogenetic Cultures

Our proteomic and untargeted metabolomic data suggested the activation of the phenylpropanoid pathway in avocado embryogenic cultures. Therefore, we carried out dose-response experiments by exposing EC to exponential concentrations (1, 10, 100, and 1000 μm) of specific polyphenolic compounds for 14 or 28 days. In our experiments, we included *trans*-ferulic acid (*t*-FA) due to its essential role in cell wall rigidity and the formation of other important organic compounds. We also selected *p*-coumaric acid (*p*-CA), a direct precursor of *p*-coumaroyl-CoA, and a branching point for the biosynthesis of flavonoids, monolignols and several other compounds [33]. We selected *p*-hydroxybenzoic acid (PHBA), associated with the final product of the β-oxidative pathway [34].

Criollo EC treated with 10 µM PHBA exhibited a significant improvement in embryo production after 28 days (Figure 3A). In contrast, PHBA did not show a significant effect on the production of embryos in Hass EC (Figure 3B). In addition, *p*-CA enhanced the production of embryos in both Criollo and Hass cultures treated for 14 and 28 days with a concentration of 1 and 10 µm, respectively (Figure 3B,E). In both Criollo and Hass EC, the improvement in embryo production was significant after 28 days of treatment. Cultures treated with *t*-FA exhibited a similar boost of embryo production with a concentration of 1 and 100 µm in both cultures, observing significant values at 28 days (Figure 3C,F). A visual analysis of the appearance of the cultures corroborates the higher number of embryos in EC treated with the compounds, as mentioned earlier. In some cases, we could observe dark regions in treated tissues, especially those treated with *t*-FA, which could suggest phenolization of the tissues (Figure S5A,B). Surprisingly, embryos treated with *p*-CA and *t*-FA exhibited minimal *PAL1* gene expression after 24 h (Figure S6).

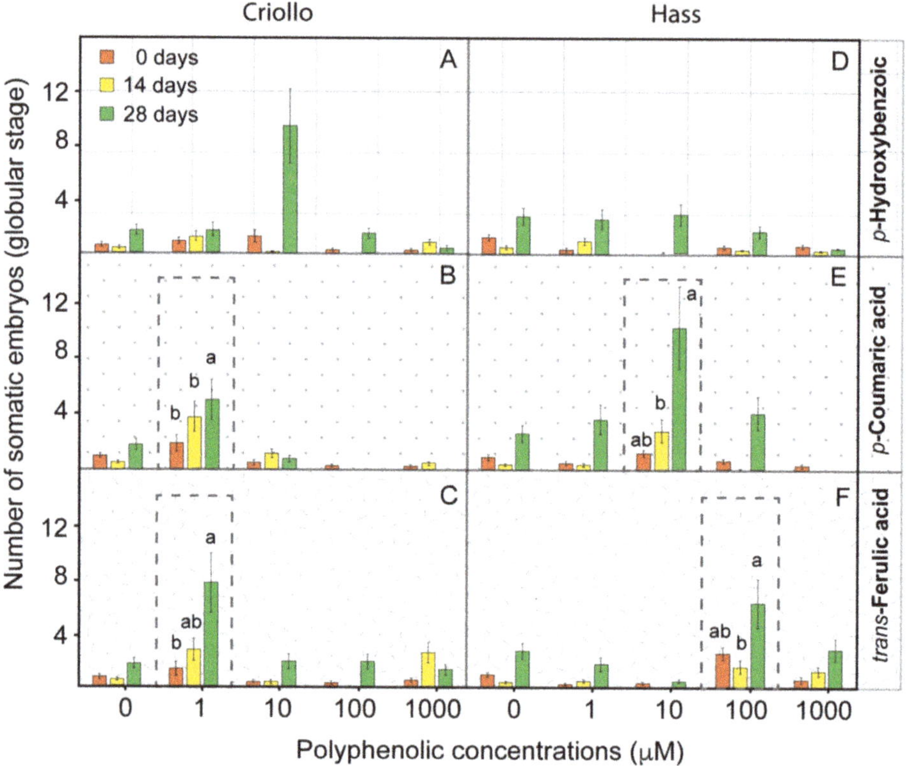

Figure 3. Effect of polyphenolics in embryogenic cultures during 28 days of dose-response treatments. The Criollo (**A–C**) and Hass (**D–F**) cultures were treated with 1, 10, 100, and 1,000 µm of *p*-hydroxybenzoic acid (**A,D**), *p*-coumaric acid (**B,E**), and *trans*-ferulic acid (**C,F**). The effect of polyphenolics was determined based on the number of newly formed globular embryos per plate. A general linear model (GLM) was applied with default settings to determine the statistical significance between treatments. Different letters (a, b and ab) represent a significant difference among treatments ($p < 0.05$) marked with dash lines.

2.4. The Improvement of the Production of Embryos in Avocado EC Is Associated with the Alteration of the Endogenous Content of Polyphenolic Compounds

The PP is at the crossroad of plant growth, structural support, biotic, and abiotic stress. Stress stimuli are among the main driving force of the induction of SE [21]. Consequently, exposing EC to these metabolites might alter the metabolome during the overproduction of somatic embryos. To corroborate our hypothesis, we carried out a target metabolomics approach. We analyzed a short time of treatments, including 6 and 12 h, as well as extended periods of 14 (335 h) and 28 days (672 h, Figure S3).

We could determine a complete alteration of the endogenous content of several polyphenolics in EC treated with *p*-CA and *t*-FA for short times (Table 1, Table S2). The most noticeable change was associated with the sharp reduction in the content of *p*-CA, quercetin 3,4′-di-*O*-glucoside, *t*-FA, and vanillin (VA) in Criollo EC treated with 1 μM *p*-CA after 6 h compared to the control (Table 1). From 6 to 12 h, we could observe a general reduction in all polyphenolic compounds analyzed. However, after 14 days, the content of *p*-CA and PHBA was higher in control than treated EC with 1 μM *p*-CA and t-FA (Table 1). In contrast, PHBA acid showed an exponential overaccumulation after 28 days in samples exposed to 1 μM *p*-CA (Table 1). The PHBA and sinapic acid (SA) exhibited a slight increase in samples treated with 1 μM *t*-FA after 28 days of exposition.

The early effect of 10 μM *p*-CA in Hass EC included the drastic reduction in the content of quercetin 3,4′-di-*O*-glucoside, *t*-FA, and SA (Table 1). In contrast, *trans*-cinnamic acid (t-CA) exhibited a substantial increase in its content after 6 h of exposition. After that, quercetin 3, 4′-di-*O*-glucoside, *p*-CA, *t*-FA, and SA were detected in higher concentrations in control samples than EC treated with 10 μM *p*-CA after 12 h. The *t*-FA was the major polyphenolic detected at the last time of analysis in samples treated with 10 μM *p*-CA. The early effect of 100 μM *t*-FA in Hass EC was in contrast with previous analysis comprising the marked overaccumulation of VA, *p*-CA, quercetin 3,4′-di-*O*-glucoside, naringin, and *t*-CA compared to control sample (Table 1). The content of these polyphenolics exhibited a continuous reduction during the time of exposition while, *t*-FA showed a sharp increase after 14 days of analysis, and the content PHBA showed a slight increase after 28 days in tissues exposed to 100 μM *t*-FA in comparison with the control sample.

2.5. Untargeted Metabolomics Provides New Clues Related to the Effect of P-Coumaric and Trans-Ferulic Acid on the Improvement of Embryo Production in Avocado

The principal component analysis (PCA) of our results exhibited a significant effect of 1 μM *p*-CA and *t*-FA on Criollo EC metabolome profile compared to control samples, throughout the time of exposition (Figure 4). After 12 h, the metabolic profile of Criollo EC treated with both polyphenolics was different compared with the control but similar between each treatment. After 14 and 28 days, the callus treated with 1 μM *p*-CA or *t*-FA and control samples exhibited different metabolic signatures (Figure 4C,D). Hass EC treated with 10 μM *p*-CA and 100 μM *t*-FA exhibited different metabolome profiles versus the control within 6 h (Figure 4A). After 12 h, only Hass EC treated with 10 μM *p*-CA showed a different profile in comparison to control samples (Figure 4B). In contrast, after 14 days of exposition samples treated with 100 μM, *t*-FA showed an entirely different metabolome signature compared to the control (Figure 4C). After 28 days of exposition, neither 10 μM *p*-CA nor 100 μM *t*-FA affected the metabolome profile of Hass EC, while Criollo EC treated with 1 μM *t*-FA exhibited an entirely different metabolome profiled compared to control sample (Figure 4D).

To determine possible molecular markers associated with the improvement of the proliferation of somatic embryos in EC, we focused on scrutinizing which metabolites are overaccumulated after 12 h and 28 days of treatment with *t*-FA 100 μM in Hass, and *t*-FA 1 μM in Criollo EC.

Table 1. The endogenous concentration of phenolic compounds (mg g^{-1} fresh weight) in embryogenic cultures of Criollo and Hass avocado treated with different concentrations of *p*-coumaric (1 and 10 µM) and *trans*-ferulic acid (10 and 100 µM) during 6 and 12 h, and 14 and 28 days. The analysis was carried out with dynamic multiple reaction monitoring (dMRM) study. Replicate values, error standard and statistic values are presented in Table S2. *p*-hydroxybenzoic acid (PHBA), vanillin (VA), *p*-coumaric acid (*p*-CA), *trans*-ferulic acid (*t*-FA), SA (Sinapic acid) and *trans*-cinnamic acid (*t*-CA). We indicated in bold and italic the values described in the text.

Cultivar Time	Treatment	PHBA	VA	*p*-CA	Quercetin 3,4'-di-*O*-glucoside	*t*-FA	SA	Naringin	*t*-CA
Criollo 0 h	0 µM Control	0.11 ± 0.01	0.11 ± 0.01	0.42 ± 0.01	0.02 ± 0.01	0.23 ± 0.01	0	0.07 ± 0.01	0.02 ± 0.01
	1 µM *p*-CA	0.11 ± 0.01	**0.11 ± 0.01**	**0.42 ± 0.01**	**0.02 ± 0.01**	**0.23 ± 0.01**	0	0.07 ± 0.01	0.02 ± 0.01
	1 µM *t*-FA	0.11 ± 0.01	0.11 ± 0.01	0.42 ± 0.01	0.02 ± 0.01	0.23 ± 0.01	0	0.07 ± 0.01	0.02 ± 0.01
Hass 0 h	0 µM Control	0.21 ± 0.01	0.03 ± 0.03	0.66 ± 0.02	0.01 ± 0.01	0.09 ± 0.01	0	0	0.04 ± 0.01
	10 µM *p*-CA	0.21 ± 0.01	0.03 ± 0.03	0.66 ± 0.02	0.01 ± 0.01	0.09 ± 0.01	0	0	0.04 ± 0.01
	100 µM *t*-FA	0.21 ± 0.01	0.03 ± 0.03	0.66 ± 0.02	0.01 ± 0.01	0.09 ± 0.01	0	0	0.04 ± 0.01
Criollo 6 h	0 µM Control	0.58 ± 0.01 ***	0.22 ± 0.01 ***	0.85 ± 0.01 ***	1.66 ± 0.07 ***	1.12 ± 0.01 ***	0.7 ± 0.01 ***	0.34 ± 0.01 ***	0.06 ± 0.01 **
	1 µM *p*-CA	0.64 ± 0.05 ***	**0.06 ± 0.01 **	**0.29 ± 0.01 **	**0 *	**0.12 ± 0.01 **	0 **	0 **	0.4 ± 0.01 ***
	1 µM *t*-FA	0.38 ± 0.01 **	0.05 ± 0.01 **	0.18 ± 0.01 **	0.03 ± 0.04 **	0.28 ± 0.01 **	0 **	0 **	0.11 ± 0.01 ***
Hass 6 h	0 µM Control	0.29 ± 0.01 *	0.06 ± 0.01 *	0.62 ± 0.01 *	0.33 ± 0.05 *	1 ± 0.01 **	0.86 ± 0.02 ***	0.06 ± 0.01 *	0.03 ± 0 *
	10 µM *p*-CA	0.25 ± 0.01 *	0.05 ± 0.001 *	0.7 ± 0.01 **	**0.02 ± 0.03 **	**0.13 ± 0.01 *	**0 *	0 *	**0.29 ± 0.01 *
	100 µM *t*-FA	0.32 ± 0.01 *	**0.24 ± 0.01 ***	**1.39 ± 0.01 ***	**0.62 ± 0.02 ***	10.6 ± 0.22 ***	0.64 ± 0.02 **	**0.18 ± 0.02 ***	**0.98 ± 0.02 ***
Criollo 12 h	0 µM Control	0.21 ± 0.01 **	0.15 ± 0.01 ***	0.61 ± 0.01 ***	0.62 ± 0.04 ***	0.52 ± 0.01 ***	0.72 ± 0.02 ***	0.05 ± 0.01 ***	0.02 ± 0.01 **
	1 µM *p*-CA	0.25 ± 0.01 ***	0.06 ± 0.01 **	0.46 ± 0.01 **	0 **	0.26 ± 0.01 **	0 **	0 ***	0.15 ± 0.01 ***
	1 µM *t*-FA	0.18 ± 0 *	0.04 ± 0.01 *	0.18 ± 0.01 *	0 **	0.12 ± 0.01 *	0 **	0 ***	0.03 ± 0.01 **
Hass 12 h	0 µM Control	0.6 ± 0.01 ***	0.11 ± 0.01 *	**0.94 ± 0.02 ***	**0.18 ± 0.02 ***	0.86 ± 0.07 ***	**0.42 ± 0.01 ***	0.21 ± 0.01 ***	0.12 ± 0.01 **
	10 µM *p*-CA	0.2 ± 0 *	0.03 ± 0.001 *	0.24 ± 0.01 *	0.04 ± 0.01 **	0.13 ± 0.01 *	0 *	0 *	0.08 ± 0.01 *
	100 µM *t*-FA	0.4 ± 0 *	0.17 ± 0.01 ***	0.48 ± 0.01 *	0.05 ± 0.04 **	0.39 ± 0.01 *	0 *	0.22 ± 0.01 ***	0.08 ± 0.01 *
Criollo 14 days	0 µM Control	**1.12 ± 0.02 ***	0.06 ± 0.001 ***	**1.28 ± 0.02 ***	0.3 ± 0.01 ***	0.23 ± 0.01 *	0.03 ± 0.02 ***	0 ***	0.01 ± 0.02 ***
	1 µM *p*-CA	0.09 ± 0.001 *	0.05 ± 0.01 **	0.39 ± 0.01 *	0 **	0.28 ± 0.02 **	0 **	0 ***	0.01 ± 0.02 ***
	1 µM *t*-FA	0.12 ± 0.01 **	0.02 ± 0.01 *	1.18 ± 0.08 **	0 **	0.56 ± 0.03 ***	0.02 ± 0.01 *****	0 ***	0.02 ± 0.01 ***

Table 1. *Cont.*

Cultivar Time	Treatment	PHBA	VA	*p*-CA	Quercetin 3,4′-di-*O*-glucoside	*t*-FA	SA	Naringin	*t*-CA
Hass 14 days	0 μM Control	0.09 ± 0.01 *	0.03 ± 0.01 ***	0.69 ± 0.03 ***	0 *	0.4 ± 0.01 **	0.01 ± 0.01 **	0 ***	0.02 ± 0.01 ***
	10 μM *p*-CA	0.11 ± 0.01 **	0.03 ± 0.01 ***	0.24 ± 0.02 *	0.06 ± 0.01 **	0.55 ± 0.03 **	0.1 ± 0.02 ***	0 ***	0.01 ± 0.01 **
	100 μM *t*-FA	0 *	0 *	0 *	0 *	**7.06 ± 3.63 *****	0 **	0 ***	0 *
Criollo 28 days	0 μM Control	0.62 ± 0.01 *	0.05 ± 0.01 **	0.96 ± 0 ***	0 ***	0.72 ± 0.01 ***	0.23 ± 0.01 **	0 ***	0.5 ± 0 ***
	1 μM *p*-CA	**3.56 ± 0.06 *****	0 *	0.05 ± 0.04 *	0 ***	0 *	0 *	0 ***	0 *
	1 μM *t*-FA	**1.35 ± 0.02 **	0.08 ± 0 ***	0.74 ± 0.01 **	0.02 ± 0.04 ***	0.12 ± 0.10 **	**1.55 ± 0.12 *****	0 ***	0.03 ± 0 **
Hass 28 days	0 μM Control	0.4 ± 0 *	0.05 ± 0 **	1.14 ± 0.01 ***	0 ***	0.78 ± 0.01 **	0.04 ± 0.01 ***	0 ***	0.01 ± 0 ***
	10 μM *p*-CA	1.36 ± 0.03 ***	0 *	0.02 ± 0.02 *	0 ***	**7.75 ± 4.43 *****	0 *	0 ***	0 **
	100 μM *t*-FA	**1.12 ± 0.01 **	0.03 ± 0 *	0.16 ± 0.01 *	0 ***	0.08 ± 0 **	0.01 ± 0.01 **	0 ***	0 **

1. A general linear model (GLM) for metabolomics data was applied with default settings to determine. the statistical significance between treatments. 2. Data are mean ± standard deviation reported in mμ g^{-1} of dried EC. 3. *, **, ***, ***** represent a significant difference among treatments ($p < 0.05$).

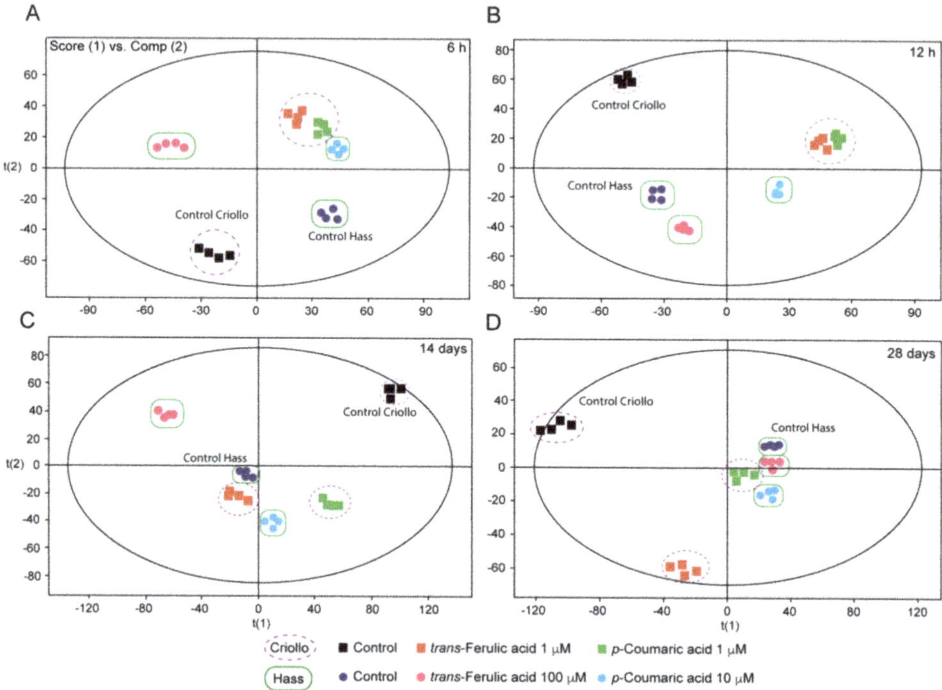

Figure 4. Principal component analysis (PCA)-mediated grouping of Criollo and Hass EC treated with *p*-coumaric acid and *trans*-ferulic acid over 6 h (**A**), 12 h (**B**), 14 days (**C**) and 28 days (**D**). We used the exact mass spectrum fingerprints to carry out the principal component analysis.

We were able to determine the overaccumulation of a putative auxin-a (FDB017851) and adenosine (FDB003554) after 12 h of treatment with both polyphenolics in the two avocado varieties (Table S3). Furthermore, after 28 days, Hass cultivar exhibited the overaccumulation of putative cellulose (FDB001182), while in Criollo, we could determine the putative 1,26-hexacosanediol-diferulate (FDB002683). Afterward, we carried out a microscopy study based on super-resolution confocal images to identify some modification at the cell wall reinforcement level. Our approach was associated with calcofluor labeling to highlight the cell wall, which allowed us to measure the thickness of it. Our study suggests that there is a slight thickening of the cell wall of embryos of avocado Hass varieties after 12 h of treatment with 100 µM t-FA (Figure 5A,C). Cell wall embryos treated with 1 µM *t*-FA did not exhibit significant differences compared to the control sample (Figure 5B,D). Furthermore, statistical analysis verified the significant thickening of the cell wall compared with the control samples (Figure 5C).

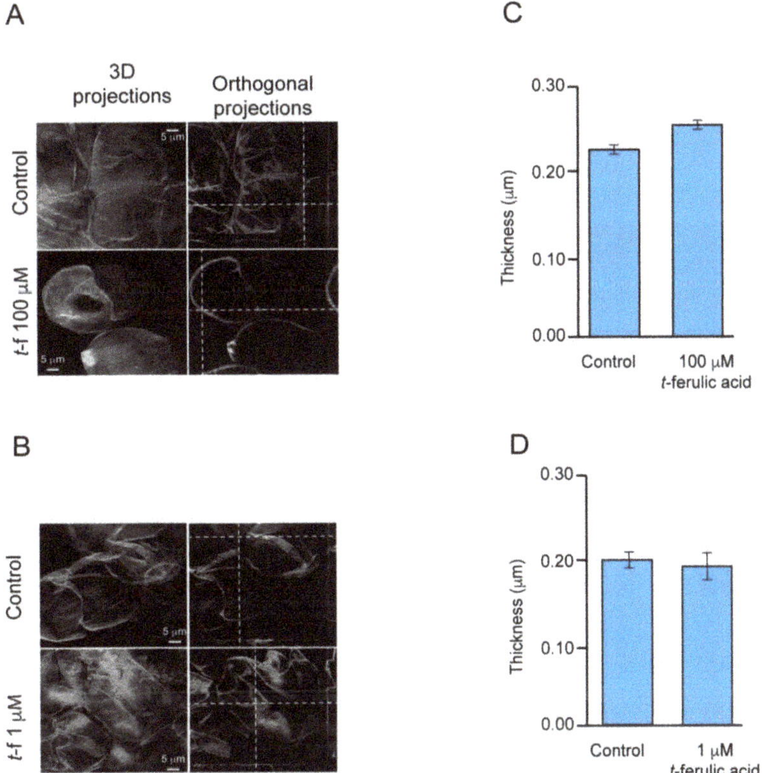

Figure 5. Determination of cell wall thickness during 12 h of treatment with ferulic acid in avocado Hass embryos. Three-dimensional projection base confocal microscopy analysis, intersections of dotted lines indicate the angle of analysis (**A**,**B**). The thicknesses of the cell wall were measured considering 30 orthogonal projections and the statistical analysis was made by R (**C**,**D**).

3. Discussion

3.1. Stress Regulation and Cell Wall Fortification a Common Feature of Embryogenic Callus and Possible Association with the Improvement of Embryogenesis

Previous studies have suggested that in vitro cultures face multiple types of stress during the acquisition of EP [21,35,36]. Our results indicate that NEC may face even more adverse conditions than EC. The cellular and molecular response to stress differ between avocado varieties (Table S1, Figures S2 and S3). To make our analysis more precise, we mainly focused on differential proteins that were differentially accumulated in the EC and NEC of both Criollo and Hass varieties in two independent studies (Core proteome, Figure 2, Figure S1). The core proteome in this study could be related to responses to stimuli, including responses to cold (GO: 0009409), water deprivation (GO: 0009414), light stimulus, metal ion (GO: 0010038), salt stress (GO: 0009651), heat (GO: 0009408), cytokinin (GO: 0009735), hormone (GO: 0009725) and other organisms (GO: 0051707, Figure 2B, Data S2).

There are several documented strategies of plant cells to overcome adverse external conditions, including the reinforcement of their cell walls and over-production of antioxidant metabolites [37,38]. Some of the aforementioned strategies may prevail in avocado embryogenic cultures (Figure 6). Evidence suggests that such stress response is critical in the establishment of SE. For example, peroxidases (POXs) are potential markers of EP acquisition across species [39,40]. The ROS-induced activity of POXs is the primary mechanism involved in wall remodeling during stress. POXs are proposed to crosslink cell wall glycoproteins such as extensins, facilitating arabinoxylans crosslinking through ferulic acid and promoting the formation of diferulic acid crosslinks between lignin molecules [41]. The higher number of extracellular POXs (Table S1) identified in ECs as compared to NEC points to a better biochemical response to oxidative stress, perhaps in part through cell wall reinforcement (Figure 1F,L). In addition, across species, the EMS is known to associate with arabinogalactans-proteins, hydroxyproline-rich glycoproteins, and pectin epitopes [42]. The presence of an EMSN and well-fortified cell wall in ECs (Figure 1D,J) suggests that these cell wall-associated proteins may play a vital role during the induction and establishment of SE in avocado cultures. Reinforcement of the cell wall with phenolic acid metabolites and lignin deposition has been indicated as a common feature of EC and organogenesis [16,27].

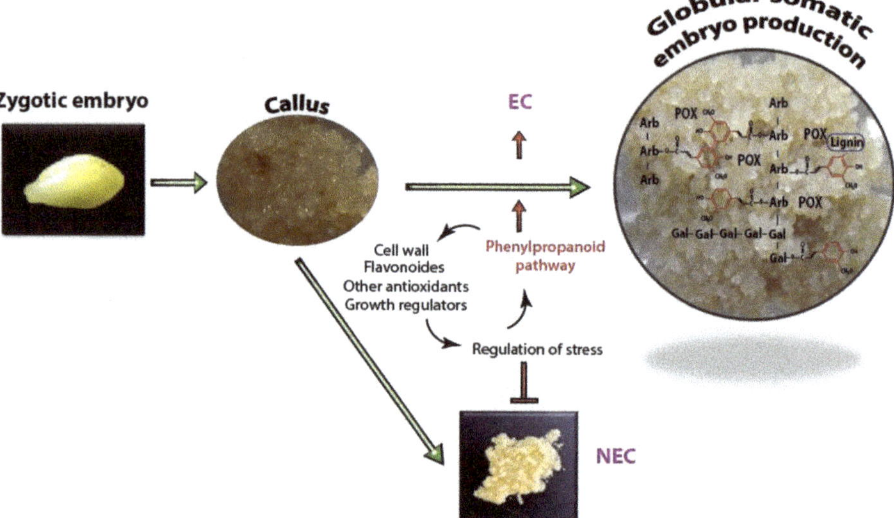

Figure 6. A model of cell wall reinforcement and regulation of stress condition mediated by the phenylpropane pathways and growth regulators in avocado somatic embryogenesis. In dicots, ferulic acid is bound to pectic polysaccharides, including the C-2 hydroxyl group of arabinofuranose (Arb) and C-6 hydroxyl group of galactopyranose (Gal). Peroxidases (POX) catalyze the oxidation of ferulic acid, leading the formation of ferulic acid dehydrodimers. The depiction of ferulic acid in the cell wall modified from Mathew and Abrahan [43].

In EC, the up-regulation of cinnamic acid 4-hydroxylases (C4H, CYP73A5), which catalyzes the conversion of t-CA to *p*-CA, provides the first step toward the biosynthesis lignin monomers (Table S1). In addition, CCoAOMT1 identified in higher proportion in EC than NEC contributes to the production of feruloyl-CoA and sinapoyl-CoA, which are related to the synthesis of feruloylated polysaccharides and was directly implicated in cell wall strengthening [44]. In addition, C3'H (CYP98A3) overaccumulation in Hass EC (Table S1) is associated with the 3'-hydroxylation of *p*-coumaric esters of shikimic/quinic acids forming lignin monomers [45–47]. The complete repression of C3'H leads to cell wall alteration and reduction in lignin deposition, which in turn affects the cell wall expansion and plant growth [47]. This shows that C3'H could contribute to cell wall reinforcement during the formation of embryos in the early stage of SE. Furthermore, overaccumulation of POX72 and other related lignin proteins underpin the lignin monomer biosynthesis in Criollo EC (Figure 6).

The up-regulation of flavonoid-related enzymes in EC, including chalcone-flavanone isomerases, naringenin, 2-oxoglutarate 3-dioxygenase and UDP-glycosyltransferase superfamily provides evidence of the activation of this branch of the phenylpropanoids in EC. As flavonoids are ROS scavengers, the overaccumulation of proteins linked to the biosynthesis of these molecules, for example, the overproduction of naringenin, flavanone, and epicatechin 3'O-glucuronide in EC (Figure S4), could indicate the contribution of these ROS scavengers to the maintenance of the ROS balance in EC [48,49].

3.2. The Improvement of the Production of Somatic Embryos by Polyphenolics Is Linked to the General Reduction in the Level of Endogenous Phenolic Compounds

Our proteomic data strongly suggest the activation of the PP in avocado EC. As previously mentioned, our analysis showed the activation of two main branches of the PP, including lignin and flavonoid biosynthesis. We hypothesized that exposing specific polyphenolics might increase the rate of embryo generation in EC based on our results of comparative proteomics between EC and NEC Criollo and Hass avocado varieties. We could determine that a lower concentration of 1 μM *p*-CA and *t*-FA significantly improved the rate of production of somatic embryos in Criollo EC. In comparison, 10 μM *p*-CA and 100 μM *t*-FA are better for Hass EC (Table 1).

We could determine that treating Criollo and Hass EC with 1–10 μM *p*-CA and 1 μM *t*-FA reduced the production of several polyphenolic metabolites after 6 h of treatment. In contrast, exposing Hass EC to 100 μM *t*-FA exhibited contrasting patterns, as observed with the above-mentioned treatments after 6 h. Furthermore, Criollo and Hass EC presented the overaccumulation of PHBA and *t*-FA after 28 days of treatment, respectively. We should keep in mind that the dynamic of the endogenous content of polyphenolics in tissue cultures depends on several factors, such as the concentration and ratio of growth regulators, response to stress factors and grade of differentiation of tissues and organs [50–53].

In a previous study, a callus treated with cytokinin displayed a drastic reduction in *p*-CA, sinapic, *t*-FA, *t*-CA, and CA. In addition, a significant induction of hydroxybenzoic acid derivatives such as PHBA, MHBA, and vanillic acids was observed in callus cultures [52]. Cytokinin negatively regulates the expression of *PAL* genes [49]. On the other hand, the cyclic AMP molecule is involved in the activation of PAL [54]. The cyclic AMP was drastically accumulated in Criollo EC treated with *t*-FA 1 μM for 12 h (Table S2). The studies above strongly suggest an interconnection between growth regulators and the phenylpropanoid pathway. This assumption is supported by a previous report where naringenin and *cis*-cinnamic acid (c-CA, conversion of *t*-CA by light) were suggested as negative regulators of auxin transport [55]. Recent studies indicated that auxin-regulated plant growth is fine-tuned by early steps in phenylpropanoid biosynthesis, particularly *t*-CA, and its derivative enhances auxin signaling and promotes auxin-dependent leaf expansion in Arabidopsis [56]. In our targeted metabolomics study, we putatively detected the overaccumulation of Auxin A and purines in EC exposed with *t*-FA during the first 12 h in both Criollo and Hass varieties. We do not know the molecular function of the putative auxin-a in the SE of avocado. However, auxin signals are probably transduced through cAMP, but further evidence is needed to corroborate this assumption [57]. Auxin is an essential plant hormone for SE induction and the cellular differentiation process [58,59]. Ashinara et al. [59] noticed that an active utilization of purines during the early phases of SE might be required for the proliferation and cell division of the embryogenic tissue in the presence of auxin. The possible presence of these compounds could explain the proliferation of somatic embryos during the treatments with polyphenolic compounds. Further study may provide clues associated with the functionality of these molecules in avocado SE.

The PP, as a crossroad of several biological processes, is tightly regulated at several levels. For example, PAL, the first committed enzyme in the PP, is regulated at several levels, including the transcriptional regulation of *PAL* genes and posttranscriptional modifications. The activity of *PAL* is metabolically feedback regulated by particular biosynthetic intermediates or chemical signals [49]. In our study, we visualize the significant upregulation of *PAL1* gene expression in both Criollo and

Hass EC compared to NEC (Figure 2C). However, in EC treated with *p*-CA and *t*-FA, the PAL1 gene expression was barely observed after 24 h (Figure S6). Previous studies in *Phaseolus vulgaris* cell suspension cultures showed that the exogenous application of t-CA negatively regulated the enzymatic activity of PAL activity and *PAL* gene transcription [60,61]. Our result and others suggest a negative regulation of *PAL* gene expression. Our target metabolomic analysis, showing a general reduction in endogenous content of polyphenolics, is supported by studies from different plant species where PAL activity was shown to be inhibited by *t*-CA, *p*-CA, PHBA, *O*-chlorocinnamate and other related compounds [62–64]. The negative feedback regulation of *PAL* might redirect carbon flow, as observed after *C4H* inhibition, which was associated with the production of namoylmalate and overproduction of salicylic acid in Arabidopsis plants and elicited *Nicotiana tabacum* cv Bright Yellow cell suspension culture, respectively [65,66]. The negative regulation of *PAL* gene expression in Criollo EC and Hass EC was associated with the significant overaccumulation of PHBA and *t*-FA, respectively, after 28 days of treatment with *p*-CA (Table 1).

3.3. Change in Metabolic Flow May Explains the Positive Association between Polyphenolics and the Overproliferation of Avocado Somatic Embryos

It is noteworthy to mention that previous studies have indicated that phenolic compounds including *t*-CA, PHBA, vanillyl benzyl ether (VBE), 4-[(phenyl methoxy) methyl] phenol, caffeic acid, and chlorogenic acid, inhibit SE [22,31]. Although the precise inhibition mechanism of these secondary metabolites is unknown, overaccumulation of phenolic metabolites in the culture media of dense in vitro cultures abolished the induction and establishment of SE. Cvikrová et al. [67] reported a negative correlation between the mitotic activity and the content of hydroxycinnamic acids in the alfalfa cell suspension. PHBA, which was overaccumulated in Criollo EC treated with 1 µM *p*-CA after 28 days, has been associated with the repression of SE induction in *Larix leptolepis*, *Coffea canephora* and *Daucus carota* [22,31]. However, it is possible that PHBA and other related phenolic metabolites became inactive while associated with the cell wall. In avocado culture, the induction of somatic embryos by *p*-CA and *t*-FA may comprise redirection of the metabolic flow from the synthesis of polyphenolics to the production of the building blocks of lignin and flavonoid compounds having a role in cell wall reinforcement and ROS scavenging, respectively. The general reduction in endogenous content of polyphenolics in EC treated with *p*-CA and *t*-FA during the first 12 h supports our assumption (Table 1). Previous studies have suggested that an adverse effect of *p*-CA on PAL1 might change the metabolic flow toward the production of well-known lignin constituents [68,69].

Previous studies suggested the crucial role of cell wall reinforcement in the early stages of SE [70,71]. In addition, TEM showed a well-defined cell wall of EC compared to NEC in avocado cultures (Figure 1E,F,K,L), the reinforcement of which can be associated with both the cross-linking of feruloyl-polysaccharides or oxidative coupling of lignin precursors [72,73], which in part may explain the slight increase in cell wall thickness detected with confocal microscopy in Hass avocado treated with *t*-FA 100 µM for 12 h (Figure 5).

4. Materials and Methods

4.1. Chemical Reagents

Solvents used for the extraction and analysis of phytochemicals (methanol, isopropanol, acetonitrile, water and formic acid) were LC-MS grade (67-56-1, 67-63-0, 75-05-8, 7732-18-5, 85178, respectively, Sigma-Aldrich, (St. Louis, MO, USA). Authentic standards for mangiferin, (+)-catechin, quercetin-3-D-galactoside, quercetin, gallic acid, (-)-epicatechin and quercetin-3-glucoside (M3547, 43412, 83388, Q4951, G7384, 1753, 16654, respectively, Sigma-Aldrich, St. Louis, MO, USA) were used. Kaempferol-3-*O*-glucoside (90242), 4-hydroxy-benzoic acid (99-96-7), caffeic acid (6034 S), 4-coumaric acid (6031 A), ferulic acid (6077 A), quercetin 3, 4-di-*O*-glucoside (1347 S), quercetin 3-D-*O*-galactoside (1027 S) standards were purchased from Extrasynthese (https://www.extrasynthese.com/). All reagents used for proteomic analysis were purchased from Sigma-Aldrich (St. Louis, MO, USA), except as otherwise specified in the corresponding section.

4.2. Establishment of Embryogenic Cultures

Immature fruits (5–10 mm) of avocado varieties "Hass" and "Criollo" were collected and ECs of both varieties were established from immature zygotic embryos [9]. A small proportion of the resulting ECs lost embryogenic competence after two cycles of the subculture; these cultures were labeled as NEC for this study. The media used to subculture both kinds of callus included MS (Murashige and Skoog) major and minor salt [74], sucrose 30 g L^{-1}, thiamine HCl 4 mg L^{-1}, myo-inositol 100 mg L^{-1} and picloram 0.41 µM; this medium is named MSP [10]. The pH of the culture medium was adjusted to 5.7 before adding gellan gum 3 g L^{-1} (Cat. No. 71010-52-1, Caisson Labs, Smithfield, UT, USA) (https://caissonlabs.com/), and autoclaving at 1.5 kg cm^{-2} at 121 °C for 15 min. Under aseptic conditions, aliquots of media (20 mL) were dispensed onto sterile plastic Petri dishes (100 × 15 mm). The cultures were incubated in darkness at 25 ± 1 °C. The explants were subcultured every two weeks on the same culture medium over seven months.

4.3. Protein Extraction

Proteins were obtained from 0.5 g of frozen NEC and EC collected from seven-month-old cultures and after two weeks of subculturing. The in vitro cultures were first ground with liquid nitrogen in a mortar with a pestle. The powder was suspended in three volumes of phosphate buffer 100 mM (pH 7.0) containing SDS 2%, NaCl 150 mM, and 20 µL g^{-1} of protease inhibitor cocktail (L P8849, Sigma-Aldrich). The mixtures were homogenized with a tissue homogenizer (Tissue-Tearor™, BioSpec Products, Inc., Bartlesville, OK, USA) and centrifuged at 10,000× *g* for 45 min at 25 °C. Afterwards, the supernatants were recovered and kept at −80 °C for future proteomic analysis. The protein assay was carried out with the BCA™ Assay Kit (23225, Pierce, Rockford, IL, USA) using bovine serum albumin (BSA) as a standard.

4.4. Proteomics Analysis

The general procedure of proteomic analysis, including protein digestion and isobaric labeling, has been published elsewhere [75]. Detailed information on proteomic analysis is presented in Appendix A, and particular modifications in the general procedures are summarized in this section.

4.5. Protein Digestion and Tandem Mass Tags (TMT) Labeling

We started with 100 µg of protein by reducing it with tris (2-carboxyethyl) phosphine (TCEP; 10 mM, C4706, Sigma Aldrich) and alkylating with iodoacetamide (IA, A3221 Sigma Aldrich). Then, proteins were digested with trypsin (V528A, Trypsin Gold, Promega, Madison, WI, USA) at a 1:30 (*w/w*) trypsin protein ratio for 16 h at 37 °C. Afterwards, additional freshly made trypsin was added in 1:60 (*w/w*) trypsin protein ratio for 4 h at 37 °C. Then, peptides were labeled with TMT6-plex reagents according to the manufacturer's instructions (90066, Thermo Fisher Scientific, Rockford, IL, USA). The labels 126, 127 and 128 were used for NECs, while labels 129, 130 and 131 were used for ECs. Then, samples were pooled and fractionated using strong cation exchange (SCX) cartridges (60108-421, Thermo Scientific). Fractions were desalted with C_{18} cartridges and dried using a CentriVap (Labconco Kansas, MO, USA).

4.6. NanoLC-MS/MS Analysis and Synchronous Precursor Selection (SPS)-MS3 for TMT Analysis

Each reconstituted sample (5 µL) was injected into a nanoviper C_{18} trap column (3 µm, 75 µm × 2 cm, Dionex) at a flow rate (FR) of 3 µL min^{-1}, and fractionated on an EASY spray C18 RSLC column (2 µm, 75 µm × 25 cm) adapted to a nanoLC (UltiMate 3000 RSLC system, Dionex). A 100 min gradient was used with an FR of 300 nL min^{-1} and two solvents (solvent A: 0.1% formic acid in water and solvent

B: 0.1% formic acid in 90% acetonitrile). The gradient was set as follows: 10 min solvent A, 7–20% solvent B for 25 min, 20% solvent B for 15 min, 20–25% solvent B for 15 min, 25–95% solvent B for 20 min, and 8 min solvent A. Full MS scans in the Orbitrap analyzer (Orbitrap FusionTM TribidTM, Thermo-Fisher Scientific, San Jose, CA, USA) were carried out with: 120,000 of resolution (FWHM), scan range 350–1500 m/z, AGC of 2.0e5, maximum injection time of 50 ms, intensity threshold of 5×10^3, dynamic exclusion 1 at 70 s, and 10 ppm mass tolerance. For MS2 analysis, the 20 most abundant MS1s were isolated with charge states set to 2–7. A precursor selection mass range of 400–1200 m/z was used, with a precursor ion exclusion width range of 18 to 5 m/z, and an isobaric tag loss TMT. MS3 spectra were acquired using synchronous precursor selection (SPS) with ten isolation notches, as previously described [75].

4.7. Data Analysis and Interpretation

Raw data were processed with Proteome Discoverer 2.1 (PD, Thermo Fisher Scientific, USA). The subsequent search was carried out with SEQUEST HT, MASCOT (version 2.4.1, Matrix Science), and AMANDA against the avocado proteins databases. Parameters in the search included full-tryptic protease specificity, two missed cleavage allowed. Static modifications covered carbamidomethylation of cysteine (+57.021 Da) and TMT 6-plex *N*-terminal/lysine residues (+229.163 Da). Dynamic modifications comprised methionine oxidation (+15.995 Da) and deamidation in asparagine/glutamine (+0.984 Da). We used for the TMT6-plex quantification method ± 10 ppm mass tolerance, highest confidence centroid, and a precursor co-isolation filter of 45%. Protein identification was carried out with tolerances of ±10 ppm and ±0.6 Da. Peptide hits were filtered for a maximum of 1% FDR using the Percolator algorithm. Functional annotation of proteins was carried out by Blast2Go software (https://www.blast2go.com/), and gene ontology (GO) enrichment was carried out by David bioinformatic source (https://david.ncifcrf.gov/) using *Arabidopsis* homologs. We used the REVIGO web server (http://revigo. irb.hr/) for GO clustering and visual representation of biological processes (Data S2).

4.8. Phytochemical Extraction

Fifty milligrams of EC and NEC samples were lyophilized in a freeze dryer (Freezone1, Labconco, Kansas, Missouri, USA) and suspended in 1 mL of methanol containing 0.1% formic acid and then homogenized with a tissue homogenizer (Tissue-Tearor™, BioSpec Products, Bartlesville, OK, USA) in a 1.5 mL centrifuge tube. Afterwards, the samples were placed in an ultrasonic bath (Cole-Parmer, Vernon Hills, IL, USA) for 45 min at 4 °C. Then, the mixtures were centrifuged at 3000× g for 15 min at 4 °C. The supernatant was split off for further analysis. Three biological replicates were analyzed for NEC and EC.

4.9. Determination and Quantification of Phenolic Compounds and Untargeted Metabolomics Analysis

Phenolics were identified and quantified using a UPLC system (Agilent, 1290, Santa Clara, CA, USA) coupled to the QqQ mass spectrometer (Agilent, 6460, Santa Clara, CA, USA) with a dynamic multiple reaction monitoring (dMRM) method for the searching for up to 60 compounds, as previously described by our research group [76]. In addition, untargeted metabolomics analysis using liquid chromatography and high-resolution mass spectrometry and orthogonal partial least square discriminant analysis (OPLS-DA) was carried out as previously reported [75]. Detailed information is shown in Appendix A.

4.10. Stereoscope Analysis

Fresh tissue samples of CE and NEC of Hass and Criollo varieties were observed in a light stereoscope Leica S6D (Schweiz) with a 63× objective. The images were processed using LAS V4.12 software (http://www.leicamicrosystems.com).

4.11. Scanning Electron. Microscopy Preparation (SEM)

The samples from 7-month-old EC and NECs were fixed in glutaraldehyde 2.5% (pH 7.2) for 12 h. After fixation, the cultures were rinsed thrice in phosphate buffer (pH 7.2) and dehydrated in an ethanol series for 60 min each (40%, 50%, 60%, 70%, 80%, 90% to 100%). After ethanol dehydration, the samples were dried in a critical point drier (K 850, Quorum, West Chester, PA, USA) using liquid CO_2. Samples were then attached to an aluminum stub with double-stick tape. The cultures were then gold coated in a sputter coater (Q150R, Quorum, Quorum Technologies Ltd., Lewes, UK) and observations were carried out in an FEI-Quanta250 FEG microscope (Czech Republic), operated at 5 Kv acceleration voltage.

4.12. Transmission Electron. Microscopy (TEM)

The samples were fixed overnight with 0.1 M 'Sorensens's buffer (pH 7.2) containing paraformaldehyde 2%, glutaraldehyde 1% and sucrose 0.8%. Afterwards, the samples were rinsed with the same buffer and post-fixed with OsO4 1% for 2 h. Thereafter, samples were dehydrated with different concentrations of ethanol for 10 min each (30, 50, 70, 96 and 100%). Then the samples were embedded in 'Spurr's resin (14300, Electron Microcopy Science, Hatfield, PA, USA) and polymerized at 60 °C for 24 h. Afterward, thin cross-sections of 70 nm were obtained with an Ultracut ultramicrotome (EM UC7, Leica Microsystem, Wetzlar, Germany) (http://www.leicamicrosystems.com) and mounted on copper grids and studied with a JEM 1400 Plus Transmission Electron Microscope (JEOL, Akishima, Tokyo, Japan).

4.13. Confocal Microscopy

Avocado somatic embryos for Confocal Laser Scanning Microscopy (CLSM, Mannheim, Germany) were fixed in 4% paraformaldehyde in phosphate-buffered saline, 0.2 M (pH 7.2; PBS). Samples were stained with 10 µL of calcofluor white (18909 Sigma-Aldrich, (St. Louis, MO, USA) for 10 min and rinsed in PBS for 5 min. Images were acquired with a Leica TCS-SP8+STED microscope (Leica Microsystems, Mannheim, Germany) using a plan-apochromat 63x (NA 1.40, oil) objective. Calcofluor-stained samples (cell wall) were recorded in the grey channel (425–500 nm emission; excitation 405 nm). Deconvolution images were processed with SVI Hygens professional software v.18.04.1 (Hilversun, The Netherlands).

4.14. Embryogenic Callus Treatments with Phenolic Compounds

Criollo EC was cultured in Petri dishes (100 × 15 mm) containing 25 mL of MSP medium supplements with *p*-hydroxybenzoic acid, *p*-coumaric acid (1 µM) or *trans*-ferulic acid for 28 d. At the same time, Hass EC was cultured for the same time in MSP medium containing either *p*CA 10 µM or *t*FA 100 µM. We sampled EC for 0, 6, 12, 24, 336 (14 d) and 672 h (28 d). The number of somatic globular embryos was visually recorded at 14 and 28 d of each treatment. The experiment consisted of five repetitions per treatment.

4.15. RNA Extraction and Quantitative PCR Assays (qPCR)

Total RNA was extracted using "Mini Kit Plant RNeasy" following the manufacturer's instructions (74104, Qiagen, Hilden, Germany). Sequence-specific primers were designed for each gene. Arabidopsis genes homologs in avocado, including phenylalanine ammonia-lyase 1 (*PAL1*, AT2G37040), chalcone-flavanone isomerases (*CHI1*, AT5G05270 and *CHI3*, AT3G55120) and flavanone 3-hydroxylase (*F3H*, AT3G51240) were chosen. The primers were designed using Primer Express Software (version 3.0.1, IDT). Primer or probe sequences were selected to be complementary to an exon-exon junction, to ensure amplification from the cDNA template and not from genomic DNA. The transcript abundance of the genes in this study was analyzed by quantitative real-time PCR (qRT-PCR) with the StepOne™ Real-Time PCR System (Applied Biosystems, Foster City, CA, USA) using TaqMan assays. The gene expression values were normalized to Rubisco (accession No. AY337727) expression. The RT-PCR reaction was performed under the following conditions: 50 °C for 2 min, 10 min at 95 °C, followed by 40 cycles at 95 °C each 10 s, 60 °C for 1 min for denaturation, primer alignment, and amplification, respectively. Finally, qRT-PCR data were analyzed by the $2^{-\Delta\Delta Ct}$ method [77]. Analyses were performed per triplicate. The statistical analysis was performed using one-way ANOVA ($p < 0.05$), followed by Tukey's honestly significant test calculated at 5% levels of probability using the Statistical Software R®.

4.16. Statistical Analysis

Protein abundances were quantile, and log2 normalized. A linear model for microarrays data [78] approach with default setting was used to determine the statistical significance of differential protein expression between ECs and NECs for all the quantified proteins. Significantly differentially expressed proteins were defined as proteins identified with at least two peptides, which exhibited a 2-fold-change in EC to NEC ratio with a *p*-value smaller than 0.05. The statistical analysis of the endogenous content of phenolic compounds was carried out with a Mann–Whitney U-test ($n = 3$, $p < 0.05$).

5. Conclusions

Our proteomic-metabolomic approach provides insight into the improvement of SE in avocado. Our findings highlight the critical features of EC, including an active PP metabolism and cell wall fortification that might be linked with molecular features associated with embryogenic competence. Both ECs and NEC experience stress due to culture conditions, proteins associated with oxidation-reduction processes, and other stress response mechanisms were, however, present in a higher proportion in EC than NEC. This suggests that the ability to respond to stress is a hallmark of SE establishment in avocado. Our obtained evidence indicates that specific polyphenolics can induce the proliferation of embryos, which could be useful for the obtention of starting biological material during the establishment of an efficient pipeline for the improvement of embryo maturation/germination in the particular case of avocado SE.

Supplementary Materials: Supplementary materials can be found at http://www.mdpi.com/1422-0067/21/16/5679/s1. Figure S1: Workflow of proteomic studies carried out in avocado in vitro cultures, Figure S2: Volcano plots of differential proteins identified in Criollo (a) and Hass (b) avocado embryogenic (EC) and non-embryogenic cultures (NEC), Figure S3: Gene ontology enrichment and clustering of biological processes annotation of proteins accurately identified as differential either in Criollo or Hass cultures, Figure S4: S-Plot (Hass NE_EC) of the OPLS-DA comparison between Hass EC and NEC, Figure S5: Visual analysis of Criollo (a) and Hass (b) embryogenic cultures treated with one, ten, 100 and 1000 µM of 4-hydroxybenzoic acid, *p*-coumaric acid, and *trans*-ferulic acid for 28 days, Figure S6: Quantitative real-time PCR validation of for genes involved in phenylpropanoids pathway in treatments with polyphenolic compounds, Table S1: Core proteome differentially identify in both Hass and Criollo EC and NEC (-), based on LIMMA (linear model for microarray data), Table S2: Statistical analysis of values associated with the endogenous content of polyphenolics (mg g−1 fresh weight) in embryogenic Criollo and Hass cultures treated with different concentrations of *p*-coumaric acid and *trans*-ferulic acid during six h, twelve h, 14 and 28 days, Table S3: Putative identification of compounds detected in embryogenic Criollo and Hass cultures treated with *p*-coumaric acid (1 µM) and *trans*-ferulic acid (100 µM) after 12 h and 28 days, Table S4: Primers designed for quantification of relative expression by qPCR, Data S1: Differential proteins identified in comparative analysis, using TMT 6-plex label reagents and synchronous precursor selection (SPS) MS3, in embryogenic (EC) and non-embryogenic (NEC) Hass avocado cultures, Data S2: Gene ontology (GO) enrichment of the core proteome, proteins particularly identified in Hass and Criollo varieties-based David Functional annotation Bioinformatics Microarray analysis (https://david.ncifcrf.gov/) using Arabidopsis homologs. We used the REVIGO web server (http://revigo.irb.hr/) for GO clustering and visual representation of biological processes.

Author Contributions: C.A.O.-G., M.M.-R., C.P.-M. and E.R.-M. designed the experiments; C.A.O.-G., J.L.M.-V., M.R.-V., J.M.E.-C., E.I.-L., C.P.-M., F.Q.-F., A.S.-C., J.A.G.-A., L.M.S. and V.M.L.-V. conducted the experiments, and analyzed the data; E.R.-M., C.A.O.-G., M.M.-R. and V.M.L.-V. wrote the manuscript with the collaboration of all co-authors. All authors have read and agreed to the published version of the manuscript.

Funding: This research was funded by CONACYT, grant number SAGARPA-SENASICA 259915, and FORDECYT 29239. CONACYT provided the scholarship (701571) for the Ph.D. student CAOG.

Acknowledgments: We greatly appreciate the invaluable support provided by SAGARPA-SENASICA and Martín R. Aluja Schuneman in this study. We thank the technical assistance offered by Luis A. Cruz-Silva, Jiovanny Arellano de los Santos, Olinda E. Velázquez, Betsabé Ruíz Guerra and Irene Perea-Arango. We want to thank Luis Herrera-Estrella for providing the genomic information of Avocado.

Conflicts of Interest: The authors declare no conflict of interest. The funders had no role in the design of the study; in the collection, analyses, or interpretation of data; in the writing of the manuscript, or in the decision to publish the results.

Abbreviations

ALH2C4	Aldehyde dehydrogenase family 2-member C4
c-CA	*cis*-cinnamic acid
p-CA	*p* coumaric acid
t-CA	*trans*-cinnamic acid
CAD9	Cinnamyl alcohol dehydrogenase 9
BSA	Bovine serum albumin
CCoAOMT	Caffeoyl-CoA *O*-methyltransferase
CCoAOMT1	Caffeoyl-CoA *O*-methyltransferase 1
C4H/CYP73A5	Cinnamic acid 4-hydroxylases
4CL1	4-coumarate: CoA ligase 1
EC	Embryogenic callus
EMSN	Extracellular matrix surface Network
EP	Embryogenic potency
t-FA	*trans*-ferulic acid
GLM	General linear model
LDOX	Leucoanthocyanidin dioxygenase
MS	Murashige and Skoog
MSP	Murashige and Skoog medium supplemented with 0.1 mg L^{-1} picloram
NEC	Non-embryogenic callus
OMT1	Flavone 3'-O-methyltransferase 1
PAL1	Phenylalanine ammonia-lyase 1
PCA	Principal component analysis
PHBA	*p*-hydroxybenzoic acid
POX52	Peroxidase 52
PP	Phenylpropanoid pathway
ROS	Reactive oxygen species
SA	Sinapic acid
SE	Somatic embryogenesis
TCEP	tris (2-carboxyethyl) phosphine
TMT	Tandem mass tags
VBE	Vanillyl benzyl ether
VA	Vanillin

Appendix A

Detailed information of methods used for comparative proteomic analysis, targeted study of phenolics (dynamic multiple reaction monitoring, dMRM) and untargeted metabolomics using liquid chromatography and high-resolution mass spectrometry and orthogonal partial least square discriminant analysis (OPLS-DA).

References

1. Smith, S.M.; Gomez, D.F.; Beaver, R.A.; Hulcr, J.; Cognato, A.I. Reassessment of the species in the *Euwallacea fornicatus* (Coleoptera: Curculionidae: Scolytinae) complex after the rediscovery of the "lost" type specimen. *Insects* **2019**, *10*, 261. [CrossRef]

2. Encina, C.L.; Parisi, A.; O'Brien, C.; Mitter, N. Enhancing somatic embryogenesis in avocado (*Persea americana* Mill.) using a two-step culture system and including glutamine in the culture medium. *Sci. Hortic.* **2014**, *165*, 44–50. [CrossRef]

3. Lavi, U.; Lahav, E.; Genizi, A.; Degani, C.; Gazit, S.; Hillel, J. Quantitative genetic analysis of traits in Avocado cultivars. *Plant Breed.* **1991**, *106*, 149–160. [CrossRef]

4. Van Nocker, S.; Gardiner, S.E. Breeding better cultivars, faster: Applications of new technologies for the rapid deployment of superior horticultural tree crops. *Hortic. Res.* **2014**, *1*, 14022. [CrossRef] [PubMed]

5. Pliego-Alfaro, F.; Barceló-Muñoz, A.; Simón-Pérez, E.; de la Viña-Nieto, G.; Sánchez-Romero, C.; Perán-Quesada, R. La micropropagación en la mejora de patrones de aguacate (*Persea americana* Mill.): Problemas y limitaciones. *Rev. Chapingo Ser. Hortic.* **1999**, *5*, 239–244.

6. Palomo-Ríos, E.; Perez, C.; Mercado, J.A.; Pliego-Alfaro, F. Enhancing frequency of regeneration of somatic embryos of avocado (*Persea americana* Mill.) using semi-permeable cellulose acetate membranes. *Plant Cell Tissue Organ Cult.* **2013**, *115*, 199–207. [CrossRef]

7. Pliego-Alfaro, F.; Barceló-Muñoz, A.; Lopez-Gomez, R.; Ibarra-Laclette, E.; Herrera-Estrella, L.; Palomo-Ríos, E.; Mercado, J.A.; Litz, R.E. Biotechnology. In *The Avocado: Botany, Production and Uses*; CABI Publishing: Wallingford, Oxfordshire, UK, 2013; pp. 268–300.

8. O'Brien, C.; Hiti-Bandaralage, J.C.A.; Hayward, A.; Mitter, N. Avocado (*Persea americana* Mill.). In *Biological Nitrogen Fixation in Forest Ecosystems: Foundations and Applications*; Springer Science and Business Media LLC: Berlin, Germany, 2018; Volume II, pp. 305–328.

9. Pliego-Alfaro, F.; Murashige, T. Somatic embryogenesis in avocado (*Persea americana* Mill.) in vitro. *Plant Cell Tissue Organ Cult.* **1988**, *12*, 61–66. [CrossRef]

10. Litz, R.E.; Litz, W. Maturation of avocado somatic embryos and plant recovery. *Plant Cell Tissue Organ Cult.* **1999**, *58*, 141–148. [CrossRef]

11. Sánchez-Romero, C.; Márquez-Martín, B.; Pliego-Alfaro, F. Somatic and zygotic embryogenesis in avocado. In *The Plant Plasma Membrane*; Springer Science and Business Media LLC: Berlin, Germany, 2005; Volume 2, pp. 271–284.

12. Perán-Quesada, R.; Sánchez-Romero, C.; Barceló-Muñoz, A.; Pliego-Alfaro, F. Factors affecting maturation of avocado somatic embryos. *Sci. Hortic.* **2004**, *102*, 61–73. [CrossRef]

13. Márquez-Martín, B.; Sesmero, R.; Quesada, M.A.; Pliego-Alfaro, F.; Sánchez-Romero, C. Water relations in culture media influence maturation of avocado somatic embryos. *J. Plant Physiol.* **2011**, *168*, 2028–2034. [CrossRef] [PubMed]

14. Guzmán-García, E.; Sánchez-Romero, C.; Panis, B.; Carpentier, S. The use of 2D-DIGE to understand the regeneration of somatic embryos in avocado. *Proteomics* **2013**, *13*, 3498–3507. [CrossRef] [PubMed]

15. Zhou, T.; Yang, X.; Guo, K.; Deng, J.; Xu, J.; Gao, W.; Lindsey, K.; Zhang, X. ROS homeostasis regulates somatic embryogenesis via the regulation of auxin signaling in cotton. *Mol. Cell. Proteom.* **2016**, *15*, 2108–2124. [CrossRef] [PubMed]

16. Causevic, A.; Delaunay, A.; Ounnar, S.; Righezza, M.; Delmotte, F.; Brignolas, F.; Hagège, D.; Maury, S. DNA methylating and demethylating treatments modify phenotype and cell wall differentiation state in sugarbeet cell lines. *Plant Physiol. Biochem.* **2005**, *43*, 681–691. [CrossRef] [PubMed]

17. Yamamoto, N.; Kobayashi, H.; Togashi, T.; Mori, Y.; Kikuchi, K.; Kuriyama, K.; Tokuji, Y. Formation of embryogenic cell clumps from carrot epidermal cells is suppressed by 5-azacytidine, a DNA methylation inhibitor. *J. Plant Physiol.* **2005**, *162*, 47–54. [CrossRef] [PubMed]

18. Xiao, W.; Custard, K.D.; Brown, R.C.; Lemmon, B.E.; Harada, J.J.; Goldberg, R.B.; Fischer, R.L. DNA Methylation is critical for Arabidopsis embryogenesis and seed viability. *Plant Cell* **2006**, *18*, 805–814. [CrossRef]

19. Valledor, L.; Meijón, M.; Hasbun, R.; Cañal, M.J.; Rodríguez, R. Variations in DNA methylation, acetylated histone H4, and methylated histone H3 during Pinus radiata needle maturation in relation to the loss of in vitro organogenic capability. *J. Plant Physiol.* **2010**, *167*, 351–357. [CrossRef]

20. Viejo, M.; Rodríguez, R.; Valledor, L.; Pérez, M.; Cañal, M.-J.; Hasbún, R. DNA methylation during sexual embryogenesis and implications on the induction of somatic embryogenesis in *Castanea sativa* Miller. *Sex. Plant Reprod.* **2010**, *23*, 315–323. [CrossRef] [PubMed]

21. Fehér, A. Somatic embryogenesis — Stress-induced remodeling of plant cell fate. *Biochim. Biophys. Acta (BBA) Bioenerg.* **2015**, *1849*, 385–402. [CrossRef]

22. Nic-Can, G.; Galaz-Ávalos, R.M.; De-La-Peña, C.; Magaña, A.A.; Wróbel, K.; Loyola-Vargas, V.M. Somatic embryogenesis: Identified factors that lead to embryogenic repression. A case of species of the same genus. *PLoS ONE* **2015**, *10*, e0126414. [CrossRef]

23. Us-Camas, R.; Rivera-Solis, G.; Duarte-Aké, F.; De-La-Peña, C. In vitro culture: An epigenetic challenge for plants. *Plant Cell Tissue Organ Cult.* **2014**, *118*, 187–201. [CrossRef]

24. De-La-Peña, C.; Nic-Can, G.; Galaz-Ávalos, R.M.; Avilez-Montalvo, R.; Loyola-Vargas, V.M. The role of chromatin modifications in somatic embryogenesis in plants. *Front. Plant Sci.* **2015**, *6*. [CrossRef] [PubMed]

25. Duarte-Aké, F.; De-La-Peña, C. Epigenetic advances in somatic embryogenesis in sequenced genome crops. In *Somatic Embryogenesis: Fundamental Aspects and Applications*; Springer Science and Business Media LLC: Berlin, Germany, 2016; pp. 81–102.

26. Cvikrová, M.; Malá, J.; Hrubcová, M.; Eder, J.; Zoń, J.; Macháčková, I. Effect of inhibition of biosynthesis of phenylpropanoids on sessile oak somatic embryogenesis. *Plant Physiol. Biochem.* **2003**, *41*, 251–259. [CrossRef]

27. Misra, B.B.; Dey, S. Accumulation patterns of phenylpropanoids and enzymes in East Indian sandalwood tree undergoing developmental progression 'in vitro'. *Aust. J. Crop. Sci.* **2013**, *7*, 681–690.

28. Kobayashi, T.; Higashi, K.; Sasaki, K.; Asami, T.; Yoshida, S.; Kamada, H. Purification from conditioned medium and chemical identification of a factor that inhibits somatic embryogenesis in carrot. *Plant Cell Physiol.* **2000**, *41*, 268–273. [CrossRef]

29. Alemanno, L.; Ramos, T.; Gargadenec, A.; Andary, C.; Ferriere, N. Localization and identification of phenolic compounds in *Theobroma cacao* L. somatic embryogenesis. *Ann. Bot.* **2003**, *92*, 613–623. [CrossRef]

30. Umehara, M.; Ogita, S.; Sasamoto, H.; Koshino, H.; Asami, T.; Fujioka, S.; Yoshida, S.; Kamada, H. Identification of a Novel factor, Vanillyl Benzyl Ether, Which Inhibits Somatic Embryogenesis of Japanese Larch (*Larix leptolepis* Gordon). *Plant Cell Physiol.* **2005**, *46*, 445–453. [CrossRef]

31. Umehara, M.; Ogita, S.; Sasamoto, H.; Koshino, H.; Nakamura, T.; Asami, T.; Yoshida, S.; Kamada, H. Identification of a factor that complementarily inhibits somatic embryogenesis with vanillyl benzyl ether in Japanese larch. *Vitr. Cell. Dev. Boil. Anim.* **2007**, *43*, 203–208. [CrossRef]

32. Rendón-Anaya, M.; Ibarra-Laclette, E.; Méndez-Bravo, A.; Lan, T.; Zheng, C.; Carretero-Paulet, L.; Perez-Torres, C.A.; Chacón-López, A.; Hernandez-Guzmán, G.; Chang, T.-H.; et al. The avocado genome informs deep angiosperm phylogeny, highlights introgressive hybridization, and reveals pathogen-influenced gene space adaptation. *Proc. Natl. Acad. Sci. USA* **2019**, *116*, 17081–17089. [CrossRef]

33. Jiang, H.; Wood, K.V.; A Morgan, J. Metabolic engineering of the phenylpropanoid pathway in *Saccharomyces cerevisiae*. *Appl. Environ. Microbiol.* **2005**, *71*, 2962–2969. [CrossRef]

34. Qualley, A.V.; Widhalm, J.R.; Adebesin, F.; Kish, C.M.; Dudareva, N. Completion of the core -oxidative pathway of benzoic acid biosynthesis in plants. *Proc. Natl. Acad. Sci. USA* **2012**, *109*, 16383–16388. [CrossRef]

35. Smertenko, A.; Bozhkov, P.V. Somatic embryogenesis: Life and death processes during apical–basal patterning. *J. Exp. Bot.* **2014**, *65*, 1343–1360. [CrossRef] [PubMed]

36. Nic-Can, G.; Avilez-Montalvo, J.R.; Aviles-Montalvo, R.N.; Márquez-López, R.E.; Mellado-Mojica, E.; Galaz-Ávalos, R.M.; Loyola-Vargas, V.M. The Relationship between stress and somatic embryogenesis. In *Somatic Embryogenesis: Fundamental Aspects and Applications*; Springer Science and Business Media LLC: Berlin, Germany, 2016; pp. 151–170.

37. Agati, G.; Azzarello, E.; Pollastri, S.; Tattini, M. Flavonoids as antioxidants in plants: Location and functional significance. *Plant Sci.* **2012**, *196*, 67–76. [CrossRef] [PubMed]

38. Mouradov, A.; Spangenberg, G. Flavonoids: A metabolic network mediating plants adaptation to their real estate. *Front. Plant Sci.* **2014**, *5*, 620. [CrossRef] [PubMed]

39. Takeda, H.; Kotake, T.; Nakagawa, N.; Sakurai, N.; Nevins, D.J. Expression and function of cell wall-bound cationic peroxidase in asparagus somatic embryogenesis. *Plant Physiol.* **2003**, *131*, 1765–1774. [CrossRef] [PubMed]

40. Gallego, P.; Martin, L.; Blázquez, A.; Guerra, H.; Villalobos, N. Involvement of peroxidase activity in developing somatic embryos of *Medicago arborea* L. Identification of an isozyme peroxidase as biochemical marker of somatic embryogenesis. *J. Plant Physiol.* **2014**, *171*, 78–84. [CrossRef]

41. Novaković, L.; Guo, T.; Bacic, A.; Sampathkumar, A.; Johnson, K.L. Hitting the Wall—sensing and signaling pathways involved in plant cell wall remodeling in response to abiotic stress. *Plants* **2018**, *7*, 89. [CrossRef]

42. Pilarska, M.; Knox, J.; Konieczny, R. Arabinogalactan-protein and pectin epitopes in relation to an extracellular matrix surface network and somatic embryogenesis and callogenesis in *Trifolium nigrescens* Viv. *Plant Cell Tissue Organ Cult.* **2013**, *115*, 35–44. [CrossRef]

43. Mathew, S.; Abraham, T.E. Ferulic acid: An antioxidant found naturally in plant cell walls and feruloyl esterases involved in its release and their applications. *Crit. Rev. Biotechnol.* **2004**, *24*, 59–83. [CrossRef]

44. Lenucci, M.S.; Piro, G.; Dalessandro, G. In muro feruloylation and oxidative coupling in monocots. *Plant Signal. Behav.* **2009**, *4*, 228–230. [CrossRef]

45. Schoch, G.; Goepfert, S.; Morant, M.; Hehn, A.; Meyer, D.; Ullmann, P.; Werck-Reichhart, D. CYP98A3 from *Arabidopsis thaliana* is a 3′-Hydroxylase of phenolic esters, a missing link in the phenylpropanoid pathway. *J. Biol. Chem.* **2001**, *276*, 36566–36574. [CrossRef]

46. Franke, R.; Humphreys, J.M.; Hemm, M.R.; Denault, J.W.; Ruegger, M.O.; Cusumano, J.C.; Chapple, C.C.S. The Arabidopsis REF8 gene encodes the 3-hydroxylase of phenylpropanoid metabolism. *Plant J.* **2002**, *30*, 33–45. [CrossRef] [PubMed]

47. Tahir, N.; Pollet, B.; Ehlting, J.; Larsen, K.; Asnaghi, C.; Ronseau, S.; Proux, C.; Erhardt, M.; Seltzer, V.; Renou, J.-P.; et al. A coumaroyl-ester-3-hydroxylase insertion mutant reveals the existence of nonredundant meta-hydroxylation pathways and essential roles forphenolic precursors in cell expansion and plant growth. *Plant Physiol.* **2005**, *140*, 30–48. [CrossRef]

48. Zhang, Y.; Butelli, E.; De Stefano, R.; Schoonbeek, H.-J.; Magusin, A.; Pagliarani, C.; Wellner, N.; Hill, L.; Orzaez, D.; Granell, A.; et al. Anthocyanins double the shelf life of tomatoes by delaying overripening and reducing susceptibility to gray mold. *Curr. Boil.* **2013**, *23*, 1094–1100. [CrossRef] [PubMed]

49. Zhang, X.; Liu, C.-J. Multifaceted Regulations of gateway enzyme phenylalanine ammonia-lyase in the biosynthesis of phenylpropanoids. *Mol. Plant* **2015**, *8*, 17–27. [CrossRef]

50. Palacio, L.; Cantero, J.J.; Cusido, R.M.; Goleniowski, M.E. Phenolic compound production in relation to differentiation in cell and tissue cultures of *Larrea divaricata* (Cav.). *Plant Sci.* **2012**, *193*, 1–7. [CrossRef]

51. Szopa, A.; Ekiert, H. Production of biologically active phenolic acids in Aronia melanocarpa (Michx.) Elliott in vitro cultures cultivated on different variants of the Murashige and Skoog medium. *Plant Growth Regul.* **2013**, *72*, 51–58. [CrossRef]

52. Moyo, M.; Amoo, S.O.; Aremu, A.O.; Gruz, J.; Šubrtová, M.; Dolezal, K.; Van Staden, J. Plant regeneration and biochemical accumulation of hydroxybenzoic and hydroxycinnamic acid derivatives in *Hypoxis hemerocallidea* organ and callus cultures. *Plant Sci.* **2014**, *227*, 157–164. [CrossRef]

53. Yu, Y.; Wang, T.; Wu, Y.; Zhou, Y.; Jiang, Y.; Zhang, L. Effect of elicitors on the metabolites in the suspension cell culture of Salvia miltiorrhiza Bunge. *Physiol. Mol. Boil. Plants* **2018**, *25*, 229–242. [CrossRef]

54. Newton, R.P.; Smith, C.J. Cyclic nucleotides. *Phytochemistry* **2004**, *65*, 2423–2437. [CrossRef]

55. Steenackers, W.J.; Klíma, P.; Quareshy, M.; Cesarino, I.; Kumpf, R.P.; Corneillie, S.; Araújo, P.; Viaene, T.; Goeminne, G.; Nowack, M.K.; et al. cis-Cinnamic acid is a novel, natural auxin efflux inhibitor that promotes lateral root formation. *Plant Physiol.* **2016**, *173*, 552–565. [CrossRef]

56. Kurepa, J.; Shull, T.E.; Karunadasa, S.S.; Smalle, J.A. Modulation of auxin and cytokinin responses by early steps of the phenylpropanoid pathway. *BMC Plant Biol.* **2018**, *18*, 278. [CrossRef] [PubMed]

57. Trewavas, A. Plant cyclic AMP comes in from the cold. *Nature* **1997**, *390*, 657–658. [CrossRef] [PubMed]

58. Nic-Can, G.; Loyola-Vargas, V.M. The role of the auxins during somatic embryogenesis. In *Somatic Embryogenesis: Fundamental Aspects and Applications*; Springer Science and Business Media LLC: Berlin, Germany, 2016; pp. 171–182.

59. Ashihara, H.; Stasolla, C.; Loukanina, N.; Thorpe, T.A. Purine metabolism during white spruce somatic embryo development: Salvage of adenine, adenosine, and inosine. *Plant Sci.* **2001**, *160*, 647–657. [CrossRef]

60. Bolwell, G.P.; Cramer, C.L.; Lamb, C.J.; Schuch, W.; Dixon, R.A. L-Phenylalanine ammonia-lyase from *Phaseolus vulgaris*: Modulation of the levels of active enzyme by trans-cinnamic acid. *Planta* **1986**, *169*, 97–107. [CrossRef] [PubMed]

61. Mavandad, M.; Edwards, R.; Liang, X.; Lamb, C.J.; Dixon, R.A. Effects of trans-cinnamic acid on expression of the bean phenylalanine ammonia-lyase gene family. *Plant Physiol.* **1990**, *94*, 671–680. [CrossRef]

62. O'Neal, D.; Keller, C. Partial purification and some properties of phenylalanine ammonia-lyase of tobacco (*Nicotiana tabacum*). *Phytochemistry* **1970**, *9*, 1373–1383. [CrossRef]

63. Sato, T.; Kiuchi, F.; Sankawa, U. Inhibition of phenylalanine ammonia-lyase by cinnamic acid derivatives and related compounds. *Phytochemistry* **1982**, *21*, 845–850. [CrossRef]

64. Appert, C.; Logemann, E.; Hahlbrock, K.; Schmid, J.; Amrhein, N. Structural and catalytic properties of the four phenylalanine ammonia-lyase isoenzymes from parsley (*Petroselinum Crispum* Nym.). *JBIC J. Boil. Inorg. Chem.* **1994**, *225*, 491–499. [CrossRef]

65. Schoch, G.A.; Nikov, G.N.; Alworth, W.L.; Werck-Reichhart, D. Chemical inactivation of the cinnamate 4-hydroxylase allows for the accumulation of salicylic acid in elicited cells. *Plant Physiol.* **2002**, *130*, 1022–1031. [CrossRef]

66. Schilmiller, A.L.; Stout, J.; Weng, J.-K.; Humphreys, J.; Ruegger, M.O.; Chapple, C.C.S. Mutations in the cinnamate 4-hydroxylase gene impact metabolism, growth and development in Arabidopsis. *Plant J.* **2009**, *60*, 771–782. [CrossRef] [PubMed]

67. Cvikrová, M.; Malá, J.; Eder, J.; Hrubcová, M.; Vágner, M. Abscisic acid, polyamines and phenolic acids in sessile oak somatic embryos in relation to their conversion potential. *Plant Physiol. Biochem.* **1998**, *36*, 247–255. [CrossRef]

68. Widhalm, J.R.; Dudareva, N. A familiar ring to it: Biosynthesis of plant benzoic acids. *Mol. Plant* **2015**, *8*, 83–97. [CrossRef]

69. Lu, F.; Karlen, S.D.; Regner, M.; Kim, H.; Ralph, S.A.; Sun, R.-C.; Kuroda, K.-I.; Augustin, M.A.; Mawson, R.; Sabarez, H.; et al. Naturally p-hydroxybenzoylated lignins in palms. *BioEnergy Res.* **2015**, *8*, 934–952. [CrossRef]

70. Xu, C.; Zhao, L.; Pan, X.; Šamaj, J. Developmental localization and methylesterification of pectin epitopes during somatic embryogenesis of banana (Musa spp. AAA). *PLoS ONE* **2011**, *6*, e22992. [CrossRef] [PubMed]

71. Dobrowolska, I.; Majchrzak, O.; Baldwin, T.C.; Kurczyńska, E. Differences in protodermal cell wall structure in zygotic and somatic embryos of *Daucus carota* (L.) cultured on solid and in liquid media. *Protoplasma* **2011**, *249*, 117–129. [CrossRef]

72. Burr, S.J.; Fry, S.C. Feruloylated arabinoxylans are oxidatively cross-Linked by extracellular maize peroxidase but not by horseradish peroxidase. *Mol. Plant* **2009**, *2*, 883–892. [CrossRef]

73. Shigeto, J.; Nagano, M.; Fujita, K.; Tsutsumi, Y. Catalytic profile of Arabidopsis peroxidases, AtPrx-2, 25 and 71, contributing to stem lignification. *PLoS ONE* **2014**, *9*, e105332. [CrossRef]

74. Murashige, T.; Skoog, F. A revised medium for rapid growth and bioassays with tobacco tissue cultures. *Physiol. Plant* **1962**, *15*, 473–497. [CrossRef]

75. Monribot-Villanueva, J.L.; Elizalde-Contreras, J.M.; Aluja, M.; Segura-Cabrera, A.; Birke, A.; Guerrero-Analco, J.A.; Ruiz-May, E. Endorsing and extending the repertory of nutraceutical and antioxidant sources in mangoes during postharvest shelf life. *Food Chem.* **2019**, *285*, 119–129. [CrossRef]

76. Juárez-Trujillo, N.; Monribot-Villanueva, J.L.; Jiménez-Fernández, V.M.; Suárez-Montaño, R.; Aguilar-Colorado, Á.S.; Guerrero-Analco, J.A.; Jiménez, M. Phytochemical characterization of Izote (*Yucca elephantipes*) flowers. *J. Appl. Bot. Food Qual.* **2018**, *91*, 202–210.

77. Livak, K.J.; Schmittgen, T.D. Analysis of relative gene expression data using real-time quantitative PCR and the 2-ΔΔCT method. *Methods* **2001**, *25*, 402–408. [CrossRef] [PubMed]

78. Smyth, G.K. limma: Linear Models for Microarray Data. In *Nonclinical Statistics for Pharmaceutical and Biotechnology Industries*; Springer Science and Business Media LLC: Berlin, Germany, 2005; pp. 397–420.

International Journal of
Molecular Sciences

Article

Comparative Proteomics Profiling Illuminates the Fruitlet Abscission Mechanism of Sweet Cherry as Induced by Embryo Abortion

Zhi-Lang Qiu [1], Zhuang Wen [1], Kun Yang [1], Tian Tian [1,2], Guang Qiao [1], Yi Hong [1] and Xiao-Peng Wen [1,2,*]

[1] Key Laboratory of Plant Resources Conservation and Germplasm Innovation in Mountainous Region (Guizhou University), Ministry of Education, Institute of Agro-bioengineering/College of Life Sciences, Guizhou University, Guiyang 550025, China; 18786621377@163.com (Z.-L.Q.); gzu_zwen@163.com (Z.W.); kyanggz@163.com (K.Y.); 13518504594@163.com (G.Q.); hongyi715@163.com (Y.H.)
[2] Institute for Forest Resources & Environment of Guizhou, College of Forestry, Guizhou University, Guiyang 550025, China; tiantiangzu@163.com
* Correspondence: xpwensc@hotmail.com; Tel.: +86-851-88290212

Received: 18 January 2020; Accepted: 8 February 2020; Published: 11 February 2020

Abstract: Sweet cherry (*Prunus avium* L.) is a delicious nutrient-rich fruit widely cultivated in countries such as China, America, Chile, and Italy. However, the yield often drops severely due to the frequently-abnormal fruitlet abscission, and few studies on the metabolism during its ripening process at the proteomic level have been executed so far. To get a better understanding regarding the sweet cherry abscission mechanism, proteomic analysis between the abscising carpopodium and non-abscising carpopodium of sweet cherry was accomplished using a newly developed Liquid chromatography-mass spectrometry/mass spectrometry with Tandem Mass Tag (TMT-LC-MS/MS) methodology. The embryo viability experiments showed that the vigor of the abscission embryos was significantly lower than that of retention embryo. The activity of cell wall degrading enzymes in abscising carpopodium was significantly higher than that in non-abscising carpopodium. The anatomy results suggested that cells in the abscission zone were small and separated. In total, 6280 proteins were identified, among which 5681 were quantified. It has been observed that differentially accumulated proteins (DAPs) influenced several biological functions and various subcellular localizations. The Kyoto Encyclopedia of Genes and Genomes (KEGG) enrichment analysis showed that plenty of metabolic pathways were notably enriched, particularly those involved in phytohormone biosynthesis, cell wall metabolism, and cytoskeletal metabolism, including 1-aminocyclopropane-1-carboxylate oxidase proteins which promote ethylene synthesis, and proteins promoting cell wall degradation, such as endoglucanases, pectinase, and polygalacturonase. Differential expression of proteins concerning phytohormone biosynthesis might activate the shedding regulation signals. Up-regulation of several cell wall degradation-related proteins possibly regulated the shedding of plant organs. Variations of the phytohormone biosynthesis and cell wall degradation-related proteins were explored during the abscission process. Furthermore, changes in cytoskeleton-associated proteins might contribute to the abscission of carpopodium. The current work represented the first study using comparative proteomics between abscising carpopodium and non-abscising carpopodium. These results indicated that embryo abortion might lead to phytohormone synthesis disorder, which effected signal transduction pathways, and hereby controlled genes involved in cell wall degradation and then caused the abscission of fruitlet. Overall, our data may give an intrinsic explanation of the variations in metabolism during the abscission of carpopodium.

Keywords: sweet cherry; embryo abortion; fruitlet abscission; fruit drop; mechanism

1. Introduction

Sweet cherry (*Prunus avium* L.), widely cultivated in countries such as China, America, and Japan, is an important fruit crop known for its appealing color, delicious taste, and nutritional value [1]. However, abnormal fruit abscission can often reduce crop yield greatly. Previous research shows that fruitlet shedding is insufficient to obtain better economic benefits [2]. Fruit abscission is a highly regulated developmental process is effected by both internal and environmental causes [3]. Its regulatory mechanism is complicated and concerns multiple reasons. Therefore, efforts on the unveiling molecular mechanism of fruit abscission in sweet cherry plays vital role in increasing its yield.

Abscission is a fundamental process in plant biology and represents an evolutionary adaptation of plants, it allows to discard senescent or physiologically damaged organs, e.g., leaves, petals, and fruit [4,5] for better adaptation and for efficient seed dispersal [6]. Abscission is precisely regulated by structural, physiological, biochemical, and molecular changes that ultimately lead to the shedding of plant organs [7]. This event takes place in a special cell layer called as abscission zone (AZ), which consists of cell separation enabled by hydrolytic enzymes [5,8]. However, frequently severe abscission is a hard nut for fruit productivity [9]. Recently, more efforts have been leveraged on illuminating the regulation of abnormal abscission at the molecular level [10–12], which help understand the mechanisms underlying abscission along with getting bumper harvest [4,6]. Also, molecular studies on abscission can help improve current agricultural management practices, such as flower and fruit thinning, mechanical picking of fruit [4,11]. To date, the abundant studies on organ abscission had been described in the model plants, e.g., *Arabidopsis thaliana* [13] and *Solanum lycopersicum* [12,14,15], and some molecular knowledge related to fruit abscission had also been acquired from fruit tree crops (e.g., apple, citrus, lichi) [6,11,16]. However, the current information about the molecular mechanisms underlying severe fruit abscission in sweet cherry has not yet been unraveled.

According to the causes, abscission can be divided into three types, namely, normal abscission (such as abscission of ripened fruit and seed), metabolic abscission due to the completion between the reproductive growth and vegetative growth (such as premature shedding of fruit and unpollinated flowers), and abnormal abscission owing to environmental stresses (such as cold, heat, light, and pathogen) [17]. The abscission of plant organs is associated with a balance between the levels of auxin and ethylene in AZ [18–20]. It has also been observed that ethylene can induce the synthesis and secretion of various cell wall and middle hydrolases, which are accompanying to plant organ abscission [9,21], while auxin inhibits abscission by rendering AZ cells insensitive to ethylene [10,12]. Simply, abscission can be divided into four major steps: (a) Differentiation and formation of the AZ; (b) acquisition of the competence to respond to abscission signals; (c) execution of organ abscission; and (d) differentiation of a protective layer [8,14]. It has been found that the expression of multiple regulatory genes varied before and during peduncle shedding [22], and this variation affected the differential expression of transcription factors associated with the auxin and ethylene pathways [23].

Typical components of the cell wall containing cellulose, hemicellulose, pectic polysaccharides, proteins, and phenolic compounds. During the process of plant organ shedding, cell wall hydrolases are synthesized in large quantities, and enzyme activity is also elevated, which conceivably the origin of the degradation of the middle lamella and the loosening of the primary cell wall of the separation layers [21]. Cellulase (CEL) and polygalacturonase (PG), two major cell wall hydrolase enzymes, had been extensively studied in different plants and played an important role in plant organ abscission [24]. Additionally, expansin protein (EXP), xyloglucan endotransglucosylases/hydrolases (XTH), peroxidase (POD) [15,25] also play an essential function during plant organ abscission. In the process of ethylene induction or low level auxin initiation, the degrading enzyme genes of plant cell walls were also up-regulated, resulting in the abscission of plant organs [26].

Comparative proteomic analysis is a powerful tool for systematically understanding of a biological event at the molecular level [14]. Recently, proteomics has been widely used in the study of citrus [27], apple [28], pear [29] and sweet cherry [30], etc. However, there had not been reports on the mechanism of cherry fruit drop by proteomic analysis. Besides, Parallel Reaction Monitoring

(PRM) was a recently developed methodology in targeted mass spectrometry, which involves in the use of a quadrupole-equipped orbitrap [31] and has been widely used to quantify and detect target proteins [32,33]. Also, PRM has been used to validate the reliability of proteomic data, which provides a reliable guarantee for accurate and reliable proteomics data [34].

In the present study, the vigor of abscission sweet cherry embryos was inspected, the enzyme activity assay of the abscising acropodium and non-abscising carpopodium, the anatomical structure observations of the abscising acropodium and non-abscising carpopodium, then we used TMT and PRM to analyze the changes in the proteome during the embryo-induced abortion. The study found that (1) the degree of embryo abortion was significantly higher in normal fruits than in normal fruits; (2) the cell wall degrading enzyme activity was significantly higher than retention fruit; (3) the cells in the abscission zone were small and separated; (4) A total of 6280 proteins were identified, among which 5681 were quantified. A total of 1957 DAPs, including 1056 up- and 901 down-regulated proteins, were identified. Mostly, preceding proteins involved in cell wall hydrolysis-related, lignin synthesis-related, plant hormone synthesis and signaling-related enzymes, cytoskeleton-related proteins, transporters, and transcription factors. Existing study can provide data for scrutinizing the shedding of plant organs from the aspects of morphology, anatomy and molecular biology, also establishes a foundation revealing the molecular mechanism of sweet cherry fruit abscission and breeding high yield sweet cherry germplasm.

2. Results

2.1. Embryo Vigor of Shedding Fruit

After anatomical observation of embryos during the development, they might be divided into three types according to the plumpness, namely 0 < plumpness < 50% (Figure 1(a1)), 50% < plumpness < 100% (Figure 1(a2)) and plumpness = 100% (Figure 1(a3)); according to the statistics, it is shown that in the abscission fruit, the fruit with plumpness = 100% accounts for 3%, while in the normal fruit, the fruit with plumpness = 100% accounts for only 90% (Figure 1b). Based on the results of the staining experiment, it can be divided into three categories, namely 0 < coloring degree < 50% (Figure 1(a4)), 50% < coloring degree < 100% (Figure 1(a5)) and coloring degree = 100% (Figure 1(a6)). In the abscission fruits, coloring degree = 100% accounted for 3%, while the non-abscission fruits accounted for 94% (Figure 1c). The result showed that the fruit plumpness and the coloring degree of embryos are related to fruit abscission. In other words, there was a greater correlation between embryo abortion and fruit shedding.

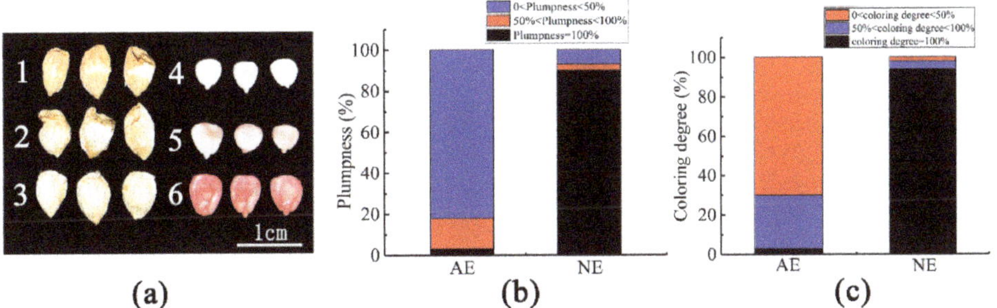

Figure 1. The embryos vigor of abscission and non-abscission fruit. (**a**) 1, 2, and 3 represent different plumpness, and 4, 5, and 6 represent the coloring degree corresponding to different plumpness. (**b**), The stacked figure of different plumpness. (**c**), The stacked figure of different coloring degree. AE, abscission embryo; NE, non-abscission embryo.

2.2. Enzyme Activity and Anatomical Structure of Carpopodium Abscission Zone

To explore whether cell wall hydrolysis and activity of antioxidant activity-related enzymes peroxidase in the abscission zone is related to the shedding of sweet cherry fruitlet. Enzyme activity assay showed that cellulase (CEL), polygalacturonase (PG), pectinase (PE), and peroxidase (POD) activities were significantly higher than retention carpopodium (Figure 2a). This result indicates that the shedding of sweet cherry fruit may be due to the hydrolysis of the cell wall in the abscission zone. In addition, the anatomical structure of carpopodium abscission zone suggested that abscising carpopodium abscission zone cell was small and dense (Figure 2b), while the non-abscising carpopodium abscission zone cell was sizeable and lean (Figure 2c). Moreover, the cells in the abscission zone (AZ) also separated. These outcomes indicate that the fruit abscission of the sweet cherry has an immense connection with the physiological, biochemical metabolism and cell structure of the abscission zone.

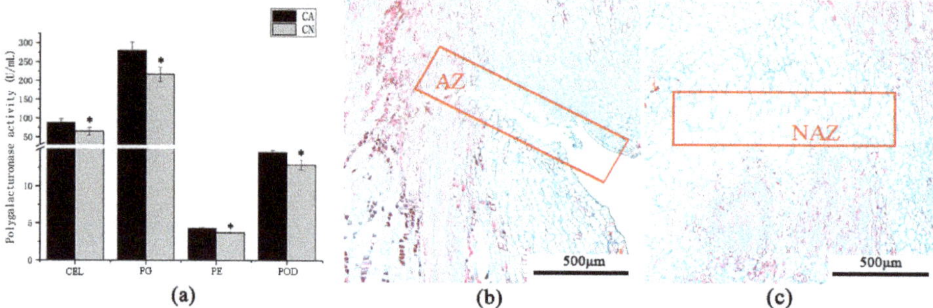

Figure 2. Enzyme activity and anatomical structure of carpopodium abscission zone. (**a**) The activities of cellulase (CEL), polygalacturonase (PG), pectinase (PE), and peroxidase (POD) in the abscising carpopodium (CA) and non-abscising carpopodium (CN); (**b**) The anatomical structure of abscising carpopodium abscission zone. (**c**) The anatomical structure of the non-abscising carpopodium abscission zone. AZ, abscission zone. NAZ, non-abscising abscission zone.

2.3. Quality Control and Quantitative Proteomic Analysis

An integrated approach involving LC-MS/MS and TMT labeling was applied to analyze the proteomic changes between abscising carpopodium and non-abscising carpopodium. The general workflow is demonstrated in Figure 3a. The satisfactory reproducibility for the current experiment has been proven via Pair-wise Pearson's correlation coefficients (Figure 3b). Overall, 34,432 peptides were revealed. Following the quality confirmation, along with average mass error < 0.02 Da, signifying an immense validity for data regarding MS (Figure 3c). Classified peptides lengths were recorded among 7 to 20 amino acids, which demonstrating fulfilled standard criteria of our sampling (Figure 3d). The number of proteins analyzed during the experiment was 6280, where 5681 were quantified. All discovered proteins were categorized to understand their function properly e.g., GO terms, represent the functional domains, KEGG pathways, and subcellular localization. The identified protein's detailed information is listed in Table S1.

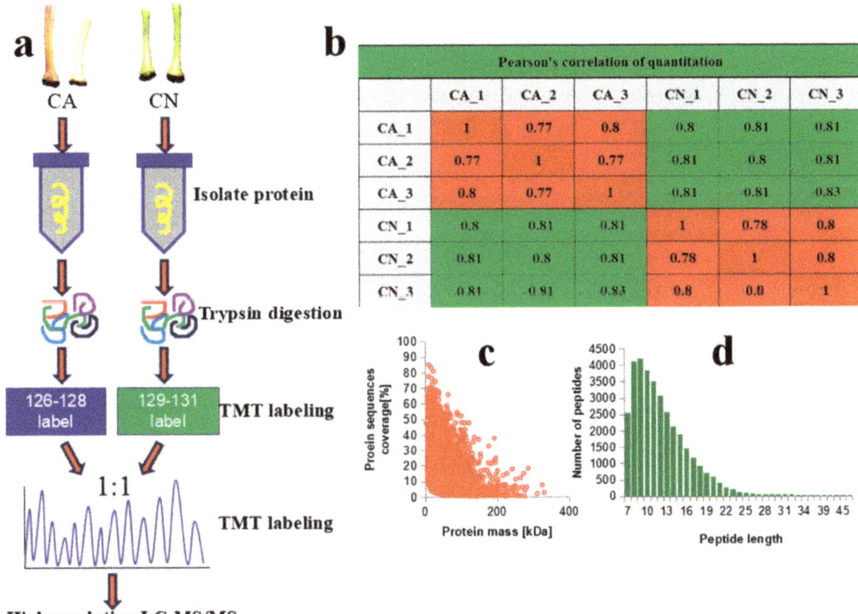

	CA_1	CA_2	CA_3	CN_1	CN_2	CN_3
CA_1	1	0.77	0.8	0.8	0.81	0.81
CA_2	0.77	1	0.77	0.81	0.8	0.81
CA_3	0.8	0.77	1	0.81	0.81	0.83
CN_1	0.8	0.81	0.81	1	0.78	0.8
CN_2	0.81	0.8	0.81	0.78	1	0.8
CN_3	0.81	0.81	0.83	0.8	0.8	1

Figure 3. The trial technique for quantitative proteome investigation and quality control approval of MS information. (**a**) Protein was extricated in three natural imitates for each sample gathering. Entire protein samples were trypsin digested and dissected by HPLC-MS/MS. (**b**) Pearson s correlation of protein quantitation. (**c**) Mass delta of all identified peptides. (**d**) Length distribution of every single distinguished peptide.

2.4. Identification of DAPs During Carpopodium Abscission

A total of 1957 DAPs, including 1056 up- and 901 down-regulated proteins, were recognized (Table S2 and Figure S1). Among DAPs, the top five up-regulated proteins were a 7-deoxyloganetin glucosyltransferase-like (4.39 fold), followed by a plasma membrane-associated cation-binding protein 1 (3.70), an UDP-glycosyltransferase 73C1-like (3.66 fold), a putative Beta-D-xylosidase (3.59 fold), an extracellular ribonuclease LE-like (3.40 fold). the top five down-regulated proteins were a glucuronoxylan 4-*O*-methyltransferase 1 (4.81 fold), anthocyanidin reductase ((2*S*)-flavan-3-ol-forming, 4.00 fold), probable auxin efflux carrier component 1c (3.86 fold), cytochrome P450 98A2 (3.57 fold), dCTP pyrophosphatase 1-like (3.40 fold) (Table S2). Subcellular locations of the DAPs were predicted (Table S3). For the up-regulated proteins, a total of 14 groups were identified, such as chloroplast- (404 proteins), cytoplasm- (201), nucleus- (168), plasma membrane- (99), extracellular- (82), vacuolar membrane- (42), mitochondria- (30), and endoplasmic reticulum- (12 proteins) (Figure S1). For the down-regulated proteins, 15 components were identified, including chloroplast- (309 proteins), cytoplasm- (267), nucleus- (189), and plasma membrane-localized protein (58) (Figure S1).

2.5. Enrichment Analysis of DAPs During Carpopodium Abscission

In total, 332 DAPs were assigned to at least one GO term. For up-regulated proteins, the highly enriched 'Biological Process' GO terms were 'inorganic anion transport', 'cell wall macromolecule catabolic process', 'peptide transport'; within the 'Molecular Function', the most significantly enriched terms were 'hydrolase activity'; and the most enriched terms in the 'Cellular Component' were 'nucleosome' and 'chromatin' (Figure 4a). For down-regulated proteins, the highly enriched 'Biological Process' GO terms were related to 'microtubule nucleation', 'alpha-amino acid metabolic process', 'chlorophyll biosynthetic process', and 'protein polymerization'. Within the 'Molecular Function', the most significantly enriched terms were 'structural molecule activity', 'structural constituent of cytoskeleton'; and the most enriched terms in the 'Cellular Component' were 'polymeric cytoskeletal fiber', 'microtubule cytoskeleton', 'cytoskeletal part', and 'cytoskeleton' (Figure 4b).

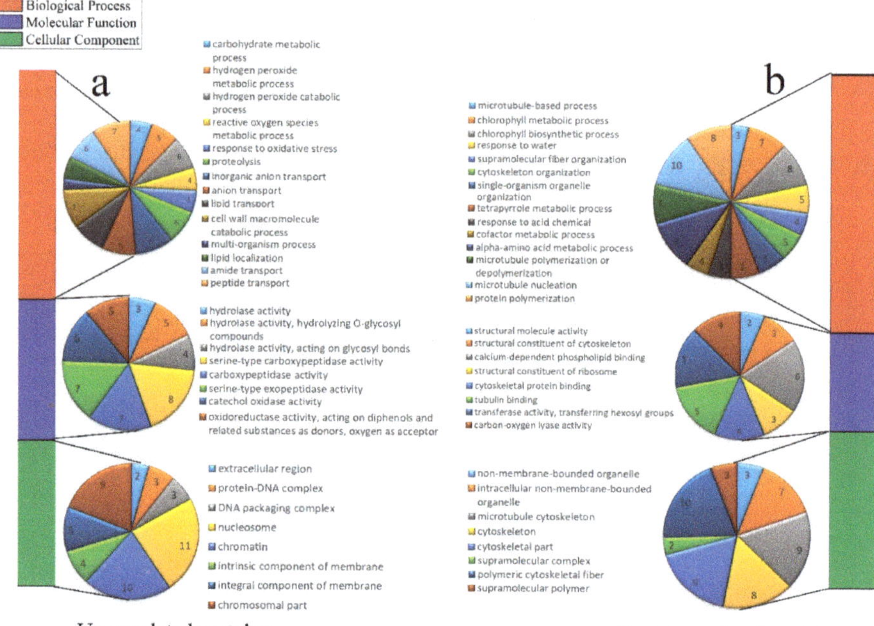

Figure 4. Gene Ontology (GO) enrichment analysis of DAPs. (**a**) Dispersion of the up-regulated proteins with GO annotation. Diverse shading squares express to various terms, including cellular component, molecular function, and biological process. The number, some of the up-regulated proteins in each second-level term, was appeared in a pie chart. (**b**) Distribution of the down-regulated proteins with GO annotation. Diverse colors squares symbolize to various terms, including cellular component, molecular function, and biological process. Number, the quantity of the down-regulated proteins in each second-level term has appeared in a pie chat.

The up-regulated proteins were mostly linked with 'phenylpropanoid biosynthesis' and 'galactose metabolism'; and the down-regulated proteins were usually engaged in 'flavonoid biosynthesis' and 'amino acids' (Figure S2).

The up-regulated proteins generally contained a Glycoside hydrolase superfamily and down-regulated proteins mainly comprised a TCP-1-like chaperonin intermediate domain (Figure S3).

2.6. Biosynthesis of Cell Wall Modifying Proteins and Lignin

Among all the DAPs, 101 proteins were predicted to be associated with cell wall metabolism and lignin biosynthesis, of which 73 were up-regulated and 28 down-regulated in the abscising carpopodium (Table S4). These included, cell wall hydrolytic enzymes such as endoglucanase CX (CEL), pectinesterase (PE), polygalacturonase (PG), pectin acetylesterase 12-like (PAE), β-galactosidase (GBAL), beta-glucosidase 45-like isoform X6 (BGLU) significantly up-regulated in the abscising carpopodium; however, the biosynthesis of cell wall modifying proteins, namely, galacturonosyltransferase (GalAT) and UDP-glucose 6-dehydrogenase (UGDH) were significantly down-regulated in the abscising carpopodium. Interestingly, the extension-associated proteins cell wall, e.g., xyloglucan endotransglucosylase/hydrolase protein (XTH) and expansin-like B1 (EXP), were significantly up-regulated in the abscising carpopodium. Excluding, the related proteins of lignin synthesis, e.g., peroxidase 16-like (POD), were also significantly up-regulated, while anthocyanidin 3-*O*-glucosyltransferase 5-like (3GT) significantly down-regulated in abscising carpopodium (Table S4, Figure 5).

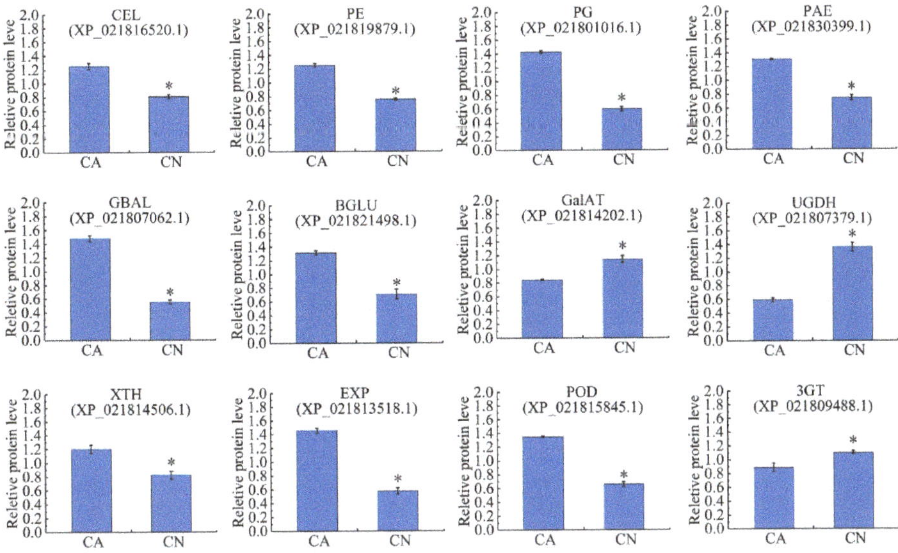

Figure 5. Relative expression levels of the proteins related to cell wall metabolism and lignin biosynthesis. Significant differences in expression level were indicated by "*".

2.7. Plant Hormone Biosynthesis and Signal Transduction

Totally, 105 proteins were annotated to be associated with the phytohormone biosynthesis and signal transduction pathways, of which 51 were up-regulated and 54 down-regulated. In the ethylene biosynthesis pathway, four 1-aminocyclopropane-1-carboxylate oxidase (ACO) proteins were significantly up-regulated in the abscising carpopodium. In the abscisic acid biosynthesis pathway, the zeaxanthin epoxidase (ZEP) and 9-cis-epoxycarotenoid dioxygenase (NCED) were significantly up-regulated in the abscising carpopodium. In the abscisic acid signal transduction pathway, ten protein phosphatase 2C (PP2C) proteins and one SnRK2 were significantly up-regulated. In the salicylic acid signal transduction pathway, one NPR5 and one pathogenesis-related protein 1 (PR1) were significantly up-regulated. In the auxin biosynthesis pathway, two tryptophan synthase alpha chain, the enzyme necessary for the synthesis of tryptophan was significantly down-regulated in the abscising carpopodium. In the auxin signal transduction pathway, the auxin transporter-like protein 2 (AUX1) and transport inhibitor response 1 (TIR1) were significantly down-regulated in abscising carpopodium. furthermore, one 2-oxoglutarate-dependent dioxygenase (DAO), which is essential for auxin catabolism and the maintenance of auxin homeostasis in reproductive organs was down-regulated in abscising carpopodium. It is worth mentioning that one polyamine oxidase was down-regulated in abscising carpopodium (Table S5, Figure 6).

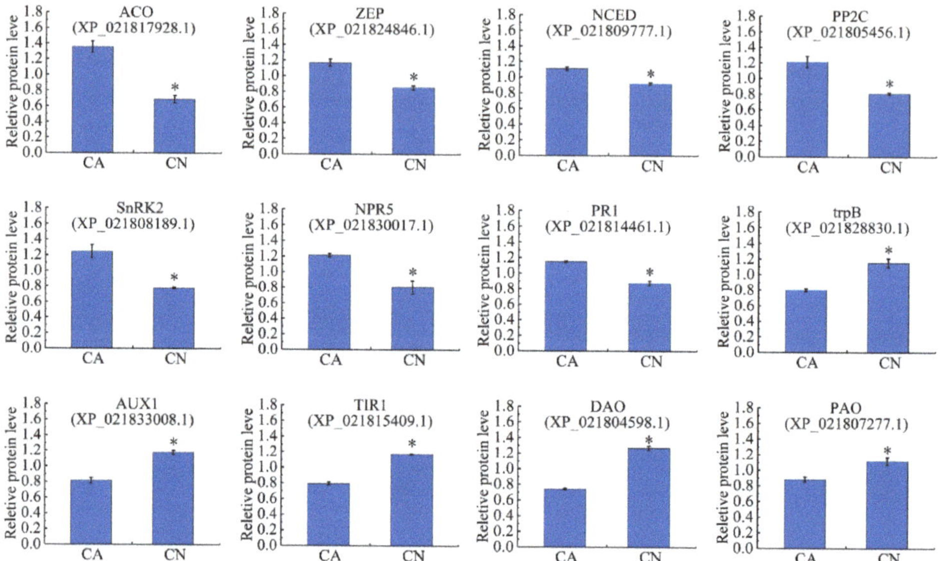

Figure 6. Relative expression levels of the proteins related to plant hormone and signal transduction. Significant differences in expression level were indicated by "*".

2.8. Cytoskeleton and Transport Proteins

Totally, 49 proteins were related with the cytoskeleton, of which 23 were up-regulated and 26 were down-regulated. These proteins mainly concerned tubulin family proteins, microtubule-associated proteins, and actin regulated proteins. These proteins mainly involved in the formation of cell wall. Therefore, these proteins may regulate the detachment of the carpopodium by regulating the formation of cell walls. Among the transport proteins, of which 27 were up-regulated and 14 were down-regulated prior proteins holding ABC transporters, lipid-transfer proteins, calcium-transporting ATPase, auxin transport protein, etc. These proteins also regulate cell wall hydrolase changes by transporting ATP, lipids, calcium, and auxins.

2.9. Transcription Factor

Totally, 17 transcription factors were accumulated, involved different family TFs including homeobox-leucine zipper, bZIP and ARF, bHLH, suggesting a complex regulation of organ separation. In particular, homeobox-leucine zipper protein ATHB-12-like was up-regulated 2.62 fold, bZIP was up-regulated 1.21 fold, bHLH3 was down-regulated 1.30 fold and ARF was down-regulated 1.38 fold, which may play a critical role in the abscission of plant organs.

2.10. Validation of DAPs by PRM

Totally, 22 DAPs significantly involved in these GO terms and pathways were selected for PRM analysis. Additionally, two particular peptides along anticipated chemical stability were selected of separate protein, and the relative protein abundance was indicated as the average of the two standardized peptide top regions (Table S6). The expression values of the up-regulated proteins were higher and those of the down-regulated proteins were lower in the CA group, in comparison to the CN group. The fold changes for these proteins were significantly different between the CA and CN groups at $p < 0.01$, in agreement with the findings from TMT analysis (Figure 7).

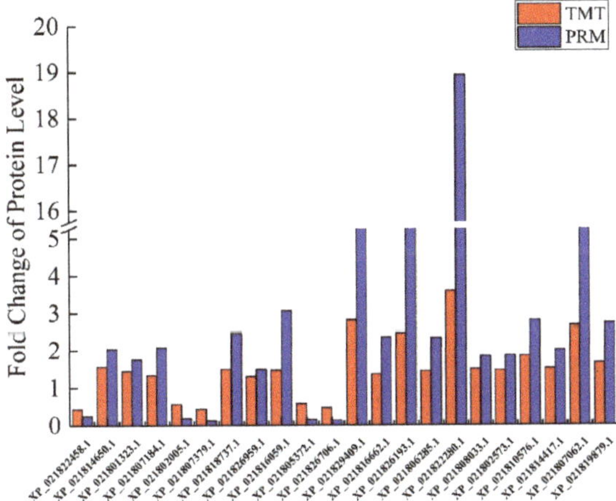

Figure 7. Confirmation of 22 selected differentially produced proteins detected by the Parallel Reaction Monitoring (PRM) technique. The ordinate represents the CA/CN ratio value, ratio > 1 represents the up-regulation, and ratio < 1 represents the down-regulation.

3. Discussion

Fruit abscission cause by embryo abortion presents in a variety of plants, such as apple [35] and mango [36]. However, there are few studies on embryo abortion-induced sweet cherry fruit shedding. It was not until the early 21st century that people began to study the physiological mechanism of sweet cherry fruit abscission, and these studies have shown that the fruit abscission has an excellent at correlation with the polar transport of auxin, carbohydrate, and abscisic acid [37,38]. Recently, a large number of evidences on fruit shedding have been published about other species, such as tomato [23], citrus [6], and litchi [16]. It is noteworthy that high-throughput proteomic analysis has been developed to reveal the mechanism of plant organ abscission at the protein level [14]. Abscission is a process that works together by external and internal factors [3]. The process has a complex mechanism or modification processes [14], including cell wall modifications [6], plant hormonal biosynthesis, signal transduction pathways [3], and pathogen defense-regulated [4]. To study the fruitlet abscission mechanism of sweet cherry as induced by embryo abortion, comparative proteomics was used to study alterations in protein between the abscising carpopodium and non-abscising carpopodium.

3.1. Embryo Abortion Leads to Fruit Abscission

In the present case, experiments from embryo vigor and fruitlet abscission demonstrated that there was a positive correlation between embryo vigor and fruitlet abscission. It can be inferred that the abortion of an embryo may cause fruit shedding. Both mango [36] and citrus [39] have proved that embryo abortion can lead to fruit abscission. Additionally, our group also found that the pollination trees have low pollen vigor and pollen deformity under an electron microscope (Unpublished). In the process of fruit setting and development, endogenous hormones play a continuous coordinating role, and embryo abortion is related to the content and balance of endogenous hormones [40]. In turn, numerous studies have found that the regulation of endogenous hormones is closely related to the embryonic development of plants [41]; in other words, the plant hormone imbalance can also cause embryo abortion. There were many reasons for embryo abortion, such as male sterility [42], pollination and fertilization [43], and endogenous hormones [41]. Therefore, by adjusting the hormone balance and improving the pollen vigor of pollination tree, the abortion rate of the embryo can be reduced and the fruit setting rate can be increased. If not, after finding the molecular mechanism of fruit shedding, it can also increase fruit yield by directly regulating the expression of genes associated with fruit shedding.

3.2. Cell Wall Metabolism and Abscission

Abscission is an active physiological process that occurs through the dissolution of cell walls at predetermined locations, the abscission zones (AZs) [3]. The most direct cause of plant organ is due to changes in cell wall hydrolase activity, resulting in the degradation of the cell wall. Several genes regulate the functioning of the plant cell wall. Changes in their expression are associated with aging [5], organ growth and development [5,44], maturation of fruits [45], and organ abscission [46].

Our proteomics profiling of sweet cherry abscising carpopodium showed bidirectional changes in the expression of the cell wall-related proteins, probably associated not only with the ongoing process of abscission but also with the progressive development of the organs that are not dropped (Table S4). However, it is worth noting that most of the enzymes involved in cell wall degradation showed a trend of up-regulation, including cellulose [3], pectinase [47], polygalacturonase [48], β-galactosidase [49], which play a major role in cell wall degradation. Moreover, the enzymatic activity of cellulose, pectinase and polygalacturonase were significantly higher than non-abscising carpopodium (Figure 2a). Besides, expansin was proven to be associated with the process of wall extension during cell growth [15]. It has, however, become clear that expansins also make a significant contribution to the process of fruit softening, which involves wall breakdown, rather than expansion. It has been observed that expansins play an important role with ethylene-mediated abscission. The function of expansin may increase the disorder of cellulose crystals, making the glucan chains easier to hydrolyze [50]. In the current case, significant accumulation of the expansin in the abscising carpopodium, the accumulation of expansin may play an important role in the shedding of sweet cherry carpopodium. Xyloglucan is one of the major hemicelluloses of primary cell walls, in dicot plants, and may account for up to 10%–20% of cell wall components [15]. XTHs belong to a multigene family which plays important role in several different processes during cell wall modification. These include root hair initiation [51], hypocotyl elongation [52], hydrolysis of seed storage carbohydrates, leaf growth and expansion, fruit softening tension, wood formation, and petal abscission. Currently, two XTHs were significantly up-regulated in abscising carpopodium (Figure 6), suggesting that It is likely that changes mediated by XTHs action may allow easier accessibility of the cell wall to other cell wall hydrolytic enzymes, thus accelerating abscission [32].

3.3. Lignin Biosynthesis and Abscission

Despite the fact that the role of lignin and lignified tissues in abscission has not yet been explained, it had been notable that during leaf abscission of woody species, ligno-suberization of the protective layers is exceptionally normal [9]. Moreover, in the process of plant organ shedding, it was usually accompanied by lignin deposition [6]. The function of lignin discharge has been related to the generation of defensive layers at the tissues staying in the plant during the last step of the abscission procedure [9,53]. Moreover, it has been suggested that lignification could also facilitate the mechanical cell wall breakage during cell separation processes [54]. Therefore, lignin deposition might be considered as a marker of abscission activation. In the present study, there were several proteins significantly up-regulated in 'Phenylpropanoid biosynthesis' pathway. It is noteworthy that 7 peroxidases were up-regulated over 1.5 times. Additionally, the enzymatic activity of peroxidases was significantly higher than the non-abscising carpopodium (Figure 2a), suggesting an increase in lignin biosynthesis. This result indicated that 'peroxidase' plays an important role in the process of lignin biosynthesis. However, lignin plays an important role in the shedding of carpopodium and the formation of protective layers.

3.4. Plant Hormone-Related Abscission

In the overall process of abscission, regulatory effects of plant hormones are of major relevance since they mediate responses of plant organs to stress [4,55]. Depending on their concentration in different tissues, their receptor concentration and affinity, their homeostasis, their transport or their interaction, hormones can act as a signal to accelerate or inhibit the effects of abscission, and the responses are complex [3]. The ethylene and abscisic acid (ABA) act as abscission-accelerating signals [3,18], while auxin was considered as abscission inhibitors [13,56]. The plant hormone ethylene plays an important role as a positive regulator in process of plant organ abscission since it can induce differential gene expression, including cell wall hydrolytic enzymes, lipid-transfer proteins, pathogen-related proteins, hormone biosynthesis, etc. [9]. However, it remains unclear whether ABA induces abscission directly via hormone activity or indirectly by generating a mild carbohydrate deficit [57]. Auxin inhibits the plant organ abscission because it can prevent the expression of some cell wall degrading enzymes [13]. Before auxin worked, it was transported by the auxin influx carriers (AUX1/LAX) and transport inhibitor response 1 (TIR1) [58]. Therefore, AUX1 and TIR1 also play an important role in plant organ abscission. Recently, four key proteins for ethylene biosynthesis, 1-aminocyclopropane-1-carboxylate oxidase (ACO) proteins, were up-regulated significantly in the abscising carpopodium. However, the S-adenosylmethionine synthetase (SAMS) were down-regulated. The down regulation of SAMS could increase methionine accumulation [59], while the up-regulated of ACO could increase ethylene biosynthesis [60]. These results indicate that ethylene biosynthesis increases lead to the carpopodium abscission. In the 'carotenoid biosynthesis' pathway, the ABA biosynthesis key protein 9-cis-epoxycarotenoid dioxygenase (NCED) was up-regulated significantly in the abscising carpopodium. In addition, in the ABA signal transduction pathway, one PP2C and two SnRK2 was significantly up-regulated, the evidence suggests that abscisic acid may play an important role in the carpopodium abscission. It is noteworthy that tryptophan synthase was down-regulated in the abscising carpopodium. This result led to the decrease in tryptophan synthesis, thereby causing a decrease in the synthesis of auxin. Additionally, the AUX1 and TIR1 protein down-regulated in the abscising carpopodium. These results indicate the precursor of auxin synthesis and the transport vector of auxin are reduced, resulting in a decrease in auxin entering the cell. In summary, the synthesis of ethylene and abscisic acid-related proteins were significantly up-regulated, while the synthesis of auxin-related proteins was significantly down-regulated, leading to imbalance of hormones, leading to the shedding of the carpopodium.

3.5. Cytoskeleton and Abscission

The cytoskeleton is a fundamental component of the constituent cells, including the actin cytoskeleton and the microtubule cytoskeleton. Studies have shown that adhesion maintenance of requires modification of the cell wall, depending on the actin cytoskeleton [61]. The reason why is that the delivery of pectin and its modifying proteins occur mainly via the actin cytoskeleton. To regulate branching and nucleation of the actin filament, the Actin-associated protein2/3 complex (Arp2/3) is highly conserved and is the critical component [62]. In addition, the microtubule cytoskeleton is also the basic structure of cellulose. There have been reports that there is a great correlation between the arrangement of microtubules and shedding [63]. Besides, microtubule-associated proteins have an important role in cellulose biosynthesis [64]. Currently, one actin-related protein 2/3 complex is significantly down-regulated in the abscising carpopodium. This result indicates that the ability to adhere to cells is limited, which increases the possibility of shedding. On the other hand, a total of seven microtubule-associated proteins are significantly down-regulated. These results indicate that the synthesis of microtubules has been inhibited. These results indicate that cell wall synthesis decreases, leading to decreased adhesion between cells, causing cell separation (Figure 2b) and carpopodium abscission.

3.6. Transcription Factor Raleted Abscission

During the plant organ abscission, a large number of transcription factors are also involved. Including ERF family [49], ARF family [13], MADS-box family [65,66], and HD-ZIP family [16,67]. In the present study, ERF and MADS-box did not accumulate, However, two HD-ZIP family proteins were significantly up-regulated. In litchi, two cellulase genes, *LcCEL2* and *LcCEL8*, can be directly activated by the HD-ZIP family transcription factor HB2, synthesize cellulase, and then hydrolyze cellulose, causing the loss of litchi fruits [67]. Besides, when the fruitlets sense the abscission signal, LcHB2/3 is induced, which stimulates the biosynthesis of ethylene and ABA through direct binding to the promoters of *LcACO2/3*, *LcACS1/4/7*, and *LcNCED3* genes., fruitlets sense the abscission signals. Afterward, *LcPG1/2* gene expression and polygalacturonase action are expanded. Additionally, ethylene and ABA may also boost the expression of LcHB2/3 by positive feedback regulation. The final breakdown of homogalacturonan in the cell walls of FAZ leads to the occurrence of fruitlet abscission [16].

4. Materials and Methods

4.1. Plant Materials

The sweet cherry 'Santina', grown in Weining County, Guizhou Province, China (E: 104.12, N: 27.25) was used as the material, and the abscising carpopodium and the non-abscising carpopodium were taken during the young fruit period. Three plants with similar growth vigor were selected, and abscising carpopodium (CA) and non-abscising carpopodium (CN) were taken at 20 d after flowering. These carpopodiums were quickly frozen in liquid nitrogen and brought back to the laboratory for storage in a −80 °C refrigerator. Also, to explore the relationship between fruit shedding and embryo development, embryos of abscission and retention fruits were used to detect vitality.

4.2. Detection of Embryo Activity

In this study, 100 retention fruits and 100 abscission fruits were randomly selected, dissected, and the appearance was observed. According to the size of the embryo, we define the percentage of embryos in the seed coat called plumpness. Afterward, embryo activity was examined. By 2,3,5-Triphenyte-trazoliumchloride (TTC). According to how much the embryo is colored, the percentage of the embryo colored as coloring degree.

4.3. Measurement of Enzyme Activity

The abscising carpopodium (CA) and non-abscising carpopodium (CN) were taken, and their activities of cellulase, pectinase, peroxidase and polygalacturonase were determined by the kit (solabio, Beijing, China) according to the description, and each measurement index was repeated three times.

4.4. Anatomical Observation of Carpopodium Abscission Zones.

The carpopodium of 'Santina' retention and abscission fruit were sampled, then the optimized glycerol alcohol mixture (50% glycerol: 70% alcohol in a volume ratio of 1:1) was used to soften carpopodium for at least 24 h. Subsequently, cross sections containing fruit stalk abscission layer with 1–2 cm in length were taken from the fruit, and were fixed immediately in FAA (Formalin-acetic acid-alcohol, 70%) for at least 48 h. Then, pretreated carpopodium were dehydrated in an ethanol series (70%, 85%, 95% (v/v)) and absolute ethanol, embedded in paraffin, sectioned at a thickness of 8–10 μm cross-sections by a rotary microtome. and dyed with safranin-fast green and slice was sealed by Canada balsam. All prepared slides were observed with a CX41RF light microscope (Olympus, Tokyo, Japan). Photographs were taken with a digital camera. Digital images in JPEG format were again processed with Photoshop CS6 (Adobe Systems Incorporated, San Jose, CA, US).

4.5. Protein Extraction and Trypsin Digestion

The extraction of plant proteins was performed based on the published methods with a slight modification. Briefly, the sample was ground with liquid N_2 into cell powder and transferred to a 5-mL tube. Next the sample was sonicated three times on ice using a high intensity ultrasonic processor (Scientz) in four volumes of lysis buffer (8 M urea, 10 mM dithiothreitol, 1% Protease Inhibitor Cocktail). The remaining debris was removed by centrifugation at 20,000 g at 4 °C for 10 min. Finally, the protein was precipitated with cold 20% TCA for 2 h at −20 °C. After centrifugation at 12,000 *g* 4 °C for 10 min, the supernatant was discarded. The remaining precipitate was washed with cold acetone for three times. The protein was redissolved in 8 M urea and the protein concentration was determined with BCA kit according to the manufacturer's instructions.

For digestion, the protein solution was reduced with 5 mM dithiothreitol for 30 min at 56 °C and alkylated with 11 mM iodoacetamide for 15 min at room temperature in darkness. The protein sample was later diluted by adding 100 mM TEAB to urea concentration < 2M. Finally, trypsin was added into the protein sample in a 1:50 trypsin-to-protein mass ratio for the first digestion overnight, and 1:100 trypsin-to-protein mass ratio for a second 4 h-digestion.

4.6. TMT Labeling and HPLC Fractionation

After trypsin digestion, sample peptides were desalted by Strata X C18 SPE column (Phenomenex, Torrance, CA, US) and vacuum-dried. Resulting peptides were reconstituted using a six-plex TMT kit (ThermoFisher, Shanghai, China) according to its operation manual. Briefly, one unit of TMT reagent was thawed and reconstituted in acetonitrile. The peptides mixtures were then incubated for 2 h at room temperature and pooled, desalted and dried by vacuum centrifugation.

The sample peptides were fractionated into fractions by high pH reverse-phase HPLC using the Agilent 300Extend C18 column (Agilent, Shanghai, China). Briefly, sample peptides were fractionated with a gradient of 8%–32% acetonitrile (pH 9.0) over 60 min into 60 fractions. Then, all fractions were combined into 18 fractions and vacuum dried by centrifuging.

4.7. LC-MS/MS Analysis

Peptides were dissolved in 0.1% formic acid and directly loaded onto a homemade reversed-phase analytical column (15-cm length, 75 μm i.d.). The gradient was comprised of an increase from 6% to 23% solvent B (0.1% formic acid in 98% acetonitrile) over 26 min, 23% to 35% in 8 min and climbing to 80% in 3 min then holding at 80% for the last 3 min, all at a constant flow rate of 400 nL/min on an EASY-nLC 1000 UPLC system (Thermo, Shanghai, China).

The peptides were subjected to NSI source followed by tandem mass spectrometry (MS/MS) in Q ExactiveTM Plus (Thermo, Shanghai, China) coupled online to the UPLC. The electrospray voltage applied was 2.0 kV. The m/z scan range was 350 to 1800 for full scan, and intact peptides were detected in the Orbitrap at a resolution of 70,000. Peptides were then selected for MS/MS using the NCE setting as 28 and the fragments were detected in the Orbitrap at a resolution of 17,500. A data-dependent procedure that alternated between one MS scan followed by 20 MS/MS scans with 15.0s dynamic exclusion. Automatic gain control (AGC) was set at 5E4. The fixed first mass was set as 100 m/z.

4.8. Database Search

The resulting MS/MS data were processed using the Maxquant with integrated Andromeda search engine v.1.5.2.8 (Matthias Mann Lab, Germany). Tandem mass spectra were searched against a *Prunus avium* database concatenated with reverse decoy database. Trypsin/P was specified as cleavage enzyme allowing up to 2 missing cleavages. The mass tolerance for precursor ions was set as 20 ppm in the first search and 5 ppm in the main search, and the mass tolerance for fragment ions was set as 0.02 Da. Carbamidomethyl on Cys was specified as fixed modification and oxidation on Met was set as variable modifications. False discovery rate (FDR) was adjusted to < 1% and minimum score for peptides was set > 40 A TMT-6-plex kit was used for quantification of the resulting peptides. The quantitative level of the peptide was measured according to its ion signal intensity ratio in the secondary spectrum.

4.9. Bioinformatics Analysis

The Gene Ontology (GO) annotation proteome was derived from the UniProt-GOA database (http://www.ebi.ac.uk/GOA/), first converting the identified protein ID to a UniProt ID and then mapping to GO IDs by protein ID. If some identified proteins are not annotated by UniProt-GOA database, the InterProScan soft are used to annotated protein's GO functional based on protein sequence alignment method. GO items can be divided into three categories, namely, biological process (BP), cellular component (CC), molecular function (MF). In this study, we mapped the differentially displayed proteins (fold changes > 1.2, $p < 0.05$) into the GO database (http://www.geneontology.org/). It was computable for the amount of proteins at each GO term and the target list used for results, which came from TMT data. The list was constructed by downloading the data on the GO database. The Kyoto Encyclopedia of Genes and Genomes (KEGG) database (https://www.kegg.jp/) was used to annotate the protein pathway, first using KEGG online service tools KAAS to annotate the protein s KEGG database description and then mapping the annotation result on the KEGG pathway database using KEGG online service tools KEGG mapper.

5. Conclusions

In this study, through protein sequencing and related physiological indicators, it was found that the mechanism of sweet cherry fruit drop may be caused by pollen abortion. Pollen abortion caused embryo abortion, then it might lead to phytohormone synthesis disorder, which effected signal transduction pathways, and hereby controlled genes involved in cell wall degradation and then caused the abscission of fruitlet. Overall, our data may give an intrinsic explanation of the variations in metabolism during the abscission of carpopodium (Figure 8).

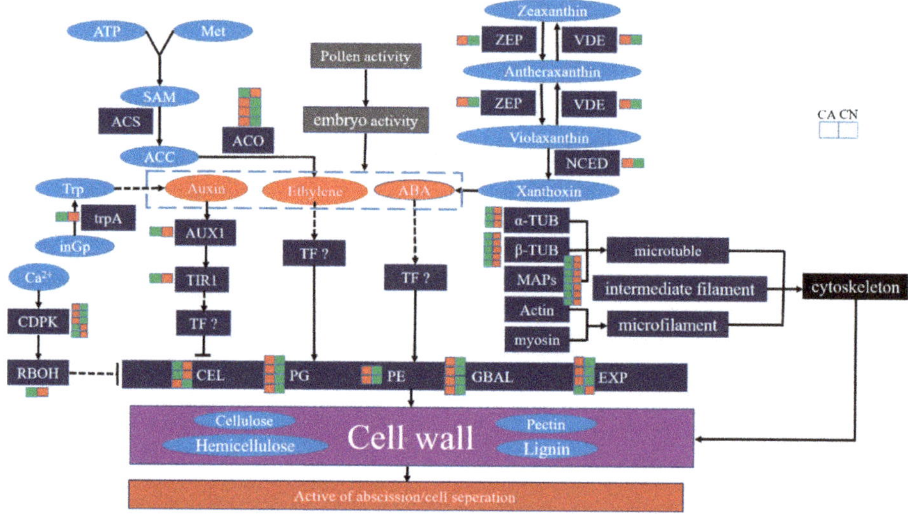

Figure 8. Model of abscission regulation. Red ovals represent hormones, dark blue represents enzymes or proteins, light blue ellipse represents compounds or ions, small red boxes represent up-regulated proteins, and small green squares represent down-regulated proteins.

Supplementary Materials: Supplementary materials can be found at http://www.mdpi.com/1422-0067/21/4/1200/s1.

Author Contributions: X.-P.W. acquired funding and revised the manuscript; Z.-L.Q. performed most of the experiments and the data analysis, and also wrote the original draft preparation; Z.W. and K.Y. contributed analysis tools and offered technical support; Z.-L.Q. carried out the experiments with the help of T.T., G.Q., and Y.H. All authors have read and approved the final version of the manuscript.

Funding: The project was supported by grants from Core Program of Guizhou Province, P. R. China (2016-2520), the Innovation Talent Program of Guizhou Province, P. R. China (2016-4010), Guizhou Graduate Research Funding (Qianjiaohe YJSCXJH [2019]027).

Acknowledgments: The authors are grateful for assistance with the manuscript provided by Hui-Min Zhang and Ju-Kun Song from Guizhou University.

Conflicts of Interest: The authors declare no conflict of interest.

Abbreviations

3GT	anthocyanidin 3-*O*-glucosyltransferase 5-like
ABA	abscisic acid
ACO	1-aminocyclopropane-1-carboxylate oxidase
AE	Abscission embryo
AGC	Automatic gain control
ARF	Auxin response factor
Arp2/3	Actin-related protein2/3 complex
AUX1	auxin transporter-like protein 2
AZ	Abscission Zone
BGLU	beta-glucosidase
bHLH	basic-helix-loop-helix
BP	biological process
bZIP	basic region/leucine zipper motif
CA	abscising carpopodium
CC	cellular component
CEL	cellulase
CN	non-abscising carpopodium
DAO	2-oxoglutarate-dependent dioxygenase DAO-like
DAPs	Differential accumulated proteins
EXP	Expansin
FAA	Formalin-acetic acid-alcohol
FDR	False discovery rate
GalAT	galacturonosyltransferase
GBAL	β-galactosidase
GO	gene ontology
HPLC	High-Performance Liquid Chromatography
KEGG	The Kyoto Encyclopedia of Genes and Genomes
MF	molecular function
NE	Non-abscission embryo
NPR5	regulatory protein NPR5-like
PAE	pectin acetylesterase
PE	pectinesterase
PG	polygalacturonase
POD	peroxidase
PP2C	protein phosphatase 2C
PR1	pathogenesis-related protein 1
PRM	Parallel Reaction Monitoring
PRM	Parallel Reaction Monitoring
SAMS	S-adenosylmethionine synthetase
SnRK2	SNF1-related protein kinase 2
TIR1	transport inhibitor response 1
TMT	Tandem Mass Tag
TMT-LC/MS	Tandem Mass Tag-liquid chromatograph-mass spectrometer
TTC	Triphenyte-trazoliumchloride
UGDH	UDP-glucose 6-dehydrogenase
XTH	xyloglucan endotransglucosylase/hydrolase protein
ZEF	zeaxanthin epoxidase

References

1. Wei, H.; Chen, X.; Zong, X.; Shu, H.; Gao, D.; Liu, Q. Comparative transcriptome analysis of genes involved in anthocyanin biosynthesis in the red and yellow fruits of sweet cherry (*Prunus avium* L.). *PLoS ONE* **2015**, *10*, 1–20. [CrossRef] [PubMed]

2. Kühn, N.; Serrano, A.; Abello, C.; Arce, A.; Espinoza, C.; Gouthu, S.; Deluc, L.; Arce-Johnson, P. Regulation of polar auxin transport in grapevine fruitlets (*Vitis vinifera* L.) and the proposed role of auxin homeostasis during fruit abscission. *BMC Plant Biol.* **2016**, *16*, 1–17. [CrossRef] [PubMed]

3. Sawicki, M.; Aït Barka, E.; Clément, C.; Vaillant-Gaveau, N.; Jacquard, C. Cross-talk between environmental stresses and plant metabolism during reproductive organ abscission. *J. Exp. Bot.* **2015**, *66*, 1707–1719. [CrossRef] [PubMed]

4. Estornell, L.H.; Agustí, J.; Merelo, P.; Talón, M.; Tadeo, F.R. Elucidating mechanisms underlying organ abscission. *Plant Sci.* **2013**, *199–200*, 48–60. [CrossRef]

5. Glazinska, P.; Wojciechowski, W.; Kulasek, M.; Glinkowski, W.; Marciniak, K.; Klajn, N.; Kesy, J.; Kopcewicz, J. De novo transcriptome profiling of flowers, flower pedicels and pods of lupinus luteus (*Yellow lupine*) reveals complex expression changes during organ abscission. *Front. Plant Sci.* **2017**, *8*. [CrossRef]

6. Merelo, P.; Agustí, J.; Arbona, V.; Costa, M.L.; Estornell, L.H.; Gómez-Cadenas, A.; Coimbra, S.; Gómez, M.D.; Pérez-Amador, M.A.; Domingo, C.; et al. Cell wall remodeling in abscission zone cells during ethylene-promoted fruit abscission in citrus. *Front. Plant Sci.* **2017**, *8*. [CrossRef]

7. Parra-Lobato, M.C.; Gomez-Jimenez, M.C. Polyamine-induced modulation of genes involved in ethylene biosynthesis and signalling pathways and nitric oxide production during olive mature fruit abscission. *J. Exp. Bot.* **2011**, *62*, 4447–4465. [CrossRef]

8. Patterson, S.E. Cutting loose. Abscission and dehiscence in Arabidopsis. *Plant Physiol.* **2001**, *126*, 494–500. [CrossRef]

9. Agustí, J.; Merelo, P.; Cercós, M.; Tadeo, F.R.; Talón, M. Ethylene-induced differential gene expression during abscission of citrus leaves. *J. Exp. Bot.* **2008**, *59*, 2717–2733. [CrossRef]

10. Nakano, T.; Fujisawa, M.; Shima, Y.; Ito, Y. Expression profiling of tomato pre-abscission pedicels provides insights into abscission zone properties including competence to respond to abscission signals. *BMC Plant Biol.* **2013**, *13*, 1. [CrossRef]

11. Nakano, T.; Kato, H.; Shima, Y.; Ito, Y. Apple SVP family MADS-box proteins and the tomato pedicel abscission zone regulator JOINTLESS have similar molecular activities. *Plant Cell Physiol.* **2015**, *56*, 1097–1106. [CrossRef]

12. Fu, X.; Shi, Z.; Jiang, Y.; Jiang, L.; Qi, M.; Xu, T.; Li, T. A family of auxin conjugate hydrolases from Solanum lycopersicum and analysis of their roles in flower pedicel abscission. *BMC Plant Biol.* **2019**, *19*, 1–17. [CrossRef] [PubMed]

13. Ellis, C.M.; Nagpal, P.; Young, J.C.; Hagen, G.; Guilfoyle, T.J.; Reed, J.W. AUXIN RESPONSE FACTOR1 and AUXIN RESPONSE FACTOR2 regulate senescence and floral organ abscission in *Arabidopsis thaliana*. *Development* **2005**, *132*, 4563–4574. [CrossRef] [PubMed]

14. Zhang, X.L.; Qi, M.F.; Xu, T.; Lu, X.J.; Li, T.L. Proteomics profiling of ethylene-induced tomato flower pedicel abscission. *J. Proteom.* **2015**, *121*, 67–87. [CrossRef] [PubMed]

15. Tsuchiya, M.; Satoh, S.; Iwai, H. Distribution of XTH, expansin, and secondary-wall-related CesA in floral and fruit abscission zones during fruit development in tomato (*Solanum lycopersicum*). *Front. Plant Sci.* **2015**, *6*, 1–9. [CrossRef]

16. Li, C.; Ma, X.; Huang, X.; Wang, H.; Wu, H.; Zhao, M.; Li, J. Involvement of HD-ZIP I transcription factors LcHB2 and LcHB3 in fruitlet abscission by promoting transcription of genes related to the biosynthesis of ethylene and ABA in litchi. *Tree Physiol.* **2019**, 510642. [CrossRef]

17. Qi, M.F.; Xu, T.; Chen, W.Z.; Li, T.L. Ultrastructural localization of polygalacturonase in ethylene-stimulated abscission of tomato pedicel explants. *Sci. World J.* **2014**, *2014*. [CrossRef]

18. Taylor, J.E.; Whitelaw, C.A. Signals in abscission. *New Phytol.* **2001**, *151*, 323–340. [CrossRef]

19. Meir, S.; Hunter, D.A.; Chen, J.; Halaly, V.; Reid, M.S. Molecular Changes Occurring during Acquisition of Abscission Competence following Auxin Depletion in *Mirabilis jalapa*. *Plant Physiol.* **2006**, *141*, 1604–1616. [CrossRef]

20. Sundaresan, S.; Philosoph-Hadas, S.; Riov, J.; Belausov, E.; Kochanek, B.; Tucker, M.L.; Meir, S. Abscission of flowers and floral organs is closely associated with alkalization of the cytosol in abscission zone cells. *J. Exp. Bot.* **2015**, *66*, 1355–1368. [CrossRef]

21. Mishra, A.; Khare, S.; Trivedi, P.K.; Nath, P. Effect of ethylene, 1-MCP, ABA and IAA on break strength, cellulase and polygalacturonase activities during cotton leaf abscission. *South African J. Bot.* **2008**, *74*, 282–287. [CrossRef]

22. Gao, Y.; Liu, C.; Li, X.; Xu, H.; Liang, Y.; Ma, N.; Fei, Z.; Gao, J.; Jiang, C.Z.; Ma, C. Transcriptome profiling of petal abscission zone and functional analysis of an Aux/IAA family gene RhiAA16 involved in petal shedding in rose. *Front. Plant Sci.* **2016**, *7*, 1–13. [CrossRef] [PubMed]

23. Sundaresan, S.; Philosoph-Hadas, S.; Riov, J.; Mugasimangalam, R.; Kuravadi, N.A.; Kochanek, B.; Salim, S.; Tucker, M.L.; Meir, S. De novo transcriptome sequencing and development of abscission zone-specific microarray as a new molecular tool for analysis of tomato organ abscission. *Front. Plant Sci.* **2016**, *6*. [CrossRef]

24. Roberts, J.A.; Gonzalez-Carranza, Z.H. Pectinase functions in abscission. *Stewart Postharvest Rev.* **2009**, *5*, 1–4. [CrossRef]

25. Poovaiah, B.W.; Rasmussen, H.P. Calcium Distribution in the Abscission Zone of Bean Leaves. *Plant Physiol.* **1973**, *52*, 683–684. [CrossRef] [PubMed]

26. Patterson, S.E.; Bleecker, A.B. Ethylene-Dependent and -Independent Processes Associated with Floral Organ Abscission in Arabidopsis. *Plant Physiol.* **2004**, *134*, 194–203. [CrossRef] [PubMed]

27. Wang, J.H.; Liu, J.J.; Chen, K.L.; Li, H.W.; He, J.; Guan, B.; He, L. Comparative transcriptome and proteome profiling of two Citrus sinensis cultivars during fruit development and ripening. *BMC Genom.* **2017**, *18*, 1–13. [CrossRef]

28. Zhang, S.; Zhang, D.; Fan, S.; Du, L.; Shen, Y.; Xing, L.; Li, Y.; Ma, J.; Han, M. Effect of exogenous GA3 and its inhibitor paclobutrazol on floral formation, endogenous hormones, and flowering-associated genes in 'Fuji' apple (*Malus domestica* Borkh.). *Plant Physiol. Biochem.* **2016**, *107*, 178–186. [CrossRef]

29. Li, J.M.; Huang, X.S.; Li, L.T.; Zheng, D.M.; Xue, C.; Zhang, S.L.; Wu, J. Proteome analysis of pear reveals key genes associated with fruit development and quality. *Planta* **2015**, *241*, 1363–1379. [CrossRef]

30. Chan, Z.; Wang, Q.; Xu, X.; Meng, X.; Qin, G.; Li, B.; Tian, S. Functions of defense-related proteins and dehydrogenases in resistance response induced by salicylic acid in sweet cherry fruits at different maturity stages. *Proteomics* **2008**, *8*, 4791–4807. [CrossRef]

31. Bargiela, R.; Herbst, F.A.; Martínez-Martínez, M.; Seifert, J.; Rojo, D.; Cappello, S.; Genovese, M.; Crisafi, F.; Denaro, R.; Chernikova, T.N.; et al. Metaproteomics and metabolomics analyses of chronically petroleum-polluted sites reveal the importance of general anaerobic processes uncoupled with degradation. *Proteomics* **2015**, *15*, 3508–3520. [CrossRef] [PubMed]

32. Tsuchiya, H.; Tanaka, K.; Saeki, Y. The parallel reaction monitoring method contributes to a highly sensitive polyubiquitin chain quantification. *Biochem. Biophys. Res. Commun.* **2013**, *436*, 223–229. [CrossRef] [PubMed]

33. Yu, Q.; Liu, B.; Ruan, D.; Niu, C.; Shen, J.; Ni, M.; Cong, W.; Lu, X.; Jin, L. A novel targeted proteomics method for identification and relative quantitation of difference in nitration degree of OGDH between healthy and diabetic mouse. *Proteomics* **2014**, *14*, 2417–2426. [CrossRef] [PubMed]

34. Dong, H.; Li, Y.; Fan, H.; Zhou, D.; Li, H. Quantitative proteomics analysis reveals resistance differences of banana cultivar 'Brazilian' to Fusarium oxysporum f. sp. cubense races 1 and 4. *J. Proteom.* **2019**, *203*, 103376. [CrossRef]

35. Ferrero, S.; Carretero-Paulet, L.; Mendes, M.A.; Botton, A.; Eccher, G.; Masiero, S.; Colombo, L. Transcriptomic signatures in seeds of apple (*Malus domestica* L. Borkh) during fruitlet abscission. *PLoS ONE* **2015**, *10*, 1–15. [CrossRef]

36. He, J.H.; Ma, F.W.; Chen, Y.Y.; Shu, H.R. Differentially expressed genes implicated in embryo abortion of mango identified by suppression subtractive hybridization. *Genet. Mol. Res.* **2012**, *11*, 3966–3974. [CrossRef]

37. Blanusa, T.; Else, M.A.; Atkinson, C.J.; Davies, W.J. The regulation of sweet cherry fruit abscission by polar auxin transport. *Plant Growth Regul.* **2005**, *45*, 189–198. [CrossRef]

38. Blanusa, T.; Else, M.A.; Davies, W.J.; Atkinson, C.J. Regulation of sweet cherry fruit abscission: The role of photo-assimilation, sugars and abscisic acid. *J. Horticultural Sci. Biotechnol.* **2006**, *81*, 613–620. [CrossRef]

39. Mesejo, C.; Muñoz-Fambuena, N.; Reig, C.; Martínez-Fuentes, A.; Agustí, M. Cell division interference in newly fertilized ovules induces stenospermocarpy in cross-pollinated citrus fruit. *Plant Sci.* **2014**, *225*, 86–94. [CrossRef]

40. Ben-Cheikh, W.; Perez-Botella, J.; Tadeo, F.R.; Talon, M.; Primo-Millo, E. Pollination increases gibberellin levels in developing ovaries of seeded varieties of citrus. *Plant Physiol.* **1997**, *114*, 557–564. [CrossRef]

41. Tokuji, Y.; Kuriyama, K. Involvement of gibberellin and cytokinin in the formation of embryogenic cell clumps in carrot (*Daucus carota*). *J. Plant Physiol.* **2003**, *160*, 133–141. [CrossRef] [PubMed]

42. Pellan-Delourme, R.; Renard, M. Cytoplasmic male sterility in rapeseed (*Brassica napus* L.): Female fertility of restored rapeseed with "Ogura" and cybrids cytoplasms. *Genome* **1988**, *30*, 234–238. [CrossRef]

43. Lal, N.; Kumar Gupta, A.; Nath, V. Fruit Retention in Different Litchi Germplasm Influenced by Temperature. *Int. J. Current Microbiol. Appl. Sci.* **2017**, *6*, 1189–1194. [CrossRef]

44. Gunawardena, A.H.L.A.N.; Greenwood, J.S.; Dengler, N.G. Cell wall degradation and modification during programmed cell death in lace plant, Aponogeton madagascariensis (Aponogetonaceae). *Amer. J. Bot.* **2007**, *94*, 1116–1128. [CrossRef] [PubMed]

45. Giné-Bordonaba, J.; Echeverria, G.; Ubach, D.; Aguiló-Aguayo, I.; López, M.L.; Larrigaudière, C. Biochemical and physiological changes during fruit development and ripening of two sweet cherry varieties with different levels of cracking tolerance. *Plant Physiol. Biochem.* **2017**, *111*, 216–225. [CrossRef]

46. Kim, J.; Sundaresan, S.; Philosoph-Hadas, S.; Yang, R.; Meir, S.; Tucker, M.L. Examination of the abscission-associated transcriptomes for soybean, tomato, and arabidopsis highlights the conserved biosynthesis of an extensible extracellular matrix and boundary layer. *Front. Plant Sci.* **2015**, *6*, 1–15. [CrossRef]

47. Goldental-Cohen, S.; Burstein, C.; Biton, I.; Ben Sasson, S.; Sadeh, A.; Many, Y.; Doron-Faigenboim, A.; Zemach, H.; Mugira, Y.; Schneider, D.; et al. Ethephon induced oxidative stress in the olive leaf abscission zone enables development of a selective abscission compound. *BMC Plant Biol.* **2017**, *17*, 1–17. [CrossRef]

48. Ke, X.; Wang, H.; Li, Y.; Zhu, B.; Zang, Y.; He, Y.; Cao, J.; Zhu, Z.; Yu, Y. Genome-wide identification and analysis of polygalacturonase genes in solanum lycopersicum. *Int. J. Mol. Sci.* **2018**, *19*, 2290. [CrossRef]

49. Gao, Y.; Liu, Y.; Liang, Y.; Lu, J.; Jiang, C.; Fei, Z.; Jiang, C.Z.; Ma, C.; Gao, J. Rosa hybrida RhERF1 and RhERF4 mediate ethylene- and auxin-regulated petal abscission by influencing pectin degradation. *Plant J.* **2019**, *99*, 1159–1171. [CrossRef]

50. Cosgrove, D.J. Enzymes and Other Agents That Enhance Cell Wall Extensibility. *Annu. Rev. Plant Physiol. Plant Mol. Biol.* **1999**, *50*, 391–417. [CrossRef]

51. Vissenberg, K.; Fry, S.C.; Verbelen, J.P. Root hair initiation is coupled to a highly localized increase of xyloglucan endotransglycosylase action in arabidopsis roots. *Plant Physiol.* **2001**, *127*, 1125–1135. [CrossRef] [PubMed]

52. Carmen, C.; Jocelyn, R.K.C.; Alan, B.B. Auxin regulation and spatial localization of an endo-1,4-beta-D-glucanase and a xyloglucan endotransglycosylase in expanding tomato hypocotyls. *Plant J.* **1997**, *12*, 417–426. [CrossRef]

53. Van Nocker, S. Development of the abscission zone. *Stewart Postharvest Rev.* **2009**, *5*. [CrossRef]

54. Liljegren, S.J.; Ditta, G.S.; Eshed, Y.; Savidge, B.; Bowmant, J.L.; Yanofsky, M.F. SHATTERPROOF MADS-box genes control dispersal in Arabidopsis. *Nature* **2000**, *404*, 766–770. [CrossRef]

55. Smékalová, V.; Doskočilová, A.; Komis, G.; Šamaj, J. Crosstalk between secondary messengers, hormones and MAPK modules during abiotic stress signalling in plants. *Biotechnol. Adv.* **2014**, *32*, 2–11. [CrossRef]

56. Nakano, T.; Ito, Y. Molecular mechanisms controlling plant organ abscission. *Plant Biotechnol.* **2013**, *30*, 209–216. [CrossRef]

57. Einhorn, T.C.; Arrington, M. ABA and Shading Induce 'Bartlett' Pear Abscission and Inhibit Photosynthesis but Are Not Additive. *J. Plant Growth Regul.* **2018**, *37*, 300–308. [CrossRef]

58. Shen, C.J.; Bai, Y.H.; Wang, S.K.; Zhang, S.N.; Wu, Y.R.; Chen, M.; Jiang, D.A.; Qi, Y.H. Expression profile of PIN, AUX/LAX and PGP auxin transporter gene families in Sorghum bicolor under phytohormone and abiotic stress. *FEBS J.* **2010**, *277*, 2954–2969. [CrossRef]

59. Goto, D.B.; Ogi, M.; Kijima, F.; Kumagai, T.; Van Werven, F.; Onouchi, H.; Naito, S. A single-nucleotide mutation in a gene encoding S-adenosylmethionine synthetase is associated with methionine over-accumulation phenotype in Arabidopsis thaliana. *Genes Genet. Syst.* **2002**, *77*, 89–95. [CrossRef]

60. Chersicola, M.; Kladnik, A.; Žnidarič, M.T.; Mrak, T.; Gruden, K.; Dermastia, M. 1-Aminocyclopropane-1-Carboxylate Oxidase Induction in Tomato Flower Pedicel Phloem and Abscission Related Processes Are Differentially Sensitive To Ethylene. *Front. Plant Sci.* **2017**, *8*, 1–14. [CrossRef]

61. Daher, F.B.; Braybrook, S.A. How to let go: Pectin and plant cell adhesion. *Front. Plant Sci.* **2015**, *6*, 1–8. [CrossRef] [PubMed]

62. Higgs, H.N.; Pollard, T.D. Regulation of Actin Filament Network Formation Through ARP2/3 Complex: Activation by a Diverse Array of Proteins. *Annual Rev. Biochem.* **2001**, *70*, 649–676. [CrossRef] [PubMed]

63. Li, Y.; Du, M.; Tian, X.; Xu, D.; Li, Z. Study on the changes of micr otubule cytoskeleton of abscission zone dur ing leaf abscission in cotton (in chinese). *J. Shihezi Univ.* **2016**, *34*, 1–7. [CrossRef]

64. Rajangam, A.S.; Kumar, M.; Aspeborg, H.; Guerriero, G.; Arvestad, L.; Pansri, P.; Brown, C.J.L.; Hober, S.; Blomqvist, K.; Divne, C.; et al. MAP20, a microtubule-associated protein in the secondary cell walls of hybrid aspen, is a target of the cellulose synthesis inhibitor 2,6-dichlorobenzonitrile. *Plant Physiol.* **2008**, *148*, 1283–1294. [CrossRef]

65. Wing, R.A.; Mao, L.; Begum, D.; Chuang, H.; Budiman, M.A.; Szymkowiak, E.J.; Irish, E.E. JOINTLESS is a MADS-box gene controlling tomato flower abscission zone development. *Nature* **2000**, *406*, 910–913. [CrossRef]

66. Xie, Q.; Hu, Z.; Zhu, Z.; Dong, T.; Zhao, Z.; Cui, B.; Chen, G. Overexpression of a novel MADS-box gene SlFYFL delays senescence, fruit ripening and abscission in tomato. *Sci. Rep.* **2014**, *4*, 1–10. [CrossRef]

67. Li, C.; Zhao, M.; Ma, X.; Wen, Z.; Ying, P.; Peng, M.; Ning, X.; Xia, R.; Wu, H.; Li, J. The HD-Zip transcription factor LcHB2 regulates litchi fruit abscission through the activation of two cellulase genes. *J. Exp. Bot.* **2019**, *70*, 5189–5203. [CrossRef]

International Journal of
Molecular Sciences

Article

Comparative Proteomic Analysis by iTRAQ Reveals that Plastid Pigment Metabolism Contributes to Leaf Color Changes in Tobacco (*Nicotiana tabacum*) during Curing

Shengjiang Wu [1,2], Yushuang Guo [2], Muhammad Faheem Adil [3], Shafaque Sehar [3], Bin Cai [2], Zhangmin Xiang [2], Yonggao Tu [2], Degang Zhao [1,4,*] and Imran Haider Shamsi [3,*]

[1] State Key Laboratory Breeding Base of Green Pesticide and Agricultural Bioengineering, The Key Laboratory of Plant Resources Conservation and Germplasm Innovation in Mountainous Region (Ministry of Education), Guizhou University, Guiyang 550025, China; wushengjiang1210@163.com

[2] Key Laboratory of Molecular Genetics/Upland Flue-cured Tobacco Quality and Ecology Key Laboratory, Guizhou Academy of Tobacco Science, CNTC, Guiyang 550081, China; yshguo@126.com (Y.G.); bincaiuk@gmail.com (B.C.); xiangzhangmin@126.com (Z.X.); yc12101875@126.com (Y.T.)

[3] Department of Agronomy, College of Agriculture and Biotechnology, Zijingang Campus, Zhejiang University, Hangzhou 310058, China; 11516093@zju.edu.cn (M.F.A.); 11816126@zju.edu.cn (S.S.)

[4] Guizhou Academy of Agricultural Sciences, Guiyang 550006, China

[*] Correspondence: dgzhao@gzu.edu.cn (D.Z.); drimran@zju.edu.cn (I.H.S.)

Received: 17 February 2020; Accepted: 30 March 2020; Published: 31 March 2020

Abstract: Tobacco (*Nicotiana tabacum*), is a world's major non-food agricultural crop widely cultivated for its economic value. Among several color change associated biological processes, plastid pigment metabolism is of trivial importance in postharvest plant organs during curing and storage. However, the molecular mechanisms involved in carotenoid and chlorophyll metabolism, as well as color change in tobacco leaves during curing, need further elaboration. Here, proteomic analysis at different curing stages (0 h, 48 h, 72 h) was performed in tobacco cv. Bi'na1 with an aim to investigate the molecular mechanisms of pigment metabolism in tobacco leaves as revealed by the iTRAQ proteomic approach. Our results displayed significant differences in leaf color parameters and ultrastructural fingerprints that indicate an acceleration of chloroplast disintegration and promotion of pigment degradation in tobacco leaves due to curing. In total, 5931 proteins were identified, of which 923 (450 up-regulated, 452 down-regulated, and 21 common) differentially expressed proteins (DEPs) were obtained from tobacco leaves. To elucidate the molecular mechanisms of pigment metabolism and color change, 19 DEPs involved in carotenoid metabolism and 12 DEPs related to chlorophyll metabolism were screened. The results exhibited the complex regulation of DEPs in carotenoid metabolism, a negative regulation in chlorophyll biosynthesis, and a positive regulation in chlorophyll breakdown, which delayed the degradation of xanthophylls and accelerated the breakdown of chlorophylls, promoting the formation of yellow color during curing. Particularly, the up-regulation of the chlorophyllase-1-like isoform X2 was the key protein regulatory mechanism responsible for chlorophyll metabolism and color change. The expression pattern of 8 genes was consistent with the iTRAQ data. These results not only provide new insights into pigment metabolism and color change underlying the postharvest physiological regulatory networks in plants, but also a broader perspective, which prompts us to pay attention to further screen key proteins in tobacco leaves during curing.

Keywords: *Nicotiana tabacum*; ultrastructure; postharvest physiology; pigment metabolism; iTRAQ

1. Introduction

Tobacco (*Nicotiana tabacum*) is an extensively investigated model plant and one of the most widely cultivated non-food crops. Given its agricultural importance, tobacco is grown in more than 100 countries for its foliage, mainly consumed as cigarettes, cigars, snus, snuff, etc. The plant organs, including fruit, flowers, and leaves, undergo a series of complex physiological and biochemical changes when they detach from their mother plant [1–3]. Curing is the process of transforming raw materials into target requirements through certain processes. Color is one of the quality factors of cash crops and agricultural products, and color change associated with carotenoid and chlorophyll metabolism is one of the most obvious phenomena in postharvest vegetative organs during curing and storage. A substantial amount of research has been carried out to comprehend this fundamental postharvest physiological process to improve the commercial value of agricultural goods during the past few decades [4–6].

The carotenoids represent the most widespread group of pigments in nature, with over 750 members and an estimated yield of 100 million tons per year [6–8], and many carotenoid compounds have been examined, such as β-carotene, lutein, violaxanthin, and neoxanthin [9–11]. Certain plant proteins, such as carotenoid cleavage dioxygenases (CCDs) and violaxanthin de-epoxidase (VDE), significantly participate in the regulation of the carotenoid and degradation products content in plants [6,10,12]. Additionally, lipoxygenase (LOX) is an important enzyme that catalyzes the co-oxidation of β-carotene and plays a significant part in the deterioration of β-carotene levels [13,14], while peroxidase (POD) is involved in the cleavage of various carotenes, such as xanthophylls and apocarotenals, to flavor compounds [15]. Carotenoid degradation products are important volatile flavor components and precursors for plant growth regulators such as the phytohormone abscisic acid (ABA) and strigolactones in a range of plant species [10,12,16].

In addition, chlorophyll metabolism is an important biological phenomenon, and it has been estimated that about one billion tons of chlorophyll are destroyed on a global scale each year [17,18]. Chlorophyll compounds mainly include chlorophyll *a* and chlorophyll *b* in plants [17–19]. Proteins, such as chlorophyllide-a oxygenase (CAO) and chlorophyllase (Chlase), are involved in the chlorophyll biosynthetic pathway and the chlorophyll breakdown pathway [17,18,20]. Chlorophyll degradation is a highly controlled sequential process that converts the fluorescent chlorophyll molecules into non-fluorescent chlorophyll catabolites (NCCs), which are stored within the vacuole in a range of plant species [18,21,22]. Furthermore, chlorophyll serves as a precursor for important volatile flavor components such as phytol and neophytadiene [16].

Parameters of the CIEL**a***b* color coordinate include lightness L^* (positive white and negative black), and two chromatic components a^* (positive red and negative green) and b^* (positive yellow and negative blue) [2,11]. Color change is one of the most dramatic events occurring in plant postharvest organs during curing and storage [2,5,23]. Pigments are compounds that absorb subsets of the visible spectrum, transmitting and reflecting back only what they do not absorb, and causing the tissue to be perceived as the reflected colors [24]. Carotenoids are pigments that range in color from yellow through orange to red, resulting from their C_{40} polyene backbone [6,14]. The green color changes to orange and red due to the breakdown of chlorophylls and the accumulation of the orange β-carotene and the red lycopene in plants [6,25]. The plants seem intensely yellow due to the accumulation of the xanthophylls, namely lutein, neoxanthin, and violaxanthin [11,25,26]. The color change is determined by a dynamic shift in pigment composition and their contents in plants, which is associated with the regulation of differentially expressed proteins (DEPs). Proteins involved in photosynthesis, glyoxylate metabolism, carbon and nitrogen metabolism, anthocyanin biosynthesis, protein processing, and redox homeostasis are crucial for color regulation in plants [8,27–29]. Moreover, leaf color change is a complex programmed process that is closely related to pigment metabolism and is regulated by fine-tuned molecular mechanisms [8,27,30].

Tobacco is the most important non-food agricultural economic crop and serves as a model plant organism to study fundamental biological processes [31,32]. In tobacco, leaf senescence during postharvest processing is different from natural senescence, and rather an accelerated one [16,23]. It is worth emphasizing that energy metabolism, photosynthesis, jasmonic acid biosynthesis, cell rescue, and reactive oxygen species scavenging are crucial for leaf senescence, and induced leaf senescence may be involved in nutrient remobilization and the cell viability maintenance [33–36]. Strikingly, carotenoid, and chlorophyll metabolism associated with color change was one of the most important biological processes in tobacco leaves during curing and senescence [23,36,37]. Fresh tobacco leaves are harvested and processed into flue-cured tobacco raw material in a bulk barn. This curing process of tobacco leaves can be divided into the yellowing stage, the leaf-drying stage, and the stem-drying stage. The yellowing stage is the first key step associated with carotenoid and chlorophyll metabolic and color changes in tobacco leaves [16,23]. Thus, studying plastid pigment metabolic and color changes in postharvest tobacco leaves during curing will provide more information for enhancing the understanding of this biological process and improving crop quality and reducing losses.

iTRAQ (isobaric tags for relative and absolute quantification), a high-throughput proteomic technology, is one platform for comparing changes in the abundance of specific proteins among different samples [8,29,38]. Although new advances have been made in our understanding of pigment metabolism and color change in plant organs [4–6,13,23], fewer studies have focused specifically on their molecular mechanisms in postharvest tobacco leaves during curing. In this study, iTRAQ-based proteomic analysis was employed to identify important regulators in pigment metabolism pathways and elucidate the molecular mechanism of pigment metabolism and color change in tobacco leaves during the yellowing stage (0 h, 48 h, 72 h). The results herein provide new insights into the molecular mechanisms involved in pigment metabolism and color change for the future study of postharvest physiological regulatory networks in plants.

2. Results

2.1. Color and Phenotypic Changes of Tobacco Leaves during Curing

During the curing process, significant differences in the leaf color parameters L^*, a^* and b^* were observed (Figure 1A). Strikingly, the L^*, a^*, and b^* values gradually increased, which were consistent with the changes in the tobacco leaf phenotypes from green to yellow during 0–72 h (Figure 1B).

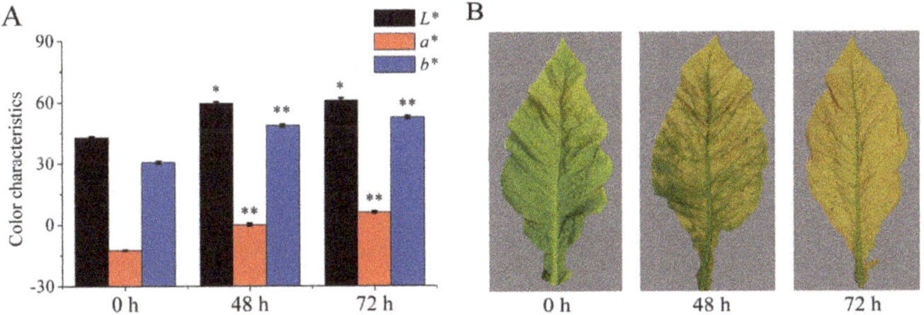

Figure 1. Color and phenotypic changes in tobacco leaves during curing. (**A**) The leaf color values of L^*, a^*, and b^* were determined during curing. Data are shown as the means ± SE, n = 30. Asterisks indicate significant differences between the values at 0 h and 48 h or 72 h based on Duncan's multiple range test in SPSS (* $p < 0.05$, ** $p < 0.01$). (**B**) Representative tobacco leaf phenotypes were documented for each flue curing stage.

2.2. Ultrastructural Observations of Tobacco Leaves during Curing

Ultrastructural observations indicated that the cell contained relatively intact chloroplast, grana thylakoids, and starch granules at 0 h; however, at 48 h, the chloroplast membranes and grana thylakoid lamellae were severely disrupted (Figure 2). At 72 h, only a few of the chloroplast and grana thylakoid lamellae remained. Ultrastructural observations of the cells showed that the curing process accelerated the chloroplast structural breakdown and promoted the degradation of the pigments in tobacco leaves during curing.

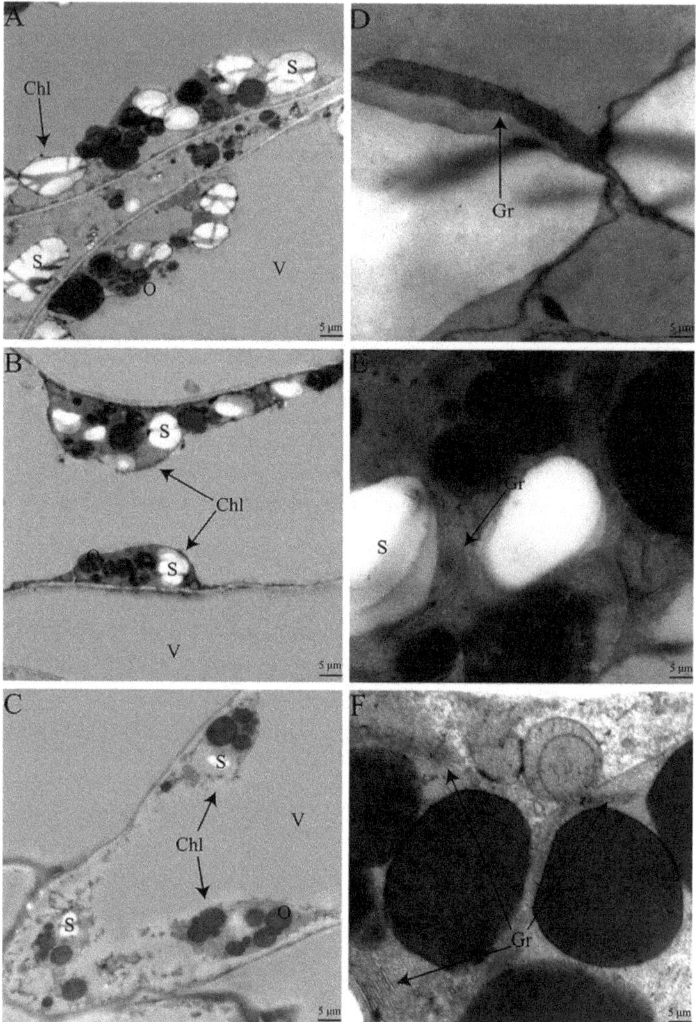

Figure 2. Ultrastructural changes in tobacco leaves during curing. Chloroplasts were gradually disrupted in tobacco leaves cells during curing (**A**, 0 h; **B**, 48 h and **C**, 72 h). Grana thylakoid lamellae were disrupted in tobacco leaves during curing (**D**, 0 h; **E**, 48 h and **F**, 72 h). Chl, chloroplast; Gr, grana thylakoid lamellae; S, starch granule; O, osmiophilic granule; V, vacuole.

2.3. Physiological Attributes of Tobacco Leaves during Curing

The plastid pigment concentrations in tobacco leaves during curing were analyzed, as presented in Table 1. Among several carotenoids, the highest levels were displayed by lutein followed by β-carotene, while violaxanthin concentration was found higher than that of neoxanthin during 0–72 h. However, the chlorophylls were found to be the most abundant plastid pigments in tobacco leaves at 0 h, followed by the carotenoids. Although the carotenoid and chlorophyll concentrations decreased significantly, the ratio between the carotenoids and chlorophylls significantly increased during curing. This difference may be explained by the observation that the chlorophyll *a* (94.05%) and chlorophyll *b* (87.53%) concentrations and SPAD value (93.37%) decreased at a greater rate than carotenoids, including β-carotene (74.35%), lutein (77.56%), violaxanthin (73.71%), and neoxanthin (79.94%) during curing. These results indicated that pigments, particularly chlorophylls, degrade at high levels in tobacco leaves during curing. To confirm that proteins involved in pigment metabolism were modulated in tobacco leaves during 0–72 h, we selected physiological parameters that can be measured using established assays (Figure 3). The Chlase activities and MDA content significantly increased in tobacco leaves during curing. However, ascorbate peroxidase (APX) activities and ascorbic acid (ASA) content significantly decreased in tobacco leaves during 0–72 h. In addition, the LOX activities showed higher at 48 h than that at 0 h and 72 h, and the POD activities in leaves at 0 h was significantly higher than that at 72 h. These findings indicate that change in physiological parameters associated with pigment metabolism and color change is significant in tobacco leaves during different curing stages.

Table 1. Pigment content changes in tobacco leaves during curing.

Concentration	Curing Time (h)		
	0	48	72
β-carotene (μg·g$_{DM}^{-1}$)	275.37 ± 7.32	151.19 ± 3.92**	70.63 ± 1.20**
Lutein (μg·g$_{DM}^{-1}$)	384.29 ± 5.02	213.85 ±6.64**	86.23 ± 1.93**
Violaxanthin (μg·g$_{DM}^{-1}$)	99.73 ± 2.98	46.49 ± 2.55**	26.22 ± 0.62**
Neoxanthin (μg·g$_{DM}^{-1}$)	39.72 ± 1.53	14.85 ± 0.44**	7.97 ± 0.28**
Chlorophyll a (mg·g$_{FM}^{-1}$)	0.71 ± 0.02	0.10 ± 0.01**	0.04 ± 0.01**
Chlorophyll b (mg·g$_{FM}^{-1}$)	0.33 ± 0.01	0.07 ± 0.01**	0.04 ± 0.01**
SPAD value	21.87 ± 0.63	5.36 ± 0.39**	1.45 ± 0.23**
Carotenoids/Chlorophylls	0.29 ± 0.01	1.30 ± 0.06**	2.22 ± 0.14**
Xanthophylls/β-carotene	1.91 ± 0.06	1.84 ± 0.07	1.71 ± 0.03*

Data are shown as the means ± SEs (n = 7), except for the relative chlorophyll content (SPAD; Soil Plant Analysis Development) value (n = 30). Asterisks indicate significant differences between the values at 0 h and 48 h or 72 h based on Duncan's multiple range test in SPSS (* p < 0.05, ** p < 0.01). Xanthophylls include neoxanthin, violaxanthin, and lutein. DM, dry mass; FM, fresh mass.

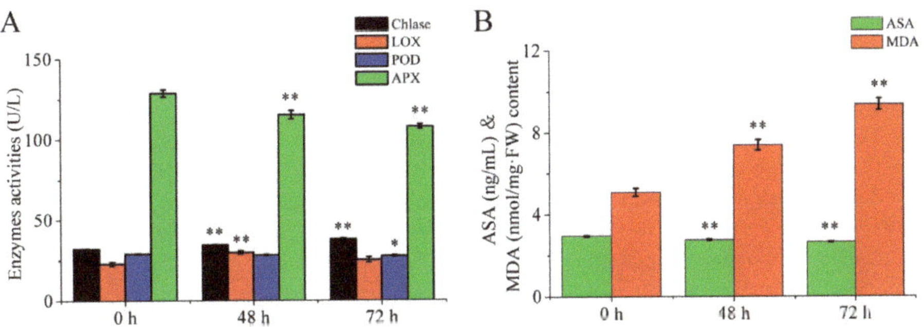

Figure 3. Changes of enzyme activities (**A**, Chlase, LOX, POD, APX) and chemical components (**B**, ASA and MDA) in tobacco leaves during curing. Data are shown as the means ± SE (n = 6). Asterisks indicate significant differences between the values at 0 h and 48 h or 72 h based on Duncan's multiple range test in SPSS (* p < 0.05, ** p < 0.01). Chlase, chlorophyllase; LOX, lipoxygenase; POD, peroxidase; APX, ascorbate peroxidase; ASA, ascorbic acid; MDA, malondialdehyde.

2.4. Pigment Degradation Products Analysis in Tobacco Leaves during Curing

For chemometric analysis, the relative concentrations of the 82 volatile components were analyzed using comprehensive two-dimensional gas chromatography time-of-flight mass spectrometry (GC×GC-TOF-MS) system (Supplementary Table S1), including 19 carotenoid metabolites and 1 chlorophyll catabolite (Supplementary Table S2). The total concentration of carotenoid and chlorophyll degradation products in tobacco leaves decreased during 0–48 h and 0–72 h and increased during 48–72 h. It is worth noting that the 6-methyl-5-hepten-2-ol, β-ionol, β-ionone, and solavetivone detected in the tobacco leaf samples during 0–72 h and the 3-oxo-α-ionol detected in the mature fresh leaves have not been previously reported to be the components in tobacco headspace volatiles. Strikingly, isophorone was found to be the most abundant carotenoid volatile metabolite in tobacco leaves followed by geranylacetone and dihydroactinidiolide during 0–72 h. The levels of six carotenoid metabolites, including 6-methyl-5-hepten-2-ol, linalool, isophorone, megastigmatrienone A, megastigmatrienone B, and solavetivone were decreased during 0–72 h. Conversely, the levels of 3-oxo-α-ionol and 3-hydroxy-β-damascone increased during 0–72 h, whereas the remaining metabolites of carotenoid and chlorophyll did not persistently increase or decrease during 0–72 h. These findings indicate that the postharvest tobacco leaves underwent a series of complex physiological and biochemical changes involving pigment metabolism during curing.

2.5. Protein Profile Analysis of Tobacco Leaves Using iTRAQ

In order to clarify molecular mechanisms involved in carotenoid and chlorophyll metabolism and color change, data analysis based on the phenotypic, physiological, and chemical changes in tobacco leaves during 0–72 h using iTRAQ was found credible. In total, 1,043,678 spectra were identified from the iTRAQ analysis using the leaf samples at different curing stages as the materials. MASCOT (Modular Approach to Software Construction Operation and Test), a powerful database retrieval software, which can realize the identification from mass spectrometry data to protein, generated a total of 372,574 spectra matched to in silico peptide spectra, 218,381 unique spectra, 30,498 peptides, 22,993 unique peptides, and 5931 proteins from the iTRAQ experiments Run1, Run2, and Run3 (Figure 4A and Supplementary Table S3). In order to obtain the relationship between the spectrum and the peptide segment, the mass spectrum was matched with the theoretical spectrum, where peptide segments were used as the dimension for data processing and calculation (a peptide segment may correspond to more than one spectrum). Among the identified proteins in three technical duplicate experiments, 1296 proteins had 1 identified unique peptide, 3715 had 2, 2564 had 3, 516 had more than 11, and the remainder had 4–10 (Figure 4B). The peptide information validated that many unique peptides are shared with different proteins (the identified peptides were compared with protein databases). The relative molecular mass of identified proteins was mainly distributed at 10~80 kDa, and the proportion (18.06%) of proteins with a relative molecular mass of 30~40 kDa was the highest (Supplementary Figure S1A). A total of 5488 proteins were identified with 0~10% sequence coverage. However, only 1.18% of the proteins were identified with sequence coverage > 20% (Supplementary Figure S1B). The coefficient of variation (CV), defined as the ratio of the standard deviation (SD) to the mean (CV = SD/mean) was used to evaluate the reproducibility of protein quantification. The lower the CV, the better the reproducibility. The CV distribution (mean CV: 0.16) in three replicates showed good reproducibility (Supplementary Figure S2).

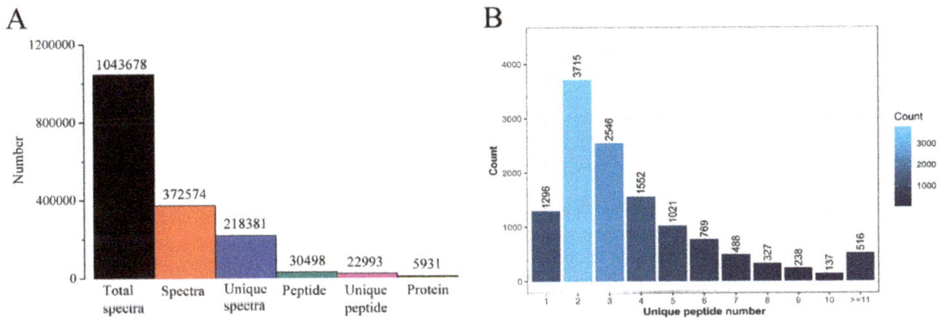

Figure 4. Basic information of iTRAQ output. (**A**) Spectra, peptides, and proteins identified in tobacco leaves. (**B**) Number of peptides that matched proteins.

2.6. DEPs Identified and Functional Analysis

The proteins were screened with a fold-change value > 1.5 or < 0.67 and a Q-value of < 0.05. Based on these criteria, 450 (319, 348 and 36) up-regulated proteins, 452 (355, 376 and 59) down-regulated proteins, and 21 commonly expressed (up/down-regulated) proteins with a total of 923 (674, 724 and 95) DEPs in the comparisons of leaves at 48 h and 0 h ("48 h vs. 0 h" hereafter), leaves at 72 h and 0 h ("72 h vs. 0 h" hereafter), and leaves at 72 h and 48 h ("72 h vs. 48 h" hereafter) were identified, respectively (Figure 5A,C and Supplementary Figure S3). The heatmap/hierarchical clustering analysis was conducted for all the identified proteins and DEPs using the pheatmap package in R language. As shown in Figure 5B,C, the identified proteins and DEPs in different leaf samples were easily discriminated, which demonstrated the significant differences in protein levels in tobacco leaves during different curing stages. To further understand their functions, 837 DEPs were annotated on the basis of gene ontology (GO) terms in three categories: Biological process, cellular component, and molecular function (Supplementary Figure S4). In particular, the catalytic activity involved 375, 409, and 48 DEPs in comparisons of 48 h vs. 0 h, 72 h vs. 0 h and 72 h vs. 48 h, and was the most commonly annotated category under the biological process term. In contrast, 389/387, 402/401, and 49/49 DEPs were annotated under cell/cell part in the cellular components term in different comparisons. In the molecular function category, 365/349, 375/357, and 42/40 DEPs were annotated under the cellular process/metabolic process in different comparisons.

According to the biological functional properties, the eukaryotic orthologous groups (KOGs) categories of DEPs are shown in Supplementary Figure S5A–C. Post-translational modification, protein turnover and chaperones (19.90%/16.12%), general function prediction only (13.10%/13.84%), translation, ribosomal structure and biogenesis (11.22%/10.10%), carbohydrate transport and metabolism (9.35%/9.61%), and energy production and conversion (6.12%/8.14%) were the main functional categories identified from the comparisons of 48 h vs. 0 h and 72 h vs. 0 h in tobacco leaves during curing, whereas the main KOGs categories obtained from the comparison of 72 h vs. 48 h were posttranslational modification, protein turnover and chaperones (18.92%), general function prediction only (13.51%), translation, ribosomal structure and biogenesis (12.16%), amino acid transport and metabolism (6.76%), and energy production and conversion (6.76%). Furthermore, the Kyoto Encyclopedia of Genes and Genomes (KEGG) pathway annotation of DEPs are shown in Supplementary Figure S6A–C. The global and overview maps (255, 263, and 28) and carbohydrate metabolism (94, 112, and 14) pathways exhibited more annotated DEPs in different comparisons.

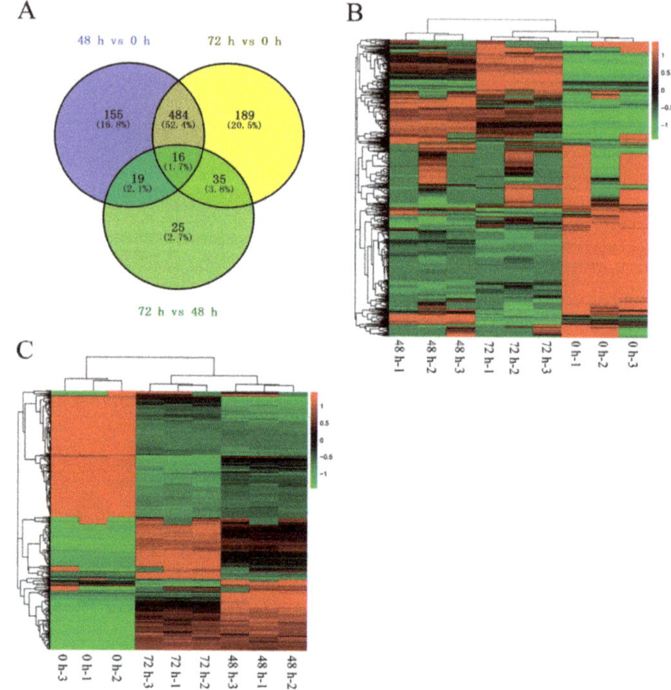

Figure 5. Venn diagram and heatmap representing identified proteins and DEPs from different comparison groups in tobacco leaf samples. (**A**) Total DEPs identified in all tobacco leaf samples; (**B**) Heatmap/hierarchical clustering of all identified proteins; and (**C**) Heatmap/hierarchical clustering of DEPs identified in all tobacco leaf samples. The numbers of DEPs identified from three biological replicates are shown in the different segments (Figure 5A). Red and green indicate higher expression and lower expression, respectively (Figure 5B,C).

2.7. DEPs Involved in Carotenoid and Chlorophyll Metabolism

At the post-transcriptional level, 31 DEPs involved in carotenoid and chlorophyll metabolism were identified in tobacco leaves during curing, and the detail of these DEPs and BLAST data are listed in Supplementary Table S4. iTRAQ analysis revealed that among five DEPs in the carotenoid biosynthetic pathway, two were up-regulated during 0–48 h and/or 0–72 h, and three were down-regulated during 0–48 h and/or 0–72 h (Figure 6A). Alternatively, 14 proteins involved in carotenoid degradation were differentially expressed in tobacco leaves during curing. These proteins included 2 LOX proteins and 12 POD proteins. Two LOX proteins were both up-regulated during 0–48 h and 0–72 h. Of the 12 POD proteins, 3 were up-regulated during 0–48 h and/or 0–72 h, and 9 were down-regulated during 0–48 h and/or 0–72 h and/or 48–72 h. The differences in the abundance of these proteins indicated the complex regulatory network involved in carotenoid metabolism in tobacco leaves during curing (Figure 7). A total of 12 DEPs were involved in chlorophyll metabolism in tobacco leaves during curing, including 11 DEPs in the chlorophyll biosynthetic pathway and 1 DEP in the chlorophyll breakdown pathway (Figure 6B). Ten DEPs in the chlorophyll biosynthetic pathway were significantly down-regulated during 0–48 h and/or 0–72 h and/or 48–72 h. In contrast, ferrochelatase isoform I was up-regulated during 0–48 h and 0–72 h in the chlorophyll biosynthetic pathway, and chlorophyllase-1-like isoform X2 (Chlase-1-X2) was significantly up-regulated during 0–72 h and 48–72 h in the chlorophyll breakdown pathway. The results indicated that these DEPs potentially play important roles in chlorophyll metabolism (Figure 7).

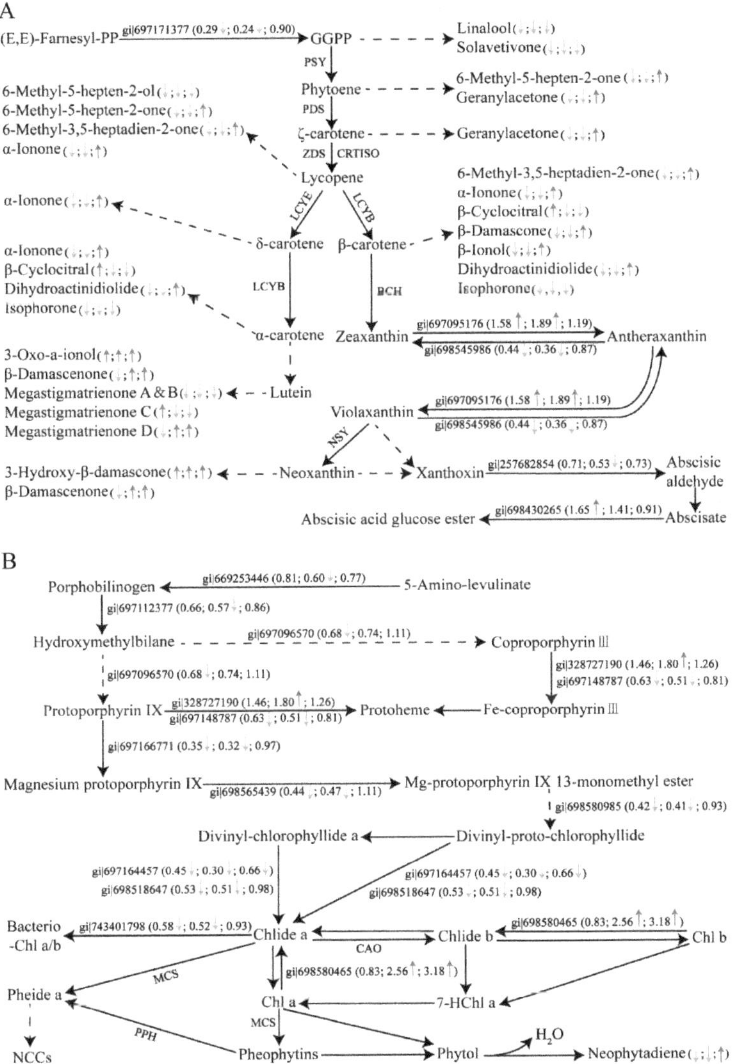

Figure 6. The carotenoid (**A**) and chlorophyll (**B**) metabolic pathway in tobacco leaves during curing. Up-regulated proteins or increased metabolites are marked with upward red arrows, while down-regulated proteins or decreased metabolites are marked with downward green arrows. The numbers represent the fold change. The left arrows or numbers represented the difference of proteins or metabolites during 0–48 h, the middle arrows or numbers indicated the difference during 0–72 h, and the right indicated the difference during 48–72 h. Gi numbers and ratios of the DEPs are shown in Supplementary Table S4. (**A**) GGPP, geranylgeranyl diphosphate; PSY, phytoene synthase; PDS, phytoene desaturase; ZDS, ζ-carotene desaturase; CRTISO, carotenoid isomerase; LCYB, lycopene β-cyclase; LCYE, lycopene ε-cyclase; BCH, β-carotene hydroxylase; NSY, neoxanthin synthase. (**B**) Chlide a/b, chlorophyllide a/b; MCS, metal-chelating substance; CAO, chlorophyllide-a oxygenase; Pheide a, pheophorbide a; Chl a/b, chlorophyll a/b; 7HChl a, 7-Hydroxy-chlorophyll a; NCCs, non-fluorescent chlorophyll catabolites.

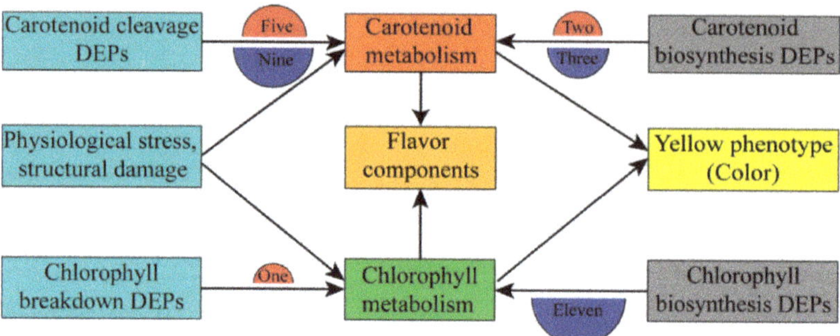

Figure 7. Branching program related to pigment metabolism and color change in tobacco leaves during curing. The semicircle filled with the red color indicated that the DEPs were positive regulators, and the blue color suggested that the DEPs were negative regulators in the pigment biosynthetic and breakdown pathway. The numbers represented the numbers of the total up- and down-regulated proteins in different pathways.

2.8. Validation of iTRAQ Data by qRT-PCR

To provide the accurate data for the molecular mechanisms related to the pigment metabolism and color change in postharvest tobacco leaves during curing, the mRNA expression levels of eight key DEPs were detected by quantitative real-time polymerase chain reaction (qRT-PCR). The results exhibited that qRT-PCR data of eight genes aligned with the iTRAQ results (Figure 8).

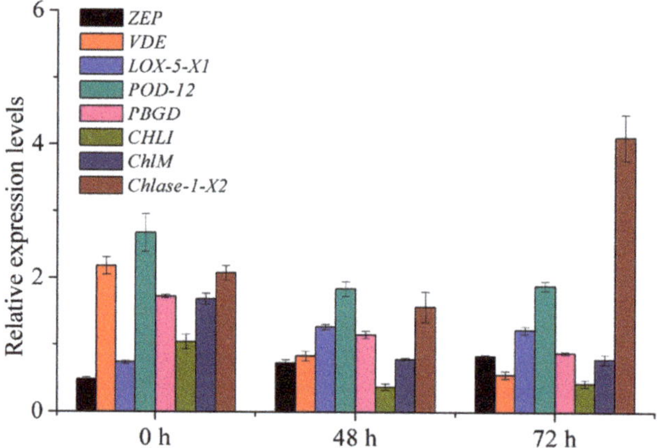

Figure 8. Verification of iTRAQ results by qRT-PCR. Values represent the means ± SE (*n* = 3). *ZEP*, zeaxanthin epoxidase, chloroplastic-like; *VDE*, violaxanthin de-epoxidase, chloroplastic; *LOX-5-X1*, probable linoleate 9S-lipoxygenase 5 isoform X1; *POD-12*, peroxidase 12-like; *PBGD*, porphobilinogen deaminase, chloroplastic-like; *CHLI*, magnesium-chelatase subunit ChlI, chloroplastic; *ChlM*, magnesium protoporphyrin IX methyltransferase, chloroplastic; *Chlase-1-X2*, chlorophyllase-1-like isoform X2.

3. Discussion

Proteomics analysis provides a broad perspective on the process of leaf color change, which prompts us not only to pay attention to the pigment metabolism pathway but also to further screen some key proteins related to the pigment metabolism and color change in tobacco leaves during curing [27–29]. Quantitative proteome analysis revealed that hundreds of DEPs were identified in all leaf samples during curing. Although, many DEPs might be associated with pigment metabolism and color change in tobacco leaves during curing, 19 DEPs related to carotenoid metabolism, and 12 DEPs involved in chlorophyll metabolism were selected based on bioinformatics analysis. This analysis helped us to clearly identify DEPs associated with pigment metabolism and color change and the postharvest physiological regulatory networks in tobacco leaves during curing (Figure 7).

3.1. Leaf Color Change is Determined by the Carotenoid and Chlorophyll Content

Color change in plants is determined by the content of various plastid pigments, and plant organs are intensely yellow due to the accumulation of the xanthophylls, including lutein, neoxanthin, and violaxanthin [11,25,26]. The green color changes to orange due to the breakdown of chlorophylls and the accumulation of the β-carotene in plants [6,25]. Regardless, the concentrations of both carotenoid and chlorophyll significantly decreased in tobacco leaves during curing. Chlorophyll was found to be the most abundant plastid pigment in tobacco leaves at 0 h, but the ratio of carotenoid/chlorophyll of different samples was all larger than or equal to 1.30 during 48–72 h. In addition, the ratios between xanthophylls and β-carotene of different samples were all larger than or equal to 1.71 during 0–72 h. Thus, the tobacco leaves showed a green phenotype at 0 h and a yellow phenotype at 48 h and 72 h. Pigment metabolism and their relative contents were responsible for the formation of the yellow phenotype, which was consistent with previous reports [11,30,39]. The data were expressed as the color values of lightness L^*, greenness a^*, and yellowness b^* [2,11,27]. In this study, the change in color was quantified as the increment in the values of L^*, a^*, and b^*, which is associated with the pigment degradation and the increase in the relative concentrations of the carotenoid and the phenotypic change in tobacco leaves during curing.

3.2. Effect of Cell Ultrastructure Damage on Pigment Metabolism and Leaf Color Change

The metabolism of chlorophyll and carotenoid occurs in the chloroplast (complex organelle with several distinct sub-organellar compartments to internally sort the proteins) and chromoplast membranes [8,40]. Compared with 0 h, chloroplasts and grana thylakoid lamellae appeared to be more severely damaged in tobacco leaves at 48 h and 72 h, especially at 72 h. Chloroplast structure and functions are plausibly linked to pigment metabolism and leaf color change [8,30]. Structural damage of the chloroplast might accelerate the degradation of the chlorophyll and carotenoid, and alter the proportion of pigment compositions and promote the formation of yellow color in tobacco leaves during curing.

3.3. Role of Physiological Parameters in Pigment Metabolism and Leaf Color Change

Chlase, POD, and LOX are all important enzymes involved in pigment metabolism [14,15,19], and APX, ASA, and MDA are significant parameters to deduce the physiological state in plants [41,42]. The increased Chlase activities in tobacco leaves during curing might accelerate the degradation of chlorophyll [17,18]. In contrast, the increased LOX activities in tobacco leaves during 0–48 h might promote the degradation of carotenoid [13,14], but the decreased POD activities during 0–72 h might result in the delayed degradation of carotenoid in tobacco leaves during curing [15]. APX is an important enzyme for detoxification of H_2O_2 in plants [43], and ASA is a key substrate for the detoxification of reactive oxygen entities [36]. The decreased APX activities and ASA content in tobacco leaves during 0–72 h might accelerate leaf senescence and promote the degradation of pigment. MDA is a marker for lipid peroxidation and a characteristic of senescence in plants, and the increased MDA content in tobacco leaves during curing might also accelerate leaf senescence and promote the degradation of pigment, which is consistent with the results of leaf senescence in tomato plants [42]. Following the progressive stress of physiology and catalysis of the enzymes, the pigment content remarkably decreased during curing, especially chlorophyll content, and the leaves kept a yellow phenotype for 72 h.

3.4. Role of DEPs in Carotenoid and Chlorophyll Metabolism and Color Change

In this study, 31 DEPs involved in carotenoid and chlorophyll metabolism were identified in tobacco leaves during curing. Although, these DEPs in the pigment metabolic pathway were mainly responsible for the change of pigment content and leaf color, the carotenoid cleavage, and chlorophyll breakdown were the primary biological processes in postharvest plant leaves during curing [16,23,44]. In the present study, 5 DEPs involved in carotenoid biosynthesis and 14 DEPs related to carotenoid cleavage were identified in tobacco leaves during curing. In the carotenoid biosynthetic pathway, geranylgeranyl pyrophosphate synthase, chloroplastic-like (GGPPS) was down-regulated in tobacco leaves during curing, which catalyzes the conversion of (E, E)-farnesyl diphosphate (FPP) into geranylgeranyl diphosphate (GGPP). In plants, FPP and GGPP are isoprenoid precursors necessary for carotenoid biosynthesis. The down-regulated GGPPS in tobacco leaves during 0–48 h and 0–72 h might reduce the content of carotenoids, and it ultimately leads to the decreased content of their degradation products, such as linalool, solavetivone, 6-methyl-5-hepten-2-ol, 6-methyl-5-hepten-2-one, 6-methyl-3,5-heptadien-2-one, β-ionone, and isophorone.

In addition, zeaxanthin epoxidase, chloroplastic-like (ZEP) was significantly up-regulated, and VDE was significantly down-regulated in tobacco leaves during 0–48 h and 0–72 h. VDE is a member of a group of proteins known as lipocalins that bind and transport small hydrophobic molecules [7]. Zeaxanthin is epoxidized by ZEP, finally yielding violaxanthin, however, this epoxidation is reversible with the effect of VDE [11]. When zeaxanthin synthesis is inhibited by VDE, violaxanthin might be catalyzed by neoxanthin synthase (NSY) to yield neoxanthin. Then neoxanthin could be converted into 3-hydroxy-β-damascone and β-damascenone under the catalysis of carotenoid cleavage enzymes, resulting in the increase in their concentrations in tobacco leaves during 0–72 h. Furthermore, the violaxanthin concentration was higher than that of neoxanthin during 0–72 h, which might be closely associated with the up-regulated ZEP and down-regulated VDE.

In the carotenoid biosynthetic pathway, unnamed protein product protein was significantly down-regulated in tobacco leaves during 0–72 h, while scopoletin glucosyltransferase-like protein was significantly up-regulated during 0–48 h. Both of them were involved in ABA biosynthesis in tobacco leaves during curing. The down-regulated unnamed protein product protein might not be helpful for the biosynthetic conversion of xanthoxin into ABA, whereas the up-regulated scopoletin glucosyltransferase-like protein might be conducive to ABA biosynthesis and finally lead to the decreased content of carotenes. Alternatively, it is worth emphasizing that CCDs are known to be important for cleaving carotenoid compounds and forming important flavor and fragrance volatiles or their apocarotenoids [14,39,45]. However, none of them showed a significant up- or down-regulation in different comparisons. Thereby, we speculate that CCDs might not be crucial for carotenoid metabolism in tobacco leaves during 0–72 h. Moreover, LOX and POD are important enzymes involved in the cleavage of carotenoids [13–15]. Two up-regulated LOX proteins might accelerate the degradation of carotenoids and keep them at low levels in tobacco leaves during curing. Three POD proteins were up-regulated, but the other nine POD-related proteins were down-regulated in tobacco leaves during curing. These up-regulated POD proteins might accelerate the degradation of carotenoids, but down-regulated POD proteins might not be conducive to carotenoid cleavage and forming important flavor. In addition, 19 carotenoid metabolites showed 6 changing trends with increased and/or decreased relative concentrations in different curing stages. These results suggested that carotenoid metabolism was involved in a complex regulatory network.

In contrast, a total of 12 DEPs were involved in chlorophyll metabolism in tobacco leaves during curing, including 1 up-regulated and 10 down-regulated DEPs in the chlorophyll biosynthetic pathway and 1 up-regulated DEP in the chlorophyll breakdown pathway. The down-regulated delta-aminolevulinic acid dehydratase, porphobilinogen deaminase, chloroplastic-like, protoporphyrinogen oxidase, chloroplastic, magnesium-chelatase subunit ChlI, chloroplastic isoform X1, magnesium protoporphyrin IX methyltransferase, chloroplastic, magnesium-protoporphyrin IX monomethyl ester [oxidative] cyclase, chloroplastic, uncharacterized protein ycf39 isoform X2, protein

TIC 62, chloroplastic isoform X3 and geranylgeranyl reductase in the chlorophyll biosynthetic pathway inhibited chlorophyll biosynthesis and indirectly reduced chlorophyll content in tobacco leaves during curing, which is consistent with previous reports related to color regulation in plants [8,27,29]. Additionally, two DEPs were identified in the chlorophyll biosynthesis shunt related to protoheme. The ferrochelatase isoform I protein was significantly up-regulated during 0–72 h, while ferrochelatase-2, chloroplastic isoform X1 protein was significantly down-regulated during 0–48 h and 0–72 h, which might indirectly regulate chlorophyll metabolism via an effect targeted on the coproporphyrin III and protoporphyrin IX. The results suggested that these DEPs were negative regulators of chlorophyll biosynthesis in tobacco leaves during curing.

Alternatively, it is worth emphasizing that Chlase-1-X2 catalyzes the conversion of chlorophyll into chlorophyllide and phytol and is thought to be a key rate-limiting step in the chlorophyll breakdown pathway [19,46]. Subsequently, neophytadiene is formed via the dehydration of phytol. The Chlase-1-X2 was significantly up-regulated during 0–72 h and 48–72 h, which accelerated chlorophyll degradation in tobacco leaves during curing. However, the Chlase-1-X2 was down-regulated (mean ratio = 0.83) in tobacco leaves during 0–48 h. The relative concentrations of neophytadiene decreased markedly during 0–48 h and increased significantly during 48–72 h, which indicated a positive correlation between neophytadiene and Chlase-1-X2 protein. Thus, we inferred that Chlase-1-X2 played a leading role in chlorophyll breakdown in tobacco leaves during curing.

4. Materials and Methods

4.1. Plant Material and Sampling

Tobacco cultivar "Bi'na1" was obtained from the Guizhou Academy of Tobacco Science, China. Tobacco plants were grown in Fuquan City, Guizhou Province, Southwest China. Plants were cultivated based on the local production standard to produce high-quality tobacco leaves. For the flue-curing experiment, uniform mature leaves from the middle parts of the tobacco plants (at the 10th leaf position; from the bottom to approximately 60 cm in height) in individually labeled plants were harvested 80 days after field transplantation. The flue-curing barns were designed by the Guizhou Academy of Tobacco Science based on the Technique Standard of Bulk Curing Barns (Probative) Castigatory Version No. [2008]575. The curing schedule was executed following the Code of Practice for Tobacco Curing by Loose-leaves (YC/T 457-2013) (Supplementary Figure S7). It is worth noting that an intelligent curing system (DDMB06YS) was adopted, which can automatically control the dry bulb and wet bulb temperatures during curing.

For each experiment, tobacco leaves were collected at 3 phases of flue-curing, i.e., 0 h, 48 h, and 72 h during the yellowing stage. Leaf samples of 0 h were collected from labeled plants in the fields before flue-curing. At 48 h (from the beginning of the curing process), approximately 80% of the leaf area turned yellow in the middle of the yellowing stage (dry bulb temperature 38 °C and wet bulb temperature 35~36 °C). At 72 h, the tobacco leaf had the yellow laminae and green midribs indicating the end of the yellowing stage (dry bulb temperature 42 °C and wet bulb temperature 33~34 °C). The samples were divided into two duplicates at each stage during curing; one was used directly to determine the color parameters, cell ultrastructure, and pigment content, and the other one was frozen in liquid nitrogen immediately to further measure the carotenoid composition and degradation products, physiological parameters and the DEPs involved in pigment metabolism. Three independent curing experiments were performed.

4.2. Color Analysis

The leaf color was determined at 0 h, 48 h, and 72 h using a Minolta Chroma Meter CR-10 (Konica Minolta Sensing, Inc., Japan) calibrated previously with a white standard tile by taking 6 measurements per leaf in the equatorial region. The data were expressed as the color values of lightness (L^* = measures light reflected), redness (a^* = measures positive red and negative green), yellowness (b^* = measures positive yellow and negative blue). Thirty leaves were used for these determinations at each yellowing stage during curing.

4.3. Ultrastructural Observation

Sample sections of 1 mm² (2 cm distance from the midrib) were excised from the middle portion of the labeled leaves. Ultrastructural changes were studied by observing ultrathin sections of leaf palisade tissue at 0 h, 48 h, and 72 h using a Hitachi H-600 electron microscope (Kyoto, Japan) [47].

4.4. Physiological Measurements

Approximately 0.1 g of fresh tissue was immersed in 95% ethanol for 24 h in the absence of light. The absorbance of the extracts was measured using a UV-1800 ultraviolet/visible spectrophotometer (Shimadzu, Kyoto, Japan) at wavelengths of 470, 649, and 665 nm. Chlorophyll *a*, chlorophyll *b*, and total carotenoid concentrations were calculated as previously described [48]. The frozen leaf samples were ground into a fine powder using liquid nitrogen with a mortar and pestle, then freeze-dried. The β-carotene, lutein, neoxanthin, and violaxanthin were quantified via HPLC, as described previously [49]. Moreover, the chlorophyll content, as expressed by the SPAD value, was measured using a Chlorophyll Meter (Model SPAD-502, Tokyo, Japan). Thirty leaves were measured by taking 6 measurements per leaf in the equatorial region. The leaf samples were analyzed for Chlase, LOX, POD, APX, ASA activities, and malondialdehyde (MDA) content. Plant enzyme-linked immunosorbent assay kits were purchased from the Shanghai Jianglai Bio-Technology Co., Ltd. (Shanghai, China) to measure Chlase and LOX activity, and from the Nanjing Jiancheng Bioengineering Institute (Nanjing, China) to determine the APX and POD activities, along with ASA and MDA content.

4.5. Pigment Degradation Products Analysis

Carotenoid and chlorophyll degradation products were determined using qualitative and quantitative methods in postharvest tobacco leaves during curing. Freeze-dried tobacco leaf samples were previously treated using headspace solid-phase micro-extraction (HS-SPME) and analyzed using a GC×GC-TOF-MS [50].

4.6. Protein Extraction

Total proteins were extracted from leaf tissue at 0 h, 48 h, and 72 h during curing, as previously described [51]. The samples were transferred to a 2 mL centrifuge tube, and 5% cross-linked polyvinylpyrrolidone (PVPP) powder and homogenization lysis buffer (7 M urea, 2 M thiourea, 4% 3-[(3-cholamidopropyl) dimethylammonio]-1-propanesulfonate [CHAPS], 40 mM Tris-HCl, pH 8.5) were added. A grinder (power is 60 HZ, time is 2 min) was used to break the tissues, then 2 × volume of Tris-saturated phenol was added and shaken for 15 min. After centrifugation (25,000× *g* for 15 min at 4 °C), the upper phenol phase was taken into a 10 mL centrifuge tube, and 5 × volume of 0.1 M cold ammonium acetate/methanol and 10 mM dithiothreitol (DTT; final concentration) were added, then placed at −20 °C for 2 h. These steps were repeated twice. Then, 1 mL of cold acetone was added and again placed at −20 °C for 30 min. The supernatant was discarded after centrifugation, and this step was repeated once. Air-dry precipitation, 1XCocktail was added with SDS L3 and ethylene diamine tetra acetic acid (EDTA), then 10 mM DTT was added after putting on ice for 5 min. Protein was solubilized using a grinder and centrifuged, and then the supernatant was discarded and put into a water bath for 1 h at 56 °C after adding 10 mM DTT. Afterward, 55 mM iodoacetamide (IAM) was added and placed in a dark room for 45 min. 1 mL cold acetone was added, and placed at −20 °C for 2 h, then centrifuged. The steps of protein solubilization and centrifugation were repeated. The protein concentration was determined by the Bradford assay using bovine serum albumin (BSA) as a standard [52]. The samples were kept at −80 °C for further analysis.

4.7. iTRAQ Labeling and Strong Cation Exchange (SCX) Fractionation

Proteins were digested using trypsin gold (Promega, Madison, WI, USA) with the ratio of protein:trypsin = 30:1 at 37 °C for 16 h. Peptides were processed according to the manufacturer's protocol for an 8-plex iTRAQ reagent (Applied Biosystems, Foster City, CA, USA). Three protein samples were labeled with the iTRAQ tags as follows: 0 h (115 tag), 48 h (119 tag), and 72 h (121 tag). SCX chromatography was performed using an LC-20AB HPLC pump system (Shimadzu, Kyoto, Japan). The iTRAQ-labeled peptide mixtures were reconstituted with 4 mL buffer A (25 mM NaH_2PO_4 in 25% ACN, pH 2.7) and loaded onto a 4.6 mm × 250 mm Ultremex SCX column containing 5 μm particles (Phenomenex). The peptides were eluted at a flow rate of 1 mL/min with a gradient of buffer A for 10 min, 5–60% buffer B (25 mM NaH_2PO_4, 1 M KCl in 25% ACN, pH 2.7) for 27 min, 60%\ 100% buffer B for 1 min.

4.8. LC-ESI-MS/MS Analysis

Each fraction was resuspended in buffer A (2% acetonitrile, 0.1% formic acid) and centrifuged at 20,000× *g* for 10 min. The samples were loaded at 8 μL min^{-1} for 4 min, and the 44 min gradient was then run at 300 nL min^{-1} starting from 2% to 35% B (98% acetonitrile, 0.1% formic acid), followed by a 2 min linear gradient to 80%, and maintenance at 80% B for 4 min, and finally a return to 5% in 1 min. The peptides were subjected to nanoelectrospray ionization followed by tandem mass spectrometry (MS/MS) in a QEXACTIVE (Thermo Fisher Scientific, San Jose, CA, USA) coupled online to the HPLC for data-dependent acquisition (DDA) mode detection. The main parameters were set: The ion source voltage was set to 1.6 kV; the MS1 scan range was 350~1600 m/z; the resolution was set to 70,000; the MS2 starting m/z was fixed at 100; the resolution was 17,500. The screening conditions for the secondary fragmentation were: Charge 2+ to 7+, and the top 20 parent ions with the peak intensity exceeding 10,000. The ion fragmentation mode was the high-energy collision dissociation (HCD), and the fragment ions were detected in Orbitrap. The dynamic exclusion time was set to 15 s. Automatic gain control (AGC) was set to: MS1 3E6, MS2 1E5.

4.9. iTRAQ protein Identification and Quantification

The MASCOT search engine (Matrix Science, London, UK; version 2.3.02) was used to simultaneously identify and quantify proteins against the *Nicotiana tabacum* database (http://www. ncbi.nlm.nih.gov/protein?term=txid4085[Organism]; 85,194 entries). For protein identification, a mass tolerance of 20 Da (ppm) was permitted for intact peptide masses, and a mass tolerance of 0.05 Da was permitted for fragmented ions; there was an allowance for one missed cleavage in the trypsin digests. Oxidation (M) and iTRAQ8plex (Y) represent variable modifications, and carbamidomethyl (C), iTRAQ8plex (N-term) and iTRAQ8plex (K) represent fixed modifications. All unique peptides (at least one unique spectrum) were permitted for protein quantitation. An automated software called IQuant [53] was employed for quantitatively analyzing the labeled peptides with isobaric tags. It integrates Mascot Percolator [54], a well-performing machine learning method for rescoring database search results, to provide reliable significance measures. To assess the confidence of peptides, the peptide spectral matches (PSMs) were pre-filtered at 1% PSM-level false discovery rate (FDR). In order to control the rate of false-positive at the protein level, a protein FDR at 1%, which was based on "picked" protein FDR strategy [55], would also be estimated after protein inference (protein-level FDR ≤ 0.01). DEPs were required to satisfy these conditions for identification: Confident protein identification involved at least one unique peptide, changes of greater than 1.5-fold or less than 0.67-fold, and Q-values less than 0.05 in at least 2 replicate experiments. The quantitative protein ratios were then weighted and normalized by the median ratio in MASCOT.

4.10. Bioinformatics Analysis

The GO database (http://en.wikipedia.org/wiki/Gene_Ontology) represents an international standardization of gene functional classification systems. The KOGs database (https://www.ncbi.nlm.nih.gov/pubmed/14759257) was used for orthologous protein classification. The pathways were used as queries to search the KEGG pathway database (http://www.genome.jp/kegg/pathway.html). Heatmap/hierarchical clustering of DEPs was conducted by pheatmap package in R language. The mass spectrometry proteomics data have been deposited to the iProX data repository (National Center for Protein Sciences, Beijing, China) with the dataset identifier IPX0001410001 (https://www.iprox.org/).

4.11. RNA Extraction and qRT-PCR Analysis

Total RNA was extracted from tobacco leaf samples by TRIzol reagent (Invitrogen), and cDNA was reverse transcribed from 1 µg of total RNA using PrimeScript™ RT Reagent Kit (TaKaRa), according to the manufacturer's instructions. qRT-PCR was performed using the iQ™5 real-time PCR detection system (Bio-Rad, USA) with the following conditions: 95 °C for 15 s, followed by 40 cycles of 95 °C for 15 s, 60 °C for 30 s, and 72 °C for 30 s. The tobacco *β-actin* gene was used as an endogenous control. The transcript levels of genes were calculated according to the $2^{-\Delta\Delta Ct}$ method [56]. Experiments were performed in triplicate for each treatment. Primer sequences are listed in Supplementary Table S5.

4.12. Data Analysis

Data were analyzed statistically using Duncan's Multiple Range Test with SPSS version 16.0 (SPSS, Chicago, IL, USA). All the photographs and figures were processed and analyzed using Adobe Illustrator CS5 (Adobe Systems Inc., San Francisco, CA, USA) or Origin 8.0 software (Origin lab, Corp., Northampton, MA, USA).

5. Conclusions

There was a significant decrease in the content of chlorophyll than carotenoid in tobacco leaves during curing, which was not only associated with the complex regulation of DEPs in carotenoid metabolism, but also correlated with DEPs playing a negative role in chlorophyll biosynthesis and a positive role in chlorophyll breakdown. The total concentration of carotenoid and chlorophyll degradation products in tobacco leaves decreased during 0–48 h and 0–72 h and increased during 48–72 h, which was the result of the combined action of DEPs in the pigment metabolic pathway, especially in the breakdown pathway. These DEPs delayed the degradation of xanthophylls and accelerated the breakdown of chlorophylls, promoting the formation of yellow color during 0–72 h. In particular, the up-regulation of the Chlase-1-X2 was the key protein regulatory mechanism responsible for chlorophyll metabolism and color change. In the future, we will attempt to carry out further research to elucidate the regulatory factors (e.g., environmental and genetic factors) that regulate the pigment metabolic flow and color change in postharvest tobacco leaves during curing. All these findings provide useful molecular information for a better understanding of the complicated postharvest physiological regulatory networks and the molecular mechanisms involved in pigment metabolism and color change in plants.

Supplementary Materials: Supplementary materials can be found at http://www.mdpi.com/1422-0067/21/7/2394/s1. Supplementary Figure S1. Protein mass distribution in tobacco leaves (**A**) and coverage of the proteins by the identified peptides (**B**). Supplementary Figure S2. The distribution of coefficient of variation in three replicates. (**A**) Replicate 1 of different leaf samples; (**B**) replicate 2 of different leaf samples; (**C**) replicate 3 of different leaf samples. Supplementary Figure S3. Volcano plots and Venn diagram of differentially expressed proteins in three replicates. (**A**) 48 h vs 0 h; (**B**) 72 h vs 0 h; (**C**) 72 h vs 48 h; (**D**) up-regulated proteins; (**E**) down-regulated proteins Supplementary Figure S4. Barplot of the Gene Ontology analysis of DEPs obtained from different comparisons in tobacco leaf samples. (**A**) 48 h vs 0 h; (**B**) 72 h vs 0 h; (**C**) 72 h vs 48 h. Supplementary Figure S5. KOGs categories of DEPs obtained from different comparisons in tobacco leaf samples. (**A**) 48 h vs 0 h; (**B**) 72 h vs 0 h; (**C**) 72 h vs 48 h. Supplementary Figure S6. Barplots of KEGG pathway analysis of DEPs obtained from different comparisons in tobacco leaf samples. (**A**) 48 h vs 0 h; (**B**) 72 h vs 0 h; (**C**) 72 h vs 48 h. Supplementary Figure S7. Curing schedule of tobacco leaves. Supplementary Table S1. The relative concentrations of the 82 volatile components in tobacco leaves during curing (ng g^{-1}). Supplementary Table S2. Plastid pigment metabolites in tobacco leaves during curing. Supplementary Table S3. An overview of protein identification in tobacco leaves during curing. Supplementary Table S4. DEPs involved in pigment metabolism and color change in tobacco leaves during curing. Supplementary Table S5. Primers used for qRT-PCR.

Author Contributions: Conceptualization, D.Z.; I.H.S., and S.W.; data curation, S.W.; funding acquisition, D.Z.; investigation, B.C.; M.F.A., and S.S.; methodology, D.Z. and S.W.; project administration, D.Z.; supervision, D.Z.; validation, S.W., Z.X.; S.S., M.F.A., and I.H.S.; writing—original draft, S.W. and M.F.A.; writing—review and editing, Y.G., Y.T.; D.Z., and I.H.S. All authors have read and agreed to the published version of the manuscript.

Funding: This work was financially supported by the National Natural Science Foundation of China, International (Regional) Cooperation and Exchange Program, Research fund for International young scientists (517102-N11808ZJ), Sino Pakistan Project (31961143008), Jiangsu Collaborative Innovation Center for Modern Crop Production (JCICMCP) China and by the Training Program Project of High Level Innovative Talents in Guizhou Province of China [(2016)4003], the Center Construction Project of Collaborative Innovation in Guizhou Province in 2011 [(2014)01], Guizhou Key Program from the Guizhou Provincial Tobacco Company (201315, 201717 and 201813), and the Science and Technology support Program of Guizhou Province [(2016)2536].

Conflicts of Interest: The authors declare no conflict of interest.

References

1.   Ma, C.; Liang, B.; Chang, B.; Liu, L.; Yan, J.; Yang, Y.; Zhao, Z. Transcriptome Profiling Reveals Transcriptional Regulation by DNA Methyltransferase Inhibitor 5-Aza-2′-Deoxycytidine Enhancing Red Pigmentation in Bagged "Granny Smith" Apples (*Malus domestica*). *Int. J. Mol. Sci.* **2018**, *19*, 3133. [CrossRef]

2.   Chen, J.; Funnell, K.A.; Lewis, D.H.; Eason, J.R.; Woolley, D.J. Relationship between changes in colour and pigment content during spathe regreening of *Zantedeschia* 'Best Gold'. *Postharvest Biol. Technol.* **2012**, *67*, 124–129. [CrossRef]

3.   Wei, X.; Deng, X.; Cai, D.; Ji, Z.; Wang, C.; Yu, J.; Li, J.; Chen, S. Decreased tobacco-specific nitrosamines by microbial treatment with Bacillus amyloliquefaciens DA9 during the air-curing process of burley tobacco. *J. Agric. Food Chem.* **2014**, *62*, 12701–12706. [CrossRef] [PubMed]

4.   Li, D.; Zhang, X.; Li, L.; Aghdam, M.S.; Wei, X.; Liu, J.; Xu, Y.; Luo, Z. Elevated CO$_2$ delayed the chlorophyll degradation and anthocyanin accumulation in postharvest strawberry fruit. *Food Chem.* **2019**, *285*, 163–170. [CrossRef] [PubMed]

5.   Yuan, Z.; Deng, L.; Yin, B.; Yao, S.; Zeng, K. Effects of blue LED light irradiation on pigment metabolism of ethephon-degreened mandarin fruit. *Postharvest Biol. Technol.* **2017**, *134*, 45–54. [CrossRef]

6.   Yuan, H.; Zhang, J.; Nageswaran, D.; Li, L. Carotenoid metabolism and regulation in horticultural crops. *Hortic. Res.* **2015**, *2*, 15036. [CrossRef]

7.   Fraser, P.D.; Bramley, P.M. The biosynthesis and nutritional uses of carotenoids. *Prog. Lipid Res.* **2004**, *43*, 228–265. [CrossRef]

8.   Ma, C.; Cao, J.; Li, J.; Zhou, B.; Tang, J.; Miao, A. Phenotypic, histological and proteomic analyses reveal multiple differences associated with chloroplast development in yellow and variegated variants from *Camellia sinensis*. *Sci. Rep.* **2016**, *6*, 33369. [CrossRef]

9.   Davison, P.A.; Hunter, C.H.; Horton, P. Overexpression of β-carotene hydroxylase enhances stress tolerance in *Arabidopsis*. *Nature* **2002**, *418*, 203–206. [CrossRef]

10.   Karppinen, K.; Zoratti, L.; Sarala, M.; Carvalho, E.; Hirsimäki, J.; Mentula, H.; Martens, S.; Häggman, H.; Jaakola, L. Carotenoid metabolism during bilberry (*Vaccinium myrtillus* L.) fruit development under different light conditions is regulated by biosynthesis and degradation. *BMC Plant Biol.* **2016**, *16*, 95. [CrossRef]

11.   Wang, Y.; Zhang, C.; Dong, B.; Fu, J.; Hu, S.; Zhao, H. Carotenoid Accumulation and Its Contribution to Flower Coloration of *Osmanthus fragrans*. *Front. Plant Sci.* **2018**, *9*, 1499. [CrossRef] [PubMed]

12.   Leng, X.; Wang, P.; Wang, C.; Zhu, X.; Li, X.; Li, H.; Mu, Q.; Li, A.; Liu, Z.; Fang, J. Genome-wide identification and characterization of genes involved in carotenoid metabolic in three stages of grapevine fruit development. *Sci. Rep.* **2017**, *7*, 4216. [CrossRef] [PubMed]

13.   Gayen, D.; Ali, N.; Sarkar, S.N.; Datta, S.K.; Datta, K. Down-regulation of *lipoxygenase* gene reduces degradation of carotenoids of golden rice during storage. *Planta* **2015**, *242*, 353–363. [CrossRef] [PubMed]

14.   Zhai, S.; Xia, X.; He, Z. Carotenoids in Staple Cereals: Metabolism, Regulation, and Genetic Manipulation. *Front. Plant Sci.* **2016**, *7*, 1197. [CrossRef] [PubMed]

15. Zelena, K.; Hardebusch, B.; Hülsdau, B.; Berger, R.G.; Zorn, H. Generation of Norisoprenoid Flavors from Carotenoids by Fungal Peroxidases. *J. Agric. Food Chem.* **2009**, *57*, 9951–9955. [CrossRef]

16. Wahlberg, I.; Karlsson, K.; Austin, D.J.; Junker, N.; Roeraade, J.; Enzell, C.R.; Johnson, W.H. Effects of flue-curing and ageing on the volatile, neutral and acidic constituents of Virginia tobacco. *Phytochemistry* **1977**, *16*, 1217–1231. [CrossRef]

17. Hauenstein, M.; Christ, B.; Das, A.; Aubry, S.; Hörtensteiner, S. A Role for TIC55 as a Hydroxylase of Phyllobilins, the Products of Chlorophyll Breakdown during Plant Senescence. *Plant Cell* **2016**, *28*, 2510–2527. [CrossRef]

18. Kräutler, B. Breakdown of Chlorophyll in Higher Plants—Phyllobilins as Abundant, Yet Hardly Visible Signs of Ripening, Senescence, and Cell Death. *Angew. Chem. Int. Ed.* **2016**, *55*, 4882–4907. [CrossRef]

19. Barry, C.S.; McQuinn, R.P.; Chung, M.Y.; Besuden, A.; Giovannoni, J.J. Amino acid substitutions in homologs of the STAY-GREEN protein are responsible for the *green-flesh* and *chlorophyll retainer* mutations of tomato and pepper. *Plant Physiol.* **2008**, *147*, 179–187. [CrossRef]

20. Moser, S.; Müller, T.; Holzinger, A.; Lutz, C.; Jockusch, S.; Turro, N.J.; Kräutler, B. Fluorescent chlorophyll catabolites in bananas light up blue halos of cell death. *Proc. Natl. Acad. Sci. USA* **2009**, *106*, 15538–15543. [CrossRef]

21. Banala, S.; Moser, S.; Müller, T.; Kreutz, C.; Holzinger, A.; Lütz, C.; Kräutler, B. Hypermodified fluorescent chlorophyll catabolites: Source of blue luminescence in senescent leaves. *Angew. Chem. Int. Ed.* **2010**, *49*, 5174–5177. [CrossRef] [PubMed]

22. Li, C.; Wurst, K.; Jockusch, S.; Gruber, K.; Podewitz, M.; Liedl, K.R.; Kräutler, B. Chlorophyll-Derived Yellow Phyllobilins of Higher Plants as Medium-Responsive Chiral Photo switches. *Angew. Chem. Int. Ed.* **2016**, *55*, 15760–15765. [CrossRef] [PubMed]

23. Weston, T.J. Biochemical characteristics of tobacco leaves during flue-curing. *Phytochemistry* **1968**, *7*, 921–930. [CrossRef]

24. Glover, B.J.; Whitney, H.M. Structural colour and iridescence in plants: The poorly studied relations of pigment colour. *Ann. Bot.* **2010**, *105*, 505–511. [CrossRef]

25. Iorizzo, M.; Ellison, S.; Senalik, D.; Zeng, P.; Satapoomin, P.; Huang, J.; Bowman, M.; Iovene, M.; Sanseverino, W.; Cavagnaro, P.; et al. A high-quality carrot genome assembly provides new insights into carotenoid accumulation and asterid genome evolution. *Nat. Genet.* **2016**, *48*, 657–666. [CrossRef]

26. Galpaz, N.; Ronen, G.; Khalfa, Z.; Zamir, D.; Hirschberg, J.A. Chromoplast-specific carotenoid biosynthesis pathway is revealed by cloning of the tomato white-flower locus. *Plant Cell* **2006**, *18*, 1947–1960. [CrossRef]

27. Yu, J.J.; Zhang, J.Z.; Zhao, Q.; Liu, Y.L.; Chen, S.X.; Guo, H.L.; Shi, L.; Dai, S.J. Proteomic Analysis Reveals the Leaf Color Regulation Mechanism in Chimera *Hosta* "Gold Standard" Leaves. *Int. J. Mol. Sci.* **2016**, *17*, 346. [CrossRef]

28. Chu, P.; Yan, G.X.; Yang, Q.; Zhai, L.N.; Zhang, C.; Zhang, F.Q.; Guan, R.Z. iTRAQ-based quantitative proteomics analysis of *Brassica napus* leaves reveals pathways associated with chlorophyll deficiency. *J. Proteom.* **2015**, *113*, 244–259. [CrossRef]

29. Lin, M.; Fang, J.; Qi, X.; Li, Y.; Chen, J.; Sun, L.; Zhong, Y. iTRAQ-based quantitative proteomic analysis reveals alterations in the metabolism of *Actinidia arguta*. *Sci. Rep.* **2017**, *7*, 5670. [CrossRef]

30. Wu, H.; Shi, N.; An, X.; Liu, C.; Fu, H.; Cao, L.; Feng, Y.; Sun, D.; Zhang, L. Candidate Genes for Yellow Leaf Color in Common Wheat (*Triticum aestivum* L.) and Major Related Metabolic Pathways according to Transcriptome Profiling. *Int. J. Mol. Sci.* **2018**, *19*, 1594. [CrossRef]

31. Yang, H.; Zhao, L.; Zhao, S.; Wang, J.; Shi, H. Biochemical and transcriptomic analyses of drought stress responses of LY1306 tobacco strain. *Sci. Rep.* **2017**, *7*, 17442. [CrossRef] [PubMed]

32. Sierro, N.; Battey, J.N.; Ouadi, S.; Bakaher, N.; Bovet, L.; Willig, A.; Goepfert, S.; Peitsch, M.C.; Ivanov, N.V. The tobacco genome sequence and its comparison with those of tomato and potato. *Nat. Commun.* **2014**, *5*, 3833. [CrossRef] [PubMed]

33. Guo, Y. Towards systems biological understanding of leaf senescence. *Plant Mol. Biol.* **2013**, *82*, 519–528. [CrossRef] [PubMed]

34. Gupta, R.; Lee, S.J.; Min, C.W.; Kim, S.W.; Park, K.H.; Bae, D.W.; Lee, B.W.; Agrawal, G.K.; Rakwal, R.; Kim, S.T. Coupling of gel-based 2-DE and 1-DE shotgun proteomics approaches to dig deep into the leaf senescence proteome of *Glycine max*. *J. Proteom.* **2016**, *148*, 65–74. [CrossRef] [PubMed]

35. Wei, S.; Wang, X.; Zhang, J.; Liu, P.; Zhao, B.; Li, G.; Dong, S. The role of nitrogen in leaf senescence of summer maize and analysis of underlying mechanisms using comparative proteomics. *Plant Sci.* **2015**, *233*, 72–81. [CrossRef]

36. Li, L.; Zhao, J.; Zhao, Y.; Lu, X.; Zhou, Z.; Zhao, C.; Xu, G. Comprehensive investigation of tobacco leaves during natural early senescence via multi-platform metabolomics analyses. *Sci. Rep.* **2016**, *6*, 37976. [CrossRef]

37. Burton, H.R.; Kasperbauer, M.J. Changes in chemical composition of tobacco lamina during senescence and curing. 1. Plastid pigments. *J. Agric. Food Chem.* **1985**, *33*, 879–883. [CrossRef]

38. Li, S.; Su, X.; Jin, Q.; Li, G.; Sun, Y.; Abdullah, M.; Cai, Y.; Lin, Y. iTRAQ-Based Identification of Proteins Related to Lignin Synthesis in the Pear Pollinated with Pollen from Different Varieties. *Molecules* **2018**, *23*, 548. [CrossRef]

39. Zhang, B.; Liu, C.; Yao, X.; Wang, F.; Wu, J.; King, G.J.; Liu, K. Disruption of a *CAROTENOID CLEAVAGE DIOXYGENASE 4* gene converts flower colour from white to yellow in *Brassica* species. *New Phytol.* **2015**, *206*, 1513–1526. [CrossRef]

40. Wang, Y.L.; Cao, C.; Zhou, H.; Zeng, Y.; Yang, J.; Wang, Y.X. Biochemical and transcriptome analyses of a novel chlorophyll-deficient chlorina tea plant cultivar. *BMC Plant Biol.* **2014**, *14*, 352. [CrossRef]

41. Akram, N.A.; Shafiq, F.; Ashraf, M. Ascorbic Acid-A Potential Oxidant Scavenger and Its Role in Plant Development and Abiotic Stress Tolerance. *Front. Plant Sci.* **2017**, *8*, 613. [CrossRef]

42. Yarmolinsky, D.; Brychkova, G.; Kurmanbayeva, A.; Bekturova, A.; Ventura, Y.; Khozin-Goldberg, I.; Eppel, A.; Fluhr, R.; Sagi, M. Impairment in Sulfite Reductase Leads to Early Leaf Senescence in Tomato Plants. *Plant Physiol.* **2014**, *165*, 1505–1520. [CrossRef] [PubMed]

43. Qin, Y.; Djabou, A.S.M.; An, F.; Li, K.; Li, Z.; Yang, L.; Wang, X.; Chen, S. Proteomic analysis of injured storage roots in cassava (*Manihot esculenta* Crantz) under postharvest physiological deterioration. *PLoS ONE* **2017**, *12*, e0174238. [CrossRef] [PubMed]

44. Wu, Z.J.; Ma, H.Y.; Zhuang, J. iTRAQ-based proteomics monitors the withering dynamics in postharvest leaves of tea plant (*Camellia sinensis*). *Mol. Genet. Genom.* **2018**, *293*, 45–59. [CrossRef] [PubMed]

45. Lätari, K.; Wüst, F.; Hübner, M.; Schaub, P.; Beisel, K.G.; Matsubara, S.; Beyer, P.; Welsch, R. Tissue-Specific Apocarotenoid Glycosylation Contributes to Carotenoid Homeostasis in *Arabidopsis* Leaves. *Plant Physiol.* **2015**, *168*, 1550–1562. [CrossRef] [PubMed]

46. Hörtensteiner, S. Chlorophyll degradation during senescence. *Annu. Rev. Plant. Biol.* **2006**, *57*, 55–77. [CrossRef] [PubMed]

47. Bollenbach, T.J.; Lange, H.; Gutierrez, R.; Erhardt, M.; Stern, D.B.; Gagliardi, D. RNR1, a 3′-5′ exoribonuclease belonging to the RNR superfamily, catalyzes 3′ maturation of chloroplast ribosomal RNAs in *Arabidopsis thaliana*. *Nucleic Acids Res.* **2005**, *33*, 2751–2763. [CrossRef]

48. Fargašová, A.; Molnárová, M. Assessment of Cr and Ni phytotoxicity from cutlery-washing waste-waters using biomass and chlorophyll production tests on mustard *Sinapis alba* L. seedlings. *Environ. Sci. Pollut. Res.* **2010**, *17*, 187–194. [CrossRef]

49. Aruna, G.; Baskaran, V. Comparative study on the levels of carotenoids lutein, zeaxanthin and β-carotene in Indian spices of nutritional and medicinal importance. *Food Chem.* **2010**, *123*, 404–409. [CrossRef]

50. Xiang, Z.; Cai, K.; Liang, G.; Zhou, S.; Ge, Y.; Zhang, J.; Geng, Z. Analysis of volatile flavour components in flue-cured tobacco by headspace solid-phase microextraction combined with GC×GC-TOFMS. *Anal. Methods* **2014**, *6*, 3300. [CrossRef]

51. Isaacson, T.; Damasceno, C.M.; Saravanan, R.S.; He, Y.; Catalá, C.; Saladié, M.; Rose, J.K. Sample extraction techniques for enhanced proteomic analysis of plant tissues. *Nat. Protoc.* **2006**, *1*, 769–774. [CrossRef] [PubMed]

52. Bradford, M.M. A Rapid and Sensitive Method for the Quantitation of Microgram Quantities of Protein Utilizing the Principle of Protein-Dye Binding. *Anal. Biochem.* **1976**, *722*, 48–254. [CrossRef]

53. Wen, B.; Zhou, R.; Feng, Q.; Wang, Q.; Wang, J.; Liu, S. IQuant: An automated pipeline for quantitative proteomics based upon isobaric tags. *Proteomics* **2014**, *14*, 2280–2285. [CrossRef] [PubMed]

54. Brosch, M.; Yu, L.; Hubbard, T.; Choudhary, J. Accurate and Sensitive Peptide Identification with Mascot Percolator. *J. Proteome Res.* **2009**, *8*, 3176–3181. [CrossRef] [PubMed]

55. Savitski, M.M.; Wilhelm, M.; Hahne, H.; Kuster, B.; Bantscheff, M. A Scalable Approach for Protein False Discovery Rate Estimation in Large Proteomic Data Sets. *Mol. Cell. Proteom.* **2015**, *14*, 2394–2404. [CrossRef] [PubMed]
56. Livak, K.J.; Schmittgen, T.D. Analysis of relative gene expression data using real-time quantitative PCR and the 2(-Delta C(T)) Method. *Methods* **2001**, *25*, 402–408. [CrossRef]

International Journal of
Molecular Sciences

Article

Proteome-Wide Analyses Provide New Insights into the Compatible Interaction of Rice with the Root-Knot Nematode *Meloidogyne graminicola*

Chao Xiang [1,†], Xiaoping Yang [2,†], Deliang Peng [1], Houxiang Kang [1], Maoyan Liu [1,3], Wei Li [3], Wenkun Huang [1,*] and Shiming Liu [1,*]

[1] State Key Laboratory for Biology of Plant Diseases and Insect Pests, Institute of Plant Protection, Chinese Academy of Agricultural Sciences, Beijing 100193, China; xiangchao2018@163.com (C.X.); pengdeliang@caas.cn (D.P.); Kanghouxiangcaas@163.com (H.K.); liu-mao-yan@foxmail.com (M.L.)

[2] Hunan Biological and Electromechanical Polytechnic, Changsha 410127, China; xiaopingyang75@163.com

[3] College of Plant Protection, Hunan Agricultural University, Changsha 410128, China; liwei350551@163.com

* Correspondence: wkhuang@ippcaas.cn (W.H.); smliuhn@yahoo.com (S.L.)

† These authors equally contributed to this work.

Received: 20 July 2020; Accepted: 4 August 2020; Published: 6 August 2020

Abstract: The root-knot nematode *Meloidogyne graminicola* is an important pathogen in rice, causing huge yield losses annually worldwide. Details of the interaction between rice and *M. graminicola* and the resistance genes in rice still remain unclear. In this study, proteome-wide analyses of the compatible interaction of the *japonica* rice cultivar "Nipponbare" (NPB) with *M. graminicola* were performed. In total, 6072 proteins were identified in NPB roots with and without infection of *M. graminicola* by label-free quantitative mass spectrometry. Of these, 513 specifically or significantly differentially expressed proteins were identified to be uniquely caused by nematode infection. Among these unique proteins, 99 proteins were enriched on seven Kyoto Encyclopedia of Genes and Genomes (KEGG) pathways. By comparison of protein expression and gene transcription, LOC_Os01g06600 (ACX, a glutaryl-CoA dehydrogenase), LOC_Os09g23560 (CAD, a cinnamyl-alcohol dehydrogenase), LOC_Os03g39850 (GST, a glutathione S-transferase) and LOC_Os11g11960 (RPM1, a disease resistance protein) on the alpha-linolenic acid metabolism, phenylpropanoid biosynthesis, glutathione metabolism and plant–pathogen interaction pathways, respectively, were all associated with disease defense and identified to be significantly down-regulated in the compatible interaction of NPB with nematodes, while the corresponding genes were remarkably up-regulated in the roots of a resistant rice accession "Khao Pahk Maw" with infection of nematodes. These four genes likely played important roles in the compatible interaction of rice with *M. graminicola*. Conversely, these disease defense-related genes were hypothesized to be likely involved in the resistance of resistant rice lines to this nematode. The proteome-wide analyses provided many new insights into the interaction of rice with *M. graminicola*.

Keywords: rice; root-knot nematode; *Meloidogyne graminicola*; compatible interaction; proteome-wide analyses

---

1. Introduction

As one of the top ten plant-parasitic nematodes [1], the root-knot nematode (RKN) *Meloidogyne graminicola* is a tremendous threat to rice (*Oryza sativa* L.) worldwide. Recent cultivation pattern alterations and climate changes have further aggravated the damaging effect of this nematode [2–4]. This nematode mainly infects rice root tips and induces root cells to form giant cells as a nutrition resource with a characteristic symptom of hook-shaped galls on the roots, resulting in deficient growth and severe rice yield losses of up to 87% [5,6]. In well-drained soil, at 22–29 °C, the swelling of root tips can be vaguely observed at 1 day post inoculation (dpi), hook-shaped galls can be clearly visible on the root tips at 3 dpi, and at 7 to 15 dpi, following 1–2 times of molting, the nematodes eventually develop into females, which then lay eggs in the galls [7,8]. The short life cycle makes *M. graminicola* difficultly controlled [2]. Planting resistant rice cultivars is an eco-friendly and economical means to prevent and control this pathogen. Analyses of details of the interaction of rice with *M. graminicola* and identification of the *M. graminicola*-resistant proteins (genes) are of significance for the management of this nematode.

The resistance of rice to *M. graminicola* is controlled by quantitative trait loci (QTL). Galeng-Lawilao et al. (2018) [9] identified a total of 12 QTL underlying resistance and tolerance to *M. graminicola*. The identification of candidate genes underlying resistance to *M. graminicola* is mostly performed using transcriptomes obtained by RNA sequencing (RNA-seq). Kyndt et al. (2012) [10] identified 382 loci that were significantly differentially expressed in *M. graminicola*-infected galls following study of the transcriptional reprogramming patterns of galls induced in the compatible interaction between the *japonica* rice cultivar "Nipponbare"(NPB) and *M. graminicola* at 3 and 7 dpi using RNA-seq. Analyses of the transcriptome of NPB giant cells at 7 and 14 dpi by using both laser-capture microdissection and RNA-seq indicated that the expression of genes associated with chloroplast biogenesis and tetrapyrrole synthesis was notably changed, while the majority of defense-related genes were extremely suppressed by infection of *M. graminicola* [11]. Through comparison of the transcriptomes of rice accessions compatibly and incompatibly interacting with *M. graminicola* at 2, 4 and 8 dpi, 32 potential *M. graminicola*-resistance-associated genes such as phenylalanine ammonia lyase, thionin and chalcone synthase genes were identified [12]. However, the rice genes underlying resistance to *M. graminicola* are not yet clear.

Many genes (regulators) impact traits at translation level rather than at transcription level. For example, in the induction of immunity triggered by the microbe-associated molecular pattern elf18, translational efficiency of Arabidopsis was mediated by the R-motif in a highly enriched messenger RNA through interaction with poly(A)-binding proteins [13]. The proteome can provide accurate and direct bioinformation and may be better than the transcriptome for the identification of many genes (regulators) underlying traits. Recently, sole proteomics analyses or combination analyses with other omics have been applied to the interaction of rice with pathogens, such as the regulation by chitosan oligosaccharide of the expression of plant defense-related proteins and virus proteins in the interaction of rice with southern rice black-streaked dwarf virus [14], the interaction of rice with *Fusarium fujikuroi* [15], and the responses of diverse genotypes of rice to *Magnaporthe oryzae* [16]. In this study, on one hand, to gain new insights into the interaction of rice with the RKN *M. graminicola*, the susceptible rice cultivar NPB was selected to perform proteome-wide analyses of the compatible interaction with *M. graminicola* by label-free quantitative mass spectrometry. Through comparison between proteins with and without infection of nematodes, the specifically expressed proteins and the significantly differentially expressed proteins (SDEPs), and then the proteins uniquely caused by the infection of nematodes in the roots of NPB, were identified. On the other hand, because the *M. graminicola*-resistant genes are not yet identified in rice, we also tried to find some useful bioinformation on the resistance of rice against this nematode through the proteome-wide analyses. So far, most of the identified RKN-resistant genes such as *Mi-1*, *Mi-9* and *Ma* all belong to the nucleotide binding site-leucine-rich repeat (NBS-LRR) family [17–21]. The expression of this type of proteins was one of our primary targets to be studied. Rice genes/proteins on certain defense-related and plant–pathogen interaction pathways may be down-regulated in the compatible interaction of rice with nematodes that the defense of rice against nematodes is likely suppressed, resulting in susceptibility, while these identical genes/proteins are up-regulated in the incompatible interaction of rice with nematodes. Therefore, we paid more attention to this type of proteins and tried to identify proteins with opposite expression patterns in susceptible and resistant rice accessions using the unique proteins obtained by the proteome-wide analyses. For this, we confirmed a previously reported *M. graminicola*-resistant rice accession "Khao Pahk Maw" [22] and used it to analyze the transcriptional abundance of corresponding genes. Meanwhile, during the analyses, we noticed that the expressions of a vesicle-associated membrane protein (VAMP, LOC_Os10g06540), an aminotransferase domain-containing protein (AT, LOC_Os05g15530) and the PEX3 (Loc_Os09g14510) were down-regulated in the compatible interaction of NPB with nematodes. They are associated with SNARE interactions in vesicular transport, folate biosynthesis and peroxisome pathway, respectively. The vesicular transport-related GmSNAP18 and the folate biosynthesis-related GmSHMT08 have been cloned and functionally identified to play very important roles in the resistance of soybean to soybean cyst nematode, which is one of the devastating pathogens in soybean [23–25], and PEX3 is involved in the cleavage of hydrogen peroxide and ß-oxidation of very long chain fatty acids to short chain fatty acids, playing key roles in the conversion of bioactive oxygen species and lipid metabolism [26]. Therefore, these three proteins/genes were further analyzed in this study. According to the analysis results, besides VAMP, AT and PEX3, another five proteins, namely, LOC_Os01g06600 (ACX, a glutaryl-CoA dehydrogenase) on the alpha-linolenic acid metabolism pathway, LOC_Os09g23560 (CAD, a cinnamyl-alcohol dehydrogenase) on the phenylpropanoid biosynthesis pathway, LOC_Os03g39850 (GST, a glutathione S-transferase) on the glutathione metabolism pathway, and LOC_Os11g11960 (RPM1, a disease resistance protein) and LOC_Os02g46090 (CDPK, a calcium/calmodulin dependent protein kinase) on the plant–pathogen interaction pathway, were also studied in detail. The findings obtained in this study provide many new insights into the compatible interaction of rice with the RKN *M. graminicola*.

2. Results

2.1. Proteome Measurement of the Roots of Rice Cultivar NPB with or without Infection of M. graminicola

For the proteome measurement by label-free quantitative mass spectrometry (MS) technology, 14-day-old seedlings of the rice cultivar NPB were divided into two groups; one group ("MG") was inoculated and the other group ("CK") was not inoculated with *M. graminicola*. The roots were collected at four time-points: before inoculation, and 1, 3 and 7 dpi of *M. graminicola*. The samples collected before inoculation were used as the control (CK0), and the other collected samples were correspondingly used as MG1DPI and CK1DPI (1 dpi), MG3DPI and CK3DPI (3 dpi), and MG7DPI and CK7DPI (7 dpi). Because the galls started to be obviously visible at 3 dpi [7,8], the root tips were collected at two time-points (before inoculation and 1 dpi), while the root parts mainly containing the galls were selected under stereomicroscopy and collected at the other two time-points (3 and 7 dpi). In this study, three biological replicates were set for each sample; each replicate was measured by label-free quantitative mass spectrometry independently, and their obtained proteome data were analyzed individually. A large number of proteins were quantitatively identified in each sample. In total, 6072 proteins were identified in the roots of NPB with and without infection of nematodes. Among them, 5821, 5862, 5860, 5773, 5822, 5844 and 5784 proteins were identified in CK0, CK1DPI, CK3DPI, CK7DPI, MG1DPI, MG3DPI and MG7DPI, respectively (Table 1). Compared to the total numbers of proteins in CK0, no significant difference was shown in the identified protein numbers among all the other measured samples.

The infection of nematodes causes not only the expression of new proteins but also changes in the expression levels of proteins (up- or down-regulated) in the host plants. The Volcano plot is a type of scatterplot that enables the identification of genes/proteins with large fold changes that are also of statistical significance. Because all the comparisons later on were carried out based on the specifically expressed proteins and the significantly differentially expressed proteins (SDEPs, up- and down-regulated) in the roots of each sample compared to CK0, we first conducted Volcano plot analyses to identify SDEPs in each sample using all the identified proteins. The results showed different patterns of SDEPs in the roots at various days with or without infection of *M. graminicola* as per the Volcano plots (Figure S1). Subsequently, to compare the similarity of proteome data among replicates, all the acquired SDEPs of each replicate were employed to perform hierarchical clustering analyses. The results showed that the three biological replicates in each sample had highly similar expression profiles (Figure S2), suggesting the reliability of the proteome data.

Table 1. General information of proteome measurement of the seedling roots of NPB by label-free quantitative mass spectrometry. The numbers represent the numbers of proteins quantitatively identified. CK0; MG1DPI and CK1DPI; MG3DPI and CK3DPI; and MG7DPI and CK7DPI denote the root samples just before infection and at 1, 3 and 7 dpi with ('MG') and without ('CK') infection of *M. graminicola*, respectively.

Samples	Repeat 1	Repeat 2	Repeat 3	Total	Common in All 3 Repeats	Common Only in 2 Repeats
CK0	5801	5819	5683	5821	5473	348
CK1DPI	5784	5838	5824	5862	5568	294
CK3DPI	5861	5774	5771	5860	5503	357
CK7DPI	5702	5763	5692	5773	5379	394
MG1DPI	5795	5760	5797	5822	5538	284
MG3DPI	5821	5769	5772	5844	5511	333
MG7DPI	5734	5757	5744	5784	5482	302
Average				5824		
Total				6072		

2.2. Proteome-wide Analyses of the Roots of NPB with Infection of M. graminicola

To gain insights into the compatible interaction of rice NPB with *M. graminicola*, proteome-wide comparisons were carried out between the roots of NPB with and without infection of *M. graminicola* (MG_versus_CK). Compared to the corresponding samples of the "CK" group (without nematode infection), the numbers of SDEPs, which are summarized from the details as shown in Figure S1A–C, and the specifically expressed proteins of the samples in the "MG" group (with nematode infection) are presented in Figure 1. The results showed that compared to the corresponding CK1DPI, CK3DPI and CK7DPI, in total, 111 proteins were newly emerged (Figure 1A), 247 SDEPs were up-regulated (Figure 1B) and 334 SDEPs were down-regulated (Figure 1C) among MG1DPI, MG3DPI and MG7DPI. Among these, 32, 50 and 41 specifically expressed proteins; 42, 181 and 27 up-regulated SDEPs; and 83, 200 and 58 down-regulated SDEPs were identified at 1, 3 and 7 dpi, respectively (Figure 1). Obviously, the numbers of both newly emerged specifically expressed proteins and SDEPs were higher at 3 dpi than at both 1 dpi and 7 dpi. In particular, the numbers of SDEPs at 3 dpi were much higher than those at both 1 and 7 dpi (Figure 1). In addition, the root materials used to perform MS measurement mainly contained the galls from 3 dpi onwards. All these results indicated that the gall cells might be most active at 3 dpi. Meanwhile, these results also indicated that the infection of nematodes might considerably induce or suppress the expression of proteins in the roots of NPB.

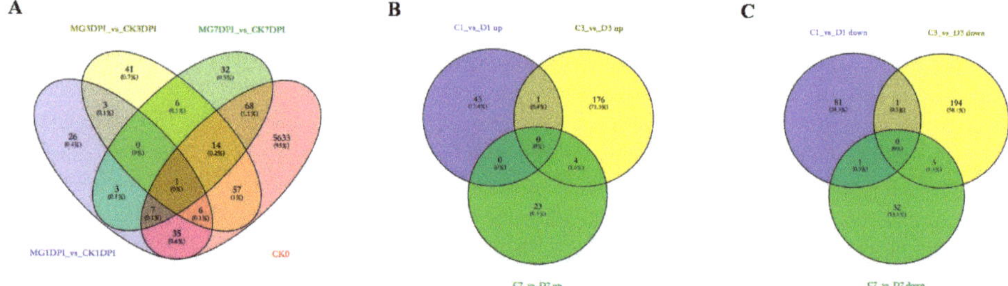

Figure 1. Proteomic profiles of roots of rice NPB with infection of *M. graminicola* during 7 days of growth: (**A**) specifically expressed proteins, (**B**) up-regulated significantly differentially expressed proteins (SDEPs), and (**C**) down-regulated SDEPs.

Successively, using all the specifically expressed proteins and SDEPs from MG1DPI_versus_CK1DPI, MG3DPI_versus_CK3DPI or MG7DPI_versus_CK7DPI, the enrichment of Gene Ontology (GO) and Kyoto Encyclopedia of Genes and Genomes (KEGG) pathways was analyzed separately. The results (only the 20 most significantly enriched GO terms and KEGG pathways in each comparison are shown, Figure 2) showed that the significance and quantity of proteins enriched in GO and KEGG terms were different at various dpi. In particular, the proteins in the terms "vacuole", "response to stress", "response to stimulus" and "peroxisome" were significantly enriched at 3 dpi (Figure 2B), while no proteins in these terms were enriched at 7 dpi (Figure 2C). These terms are all relevant to the defense of plants against stress including pathogens. These results indicated that the host rice most strongly responded to the infection of nematodes at 3 dpi. Regarding the KEGG enrichment analyses, the proteins in the terms "metabolic pathways" and "biosynthesis of secondary metabolites" were significantly enriched at both 1 and 3 dpi. However, at 7 dpi, the proteins were not significantly enriched in the term "biosynthesis of secondary metabolites", despite the proteins being significantly enriched in "metabolic pathways" (Figure 2D–F).

Subsequently, all the specifically expressed proteins and SDEPs in the entire MG_versus_CK group were put together to perform GO and KEGG enrichment analyses (Figure 3). The results revealed that 289, 263 and 460 proteins associated with the "cytoplasm", "cytoplasmic part" and "cell" GO terms (Figure 3A), and 123, 67 and 19 proteins involved in "metabolic pathways", "biosynthesis of secondary metabolites" and "phenylpropanoid biosynthesis" out of the 11 KEGG pathways (Figure 3B), respectively, were most significantly enriched. These results suggested that the infection of *M. graminicola* mainly induced significant changes in cellular components as well as molecular functions including many types of metabolic biosynthesis pathways in the host rice NPB.

2.3. Proteins in the Roots of NPB Uniquely Caused by Infection of M. graminicola

Firstly, compared to the proteins in CK0, the specifically expressed proteins and SDEPs were analyzed using the data of samples in the "CK" group (CK1DPI, CK3DPI and CK7DPI) to show the proteome changes of the NPB seedling roots without infection of *M. graminicola* (CK_versus_CK0). The numbers of specifically expressed proteins are shown in Figure 4A, and the numbers of SDEPs summarized from Figure S1D–F are exhibited in Figure 4B,C. In total, 216 proteins were specifically expressed in the "CK" group, including 137 proteins in CK1DPI, 159 proteins in CK3DPI and 152 proteins in CK7DPI, with 77 proteins shared among CK1DPI, CK3DPI and CK7DPI (Figure 4A). A total of 349 SDEPs were up-regulated including 92 SDEPs in CK1DPI, 164 SDEPs in CK3DPI and 183 SDEPs in CK7DPI, with 16 SDEPs shared among CK1DPI, CK3DPI and CK7DPI (Figure 4B). A total of 379 SDEPs were down-regulated including 55 SDEPs in CK1DPI, 154 SDEPs in CK3DPI and 278 SDEPs in CK7DPI, with 22 SDEPs shared among CK1DPI, CK3DPI and CK7DPI (Figure 4C).

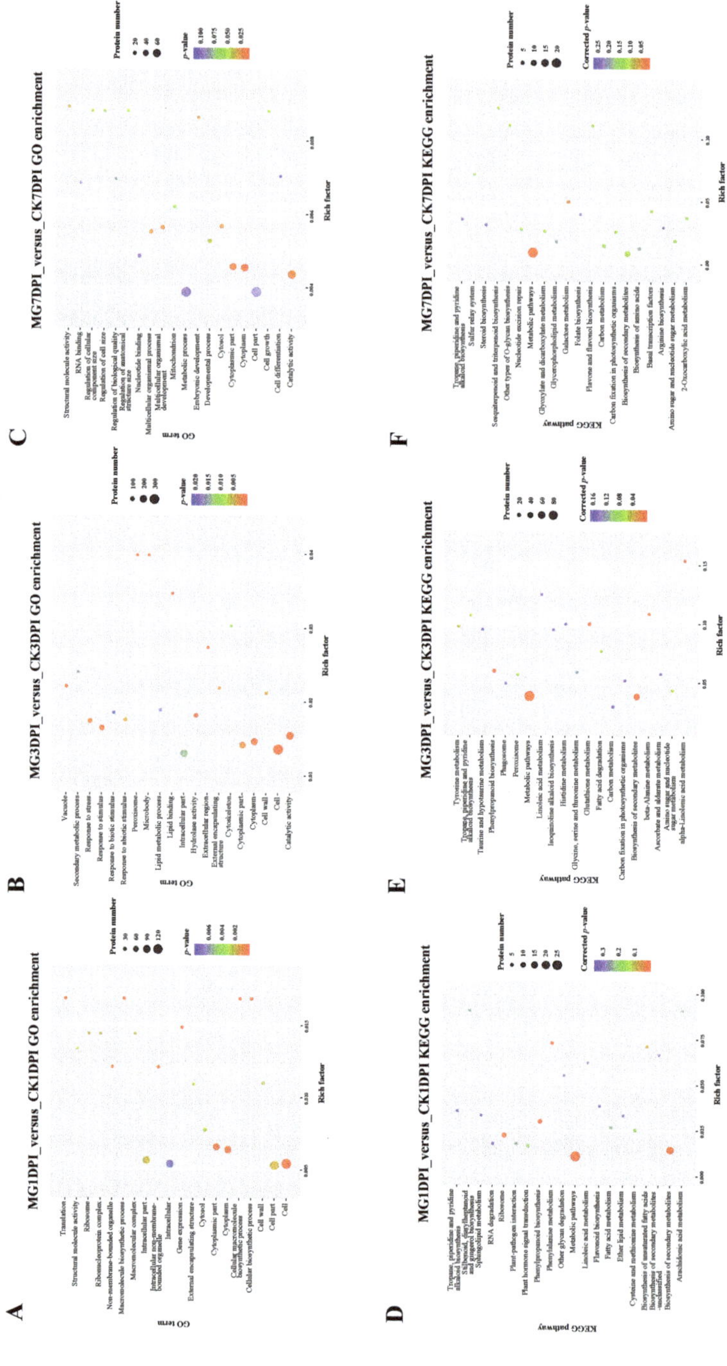

Figure 2. GO and KEGG enrichment analyses using all the specifically expressed proteins and SDEPs of NPB roots in MG1DPI_versus_CK1DPI, MG3DPI_versus_CK3DPI, or MG7DPI_versus_CK7DPI. (A–C) GO enrichment analyses in MG1DPI_versus_CK1DPI, MG3DPI_versus_CK3DPI, and MG7DPI_versus_CK7DPI, respectively. (D–F) KEGG enrichment analyses in MG1DPI_versus_CK1DPI, MG3DPI_versus_CK3DPI, and MG7DPI_versus_CK7DPI, respectively.

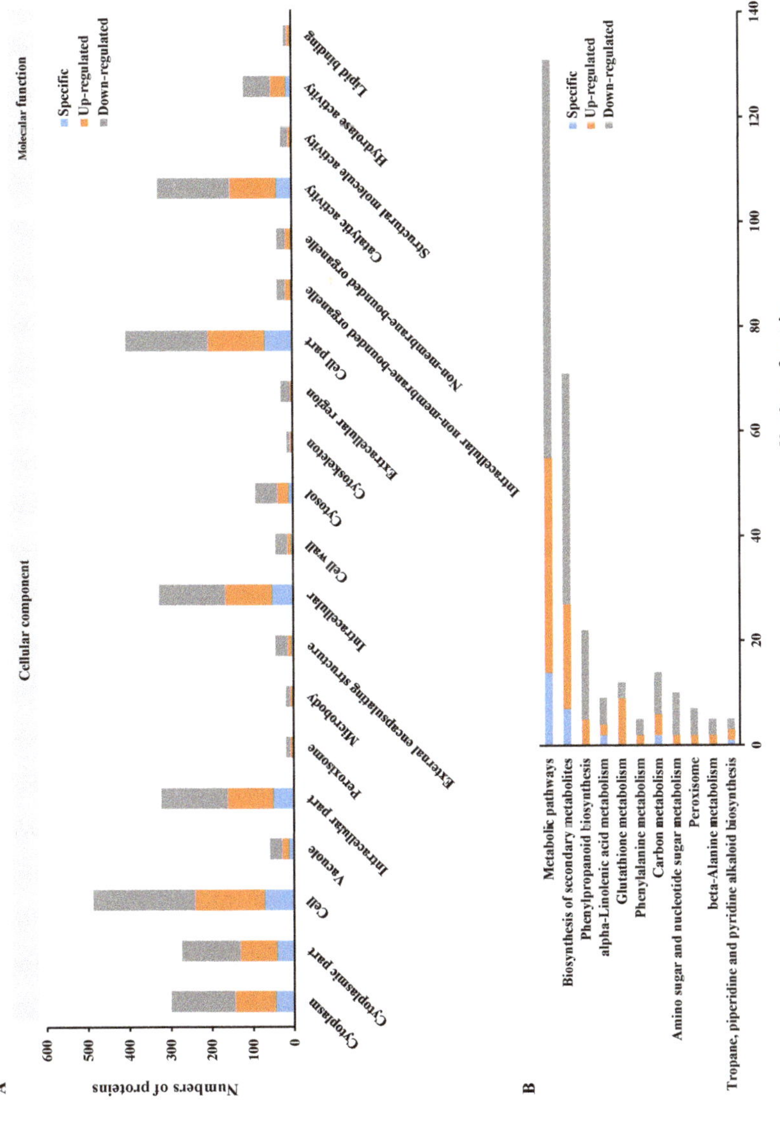

Figure 3. GO and KEGG enrichment analyses using all the specifically expressed proteins and SDEPs of NPB roots in the whole MG_versus_CK group: (**A**) GO enrichment analyses, and (**B**) KEGG pathway enrichment analyses.

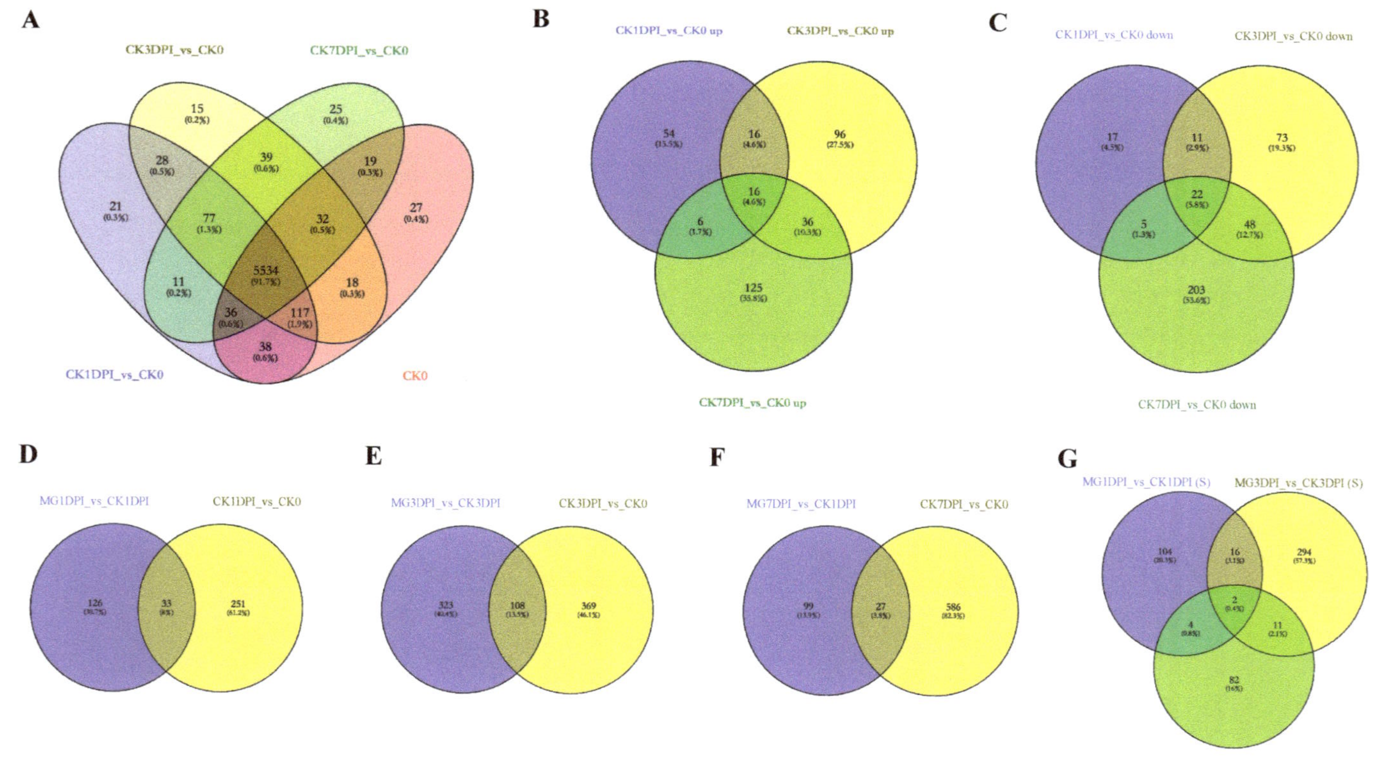

Figure 4. Proteins in the roots of NPB uniquely caused by infection of *M. graminicola*. (**A–C**) Specifically expressed proteins, up-regulated significantly differentially expressed proteins (SDEPs) and down-regulated SDEPs in the roots of NPB without infection of *M. graminicola* at 1, 3 and 7 days of growth, respectively (CK_versus_CK0). (**D–F**) Comparison of the proteins (all the specifically expressed proteins and up- and down-regulated SDEPs) in the roots of NPB between samples with and without infection of *M. graminicola* at 1, 3 and 7 dpi, respectively (MG_versus_CK and CK_versus_CK0). (**G**) All the unique proteins in the roots of NPB with infection of *M. graminicola* at 1, 3 and 7 dpi (the blue parts in (**D–F**)).

Secondly, all the obtained specifically expressed proteins and SDEPs in the "CK" group compared to CK0 were also employed to perform GO and KEGG enrichment analyses. The detailed results of GO and KEGG enrichment analyses in CK1DPI_versus_CK0, CK3DPI_versus_CK0 or CK7DPI_versus_CK0 are shown in Figure S3 (just the 20 most significantly enriched terms are shown), while the summarized enrichment results using all the specifically expressed proteins and SDEPs in the whole CK_versus_CK0 group are displayed in Figure S4. The GO enrichment analyses indicated that many proteins associated with "biological process", "cellular component" and "molecular function" were significantly enriched. In particular, 63, 64 and 69 proteins involved in "ribosome", "structural molecule activity" and "cellular biosynthetic process", respectively, were most significantly enriched (Figure S4A). The KEGG enrichment analyses indicated that of the 25 significantly enriched KEGG pathways, 180, 55 and 85 proteins involved in "metabolic pathways", "ribosome" and "biosynthesis of secondary metabolites", respectively, were most significantly enriched (Figure S4B). All these results indicated that the expressed proteins in NPB seedling roots without infection of *M. graminicola* were being dynamically as well as dramatically changed along with the 7 days of growth at this seedling stage.

Thirdly, we compared the proteins between MG_versus_CK (Figure 1) and CK_versus_CK0 (Figure 4A–C). There were some common proteins shared between the two groups at each time-point: 33, 108 and 27 proteins at 1, 3 and 7 dpi, respectively (Figure 4D–F). These results suggested that infection of nematodes not only caused dramatic changes in feeding site giant cells (galls) but also might impact the growth of rice roots and even of the whole seedlings. To concentrate on the changes in the galls at protein level and to gain insights into the accurate compatible interaction of rice with nematodes, we mainly focused on the proteins after removal of the common proteins shared between MG_versus_CK and CK_versus_CK0 at each time-point. Of the 656 proteins in the MG_versus_CK group, 513 unique proteins were obtained (Figure 4G), while 143 proteins were shared with the CK_versus_CK0 group. The detailed unique protein information is listed in Table S1. Obviously, these proteins in the roots of NPB seem to be uniquely caused by the infection of *M. graminicola*.

2.4. Important Proteins in the Roots Associated with the Compatible Interaction of NPB with M. graminicola

All the 513 unique proteins identified above were then used in GO and KEGG enrichment analyses. The GO enrichment analyses showed that 402 proteins were involved in 14 significantly enriched GO terms related to cellular components and molecular functions. Of these, the proteins associated with "cytoplasmic part", "cytoplasm" and "catalytic activity" were most significantly enriched (Figure 5A). The KEGG enrichment analyses indicated that only 99 proteins involved in 7 pathways (there were 62 proteins involved in 2–4 pathways, Table S1) were significantly enriched (Figure 5B). Obviously, the numbers of enriched proteins and the quantity of enrichment GO terms and KEGG pathways (Figure 5A,B) were both less than those without removal of the common proteins shared with the CK_versus_CK0 group (Figure 3), simplifying the subsequent analyses.

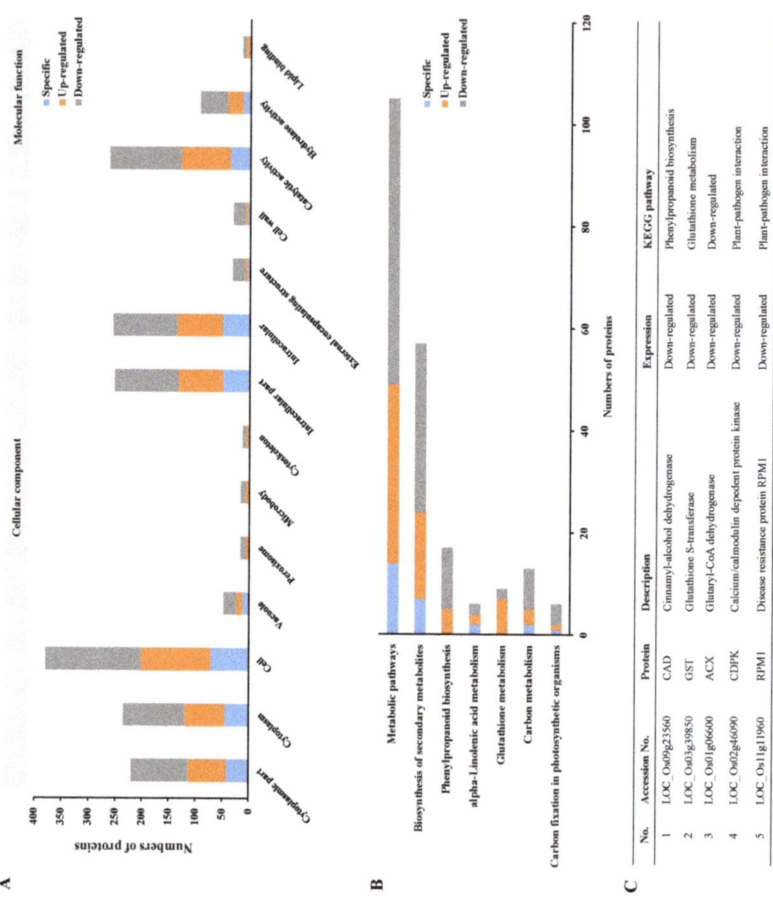

Figure 5. Proteins in the roots associated with the interaction of NPB with *M. graminicola*: (**A**) GO enrichment analyses using all the unique proteins in Figure 4G, (**B**) KEGG pathway enrichment analyses using all the unique proteins in Figure 4G, and (**C**) information on the selected proteins in the roots associated with the interaction of rice with *M. graminicola*.

Of the 7 KEGG pathways (Figure 5B), 3 pathways, namely, phenylpropanoid biosynthesis, glutathione metabolism and alpha-linolenic acid metabolism, have been reported to be involved in the defense against stresses. There were 28 enriched proteins on these 3 pathways: 14 proteins on phenylpropanoid biosynthesis, 8 proteins on glutathione metabolism and 6 proteins on alpha-linolenic acid metabolism (Figure 5B). Of these, 3 proteins were down-regulated, namely, LOC_Os09g23560 (CAD, a cinnamyl-alcohol dehydrogenase) on the phenylpropanoid biosynthesis pathway, LOC_Os03g39850 (GST, a glutathione S-transferase) on the glutathione metabolism pathway and LOC_Os01g06600 (ACX, a glutaryl-CoA dehydrogenase) on the alpha-linolenic acid metabolism pathway (Figure 5C, Table S1). The down-regulation of the defense genes on these 3 pathways might contribute to the compatible interaction of NPB with *M. graminicola*. In addition, there were 2 down-regulated proteins, LOC_Os02g46090 (CDPK, a calcium/calmodulin dependent protein kinase) and LOC_Os11g11960 (RPM1, a disease resistance protein), on the plant–pathogen interaction pathway in the compatible interaction of NPB with *M. graminicola* (Figure 5C, Table S1). These 2 proteins might also be involved in the compatible interaction. Thus, these 5 proteins were identified as the important proteins in the roots associated with the compatible interaction of NPB with *M. graminicola*.

2.5. Proteins with Opposite Expression Patterns in Susceptible and Resistant Rice Lines in Response to the Infection of M. graminicola

We first analyzed the reliability of the proteome data by comparison of the expression trends of proteins and transcripts of the corresponding genes. Besides the five above-mentioned proteins' genes, three genes, namely, *VAMP* (*LOC_Os10g06540*) on the SNARE interactions in vesicular transport pathway, *AT* (*LOC_Os05g15530*) on the folate biosynthesis pathway and *PEX3* (*LOC_Os09g14510*) on the peroxisome pathway, were also selected to analyze their transcriptional abundance in the roots of rice before and after infection of *M. graminicola*. As per the proteome data, all the eight selected proteins were down-regulated in *M. graminicola*-infected NPB at one to three time-points (1, 3 and 7 dpi) (Figure 5C, Table S1). The transcriptional abundance results indicated that *AT* was significantly up-regulated in NPB at 7 dpi (Figure 6A), and its encoded protein was slightly up-regulated (0.25 fold changes, Figure 6B) at 7 dpi. In contrast, all the other seven genes were down-regulated in the NPB roots at most of or all of the time-points (Figure 6A), almost matching the trends of expression of the corresponding proteins (Figure 6B). These results suggested that the obtained quantitative proteome data were reliable.

Subsequently, we confirmed one *M. graminicola*- highly resistant rice accession, Khao Pahk Maw, which was identified to be resistant to *M. graminicola* [22]. In the scoring at 14 dpi, only one gall was observed on 189 roots of Khao Pahk Maw with the gall index (GI) of 1.1, while the GI of the susceptible cultivar NPB was 52.1.

The Khao Pahk Maw accession was then used to analyze the transcriptional abundance of the eight selected genes in the roots before and after inoculation at different time-points. The results showed that among the eight selected genes, the transcriptional abundance of *ACX*, *CAD*, *GST*, *RPM1* and *AT* was all significantly up-regulated (relative expression value > 2), while the transcription abundance of *PEX3*, *CDPK* and *VAMP* was not remarkably changed (relative expression value ≤ 2), in the nematode-infected roots compared to the non-infected roots of Khao Pahk Maw (Figure 6C). Comparing the transcription and proteomics data of the roots in susceptible NPB and resistant Khao Pahk Maw for the eight proteins/genes, clearly only ACX, CAD, GST and RPM1 showed opposite reactions in the susceptible and resistant rice lines that all of them were significantly down-regulated in susceptible NPB, whereas they were remarkably up-regulated in resistant Khao Pahk Maw, in response to the infection of *M. graminicola* (Figure 6A–C).

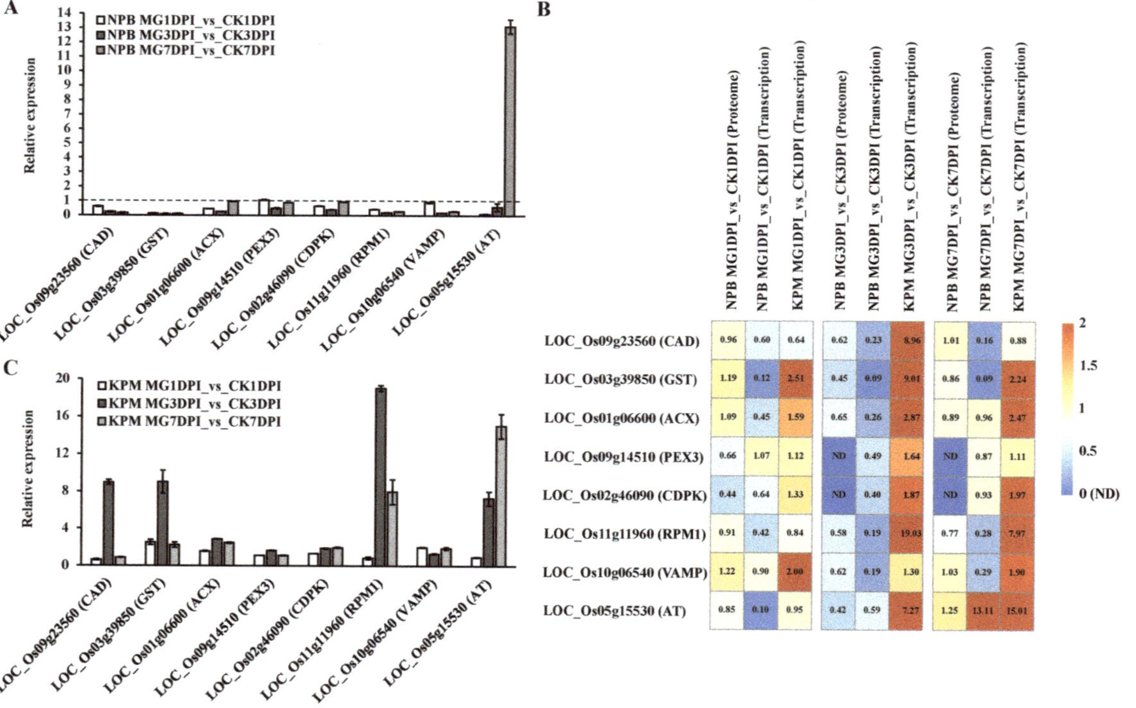

	NPB MG1DPI_vs_CK1DPI (Proteome)	NPB MG1DPI_vs_CK1DPI (Transcription)	KPM MG1DPI_vs_CK1DPI (Transcription)	NPB MG3DPI_vs_CK3DPI (Proteome)	NPB MG3DPI_vs_CK3DPI (Transcription)	KPM MG3DPI_vs_CK3DPI (Transcription)	NPB MG7DPI_vs_CK7DPI (Proteome)	NPB MG7DPI_vs_CK7DPI (Transcription)	KPM MG7DPI_vs_CK7DPI (Transcription)
LOC_Os09g23560 (CAD)	0.96	0.60	0.64	0.62	0.23	8.96	1.01	0.16	0.88
LOC_Os03g39850 (GST)	1.19	0.12	2.51	0.45	0.09	9.01	0.86	0.09	2.24
LOC_Os01g06600 (ACX)	1.09	0.45	1.59	0.65	0.26	2.87	0.89	0.96	2.47
LOC_Os09g14510 (PEX3)	0.66	1.07	1.12	ND	0.49	1.64	ND	0.87	1.11
LOC_Os02g46090 (CDPK)	0.44	0.64	1.33	ND	0.40	1.87	ND	0.93	1.97
LOC_Os11g11960 (RPM1)	0.91	0.42	0.84	0.58	0.19	19.03	0.77	0.28	7.97
LOC_Os10g06540 (VAMP)	1.22	0.90	2.00	0.62	0.19	1.30	1.03	0.29	1.90
LOC_Os05g15530 (AT)	0.85	0.10	0.95	0.42	0.59	7.27	1.25	13.11	15.01

Figure 6. Identification of proteins with opposite expression patterns in resistant and susceptible rice lines in response to the infection of *M. graminicola*. (**A**) Transcriptional abundance of genes encoding 8 selected proteins in the roots of NPB before and 1, 3 and 7 dpi of *M. graminicola*. (**B**) Detailed data of proteome and transcription of the 8 selected proteins/genes in the roots of NPB and Khao Pahk Maw. KPM, Khao Pahk Maw; ND, not detectable. All the numbers denote fold changes compared to CK0. (**C**) Transcriptional abundance of genes encoding 8 selected proteins in the roots of the resistant rice accession Khao Pahk Maw before and 1, 3 and 7 dpi of *M. graminicola*. KPM, Khao Pahk Maw.

3. Discussion

The interaction of plants with plant-parasitic root-knot nematodes is usually studied by transcriptomic and/or histocytological analyses, but proteome-wide analyses have also been applied in this field and have provided promising results. Recently, combination analyses of quantitative proteomics with RNA-seq on a resistant cucumber cultivar "IL10-1" and a susceptible cucumber cultivar "CC3" revealed the important roles of the MAPK signaling and flavonoid metabolic pathway in the resistance of cucumber to *M. incognita* [27]. In this study, we conducted proteome-wide analyses of the compatible interaction of NPB rice with *M. graminicola* (Figure S5A,B). A large number of proteins were quantitatively identified in each sample by label-free quantitative MS measurement (Table 1). There were more newly emerged proteins and up- and down-regulated SDEPs in the roots of NPB at 3 dpi than at both 1 and 7 dpi (Figure 1). In addition, the galls started to be noticeably visible at 3 dpi [7,8]. Therefore, it may be concluded that the feeding site giant cells in roots of rice were most active at 3 dpi of nematodes, they needed to biosynthesize much more substances for the growth of feeding site giant cells (galls) in the interaction. Further, compared to before inoculation (CK0), at 3 dpi, there were 181 up-regulated SDEPs and 200 down-regulated SDEPs, the quantity of SDEPs was much more than that of the newly emerged proteins (50) in the infected roots (Figure 1). Therefore, the infection of nematodes might largely impact (stimulate or suppress) the expression of proteins in the rice roots. Previously, transcriptomic analyses of NPB giant cells indicated that the majority of defense-related genes were enormously suppressed after infection of *M. graminicola* [11]. In this study, many proteins were also significantly enriched in several defense-related GO terms such as "vacuole", "response to stress", "response to stimulus" and "peroxisome" at 3 dpi, while no proteins were enriched in these terms at both 1 and 7 dpi (Figure 2A–C). Moreover, the proteins were significantly enriched on the KEGG pathway "biosynthesis of secondary metabolites" at 1 and 3 dpi rather than at 7 dpi (Figure 2D–F). Taken together, these results suggest that the rice roots were likely to still be strongly establishing the compatible interaction with nematodes before and at 3 dpi, until 7 dpi such an interaction might already be completely stable.

While comparing the proteome data between MG_versus_CK (Figure 1) and CK_versus_CK0 (Figure 4A–C) at identical time points, there were still a partiality of common proteins at each time-point (Figure 4D–F). In total, 143 proteins were shared between these two groups, a decrease of around 21.8% in quantity relative to the whole protein numbers (656) in these groups, suggesting that the changes in these common proteins might not be directly related to the compatible interaction of rice with *M. graminicola*. Of the 513 uniquely changed proteins (Figure 4G, detailed in Table S1), 402 and 99 proteins were enriched by GO and KEGG analyses, respectively (Figure 5A,B). These proteins are involved in the biosynthesis of many cellular components and are associated with many pathways, with the largest numbers of proteins in "cell" and "catalytic activity" in the GO terms and "metabolic pathways" and "biosynthesis of secondary metabolites" on the KEGG pathways (Figure 5A,B). All these results suggest that complicated but dramatic changes in cells took place in the compatible interaction of rice with *M. graminicola*. The changes in the abundance of the uniquely enriched proteins are therefore identified as being likely associated with the susceptibility of NPB to *M. graminicola*.

PEX3 on the peroxisome pathway for cleavage of hydrogen peroxide and ß-oxidation of very long chain fatty acids to short chain fatty acids was reported to be remarkably up-regulated by 4.13 fold changes in NPB at 7 dpi of *M. graminicola* by RNA-seq [11,26]. In contrast, in this study, the protein PEX3 was down-regulated at 1 dpi but not detectable at 3 and 7 dpi (Figure 6B), while the transcriptional abundance of *PEX3* was significantly decreased in NPB at 3 dpi but not significantly changed at 1 and 7 dpi (Figure 6A,B). The expression of VAMP involved in SNARE interactions in vesicular transport was not significantly changed in proteome data, whereas it was considerably decreased in transcription data at 7 dpi (Figure 6B). The expression of AT involved in folate biosynthesis was not changed much at 1 dpi but was slightly increased at 7 dpi in proteome data, while it was remarkably reduced at 1 dpi and significantly increased at 7 dpi in transcription data (Figure 6B). All these results obtained from this study and RNA-seq analyses by Ji et al. (2013) [11] obviously suggest different patterns of the same proteins/genes between protein and gene transcription expression in some cases. Multiple omics analyses will be much better to dissect the mechanism of compatible interaction of rice with nematodes. However, the substantial bioinformation acquired by proteome-wide analyses in this study provides many new insights into the compatible interaction of rice with *M. graminicola*.

So far, sources of *M. graminicola*-resistant rice are very limited [2,22,28,29] and the *M. graminicola*-resistant genes are not yet clear in rice, despite the fact that some resistance candidate genes were identified mainly through analyses of the transcriptomes by RNA-seq [10–12]. In this study, we confirmed the high resistance of the rice accession Khao Pahk Maw [22]. To conduct the proteome measurement for the interaction analyses, we chose to perform proteome-wide analyses of the compatible interaction of NPB with *M. graminicola*. Because the previously identified RKN-resistant genes *Mi-1*, *Mi-9* and *Ma* all belong to the NBS-LRR gene family [17–21], in this study, we first searched for significantly differentially expressed NBS-LRR proteins. However, no NBS-LRR proteins were enriched in the compatible interaction by proteome-wide analyses (Table S1). Analyses of the seven pathways carrying the 99 uniquely enriched proteins (Figure 5B, Table S1) indicated that three pathways, namely, alpha-linolenic acid metabolism, glutathione metabolism and phenylpropanoid biosynthesis, were noticeably reported to be involved in the defense of the plant against stresses including pathogens. For example, a glutathione S-transferase Fhb7 on the glutathione metabolism pathway was recently identified to confer resistance to Fusarium head blight in wheat [30]. Thus, these three pathways were most likely associated with the compatible interaction of NPB with nematodes. Of the 28 proteins associated with these three pathways, an ACX on the alpha-linolenic acid metabolism pathway, a CAD on the phenylpropanoid biosynthesis pathway and a GST on the glutathione metabolism pathway were all down-regulated during the compatible interaction with *M. graminicola* (Figure 5C, Table S1). In addition, an RPM1 and CDPK on the plant–pathogen interaction pathway were also down-regulated in the roots of NPB by infection of *M. graminicola* (Table S1). Among the genes encoding these five important proteins, four genes, *ACX*, *CAD*, *GST* and *RPM1*, were significantly up-regulated in the *M. graminicola*-highly resistant rice accession Khao Pahk Maw (Figure 6C) after infection of nematodes (Figure 6B,C), in contrast to the expression in the roots of susceptible rice NPB under infection conditions (Figure 6A,B). The down-regulation of these four genes might suppress the defense of rice against nematodes, so they likely played important roles in the compatible interaction of rice with *M. graminicola*. Conversely, these four genes are all associated with disease defense and are therefore hypothesized to be likely involved in the resistance of resistant rice lines to *M. graminicola*, all of which are different from the candidates previously reported to be involved in the resistance of rice to *M. graminicola*, to the best of our knowledge. However, due to no significant increase being found in the expression of *CDPK*, which is a calcium/calmodulin dependent protein kinase responsible for propagating immune signaling for the resistance of plants to pathogens [31], in resistant Khao Pahk Maw (Figure 6C), it may not be involved in the interaction of rice with *M. graminicola*. Regarding these four identified genes, the transcription expression of *RPM1* has been recorded by RNA-seq data, whereas RNA-seq data were not available for the other three genes [11]. The expression results of *RPM1* obtained in this study (Figure 6A,B) are consistent with the expression trend in RNA-seq data showing that *RPM1* was significantly down-regulated by 6.0 fold changes in NPB at 7 dpi of *M. graminicola* [11]. *RPM1* was identified as a resistant gene against the bacterium *Pseudomonas syringae* many years ago [32]. We strongly hypothesize that *RPM1* may confer resistance to *M. graminicola* in rice, and that it may possess broad-spectrum resistance against pathogens including the parasitic nematode *M. graminicola*. However, this needs to be extensively investigated.

In addition, there were 12 specifically expressed proteins, many up-regulated proteins and some other down-regulated proteins in NPB within the 99 enriched proteins identified on the seven KEGG pathways (Table S1). Of the 513 proteins uniquely caused by the infection of nematodes (Table S1), 402 proteins were enriched in 14 GO terms associated with cellular components and molecular functions (Figure 5A), some of which overlapped with the 99 proteins on the seven identified KEGG pathways (Table S1). These proteins may also be involved in the compatible interaction of NPB with nematodes. More detailed study of the compatible interaction of rice with *M. graminicola* is recommended in combination with other omics analyses on the basis of the proteomic data acquired in this study.

4. Materials and Methods

4.1. Rice Planting, Nematode Inoculation and Resistance Evaluation

The rice cultivar NPB and the *O. sativa* L. subsp. *Aus* accession Khao Pahk Maw that originated from Japan and Thailand and were reported to be susceptible and resistant to *M. graminicola*, respectively [11,22], were used as the experimental materials in this study. Their seeds were obtained from the International Rice Germplasm Center. The rice seeds germinated and were transplanted as described by Zhan et al. (2018) [33] with slight modifications. Briefly, the rice seeds were first soaked in 5.25% sodium hypochlorite (NaClO) for 10 min and germinated in the dark for 6 d at 28 °C; afterwards, each germinated seed was transferred into a polyvinylchloride tube containing SAP (sand and water-absorbent synthetic polymer) [34], growing under a 16 h/8 h light/dark regime at 75% relative humidity at 26 °C. Extra care was taken to irrigate each tube with 20 mL of Hoagland's solution every three days.

M. graminicola was propagated and extracted as described by Huang et al. (2015) [35]. Fourteen-day-old plants were inoculated with 300 stage juveniles (J2s) of *M. graminicola* per plant. The control plants were inoculated with water. Resistance/susceptibility scoring was conducted using the methods described by Zhan et al. (2018) [36]. The gall index (GI) at 14 dpi was used to score resistance/sensitivity. The evaluation criteria were as follows: immune (I) GI = 0; highly resistant (HR) $0.1 \leq GI \leq 5.0$; resistant (R) $5.1 \leq GI \leq 25.0$; moderately susceptible (MS) $25.1 \leq GI \leq 50.0$; susceptible (S) $50.1 \leq GI \leq 75.0$; highly susceptible (HS) $GI > 75.0$.

Before the protein MS measurement, we observed the phenotype of the NPB with the infection of *M. graminicola*. After inoculation of *M. graminicola* on the roots of rice cultivar NPB, galls started to be clearly visible on the roots at 3 dpi [7,8]; many galls emerged on the roots at 7 dpi (Figure S5A). The galls were stained using the method described by Bybd et al. (1983) [37] with minor modifications. Briefly, the roots of the seedlings were successively treated with 5% NaClO for 10 m, rinsed thoroughly with water, and boiled in an acid fuchsin solution (5% acid fuchsin in 25% acetic acid) for 1 m. Subsequently, the seedlings were transferred to glycerin for decolorization. After staining, the nematodes inside the roots were observed under a stereoscope (Leica S9E, Leica, Wetzlar, Germany). Several nematodes were observed in each gall at 7 dpi (Figure S5B). These results indicated that NPB showed compatible interaction with *M. graminicola*, consistent with the results reported previously [11].

4.2. Sample Collection

For both NPB and Khao Pahk Maw, three samples were collected before inoculation as the controls (CK0). Additionally, regarding NPB, the samples of *M. graminicola*-infected root tips or the root parts mainly containing visible galls were collected from 3 seedlings at 1, 3 and 7 dpi. About 10–15 rice galls were mixed as one sample up to 500 mg at each time-point. All the samples were immediately frozen in liquid nitrogen and stored at −80 °C for further usage. The NPB samples were used for both proteome measurement and quantitative real-time PCR (qRT-PCR) analyses, while the samples of Khao Pahk Maw were used for qRT-PCR analyses only. Three biological replicates were set for each control or treatment, and each replicate was used as an independent sample.

4.3. Protein Extraction and Digestion

The proteins were extracted as described previously with slightly modifications [38]. Briefly, powder ground in liquid nitrogen was suspended in precooled trichloroacetic acid/acetone (*v:v* = 1:9) overnight and then centrifuged at 14,000× *g* for 15 min, and the precipitate was washed thrice with acetone following each time of centrifuge. White powder was re-suspended in SDT lysis buffer (4% SDS, 100 mM Tris-HCl, 1 mM DTT, 1 mM PMSF, pH7.6, including one-fold PhosSTOP phosphatase inhibitor mixture from Roche). The protein digestion was primarily performed as per the Filter Aided Sample Preparation (FASP) protocol [39]. Protein concentration was determined using the bicinchoninic acid (BCA) method [40].

4.4. High Performance Liquid Chromatography–Tandem Mass Spectrometry (HPLC–MS/MS)

Digested peptide mixtures were pressure-loaded onto a fused silica capillary column packed with 3 μm dionex C18 material (Phenomenex, Tianjin, China). The RP (reverse phase) sections with 100Å were 15 cm long, and the column was washed with buffer A (water containing 0.1% formic acid) and buffer B (acetonitrile containing 0.1% formic acid). After desalting, a 5 mm, 300 μm C18 capture tip was placed in line with an Agilent 1100 quaternary HPLC (Agilent, California, USA), and the samples were analyzed using a 12-step separation.

The first step consisted of a 5-min gradient from 0% to 2% buffer B, followed by a 45-min gradient to 40% buffer B, and then by gradient from 40% to 80% buffer B for 3 min and 10 min of 80% buffer B. After a 2-min buffer B gradient from 80% to 2%, approximately 100 μg of tryptic peptide mixture was loaded onto the columns and was separated at a flow rate of 0.5 μL/min by using a linear gradient. As peptides were eluted from the micro-capillary column, they were electrosprayed directly into a Bruker micrOTOF-Q II mass spectrometer (Bruker Daltonics, Billerica, MA, USA) with the application of a distal 180 °C source temperature. The mass spectrometer was operated in the MS/MS (auto) mode. Survey MS scans were acquired in the TOF-Q II with the resolution set to a value of 20,000. Each survey scan (50~2500) was followed by five data-dependent MS/MS scans at 2 Hz normalized scan speed.

4.5. Sequence Database Search and Data Analyses

Proteome Discoverer 2.1 (Thermo Fisher Scientific, Waltham, MA, USA) software was used for data analyses. Peptide identification was performed with the SEQUST search engine using the NPB proteome databases downloaded from the Rice Genome Annotation Project (http://rice.plantbiology.msu.edu/index.shtml) (access on 1 January 2019). Decoys for the database search were generated with the revert function. The following options were used to identify the proteins: Peptide mass tolerance = ±10 ppm, MS/MS tolerance = 0.02 Da, enzyme = trypsin, max missed cleavage = 2, fixed modification: Carbamidomethyl (C), variable modification: oxidation (M) and Acetyl (Protein N-term), database pattern = decoy. The false discovery rate (FDR) for peptides and proteins was set to 0.01.

The proteins only identified in one biological replicate were ignored (the same below), and only the up- or down-regulated proteins in at least 2 replicates with relative quantification p-value < 0.05 and 1.5 fold changes were selected as the SDEPs in the data using Cuffdiff [41].

4.6. Volcano Plot and Hierarchical Clustering Analyses

The Volcano plot and hierarchical clustering analyses were performed in the R environment (https://www.R-project.org) (access on 1 July 2020). The aes function in the package ggplot2 was used for Volcano plot drawing (https://CRAN.R-project.org/package=ggplot2) (access on 1 July 2020). The scale and hclust functions in the package stats were used for z-score normalization and hierarchical clustering, and the pheatmap function in the package pheatmap was employed for hierarchical clustering drawing (https://CRAN.R-project.org/package=pheatmap) (access on 1 July 2020).

4.7. Bioinformatic Analysis

The agriGO v2.0 [42] software was employed for Gene Ontology (GO) annotation and enrichment analyses, and the Kyoto Encyclopedia of Genes and Genomes (KEGG) was used for pathway annotation and enrichment analyses [43] to gain a better understanding of the functions of the selected proteins. The GO enrichment on three ontologies, including biological process, molecular function and cellular component, and the KEGG pathway enrichment were applied on the basis of Fisher's exact test, and only the GO categories and KEGG pathways with p-value < 0.05 and FDR < 0.05 or corrected p-value < 0.05 were considered as significant enrichment.

4.8. RNA Extraction and qRT-PCR

Total RNA was extracted from galls/roots using the RNeasy Plant Mini Kit (Qiagen, Dusseldorf, Germany), and cDNA was synthesized using the PrimeScript™ RT reagent Kit with gDNA Eraser (Perfect Real Time) (TaKaRa, Tokyo, Japan) as per the manufacturer's instructions. The quality and concentration of extracted RNA were measured using a Nanodrop 2000 (Thermo Scientific, Waltham, MA, USA).

The expression of the 8 selected genes in NPB and Khao Pahk Maw was measured by qRT-PCR using the designed primers listed in Table S2. The qRT-PCR was performed on an ABI 7500 Fast Real Time PCR System (Thermo Fisher Scientific, Waltham, MA, USA). The ubiquitin gene *UBQ5* (*LOC_Os01g22490*) was used as the reference gene [44]. Three replicates were set for each sample. Relative gene expression level was calculated using the $2^{-\Delta\Delta Ct}$ method [45].

4.9. Statistical Analyses of Data

The statistical significance of all the data was analyzed by ANOVA using SPSS version 25 (IBM, Armonk, NY, USA). All the data were analyzed using a paired *t*-test and Duncan's multiple comparison test.

Supplementary Materials: Supplementary materials can be found at http://www.mdpi.com/1422-0067/21/16/5640/s1. Figure S1: Volcano plots of the relative protein abundance changes of all the identified proteins in the roots of NPB with or without infection of *M. graminicola*. Figure S2: Clustering heatmaps of the three biological replicates. Figure S3: GO and KEGG enrichment analyses using all the specifically expressed proteins and SDEPs of NPB roots in CK1DPI_versus_CK0, CK3DPI_versus_CK0 or CK7DPI_versus_CK. Figure S4: GO and KEGG enrichment analyses of the specifically expressed proteins and SDEPs of the NPB roots in the whole CK_versus_CK0 group. Figure S5: Compatible interaction of the *japonica* rice cultivar NPB with *M. graminicola*. Table S1: List of the specifically expressed proteins and significantly differentially expressed proteins (SDEPs) in the roots of NPB rice uniquely caused by the infection of *M. graminicola*. Table S2: List of the primers used in this study.

Author Contributions: W.H. and S.L. conceived the project and designed the experiments. C.X., W.H. and M.L. performed the experiments. C.X., S.L., W.H., X.Y. and W.L. analyzed the data. D.P., X.Y. and H.K. provided the experimental materials. D.P. coordinated the project. S.L. and C.X. wrote the manuscript. All authors have read and agreed to the published version of the manuscript.

Funding: This work was financially supported by the National Natural Science Foundation of China (31571986, 31972248) and the Special Fund for Agro-Scientific Research in the Public Interest (201503114).

Acknowledgments: The authors thank Beijing BangFei Bioscience Co. Ltd. (http://www.bangfeibio.com) for their label-free quantitative mass spectrometry measurement and primary interpretation of the obtained proteome data. The authors also thank the International Rice Germplasm Center for providing rice seeds.

Conflicts of Interest: The authors declare that they have no conflict of interest.

References

1. Jones, J.T.; Haegeman, A.; Danchin, E.G.J.; Gaur, H.S.; Helder, J.; Jones, M.G.K.; Kikuchi, T.; Manzanilla-López, R.; Palomares-Rius, J.E.; Wesemael, W.M.L.; et al. Top 10 plant-parasitic nematodes in molecular plant pathology. *Mol. Plant Pathol.* **2013**, *14*, 946–961. [CrossRef] [PubMed]

2. De Waele, D.; Elsen, A. Challenges in tropical plant nematology. *Annu. Rev. Phytopathol.* **2007**, *45*, 457–485. [CrossRef] [PubMed]

3. Liu, Y.; Ding, Z.; Peng, D.L.; Liu, S.M.; Kong, L.A.; Peng, H.; Xiang, C.; Li, Z.C.; Huang, W.K. Evaluation of the biocontrol potential of *Aspergillus welwitschiae* against the root-knot nematode *Meloidogyne graminicola* in rice (*Oryza sativa* L.). *J. Integr. Agric.* **2019**, *18*, 1–10. [CrossRef]

4. Xiang, C.; Liu, Y.; Liu, S.M.; Huang, Y.F.; Kong, L.A.; Peng, H.; Liu, M.Y.; Liu, J.; Peng, D.L.; Huang, W.K. αβ-Dehydrocurvularin isolated from the fungus *Aspergillus welwitschiae* effectively inhibited the behaviour and development of the root-knot nematode *Meloidogyne graminicola* in rice roots. *BMC Microbiol.* **2020**, *20*, 48. [CrossRef] [PubMed]

5. Netscher, C.; Erlan, X. A root-knot nematode, *Meloidogyne graminicola*, parasitic on rice in Indonesia. *Afro-Asia J. Nematol.* **1993**, *3*, 90–95.

6. Mantelin, S.; Bellafiore, S.; Kyndt, T. *Meloidogyne graminicola*: A major threat to rice agriculture. *Mol. Plant Pathol.* **2017**, *18*, 3–15. [CrossRef]

7. Nguyễn, P.V.; Bellafiore, S.; Petitot, A.S.; Haidar, R.; Bak, A.; Abed, A.; Gantet, P.; Mezzalira, I.; Engler, J.A.; Fernandez, D. *Meloidogyne incognita*-rice (*Oryza sativa*) interaction: A new model system to study plant-root-knot nematode interactions in monocotyledons. *Rice* **2014**, *7*, 23. [CrossRef]

8. Peng, D.L.; Gaur, S.H.; Bridge, J. Nematode parasites of rice. In *Plant-Parasitic Nematodes in Subtropical and Tropical Agriculture*, 3rd ed.; Sikora, R.A., Coyne, D., Hallmann, J., Timper, P., Eds.; CABI Publishing: Wallingford, UK, 2018; pp. 120–162. ISBN 1786391244.

9. Galeng-Lawilao, J.; Kumar, A.; De Waele, D. QTL mapping for resistance to and tolerance for the rice root-knot nematode, *Meloidogyne graminicola*. *BMC Genet.* **2018**, *19*, 53. [CrossRef]

10. Kyndt, T.; Denil, S.; Haegeman, A.; Trooskens, G.; Bauters, L.; Criekinge, W.V.; De Meyer, T.; Gheysen, G. Transcriptional reprogramming by root knot and migratory nematode infection in rice. *New Phytol.* **2012**, *196*, 887–900. [CrossRef]

11. Ji, H.; Gheysen, G.; Denil, S.; Lindsey, K.; Topping, J.F.; Nahar, K.; Haegeman, A.; De Vos, W.H.; Trooskens, G.; Criekinge, W.V.; et al. Transcriptional analysis through RNA sequencing of giant cells induced by *Meloidogyne graminicola* in rice roots. *J. Exp. Bot.* **2013**, *64*, 3885–3898. [CrossRef]

12. Petitot, A.S.; Kyndt, T.; Haidar, R.; Dereeper, A.; Collin, M.; de Almeida Engler, J.; Gheysen, G.; Fernandez, D. Transcriptomic and histological responses of African rice (*Oryza glaberrima*) to *Meloidogyne graminicola* provide new insights into root-knot nematode resistance in monocots. *Ann. Bot-London* **2017**, *119*, 885–899. [CrossRef]

13. Xu, G.; Greene, H.H.; Yoo, H.; Liu, L.; Marqués, J.; Motley, J.; Dong, X. Global translational reprogramming is a fundamental layer of immune regulation in plants. *Nature* **2017**, *545*, 487–490. [CrossRef] [PubMed]

14. Yang, A.; Yu, L.; Chen, Z.; Zhang, S.; Shi, J.; Zhao, X.; Yang, Y.; Hu, D.; Song, B. Label-free quantitative proteomic analysis of chitosan oligosaccharide-treated rice infected with southern rice black-streaked dwarf virus. *Viruses* **2017**, *9*, 115. [CrossRef] [PubMed]

15. Ji, Z.; Zeng, Y.; Liang, Y.; Qian, Q.; Yang, C. Proteomic dissection of the rice-*Fusarium fujikuroi* interaction and the correlation between the proteome and transcriptome under disease stress. *BMC Genom.* **2019**, *20*, 91. [CrossRef] [PubMed]

16. Ma, Z.; Wang, L.; Zhao, M.; Gu, S.; Wang, C.; Zhao, J.; Tang, Z.; Gao, H.; Zhang, L.; Fu, L.; et al. iTRAQ proteomics reveals the regulatory response to *Magnaporthe oryzae* in durable resistant vs. susceptible rice genotypes. *PLoS ONE* **2020**, *15*, e0227470. [CrossRef] [PubMed]

17. Milligan, S.B.; Bodeau, J.; Yaghoobi, J.; Kaloshian, I.; Zabel, P.; Williamson, V.M. The root knot nematode resistance gene *Mi* from tomato is a member of the leucine zipper, nucleotide binding, leucine-rich repeat family of plant genes. *Plant Cell* **1998**, *10*, 1307–1319. [CrossRef] [PubMed]

18. Ammiraju, J.; Veremis, J.; Huang, X.; Roberts, P.; Kaloshian, I. The heat-stable root-knot nematode resistance gene *Mi-9* from *Lycopersicon peruvianum* is localized on the short arm of chromosome. *Theor. App. Genet.* **2003**, *106*, 478–484. [CrossRef]

19. Tzortzakakis, E.A.; Adam, M.A.M.; Blok, V.C.; Paraskevopoulos, C.; Bourtzis, K. Occurrence of resistance-breaking populations of root-knot nematodes on tomato in Greece. *Eur. J. Plant. Pathol.* **2005**, *113*, 101–105. [CrossRef]

20. Jablonska, B.; Ammiraju, J.S.; Bhattarai, K.K.; Mantelin, S.; Martinez, I.O.; Roberts, P.A.; Kaloshian, I. The *Mi-9* gene from *Solanum arcanum* conferring heat-stable resistance to root-knot nematodes is a homolog of *Mi-1*. *Plant Physiol.* **2007**, *143*, 1044–1054. [CrossRef]

21. Claverie, M.; Dirlewanger, E.; Bosselut, N.; Ghelder, C.V.; Voisin, R.; Kleinhentz, M.; Lafargue, B.; Abad, P.; Rosso, M.N.; Chalhoub, B.; et al. The *Ma* gene for complete-spectrum resistance to *Meloidogyne* species in *Prunus* is a TNL with a huge repeated C-terminal post-LRR region. *Plant Physiol.* **2011**, *156*, 779–792. [CrossRef]

22. Dimkpa, S.O.N.; Lahari, Z.; Shrestha, R.; Douglas, A.; Gheysen, G.; Price, A.H. A genome-wide association study of a global rice panel reveals resistance in *Oryza sativa* to root-knot nematodes. *J. Exp. Bot.* **2016**, *67*, 1191–1200. [CrossRef]

23. Cook, D.E.; Lee, T.G.; Guo, X.; Melito, S.; Wang, K.; Bayless, A.M.; Wang, J.; Hughes, T.J.; Willis, D.K.; Clemente, T.E.; et al. Copy number variation of multiple genes at *Rhg1* mediates nematode resistance in soybean. *Sci.* **2012**, *338*, 1206–1209. [CrossRef] [PubMed]

24. Liu, S.; Kandoth, P.K.; Warren, S.D.; Yeckel, G.; Heinz, R.; Alden, J.; Yang, C.; Jamai, A.; El-Mellouki, T.; Juvale, P.S.; et al. A soybean cyst nematode resistance gene points to a new mechanism of plant resistance to pathogens. *Nature* **2012**, *492*, 256–260. [CrossRef] [PubMed]

25. Liu, S.; Kandoth, P.K.; Lakhssassi, N.; Kang, J.; Colantonio, V.; Heinz, R.; Yeckel, G.; Zhou, Z.; Bekal, S.; Dapprich, J.; et al. The soybean *GmSNAP18* gene underlies two types of resistance to soybean cyst nematode. *Nat. Commun.* **2017**, *8*, 14822. [CrossRef] [PubMed]

26. Corpas, F.J.; Del Río, L.A.; Palma, J.M. Plant peroxisomes at the crossroad of NO and H_2O_2 metabolism. *J. Integrat. Plant Biol.* **2019**, *61*, 803–816.

27. Wang, X.; Cheng, C.; Li, Q.; Zhang, K.; Lou, Q.; Li, J.; Chen, J. Multi-omics analysis revealed that MAPK signaling and flavonoid metabolic pathway contributed to resistance against *Meloidogyne incognita* in the introgression line cucumber. *J. Proteom.* **2020**, *220*, 103675. [CrossRef]

28. Kyndt, T.; Fernandez, D.; Gheysen, G. Plant-parasitic nematode infection in rice: Molecular and cellular insights. *Annu. Rev. Phytopathol.* **2014**, *52*, 7–19. [CrossRef]

29. Hatzade, B.; Singh, D.; Phani, V.; Kumbhar, S.; Rao, U. Profiling of defense responsive pathway regulatory genes in Asian rice (*Oryza sativa*) against infection of *Meloidogyne graminicola* (Nematoda: Meloidogynidae). *3 Biotech* **2020**, *10*, 60. [CrossRef]

30. Wang, H.; Sun, S.; Ge, W.; Zhao, L.; Hou, B.; Wang, K.; Lyu, Z.; Chen, L.; Xu, S.; Guo, J.; et al. Horizontal gene transfer of *Fhb7* from fungus underlies *Fusarium* head blight resistance in wheat. *Science* **2020**, *368*, eaba5435. [CrossRef]

31. Bredow, M.; Monaghan, J. Regulation of plant immune signaling by calcium-dependent protein kinases. *Mol. Plant Microb. Interact.* **2019**, *32*, 6–19. [CrossRef]

32. Grant, M.R.; Godiard, L.; Straube, E.; Ashfield, T.; Dangl, J.L. Structure of the *Arabidopsis RPM1* gene enabling dual specificity disease resistance. *Science* **1995**, *269*, 843–846. [CrossRef] [PubMed]

33. Zhan, L.P.; Peng, D.L.; Wang, X.L.; Kong, L.A.; Peng, H.; Liu, S.M.; Liu, Y.; Huang, W.K. Priming effect of root-applied silicon on the enhancement of induced resistance to the root-knot nematode *Meloidogyne graminicola* in rice. *BMC Plant Biol.* **2018**, *18*, 50. [CrossRef] [PubMed]

34. Reversat, G.; Boyer, J.; Sannier, C.; Pando-Bahuon, A. Use of a mixture of sand and water-absorbent synthetic polymer as substrate for the xenic culturing of plant-parasitic nematodes in the laboratory. *Nematology* **1999**, *1*, 209–212. [CrossRef]

35. Huang, W.K.; Ji, H.L.; Gheysen, G.; Debode, J.; Kyndt, T. Biochar-amended potting medium reduces the susceptibility of rice to root-knot nematode infections. *BMC Plant Bio.* **2015**, *15*, 267. [CrossRef] [PubMed]

36. Zhan, L.P.; Ding, Z.; Peng, D.L.; Peng, H.; Kong, L.A.; Liu, S.M.; Liu, Y.; Li, Z.C.; Huang, W.K. Evaluation of Chinese rice varieties resistant to the root-knot nematode *Meloidogyne graminicola*. *J. Integr. Agric.* **2018**, *17*, 621–630. [CrossRef]

37. Bybd, D.W.; Kirkpatrick, T.; Barker, K.R. An improved technique for clearing and staining plant tissues for detection of nematodes. *J. Nematol.* **1983**, *15*, 142.

38. Wiśniewski, J.R.; Zougman, A.; Nagaraj, N.; Mann, M. Universal sample preparation method for proteome analysis. *Nat. Methods.* **2009**, *6*, 359–362. [CrossRef]

39. Wiśniewski, J.R.; Zougman, A.; Mann, M. Combination of FASP and stagetip-based fractionation allows in-depth analysis of the hippocampal membrane proteome. *J. Proteome Res.* **2009**, *8*, 5674–5678. [CrossRef]

40. Kao, S.H.; Wong, H.K.; Chiang, C.Y.; Chen, H.M. Evaluating the compatibility of three colorimetric protein assays for two-dimensional electrophoresis experiments. *Proteomics* **2008**, *8*, 2178–2184. [CrossRef]

41. Trapnell, C.; Hendricjson, D.G.; Sauvageau, M.; Goff, L.; Rinn, J.L.; Pachter, L. Differential analysis of gene regulation at transcript resolution with RNA-seq. *Nat. Biotechnol.* **2013**, *31*, 46–53. [CrossRef]

42. Tian, T.; Liu, Y.; Yan, H.Y.; You, Q.; Yi, X.; Du, Z.; Xu, W.Y.; Su, Z. agriGO v2.0: A GO analysis toolkit for the agricultural community. *Nucleic Acids Res.* **2017**, *45*, W122–W129. [CrossRef]

43. Kanehisa, M.; Sato, Y. KEGG Mapper for inferring cellular functions from protein sequences. *Protein Sci.* **2020**, *29*, 28–35. [CrossRef] [PubMed]

44. Jain, M.; Nijhawan, A.; Tyagi, A.K.; Khurana, J.P. Validation of housekeeping genes as internal control for studying gene expression in rice by quantitative real-time PCR. *Biochem. Biophys. Res. Commun.* **2006**, *345*, 646–651. [CrossRef] [PubMed]
45. Livak, K.; Schmittgen, T. Analysis of relative gene expression data using real-time quantitative PCR and the $2^{-\Delta\Delta ct}$ method. *Methods* **2000**, *25*, 402–408. [CrossRef] [PubMed]

MDPI

St. Alban-Anlage 66

4052 Basel

Switzerland

Tel. +41 61 683 77 34

Fax +41 61 302 89 18

www.mdpi.com

International Journal of Molecular Sciences Editorial Office

E-mail: ijms@mdpi.com

www.mdpi.com/journal/ijms

www.ingramcontent.com/pod-product-compliance
Lightning Source LLC
LaVergne TN
LVHW071628170726
843515LV00009B/2508